International Conference on Disaster Management and Climate Change

IOP Conference Series: Earth and Environmental Science
Volume 986

Online
4 September 2021

ISBN: 978-1-7138-4967-4
ISSN: 1755-1307

Printed from e-media with permission by:

Curran Associates, Inc.
57 Morehouse Lane
Red Hook, NY 12571

Some format issues inherent in the e-media version may also appear in this print version.

Institute of Physics
Dirac House, Temple Back
Bristol BS1 6BE UK

Phone: 44 1 17 929 7481
Fax: 44 1 17 920 0979

techtracking@iop.org

Additional copies of this publication are available from:

Curran Associates, Inc.
57 Morehouse Lane
Red Hook, NY 12571 USA
Phone: 845-758-0400
Fax: 845-758-2633
Email: curran@proceedings.com
Web: www.proceedings.com

TABLE OF CONTENTS

IOP Conf. Series: Earth and Environmental Science **986** (2022) 011001 doi:10.1088/1755-1315/986/1/011001

PREFACE

The Disaster Study Center Sebelas Maret University has successfully held an "International Conference on Disaster Management and Climate Change (ICoDMC)" via the online zoom system on September 4, 2021. ICoDMC aims to disseminate research results, various ideas on controlling environmental degradation and disaster management for a better life and providing input to the Government and other policy-making institutions in environmental and disaster management. The Theme of 2021 is Environment degradation and disaster management.

ICoDMC invited great speakers who are very competent in the field of disaster and the environment, and they are Letnan Jendral TNI Ganip Warsito (Head of the National Disaster Management Agency), Prof. JC Gaillard (The University of Auckland, New Zealand), Prof. Josaphat Tetuko Sri Sumantyo (Center for Environment Remote). Sensing, Chiba University), Associate. Prof. Dr Mohd Hairy Ibrahim (UPSI Malaysia), Dr. (Cand). Sorja Koesuma, S.Si., M.Si. (Sebelas Maret University) and Dr. Rita Noviani, S.Si., M.Sc (Sebelas Maret University)

Topic issues for this seminar are Disaster Education, Living in Harmony With Disaster Risk, Regional Planning for Disaster Management, Building and Landscape For Disaster Mitigation, Mitigation and Adaptation To Climate Change, Climate Change Policy, Land Degradation and Land Conservation, Water Quality and Quantity Problems, Air Pollution and Greenhouse Effect and Urban Heat Island (UHI). There were 135 articles submitted according to the topic above, and after being selected, there were 99 articles chosen to be presented and submitted to the IOP-EES Proceedings. Article authors consist of lecturers, practitioners, teachers and students. The participants who joined at the time of the seminar was 320 participants.

Thanks to all of the speakers, authors and participants. See you at the second ICoDMC.

Best Regards

Prof. Dr. Chatarina Muryani, M.Si

Chair of the committee

IOP Conf. Series: Earth and Environmental Science **986** (2022) 011001 doi:10.1088/1755-1315/986/1/011001

Committee
International Conference on Disaster Management and Climate Changes (ICoDMC)
2021

Advisory Board :

1. Prof. Jiro Takai, Ph.D (Nagoya University, Japan)
2. Prof. Amy Mackenzie (Southern Cross University, Australia)
3. Prof. Dr. Hafizan Juahir (Universiti Sultan Zainal Abidin, Malaysia)
4. Prof. Dr. Muh Aris Marfai, S.Si., M.Sc (Universitas Gadjah Mada, Indonesia)
5. Prof. Dr. H. Jamal Wiwoho, S.H., M.Hum (Universitas Sebelas Maret, Indonesia)

Steering Committee :

1. Prof. Dr. Okid Parama Astirin, M.S (Environmental Science, Universitas Sebelas Maret)
2. Prof. Dr. Eng. Syamsul Hadi, S.T., M.T (Universitas Sebelas Maret, Indonesia)

Conference Chair :

Prof. Dr. Chatarina Muryani, M.Si (Universitas Sebelas Maret, Indonesia)

Conference Vice Chair :

Sorja Koesuma, S.Si., M.Si (Universitas Sebelas Maret, Indonesia)

Editor

1. Associate. Prof. Dr Mohd Hairy Ibrahim (Universiti Pendidikan Sultan Indris, Malaysia) (Scopus Id : 55497320300)
2. Prof. Dr. Enok Maryani, M.S (Universitas Pendidikan Indonesia) (Scopus Id : 57193733326)
3. Prof. Dr. Eva Banowati, M.Si (Universitas Nrgeri Semarang, Indonesia) (Scopus Id : 57195296290)
4. Dr. Sidharta Adyatma, M.Si (Universitas Lambung Mangkurat, Indonesia) (Scopus Id : 57211325724)
5. Dr. Estuning Tyas Wulan Mei, M.Si (Universitas Gadjah Mada, Indonesia) (Scopus Id : 55018646400)
6. Dr. Pipit Wijayanti, S.Si., M.Sc (Universitas Sebelas Maret, Indonesia) (Scopus Id : 57222575230)
7. Dr. Rita Noviani, S.Si., M.Sc (Universitas Sebelas Maret, Indonesia)
8. Dr. Yasin Yusup, S.Si., M.Si (Universitas Sebelas Maret, Indonesia) (Scopus Id : 57202301304)
9. Ahmad Marzuki, S.Si., Ph.D (Universitas Sebelas Maret, Indonesia) (Scopus Id : 36810005100)

IOP Conf. Series: Earth and Environmental Science **986** (2022) 011001 doi:10.1088/1755-1315/986/1/011001

Local Committee :

1. Lintang Ronggowulan, S.Pd., M.Pd (Universitas Sebelas Maret, Indonesia)
2. Seno Budhi Ajar, S.Pd., M.Si (Universitas Sebelas Maret, Indonesia)
3. Farida Hidayat, S.Psi., M.Si (Universitas Sebelas Maret, Indonesia)
4. Anton Subarno, S.Pd., M.Pd., Ph.D (Universitas Sebelas Maret, Indonesia)

IOP Conf. Series: Earth and Environmental Science **986** (2022) 011002 doi:10.1088/1755-1315/986/1/011002

Peer Review Statement

All papers published in this volume have been reviewed through processes administered by the Editors. Reviews were conducted by expert referees to the professional and scientific standards expected of a proceedings journal published by IOP Publishing.

- **Type of peer review:** Single Anonymous
- **Conference submission management system:** Easychair
- **Number of submissions received:** 135
- **Number of submissions sent for review:** 135
- **Number of submissions accepted:** 99
- **Acceptance Rate (Submissions Accepted / Submissions Received × 100):** 73.3
- **Average number of reviews per paper:** 6.75
- **Total number of reviewers involved:** 20
- **Contact person for queries:**
  Name: Chatarina Muryani
  Email: chatarinamuryani@staff.uns.ac.id
  Affiliation: Universitas Sebelas Maret

Cost assessment of biomass coal fuel on air pollution and coal consumption reduction

D Marganingrum[1*], M N Khalifah[2], and P Nursetyowati[2]

[1] Research Center for Geotechnology-National Research and Innovation Agency
[2] Environmental Engineering Department-Bakrie University

*Email: dmarganingrum@yahoo.com

Abstract. The use of coal as an energy source in the industrial sector has been quite massive, but the negative impact has not been a serious concern. We realized that the use of coal had caused pollution, including air pollution from coal-burning processing and water and soil pollution from the coal solid waste (fly ash and bottom ash), which not be appropriately managed. This study aims to assess the beneficiary of utilizing the unburnt carbon as co-firing in boiler industrial compared with air emission impact—the unburnt carbon mixed with municipal solid waste treated firstly to be briquette. The briquette was called Biomass Coal Fuel (BCF), and it was used as coal substitution of 10-15%. The cost of air pollution calculation referred to the Ministry Environmental Regulation Number 13 of 2011. We have compared the cost of air pollution between before and after BCF substitution. Meanwhile, the beneficiaries were calculated from the reduction of coal consumption. The result showed that emissions cost just approximately 4-5% of the beneficiaries due to reductions in coal consumption that about 145.656 – 218.484 million rupiahs per year. The factory can use this beneficiary to improve air emission controlling and support environmental conservation.

1. Introduction

Currently, the use of fossil fuels is increasing rapidly along with increased activities in the industrial and transportation sectors [1]. One source of fossil fuel energy that is widely used in industry is coal. Coal is needed to meet energy demand and to reduce dependence on oil, whose reserves are starting to decrease, while the availability of renewable energy sources still faces several obstacles [2]. All domestic coal consumption in 2016 was 76% of which was used by PLTU. The remaining 24% is used in the cement, metallurgy, fertilizer, paper, textile, and briquette industries [3].

One of the environmental impacts of using coal is air pollution which can trigger global warming [4]. The environmental impact of the increasingly massive use of coal has become a worldwide concern. Various studies have been conducted to reduce the negative impacts of using coal. Increasing the calorific value of low-quality coal has been carried out to increase the efficiency of its use [5–7]. The research to reduce sulfur content in coal has also been done to reduce its emission level [8].

In addition to having an impact on air pollution, the process of utilizing and using coal also produces solid waste, which also has the potential to pollute the environment [9]. Coal solid waste can be in the form of fly ash or bottom ash (FABA). Fly ash has a size of 0.001–100 m, and bottom ash has 0.1-10 mm [10]. Many studies on the utilization of coal solid waste have been carried out [11–15]. FABA waste is generally used as a mixture for making concrete [16]. Biomass had been combined with bottom ash into solid fuels or briquettes [17]. However, the utilization of FABA waste, especially from industrial steam boilers, is still constrained by regulations because it is still categorized as a toxic and hazardous waste [11,16].

One textile factory in Bandung Regency has used unburnt carbon mixed with biomass from urban waste as briquettes. The factory used the briquettes as a substitute for coal in boiler machines to meet the energy demands [18,19]. This activity can form a closed-loop flow in the company's processes. Besides providing economic benefits, companies can also pay attention to environmental ethics in business processes so that the role of industry in sustainable development can be done as expected [9].

It is necessary to conduct an economic assessment and its impact to formulate the cost of ecosystem services to compensate for the increasingly massive use of coal. This assessment can help in realizing sustainable development that supports the diversification of energy sources. This paper describes the assessment of the benefits and costs of the factory that used coal to participate in supporting environmental conservation.

2. Material and Method

2.1. Material

This study was conducted at a factory that uses coal as an energy source in the textile production process. Coal is used in three boiler machines with a coal consumption of almost 30 tons per day. The waste of fly ash and bottom ash (FABA) produced is 3 tons per day. From the 3 tons of FABA, 750 kg of unburned carbon is produced per day. The factory uses the unburned carbon as a briquette mixture material used for coal substitution in one of the boiler machines, namely the Omnical boiler (steam boiler). Figure 1 shows the closed-loop flow in the production process that can provide beneficiary for the factory.

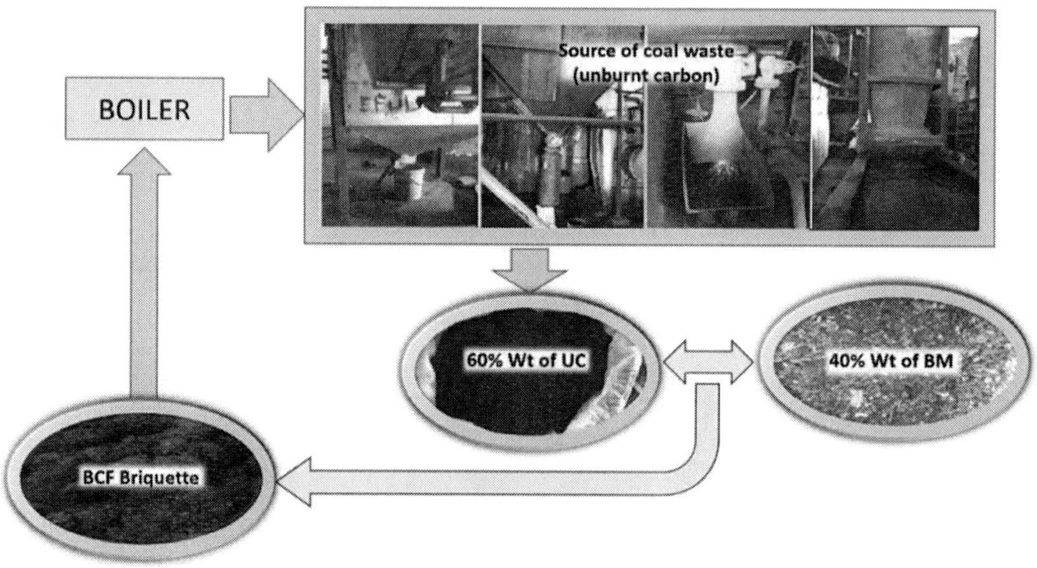

Figure 1. Flowchart of closed-loop in factory which this study conducted

2.2. Method

This paper is still at the preliminary assessment. This paper has not explained the economic assessment of all production activities in the factory. The valuation method is limited to the benefits and cost of using Biomass Coal Fuel (BCF) briquettes to substitute coal consumed. The assessment is carried out based on the accumulated value of the pollution unit based on the Regulation of the Minister of the Environment Number 13 of 2011. The accumulated compensation of air emissions is calculated in the

IOP Conf. Series: Earth and Environmental Science **986** (2022) 012001 doi:10.1088/1755-1315/986/1/012001

combustion process, between before and after substitution of BCF briquettes. The cost basis per unit of pollution according to the Minister of Environment Regulation No. 13 of 2011 is Rp 24,750.00. Table 1 shows the pollution unit values for each type of parameter. Experts determine this value based on the balance between the level of danger, toxicity, and ability of nature to degrade when released into the environment. It is smaller value means it is more dangerous and difficult to be degraded by nature. This study only focuses on three main air pollutant parameters from the coal combustion process, namely particulate, sulfur dioxide (SO_2), and nitrogen dioxide (NO_2), following applicable regulations. The total cost of accumulated pollution units from the two data is then compared. Its difference is calculated to compare with the beneficiaries obtained by the factory from reducing the coal consumption due to BCF substitution.

Table 1. Value of air emission parameter unit

Parameter	Value of unit pollution [a)]
Particulate Matter	250 Kg
Sulfur Oxide (SO_X)	200 Kg
Nitrogen Oxide (NO_X)	200 Kg

Source: Ministry of Environmental Regulation Number 13 of 2011
[a)] Experts determined

The economic equation used to assess losses from air pollution is as follows [20]:

$$C_1 = A \times B \tag{1}$$

Where C1 is the estimated value of the economic loss of air pollution based on the cost per pollution unit (Rp/year), A is the number of pollution units (kg/year), and B is the cost basis per unit of pollution (Rp/kg). The value of economic losses due to air pollution should be the responsibility of the polluter.

Another cost in this case, besides the cost of economic losses due to air pollution, is producing BCF briquettes. These costs consist of the cost of purchasing biomass and the operational costs of making BCF briquettes. Operating costs are used for electricity and water. At the same time, assets and labor are not included in the calculation because they use assets and workers already available at the factory. The cost of producing BCF (C_2) briquettes is formulated as follows:

$$C_2 = (C_{C\text{-}BCF} \times C_{BM} \times 0.4) + C_{OM} \tag{2}$$

Where $C_{C\text{-}BCF}$ is the amount of BCF consumption used to substitute coal (kg/year); C_{BM} is the unit cost per kg of biomass (=Rp 750,-), 0.4 is the number of weight percentage of biomass used in the production of BCF briquette, and C_{OM} is the operational cost of BCF briquette production which is set at 10% of gross profit.

Then the economic loss is C. This is the total cost during the use of BCF briquettes, including the unit cost due to air pollution (C1) and the cost of BCF briquettes production (C2). The total cost (C) is formulated as follows:

$$C = C_1 + C_2 \tag{3}$$

The economic loss (C) value is compared with the benefits of cost-reducing of coal consumption due to BCF briquettes substitution. The benefits obtained by the factory due to using BCF briquette substitution (B in Rp/year) can be formulated by the following equation:

$$B = P_{BCF} \times C_{C\text{-}C} \times C_C \tag{4}$$

Where P_{BCF} is the percentage of BCF used to substitute coal (%), $C_{C\text{-}C}$ is the consumption of coal (kg/year), and C_C is the cost of coal (Rp/kg).

The values of C and B are then compared. The difference is the profit or loss obtained by the factory when using BCF briquette as coal substitution. If B is less than C, the factory suffers a loss but is still required to pay the cost of air conditioning ecosystem services due to emissions from coal combustion and or the use of a mixture of coal and BCF briquettes. On the other hand, if B is more than C, the factory's profits can support the cost of air ecosystem services due to the air pollution.

3. Result and Discussion

Emissions caused by the coal combustion process are influenced by several things, including the characteristics of the coal used, the capacity and type of boiler used, operating technology, and emission control technology [21]. The factory where this study was conducted uses coal, following the availability of suppliers when it is ordered. However, most of the types of coal used are sub-bituminous types with lumpy sizes. The type of boiler used is a fluidized bed with an emission control system using a wet scrubber and a cyclone.

Figure 2. Coal type which used by factory for boiler that monitored its air emission

Monitoring of air emission and ambient is conducted by the factory usually twice a year. Table 1 shows the monitoring air emissions using whole coal and coal substituting 10- 15% of BCF briquettes from 2018 to 2020. BCF briquette substitution was only carried out in the 2020 study activity. Meanwhile, emission data from the use of whole coal feeder data from 2018. Based on Table 2, all air monitoring results still meet the quality standards. The standard for air emission refers to Attachment IV to the Regulation of the Environment Minister Number 7 of 2007. In the regulation, there are only four parameters: particulates, nitrogen dioxide, sulfur dioxide, and opacity. However, for the opacity parameter, the pollutant unit value is not regulated in the Minister of the Environment Regulation Number 13 of 2011. In this paper, only three parameters are used, as shown in Table 2.

Table 2. Air emission

Parameter	Unit	Air emission of coal feeder				Air emission by BCF substitution feeder	
		29/09/2018	18/02/2019	26/02/2020	12/10/2020	09/03/2020	25/09/2020
Particulate	mg/Nm^3	107.00	212.20	12.10	20.98	70.90	7.50
Nitrogen Dioxide (NO_2)	mg/Nm^3	64.00	154.40	174.20	23.00	370.30	9.41
Sulfur Dioxide (SO_2)	mg/Nm^3	41.00	182.10	150.00	18.00	493.90	7.85
Flow rate	m^3/s	6.7	5.2	4.3	4.9	6.5	4.8

Table 3. Standard of air emission

Parameter	Unit	Standard
Particulate	mg/Nm3	230
Nitrogen Dioxide (NO_2)	mg/Nm3	825
Sulfur Dioxide (SO_2)	mg/Nm3	750

The pollution unit load for each emission parameter is the product of the emission concentration of each parameter with the flow rate. The results of the accumulated emission unit load calculation for one year (A) are shown in Table 3. The costs that the factory must pay as an estimate of the cost of economic losses due to air pollution, as written in equation (1), are shown in Table 4. The calculation of the emission unit load written in Table 4 is the average emission unit load data for each type of feeder used. The total cost of losses due to emissions arising from the use of coal and substitution with BCF briquettes is Rp.3,588,488,-/year or Rp.7,823,501,-/year, respectively. The difference between these values indicates an increase in emission costs caused by coal waste (unburnt carbon) for coal substitution. However, the economic calculations do not stop here, and it is necessary to look for the positive side that can be obtained by the factory from the utilization of its coal waste so that a closed-loop flow that supports environmental conservation is achieved.

Table 4. Air emission loading

Parameter	**Air emission loading of coal feeder (kg/year)**				**Air emission loading by BCF substitution feeder (kg/year)**	
	29/09/2018	18/02/2019	26/02/2020	12/10/2020	09/03/2020	25/09/2020
Particulate	14,866	22,881	1,079	1,871	9,556	746
Nitrogen Dioxide (NO_2)	8,892	16,649	15,533	2,051	49,911	937
Sulfur Dioxide (SO_2)	5,696	19,635	13,375	1,605	66,570	781

Table 5. Average cost of air emission pollution

Parameter	**Unit cost (Rp/kg)**	**Value of unit pollution (kg)**	**Average air emission loading (kg/year)**		**Cost of air pollution (Rp/year)**	
			Coal Feeder	**BCF Substitution**	**Coal Feeder**	**BCF Substitution**
Particulate	24,750	250	10,174	5,151	1,007,230	509,983
Sulfur Dioxide (SO2)	24,750	200	10,078	33,676	1,247,128	4,167,352
Nitrogen Oxide (NO2)	24,750	200	10,781	25,424	1,334,131	3,146,165
Total (Rp/year)					**3,588,488**	**7,823,501**

In this study, the economic values for BCF briquette substitution are simulated by 10% and 15%. The purchase price of coal is uncertain, but the average cost to purchase coal is Rp 1,127,- per kg. Meanwhile, the average coal consumption in the boiler used as a case study is 8,500 kg/day. Based on the calculation results, the profit from substitution of 10% and 15% BCF briquettes as equation (4) is Rp 229,840,000,-/year and Rp 344,760,000,-/year, respectively (Table 5).

Table 6. Total cost of coal consumption

Coal Consumption in Boiler (kg/day)	**Average cost of coal (Rp/Kg)**	**Total cost of coal consumption (Rp/year)**	**Total cost of coal consumption due to 10% BCF substitution**	**Total cost of coal consumption due to 15% BCF substitution**
8,500	1,127	2,298,400,000	2,068,560,000	1,953,640,000
Saving cost of coal consumption (Rp/year) (Gross Benefit/GB)			**229,840,000**	**344,760,000**

This beneficiary is a gross profit because it must be deducted by the cost of BCF briquette production. The cost of purchasing biomass based on this case study is Rp. 750,-/kg. At the same time, the operational cost of making briquettes is assumed to be 10% of the gross profit obtained in the production of BCF briquettes. Calculating the difference in profit from decreasing coal consumption costs with emissions and production costs of BCF briquettes obtained a net profit of Rp 145,656,000,- and Rp 218,484,000,-respectively, for 10% and 15% substitution of BCF briquettes (Table 6).

Table 7. Net benefit of BCF substitution

Cost Explanation	with 10% Substitution of BCF (Rp)	with 15% Substitution of BCF (Rp)
Biomass purchasing cost	61,200,000	91,800,000
Operational cost (10% of GB)	22,984,000	34,476,000
Total cost (C)	84,184,000	126,276,000
Gross Benefit (B)	229,840,000	344,760,000
Net benefit (B – C)	145,656,000	218,484,000

From the calculation results, the beneficiaries' value from using BCF briquettes as a substitute for coal is much greater than the factory's costs, both economic losses due to air emissions and production costs for BCF briquettes. Meanwhile, the value of financial losses due to air emissions is only around 4-5% of the company's net profit from using BCF briquettes as a substitute for coal. Unfortunately, the Government of Indonesia has not determined the cost of climate-regulating ecosystem services if the air emissions generated are still below the quality standard.

Based on the results of the analysis and calculations that have been carried out, it is time to apply the cost of air conditioning ecosystem services even though the emissions generated are below the quality standard. This assessment is essential to anticipate the increasingly massive use of coal and its impacts. Of course, this needs support from the Government so that more industries can use their waste for their interests. Coal solid waste has a worse impact on the water and soil environment if it is not utilized, so using them can reduce the negative impact.

Coal waste is categorized as hazardous and toxic waste because it contains heavy metals, including mercury (Hg), chromium (Cr), lead (Pb), copper (Cu), and cadmium (Cd) [22–24]. When dumped in the ground, this heavy metal will not only pollute the soil but also pollute the surrounding water because it can flow with the runoff. The economic loss caused by this heavy metal to water pollution is greater than the air emissions caused by coal as fuel. The value of one unit of wastewater pollution with parameters Hg is 20 grams, Cr and Pb is 500 grams, Cu is 1000 grams, and Cd is 100 grams. The smaller the value of the pollution unit, the parameter is more dangerous to the environment. Therefore, it is necessary to review the losses and advantages of closed-loop activities in the factory as a whole so that the factory can minimize environmental damage and the factory is not lost too. The factory will not bear the loss as long as there is support from the government to utilize its coal waste. In other words, there is a harmonious relationship between the government and business actors to support environmental conservation efforts jointly.

4. Conclusions

The use of coal as an energy source impacts air pollution and water and soil pollution caused by coal solid waste that is not appropriately managed. Meanwhile, the use of coal waste classified as LB3 is still undergoing debate. The study of coal substitution using BCF briquettes made from unburnt carbon and biomass increased the air pollution load. Even the pollution value is still below the quality standard. Economically, the loss that must be paid by the factory on the cost of air pollution can still be covered from the profits of the decrease in coal consumption. The amount of environmental costs that the factory must pay is only around 4-5% of the profits obtained by the factory because it uses 10-15% BCF briquette substitution.

Acknowledgement
The authors would like to thank the Ministry of Finance of the Republic of Indonesia for providing financial assistance to conduct this research through the LPDP RISPRO program in 2019–2020. The authors also thank the management and staff members of textile factory in the Bandung Regency, where this study was conducted. The authors are grateful to the Laboratory of the Laboratory for Environmental Quality Control, PDAM Tirtawening, Bandung, for their support in testing the samples.

References
[1] Yudiartono, Anindhita, Sugiyono A, Wahid L M . and Adiarso 2018 *Outlook Energi Indonesia 2018: Energi Berkelanjutan untuk Transportasi Darat* (Jakarta: Pusat Pengkajian Industri Proses dan Energi (PPIPE), Badan)
[2] Haryadi H and Suciyanti M 2018 Analisis Perkiraan Kebutuhan Batubara Untuk Industri Domestik Tahun 2020-2035 Dalam Mendukung Kebijakan Domestic Market Obligation Dan Kebijakan Energi Nasional *J. Teknol. Miner. dan Batubara* **14** 59
[3] Setiawan A, Wibowo A P and Rosyid F A 2020 Analisis pengaruh ekspor dan konsumsi batubara terhadap pertumbuhan ekonomi Indonesia *J. Teknol. Miner. dan Batubara* **16** 109–24
[4] Tampubolon F R S, Yuwono A S, Tambunan A H and Achsani N A 2021 Penggunaan Bahan Bakar Alternatif dalam Pengelolaan Tambang Batubara sebagai Sumber Energi yang Ramah terhadap Lingkungan *J. Ilmu Lingkung.* **19** 89–97
[5] Kasim T and Probowo H 2017 Katalis Minyak Pelumas Bekas Pada Batubara Low Calorie Daerah Tanjung Belit , Kecamatan Jujuhan , *J. Sains dan Teknol.* **17** 1–9
[6] Komariah W E 2012 *Peningkatan Kualitas Batubara Indonesia Peringkat Rendah Melalui Penghilangan Moisture Dengan Pemanasa Gelombang Mikro* (Universitas Indonesia)
[7] Said M and Fanani Z 2018 Peningkatan Kualitas Batubara Sub-Bituminus Dengan Metode Froth Flotation *J. Penelit. Sain* **11** 587–97
[8] Aladin A 2012 Penentuan Rasio Optimum Campuran CPO: Batubara Dalam Desulfurisasi dan Deashing Secara Flotasi Sistem Kontinyu *J. Rekayasa Proses* **3** 50–6
[9] Abidin A, Leksono E B, Studi P, Industri T, Teknik F, Gresik U M, Korespondensi P, Loop C, Ash F and Logistics R 2021 Koagulan dengan Konsep Reverse Logistics **7** 39–44
[10] Seshadri B, Bolan N S, Naidu R and Brodie K 2010 The role of coal combustion products in managing the bioavailability of nutrients and heavy metals in soils *J. Soil Sci. Plant Nutr.* **10** 378–98
[11] Damayanti R 2018 Abu batubara dan pemanfaatannya: Tinjauan teknis karakteristik secara kimia dan toksikologinya *J. Teknol. Miner. dan Batubara* **14** 213–31
[12] Santoso I and Roy S K 2003 Pengaruh Penggunaan Bottom Ash Terhadap Karakteristik Campuran Aspal Beton *Civ. Eng. Dimens.* **5** pp.75-81
[13] Slamet S and Gunawan B 2016 Bio Briket Campuran Bottom Ash Batu Bara Limbah PLTU dan Biomassa Melalui Proses Karbonasi Sebagai Sumber Energi Terbarukan *Prosiding SNATIF ke-6 Tahun 2019* pp 43–50
[14] Syafrudin S, Zaman B, Indriyani I, Erga A S and Natalia H B 2015 The Utilization of Bottom Ash Coal for Briquette Products by Adding Teak Leaves Charcoal, Coconut Shell Charcoal, and Rice Husk Charcoal *Waste Technol.* **3** 14–21
[15] Triantoro A, Mustofa A, Kartini K and Hanafi A 2019 Studi Analisa Kualitas Biobriket Campuran Bottom Ash Batubara dan Onggok Tepung Tapioka Menggunakan Karbonisasi *J. Fis. FLUX* **1** 54
[16] Hannan N I R R, Shahidan S, Ali N and Maarof M Z 2017 A Comprehensive Review on the Properties of Coal Bottom Ash in Concrete as Sound Absorption Material *MATEC Web Conf.* **103** 1–10
[17] James A K, Thring R W, Helle S and Ghuman H S 2012 Ash management review-applications of biomass bottom ash *Energies* **5** 3856–73
[18] Marganingrum D, Estiaty L M, Irawan C and Hidawati 2020 The biomass coal fermented (Bcf)

briquette as an alternative fuel *MSCEIS 2019 Conf. Proceeding* **1** 811–9
[19] Rudianto I S 2021 Characteristics of Particulate Emissions from Co-Firing in An Industrial Boiler *Ecolab* **15** 23–9
[20] Bestari L R, Hidayat A and Yani M 2014 The Estimation of Diesel Vehicle's Tax Value in Relation to Air Pollution Loss (Case Study: Metro Mini in DKI Jakarta) *J. Agric. , Resour. , Environ. Econ.* **2** 98–111
[21] Liu F, Zhang Q, Tong D, Zheng B, Li M, Huo H and He K B 2015 High-resolution inventory of technologies, activities, and emissions of coal-fired power plants in China from 1990 to 2010 *Atmos. Chem. Phys.* **15** 13299–317
[22] Jiang X, Lu W X, Zhao H Q, Yang Q C and Yang Z P 2014 Potential ecological risk assessment and prediction of soil heavy-metal pollution around coal gangue dump *Nat. Hazards Earth Syst. Sci.* **14** 1599–610
[23] Yahya A A, Ali N, Mohd Kamal N L, Shahidan S, Beddu S, Nuruddin M F and Shafiq N 2017 Reducing Heavy Metal Element from Coal Bottom Ash by Using Citric Acid Leaching Treatment *MATEC Web Conf.* **103** 0–6
[24] Untung S R 2011 Toxicology Test on Coal Ash From Asam-Asam Coal Fired Power Plant , Tanah Laut-South Kalimantan *Indones. Min. J.* **14** 38–42

IOP Conf. Series: Earth and Environmental Science **986** (2022) 012002 doi:10.1088/1755-1315/986/1/012002

Assessment of ceramic based photocatalytic as indoor air purifier during covid-19 pandemic

H Prasetya[1*], L Agustina[2], A Rinovian[1] and F D Muttaqin[1]

[1] Laboratory for Ceramic Based Creative Industry, Agency for Assessment and Application of Technology. Jl. Bypass Ngurah Rai, Suwung Kauh, Tanah Kilap, Denpasar Selatan, Kota Denpasar, Bali

[2] Department of Agroindustrial Technology, Lambung Mangkurat University Jl. A. Yani Km 35.5, Banjarbaru, Kalimantan Selatan

*Email: hermawan.prasetya@bppt.go.id

Abstract. In reducing number people who exposed covid-19, Government of Indonesia (GOI) have been issued micro-scale people activities limitation policy. Based on the policy, most people activities should be carried out in indoor area, called Work from Home (WFH). Indoor activities are at risk of expose to air pollution if there is a pollution source and unhealthy air circulation. Air purifier need develop to degrade pollution and inhibit microorganism. One of the promising technologies in air purification is Photocatalytic. Photocatalytic is an eco-friendly technique that emerged as a promising alternative for the degradation both organic and non-organic pollutants. Ceramic material (compared to aluminum and stainless steel) composited with Titanium dioxide (TiO_2) has been investigated to have potential as an optimal photocatalytic medium. Aim of this study was to explore ability of ceramic based photocatalytic in degrade non-organic air pollutant. Earthenware ceramic was coated with photocatalyst material and examined their photocatalytic activation. A simple photo-reactor was established, to investigated performance of ceramic based photocatalyst air quality improvement which was contaminated with cigarette smoke. Result of the study was earthenware ceramic based photocatalyst able to improve air quality by reducing concentration of pollutants both physical and chemical parameters significantly.

1. Introduction

Work from Home scheme have been adopting both by governmental and private offices during Covid-19 pandemic in decreasing number people who infected by the virus [1]. Recently, most people activities conducted in room (indoor). The indoor activities have been affecting potentially to health sector [2], because indoor activities are at risk of expose to air pollution if there is a pollution source and unhealthy air circulation [3]. Air quality is strongly influenced by the size and type of existing pollutant sources such as human activities [4]. Indoor air pollutants can be emitted by various products used or activities conducted in house such as floor cleaning products, carpet cleaners, cosmetic ingredients, paint, cigarette smoke, and kitchen smoke [3].

Several technologies have been applied to reduce indoor air pollution. One of potential technology to be applied was photocatalyst [3, 4]. This technology has been widely developed since the last few years for air polution reducing. Also, Photocatalytic technology is also applied in waste treatment, water purification, self-cleaning, anti-fog, and bacterial disinfection [4].

IOP Conf. Series: Earth and Environmental Science **986** (2022) 012002 doi:10.1088/1755-1315/986/1/012002

Photocatalytic was defined as chemical transformation in the presence of catalyst and light [5]. The photoatalytic process occurs under photon conditions that is equal to or more than the amount of band gap energy (*hv*) of catalyst material, and then allowing electrons (e^-) to be excited from h^+.This process produce electron-hole pair [5, 6]. Reduction and oxidation reaction of electron-hole pair produce two radical agents, i.e: radical hydroxyl (*OH) and radical superoxide ($*O_2^-$) [5, 7, 8]. The agents will react to harmfull chemical and organic matter and convert them into harmless compounds.

Ceramic was applied potentially as medium for photocalyst. Ceramic material (compared to aluminum and stainless steel) composited with Titanium dioxide (TiO_2) photocatalyst has been investigated to have potential as an optimal photocatalytic medium. Texture of ceramic able to bind more photocatalyst composite material [6]. In other case, earthenware ceramic modified and coated by Titanium dioxide (TiO_2), can be used as photocatalic reactor for water purification [9]. Also, ceramic filter coated by photocatalyst material, both nano TiO_2 [10] and nano ZnO [11], removed effectivelly Escherichia coli from water. These researchs inspired us to develop decorative ceramics as air purifiers, therefore the products have dual functions, both aesthetic and health functions.

This research conducted to explore ability of ceramic based photocatalytic in degrade non-organic air pollutant. Earthenware ceramic was coated with photocatalyst material and examined their photocatalytic activation. A simple photo-reactor was established, to investigated performance of ceramic based photocatalyst air quality improvement which was contaminated with cigarette smoke.

2. Methods

2.1. Preparing Ceramic as Photocatalyst Medium

Ready-to-use earthenware from Darmasaba Sub District, Bali Province, pressed into a square shape with a size of 8x8x0.5 cm. Each shape is punched into more than 10 holes. The next step, the earthenware is burned using an electric furnace at a temperature of 900 °C. The burnt shrinkage of the earthenware was 15 %.

Figure 1. Photocatalyst Medium from Ceramic

2.2. Formulation of Photocatalyst Composit

There are several photocatalyst materials which be applied as waste degradation potentially, i.e: Titanium dioxide (TiO_2), Tungsten trioxide (WO_3) and Zinc dioxide (ZnO_2) [12]. TiO_2 nanoparticle Degusa P25 combined with activated carbon was applied to degrade cigarete smoke pollutan [13]. Formulation of photocatalyst composit cotain of:

- TiO_2 nanoparticle Degusa P25 as photocatalyst material,
- Activated carbon as air absorbance,
- Tetra ethyl ortho silicate (TEOS) as SiO_2 source and binder,
- Aquades (mineral free water) as solvent.

TiO_2 was dissolved into mineral-free water and sonicated for 10 minutes. Activated carbon was added into the solution and sonicated for 20 minutes. The solution coated by photocatalyst medium and burned into electric furnace at temperature 150 °C for 1 hour.

Figure 2. TiO_2 Solutin Sonication

2.3. Photocatalytic Activation Testing
Photocatalityc activation tested by synthetic dye. Most research use rhodamine B (RhB), methylene blue (MB), and methylene orange (MO) as an indicator for photocatalytic process [14]. This study was used MB ($C_{16}H_{18}N_3SCL$) to test photocatalytic activation. 1 ml of MB (concentration of 1000 ppm) was diluted with 1000 ml of distilled water. The solution was poured into glass box, stirred by magnetic stirrer, and irradiated with UV lamp for 2 hours. Every 30 minutes, a sample of the solution was gained to measure its absorbance by using spectrophotometer UV-VIS.

Absorbance measurement was conducted at certain wavelength (notation with Lambda). The optimum wavelength of light is found by measuring the absorbance value of the solution in a certain wavelength range. The wavelength for the absorption measurement of the MB 1 ppm solution used in this research was 300-700 nm.

2.4. Polutan Degradation Testing
Government Republic of Indonesia have been issued regulation on Air Sanitation Guideline at Room in Home [15]. Based on the regulation, there were three requirements of air quality in home, i.e.: physical, chemical, and biological qualities. Physical quality parameters are Particulate Matter (PM) both $PM_{2.5}$ dan PM_{10}, air temperature, humidity, lighting, and air circulation. Chemical quality parameters contain of Sulfur dioxide (SO_2), Nitrogen dioxide (NO_2), Carbon monoxide (CO), Carbon dioxide (CO_2), Plumbum, cigarette smoke (Environmental Tobacco Smoke/ ETS), Asbestos, Formaldehyde (CH_2O), and Volatile Organic Compound (VOC). Biological quality paremeters are bacteria and fungi. This study did not test all quality parameters, but only selected parameters. Particulate Matter both $PM_{1.0}$, $PM_{2.5}$ and PM_{10} to represent physical qualitay parameters, as well as Carbon monoxide (CO), Formaldehyde (CH_2O), and Volatile Organic Compound (VOC) to represent chemical quality parameters.

A simple photoreactor was established, to investigated performance of ceramic based photocatalyst air quality improvement which was contaminated with cigarette smoke. Photoreactor was made using multiplex because the materials are strong, inexpensive, weightless, easy to obtain and shape. Then the multiplex was coated with white duplex to make it reflect light well.

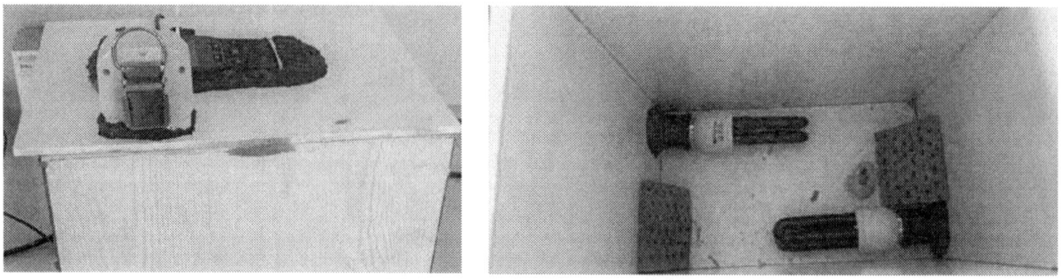

Figure 3. Simple Photoreactor for Pollutan Degradation Testing

3. Result And Analysis

3.1. Photocatalytic Activation

Photocatalyst activity was tested using a UV-Vis spectrophotometer, which is to:

- Lambda maximum measurement (is the wavelength where the maximum absorbance value in the test solution) for each type of test solution,
- Measurement of photodegradation activity for each sample variation, which is measured at the max lambda value.

To get the maximum value, used wavelengths of 400-700 nm for the measurement of absorbance values, with an interval of 10. Based on the measurement using the lambda value, the absorption value was high in the Lambda value 650–670 nm. This range of lambda (650-670 nm; interval value 2) is used to get a precise value. The results of measurements with a narrower lambda range obtained the highest absorbance value, which was 664 nm (Fig. 4). Similar study used lambda value 663 nm [16].

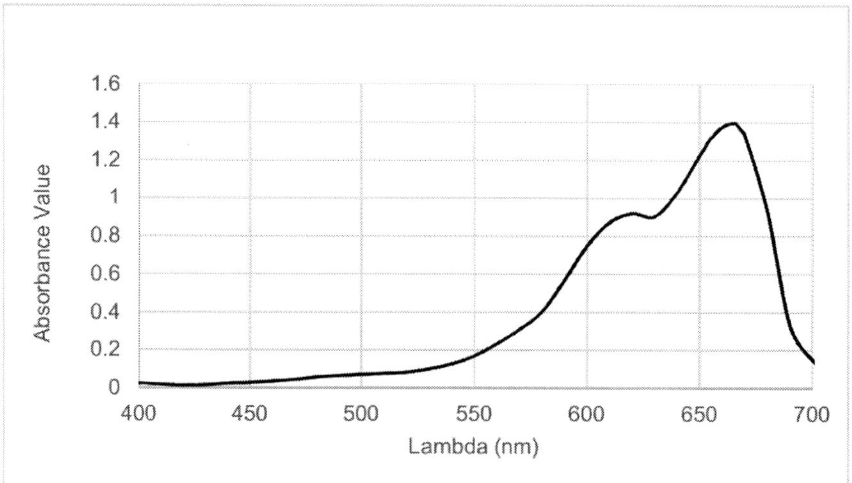

Figure 4. Measurement of MB absorbance value

The experiment was continued with the activation of photocatalyst, measurement of photocatalyst activation is carried out using a UV-Vis spectrophotometer at a Lambda value of 664 nm. These measurements were carried out on a photoreactor containing a solution of methylene blue with a ceramic membrane that had been coated with TiO_2 using two light sources, (1) sunlight, and (2) UV lamp. The 5 ppm dye solution was poured into the photoreactor then given 2 pieces of coated ceramic membrane that had been coated with TiO_2. The irradiation was carried out for 2 hours, starting from 10.30 am to

12.30 pm. Light source for photocatalysis has been envolved from sunlight to the use of UV lamp [17].The use of UV lamps as an alternative light source because of its ability to produce a large and stable photon flux. The UV lamp that is often used is UV-A with a maximum wavelength between 355-365 nm. Absorbance measurements are presented in Figure 5.

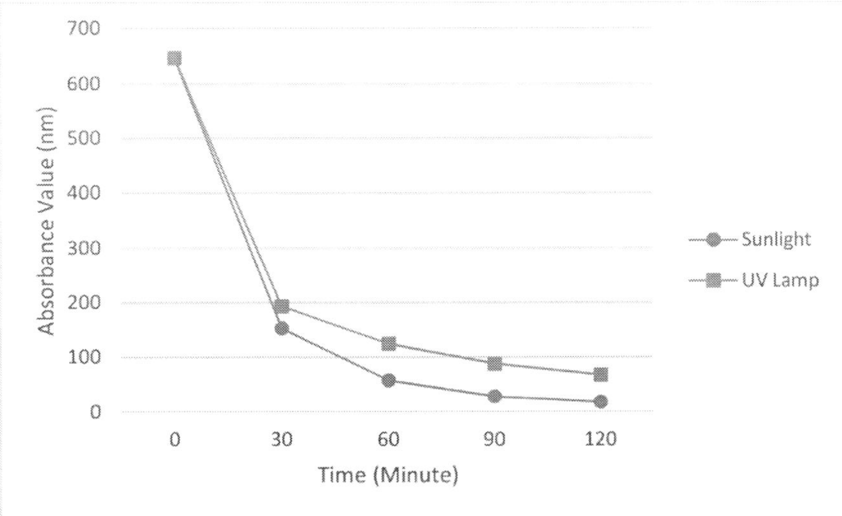

Figure 5. Comparison of photoreactor absorbance values by two light sources

Figure 5 shows that both with sunlight and UV lamps, ceramics that have been coated with TiO_2 have been able to degrade MB. This shows that photocatalysis has occurred using either sunlight or UV lamps. Photocatalytic process that takes place in photoreactors irradiated with sunlight degrades MB faster than photoreactors with UV lamps. The use of a catalyst in the photocatalytic process will increase the ability of biodegradability. TiO_2 as a photocatalyst able to absorb light (photon energy) both produced by sunlight and UV lamps to produce electrons (e^-) and holes (h^+) that will carry out chemical transformations through reduction and oxidation processes. The resulting electrons are able to reduce the dye (MB), that is diffused by an electron acceptor (e.g. O_2) on the surface to form O_2^-. While holes will oxidize organic molecules and react with OH^- or H_2O to produce $OH.O_2^-$ and OH so that it can degrade almost all dyes into mineral as end products [18].

Furthermore, to determine the MB concentration in the solution, absorbance measurements were carried out at various concentrations which would be used to make equations based on the measurement results. The measurement steps are (1) making a solution of MB 1-5 ppm and aquades as a blanko solution, (2) measuring the absorbance value for each concentration of MB using a UV-Vis Spectrophotometer, and (3) create a formula by choosing the regression formula that has the largest R^2 value. The results of the measurement and formulation of the formula are presented in Figure 6.

Ppm	Absorbance
0	0
1	0.082
2	0.183
3	0.31
4	0.452
5	0.647

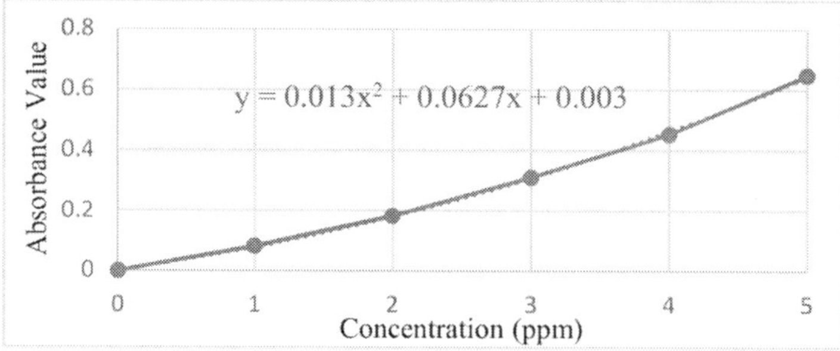

Figure 6. Measurement and result of concentration formula with absorbance value

3.2. Photocatalytic Performance

After obtaining the results of the photocatalyst activation using MB solution which showed a positive response, further testing was carried out on the ability of photocatalyst ceramics in degrading pollutants in the air. The test was carried out using a sample of cigarette smoke. Cigarette smoke was fed into the photoreactor in which there are two ceramic test objects that have been coated with TiO_2 photocatalyst. The air parameters measured were (1) pollutant particles consisting of PM_1, $PM_{2.5}$ and PM_{10}, (2) Formaldehyde (CH_2O), (3) volatile organic compounds (VOC), and (4) carbon monoxide (CO). The measurements were carried out for one hour.

3.2.1. Degradation of Physical Parameters

Reduction of concentration of particulate matters over time indicated rate of pollutant degradation. Measurement of the concentration was carried out every 10 minutes for one hour. For one-hour, particulate matter concentrations decreased significantly. Percentage of reduction of $PM_{1.0}$, $PM_{2.5}$ and PM_{10} were 99 %, 99 % and 98 % respectively. Figure 7 show degradation patterns of each particulate matter. Size of particulate seem affected to pollutan degradation in term of starting time and percentage of degradation. $PM_{1.0}$ and $PM_{2.5}$ started to degradate after 10 minutes, but PM_{10} just started after 20 minutes.

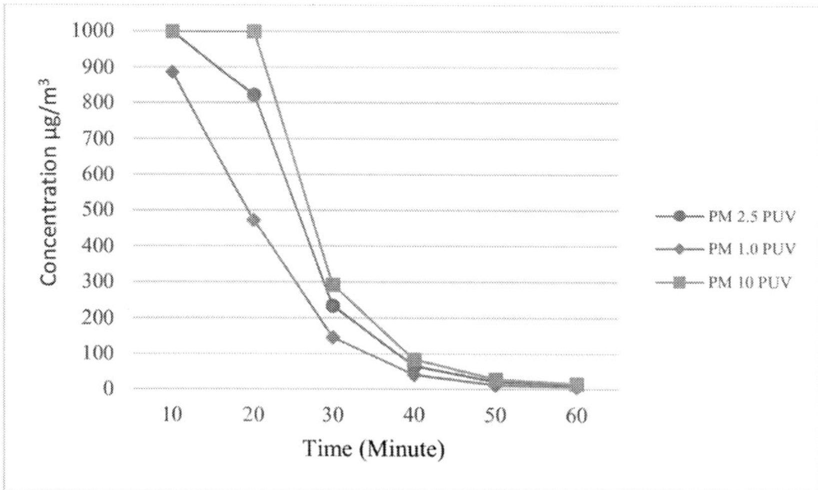

Figure 7. Physical Parameter Degradation by Photocatalyst Coated Ceramic

Figure 8 shows that in the first 10 minutes, photocatalyst-coated ceramics had the highest degradation ability against smaller pollutant particles ($PM_{2.5}$), and then the total percentage of degradation for larger particles was relatively the same in the next 10 minutes interval. These results are expected to show a positive effect if photocatalyst-coated ceramics are used in a room as an air purifier, especially in terms of its ability to degrade smaller particulates. The particulates with a diameter of <2.5μm ($PM_{2.5}$) are more damaging than particulates with a diameter of 10 μm (PM_{10}), because smaller particulates can penetrate into the lungs to the alveoli and significantly associated with increased respiratory distress [19].

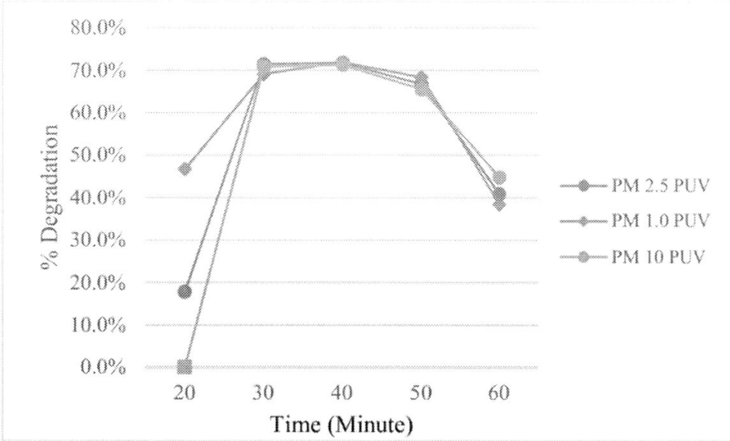

Figure 8. Degradation Pollutant Percentage

3.2.2. Degradation of Chemical Parameters

Photocatalyst coated ceramic was tested ability in degradation of three chemical parameters of air quality. These parameters were Formaldehyde (CH_2O), Volatile organic compounds (VOC), and Carbon monoxide (CO). Concentration of three chemical parameter measured every 10 minutes for one hour. Percentage of Reduction were CH_2O by 72.0 %, VOC by 92.0 % and CO by 58.4 % respectively. Ability of photocatalyst coated ceramic to chemical parameters seem less than physical parameters.

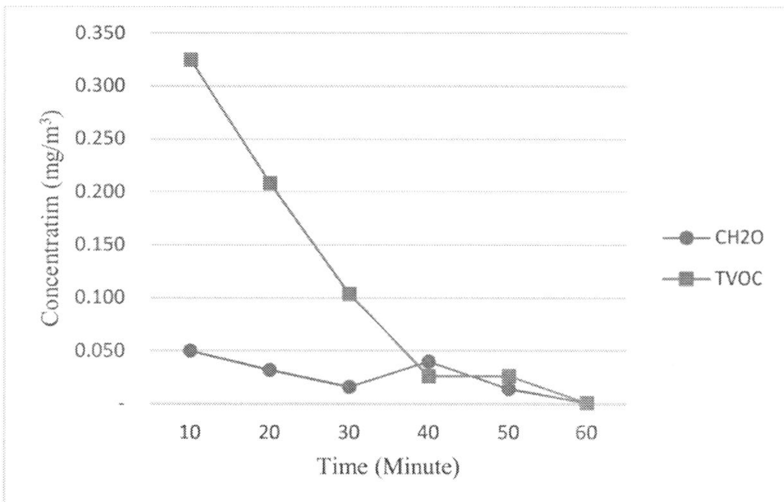

Figure 9. Concentration CH_20 and Total VOC

Figure 9 show the different degradation patterns between CH_2O and Total VOC. The slope of VOC degradation was sharper than CH_2O. The slope of CH_2O and VOC tend to change at minute 40. Figure 10 show degradation pattern of CO which tend to linear.

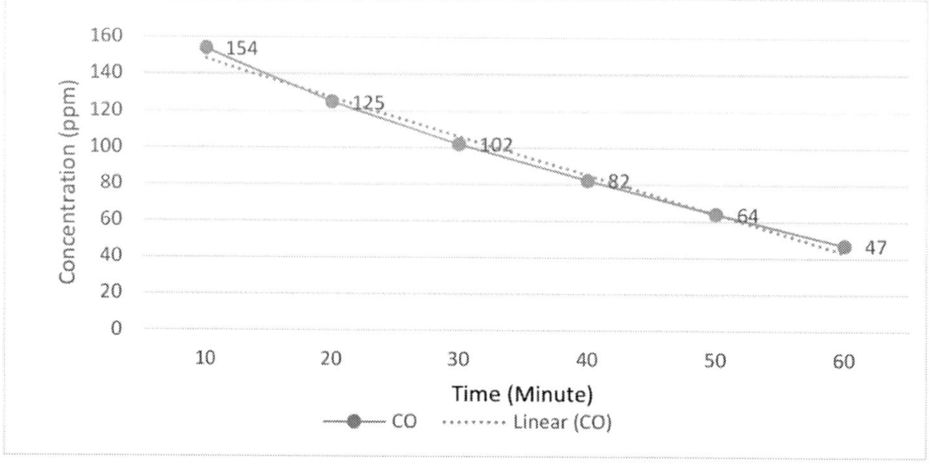

Figure 10. Concentration CO

The mechanism of degradation of TiO_2 photocatalyst through pollutant parameters in the air is basically the same as the reaction that occurs with methylene blue. The resulting electron will react with the H_2O in the air to become a radical highly reactive hydroxyl (OH^-) and then OH- will react with organic pollutant molecules in the air to break chemical bonds and convert them into simpler, harmlesser substances such as CO_2 and H_2O.

4. Conclusions

Photocatalyst coated ceramic able to degrade air pollutan both physical and chemical parameters significantly. Percentage of degradation at physical quality parameters larger than chemical quality. Percentage of reduction of $PM_{1.0}$, $PM_{2.5}$ and PM_{10} were 99 %, 99 % and 98 % respectivelly. Percentage of Reduction were CH_2O by 72.0 %, VOC by 92.0 % and CO by 58.4 % respectively.

References

[1] Mungkasa, O., *Bekerja dari Rumah (Working From Home/WFH) : Menuju Tatanan Baru Era Pandemi Covid 19.* The Indonesian Journal of Development Planning, 2020. **IV (2)** p. 126-150.

[2] Kasih, A.P., *Pakar Unair: Pandemi Covid-19 Membuat Masyarakat Cenderung Lebih Konsumtif,* in *Kompas*. 2020, Kompas Gramedia Grou[Jakarta. p. 1.

[3] Saragih, W.J., *Degradasi Polutan Udara Ruangan Menggunakan Lampu Hias Dengan Penutup Berlapis Katalis TiO2 Termodifikasi,* in *Teknik Kimia*. 2011, Universitas Indonesia: Depok. p. 62.

[4] Mudhofir, F., I. Yulianti, and Sujarwata, *Teknologi Lingkungan Penyaring Udara Sebagai Upaya* Degradasi *Polutan Asap Rokok.* Jurnal MIPA, 2018. **41 (1)**: p. 1-5.

[5] Sutanto, H. and S. Wibowo, *Semikonduktor Fotokatalis Seng Oksida dan Titania (Sintesis, Deposisi dan Aplikasi)*. 2015, Semarang: Penerbit Telescope. 154.

[6] Naimah, S., et al., *Keramik Sebagai Media Fotokatalis TiO2-Karbon Aktif Serta Aplikasinya Pada Kesehatan Lingkungan.* J Kimia Kemasan, 2015. **37 (2)** p. 123-132.

[7] Saravanan, R., F. Gracia, and A. Stephen, *Basic Principles, Mechanism, and Challenges of Photocatalysis*, in *Nanocomposites for Visible Light-induced Photocatalysis*, M.M. Khan, D. Pradhan, and Y. Sohn, Editors. 2017, Springer: Switzerland AG. p. 19-39.

[8] Shiraki, K., et al., *Improved Photocatalytic Air Cleaner with Decomposition of Aldehyde and Aerosol-Associated Influenza Virus Infectivity in Indoor Air.* Aerosol and Air Quality Research, 2017. **17**: p. 2901–2912.

[9] Ediputra, K., et al. *Modifikasi Tembikar Tanah Liat Dengan Fotokatalis TiO2 Untuk Menurunkan Kadar Polutan Organik Air Gambut.* in *Seminar Nasional Pelestarian Lingkungan* 2017. Pekanbaru: Universitas Riau.

[10] Le, T.S., et al., *Air Purification Equipment Combining A Filter Coated By Silver Nanoparticles With A Nano-TiO2 Photocatalyst For Use In Hospitals.* Adv. Nat. Sci.: Nanosci. Nanotechnol, 2015. **6** p. 1-8.

[11] Huang, J., et al., *Performance Of Ceramic Disk Filter Coated With Nano Zno For Removing Escherichia Coli From Water In Small Rural And Remote Communities Of Developing Regions.* Environmental Pollution 2018. **238**: p. 52-62.

[12] Sucahya, T.N., N. Permatasari, and A.B.D. Nandiyanto, *Review: Fotokatalis Untuk Pengolahan Limbah Cair.* Jurnal Integrasi Proses, 2016. **6 (1)** p. 1 - 15.

[13] Slamet and M. Ibadurrohman, *Purikasi Udara Dari Polutan Asap Rokok Menggunakan Kombinasi Proses Absorpsi Fotokatalis Nanokomposit TiO2 Karbon Aktif* Jurnal Riset Industri, 2009. **III (1)**: p. 1-10.

[14] Koe, W.S., et al., *An Overview Of Photocatalytic Degradation: Photocatalysts, Mechanisms, And Development Of Photocatalytic Membrane.* Environ Sci Pollut Res Int, 2020. **27(3)**: p. 2522-2565.

[15] Kesehatan, K., *Pedoman Penyehatan Udara Dalam Ruang Rumah*, in *1077/MENKES/PER/V/2011*, K.K. RI, Editor. 2011: Jakarta.

[16] Hardiana, A., et al., *Sintesis Keramik Komposit ZrO2-ZnFe2O4 sebagai Fotokatalis Magnetik untuk Degradasi Metilen Biru.* ALCHEMY Jurnal Penelitian Kimia 17(1), 2021: p. 43-53.

[17] S. Kanmani and K.P. Sundar, *Progression of Photocatalytic reactors and it's comparison: A Review.* Chemical Engineering Research and Design, 2020. **154**,: p. 135-150.

[18] Ibhadon, A.O. and P. Fitzpatrick, *Heterogeneous Photocatalysis: Recent Advances and Applications.* Catalysts 2013. **3**: p. 189-218.

[19] Kumar, R., et al., *Indoor air pollution and asthma in children at Delhi, India.* Pneumonologia Alergologia Polska, 2015. **83 (4)**: p. 275–282.

Micro household waste management and pandemic social impact relaxation; case study in DKI Jakarta

A Husen

Department of Environmental Management, Universitas Negeri Jakarta, Jalan Rawamangun Muka, Jakarta Timur, 13220, Indonesia

*Email: ach.husen59@gmail.com

Abstract. Several community-based waste management strategies in urban areas are well known. However, if examined on a micro basis, it is time to develop micro household waste management which is related to efforts to relax social impacts during the pandemic. Development research (R and D) is applied in this study. Through the Research and Development method, thisis study aims to obtain a model of household micro-waste management and social relaxation efforts due to the Corona Virus-19 pandemic in DKI Jakarta.. Until the end of the first semester of 2020, the results of the study showed that: (1) the community-based waste management model in DKI Jakarta places more emphasis on sorting and processing waste at Temporary Disposal Sites (TPS), towards the development of waste banks and waste recycling, and (2) The effectiveness of the community-based waste management model requires several strategic interventions, namely: funders (funders), direct practice, collaborators, regional organizations, and local governments.

1. Introduction

Nowadays, Floods that hit Jakarta and the surrounding areas (Bekasi, Tangerang, Depok to Bogor) in early 2020 were reported to have quite varied water levels with the peak flood water level reaching 3 meters. There are 97 flood points that occur in 5 areas of DKI Jakarta which have an impact on the decline of activities and worsening of the condition of facilities and infrastructure in Jakarta and its surroundings. Water stagnates and disrupts the operation of public facilities such as the Jabodetabek Commuter Line transportation mode, airport trains and toll roads. This incident in addition to causing damage and loss of property also caused casualties.

Flood damage is influenced by many more factors among which are flow velocity, flood duration, contamination, sediment concentration, lead time and information content of flood warning, and the quality of external response in a flood situation [1–6]. Post-flood damage and losses can actually be reduced if the factors that cause flooding can be anticipated. Some of the factors that cause flooding include natural factors and human factors where the management and utilization of natural resources and the environment are irresponsible, unprofessional, and greedy.

Human factors, one that must be taken seriously is efforts to improve people's attitudes and behavior such as still throwing garbage not in the trash, reluctance to clean drainage from garbage, and others. Flood risk reduction methods that can be applied include in the fields of flood design and flood risk prevention because it allows to evaluate the cost effectiveness of work so as to optimize investment [7–11].

Regarding the waste produced by residents, in DKI Jakarta alone, it is known that the volume of waste reaches 7,164.53 m^3 per day (only in 2018) [12]. In that year, the amount that could be transported to the final waste disposal site reached 6,872.18 m^3 per day (95.92%) and the rest (292.35 m^3 per day) was not transported. And if it is accumulated per month in a year, it will "pile up" 8,770.5 m^3 of untransported waste per day. This number is not only considered a large number but also has the potential to cause water, air, and soil pollution; increase greenhouse gases, floods, and become a source of disease and other problems.

In fact, efforts to handle waste have been carried out, but it seems that the initiation and implementation are more dominated by the government, in addition to technically it is also necessary to evaluate and find effective and efficient solutions. The old waste management system, which includes: collection, transportation, and disposal, is considered to have not been able to handle the waste problem completely. Not to mention the high cost of transporting waste (ritation), limited land for final disposal (TPA), and environmental feasibility, so the right pattern or model of waste management is a necessity. With a survey approach, the model developed through research and development efforts is expected to be in accordance with the needs for community-based waste management. A detailed description of the survey is given by *Kreibich et al.* [6] the research is aimed at obtaining a model of household micro-waste management and social relaxation efforts due to the Corona Virus-19 pandemic in DKI Jakarta.

2. Literature Review

Solid waste management (SWM) involves the collection, storage, transportation, processing, treatment, recycling and final disposal of waste [13]. Systems need to be simple, affordable, and sustainable (financially, environmentally and socially) and should be equitable, providing collection services to poor as well as wealthy households. SWM therefore, requires adequate infrastructure provision and maintenance for all activities

A review by Marine Colona and Ben Fawcett [14] explains that the efforts of community-based organizations to promote sustainable integrated waste management systems in India's major cities by local non-governmental organizations (NGOs) and based on a 'zero waste management scheme' demonstrate the success of the scheme. which are limited both in saving most of the waste generated from disposal, and in rehabilitating the local poor. However, they show that motivated individuals can successfully set up and manage waste collection systems leading to overall environmental improvements.

Environmental barriers across Europe from inefficient and unsustainable recycling systems and flows limit household waste management and jeopardize the circular economy [15]. Through an artificial intelligence (AI) approach to calculate the remaining urban waste. And by adopting the k-means clustering method with a multi-criteria taxonomic development level index (TIDL) and producing a broad impact in Europe, especially on sustainable economic value chains, leading to an environmentally friendly economy.

3. Research Methodology

With a survey approach, Kreibich et al. [6], the research uses research and development methods. This study aims to obtain a model of household micro-waste management and social relaxation efforts due to the Corona Virus-19 pandemic in DKI Jakarta.

Data collection techniques are documentation, observation and interviews through instruments containing a list of questions that refer to the concept of waste management in the community. The substance of the measurable data obtained through interviews [16,17] is information about waste management in each household used as research respondents as well as respondents' opinions regarding the efforts made to reduce pressure (social stress) during government policy on social restrictions (restrictions on interaction) due to the Corona Virus 19 pandemic. The data were analyzed descriptively quantitatively where a concept maturation was needed between confirmatory and exploratory data analysis [18].

4. Findings and Discussion

4.1. Waste Sorting Stage

The research sample was obtained at random in stages. First, it is determined the level of the region to be selected. Based on a random method, 3 (three) regions were selected, namely: East Jakarta, South Jakarta, and North Jakarta. The district level is then determined. According to a simple random method, in East Jakarta, Duren Sawit District was selected. Then in the same way, from Duren Sawit Sub district, Duren Sawit Village was chosen, and finally RW 06 was chosen. In this area, 12 (twelve) people were chosen, from Rt. 001, Rt. 004, and Rt. 010. The South Jakarta area, according to a simple random method, was selected as Setiabudi District. Then in the same way, from Setiabudi Sub district, Pasar Manggis Village was chosen, and finally RW 09 was chosen. From this area, 9 (nine) people were chosen, who came from Rt. 009, Rt. 012, and Rt. 013. In the same way, in the North Jakarta area, Pademangan Sub-district was chosen, then West Pademangan Village, and finally RW 13. From this area, 9 (nine) people were selected, who came from Rt. 009, Rt. 010, and Rt. 011.

At this stage, from 30 data sources, as many as 25 people (83.33%) stated that they did not do waste sorting, and only 5 (five) people (16.67%) answered that they did waste sorting, and the results of the sorting they put them in different place. Among the 5 (five) people, 3 (three) people throw their trash in different places, while 2 (two) other people throw it in the same place.

After sorting the waste, the next activity is disposing of waste (collection). Regarding the place where the household waste is disposed of, 24 people (80%) said they had their own garbage collection center, while 6 people (20%) said they did not have one. In line with that, the 24 people answered that they disposed of their garbage in their own shelter. As for the 6 people who do not have a garbage collection place, there are those who throw their garbage in a neighbor's shelter (1 person or 16.67%), to a garbage cart (2 people or 33.33%), to a Temporary Disposal Site (2 people or 33, 33%), and 1 other person (16.67%) to a landfill in the market.

Waste transportation is carried out in 2 (two) stages, namely first, from the collection point in the household to the Temporary Disposal Site (TPS). Second, from TPS to Final Disposal Site (TPA). In the first stage of transportation, 20 people (66.67%) answered that the waste collected in household shelters was disposed of to the TPS, 4 people (13.33%) said to the TPA, 3 (10%) people said to the garbage cart, the rest of which 3 people (10%) answered they did not know. When asked whether there was a TPS in their neighborhood, 24 people (80%) said there was, another 6 people (20%) answered no. After being observed in the field to confirm the answers of these 6 people, it turned out that it was true that in their environment there was a Temporary Disposal Site (garbage). Thus it was understandable that they did not seem to know. Who transports waste from household shelters to TPS, all sources (100%) mention cleaners. What facilities are used to transport waste from household shelters to the TPS, all of them (100%) answered that they were transported by garbage carts.

Regarding the knowledge of the community members where the TPS is located, 29 people (96.67%) answered they knew, and only 1 person (3.33%) said they did not know. Exploring further about the status of land used for Disposal (garbage) Meanwhile, 14 people (46.67%) answered state land, 10 people (33.33%) said it was individual land, the remaining 6 people (20%) said no know. Equivalent to that answer, when asked whether the use of the land obtained permission from the related party, 24 people (80%) answered that they had obtained permission from the interested party, the remaining 6 people (20%) stated that they did not know.

4.2. Waste Management

At this stage, of the 3 (three) locations that were the target of observation, 2 (two) locations, namely Duren Sawit and Pasar Kemis, did not have further waste processing. This means that after the waste arrives at the TPS, the collected waste is immediately transported and disposed of to the Final Disposal Site (TPA). This is in accordance with data obtained from community members in the two locations, which show that 100% (21 people) of sources said there was no waste processing. Only 1 (one) location, namely in West Pademangan, the waste collected at the TPS is further processed as seen from the answers from sources at that location (9 people or 100%). Based on the answers from the 9 people (100%) it is known that the waste processing carried out there is a waste bank and recycling. They said that the mechanisms implemented include: 1) waste is sorted by type, 2) organic waste is disposed of at the TPA, while cardboard/paper, and plastic and cans are recycled. From this information, it is known that the waste bank processes and recycles waste (reuse of waste) especially cardboard/paper, plastic waste, and cans). It was also known from them that the processing and recycling of waste there is carried out by cleaning staff, assisted by youth organizations, under the coordination of RW and PKK administrators.

How much is the benefit of the waste processing for the residents, from 9 people at the observation location who process waste; 6 people (66.67%) answered that they had benefited, 1 person (11.11%) said they had not, and the remaining 2 people (22.22%) said no. Meanwhile, from the observation locations where there was no waste processing (21 sources), when asked why in their environment they did not carry out waste processing, 10 people (47.62%) answered they did not know, 3 people (14.29%) said there was no initiative, 2 people (9.52%) said they did not have skilled workers, 1 person (4.76%) stated that the land was constrained, while 5 other people (23.81%) did not answer. It is further examined whether the person in charge of the area (RW administrator) has a plan to process waste in the future, only 5 people (23.81%) answered (knew) there was a plan, while 16 people (76.19%) said do not know. When asked what their attitude would be if the RW administrators run a waste bank and recycle the collected waste, all of them (100%) answered in favor.

Waste management carried out by community members will be relatively effective if they get support from the government, both at the kelurahan level, as well as at a higher level. At the research location, the community's waste management is sufficiently supported by the government. This is reflected in the answers from sources (30 people), 20 people (66.67%) said they received support, 9 people (30%) said they did not know, 1 other person (3.33%) did not answer. This support was manifested in: providing transportation facilities (9 people or 30%), 6 people (20%) said providing land, 5 people (16.67%) mentioning providing carts and trash cans, 9 people (30%) answering no know, while 1 person (3.33%) did not answer.

In waste management, in addition to efforts to reuse (either through waste banks or recycling or reuse and recycle), which is no less important and needs to be developed is the effort to reduce waste generation. Related to this, 18 people (60%) said they are trying to reduce waste, 11 people (36.67%) answered no, while 1 person (3.33%) did not answer. Of the 18 people who tried to reduce waste, 12 people (66.67%) did it by reducing the use of plastic, 3 people (16.67%) brought their own shopping bags/bags, 2 people (11.11%) by buying products without waste , and 1 person (5.55%) did not answer. When asked about the use of recycled products, only 6 (six) residents (20%) answered that they used recycled products, while the remaining 24 people (80%) said they had not. How do they treat unused items such as cardboard, newspapers, plastic bottles and so on, 15 people (50%) answered that they were collected and then sold, 9 people (30%) said they were given to scavengers, while 6 people (20%) stated thrown away.

Actually, participation in waste management takes many forms, but what is commonly done and generally understood by the community is paying a monthly fee. This is reflected in the source's answer. Of the 30 data sources, 29 people (96.67%) said there was a fee, and only 1 person (3.33%) answered that they did not know. The amount of the imposition of fees or monthly fees varies greatly, ranging from 5 (five) thousand rupiahs to 100 thousand rupiahs. People who can't afford it are not even charged a fee. Perhaps because the awareness of the community is quite high or because the pattern of determining the imposition of fees is democratic and contains the principle of cross subsidies, almost all data sources (28 people or 93.33%) say they have no objection to the imposition of fees. Only 2 people (6.67%) answered the objection.

People's expectations of waste management in the future, the following data illustrates the desire: (1) better management (spoken by 9 people/30%); (2) improved facilities and infrastructure (8 people/26.67%); (3) permanent land for waste management in RT/RW (6 people/20%); (4) skills for garbage officers are improved in carrying out their duties every day (3 people/10%); (5) community participation is increasing (3 people/10%), the rest (1 person/3.33%) did not answer.

4.3. The Model

The image below, shows that that the objective conditions of waste management in the research location show: 1) most of the people have not sorted waste; 2) residents have their own garbage collection points; 3) in the neighborhood of RT / RW there is a temporary garbage disposal site; 4) the garbage is moved by the janitor from the resident's shelter to the temporary waste disposal site (TPS). At the temporary disposal site (TPS), waste that is not processed further (there are no more waste processing activities at the TPS).

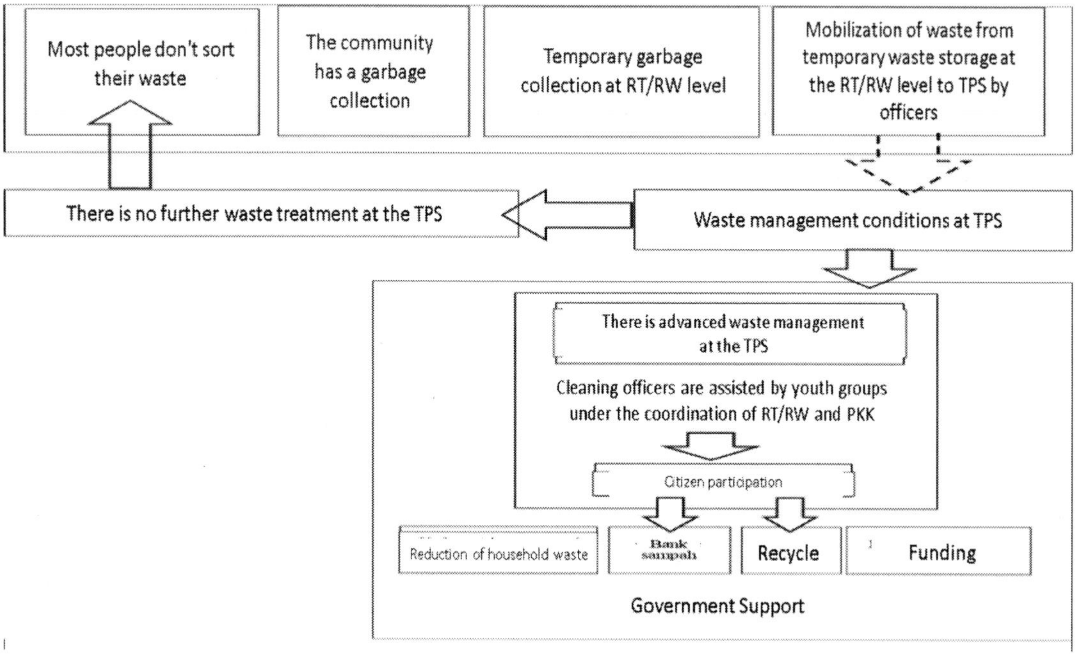

Figure 1. Household Micro Waste Management Model as a Relaxation For the Social Impact of the Covid-19 Pandemic

The advantages of the above model include the fairly good effectiveness of community-based waste management; and therefore several strategic interventions are needed, namely funders (funders), direct practice, collaborators, local organizations, and local governments.

At the research site, of the three areas studied, one of them has a fairly good waste management at the TPS. This is known from the existence of: 1) the existence of waste processing activities at the TPS; 2) carried out by cleaning staff, assisted by Karang Taruna under the coordination of RW and PKK administrators; 3) the participation of citizens through the following activities: a) efforts to reduce household waste, b) waste bank management, c) waste recycling, and d) citizen participation in waste management fees/retributions.

Therefore, what is worth considering as an effort to utilize waste both economically and ecologically is the development of a waste bank and recycling in temporary disposal sites (TPS). The existence of waste processing activities at the temporary waste disposal site will result in the community getting benefits both directly (in the form of consumer goods as a result of exchange at the waste bank) and indirectly (in the form of a cleaner, more beautiful, and comfortable environment). As a return impact, the community will be more participatory and contributive in supporting independent waste management in their environment.

5. Conclusion and Further Research

Based on the previous discussion, it can be concluded that: (1) the community-based waste management model in DKI Jakarta places more emphasis on sorting and processing waste at Temporary Disposal Sites (TPS), towards the development of waste banks and waste recycling, and (2) The effectiveness of the community-based waste management model requires several strategic interventions, namely: funders (funders), direct practice, collaborators, regional organizations, and local governments.

In the future, the concrete step of the research is a practical guide book for community-based waste management, so that in the implementation of direct practice, managers, especially community members, receive directed explanations about the stages of micro-waste management in the community.

Acknowledges
This research was funded by BLU funds from the Pascasarjana Universitas Nageri Jakarta in the 2020 fiscal year.

References

[1] McBean E A, Gorrie J, Fortin M, Ding J and Monlton R 1988 Adjustment Factors for Flood Damage Curves *J. Water Resour. Plan. Manag.* **114** 635–46
[2] Smith D I 1994 Flood damage estimation - a review of urban stage-damage curves and loss functions *Water SA* **20** 231–8
[3] Wind H G, Nierop T M, De Blois C J and De Kok J L 1999 Analysis of flood damages from the 1993 and 1995 Meuse floods *Water Resour. Res.* **35** 3459–65
[4] Penning-Rowsell E C and Green C 2000 New insights into the appraisal of flood-alleviation benefits: (1) Flood damage and flood loss information *Water Environ. J.* **14** 347–53
[5] International Commission for the Protection of the Rhine (ICPR) 2002 *Non Structural Flood Plain Management Measures and their Effectiveness* (Germany)
[6] Kreibich H, Thieken A H, Petrow T, Müller M and Merz B 2005 Flood loss reduction of private households due to building precautionary measures - Lessons learned from the Elbe flood in August 2002 *Nat. Hazards Earth Syst. Sci.* **5** 117–26
[7] U.S. Army Corps of Engineers (USACE) 1996 *Risk-based analysis for flood damage reduction studies* (Washington, D. C)
[8] Resendiz- Carrillo D and Lave L B 1990 Evaluating dam safety retrofits with uncertain benefits: The Case of Mohawk Dam (Walhonding River, Ohio) *Water Resour. Res.* **26** 1093–8
[9] Olsen J R, Beling P A, Lambert J H and Haimes Y Y 1998 Input-Output Economic Evaluation of System of Levees *J. Water Resour. Plan. Manag.* **124** 237–45
[10] Al-Futuaisi A and Stedinger J R 1999 HYDROLOGIC AND ECONOMIC UNCERTAINTIES AND FLOOD-RISK PROJECT DESIGN *J. WATER Resour. Plan. Manag.* 314–24
[11] Ganoulis J 2003 Risk-based floodplain management: A case study from Greece *Int. J. River Basin Manag.* **1** 41–7
[12] Central Bureau of Statistics 2018 Indonesian Environmental Statistics, Waste Management in Indonesia
[13] Rouse J 2002 planning for sustainable municipal solid waste management: Practical Action; the Schumacher Centre for Technology and Development Bourton-on-Dunsmore Rugby, Warwickshire, United Kingdom
[14] Colon M and Fawcett B 2006 Community-based household waste management: Lessons learnt from EXNORA's "zero waste management" scheme in two South Indian cities *Habitat Int.* **30** 916–31
[15] Papagiannis F, PatriziaGazzola and OlenaBurakcIlyaPokutsa 2021 A European household waste management approach: Intelligently clean Ukraine *J. Environ. Manage.* **294**
[16] J G and N T 2014 *Qualitative methods for health research (3rd Ed.)* (SAGE publications)
[17] M O and N K 2015 *Advanced qualitative research: A guide to using theory* (SAGE publications)
[18] Thompson C B and Panacek E A 2006 Clinical research and critical care transport: How to get started *Air Med. J.* **25** 107–11

IOP Conf. Series: Earth and Environmental Science **986** (2022) 012004 doi:10.1088/1755-1315/986/1/012004

Air Quality Classification in Urban Environment using Machine Learning Approach

Faqih Hamami[1,*] and Iqbal Ahmad Dahlan[2]

[1]School of Industrial and System Engineering, Telkom University Bandung
[2]Military Engineering Faculty, Indonesia Defense University Bogor

[*]Email: faqihhamami@telkomuniversity.ac.id

Abstract. Air pollution comes from human activities that can threaten living things. It is affected by gasses including PM_{10}, SO_2, CO, O_3, NO_2 and others. Air pollution leads to dangerous diseases even death. Monitoring air quality is important task to understand pollution concentration. Air quality monitoring is better when it can classify whether air quality is habitable or not. This research proposes air quality classification using classification algorithms such as Logistic Regression, KKN, Decision Tree, and Random Forest algorithm. Dataset was taken from Jakarta's open data for 12 months with several attributes including gas concentration and with several pre-processing steps. Based on experiment, decision tree model has the best accuracy to classify air quality level up to 100% with tuning several hyperparameters.

1. Introduction

Urban areas have a high population density with high activity. Urban population increased billions in decades. According to PBB, half of the earth's population live in cities and likely to increase by three billion by 2050. Based on BPS data, Indonesia is one of the countries with the fastest urbanization. In 2020, 56.7% of the Indonesian population live in urban areas and predicted will increase to 66.6% in 2035. Population growth in urban area give both positive and negative effects. Urbanization improve the economy. People tend to migrate to bigger cities to find better jobs and schools. Jakarta as the nation's capital become a magnet for many people from all over regions to urbanize. In 2019, millions of commuters from the Jabodetabek area were primarily connected to DKI Jakarta. 83% of these commuters are workers, while the remaining are students.

On the negative side, besides causing high population density, urbanization also causes congestion, pollution, slums and crime. Congestion occurs in big cities because many people choose to use private vehicles. It is estimated that due to traffic jams, Jakarta suffers an economic loss of 65 trillion per year. Besides Jakarta, based on data from the Inrix Global Traffic Scorecard, dozens of cities in Indonesia occupy the city with the highest traffic congestion in the world.

Pollution is also a big problem for urban areas. The high use of motorized vehicles results in uncontrolled pollution. Industrial waste also adds to air pollution level. Based on the AQ Index website, Indonesia ranks in the top 10 cities with poor air quality index. According to the Air Quality Live Index, Jakarta has a $PM_{2.5}$ concentration six times higher than the WHO safe limit. If this condition continues, the air quality will become dangerous for human health.

Air monitoring is needed to maintain safe and healthy air quality. Some harmful gases such as Particulate Matter (PM), sulfur dioxide (SO_2), carbon monoxide (CO), ozone (O_3) and carbon dioxide (CO_2) are parameters that use to determine air quality level. This research proposes a machine learning approach to classify air quality level in urban areas. Hopefully air quality can be predicted well and

IOP Conf. Series: Earth and Environmental Science **986** (2022) 012004 doi:10.1088/1755-1315/986/1/012004

become a reference for the government to take policies regarding the use of motorized vehicles and industrial waste.

Nowadays, some cities implemented some innovation to address their city's problem. [1] proposed Garuda Smart city Framework 2.0 as indicator to calculate economy, social and environment indicators. It also has an enabler indicator such as technology/ infrastructure, people and governance. This paper proposes smart environment innovation by using machine learning technology to classify the indicators from the data.

2. Methods

Since the industrial revolution in the late eighteenth century, the world has faced serious problem regarding environmental issues such as pollution. Many papers propose the method to monitor this issue with machine learning or deep learning approaches. [2] addressed this air quality with ML/DL to monitor air pollutants that containing nitrogen dioxide (NO_2), carbon monoxide (CO), ozone (O_3), sulphur dioxide (SO_2), and particulate matter (PM). This paper proposed LightGBM model to forecast the $PM_{2.5}$ concentration with the RMSE around at 1.0. and MAE evaluation indicator score dropped by about 0.25.

[3] implemented the comparison between various ML algorithms to predict classification of air quality index in city. The results show that the three best methods are random forest, Gradient Boosting, and K-Nearest Neighbours with precision, recall, and f1-score values more than 0.63. Another research is implemented by using deep learning with LSTM algorithm [4]. This paper developed LSTM network for air pollution concentrations forecasting and reveal that RMSE around at 5.58%.

Air quality monitoring is also can be monitored by using mobile microscopy and machine learning. This paper [5] used a remote data collection from a smartphone and then analyzed using machine learning to calculate the particle concentration in the air and finally classify the pollution level from the data. This paper also created a platform c-air for device control and display of results. This system provides accuracy around at 93% but it does not clearly illustrate how machine learning can reveal that accuracy only illustrate the platform, architecture and the graphic of the results. In air quality level classification, [6] proposed air quality classification using decision tree algorithm. [7] proposes to classify air quality using Naive Bayes algorithm while [8] proposes air quality using artificial neural network approach.

This research intends to classify air quality level in Jakarta with several machine learning algorithms. Some of the implemented algorithms are Logistic Regression, KKN, Decision Tree and Random Forest algorithm. Dataset is taken from Jakarta's open data with the distribution of from 5 areas of Jakarta which are, central Jakarta, north, south, west and east. The attributes from dataset are gas concentrations, timestamp, location and target class. Target class is the air quality level that determines air quality based on gas concentration. Air quality will be classified into 3 classes, 1) *Good, 2) Moderate* and *3) Unhealthy*.

2.1. Pre-Processing Phase
Even though the dataset is taken from Jakarta's open data, it still needs a pre-processing phase before executed for classification input. From nine attributes, only six attributes will be taken. They are concentration of gasses as independent variable and category level as dependent variable. Location and timestamp are ignored. Cleansing dataset is also part of pre-processing phase such as: deleting record that does not have category class, transforming different target class but having the same meaning.

2.2. Modelling and Evaluation
Several algorithms are implemented to classify multiple gasses concentration. The output air quality prediction is whether the level is good, moderate or unhealthy class. The classification algorithms are Logistic Regression, K-Nearest Neighbours (KNN), Decision Tree and Random Forest. In modelling phase, dataset is divided into train data and test data with ratios of 60:40, 70:30 and 80:20.

Besides implementing several different types of classification algorithms and simulating the distribution of datasets, tuning hyperparameter is also considered to improve the accuracy of the classification model. The classification model is then implemented on the test data and evaluation of model is obtained using metrics. To evaluate the model, we implement confusion matrix and classification report. In confusion matrix, can be obtained *true positive, true negative, false positive* and *false negative* to gain precision, recall and F1 score.

2.3. Model Deployment
In the deployment phase, the best model will be deployed using Flask. Flask is a micro web framework for creating websites in python. The web page will display prediction results predicted from classification model that has been built.

3. Result and Discussions
The implementation uses the Scikit-learn library with the Python programming language combined with Flask as web framework. In the experimental stage, the clean dataset become input of machine learning algorithms. In modelling, grid search method approach is implemented to get the best hyperparameters for each algorithm. We take two hyperparameters for each algorithm. In the hyperparameter search process, the k-fold cross validation method also carried out and divides dataset into 5 folds. The results of best hyperparameters in each algorithm can be seen in Table 1.

Table 1. the best hyperparameter each algorithm

Algorithm	**Best Hyperparameters**
Logistic Regression	{'C': 0.00026366508987303583, 'penalty': 'l2'}
KNN	{'leaf_size': 1, 'n_neighbors': 27}
Decision Tree	{'max_leaf_nodes': 10, 'min_samples_split': 2}
Random Forest	{'min_samples_leaf': 1, 'n_estimators': 500}

The best hyperparameters are implemented for each algorithm in 3 simulations with 3 ratios, 60:40, 70:30 and 80:20 as shown in Table 2.

Table 2. Comparison of classification algorithm accuracy based on split data

Data Split	**Algorithm**	**Accuracy**
60:40	Logistic Regression	0.896
	KNN	0.935
	Decision Tree	1
	Random Forest	0.998
Data Split	**Algorithm**	**Accuracy**
70:30	Logistic Regression	0.914
	KNN	0.939
	Decision Tree	1
	Random Forest	0.998
80:20	Logistic Regression	0.887
	KNN	0.918
	Decision Tree	1
	Random Forest	0.994

From Table 1, it is found that the decision tree algorithm gets perfect accuracy in every simulation of the distribution of the dataset. Meanwhile, for logistic regression the average accuracy is 0.889, the average accuracy for random forest is 0.997 and the average accuracy of KNN is 0.93. The accuracy of each algorithm is quite good above 90%. For model accuracy other than decision tree model, it can be improved by increasing the number of hyperparameters that are tuned and the values given are more varied. This approach is able to improve the accuracy of the model but it requires a longer computation time.

Based on the experimental results, the classification model from the decision tree has the highest accuracy so that this model is deployed in the system. Several researches sometimes only stop at the data modeling. This research proposes to wrap the classification model using the Flask framework to display the prediction form and its results in web application. The first page displays the input form for the dataset path. the user can directly press the predict button and the system will display the results of the water quality classification being at the good, moderate or bad level. The air quality classification page is shown in Figure 1.

Figure 1. Air Quality Classification Homepage

Users only need to input the path of new dataset from the sensing result of the gas concentration. While Figure 2 shows the results of the prediction of the air quality level. The classification result is displayed in a table with the gases concentration are being the input from user.

IOP Conf. Series: Earth and Environmental Science **986** (2022) 012004 doi:10.1088/1755-1315/986/1/012004

localhost:5001/predict

Air Quality Classification

#	PM10	SO2	CO	O3	NO2	Prediction
1	53	34	25	21	2	moderate
2	26	32	16	19	1	good
3	28	32	17	20	2	good
4	24	35	22	20	2	good
5	29	38	25	14	2	good
6	21	33	24	18	2	good

Figure 2. Prediction of Air Quality

4. Conclusions

Air quality is an important component in human life, especially in urban areas due to the large number of motorized vehicles, industry and so on. Clean air is the need of every human being. Gas concentrations that exceed the threshold can cause harm to humans. It is necessary to monitor the air quality index with the concentration of gases. Predicting the air quality index is important so that it can be seen that the environment we live in is habitable. In this research, we propose to identify and classify the concentration of gases into an air quality index of good, moderate and unhealthy with a machine learning approach. Some algorithms are implemented by tuning hyperparameters to find their best values. Based on the results of the analysis, it is found that the decision tree model is able to achieve accuracy up to 100% by tuning two hyperparameters, which are *max leaf nodes* and *min samples split*.

References

[1] Supangkat S H, Hidayat F, Dahlan I A and Hamami F 2019 The implementation of traffic analytics using deep learning and big data technology with garuda smart city framework *2019 IEEE 8th Glob. Conf. Consum. Electron. GCCE 2019* 883–7

[2] Zhang Y, Wang Y, Gao M, Ma Q, Zhao J, Zhang R, Wang Q and Huang L 2019 A Predictive Data Feature Exploration-Based Air Quality Prediction Approach *IEEE Access* **7** 30732–43

[3] Information I and Singgih I K 2020 International Journal of Informatics , Information System and Computer Engineering Air Quality Prediction in Smart City ' s Information System

[4] Hamami F and Dahlan I A 2020 Univariate Time Series Data Forecasting of Air Pollution using LSTM Neural Network *2020 Int. Conf. Adv. Data Sci. E-Learning Inf. Syst. ICADEIS 2020* 12–6

[5] Wu Y C, Shiledar A, Li Y C, Wong J, Feng S, Chen X, Chen C, Jin K, Janamian S, Yang Z, Ballard Z S, Göröcs Z, Feizi A and Ozcan A 2017 Air quality monitoring using mobile microscopy and machine learning *Light Sci. Appl.* **6**

[6] Sugiarto B and Sustika R 2017 Data classification for air quality on wireless sensor network monitoring system using decision tree algorithm *Proc. - 2016 2nd Int. Conf. Sci. Technol. ICST 2016* 172–6

[7] Kujaroentavon K, Kiattisin S, Leelasantitham A and Thammaboosadee S 2014 Air quality classification in Thailand based on decision tree *BMEiCON 2014 - 7th Biomed. Eng. Int. Conf.*

[8] Hamami F and Fithriyah I 2020 Classification of air pollution levels using artificial neural network *2020 Int. Conf. Inf. Technol. Syst. Innov. ICITSI 2020 - Proc.* 217–20

IOP Conf. Series: Earth and Environmental Science **986** (2022) 012005 doi:10.1088/1755-1315/986/1/012005

Social design as a strengthening strategy of the revitalization of the sub-communal RISHA Mojo *Kampung Kota* – Surakarta during the COVID-19 Pandemic

Kusumaningdyah N.H.[1] and P Purwandaru[2]

[1] Center of Japanese Studies, Institution of Research and Community Service (LPPM). Department of Architecture, Faculty Engineering and Urban Rural Design and Conservation Laboratory-URDC, Universitas Sebelas Maret, Surakarta Central Java, Indonesia

[2] Center of Japanese Studies, Institution of Research and Community Service (LPPM). Design Interior Study Program, Universitas Sebelas Maret, Surakarta, Central Java, Indonesia

E-mail: Kusumaningdyah_nh@staff.uns.ac.id

Abstract. *Kampung kota* is a type of Indonesian residential that is occupied by a large portion of urban dwellers. According to Pauline (1993) *Kampung kota* is still considered to have a social interaction habit as mutual assistance among the residences called *gotong royong* [1]. Despite the effort to plan the formal residential area, most *kampung kota* is categorized as slum and squatter areas. There are a lot of improvement programs initiated by the government to create a better *kampung kota*. One of the programs in Surakarta is located in neighborhood one (RW 1), Pasar Kliwon District. The residents were given temporary living space in *kampung* sub-communal RISHA Mojo before they were moved into the new area of Kampung Metal in Semanggi sub-district. *Kampung* Sub-Communal RHISA Mojo in Surakarta is a temporary housing area provided by the government to support the *kampung* improvement program in the residents' old area. Living in communal housing is not directly being guaranteed with decent hygiene environment. When the COVID-19 pandemic occurred in 2020, the residents were dealing with hygiene issues. Therefore, the goal of this paper is to show how social design at *Bangkit Berbarengan* Project that initiated by academicians of URDC Laboratory and NGO of ARKOM Solo in Surakarta becomes a strategy to improve the environmental quality in the sub-communal area, especially in the pandemic era. Social design thinking is to keep the human need approach and in formalistic approach of design that concern more about inclusivity (Sommer, 1983)[2]. This paper will discuss the approach of the social design process that applies actor-network as a method of knowledge. The case study of the social design project creates social dynamic assemblies and engages multi-stakeholders. The social design outputs are the local campaign to advocate the resident of the COVID-19 and the public facilities designed by the residents. We used qualitative research method. We investigated the residents' behaviors regarding their environment by conducting observation and in-depth interviews. Through this investigation, we found that the advocacy program for COVID-19 prevention can be easily accepted by the residents through the usage of a local icon. A creative activity of making public facilities that involved the residents also creates a bigger sense of belonging between them.

1. Introduction

At the beginning of 2020, there was a non-natural disaster, an outbreak of the COVID-19 pandemic. This pandemic is still spreading almost all over the world, including in Surakarta (Solo). Mayor of Solo, FX Hadi Rudyatmo, on March 13, 2020, determined that Solo had an Extraordinary Event Status (KLB) in the newspaper [3]. When COVID-19 occurred for the first time in Solo, the pandemic brought several disadvantages for many parties, one of which is the residents of *kampung* Sub-communal RISHA Mojo in Mojo Sub-district, Pasar Kliwon District, Surakarta. Due to this situation, the residents were asked not to do many activities outside their settlements. Nevertheless, the pandemic also gave the residents some chances to maximize the role of togetherness in overcoming the pandemic through the case study of the social design project of *Bangkit Berbarengan*. The project was initiated by the NGO's ARKOM Solo and academicians of URDC Laboratory, Department of Architecture UNS. This project was implemented during the first phase of the COVID-19 pandemic in 2020 for almost eight months.

Kampung kota is a type of Indonesian residential that is occupied by a large portion of urban dwellers. *Kampung kota* is still considered to have a social interaction habit among the residences called *gotong royong* [1]. Despite the effort to plan the formal residential area, most *kampung kota* remains to be categorized as slum and squatter areas. In order to improve the *kampung kota,* many improvement programs were initiated by the Solo government. One of the programs is implemented in neighborhood one (RW 1), *Kampung* Sub-communal RISHA, Mojo, Pasar Kliwon District. *Kampung* Sub-communal RISHA, Mojo is an area provided by the government as a temporary housing to support the *kampung* improvement program in their former area [4]. The residents were given temporary living space in *Kampung* Sub-communal RISHA, Mojo before they were moved into the new area of Kampung Metal Semanggi 23 in Mojo Sub-district.

Living in communal housing is not directly being guaranteed with decent hygiene environment. When the pandemic COVID-19 occurred in 2020, the residents were dealing with hygiene issues. They were also struggling with adjusting the working from home system during the pandemic era. Therefore, the goal of this paper is to show how the social design project of *Bangkit Berbarengan* become a strategy of the knowledge diffusion for inclusive innovation in dealing with disaster risk mitigation. This paper will discuss the approach of the social design process that applies actor-network as a method of knowledge. The case study of the social design project creates social dynamic assemblies and engages multi-stakeholders.

1.1. Purpose

The study aims to develop an adaptation model for disaster preparedness that is adapted to the locality of the *kampung*'s residents. Thus, the model can be applied as design strategies in the community to deal with the COVID-19 pandemic through the social design project in *Kampung* Sub-communal RISHA, Mojo. The design process is being observed and evaluated from (1) identification of the needs of the residents in the *Kampung* Sub-communal Mojo at the early phase of COVID-19 pandemic; (2) the actors-network in the case creates social design assemblies in a dynamic model that might perform differences on each of stage process; (3) assessment the factors affecting the social design process.

1.2. Methodology

This research is evaluation research that has an intention to evaluate the *Bangkit Bebarengan* project from the social design strategies that apply actor-network. According to Neuman (2013) [5], evaluation research as applied research tries to determine the quality of a project program's work in reaching its goals and objective. The qualitative data in this research was assembled through constant observations during the implementation process of the *Bangkit Berbarengan* project. The observation was done in every stage's project process. The method used in this research was descriptive method within its analysis. The project of *Bangkit Berbarengan* was implemented with social design thinking. Social design thinking as a practice involves co-creating products or services in sustained consultation with users [6].

There were three stages of analysis conducted. **First**, exploration. In the exploration stage, observation and shadowing were performed to know the need belonging to the *kampung* resident. Based on the observation result, the *kampung* resident needed some daily things, such as rice, sugar, and eggs. Those items were the basic item needed at the beginning of the pandemic in 2020. **Second**, the ideation stage in which the design thinking advocacy idea was created for tackling the issue of COVID-19. In this stage, a focus group discussion was organized to take the local idea that fits the resident. The constant consultation was actively organized. This approach sustained the problem to be seen through multiply lenses before finding the approach solution [6][7]. This analysis was expected to understand the network actors in the forms of social design assemblies that might differ on each stage process. **Third,** the analysis stage was done based on the theory of Human Centred Design (HCD) by IDEO (2015) [9]. The project of *Bangkit Berbarengan* is the process in which humans become the focus center. The project is concerned with the character of society and its behavior in the design process. This analysis is intended to understand techniques and how the result occurs in each stage of the HCD process (exploration, ideation, and implementation process).

2. Project *Bangkit Berbarengan Sub-communal Kampung of RISHA Mojo*

Project *Bangkit Berbarengan* in Mojo Sub-district, Pasar Kliwon District, Surakarta had been initiated since May – December 2020. The project was initiated by the NGO's ARKOM Solo and the academician of URDC Laboratory, Department of Architecture UNS. This project was done for almost 8 months at the beginning of the first phase of the COVID-19 pandemic in 2020. The settlement of *Kampung* Sub-communal RISHA, Mojo has intervened as the vulnerable settlement in the city center.

The objective of conducting this program is to create a community that has a decent understanding of the COVID-19 outbreak and its impacts, and proactively takes action on adaptation and mitigation of COVID-19 in which contributes to the achievement of sustainable national development. The various programs were implemented such as the advocating awareness of COVID-19's knowledge and the construction of communal handwashing facilities (*padasan*).

Supported by a collaboration with *Kampung kota*, this project managed to deploy 11 students of the 2020 class of Department of Architecture, Universitas Sebelas Maret to do a collaborative project. The students and ARKOM Solo were the facilitators in this project. The facilitators were responsible for assisting the focus group discussion, data collection, and data analysis.

2.1. Stage Process of Social Design Thinking

The process of designing the community participation was divided into four stages. **Stage I** was the exploration process where the students were assigned to explore the site context. During the exploration stage, observation and shadowing were done to understand the urgency faced by the residents. In this process, an information center was set up and the aspirations of residents were collected. Fast response to emergency toward the COVID-19 pandemic was also important in this stage. (See Figure 1).

Figure 1. Stage I – Exploration: Fast Response Program of COVID-19

Stage II was ideation. The facilitators were assigned to create ideas for advocating awareness of the COVID-19 pandemic. The ideation had two design thinking products. The products were poster of advocacy and the idea of communal hand washbasins (see Figure 2). In the second stage, a focus group discussion was convened to gain an understanding of the problem design. The design exploration was a diverse process in enriching alternative participatory design thinking. The students explored their interpretation on issues formulated with the residents of *Kampung* Sub-communal RISHA, Mojo. In this stage, creativity ideation had more importance than the implementation constrain. The students were expected to build their academic competence in architectural design thinking to deal with the COVID-19 pandemic. ARKOM Solo as a design studio mentor and facilitator also took a role as the representative of the resident in stating the problem and existing conditions.

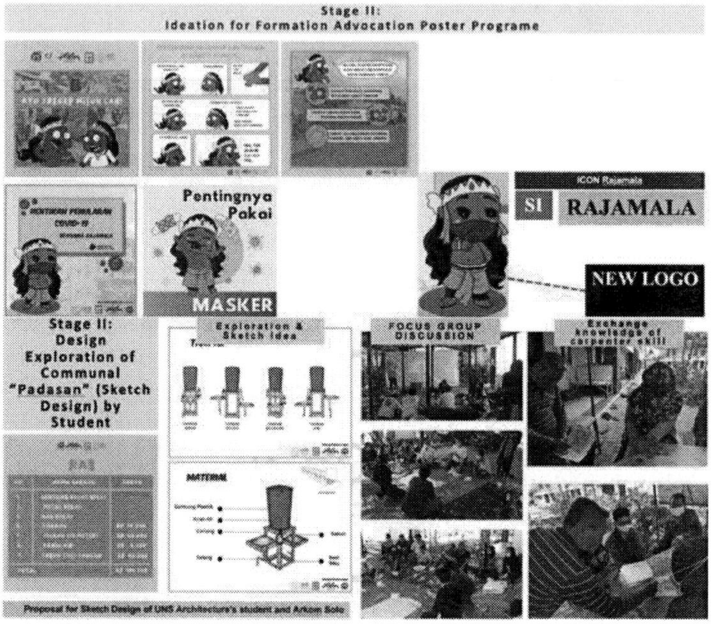

Figure 2. Stage II - Ideation for Poster Program

Stage III was project implementation. The third stage showed how the residents, facilitators, and students have implemented the idea they formulated. The advocacy poster was released to the resident of *Kampung* Sub-communal RISHA, Mojo through social media. Another design implementation was constructed the communal hand washes basins in the settlement of *Kampung* Sub-communal Mojo (See Figure 3). **Stage IV** was monitoring. The goal in performing the fourth stage was to evaluate and monitor the design action of inclusive public amenities (See Figure 3).

2.2. Actors Involved

Many actors participated in this project. Some of them came from various backgrounds (stakeholders' type). The Center of Japanese Studies at LPPM UNS took part in funding the project. The Project team with Kusumaningdyah N.H. as the team leader handled the project thoroughly. The team teaching of *Kampung Kota* subject consists of Ana Hardiana, Maya A.N, Anita Dianingrum, and Pratiwi A.S was involved actively in assisting the project team. Architecture students batch 2017 of UNS also participated in this project by elaborating design alternatives and implementing the design. ARKOM (*Arsitektur Komunitas*) Solo took part as an intermediate partner. ARKOM Solo team that consists of Muhammad Nasrullah Amin, Indra Yuda Wardiana, and Dinan Nafindro Nugroho were connected to the resident of *Kampung* Sub-communal RISHA, Mojo. Others actors, such as residents, the local leader (Head of *Rukun Tetangga*), and the head of *Ibu-Ibu PKK* were involved assiduously in this process.

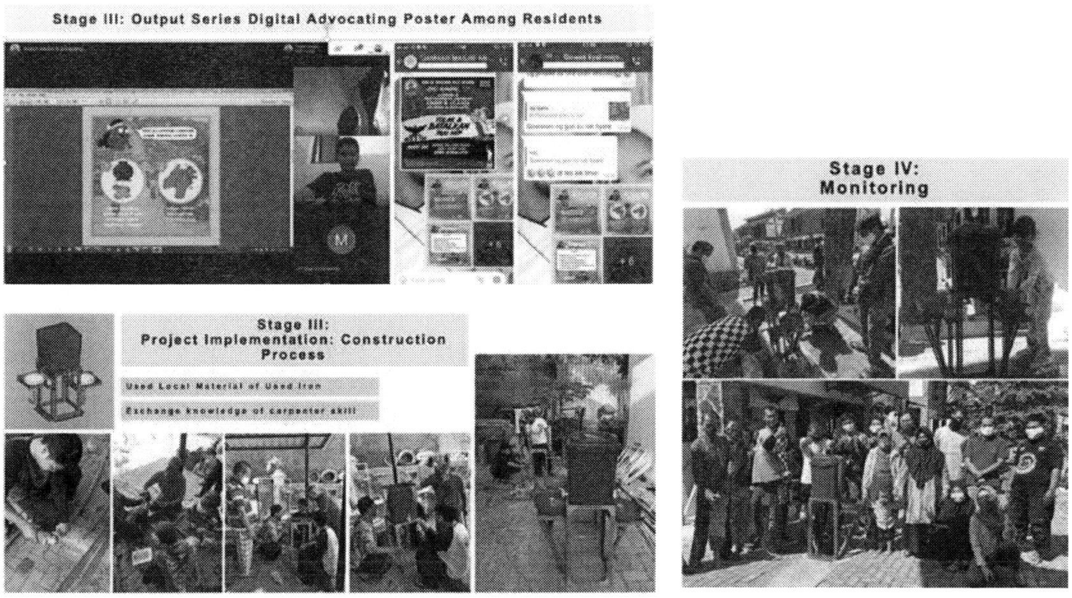

Figure 3. Stage III – The Implementation Program and Stage IV - Monitoring

3. Results and Discussion

An examination of how the participatory design in the deployed social design process reveals on the case study of *Kampung* Sub-communal RISHA, Mojo. We identified the process of social design thinking and key factors that influence these methods, and reveal the opportunities for conducting social design thinking in the urban *kampung*.

3.1. *Fast Response on COVID-19 Pandemic*

Stage I reveals that during the COVID-19 pandemic, the basic needs are highly needed by the residents. Before the project initiation was implemented in the *kampung* settlement, ARKOM Solo with the help of UNS Architecture students initiated a donation. The donation was used to buy the basic needs. The initiation was taken place to improve the risk management of the residents who were affected by the COVID-19 pandemic. Hand in hand, each of the collaborators of the *Bangkit Berbarengan* project managed to increase the resident's trust in this project.

3.2. *Design for Social Innovation*

The process of social design thinking advocacy design as a collaborative effort in which the design process is spread among various participated stakeholders and competencies (Anne, 2016) [9]. In stage II, the process was not based merely on consultation with users or residents and stakeholders, but on their active participation. This is the "participatory design" process, a design for social innovation priority principle.

It can be seen in phase II – the ideas of the facilitators (Architecture students and ARKOM Solo) through a participatory design process became a challenge to blend and engage with the residents' culture and socially innovative design. The idea of re-using the recycled-iron material was an authentic idea coming from the resident. Afterward, the facilitator added a creativity value to the recycled iron and changed it to communal hand washbasin.

Academics and practitioners highlighted that design thinking added understanding, precision, and breadth to the design process by emphasizing the importance of: - good data collection; - a clear and brief design and how to construct it; - rapid prototyping; - its function to (social) innovation; and - working in new, more interdisciplinary ways which emphasize problem-solving and systems change

IOP Conf. Series: Earth and Environmental Science **986** (2022) 012005 doi:10.1088/1755-1315/986/1/012005

through collaborative action (Anne, 2016; Westley et al 2012) [9][10][11]. Therefore, in the process stages of *Bangkit Berbarengan*, stage I and stage II were essential. A brief yet clear design and the instruction on how to construct it was important to improve the alternative design idea.

Stage III was the construction process. This process was a collaborative process between residents and UNS Architecture students. Two activities were performed in this stage. The first activity was the design translation in which the designers discussed and translated both the resident's idea and applicable design, while the second was the design realization activity that was done together with the local carpenter. The sharing of knowledge occurred in this stage process. The residents shared their knowledge of carpentry with the Architecture students. On the other hand, the Architecture students managed to add value to the recycled iron that the material was easily be found in the settlement surroundings.

3.3. *The Network Actors*

The four stages of social design thinking had been observed. The result indicates that each study performs a different result. The networks actors involved were diverse. The first stage consisted of the exploration process. The role of ARKOM Solo and the reader of the community were essential, particularly in the earlier COVID-19 pandemic. Both of them helped to give the site observation. The second stage performed the ideation activity through the participatory design process. In this stage, the Architecture students took a role as a designer who translated the resident's needs. The exploration of the design was a different process in enhancing the alternative design. In the third stage, the involvement of an expert of local carpenter took part as an essential thing. The last stage was monitoring. The role of residents was significant. The residents who used the communal washbasins every day will maintain the washbasin functions and conditions (See Figure 4).

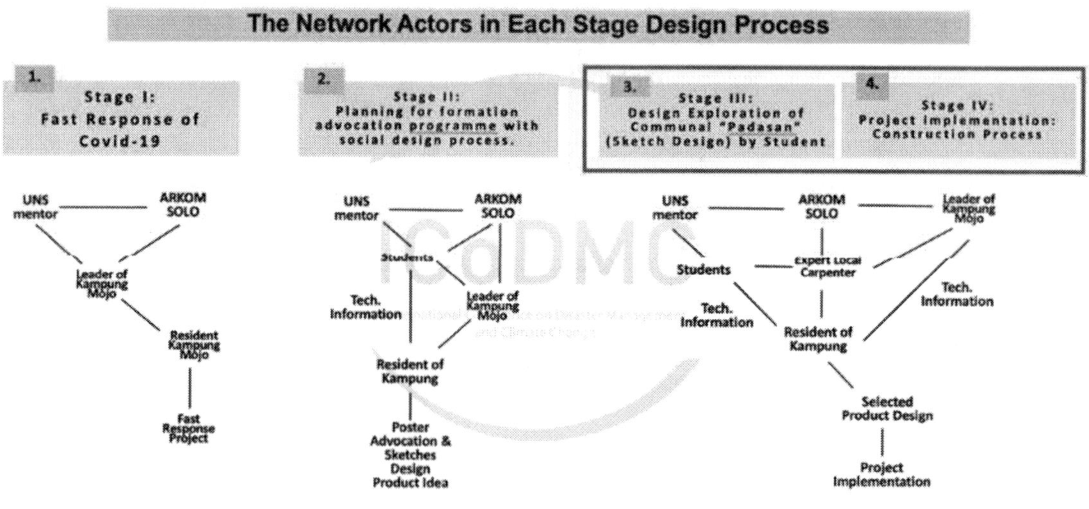

Figure 4. The Network Actors in Each Stage Design Process

3.4. *The Factors Affecting the Social Design Process*

The process was successfully being implemented. The social design thinking required further conditions to support the continuity of the *Bangkit Berbarengan* project. Some factors that affected the sustaining of the social design program were leadership, educated community, and local community initiative [9][11]. Therefore, in this project, leadership was the important element and a passionate leader is highly needed. The leader resident or head of RT - *Bapak* Drajad provided an essential factor on how the leader manages several activities in the project *Bangkit Berbarengan* during the pandemic. The communication was kept intense between the intermediate partners and the academicians (architecture students and mentor). Educated community: The COVID-19 pandemic should be enough reason to make the community understand and be aware of the dangers of the virus. Not only gaining understanding and awareness, but they also need to understand which project that brings advantage. The knowledge about the risk of COVID-19 was designed with the local icon by the academician. Therefore, the knowledge could be easily accepted by the resident of *Kampung* Sub-communal RISHA. Local community initiative the strong social capital within the neighborhood is the creative community. The residents of *Kampung Sub-communal Risha-Mojo* have a creative carpenter. This social capital enhanced the collaborative program to tackle the mitigation risk during the COVID-19 pandemic. The social bounding also fostered the involvement process of the social design thinking realization.

4. Conclusion

The proposal in this paper aims to recommend how social design thinking becomes one of a strategy as the knowledge diffusion for inclusive innovation in dealing with disaster risk mitigation. Therefore, this article shows that social design thinking is a process of developing social assemblies in several phases. The project can be done with the good relation of all stakeholders. The resident's trust become the essential part of the project.

A design that is successfully accepted by the residents is a vital success indicator of the social design program in the pandemic era. The project of *Bangkit Berbarengan* is easily accepted by the residents through the participatory design, and it creates a bigger sense of belonging to the public facilities made by the residents.

Strategies taken for the assembling group are also significant. The collaborated actors are varied in each step, and the role of actors-networks is diverse. The involved actors who come from diverse backgrounds become vital elements when the design process is oriented to inclusive social innovation. We used the *Bangkit Berbarengan* project to consider the pre-condition needed to establish using the inclusive social innovation design approach that is (1) Resident's Trust, (2) Diverse Stakeholders, (3) Network Actors, and (4) Leadership.

Acknowledgment

The author would like to thank *Arsitek Komunitas*-ARKOM (Community Architect) Solo led by Muhammad Nasrullah Amin, Indra Yuda Wardiana, and Dinan Nafindro Nugroho who have been cooperative in the collaboration project with the team teaching of *Kampung Kota* Subject of the Department of Architecture, Faculty Engineering, Universitas Sebelas Maret. ARKOM Solo as an intermediate partner has been very active in informing the communication and discussion with the resident of *Kampung* Sub-communal Mojo. The author also values the Institution of Research and Community Service (LPPM UNS) for the financial support of the Institutional Strengthens Research and Community Development's Grant Program (PNBP) 2020 through the Center of Japan Study (Pusat Studi Jepang UNS). The high appreciation goes to the Architecture's student batch 2017 of UNS of *Kampung Kota* Subject and the team-teaching lecturer for their active participation in this project.

References

[1] Pauline D M 1993 *Kampung Improvement in The Small and Medium Sized Cities of Central Java*, Review of Urban and Regional Development Studies 5.

[2] Sommer R 1983 *Social Design Creating Buildings With People in Mind.* New Jersey: Prentice Hall Inc.

[3] Labib Zaman 2020 *Dalam Sehari Kasus COVID-19 di Solo Tambah 100, Walikota Rudy" Jangan Mudik Dulu. Regional kompas.com.* https://regional.kompas.com/read/2020/11/26/09461901

[4] Aosgi 2019 *Risha, Huntara Bagi Warga Terdampak Penataan Kawasan.* https://surakarta.go.id/?p=12640

[5] Neuman, W. Lawrence. (2013). Social Research Methods : Qualitative and Quantitative Approach. Pearson.

[6] Rama KRK 2017 *Design Thingking in Social Organizations: Understanding the Role od User*

Enggagement. Journal Creating Innovaation Management 2019 (Willeyonlinelibrary) p 1-12
[7] Zahra S A 2002 *Absorptive Capacity: A review, Reconceptualizaiton, and Extension. Academu of Management Review,* **27**, *185-2003*
[8] Ideo.org 2015 *The Field Guide To Human-Centered Design* Canada
[9] Anne Chick 2016 *Design for Social Innovation: Emerging Principles and Approaches* Iridescent, **2**:*1, 78-90*
[10] Westley F, Goebey S and Robinson K 2012 *Change Lab / Design Lab for social innovation.* (Waterloo: Waterloo Institute of Social Innovation and Resilience, University of Waterloo).
[11] Kusumaningdyah N H, et al 2021 *Evaluating human-centered design methods as an approach for inclusive green design: Case study Kampung Kota Surakarta.* IOP Conf. Series: Earth and Environmental Science 780 (2021) 012030

IOP Conf. Series: Earth and Environmental Science **986** (2022) 012006 doi:10.1088/1755-1315/986/1/012006

Compulsory criteria for sustainable environmentally friendly energy

AH Ramelan[1,2], A Supriyanto[1,2*], Riyatun[1,2], Suryanto[1], F Puspitasari[3], H Krisyana[4], BS Effendi[4]

[1] Lembaga Penelitian dan Pengabdian Masyarakat (LPPM) - Sebelas Maret University-Indonesia
[2] Physics Department, Sebelas Maret University - Indonesia
[3] Kementerian Koodinator Maritim dan Investasi - Indonesia
[4] PT ThorCon Power Indonesia

*Email: agusf22@staff.uns.ac.id

Abstract. Indonesia commits the Paris Agreement COP 21 related to handling climate change and the 13th Sustainable Development Goals (SDGs) from the United Nations. The two agreements require the fulfillment energy of large, reliable, environmentally friendly and sustainable energy. The provision of environmentally friendly energy is generally associated with (1) low greenhouse gases produced, always related to renewable energy. This paper is a study of the re-definition of "environmentally friendly energy", adding that energy supply requirements must be (2) minimal in land use, (3) maintain ecosystem balance, and (4) properly manage waste. With these 4 criteria, an assessment of renewable energy is carried out, namely water, wind and solar, while new energy is nuclear. In addition to these 4 environmentally friendly criteria, energy sources must have sustainable properties. The new energy sources become transitional energy that will replace fossil energy whose sources are dwindling. Short-term sustainability is 24 hours, uninterrupted, while long-term is the availability of mining resources that ensure the sustainability of these energy sources. The study concludes that hydro/hydro energy and nuclear energy are the most environmentally friendly and sustainable combinations of energy providers compared to wind and solar energy.

1. Introduction

The Paris Agreement is closely related to the 13th Sustainable Development Goals (SDGs) achievement, namely action to tackle climate change. Indonesia has committed to COP 21 and SDGs, so real action from the government is needed to prevent climate change and its impacts. The application of environmentally friendly production and energy consumption methods is a top priority [1–5].

Fossil energy is running low, so it is necessary to choose transitional energy to cut off Indonesia's electricity supply. The criteria for transitional energy must be able to replace the fossil energy. It is affordable, large-scale, continuously supply electricity 24 hours a day, and is sustainable forever [6–8]. Energy must stop global warming and some criteria. It does not require a large land area, does not damage the ecosystem, and manages its waste correctly. Sustainable environmentally friendly energy has reliable and sustainable properties [9,10]. Indonesia has environmentally friendly energy sources in renewable energy, namely water, wind and solar. Can all three be relied upon to replace transitional energy when fossil energy runs out? Indonesia must boost the procurement of environmentally friendly energy, but this business seems to be running in place. Then does nuclear energy as a new energy in Indonesia meet the requirements of environmentally friendly energy and can be used as transitional energy?

This paper is compiled based on reviewing several articles, textbooks, and official data sources from the official web. All sources have been listed in the body text section and the reference list.

2. Discussion

Indonesia must meet the demand for energy quality. Concerning Indonesia's commitment to realizing COP 21, clean energy from GHGs must be provided. The GHG limit is not enough; what is meant by environmentally friendly energy must meet what criteria? The Big Indonesian Dictionary also states that being environmentally friendly means being safe or not damaging to the environment. Environmentally friendly, go-green related to energy claims minimal harm to humans and living things, not even endangering the ecosystem or the environment.

In obtaining and providing electrical energy, the quality of the results should still pay attention to the impact of the process on the environment. Several criteria for environmentally friendly energy need to be clearly defined to benefit all parties involved in the supply and use of energy. In addition, the provision of clean energy and environmentally friendly processes must be accompanied by sustainability. The government must ensure the availability of fuel and energy raw materials that can convert energy into electricity by a storage process, not combustion.

2.1. Environmentally Friendly Energy Criteria

The definition of environmentally friendly energy is commonly known to not damage the environment through GHG emissions, namely very little or almost no CO_2 emissions. Environmentally friendly energy is associated with energy from nature, defined as energy that will not run out or continuously renew.

This study tries to formulate criteria for environmentally friendly energy, which is expected to explain the definition of environmentally friendly energy. In this study, the requirements for environmentally friendly energy include [1,3]:

(1) Low greenhouse gas emissions,
(2) Minimum in terms of land use,
(3) The balance of the ecosystem is maintained,
(4) Handling of waste is managed properly and does not cause pollution.

Each of these criteria will be explained and will be evaluated against these criteria. In the end, a conclusion will be obtained on what kind of energy has all these criteria.

2.2. Low Greenhouse Gas (GHG) Emissions

The composition of the atmosphere that changes due to human activities causes the greenhouse effect caused by GHGs. GHGs consist of CO_2, methane (CH_4), nitrous oxide (N_2O) and fluorinated gases (HFCs, PFCs, NF_3, SF_6), and gas equivalent to carbon dioxide equivalent (CO_2(e)). The percentage of GHG in the world is shown in Figure 1. (a) and Figure 1. (b) shows the GHG sources

The main content of GHG in Figure 1(a) is CO_2, 76%, while the primary source of CO_2 (57%) comes from burning fossil fuels, namely coal, oil, and natural gas as power generation energy. Figure 2(a) The amount of CO_2 per year shows an increasing number, in 2018 reaching almost 40 Gt CO_2. The main contribution of CO_2 comes from coal, oil, and natural gas.

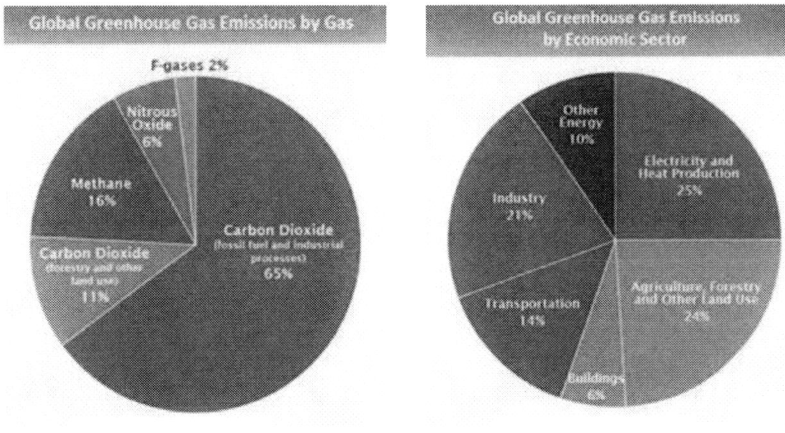

(a) (b)

Figure 1. (a) Composition of world GHG emissions and (b) producing sectors [11]

The second-largest GHG (16%) is methane (CH_4). According to the United Nations Economic Commission for Europe (UNECE), methane is a powerful GHG – about 25 times CO_2 with a 100-year global warming potential. In 20 years, methane GHG emissions have reached 84 times CO_2. Based on Figure 2(b), The three sources of electrical energy with the lowest CO_2 emissions are hydro, geothermal, and nuclear, a source of electricity.

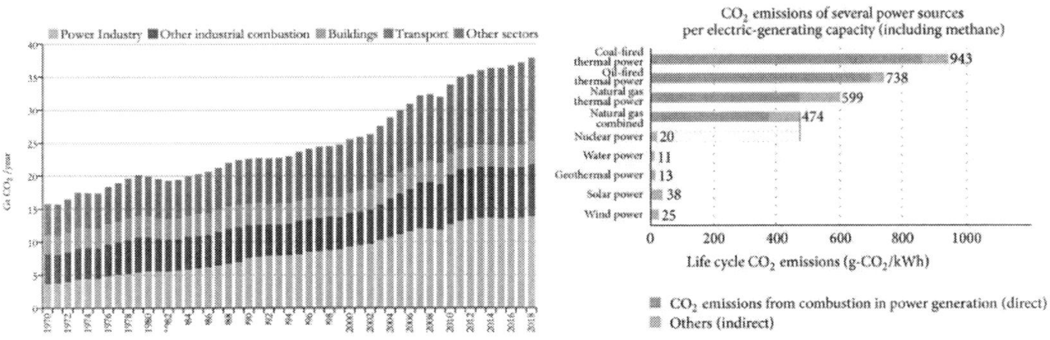

(a) (b)

Figure 2. (a) Total annual global fossil CO_2 emissions in Gt CO_2/year by sector [11], (b) GHG emissions from various energy sources [12]

2.3. Minimal in Land Use

The second criterion that environmentally friendly energy must have been minimal land use. The less land, the smaller the potential for deforestation. Almost 71% of the earth's area is water, 29% island. The human population in 2020 is 7.8 billion, Figure 3(a); even though the human population is saturated, energy needs

IOP Conf. Series: Earth and Environmental Science **986** (2022) 012006 doi:10.1088/1755-1315/986/1/012006

will still increase. It takes a power generation technology with a small footprint but has a high energy density—the more efficient the energy source, the greater the energy density, evaluated from Table 1.

Table 1. Energy density of various types of fuel [11]

Fuel type	MJ/kg	MJ/m^2	Land use m^2/GWh/yr
Coal	24		5,700
Diesel	48		
Natural Gas	55		3,200
Uranium – 235	3,900,000		1,200
Uranium – breeder	80,620,000		
Thorium – breeder	79,420,000		
Sun		0.0000015	7,500
Geothermal		0.05	900
Wind (10 mph; 5m/s)		7	1,100
Hydro		0.5 - 50	200,000
Biomass			460,000

Figure 3(b) released by Hinkley Point C (HPC), a 3,200 MWe nuclear power plant with two EPR (Experimental Power Reactor) reactors in Somerset, England. HPC needs 430 acres of land to produce the same power, while generator PVC and wind need 130,000 acres and 250,000 acres, respectively.

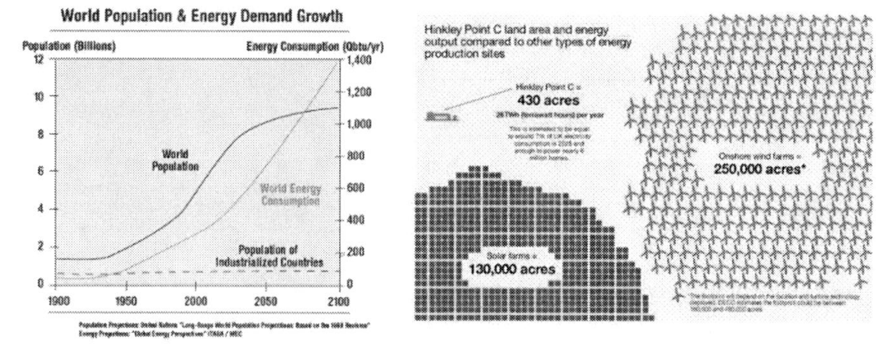

(a) (b)

Figure 3. (a) Projected human population growth and energy demand [13] (b) Comparison of footprints of the same energy producers: wind, solar and nuclear energy [11]

Figure 2 (b) shows 3 candidate power plants with the lowest GHGs, namely hydro, geothermal, and nuclear, respectively 11, 13 and 20 (g-CO_2/kWh). Nuclear energy density has the highest value, far from other sources. The fission ratio of 1 kg of Uranium-235 produces 24,000,000 kW of electricity, while 1 kg of coal produces 8 kW of electricity. Supported by the analysis of footprint area, capacity factor and installed power to produce 1 GW in Table 2, wind requires 360 times, and solar requires 75 times wider. Nuclear power plant capacity factor reaches 90% (the highest), solar and wind are very low (very dependent on location, weather, and abundance of raw materials).

Matching the energy production capacity of 1 GW from nuclear, solar energy must provide about 1.9 – 2.8 GW of installed power and provide about 19,000 ha of land, while wind power must provide 3.3 – 5.4 GW and around 93,000 ha of land. In contrast, the nuclear power plant requires around 340 ha.

Table 2. Comparison of footprint, capacity factor and power of a nuclear power plant of 1 GW [11]

Generator Type	Footprint (ha)	Capacity factor (%)	Installed power (GW)
PLTN	259 - 337	90	1,3
Sun (PV)	3.755 – 1.9425	32 – 47	1,9 – 2,8
Wind	67.340 - 93.240	17 – 28	3,3 – 5,4

2.4. Maintained Ecosystem Balance

The next criterion for environmentally friendly energy is not to harm the ecosystem. Maintaining the balance of the ecosystem is important because every ecosystem has the power to maintain the balance of the earth. For example, the potential adverse effects of wind turbines are killing birds and bats, hitting propellers, disrupting the rotating frequency of the propellers, and being electrocuted by electrical installations. The American Bird Conservancy estimates that about 366,000 birds were killed by wind turbines in America in 2012 until wind turbines were called "bird choppers" in Table 3.

Table 3. Estimated annual bird deaths due to collisions with wind turbines in the United States [14]

Study Results	**Year**	**Average Bird Mortality/year**	**Minimum – Maximum/year**
Loss and others	2013	234.000	140.000 – 328.000
Smallwood	2013	573.093	467.097 – 679.089
Erickson and others	2014	291.000	214.000 – 368.000

The American Wind Energy Association reports 60,067 MW of wind energy capacity in the United States in 2012, compared to 111,808 MW today (2021). It means that there is an increase of 86. It can be projected that approximately 681,000 birds are killed by wind turbines in the United States each year. Jianu reported from wind power plants to the environment, namely noise pollution by turbine rotation, low-frequency sound, and visual disturbances (shadow flicker effect) [15]. George C. Ledec explain the negative impacts on the surrounding environment in the form of 1) noise and visuals, 2) bird deaths, 3) soil erosion and deforestation, 4) lightning from towers, 5) electromagnetic radiation, and 6) human health and psychology. [14].

Some efforts to minimize negative effect is setting the height of the tower, setting the speed of the propeller, as well as signals that can cause birds to fly away. The arrival of a flock of birds should be predictable either by observing the bird movement period or recorded at any time using the RADAR technique. The design and shape of the propeller are aerodynamic, so it doesn't have to be big but can still rotate the turbine and materials that don't interfere with bird views so that bird deaths can be reduced.

The impact of wind energy also causes other disturbances, soil erosion and deforestation due to the need for extensive land areas. Wind turbines can cause electromagnetic interference that interferes with television signal reception or microwave transmission for communications. In addition, offshore wind turbines can also be a nuisance to sailors and ships. The construction of wind power poles can disrupt the life of the seabed surface; around 80 – 110 dB low-frequency noise interferes with the communication of whales and marine predators.

Abhishek Gawande and Pradeep Chaudhry discuss the negative impacts caused by solar power plants [16]. A California solar farm, Ivanpah, has killed hundreds of desert tortoises, put the reptiles at risk of extinction, and annually kills an estimated 6,000 birds from fires. The installation of solar panels also requires extensive land clearing, which can risk erosion and soil compaction.

The negative impact of hydroelectric power plants on the environment is that the construction of dams requires an extensive area. The effect of diverting the flow system can disrupt the reproductive migration pathways of certain fish. In some cases, the conversion of the function of the lake into a source of electrical energy causes residents to be forced to migrate, reduce their income because the working area is limited by power plants, and threaten safety if a dam occurs in a disaster.

In conventional nuclear power plants, the Nuclear Energy Institute estimates that nuclear reactors consume between 1,514 and 2,725 litres of water per MWh, coal at 1,220 to 2,270 litres per MWh, and gas at 700 to 1,200 litres per MWh. Nuclear power plants can use up to 1 billion gallons of water a day and, for this reason, nuclear power plants are built near rivers, lakes, or oceans. The cooling system used in nuclear power plants uses water from the sea or rivers. After the water is used to cool power plants, it is returned to the sea or river, with a warmer temperature, not hotter, carrying no chemicals or radioactive substances.

All electrical energy generation technologies are developing rapidly, maximizing profits and minimizing negative impacts. Academics and researchers have developed research on solar cell materials with high efficiency without high-temperature rises. The designs of wind and water turbine generators have been further refined, as have the generation of solar cells. In the nuclear reactor design, Generation IV has been developed with the inherent high safety features of passive.

From all explanations related to the impact of power plants on the balance of the ecosystem, it is known that a power plant cannot affect the surrounding environment. What is important is choosing the type of energy for the plant with the most negligible impact on the environment and finding solutions to these problems, one of which can be done through technology development.

2.5. Proper Waste Handling

Solid, liquid, and gaseous wastes harm the environment and human health. In principle, humans live surrounded by waste. The negative impact of waste depends on the level of toxicity and how it is handled. Regarding the supply of energy, environmentally friendly energy must have excellent waste treatment and management criteria. There are 3 essential keywords; minimal in volume, contained and managed.

The smaller the volume of waste, the easier it is to handle and the smaller the negative impact on the environment. FABA coal power plant waste (fly ash - bottom ash) has short and long has implications on health. However, FABA is recycled into concrete or boards. FABA is no longer called Hazardous and Toxic waste by the Indonesian government.

Waste from wind is a component of wind turbines that have passed their lifespan (lifetimes). The turbine blades, 10-15% of the turbine's weight, are made of carbon fibre and fibreglass and cannot be recycled easily or economically.

Photovoltaic (PV) solar panel waste is hazardous waste containing Pb, (Cd) and others. The International Renewable Energy Agency (IRENA) in 2016 estimated about 250,000 metric tons of solar panel waste in the world. The average panel lifespan is 25 years. Worldwide solar panel waste accounts for between 4%–14% of total generation capacity by 2030 and will increase to more than 80% (approximately 78 million metric tons) by 2050.

The amount of waste from the nuclear industry is relatively small, with 97% of low- or medium-level waste (LLW or ILW). Waste will be stored in a repository near the surface. The French state has mastered the processing of spent fuel, the volume of HLW remaining 0.2% of all radioactive waste. A large-scale

IOP Conf. Series: Earth and Environmental Science **986** (2022) 012006 doi:10.1088/1755-1315/986/1/012006

reactor (1 GWe) produces about 25–30 tonnes of spent fuel per year, with an HLW of 22,000 m^3, an LLW of about 24 million m3.

Lauren Delbridge, 2019 wrote an environmental impact study on the coal in Chesterfield. Analysis of mortality and other health effects caused by fine particles of a mixture of soot (fly ash), heavy metals, SO_2, NO_3 and NO. More than 13,000 deaths and tens of thousands of cases of chronic bronchitis, acute bronchitis, asthma, congestive heart failure, dysrhythmias, heart disease, chronic lung disease, and pneumonia every year are caused by fine particulate pollution from the PLTU emissions [17]

Renewable energy does not always contain its waste reliably; large quantities of solar panels and wind turbines end up in piles and pollute the soil with toxic metals. The use of windmills is widely used in Europe and the United States. Although they do not emit carbon emissions, the final waste of windmills pollutes the environment quite significantly. Windmill waste is buried in soil and takes decades to degrade.

Waste from PV solar power plants also pollutes the environment, and toxic chemicals will evaporate and pollute the air. The technological challenges of developing solar cells are made from non-toxic and recyclable materials. Waste from solar panels has no management arrangements. The European Union is pioneering solar panel e-waste management regulations, including specific collection, recovery, and recycling targets.

Nuclear power confines waste and manage it on a "from the cradle to the grave" principle. Used nuclear fuel is stored in a wet (cooling pool) or dry (dry cask) storage facility to reduce heat and radioactivity, then wait for it to drop to the clear level or be processed and extracted elements that are still useful. Radioactive waste is generated from many activities, including nuclear medicine, nuclear research, nuclear power generation, rare earth metal mining, and nuclear weapons reprocessing. Although nuclear waste contains radioactive substances, radioactive waste is managed very well with rigorous supervision. Nuclear waste management regulations can be said to be "uniform" or have the same standards worldwide.

2.6. Environmentally Friendly Energy (EFE) Assessment

All energy plants have advantages and disadvantages, as in Table 4. Indonesia depends on whether the power plant will be used as a large-scale baseload to replace coal power plants or used for remote areas where a small power load is required.

Table 4. Analysis of the assessment of the 4 EFE criteria

Power plants	Emission	Ecosystem Disturbance	Footprint	Waste	EFE Criteria Fulfillment	Category EFE
Wind	No (1)	Yes (0)	Large (0)	No (0)	1	Less
Sun	No (1)	Yes (0)	Large (0)	No (0)	1	Less
Geothermal	No (1)	Yes (0)	Small (1)	Yes (1)	3	Good
Hydro	No (1)	Yes (0)	Large (0)	Yes (1)	2	Enough
Nuclear	No (1)	No (1)	Small (1)	Yes (1)	4	Very good

Indonesia consists of more than 17,000 islands, with various conditions: small islands, remote and far from big islands. All types of environmentally friendly power plants can be built depending on regional interests.

2.7 Sustainable Energy

Sustainable energy provides energy supply at any time for an extended period. Countries can include several forms of sustainable energy sources in the national energy mix to stop fossil fuels. Energy sustainability reviews should be viewed from the aspect of natural resources and economics.

Energy sources that do not need combustion are called raw materials. Solar, hydro, geothermal, and wind energy are directly utilized "free". However, the process to extract it into electricity requires more complicated components and technology. Figure 4. shows the comparison of material requirements to build a power plant per unit of electrical power. Nuclear power plants require the least amount of material.

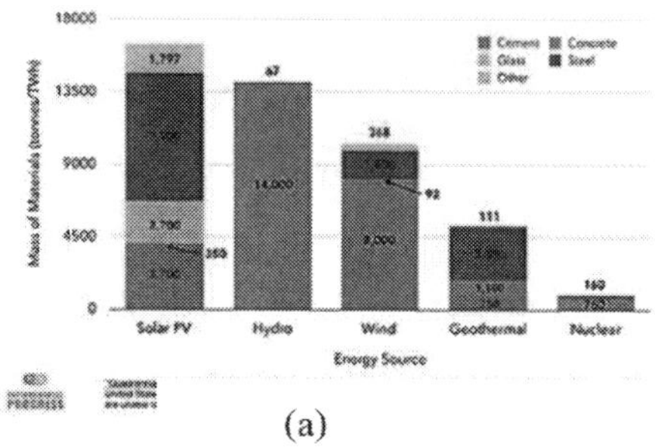

(a)

Energy Form	Resource	
	Fuel	Material Use
Wind Turbin	No	10.000
Solar PV	No	17.000
Geotermal	No	5.000
Hydro	No	14.000
Nuclear	Yes	900

(b)

Figure 4. the comparison of material requirements to build a power plant per unit of electrical power

The utilization of new and renewable energy may not provide financial benefits, but the demands of stopping global warming force a country to develop this energy. In Indonesia, the provision of subsidies, both for entrepreneurs and consumers who use electricity from NRE. The 2021 APBN is 1,006 T, and the plan is to use 53.5 T (5.32%) for electricity subsidies; the realization in 2020 for contributions will reach around 61 T. The covid pandemic is one of the factors that weigh the number of electricity subsidies.

3. Conclusion

Wind energy and solar energy do not produce CO_2 emissions in the process, more efficient than fossil energy or biomass energy. Currently, the cost of wind energy and solar energy has become a mainstream technology, which is cost-effective because the price is getting cheaper, mass production has been abundant. However, the nature of solar and wind energy is intermittent, not continuous, unlike fossil fuels which can operate 24 hours and become primary energy. Weather conditions and seasons significantly affect the yield of solar and wind energy so that they cannot be used as primary energy. Solar energy requires a storage device for solar energy in the form of a battery, limited to storage capacity, usage time and battery regeneration after some use.

Several other forms of power generation that do not produce CO_2, such as nuclear energy, geothermal and hydroelectric power plants, can also be used for decarbonization. Combining renewable energy and new energy such as nuclear energy, geothermal, and hydropower in planning the energy mix in Indonesia is very good. It can be an opportunity to eliminate CO_2 emissions or decarbonize the energy sector so that the goal of COP 21 and SDGs in the energy sector in Indonesia can be achieved.

Acknowledgements
This paper results from a joint study between LPPM UNS and PT ThorCon Power Indonesia, with PKS No. 001/PKS.THC/01/I/2021.

References

[1] Omer A M 2013 Biomass energy resources utilisation and waste management *New Dev. Blue Biotechnol. Environ. Pollut. Assess.* **3** 481–515
[2] Rogner H H, Toth F L and McDonald A 2010 Judge nuclear on its merits *ATW - Int. Zeitschrift fur Kernenergie* **55** 761–3
[3] Brecha R J 2019 Threshold electricity consumption enables multiple sustainable development goals *Sustain.* **11** 1–11
[4] Okwanya I and Abah P O 2018 Impact of Energy Consumption on Poverty Reduction in Africa *CBN J. Appl. Stat.* **9** 105–39
[5] Our World Data 2021 Research and Data
[6] Dewan Energi Nasional Republik Indonesia 2017 *Kebijakan Energi Nasional Menuju Tahun 2050: Peran Batu bara Mendukung Ketahanan Energy Naisonal* (Jakarta)
[7] Usman E 2020 *Buku Bauran Energi Nasional* (Jakarta: Dewan Energi Nasional Republik Indonesia)
[8] Mohd Thas Thaker M A, Mohd Thas Thaker H, Amin M F and Allah Pitchay A 2019 Electricity Consumption and Economic Growth: A Revisit Study of Their Causality in Malaysia *Etikonomi* **18** 1–12
[9] Takada H H, Ribeiro C O, Costa O L V and Stern J M 2020 Gini and entropy-based spread indexes for primary energy consumption efficiency and CO2 emission *Energies* **13** 1–17
[10] Pemerintah Republik Indonesia 2014 PP No. 79 Thn 2014 1–36
[11] Wiedmann T and Minx J 2007 A Definition of ' Carbon Footprint *Science (80-.).* **1** 1–11
[12] Abdallah L and El-Shennawy T 2013 Reducing carbon dioxide emissions from electricity sector using smart electric grid applications *J. Eng. (United Kingdom)* **2013**
[13] Dewan Energi Nasional Republik Indonesia 2020 *Perkembangan Bauran Energi Primer Nasional Tahun 2015 – 2019* (Jakarta)
[14] Ledec G C 2012 Annual Meeting of the International Association for Impact Assessment 27 May - 1 June 2012 *Recent Wind Power Dev. to Reduce Impacts* 1–7
[15] Jianu O, Rosen M A and Naterer G 2012 Noise pollution prevention in wind turbines: Status and recent advances *Sustainability* **4** 1104–17
[16] Gawande A and Chaudhry P 2019 Energy : a View Point With Reference To India **17** 8115–25
[17] Delbridge L 2019 Coal Ash Wastescapes: The Byproduct Of Our Coal-Fired Power Dependency *Scenar. J.* **07**

Application of simulation integrated learning model with video assisted MIKIR approach as an effort to improve understanding of earthquake mitigation

A Rusilowati[1], Supriyadi[2] and M H Pangestu[1*]

[1] Physics Education, Universitas Negeri Semarang
[2] Physics, Universitas Negeri Semarang

*Email: maulanahelmy24@gmail.com

Abstract .The research objectives (1) measure the increase in science learning outcomes and understanding of earthquake mitigation, (2) test the achievement of science learning outcomes and earthquake mitigation more than 50%,(3) describe the acceptance of the applied learning model. The research method used pre-experimental which was conducted at SMP N 2 Sedayu, Bantul Regency. The research instrument uses pretest-posttest questions and questionnaires that have been tested for validity, reliability, discriminating power and difficulty. Data analysis using N-gain, one sample t-test one party, and descriptive percentage. The results of this research are 52.5% with an average pretest value of 48.2 and a posttest of 75.4, while the results of the analysis of an increase in earthquake mitigation were 46.8%, which was obtained from an N-gain of 0.468 which calculated differently from the average value of pretest 45.5 and posttest 71.0. The results of the hypothetical science material were significantly 52.2% while for earthquake mitigation it only reached 46.8% one of the factors that influence these results, many students do not do the given simulation. The results of the student acceptance questionnaire on the learning model showed an average score of 84.72% very good category. It can be said that the applied learning model can be accepted by students.

1. Introduction

Earthquakes are natural disasters that humans cannot predict with certainty when they occur. Earthquakes according to their causes can be categorized into tectonic, volcanic, collapse, and man-made earthquakes. Tectonic earthquakes are caused by the movement of tectonic plates. Volcanic earthquakes are caused by volcanic activity. Collapse earthquakes are caused by cave collapses, and man-made earthquakes, for example, are caused by bomb blasts. Indonesia is a country located in an earthquake-prone area. This is because Indonesia's geographical location is located between three world plates, namely the Pacific Ocean plate which moves towards the West-Northwest, the Indian Ocean plate-Australian Continent (Indo-Australia) which moves to the North-Northeast, the Eurasian Continent plate which is relatively shut up (Gina, Sutomo, Awaluddin, dan Irwan, 2009; 2011) [1]. From the boundaries of these three plates form a series of world volcanoes, which encircle the Pacific Ocean called the Pacific Ring of Fire, A series of volcanoes in Indonesia meet and form volcanoes in Sumatra, Java and Nusa Tenggara [2].

From the point of view of plate tectonics, it is possible for Indonesia to have a high potential for earthquakes. The occurrence of an earthquake can cause large material losses, such as the earthquake that occurred in Kebumen, Central Java, which resulted in more than 100 buildings being severely damaged [3]. Loss of life and damage to buildings were also experienced during the 2006 earthquake in Central Java and Yogyakarta.

To reduce the large impact caused by earthquake in the form of building damage and loss of life, efforts are needed to mitigate earthquake disasters. Earthquake mitigation efforts are an effort to prepare for earthquake disasters, especially in areas with high potential for disasters that need to be instilled in the community. Disaster mitigation is a step that needs to be taken as the main starting point of disaster management [4].

One of the efforts that can be made to increase understanding and awareness of earthquake mitigation is through education. This is because formal education such as schools and universities is a strategic place to instill disaster mitigation [4].

Earthquake mitigation can be done by integrating it into learning at school in one of the subjects at the elementary, junior high or high school level. One of the subjects that can be integrated with earthquake mitigation materials is science learning. This is because science lessons take a scientific approach and require students to learn actively.

Science learning for earthquake mitigation can be done with an Simulation Integrated model with a video-assisted MIKIR approach. This model was developed by Rusilowati et al, (2020) [4]. The Simulation Integrated Model is a learning model that integrates real simulations in the implementation of learning, while the MIKIR approach is an approach that conditions children to experience, interact, communicate and reflect. The explanation of each aspect of MIKIR is as follows.

1) Experiencing (M): namely doing activities (doing) and/or observing (observing) during the learning process.
2) Interaction (I): is the process of exchanging ideas between two or more people, exchanging thoughts/ideas/ideas, discussing, responding to other people's ideas/opinions.
3) Communication (Ki): the process of conveying ideas/thoughts or feelings by one person to another. Communication can be in oral or written form, conveying ideas, conveying work results, reporting experimental results, reporting the results of group discussion.
4) Reflection (R): is an activity to look back on the learning experience and take lessons (lessons learned) in order to learn better in the future.

During the pandemic, real simulations cannot be carried out directly together in class, so it is necessary to simulate via video independently. Learning media with the help of video is one of the innovations in a learning media that is able to help deliver material in real terms. As stated by Wulandari (2018), learning media in the form of videos, accompanied by teacher explanations can make it easier for students to understand learning internally and externally [5].

Learning with the Simulation Integrated model with a MIKIR approach and video-assisted will be one of the alternatives and innovations in the science learning process and earthquake mitigation. The aims of this study are (1) to apply the Simulation Integrated learning model with a video-assisted MIKIR approach, (2) improve understanding of science material, (3) improve understanding of earthquake mitigation, (4) test the achievement of an increase of more than 50%, and (5) describe the acceptability of the applied learning model.

2. Methods

The research method using a pre-experimental design was carried out at SMPN 2 Sedayu, Bantul Regency, DIY, on February 16 to March 18, 2021 by taking samples of class VII C and VII D based on the technique using purposive sampling technique.

The stages of implementing the integrated simulation model are as follows:

1) The initial stage, this stage is carried out by giving pretest material for the structure of the earth's layer and material for earthquake mitigation. This is intended to measure students' prior knowledge about the material of the earth's layer.
2) The Core Stage, is a treatment by applying an integrated simulation learning model. Previously, students had been sent learning videos, which contained science material on the structure of the

earth's layer and earthquake mitigation with simulations, so I have observed and studied it first. Then the application of the integrated simulation model is carried out according to the learning design that has been made and uses the MIKIR approach. At the end of the lesson students are given the task of simulating how to protect themselves in the event of an earthquake, and recording it. The recording results were sent to the teacher via Google Classroom and asked to communicate it at the next meeting.

3) The final stage, carried out by giving posttest questions and student response questionnaires, which aims to determine the increase in students' knowledge on the material of the earth's layer structure and earthquake mitigation after the implementation of the integrated simulation learning model and find out student responses regarding the application of this learning model.

The data in the study were obtained from the results of the pretest-posttest scores and response questionnaires that had been given to students. The pretest and posttest questions on the material of the earth's layer structure and earthquake mitigation have been tested for validity, reliability and discriminating power and the level of difficulty of the questions. Data analysis of student knowledge was carried out using the N-gain test and the one sample t test.

The N-gain test was conducted to determine the increase in learning outcomes while the t-test was carried out to determine the achievement of the hypothesized increase. The criteria set to test the significance of increasing concept understanding is 50%. Student responses to the learning model were analyzed by percentage and categorized based on the questionnaire criteria.

3. Results and Discussion

The application of the integrated simulation model with the video-assisted MIKIR approach is applied by fulfilling the following syntax.

Table 1. Syntax of the Simulation Integrated Learning Model Approaching MIKIR

Phase / Stages	Student Activities	Approach
Engage/ Controlling	- Pay attention to the delivery of learning objectives - Responding to apperceptions from the teacher. - Pay attention to giving motivation	Interaction
Observe	- Pay attention to the delivery of subject matter - Observing the simulation (video/model)	Experience & Interaction
Doing	- Doing real simulations in groups	Experience & Interaction
Communicate	- Presenting the experience of doing simulations	Communicate
Closing	- Summarizing the subject matter with the teacher - Doing reflection	Reflection

The Simulation Integrated learning model with a video-assisted MIKIR approach can require students to perform simulations, which means that students can learn to experience, active interaction in learning discussions, communication with students and teachers, re-communicating the results of the summary and simulation practices in class, reflecting on what has been learned. studied and want to learn more. In addition, students can take meaning from learning the material of the earth's layer structure based on the interaction process that has been carried out in learning. The limitations of online learning mean that students cannot do simulations directly with their friends, but they can do it independently and record it for sharing in online classes. Students are given a stimulus to understand the problem of the impact of global warming caused by the depletion of ozone in the atmosphere. In the matter of earthquake mitigation, the implementation of the integrated simulation model was carried out in two meetings through the Wattshap group and Google meetings. Students are given learning materials about earthquake mitigation and video links to access earthquake mitigation learning videos before online learning begins.

Interactive discussions are carried out through google meetings with the teacher who leads the discussion. In the first meeting, there were interactive and communicative discussions with the topic of introducing earthquakes and their mitigation. Interactive discussion process to see student activity. At the end of the first meeting, the teacher gave the LKPD assignment in the form of a simulation which was documented by video individually at home. Students perform simulations in real terms with the help of the earthquake simulation videos that have been given.

At the last meeting, students were asked to re-explain what was done during the simulation practice based on their own understanding. In addition, the teacher also provides opportunities for other students to comment on and complement each other's results conveyed by their friends. This requires students to be active and provide opportunities for students to interact with each other and practice communicating. Increased understanding of science material is seen from the results of increased cognitive learning in the form of N-Gain scores based on changes in pretest to posttest scores. The results of the assessment can be seen in the table 2.

Table 2. Average of Pretest and Posttest scores

Subjects	Test	Average	N-gain (%)
Earth's layer structure	Pretest	48,2	52,5
	Posttest	75,4	
Earthquake Mitigation	Pretest	45,5	46,5
	Posttest	71	

The data in the form of pretest-posttest were then normalized using IBM 23 statistics using the Shapiro-wilk method with a significance level of 0.05. The results of the significance of the normality test all exceed 0.05, so it can be concluded that the data is normally distributed. The results can be seen in Table 3 below.

Table 3. Data Normality Test Results

Subject	Data	Normality Test Results	
		Significance	Criteria
Earth's layer structure	Pretest	0,336	Normal
	Posttest	0,094	Normal
Earthquake Mitigation	Pretest	0,79	Normal
	Postest	0,144	Normal

Because all data have normal distribution, the next analysis can use parametric statistics.
The results of the analysis of the significance of increasing students' knowledge on the natural science material of the earth's layer structure and earthquake mitigation can be seen in Table 4.

Table 4. Significance Test for Increasing Understanding of Earth's Structure and Earthquake Mitigation

Enhancement	t-count	t-table (df=31, α =0,05)
Earth's layer structure	4,921	2,042
Earthquake Mitigation	0,552	2,064

The results of the analysis using one sample t-test on the N-gain score obtained t count of 4.921, while t table with dk = 31 and = 0.05) of 2.042.

This means that t-count > t-table, then Ho which states an increase in science cognitive learning outcomes 50% is rejected. In other words, the increase in understanding of the material understanding of the earth's layer structure is greater than 50%.

Analysis of the significance of increasing knowledge of earthquake mitigation obtained t-count < t-table. This means that Ho which states an increase in understanding of earthquake mitigation 50% is accepted. Thus it can be concluded that the increase in understanding of earthquake mitigation is less than 50%.

The calculation results increase understanding of earthquake mitigation by 46.8%. This happened because of several influencing factors, including the fact that many students did not perform simulations in accordance with the guidelines and simulation learning videos provided. In addition, not all students take online classes through google meetings for interactive discussions due to limited signals and other supporting devices such as laptops and cellphones. This is a consequence of online learning, because limited access, quotas and internet network limitations have always been the main obstacles in the online learning process. In addition, earthquake mitigation is a new thing introduced to students in grade VII. Data on student acceptance of the simulation integrated learning model with the video-assisted MIKIR approach was obtained by providing a response scale questionnaire to students, which can be seen in figure 1.

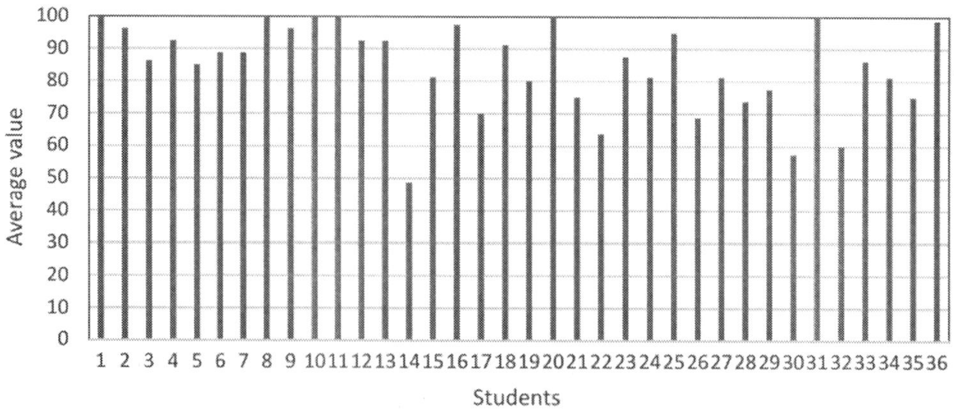

Figure 1. Learning Model Acceptance Questionnaire

Calculation of student response scores obtained an average of 84.72 in the category of high acceptability. The results of this study are in line with the research conducted by Wahyu and Rusilowati (2021) who found that learning by applying the Simulation Integrated model with the video-assisted MIKIR approach was able to increase science knowledge, flood mitigation and students' awareness of disasters [6]. This study is also in line with the research of Atmojo, et al., (2018) which found that the application of SETS-vision learning that was integrated with local wisdom was able to increase knowledge and attitudes towards natural disasters [7]. The alignment of research results is also obtained from the research of Atmojo et al., (2017) which shows that video-assisted learning is effective for increasing students' knowledge and skills about natural disaster mitigation [8] .

4. Conclusion

The Simulation Integrated learning model with the assisted video MIKIR approach that is applied boldly using the google meeting media and wattsapp group can increase the average science learning outcomes by 0.52 and understanding of earthquake mitigation by 0.46. The achievement test for increasing science learning outcomes reached more than 50%, namely 52.5%, while the understanding of earthquake mitigation did not reach 50% only reached 46.8% because one of the students did not perform the assigned simulation. The Simulation Integrated Model with the video-assisted MIKIR approach was accepted by the students, as indicated by the average percentage of the questionnaires being 84.72% in the very good category.

Acknowledgments

The researcher thanks prof. Dr. Ani Rusilowati, as a research supervisor who has supported in terms of funding and guided researchers to finish this research. In addition, the researcher would like to thank Prof. Dr. Supriyadi, as a lecturer and expert in the field of earth physics who has supported the author during the research and guided and directed the researcher.

Reference

[1] Andriyani G, Kahar S, Awaluddin M and Meilano I 2012 Kajian Regangan Selat Bali Berdasarkan Data GNSS Kontinu Tahun 2009-2011 *J. Geodesi Undip*, **1** 1.

[2] Pristanto, A I, 2010 Upaya Peningkatan Pemahaman Masyarakat Tentang Mitigasi Bencana Gempa Bumi Di Desa Tirtomartani Kecamatan Kalasan Kabupaten Sleman P rovinsi Daerah Istimewa Yogyakarta. *Skripsi. Yogyakarta: Universitas Negeri Yogyakarta.*

[3] Sarwi and Supriyadi, *Ilmu Kebumian dan Gejala Alam*, Semarang: UNNES PRESS, 2018.

[4] Rusilowati A, Hidayah I, Supriyadi, and Abidin Z 2020 *Pengembangan Model Pembelajaran Simulations Integrated Berpendekatan Mikir Mitigasi Bencana di Sekolah*, Penelitian Penugasan, no. Model Pembelajaran Untuk Mitigasi Bencana

[5] Wulandari F, 2018 Pengaruh Media Video Terhadap Kesiapsiagaan Siswa dalam Menghadapi Bencana Gempabumi di SMA Negeri 1 Gantiwarno, Klaten *J. Pend. Ilmu Pengetahuan Sos. Indo*, **3** 2, pp.18-20

[6] Rusilowati A and Jalaluddin W 2021 Implementasi Model Pembelajaran Simulations Integrated Berbantuan Video Untuk Mitigasi Bencana Banjir Di Kabupaten Brebes *J. Pend. Fisika Indo*, no. Mitigasi Banjir

[7] Atmojo S E, Anggraini D and Muhtarom T, 2017 Natural Disaster Mitigation through Integrated Social Learning Science in Primary School *Asian Social Science*, **13** 1

[8] Atmojo S E, Rusilowati A, Dwiningrum, S I A and Skotnicka M 2018 The reconstruction of disaster knowledge through thematic learning of science, environment, technology, and society integrated with local wisdom *J. Pend. IPA Indo*, **7** 2, pp.204-213.

IOP Conf. Series: Earth and Environmental Science **986** (2022) 012009 doi:10.1088/1755-1315/986/1/012009

Evaluation of safe school readiness for landslide in Tawangmangu and Matesih District, Karanganyar Regency, Indonesia, the Year 2021

C Muryani[1,2*], Sarwono[1] , G A Tjahjono[1] and S Nugraha[1]

[1] Geography Education Study Program, Universitas Sebelas Maret
[2] Disaster Research Center, Universitas Sebelas Maret

*Email: chatarinamuryani@staff.uns.ac.id

Abstract. School children are a group that is very vulnerable to disaster exposure because most of their time is spent in the classroom with their friends and teachers. To protect them, safe schools are needed both structurally, non-structurally, functionally and facilities. The purpose of this research is to find out readiness of schools located in landslide-prone areas to become landslide-safe schools. The research locations are in four elementary schools in areas prone to high landslides, namely Tengklik 1, Tengklik 2, Koripan 1 and Koripan 2. The measurement of disaster-safe school readiness used a combination of indicators from Mirzaei et al. 2019 and SPAB from the Ministry of Education and Culture. The results showed that the level of school safety from landslides was SD Tengklik 1 in the medium category, SD Tengklik 2 in the medium category, SD Koripan 2 in the low category, SD Koripan 3 in the low category.

1. Introduction

Currently, millions of children worldwide are threatened by natural and non-natural man-made disasters [1]. One of the vulnerable groups exposed to disasters is children. Children are more vulnerable than ever as there is an increasing frequency and severity of disaster risk [2,3]. Dependence on parents and their physical, emotional and mental capacities is children's level of vulnerability [4,5]. over the next decade, the number of disaster victims of children and adolescents will increase significantly [6,7].

Many students and their education are affected by disasters or emergencies. School facilities have not been designed for minimize the danger of local disasters, further increasing collapse and severe damage. The result was a series of disasters that killed tens of thousands of students [8].

The Hyogo Framework for Action 2005-2015 document was issued to consider the importance of school resilience [9,10] and the Sendai Framework for Disaster Risk Reduction 2015 - 2030 [11,12], recognize that disaster risk reduction through safety of school sites, implementation of education and access to educational content [13,14].

Disaster education is the basis for disaster risk reduction efforts. The capacity of the community to recover and facilitate himself from the impact of disasters depends on each person being can learn from insight of their ancestors and the past and how prepared they are to prepare for natural disasters that may occur in the future [15]. Children get opportunities from school to develop the skills and knowledge needed to protect themselves from disaster risk. The psychological approach of students through teachers is an effective way to prepare children for disasters [16,17]

Evi Widowati et al. (2021) [18] have developed a model for the School of Disaster Preparedness and Safety (SSSB). The developed model allows schools to increase resilience in the face of hazards by conducting self-assessment, hazards can be in the form of natural or social disasters. The factors for the formation of the SSSB include formal education curriculum, school commitment, school facilities and infrastructure, preparedness, empowerment of institutional roles, information disclosure, supervisory system, and the ability of school residents.

To increase school resilience to disasters, the Indonesian Ministry of Education and Culture, in collaboration with the National Disaster Management Agency, has established the "Satuan Pendidikan Aman Bencana (SPAB)" program as an effort to prevent and mitigate the impact of disasters in education units [16]. The objectives of SPAB in the education unit are: (a) Improving the of resources capacity; (b) Protect investment; (c) Improving the quality of education facilities and infrastructure to be safe against disasters; (d) provide protection and security to all school members (e) ensure the continuity of services; (f) providing educational services that are by the characteristics of disaster risk and the needs; (g) recovering the impact of disasters; and (h) building the independence of the education unit in running the SPAB program. The implementation of the SPAB program is regulated through the Regulation of the Minister of Education and Culture of Republic Indonesia Number 33 of 2019 concerning the Implementation of the SPAB Program. In this regulation, the implementation of the SPAB program is carried out during normal or pre-disaster situations, in emergency and post-disaster situations.

Tawangmangu and Matesih District, Karanganyar Regency are areas with a high level of landslide susceptibility. Almost every rainy season, there are reports of landslides in this area. Based on field observations, many elementary schools are located in high landslide-prone areas. For disaster risk reduction, especially for school children, schools need to be encouraged to become disaster safe schools. As an initial step, schools are prepared for disaster-safe schools. This study aims to assess the level save elementary schools in Tawangmangu and Matesih districts, Karanganyar Regency. From this research, it can be seen that disaster-safe school factors need to be improved

2. Methods

Elementary School Tengklik 1 and Tengklik 2 in Tawangmangu District and Elementary School Koripan 1 and Koripan 3 in Matesih District belong to Karanganyar Regency, Central Java Province, Indonesia. Selected as research locations because these schools are located in landslides prone areas (https://bpbd.karanganyarkab.go.id, accessed 23 September 2021)

Assessment of school level safe for landslides based on a combination of SPAB according to the Regulation of the Minister of Education and Culture Number 33 of 2019 and the Resilience School indicator from Mirzaei et al. (2019)[1].

The SPAB indicator consists of 11 indicators, there are:

a. The location is safe from disasters and easily accessible,
b. Construction of buildings safe from disasters,
c. The design and arrangement of the room safe from disasters,
d. Safe and easily accessible evacuation routes,
e. Tools and equipment for preparedness, simulation, and evacuation,
f. Conduct a periodical feasibility study of the building,
g. Ensure the facility complies with disaster safety standards,
h. Conduct disaster prevention and management education,
i. Improve the capacity of the Central and Regional Governments, education unit residents, families and communities in the implementation of the SPAB program,
j. Provide safe access to and from school,
k. Strong policies, regulations, institutions and adequate budget in the implementation of the SPAB program

To make it simpler but still measurable, the 11 indicators are classified into four hands from Mirzaei et al. (2019) [1], namely:

1. Structural (location, building construction, spatial design and arrangement, evacuation routes
2. Non-Structural (tools and equipment for preparedness, simulation, and evacuation)
3. Functional processes (a feasibility study of a building following safety standards, disaster education,
4. Facilities (accessibility, policies and regulations, institutions, budget)

They are collecting data by observation at the school and the surrounding and interviews using a questionnaire compiled based on indicators of landslide safe schools to selected informants. The informants by purposive sampling are the principal, two teachers, and the village head. For ease of analysis, the answers to each question have been directed into three options, each of which expresses a safe (score 3), less safe (score 2) and unsafe (score 1) situation. Because the questionnaire consists of 30 questions, the maximum score = 3X30 = 90 and the minimum score = 1X 30 = 30, so it can be classified as follows:

- Score 71 - 90 : Safe Schools from landslides
- Score 51 – 70 : Less safe Schools from landslides
- Score 30 – 50 : Unsafe Schools from landslides

3. Result and Discussion

Based on the results of observations and interviews with informants in four schools, the following results were obtained:

3.1 Structural Indicators

Structural indicators for landslide-safe schools are:

(a) The location of the school from the source of the disaster (2 questions)
(b) Disaster safe building construction (2 questions)
(c) Disaster safe spatial planning (2 questions)
(d) Existence of evacuation routes and gathering points (3 questions)

The location of the school (such as topography, soil type) and building construction (such as the building foundation, placement of the building on a flat or sloping area) are considered necessary in structural indicators [19]. School layout also affects student safety, for example, the size of windows and doors, doors must open outwards, designing emergency routes, also crucial in disaster-safe schools [20]

The structural indicators for Disaster Safe Schools in the four schools studied are as follows:

Table 1. Structural Indicators of Disaster Safe Schools at Primary School (PS) Tawangmangu and Matesih District

Indicators	**School**			
	TK-1	**TK-2**	**KR-2**	**KR-3**
Distance of school from threat	2	1	1	2
Length of school from resident's house	2	1	1	2
Exposure of schools to disasters	3	1	1	3
Safe building construction	3	3	2	2
Easy access school	2	2	3	2
Classroom arrangement	2	2	1	3
Evacuation route signs	2	1	1	1
Evacuation route safety	1	1	1	1
The existence of a gathering point	2	2	1	1
Total of score	**19**	**14**	**12**	**17**

Source: primary data analysis

Notes:
TK-1= Tengklik 1 PS; TK-2= Tengklik 2 PS, KR-2= Koripan 2 PS; KR-3=Koripan3 PS

The data above illustrates that structurally elementary schools in Tawang Mangu and Matesih Districts are not safe from landslides. Essential indicators that do not yet exist are evacuation route signs and safe evacuation routes. Evacuation route signs are vital to give students which direction to run if a landslide occurs. The evacuation route should be wide enough to accommodate many people who are running to save themselves.

3.2 Non-Structural Indicators

School equipment and facilities that support disaster risk reduction are included in non-structural indicators [1]. However, school equipment can cause students to be injured or even die, such as bookshelves, hanging fans, chimneys, wall hangings, as well as included in non-structural indicators [21]. In this study, non-structural indicators are facilities owned by schools that can support landslide risk reduction.

Table 2. Non-Structural Indicators of Disaster Safe Schools at Elementary School (PS) Tawangmangu and Matesih District

Indicators	**School**			
	TK-1	**TK-2**	**KR-2**	**KR-3**
Availability of fire extinguishers	1	1	1	1
The ability of Central speaker	1	3	3	1
Ability of Megaphone	3	1	1	1
The ability of School Bell	3	3	1	3
Availability of school fence	3	2	3	3
Availability of stretcher	1	1	3	1
Availability of simple medical equipment	3	3	3	2
Availability of first aid drugs	3	3	3	2
The ability of Oxygen cylinder	1	1	1	1
Total score	**17**	**18**	**17**	**13**

Source: primary data analysis
Notes:
TK-1= Tengklik 1 ES; TK-2= Tengklik 2 ES, KR-2= Koripan 2 ES; KR-3=Koripan3 ES

The data above shows that although it is not optimal, the schools studied already have several facilities that can be used for initial aid for victims. Still, specific equipment such as fire extinguishers, oxygen cylinders is not yet available in schools.

3.3 Functional Process Indicators

School resilience is also influenced by the dimensions of functional processes which include communication, management, health, and education [1]. Disaster education programs for teachers, staff, students and parents are an essential factor in building disaster safe schools [22]. In addition, the integration of disaster management into the curriculum, school regulations, syllabus, and preparation of disaster management plans is very important in school resilience to disasters [23].

The processes that should be carried out to improve school safety from landslides:

Table 3. Functional Process Indicators of Disaster Safe Elementary Schools (ES) in Tawangmangu and Matesih Districts

Indicators	School			
	TK-1	TK-2	KR-2	KR-3
Integration of disaster materials in the curriculum	3	1	2	2
Integration of disaster materials in subjects				
Integration of material in extracurricular activities	2	1	2	2
Integration of disaster materials into extracurricular activities	2	2	2	1
There is an emergency response SOP	2	2	2	2
Provided a disaster emergency simulation				
	1	1	1	1
	1	3	1	1
Total score	**11**	**10**	**10**	**9**

Source: primary data analysis
Notes:
TK-1= Tengklik 1 ES; TK-2= Tengklik 2 ES, KR-2= Koripan 2 ES; KR-3=Koripan3 ES

From the data above, the functional process conditions for disasters, especially landslides, have not been widely carried out; even in four schools, they do not have an emergency SOP. Other activities that need to be carried out are emergency simulations for students, teachers and parents.

3.4 Facility indicators

Facilities needed in school resilience are human and financial resources [24]. Different stakeholders, including internal and external, are very important in building safe schools [8]. Budgets are also needed for preparedness, emergency response, reconstruction programs, and supervision and monitoring [23]. Facility indicators in this study were measured by,

Table 4. Facility indicators of Disaster Safe Elementary Schools (ES) in Tawangmangu and Matesih Districts

Indicators	School			
	TK-1	TK-2	KR-2	KR-3
DRR Forum has been formed	1	2	1	1
There is a budget for DRR	1	1	1	1
There are student activities for DRR	1	2	1	2
There is coordination with the	1	3	1	1
community	1	2	1	1
There is coordination with the government	1	1	1	1
There is coordination with NGOs				
Total Score	**6**	**11**	**6**	**7**

Source: primary data analysis
Notes:
TK-1= Tengklik 1 ES; TK-2= Tengklik 2 ES, KR-2= Koripan 2 ES; KR-3=Koripan3 ES

From the data above, it can be seen that the facility indicators in the four schools studied are still fragile. After checking with school principals, in general, they do not understand the steps that must be taken to coordinate with various parties in disaster risk reduction. The budget for DRR is also not provided by the government.

3.5 Combined Indicators of Landslide Safe Schools

The four disaster safe indicators that have been discussed above, namely structural, non-structural, functional indicators of processes and facilities are combined to obtain the overall condition of schools in landslide-prone areas of Tawangmangu and Matesih Districts if assessed from the indicators of disaster-safe schools.

The combined results of the four indicators are,

Table 5. Combined Indicators of Disaster Safe Elementary Schools (ES) in Tawangmangu and Matesih Districts

Indicators	**School**			
	TK-1	**TK-2**	**KR-2**	**KR-3**
Structural	19	14	12	12
Non-Structural	17	18	17	13
Functional Process	11	10	10	9
Facility	6	11	6	7
Total score	**53**	**53**	**45**	**41**

Source: primary data analysis
Notes:
TK-1= Tengklik 1 ES; TK-2= Tengklik 2 ES, KR-2= Koripan 2 ES; KR-3=Koripan3 ES

The results of the combined scoring, if included in the classification of disaster safe schools the results are:

Table 6. Categories of Disaster Safe Elementary Schools (ES) in Tawangmangu and Matesih District

School	**Total Score**	**Category**
Tengklik 1 ES	53	moderate
Tengklik 2 ES	53	moderate
Koripan 2 ES	45	low
Koripan 3 ES	41	low

From the data above, it can be seen that two elementary schools in Tengklik Village, Tawangmangu District, are categorized as "moderate" disaster safe schools, while the two elementary schools located in Koripan Village, Matesih District are both in the low category. Factors that need attention and improvement are the need to determine evacuation routes and places, provision of rescue equipment for emergencies, preparation of emergency SOPs, integration of disaster materials in learning and coordination with various stakeholders.

4. Conclusion

4.1. From the results of the evaluation of Disaster Safe Schools, it can be concluded that two elementary schools in Tawangmangu District are in the medium category, two elementary schools in Matesih District are in a low category.

4.2. The limiting factors that need to be improved in quality and quantity are determining evacuation routes and places, providing rescue equipment for emergencies, preparing emergency SOP, integrating disaster materials in learning and coordinating with various stakeholders.

Acknowledgements
We acknowledge KEMENRISTEK/BRIN (Ministry of Research and Technology/National Research and Innovation Agency) for funding this research. We would like to thank LPPM Universitas Sebelas Maret for assisting researchers in managing the administration of this research. Thank you also to various parties who have supported and assisted in the completion of this research.

References
[1] Mirzaei S, Mohammadinia L, Nasiriani K, Tafti A A D, Rahaei Z, Falahzade H and Amiri H R 2019 School resilience components in disasters and emergencies: A systematic review *Trauma Mon.* **24**
[2] Tuswadi and Hayashi T 2014 Disaster Prevention Education in Merapi Volcano Area Primary Schools: Focusing on Students' Perception and Teachers' Performance *Procedia Environ. Sci.* **20** 668–77
[3] Sheffield P E, Uijttewaal S A M, Stewart J and Galvez M P 2017 Climate change and schools: Environmental hazards and resiliency *Int. J. Environ. Res. Public Health* **14**
[4] Kar N 2009 Psychological impact of disasters on children: Review of assessment and interventions *World J. Pediatr.* **5** 5–11
[5] Peek L 2008 Children and Disasters: Understanding Vulnerability, Developing Capacities, and Promoting Resilience - An Introduction *Child. Youth Environ.* **18** 1–29
[6] Kousky C 2016 Impacts of natural disasters on children *Futur. Child.* **26** 73–92
[7] Petal, M., Ireland, N., Wade, D., Robinson, G., Garaelulu, M., Nelson, T., Baker, L., Turkmen, Z., Pambid, L., & Kani S 2015 Participatory School Disaster Management Toolkit
[8] Öcal A and Topkaya Y 2011 Earthquake Preparedness in schools in seismic hazard regions in the South-East of Turkey *Disaster Prev. Manag. An Int. J.* **20** 334–48
[9] Amri A, Haynes K, Bird D K and Ronan K 2018 Bridging the divide between studies on disaster risk reduction education and child-centred disaster risk reduction: a critical review *Child. Geogr.* **16** 239–51
[10] Cui K, Han Z and Wang D 2018 Resilience of an earthquake-stricken rural community in southwest china: Correlation with disaster risk reduction efforts *Int. J. Environ. Res. Public Health* **15**
[11] Munene M B, Swartling Å G and Thomalla F 2018 Adaptive governance as a catalyst for transforming the relationship between development and disaster risk through the Sendai Framework? *Int. J. Disaster Risk Reduct.* **28** 653–63
[12] Anon 2015 The Sendai Framework for Disaster Risk Reduction: Renewing the Global Commitment to People's Resilience, Health, and Well-being *Int. J. Disaster Risk Sci.* **6** 164–76
[13] Briceño S 2015 Looking Back and Beyond Sendai: 25 Years of International Policy Experience on Disaster Risk Reduction *Int. J. Disaster Risk Sci.* **6** 1–7
[14] Tipler K, Tarrant R, Johnston D and Tuffin K 2017 Are you ready? Emergency preparedness in New Zealand schools *Int. J. Disaster Risk Reduct.* **25** 324–33
[15] The Sendai Declaration 2015 *International Forum for Promoting Education on Disaster Resilience "Developing a Resilient Community and Improving Disaster Education and Regional Disaster Preparedness"* (Japan)
[16] Elangovan A R and Kasi S 2014 Psychosocial disaster preparedness for school children by teachers *Int. J. Disaster Risk Reduct.* **12** 119–24
[17] Lam S, Zwart C, Chahal I, Lane D and Cummings H 2018 Preventing violence against children in schools: Contributions from the Be Safe program in Sri Lanka *Child Abus. Negl.* **76** 129–37
[18] Widowati E, Istiono W and Husodo A H 2021 The development of Disaster Preparedness and Safety School model: A Confirmatory Factor Analysis *Int. J. Disaster Risk Reduct.* **53** 102004
[19] Ochola S O, Eitel B and Olago D O 2010 Vulnerability of schools to floods in Nyando River

catchment, Kenya *Disasters* **34** 732–54
[20] Kisioglu A N, Demirel R and Ozturk M 2005 Assessing the indoor environment of primary schools in the southwest of Turkey *Indoor Built Environ.* **14** 141–5
[21] Grimaz S and Malisan P 2016 VISUS: A pragmatic expert-based methodology for the seismic safety triage of school facilities *Boll. di Geofis. Teor. ed Appl.* **57** 91–110
[22] Shiwaku K, Ueda Y, Oikawa Y and Shaw R 2016 *School disaster resilience assessment in the affected areas of 2011 East Japan earthquake and tsunami* vol 82 (Springer Netherlands)
[23] Tong T M T, Shaw R and Takeuchi Y 2012 Climate disaster resilience of the education sector in Thua Thien Hue Province, Central Vietnam *Nat. Hazards* **63** 685–709
[24] Lee D E, Parker G, Ward M E, Styron R A and Shelley K 2008 Katrina and the Schools of Mississippi: An Examination of Emergency and Disaster Preparedness *J. Educ. Students Placed Risk* **13** 318–34

Improving natural disater understanding of junior high school students

Ratih Puspita Dewi[1,2,*], Suharjo[1, 2], Dahroni[1, 2], Sujalwo [1,3], Ayu Fatonah [1], Yoza Wahyuningsih [1]

[1] Geography Education Department, Universitas Muhammadiyah Surakarta, Indonesia
[2] Center for Disaster Mitigation Study, Universitas Muhammadiyah Surakarta, Indonesia
[3]Informatics Engineering Education Department, Universitas Muhammadiyah Surakarta, Indonesia

[*]Email: rpd229@ums.ac.id

Abstract. Natural disaster management efforts are critical, especially in disaster-prone areas. One of them is carried out through education in schools. This study aims to implement disaster mitigation teaching media, analyze and compare the effectiveness rate of these implementations in Klaten Regency. This study was carried out quantitatively using the one-group pretest and posttest design. Respondents were determined by purposive sampling based on the location of schools in areas prone to natural disaster in the Klaten Regency. They consist of 704 students, 291 students were respondents of intracurricular learning, and 413 students were respondents of extracurricular learning. The study results showed that of all instructional media applied in disaster mitigation learning strategies, five learning media had the best effectiveness, namely volcanic mock-ups, animated videos, presentation slides, supplementary books, and posters. Furthermore, there were differences in the effectiveness of implementation, namely in intracurricular including volcanic mock-ups by 37.5%, earthquake video by 24.8%, landslide video by 25.7%, and floods slide presentation by 59.9%. Extracurricular include volcanic eruption video by 41.6%, earthquake video by 27.3%, floods supplementary book by 35.3%, hydrometeorological poster by 46.7%. An interesting finding was that the effectiveness of learning strategies with advanced digital media was not better than other simpler instructional media.

1. Introduction

Since in the early of the year 2000 the attention to disaster occurance was increased due to the disaster sequence of event and it was followed by the rising of the effort to reduce the disaster risk in the community, one of them was through school community. It was known that children are vulnerable to disaster [1]. School can be the place that represent safe and secure for the children [2]. In the global context, research that focussed to disaster was significantly growing [3]. Furthermore, research about children and disaster increased over the last decade and those research found that disaster risk reduction education and children participation had positive impact for children [4]. Research focused on children have to deal with children vulnerability and experience, the effort held by the adult to reduce children' vulnerability and improve children roles in disaster risk reduction.

Disaster Mitigation Studies Center in Universitas Muhammadiyah Surakarta, Central Java, Indonesia was established on the basis consideration of numerous disaster threat in Indonesia. The studies center vision is become information center and catalyst in disaster risk reduction through collaboration with academic community and society involving sains, technology, social, and culture as part of Muhammadiyah society. This goes along with Muhammadiyah's movement in dealing with disaster through establishment of Muhammadiyah Disaster Management Center (MDMC). The disaster mitigation studies actively provide assistance in disaster risk reduction education in Klaten Regency, Central Java since 2014 in collaboration with regional management disaster agency of Klaten Regency. The disaster mitigation center doing research about the using of teaching materials in disaster safe school through intracurricular and extracurricular activities.

Klaten regency that located in southest part of Central Java Province have a high disaster vulnerability level. It was because its location related to endogenic movement that create two landforms which are Mount Merapi landform and Bayat Hills. On the other hand, hydrometeorological phenomena such as tornado, floods, and landslide made some district in Klaten Regency have numerous disaster threat and occurance more than its surrounding regions. However, the government of Klaten Regency is one of regional government that prioritize disaster risk reduction. This is evidenced by the issuance of regent regulation about the implementation of disaster curriculum.

University take part in giving assistance to the development of disaster safe school. The university ensure the implementation of disaster safe school through giving advice of learning organization and other recommendations such as the right implementation of evacuation route sign in school and the appropriate disaster teaching materials. In disaster risk reduction, children participation are important as become part of community's prepredness and resillence so that they could involve in disaster risk reduction activities and effort [6].

2. Methods

The purposes of this study are 1) implement instructional media to improve natural disaster understanding in the intracurricullar and extracurricular activities, 2) to verify the effectiveness rate of the media 3) to compare the effectiveness rate of the media. The research was quantitative reseach with pre-experimental research design using one-group pretest and posttest design. The indicators of the test were about concept of the disaster, the cause of disaster, effect of disaster, and disaster mitigation. The test was multiple choice test. The students filled out the test for about 45 minutes both in pretest and postest. The answer then assess using scoring technique which 1 to correct answer and 0 for false answer.

2.1. Participants

The subject of this research were junior high school students in Klaten Regency. A total of 704 students from 14 selected schools participated in this study. The students were separated in to two groups. The first group consist of 291 students in the intracurricullar and the second group contained 413 students in the extracuricullar activities. The number of extracuricullar students was more than in intracurricullar student because in junior high school curriculum in Klaten Regency disaster material is delivered more in exstracurricular activities than in intracurricular activities. The extracuricullar was disaster safe school extracurricular and scout. The intracurricullar was teaching and learning activities in the classroom through integrated social science studies especially in geography related material. The basic competencies in junior high school students relate to disaster was students could describe earth structure, explain disaster phenomena, and disaster risk reduction effort. The sampling technique was using purposive sampling. The students were selected from several schools in Klaten Regency that prones to natural disaster. The disaster were floods, earthquakes, Mount Merapi eruptions, tornados, and landslides.

2.2. Teaching Strategies

Geography education department of universitas Muhammadiyah Surakarta in collaboration with regional management disaster agency of Klaten Regency take part in the implementation of school safe programme through learning activities both in intracuricullar and extracurricular activities. Teaching strategies that implemented to increase students understanding of natural disaster in their region was using both analog and digital teaching media. The analog media were book, modul, poster, flipchart and mock up, whereas digital media was video. This two-type of media were selected because instructional media hold an important role in delivering learning material to the student effectively furthermore both analog and digital media have characteristic that might appropriate to teaching and learning of certain type of disaster. The assumption was different type of disaster need proper learning media it could be the analog or digital media. The subject competency was introduction concept of disaster, mitigation, and preparedness. The distribution of the activities was adapted each hazard that school faced. The school location was separated into four categories. The first category was the school that located in the highest altitude near to the peak of Mount Merapi, the school was Public Junior High School 2 in Kemalang District. The second location was in the upperslope of Mount Merapi, the school was public junior high school 3 Maningrenggo. The third location was in the lowerslopes of Mount Merapi, the school was Public Junior High School 1 Cawas, public junior high school 1 Klaten, public junior high school 2 Klaten, public junior high school 4 Klaten, public junior high school 1 Wedi, public junior high school 1 Karangdowo, Public Junior High School 1 Ceper, Public Junior High School Wonosari, and public junior high school 1 Delanggu. The last location was in Bayat Hills, the school was Public Junior High School 1 Bayat, Public Junior High School 3 Gantiwarno, and Public Junior High School 1 Gantiwarno.

2.3. Disaster Safe School in Junior High School

Regional management disaster agency of Klaten Regency arranges disaster safe school programme for junior high school that located in the slope of Mount Merapi and Bayat Hills. In the first location, it was known that Mount Merapi eruption cause serious impact to human and environment especially in 2010 when the eruption was causing hundreds of people death. One of the victims was the caretaker of Mount Merapi. Eventhough Mount Merapi is better known cuasing impact in Special Region of Yogyakarta, its impact is also causing significant influences in Klaten Regency that located in the estern part of Mount Merapi. Regarding the second location of disaster school safe in Bayat Hills, it was known that Bayat Hills was created because of tectonic activity and structured of pretetiary rocks. *Dengkeng* fault was across this region that cuase in this location more vulnerable to earthquake.

Disaster safe school is an appropriate programme to develop disaster risk reduction education model which involve participation of the students as implementation of children right in disaster management in school. In this programme students learn to develop disaster preparedness. Students are given the opportunity to identify the most likely disaster' types that could occur in their region and disaster mitigation to prevent serious damage. For example, students identify things that could fall on them when earthquake occurs. In the more complexs learning activities students learn to create evacuation route and choose the safest gathering place. Students were expected to be inform safety behaviour to the family and the nearest community.

In creating school' policy and safety guidance, students have a chance to give their opinion. The objectives are to envolve the student confidence and creating good communication between the students and the teachers. Safe school programme prioritizes of how adults should listen to children perspective because teachers might have different backgrounds and experiences compared to the students. The important key is placing the students as part of the school and social community. The student's perspective could be one of the schools develpoment's consideration. The students give their perspective through student's organization or directly to the teachers. The principal, teachers and school staff should give positive reactions and responsible to explain school condition to the students.

Children participation in disaster risk reduction embodied in the four pillars of disaster School Safe. The first pillar is increase spiritual aspect to develop empathy and concern to natural disaster, the second pillar is disaster education, the third pillar is disaster school management and the last pillar is school' facility management. The main activities of safe school are through classroom learning activities, extracurricular, media information, and supplementary activity. The supplementary activities include seminar, discussion, contest, or other form that involve school community together with Muhammadiyah society at a wider scale. This activity aims to socialized emphaty to disaster occurance.

3. Results and Discussions

One of the goverment respond to unexpected series of numerous disaster occurance was by publishing regent regulation number 6 in the year of 2006 about guidance to disaster mitigation learning in Klaten Regency. The general purposes were to build awareness culture, preparedness and toughness to reduce loss potential and to develop emphaty and solidarity. On the other hands, the specific purposes were empowering the institutional role and school communication skills, integrating disaster risk reduction into intracurricular and extracurricular activities and built partnerships with various stakeholders to support the implementation of disaster risk reduction in schools, both structural and non-structural.

3.1. Intracurricular activities

This activitiy were done through classroom learning activities using several learning media. In this part, students learnt about Mount Merapi Eruption, landslide, earthquake, floods.

Instructional strategies for Mount Merapi Eruption

The instructional strategy was implemented using several instructional media namely supplementary books, leaflet, and mock-up. Those three insctrucional medias were implemented in three different school as describe below:

Table 1. Instructional learning programmes for Mount Merapi Eruption Materials

School Name	Instructional media	Types of disaster studied	Effectiveness rate
Public Junior High School 1 Bayat	Supplementary books	Mount Merapi eruption	Pretest: 68.7, Postest: 78.8 12.8% (very small)
Public Junior High School 3 Manisrenggo	Leaflet	Mount Merapi Eruption	Pretest: 73.4, Postest: 82.6 11.1% (very small)
Public Junior High School 1 Cawas	Mock-up	Mount Merapi Eruption	Pretest: 50.6, Postest:80.9 37.5% (Moderate high)

Sources: field experiment, 2021

The students were given the media in classes learning, however before that the pretest was held to measure student' prior knowledge to Mount Merapi Eruption and postest to know the impact of the media to student's understanding. The result showed that using supplementary book was effective to improve student understanding by 12.8%, leaflet by 11.1%, and mock-up by 37.5%. It was found that mock-up is the most effective media to impove student's understanding of Mount Merapi Eruption. Mock-up was an artificial media of device or environment in scale or full-scale model. The characterictic of mock-up was represent in three dimension (3D) that showed the shape and function of specific device or environment so that it more concrete than other media. Thus, the students have the accurate idea of the subject.

3.2. Instructional strategies for Earthquake

The media used to improve student' understanding of earthquake were video, leaflet, and poster. The experiment was done in four different school, the distribution is shown in the table 2.

Table 2. Instructional learning programmes for Earthquake Materials

School Name	Instructional media	Types of disaster studied	Effectiveness rate
Public Junior High School 2 Wonosari	Video	Earthquake	Pretest: 70.1, Postest: 90.1 22.2% (small)
Public Junior High School 3 Manisrenggo	Video	Earthquake	Pretest: 60.8, Postest: 80.9 24.8% (small)
Public Junior High School 1 Gantiwarno	Leaflet	Earthquake	Pretest: 66.7, Postest: 92.9 28.2% (low)
Public Junior High School 3 Gantiwarno	Poster	Earthquake	Pretest: 55.7, Postest: 85.2 34.6% (low)

Sources: field experiment, 2021

The table 2 show that poster inceasing student' understanding better by 34.6% compared to other media. It was followed by leaflet (28.2%), and two other video (24.8% and 22.2%). Poster was media to inform or to promote something both in text and graphic. Poster was made in apreciabel form and informative so that it could draw attention of students. The poster information was simple yet easy to understand so that students memorized the content easly.

3.3. Instructional strategies for Flood

There were two media that tested to improve student's understanding of flood materials namely powerpoint and video. The allocation is shown in table 3.

Table 3. Instructional learning programmes for Flood Materials

School Name	Instructional media	Types of disaster studied	Effectiveness rate
Public Junior High School 2 Wonosari	Video	Floods	Pretest: 64.8, Postest: 77.3 16.2% (very small)
Public Junior High School 2 Klaten	Powerpoint	Floods	Pretest: 20.4, Postest: 50.9 59.9% (very high)

Sources: field experiment, 2021

The table show the comparation of video and powerpoint in the effort to improve student' understanding. It was striking that powerpoint more effective that video. The effectiveness of powerpoint was 59.9% categorized as very high rate, however the effectiveness rate of video was 16.2% conclude as very small category. Powerpoint was one of the most widely used media in teaching and learning. It could consist of text, picture, and video. Furthermore, it can be designed with a variety of interesting animation effects so that it can avoid the student's boredom and it can give variety type of information.

3.4. Instructional strategies for Landslide
There was one media applied to tested the effectiveness to improve student' understanding of landslides the media was animated video. the video was implemented in two different school. The result show that there were slightly differences of the effectiveness rate by 25,4% and 25,7% conclude as small category. One of children participation in disaster risk reduction trough education and training was learn about disaster risk reduction [6]. In the context of prepredness and disaster mitigation there were six competencies students should achieved which were (1) types of disaster, (2) influenced factors in disaster, (3) disaster-prones areas, (4) disaster effect, (5) implementing disaster risk reduction, (6) apply preparedness habituation where students lives [7].

The result showed that the best instructional media in learning Mount Merapi Eruption was mock-up. Mock up is a prototype using affordable and easly to find materials [8]. On the other hand, the most effective instructional media to learn about earthquake is poster. the result showed that learning floods is more effective using powerpoint.

3.5. Extracurricullar activities

3.5.1. Instructional strategies for Mount Merapi Eruption
There were three instructional media implemented in extracurricular activies to enhance students' understanding of Mount Merapi Eruption materials namely poster, supplementary video and video. The allocation of those three media is described in table 4.

Table 4. Instructional learning programmes for Mount Merapi Eruption Materials

School Name	Instructional media	Types of disaster studied	Effectiveness rate
Public junior high school 2 Kemalang	Poster	Mount Merapi Eruption	Pretest: 63.2 Postest: 82.8 23.7%
Public junior high school 1 Wedi	Supplementary book	Mount Merapi Eruption	Pretest: 67.2 Postest: 89.8 25.2%
Public junior high school 2 Kemalang	Video	Mount Merapi Eruption	Pretest: 53.9 Postest: 92.3 41.6%

Sources: field experiment, 2021

Table 4 show that the highest effectiveness rate was video by (41.6%), followed by supplementary books by 25.2%, and poster by 23.7%. Video consist of text, voice, and picture. This media was more appropriate than other media in learning of Mount Merapi Eruption because it provide to types of student's learning style both auditive of visual. Furthermore, through video student can see and hear the real environment and phenomena in Mount Merapi so that they could easly undertstand the material.

3.5.2. Instructional strategies for Earthquake
There were six media implemented in extracurricular activities in order to impove student understanding of earthquake namely poster, guide book, mock-up, lealflet, and video. the media was applied in six different school in Klaten Regency.the distribution of the media is described in table 5.

Table 5. Instructional learning programmes for Earthquake Materials

School Name	Instructional media	Types of disaster studied	Effectiveness rate
Public junior high school 3 Gantiwarno	Poster	Earthquake	Pretest: 75.4 Postest: 84.7 10.9%
Public junior high Delanggu	Poster	Earthquake	Pretest: 60 Postest: 70.9 15.4%
Public junior high school 4 Klaten	Guide book	Earthquake	Pretest: 69.3 Postest: 83.7 17.2%
Public junior high school 1 Klaten	Mock-up	Earthquake	Pretest: 72.1 Postest: 91.6 21.3%
Public junior high school 2 Ceper	Leaflet	Earthquake	Pretest: 60.3 Postest: 80.9 25.5%
Public junior high school 2 Kemalang	Video	Earthquake	Pretest: 61 Postest: 84 27.3%

Sources: field experiment, 2021

The result show that video had the highest effectiveness rate by 27.3%, followed by leaflet by 25.5%, and mock-up by 21.3% those three has slightly differences. Furthermoere guide book has 17.2%, poster by 15.4%, and poster by 10.9%. Similar to learning of Mount Merapi Eruption, video was also suitable in learning of flood. Through video students can see the phenomena for real so that student can memorize better than other media.

3.5.3. *Instructional strategies for Flood*

There were four instructional media that implemented in the learning of floods related materials namely powerpoint, poster, module, and supplementary book. The distribution is described in table 6.

Table 6. Instructional learning programmes for Flood Materials

School Name	Instructional media	Types of disaster studied	Effectiveness rate
Public junior high school 2 Ceper	Powerpoint	Floods	Pretest: 70.8 Postest: 80.9 12.5%
Public junior high school 1 Karangdowo	Poster	Floods	Pretest: 74 Postest: 90.6 18.3%
Public junior high school 1 Karangdowo	Modulle	Floods	Pretest: 60.8 Postest: 80.2 24.2%
Public Junior High School 4 Klaten	Supplementary book	Floods	Pretest: 56 Postest: 86.6 35.3%

Sources: field experiment, 2021

The result showed that supplementary book had highest effectiveness rate by 35.3%, followed by module by 24.2%, poster by 18.3%, and the lowest was powerpoint by 12.5%. supplementary book was developed in order to give more information about flood where in student's book it harder to understand or there was no specific explanation about the material. Through supplementary book, student got enough information in order to understand flood.

3.5.4. Instructional strategies for hydrometeorological, tornado, and drought
In this type of disaster there were two types of media implemented they were poster and supplementary books. The distribution was described in table 7.

Table 7. Instructional learning programmes for hydrometeorological, tornado, and drought

School Name	Instructional media	Types of disaster studied	Effectiveness rate
Public junior high school 1 Karangowo	Poster	Tornado	Pretest: 60.2 Postest: 80.3 25%
Public junior high school 1 Wedi	Supplementary book	Hidrometeorological disaster	Pretest: 64 Postest: 88.8 27.9%
Public junior high school Delanggu	Poster	Drought	Pretest: 44.8 Postest: 84.1 46.7%

Sources: field experiment, 2021

The result show that the effectiveness rate of poster in order to improve students understanding of tornados was 25%, furthermore supplementary book that tested in hydrometeorological material was 27.9%, and the last was poster that implemented in drought materials was 46.7%. Eventhough, the type of disaster was different however, due to the similarity of the characteristic the three disasters were categorized as one. Overall, the highest effectiveness rate was poster. Poster give simple information in eye catching design so that students understand the hydrometeorological disaster easly than other media.

People in disaster prone areas need to enhance their disaster awareness than other in the non prones areas. The media used in informating about disaster were important. The more effective media used in disseminating disaster information the more effective action to face the disaster [9]. Furthermore people in disaster areas have higher responses to media than people in non-disaster areas [10]. The important of the media was not only in informating disaster context however it was also essential to select the appropriate media in disaster learning. It was known that in disaster risk reduction, the integration of physical science and cultural science in disaster risk reduction need to be develop. Not only science that need to be improve in disaster risk reduction however culture need to be enhance also [11,12]. One measure to improve culture of disaster awareness was through education. Through education and learning people could improve their knowledge and ability in disaster preparedness [13]. In Klaten Regency, the integration of disaster risk reduction in education has been established through various ways, one of them was integration in the teaching and learning activities both in intracurricullar and extracurricular. Furthermore, the study of disaster risk reduction for children also rising [4].

Through teaching and learning activities student can participated in disaster risk reduction directly, as stated that children participation have positive impact in disaster risk reduction management [14,15]. In the learning context the selection of instructional method was essential [16] one of it was in media selection. In this study, the teaching strategies to enhance the understanding of disaster was through the implementation of several instructional media. The result showed that in intracuricullar activities the most appropriate media in learning of Mount Merapi Eruption learning was mock-up, earthquake was poster, floods was using video, and landslide using video whereas in extracurricular activities in learning of Mount Merapi eruption the most suitable media was the video, earthquake was video, floods was supplementary book, and hydrometeorological disaster was using poster. Education was important in disaster risk reduction as stated that the more educated the more people prepared to face disaster accurately [17]. Thus, disaster risk reduction education need to be implemented in the school curriculum [18].

4. Conclusion

Through the implementation of several instructional media in junior high school it was found that the most effective instructional media for each disaster type in intracurricullar activities were mock-up for Mount Merapi Eruption (37.5%) categorized as moderate high, earthquake was poster (34.6%) categorized as low, floods was using video (59.9%) categorized as very high, and landslide using video (25,7%) categorized small category. In extracurricular activities the most effective media were Mount Merapi Eruption was video (41.6%), earthquake was video (27.3%), flood was suppelementary book (35.3%), and hydrometeorological, tornado, and drought was poster (46.7%).

Acknowledgment
Thanks to Universitas Muhammadiyah Surakarta that have funded this research

Reference

[1] Carr A, Abad Merchán M and Ullauri N 2020 Conversations about disasters: listening to Ecuadorian children *Child. Geogr.* **18** 222–33
[2] Mutch C and Gawith E 2014 The New Zealand earthquakes and the role of schools in engaging children in emotional processing of disaster experiences *Pastor. Care Educ.* **32** 54–67
[3] Mutch C 2014 The role of schools in disaster preparedness, response and recovery: what can we learn from the literature? *Pastor. Care Educ.* **32** 5–22
[4] Amri A, Haynes K, Bird D K and Ronan K 2018 Bridging the divide between studies on disaster risk reduction education and child-centred disaster risk reduction: a critical review *Child. Geogr.* **16** 239–51
[5] William A. Anderson 2013 bringing children into focus on the social science disaster research agenda *J. Chem. Inf. Model.* **53** 1689–99
[6] Pfefferbaum B, Pfefferbaum R L and Van Horn R L 2018 Involving children in disaster risk reduction: the importance of participation *Eur. J. Psychotraumatol.* **9**
[7] Hidayati D, Widayatun, Hartana P, Triyono and Kusumawati T 2015 Panduan Mengukur Tingkat Kesiapsiagaan Masyarakat dan Komunitas Sekolah 2020–4
[8] Lyly-Yrjänäinen J, Aarikka-Stenroos L and Laine T 2019 Mock-ups as a tool for assessing customer value early in the development process *Meas. Bus. Excell.* **23** 15–23
[9] Williams B D, Valero J N and Kim K 2018 Social media, trust, and disaster: Does trust in public and nonprofit organizations explain social media use during a disaster? *Qual. Quant.* **52** 537–50
[10] Geng S, Zhou Q, Li M, Song D and Wen Y 2021 Spatial–temporal differences in disaster perception and response among new media users and the influence factors: a case study of the Shouguang Flood in Shandong province *Nat. Hazards* **105** 2241–62
[11] Mercer J, Gaillard J C, Crowley K, Shannon R, Alexander B, Day S and Becker J 2012 Culture and disaster risk reduction: Lessons and opportunities *Environ. Hazards* **11** 74–95
[12] Kulatunga U 2010 Impact of culture towards disaster risk reduction *Int. J. Strateg. Prop. Manag.* **14** 304–13
[13] Hoffmann R and Blecha D 2020 Education and Disaster Vulnerability in Southeast Asia: Evidence and Policy Implications **21** 17–37
[14] Lopez Y, Hayden J, Cologon K and Hadley F 2012 Child participation and disaster risk reduction *Int. J. Early Years Educ.* **20** 300–8
[15] Martin M L 2010 Child participation in disaster risk reduction: The case of flood-affected children in Bangladesh *Third World Q.* **31** 1357–75
[16] Kitagawa K 2021 Disaster risk reduction activities as learning *Nat. Hazards* **105** 3099–118

[17] Mishra S and Suar D 2007 Do lessons people learn determine disaster cognition and preparedness? *Psychol. Dev. Soc. J.* **19** 143–59
[18] Sujata S 2010 Disaster Management Education in Indian Schools: Gaps, Challenges and a Human Resource Plan Model for School Education Sector *Indian J. Public Adm.* **56** 979–95

A systematic review of the shape of disaster education

A Subarno[1,2] and A S Dewi[3,4]

[1] Disaster Research Center, Sebelas Maret University
[2] The Department of Office Education Administration, Sebelas Maret University
[3] The Department of Management, Sebelas Maret University
[4] Doctoral Student in Economics, Sebelas Maret University

*Email: antonsubarno@fkip.uns.ac.id, dewiamina15@staff.uns.ac.id

Abstract. Researchers conducted this disaster education research to convey the importance of disaster knowledge and awareness in disaster-prone areas. The purpose of this study was to analyze the forms of disaster education in the last decade. Several previous studies on disaster education were analyzed and reviewed. This systematic review used an article selection process. It defined the purpose, conducting a literature search, pulling articles by reading abstracts, reading the full papers, abstracting data, and conducting an analysis. The disaster education system was classified into formal and non-formal education, consisting of pre-disaster, emergency responses, and post-disaster learning materials. Lastly, 95 articles have been reviewed involving a total number of participants from 29 to 430 people. The majority of research methods in this article were lectured through experimentation and integrated learning. While the research instruments used were questionnaires, documentation, observations, and interviews. In various fields of research, game-based learning enhances the understanding of disaster prevention.

1. Introduction

Disaster education should be taught to students at all levels of education [1] using diving tools and technology [2]. Various kinds of devices in residential areas can be used to learn about disasters, e.g., water, fire, rope, wet sacks, and others. The task of a disaster education unit or disaster preparedness program is to use the media to reduce the significant impact of disasters. The conceptualization of disaster education is still negligible [3]. Intergovernmental and stakeholders are still debating and negotiating disaster risk reduction and a framework for education [4]. Three possible conceptualizations for disaster education: 1) based on temporal differences between education conducted at regular or irregular times; 2) the ways of learning and teaching: formal, non-formal, and informal; and 3) establishing disaster education as a sub-discipline in the field of education [5]. [3] It is stated that disaster risk reduction cannot be separated from the important role of disaster education, this is reflected in five dimensions: knowledge, action, response, integration and participation. *Knowledge* concerns the students' understanding about the science and mechanisms of natural disaster, the answer is in the form of recognition of early warning system and the self-survival after a disaster happened, *action* motivates students to be proactive in disaster mitigation through a disaster risk formula that can be calculated from natural hazards multiplied to vulnerability and divided by the capacity of the social system, *participation* engages students to be aware on building resilience in their community, and *integration* is emphasized on the structural element of disaster education. [6] stated that the low public awareness of disaster preparedness is due to the lack of knowledge about the impact of the disaster. This awareness can be started since they are studying in elementary school. Students can become the "agents of change" in transforming and improving the understanding and preparedness of disaster education. They can share and discuss with their families what they have learned at school about disaster preparedness [1]. Furthermore, disaster education can be delivered in various forms of education by local policies [5], including lessons, training, discussions with teachers and friends, reading and listening, using hypermedia systems, games, after-class activities, and intercultural projects online that can help increase survival and trauma healing. The experience of dealing with disasters through various forms of education, although at different levels, when dealing with actual disasters can improve the control of actions and emotions [1].

On the other hand, as technology has grown significantly in the last two decades, many applications have emerged that can be used to learn about disasters—for example, disaster education games, rescue simulations, and firefighting on computers. Although many tools and applications can be used as media in disaster education, not many teachers know about them due to lack of understanding or not having enough time to teach students to use these tools and applications about disaster prevention or disaster mitigation to reduce major impacts in the event of a disaster. It is interesting to explore the way students have been taught about disaster education over the past decade based on articles published in prestigious journals.

2. Research Method

2.1. The Steps of Systematic Review

Systematic literature review means answering certain research questions by identifying, assessing, and interpreting all research evidence [7]. The method of article selection is the following [8]. Articles used in this research are all disaster education based on Scopus data. The search using title-abstract-keywords found 9985 documents, then the search was limited to the last decade and found 5502 documents. Finally, the articles were filtered using keywords, and only 380 documents were selected; then, the researchers sorted the articles that discussed the disaster education process so that 95 selected articles were obtained. The steps of a systematic review are depicted in Figure 1. The first step is to determine, create and define the topic used in the study. Then search for literacy or articles by subject. After the article is found, the first thing to do is to read the abstract of the article found and continue reading it in its entirety. That way, researchers can understand the content of the article and then analyze it.

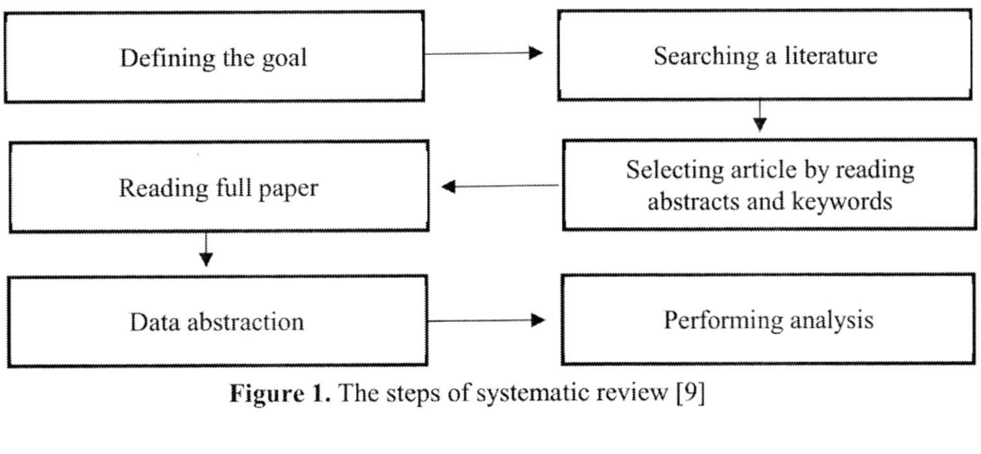

Figure 1. The steps of systematic review [9]

2.2. Article Selection Process

In reviewing the article, there are four criteria [8]: the way to select the participant, the technique to find and analyze the data, the field of the research based on disaster education, and the interpretation of the article. Disaster education in this research is categorized into five forms of education: lectures, drills, discussions, reading and listening, and games. Disaster education as a research strategy found in various research journals and relevant proceedings. The following is an example of the name of the research journal used: Advances in Intelligent Systems and Computing, AIJ Journal of Technology and Design, Applied Mechanics and Materials, ASEE Annual Conference and Exposition, Conference Proceedings, Bulletin of Earthquake Engineering, Information Systems Frontiers, IOP Conference Series: Earth and Environmental Science, Journal of Disaster Research, Journal of Physics: Conference Series, Lecture Notes in Civil Engineering, MATEC Web of Conferences, Nakahara: Journal of Environmental Design and Planning, Risk Analysis.

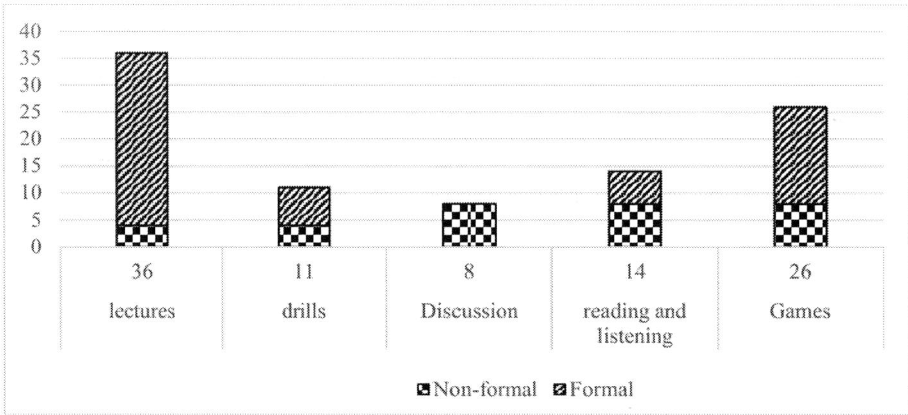

Figure 2. The form of education on disaster education articles analyzed review

Out of 95 disaster education overview articles, 36 educational lectures, 11 exercises, eight discussions, 14 reading and writing, and 26 game-style articles (See Figure 2). The researcher takes a representative sample of each form of education (see Table 1).

On the other hand, the articles reviewed are also grouped into two categories, non-formal and formal education. This refers to the sample and place of research. The number of articles' sample analyzed is 2 percent in each form of disaster education. The distribution of formal articles and non-formal articles in each state of education is presented in Table 1.

Table 1. The distribution of a formal and non-formal articles in each form of education

	Lectures	Drills	Discussion	Reading and listening	Games
Non-formal	4	4	8	8	8
Formal	32	7	0	6	18
Total number of article	36	11	8	14	26
Articles' sample analyzed review	6	2	2	3	5

In this paper, the formal education article means the research participants are students from formal schools, and the research place is located. Non-formal education article means the participants and the location of research are not in school, e.g., public places in a city, tourism destination, etc.

3. Results and Discussion

Ninety-five articles have been selected and fall into the category of disaster education. In the subject matter, detailed instrument analysis, data analysis, research methods, and methods of disaster education articles will be analyzed. The researchers study one by one of the articles. The research method commonly used in disaster education research is the quantitative method.

Table 2 shows the result of a review of disaster education articles in the form of lectures. The seven articles were taken from the Journal of Disaster Research, MATEC Web of Conferences, ASEE Annual Conference and Exposition, Engineering Geology for Society and Territory, Disaster Medicine, and Public Health Preparedness. The number of participants involved in the research was 239, 29, 266, and 430 individuals in articles no.1, no.2, no. 3, and no. 7. Participants in the three articles are not clearly identified. The number of participants in disaster education research in the form of lectures was 964 individuals.

Adiyoso and Kaneage involved 124 students of school-based disaster preparedness and 115 students of school non-based disaster preparedness in Yogyakarta. The data collection instrument used was in the form of a questionnaire to measure students' knowledge about earthquake hazards, risk perception, critical awareness, attitudes towards earthquake preparedness behavior, and earthquake preparedness behavior. A quantitative method was used to analyze the data. The analytical procedure used in this study was Multivariate Analysis of Variance (MANOVA). The role of disaster education in schools in promoting disaster preparedness behavior effectively is to increase students' knowledge of critical awareness, risk perception, and attitudes towards disaster preparedness behavior.

Table 2. An overview of the lectures form of disaster education article's sample analyzed review

No	Author(s)	Title	Participants		Research methodology	Research instruments
			Demography	**Gender**		
1	Adiyoso and Kane [10]	Effectiveness of Disaster-Based School Program on Students' Earthquake-Preparedness	Grade 11 and grade 12 of 2 junior schools from adopted and non-adopted disaster education reduction	Adopted school: 124 students, 37 males, 87 females Non-adopted school: 39 males, 76 females	Quantitative method, Multivariate Analysis of Variance (MANOVA)	A questionnaire using a 5-point scale
2	Damar and Indrayani [11]	Local wisdom based disaster education in Minangkabau society	29 people: 20 come from the ordinary group and 9 come from leaders group	No information	qualitative method using Miles and Huberman models: data reduction, displaying data, and conclusion drawing/verification	Observations, interview. 2 validation techniques: source and performing triangulation
3	Krishnan and Liao [12]	Integrating Shelter Design and Disaster Education in Architectural Curriculum	Architects and engineers	No information	Seminar and Studio design of disaster-relief shelters.	Prototype design of earthquakes, floods, mega-fires, and hurricanes, and related potential design

4	Musacchio *et al.* [13]	Long-Term Disaster-Prevention Strategies Based on Education	Children (5-10 ys) and four general public schools	No information	Comparative study	Urban disaster Prevention Strategies using MAcroseismic Faults project
5	Nagata and Kimura [14]	Proposing A Multi-Hazard Approach to Disaster Management Education to Enhance Children's "Zest for Life": Development of Disaster Management Education Programs to Be Practiced by Teachers	Elementary and junior high school students, 266 students, and 573 teachers	No information	Quantitative method, factor analysis	ADDIE survey method, 4 rating scales.
6	Pfefferbaum *et al.* [15]	The Integration of Mental and Behavioral Health Into Disaster Preparedness, Response, and Recovery	Ex - Officio of Department of Health and Human Service	No information	A qualitative method, analyzing the policy, the organizational and structural elements	Teleconferences

On the other hand, there are slight differences in article number 1. The results of the review in Article Number 2 in Table 2 use a qualitative method. In-depth interviews were conducted with 29 participants from two different community groups, namely elite groups and ordinary people groups. The local wisdom of the people of West Sumatra can assist in handling emergency response and post-earthquake recovery and reconstruction.

The results of the review of articles number 3 and number 4 in Table 2 concern the content of the curriculum. Students are introduced to disasters such as earthquakes, floods, mega-fires, and hurricanes. By getting to know them, students will see the impact of disasters, how to avoid them and how to take shelter in the event of a disaster. What is slightly different in the two articles is the participants. Article number 3 is intended for students, but article number 4 is designed for adults because adults can carry out survival measures.

Article number 5 of Table 2 aimed to educate elementary and junior high school students about natural disasters and their awareness of disaster hazards. The survey has been carried out, and the results revealed that the experience of an earthquake is the dominant aspect of being aware of disaster hazards compared to other disasters. If, for example, earthquakes and hurricane where disasters occur in their area, then they can imagine the impact of these disasters. They can imagine how to avoid them.

Article 6 explored the policy and framework of disaster education to reduce disaster risk. The integration of preparedness, response and disaster recovery must be supported by policies and administration as guidelines to implementing disaster risk reduction. In addition, disaster education can consider the 2015-2030 Sendai Framework for Disaster Risk Reduction (SFDRR). Integration of disaster education such as mitigation, prevention of preparedness to emergency response, rehabilitation, and recovery can be implemented and promoted globally.

Table 3. An overview of the drill form of disaster education article's sample analyzed review

No	Author(s)	Title	Participants		Research methodology	Research instruments
			Demography	Gender		
1	Masuzawa et al. [16]	Practice on an Education and Training Program to Development of Response Literacy to Earthquake Disaster in a Central Business District in Japan	68 employees in Shinjuku station	No information	Descriptive quantitative	Seminar and workshop followed Questionnaire survey
2	Nagata & Kimura [17]	Developing a Disaster Management Education and Training Program for Children with Intellectual Disabilities to Improve "Zest for Life" in the Event of a Disaster – A Case Study on Tochigi Prefectural Imaichi Special School for the Intellectually Disabled –	51 students	No information	quantitative measurement, paired t-test	Questionnaire

Table 3 shows the results of a review of disaster education articles in the form of drills. Both pieces are taken from the Journal of Disaster Research. This paper is grouped into formal education and non-formal education.

Research conducted by Masuzawa et al. [16] showed experimental research in the Shinjuku station area. The study began with seminars and workshops to comprehensively train employees on disaster preparedness and ended with a questionnaire consisting of five rating scales indicating participants' degrees. The four skim exercises provided are an exercise to help injured people, an exercise to respond to damaged buildings, a damage response center, and an exercise for the West exit area headquarters.

Nagata and Kimura's article concerns disaster education for children with intellectual disabilities at the Special Support School in Tochigi, Japan. Children with disabilities need special preparation and special measures to evoke their motivation to survive in a disaster. Teachers can adapt the program according to their needs, which can develop a systematic disaster management education program.

Table 4. An overview of the discussion form of disaster education article's sample analyzed review

No	Author(s)	Title	Participants		Research methodology	Research instruments
			Demography	Gender		
1	HabibiSaravi et al. [18]	Communicable diseases management in disasters: an analysis of improvement measures since 2005, Islamic Republic of Iran	131 documents	-	A qualitative document analysis, content validity assessment method	Document analysis sheet

2	Jian-Chang [19]	Community Cognition Investigation and Research on Tourists Disaster of Mountain Tourism-taking Taibai Moutain as a Case Study	390 participants, Farmer =29.3%, self-employed= 33.3%, civil servant 3.8%, student = 16.2%, migrant worker = 17.4%.	227 male and 163 female	Factor analysis	questionnaire

Table 4 shows the results of a review of disaster education articles in the form of discussions. The discussion model has a unique state because all the papers reviewed are grouped in non-formal education. This is equitable because people generally share ideas and discuss them not in a formal education setting—the two articles taken from East Mediterr Health Journal and Advance Journal of Food Science and Technology.

An article by HabibiSaravi et al. [18] Presenting research results collected from seventy-two documents from books, guidelines, reports, government documents, laws, films and newspaper articles to measure the improvement of disaster management since 2005 in the Islamic Republic of Iran. There are 4 phases of the disaster management cycle including: mitigation, prevention, response, preparedness, and recovery. There are five core disaster management covering case detection, reporting, investigation and confirmation, analysis and interpretation, and action. This article aims to build an infectious disease management system in everyday and emergency situations.

Moreover, the review of article number 2 in Table 4 and the research method described by the researcher uses community cognition with qualitative methods. The research was conducted at the tourist destination of Mount Taibai. Researchers analyzed demographic factors and awareness of disaster prevention and their impact on socio-economic conditions. Young people, higher education, local residents and those in the tourism business know more about the work of tourism disaster prevention.

Table 5. An overview of the reading and listening form of disaster education article's sample analyzed review

No	Author(s)	Title	Participants		Research methodology	Research instruments
			Demography	Gender		
1	Abunyewah, Gajendran, & Maund [20]	Conceptual Framework for Motivating Actions towards Disaster Preparedness Through Risk Communication	321 articles, books, and chapters	-	Qualitative method.	Three criteria: relevance, applied and cited and referred.
2	Sakurai & Sato [21]	Promoting Education for Disaster Resilience and the Sendai Framework for Disaster Risk Reduction	2015-2030 Sendai Framework	-	Qualitative method	-
3	Tiara et al. [22]	Event-Related Brain Potentials from Pictures Relevant to Disaster Education	Seven adults, 28-30 ys old,	Male = 7 Female = 0	Quantitative method, F test, and t-test.	Schematic view: monitors, pictures.

Table 5 revealed the results of the review of disaster education articles through reading and listening. The articles were taken from the Journal of Disaster Research and Procedia Engineering.

Research conducted by Abunyewah, Gajendran, & Maund [20] on 321 articles, books, and book chapters has been analyzed to improve disaster prevention through communication. Communication in this study means assumptions that provide more information to the community to guide disaster preparedness. Communication is mainly influenced by several variables such as the source, channel, and the messages' content. In this case, communication is a channel that combines the kinds of hazards and the recommended protective measures to prevent its impact in a simple language.

Result of article review by Sakurai and Sato [21] The educational themes in the Yokohama Strategy, HFA, and highlights in the last 20 years of disaster education experience in Japan, argues that integrating hazard knowledge into the curriculum and hazard information into the local environment is critical for the next generation. Promoting school safety and disaster education can increase public awareness of disaster hazards in Japan and around the world.

On the other hand, an article written by Tiara *et al.* [22] reported the importance of imagery or visuals in decision making on brain potential in constructing earthquake survival kits into valuable and non-useful items. The brain areas consist of frontal, parietal, central, occipital, and temporal. Each brain area has a significant difference in pairwise comparisons in response to the images of the survival kit, except for the parietal vs. temporal region. This means that a preliminary study showed that images or visuals could be integrated into disaster preparedness education through a digital environment.

Based on Table 5, it can be summarized that the importance of disaster education could be established through audio (listening) and visual (reading) media promotion incessantly. The trainee will understand the importance of disaster preparedness through education, both formal and non-formal education.

Table 6. An overview of the game form of disaster education article's sample analyzed review

No	Author(s)	Title	Participants		Research methodology	Research instruments
			Demography	Gender		
1	Hajdarevic, Konjicija, & Subasi [23]	Svxlink VOIP Implementation Using Raspberry Pi in Education and Disaster Relief Situations	educational institutions, radio amateur community,	-	An experiment	microcomputer hardware, radio amateur frequencies, radio amateur transceivers and antennas.
2	Herdiansyah et al. [6]	Disaster Awareness through Disaster Preparedness Education for Primary Schools	180 students	No information	Experiment	Training responds objects
3	Isshiki *et al.* [24]	An Offline-Based On-Demand Visualization System of Large-Scale Particle Simulation for Tsunami Disaster Prevention	No information	No information	Experiment	visualization and polygonization
4	Nouchi & Sugiura [25]	Beneficial Effects of Learning with Game-Book on Education for Disaster Prevention in Children	Thirty-four students, average 13.65 years old.	20 male, 14 female	Quantitative method, t-test.	Questionnaire
5	Babu et al., [26]	Collaborative Game-Based Learning of Post-Disaster	22 students (23-33 years), 60	No information	Quantitative method	Survey

	Management	students (23-40 years)	

Based on Table 6, five article samples were taken from the International Symposium on Telecommunications, International Conference Earth Science & Energy, the journal of Electronics and Communications in Japan, Journal of Disaster Research, and IEEE Eighth International Conference on Technology for Education. Game-based disaster education covers hardware and software technology, reading and listening skills, and motor skills. Article number 1 in Table 6 aimed to analyze the use of technology, especially in open source VOIP software (svxlink), cheap microcomputer hardware, one of the available Linux distributions, amateur radio frequencies, amateur radio transceivers, and antennas to provide education and disaster prevention. This research involved educational institutions and local communities as the object of the participants. Radio connection, including amateur radio, is a cheaper configuration that can be used as a disaster education medium without any problems in its creation and connection.

Besides that, article number 2 in Table 6 is an article entitled "Disaster awareness through disaster preparedness education for primary schools". The disaster preparedness education program that has been carried out at SDN Tanjung Anom Banten (Tanjung Anom Elementary School, Banten) involved 180 students as participants. The program, which consists of disaster mitigation and simulation, collaborates with local communities and stakeholders to acquire knowledge and students' preparedness during pre-disaster, emergency response, and post-disaster. Based on the program, it can be seen that education is an effective way to reduce disaster risk.

Article number 3 in Table 6 is still exploring the simulation method. The simulation uses a visualization system and polygonization with the particle method. improve understanding of these phenomena by means of visualizations that are used together to display the stresses on the structure and path lines of water particles. The polygonization can interactively adjust via the screen according to the simulation objectives. Polygonization and visualization systems can be used effectively for Tsunami disaster prevention.

Article number 4 in Table 6 is an article entitled "Beneficial effects of learning with game-book on education for disaster prevention in children". This article investigates the positive impact of disaster simulation game books on disaster education. A total of 34 junior high school students were involved in this study, 20 male and 14 female students. Disaster education tools must be provided because schooling is to acquire knowledge and turn knowledge into practice. Practical learning, capacity building, and experiential learning are the transformations of understanding disaster prevention education.

Article number 5 in Table 6 explains the importance of the intelligence aspect in disaster rescue or prevention. This is in line with article number 4 regarding the need to practice disaster education knowledge; this article explores the actions required during playing simulation games, such as emergency and post-disaster response. In this article, although skills training is only in games or simulations, it requires good communication and cooperation during post-disaster operations. There are three levels of play in the simulation: the first level is about gaining situational awareness of Safety, Facilities, and Lifeline; the second level is how to prioritize incident objectives and assigned tasks, and the third level requires the ability to complete some tasks and practice response activities in collaboration with other participants. This game is very effective in providing students with basic knowledge of post-disaster handling. Assessments before and after playing this simulation game are given to measure the understanding of participants (students) in dealing with and disaster situations.

4. Conclusion

Education is the fastest medium to increase awareness of disaster management. Disaster education can be applied formally or non-formally. In general, the types of disaster education can be grouped into five types, namely, through teaching, training, discussion forums, reading and listening, and in the form of games. Although this article only groups the states of disaster education through 5 states of teaching, it cannot be justified that there are only those five forms because other forms of disaster education are not covered in this article.

Disaster education is not just about the dangers of disasters and how to reduce them, but it needs to be practiced physically. There are already many technologies that can facilitate this so that students do not have to come directly to locations that might endanger themselves. Still, it can be done by simulation in an application system.

The younger generation group needs more disaster education training than the older generation group, so it is appropriate that disaster education is given to the younger generation. Besides, they are easy to understand and practice their knowledge; the younger generation is usually easier to adapt to technology and information development. In addition, there is no difference in ability by gender in training attended. Men's groups and women's groups have the exact educational and training needs for disaster.

References

[1] Yeon D H, Chung J B and I'm D H 2020 The effects of earthquake experience on disaster education for children and teens *Int. J. Environ. Res. Public Health* **17** 1–14
[2] Mitsuhara H, Iwaki K, Kozuki Y, Shishibori M, Inoue T, Yamaguchi K, Takechi Y and Morimoto M 2016 *Penumbral tourism: Place-based disaster education via real-world disaster simulation*
[3] Gong Q, Duan Y and Guo F 2021 Disaster risk reduction education in school geography curriculum: Review and outlook from a perspective of china *Sustain.* **13** 1–16
[4] Sakurai A and Sato T 2016 Promoting education for disaster resilience and the Sendai framework for disaster risk reduction *J. Disaster Res.* **11** 402–12
[5] Septima M E, Wardoyo M A I, Praptiwi N Y, Ashari A N S, Ashari A, Susanti N I, Jainudin J, Latifah F and Nugrahagung P P 2019 Disaster Education Through Local Knowledge in Some Area of Merapi Volcano vol 271 (Departement of Geography Education, Faculty of Social Science Universitas Negeri Yogyakarta, Jl. Colombo No.1 Yogyakarta55281, Indonesia: Institute of Physics Publishing)
[6] Herdiansyah H, Husein S I, Asrofani F W, Simamora P A R and Kholila B N 2020 Disaster Awareness through Disaster Preparedness Education for Primary Schools *IOP Conf. Ser. Earth Environ. Sci.* **519**
[7] Wahono R S 2007 A Systematic Literature Review of Software Defect Prediction: Research Trends, Datasets, Methods and Frameworks *J. Softw. Eng.* **1** 1–16
[8] Suprapto N and Chang T 2015 Research on University Student ' s self-efficacy scale in science education : A systematic review 120–9
[9] Suprapto N, Sunarti T, Suliyanah, Wulandari D, Hidayaatullaah H N, Adam A S and Mubarok H 2020 A systematic review of photovoice as participatory action research strategies *Int. J. Eval. Res. Educ.* **9** 675–83
[10] Adiyoso W and Kanegae H 2013 Effectiveness of disaster-based school program on students' earthquake-preparedness *J. Disaster Res.* **8** 1009–17
[11] Damar D and Indrayani I 2018 Local wisdom based disaster education in Minangkabau society *MATEC Web Conf.* **229** 1–6
[12] Krishnan S and Liao Y 2019 Integrating shelter design and disaster education in architectural curriculum (University of Illinois, Urbana-Champaign, United States: American Society for Engineering Education)
[13] Musacchio G, Bernhardsdóttir A E, Ferreira M A and Falsaperla S 2014 Long-term disaster-prevention strategies based on education *Engineering Geology for Society and Territory - Volume 7: Education, Professional Ethics and Public Recognition of Engineering Geology* (Istituto Nazionale di Geofisica e Vulcanologia, Rome, Italy: Springer International Publishing) pp 77–80
[14] Nagata T and Kimura R 2017 Proposing a multi-hazard approach to disaster management

education to enhance children's "Zest for life": Development of disaster management education programs to be practiced by teachers *J. Disaster Res.* **12** 17–41

[15] Pfefferbaum B, Flynn B W, Schonfeld D, Brown LM, Jacobs G A, Dodgen D, Donato D, Kaul R E, Stone B, Norwood A E, Reissman D B, Herrmann J, Hobfoll S E, Jones R T, Ruzek J I, Ursano R J, Taylor R J and Lindley D 2012 The integration of mental and behavioral health into disaster preparedness, response, and recovery. *Disaster Med. Public Health Prep.* **6** 60–6

[16] Masuzawa Y, Hisada Y, Murakami M, Shindo J, Miyamura M, Suwa H, Tanaka S, Mizukoshi K and Nakajima Y 2014 Practice on an education and training program to development of response literacy to earthquake disaster in a central business district in Japan *J. Disaster Res.* **9** 216–36

[17] Nagata T and Kimura R 2020 Developing a disaster management education and training program for children with intellectual disabilities to improve "zest for life" in the event of a disaster – a case study on Tochigi prefectural imaichi special school for the intellectually disabled *J. Disaster Res.* **15** 20–40

[18] Habibisaravi R, Seyedin H, Rad A M and Gouya M M 2019 Communicable diseases management in disasters: An analysis of improvement measures since 2005, Islamic Republic of Iran *East. Mediterr. Heal. J.* **25** 269–81

[19] Jian-Chang Z 2014 Community cognition investigation and research on tourists disaster of mountain tourism-taking Taibai Moutain as a case study *Adv. J. Food Sci. Technol.* **6** 812–8

[20] Abunyewah M, Gajendran T and Maund K 2018 Conceptual Framework for Motivating Actions towards Disaster Preparedness Through Risk Communication *Procedia Eng.* **212** 246–53

[21] Sakurai A and Sato T 2016 Promoting education for disaster resilience and the Sendai framework for disaster risk reduction *J. Disaster Res.* **11** 402–12

[22] Tiara A, Mikropoulos T A, Mavridis D and Mercier J 2017 Event-related brain potentials from pictures relevant to disaster education *Lect. Notes Comput. Sci. (including Subsea. Lect. Notes Artif. Intell. Lect. Notes Bioinformatics)* **10512 LNAI** 144–58

[23] Hajdarevic K, Konjicija S and Subasi A 2014 Svxlink VOIP implementation using raspberry Pi in education and disaster relief situations *BIHTEL 2014 - 10th Int. Symp. Telecommun. Proc.*

[24] Isshiki M, Asai M, Eguchi S and Miyagawa Y 2018 An Offline-Based On-Demand Visualization System of Large-Scale Particle Simulation for Tsunami Disaster Prevention *Electron. Commun. Japan* **101** 55–62

[25] Nouchi R and Sugiura M 2014 Beneficial effects of learning with game-book on education for disaster prevention in children *J. Disaster Res.* **9** 1079–87

[26] Mclain ML, Babu S K, Bijlani K, Jayakhrishnan R and Bhavani R R 2016 Collaborative Game-Based Learning of Post-Disaster Management *2016 IEEE 8th Int. Conf. Technol. Educ. Collab.* 80–7

IOP Conf. Series: Earth and Environmental Science **986** (2022) 012012 doi:10.1088/1755-1315/986/1/012012

Issues and facts on nuclear disaster

Riyatun [1*], Suharyana [1], D T Kartono[1], P K Wijayanto[2] , H Krisyana[3] and B S Effendi[3]

[1] LPPM - Sebelas Maret University, Jl Ir Sutami 36 A Surakarta Indonesia
[2] Slamet Riyadi University, Jl. Sumpah Pemuda No.18, Surakarta Indonesia
[3] PT. ThorCon Power Indonesia, World Trade Centre 3 Jl. Jenderal Sudirman Kav. 29-31 Karet, Setiabudi, Jakarta Selatan

*Email: riyatun@staff.uns.ac.id

Abstract. Concerns and fears about nuclear, both those originating from public and government perceptions, are examined in this paper. The NPP (NPP or PLTN) is a new energy in Indonesia; therefore, rejection is bound to happen. An overview of the work of NPP and the development of the latest fourth-generation nuclear reactor technology shows high security and safety features. A balanced and scientific study is carried out on issues that arise from the general public's perception of NPPs, related to accident issues, radiation issues, waste issues, and social issues. Social issues related to nuclear issues are sunset technology and the NIMBY (not in my backyard) syndrome. There is no reason for people to be afraid of nuclear. The measurement of public acceptance is carried out by analyzing data from various countries, supported by the results of a questionnaire in Indonesia, showing that there is always resistance to NPPs Indonesian are ready to accept NPPs because more than 60% agree with NPPs.

1. Introduction

Nuclear energy is a new renewable energy source (RNE) that can produce electricity continuously with large and reliable power [1]. NPPs do not produce carbon emissions directly in the energy conversion process but come from mining and transportation processes. Evidence of this process needs to be properly explained.

The nuclear issue is sunset technology. It is obsolete and has been abandoned by humans. It's true, the trend of NPPs development after March 11, 2011, is declining, but NPPs construction is up 2.3% per year and decreased in 2019 due to the covid pandemic. There is also the NIMBY syndrome, which is evident in developing all types of energy sources. This perception needs to be straightened out so that humans do not make wrong decisions.

2. Methods

This paper is compiled based on reviewing several papers, textbooks, and official data sources from the web. All sources have been listed in the body text section and are also listed in the reference list.

3. Results and Discussion

3.1. NPP in General

Nuclear reactions can occur due to fission (break) or fusion (joint) atomic nuclei. To make nuclear fission reactions, it takes fuel elements Uranium-235 or Plutonium-241. The energy produced by nuclear reactions is used for NPPs, the heat of reaction to heat water into water vapor to turn turbines for electric generators. There is no difference in the principle of NPPs with other generators, and the difference is in the fuel [2]

The basic principles of nuclear fission-based NPPs are described in Figure 1 [3,4] : a nuclear reactor, primary and secondary coolant flow in a closed cycle, a heat exchanger system or steam generator, a condenser, and a turbine connected to an electric generator. The nuclear reactor consists of a reactor core (1) and a containment building (2). The reactor core is a space where the fission chain reaction is controlled, producing energy and **radioactive substances** (RAS). The containment building (2) prevents RAS from escaping into the environment.

A closed primary cooling system takes the heat energy. The energy is brought to the heat exchanger (3), after which the water is returned to the core. The secondary water in the heat exchanger tube (3) becomes high-pressure steam and is sent to the turbine (4), the turbine moves, and the electric generator functions.

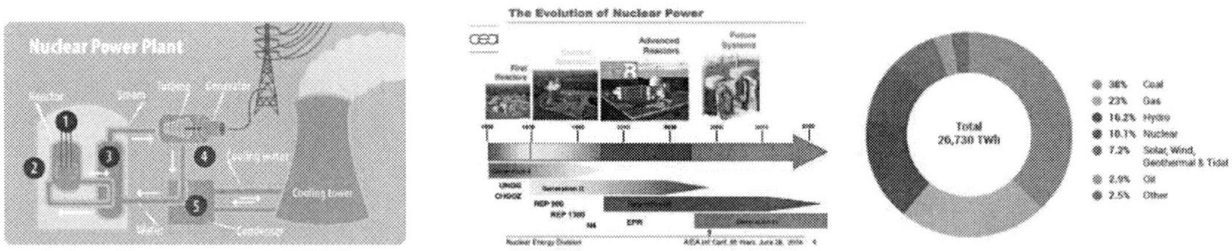

(a) (b) (c)

Figure. 1 (a) NPP basic principles [5] (b) Development of power reactors or NPPs from Generations I – IV [6] (c) The contribution of NPPs: the 4th order in the world's energy fulfillment [7]

Commercial NPPs were first put into operation in the world in 1954, grew significantly until 1985, and now have a growth rate of 2.3%. Until 2020, it was reported that 444 NPPs spread across 32 countries, producing a total of 10% of the world's electricity needs. This number will increase with 66 NPPs are under construction [8]

The technological and economical levels of NPPs are grouped by generation. **Figure 1. (b)** is the evolution of the design innovations of generation I, II, III, III+, and IV NPPs [6]. Each NPP model's characteristics or design descriptions are written in the Advanced Reactor Information System (ARIS), published annually by the International Atomic Energy Agency (IAEA) [3].

3.2. Nuclear Accident Issues

The NPPs and nuclear bombs work using the same concept, a fission chain reaction. However, NPPs cannot explode like a nuclear bomb. The concentration of uranium-235 in a nuclear bomb must be high, around 90%, while at an NPP, it is only 20%. The fission reaction in a nuclear reactor takes place under control, while in an atomic bomb, it is uncontrolled [2–4].

Several nuclear accidents since 1942 until now have occurred, well documented, all causes can be identified and analyzed to prevent similar accidents from happening again. For standardization purposes, the IAEA's International Nuclear and Radiological Event Scale (INES) divides into 7 disaster scales, levels (1–3) are called incidents, and (4–7) are called accidents. Level 0 is a non-hazardous deviation/event [9]. Three NPP accidents that have received great attention to date are Three Mile Island (March 28, 1979), Chernobyl (April 26, 1986), and Fukushima Daiichi (March 11, 2011). The results of the analysis of accident data and their impacts are summarized in **Table 1.**

TMI accidents are caused by the failure of monitoring instruments and untrained operators. Even though the reactor core melted, there was no radioactive substance that polluted the environment. Everything was enclosed in a very strong reactor building, made of thick concrete. Although there was no radiation pollution and no fatalities because it happened in the United States, the most advanced NPP technology, this accident took the public's attention [10].

The Chernobyl NPP Unit 4 accident was caused by an imperfect core design and incompetent operators. The type of Chernobyl NPP is RMBK1000, Russian technology, and only operated in countries within the Soviet Union. After the accident, all RBMK1000 units were redesigned and put into operation safely until they were permanently extinguished when they reached their planned service life. After 20 years of the Chernobyl accident, UNSCEAR reports no evidence of increased solid cancer or leukemia rates among the general population. However, there is widespread psychological concern about the effects of radiation. The total number of deaths reliably caused by the radiation accident was 62 people [11].

Table 1. Comparison of nuclear accidents that have caught the world's attention [12,13]

Accident	Three Mile Island	Chernobyl	Fukushima Daiichi
Date	28 March 1979	26 March 1986	11 March 2011
Location	USA	Ukraine	Japan
INES Scale	5	7	7
Cause	Human Error	human error and imperfect terrace design	Emergency cooling system not working
Victim	0	30	0
Evacuated Victim	140,000	166,000 in 1986 and 230,000 in subsequent years	100,000
Evacuation Zone (km)	8	30	20
Diffuse radiation (becquerels map)	0.062	5,200	770
Long term health effects	None	There were > 6,000 cases of thyroid cancer in children with 9 deaths; 4,000 – 25,000 deaths due to radiation exposure.	None

The Fukushima accident was preceded by the 9 R (Richter scale) Tohoku earthquake and was followed by a tsunami. The study of documentation and earthquake forecasts around Fukushima is 7 R, so appropriate dikes are built along the coast of the NPP site. It turned out that the earthquake was 9 R, so the barrier could not withstand the tsunami. Seawater enters the NPP complex and inundates the backup pump generator, which drains the reactor cooling water when an emergency occurs.

Figure 2 (a) shows the location of NPPs in Japan. Around the Fukushima Daiichi NPP complex, there are Fukushima Dainii and Onagawa NPP. The location of the PTLN Onagawa complex is the closest to the epicenter of the earthquake, of course, the higher the tsunami waves. The tsunami height at Onagawa was 14.3 m, while that at Fukushima Daiichi was 13.1 m.

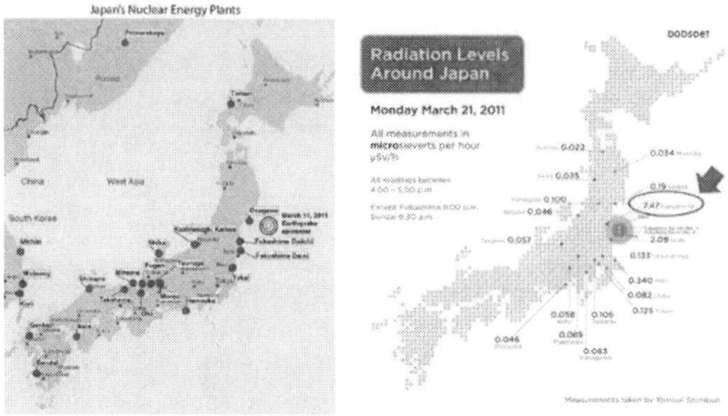

(a) (b)

Figure 2. (a) Map of NPPs in Japan [14]; (b) The results of measuring the dose rate at 10 days after the accident of the Fukushima Daiichi NPP [15]

There were 41 NPP accidents with an INES 3 scale category. **Table 2** shows serious accidents but only 3 accidents at commercial NPPs, the other being research reactors.

The National Resources Defense Council reported that between 1954 and 2011, an average of 1 NPP accident occurred per 1,309 reactor-years with a probability of 1 per 3,704 reactor-years [16] Thus, reactor accidents in many countries are well documented, their causes and impacts have been analyzed. All these experiences are used to improve the safety features of nuclear reactor technology so that generation III and III+ and IV reactors appear to have passive safety features.

Table 2. List of meltdown reactor core accidents [17,18]

Year	Location	Unit	Reactor Type
1959	California, USA	Sodium reactor exp	Sodium-cooled power reactor
1961	Idaho, USA	Stationary Low Reactor	Exp gas-cooled, water-moderated
1966	Michigan, USA	Enrico Fermi Unit 1	Liquid metal fast breeder reactor
1967	Dumfreshire, Scotland	Chapelcross Unit 2	Gas-cooled, graphite-moderated
1969	Loir-et-Chaire, France	Saint-Laureant A-1	Gas-cooled, graphite-moderated
1979	Pennsylvania, USA	Three Mile Island	Pressurized Water Reactor (PWR)
1980	Loir-et-Chaire, France	Saint-Laureant A-1	Gas-cooled graphite moderated
1986	Pripyat, Ukrania	Chernobyl Unit 4	RBKM-1000
1989	Lubmin, Germany	Greifswald Unit 5	Pressurized Water Reactor (PWR)
2011	F. Daiichi, Japan	F. Daiichi Unit 1,2,3	Boiling Water Reactor (BWR)

Nuclear accidents need to be anticipated early if nuclear power plants are to be built in Indonesia. Because there is a technology with a generation category that is increasingly sophisticated in its safety level, the nuclear power plant built in Indonesia should be the 4th generation. Do not let Indonesia build nuclear power plants with outdated technology.

3.3. Dangerous Radiation Issue

The issue of the impact of nuclear radiation on health creates fear in some people. Nuclear radiation has been used for various purposes, such as medicine for diagnosis and therapy, industrial logging, customs, and product quality control. Everyone everywhere always receives radiation. The higher the location, the greater the intensity of radiation from cosmic rays. All passengers and crew on an airplane are exposed to more radiation than people on the ground. Radioactive substances (RAS) exist in the body and enter through breathing or food, such as potassium-40 in bananas and carbon-14 in carbohydrates. Radiation also comes from the earth (terrestrial) does not endanger health because the dose is very small [3]

BAPETEN Regulation No. 4/2013 regulates the dose limit value (DLV) for radiation workers and the public. The DLV of the general population is 1 mSv/year; for radiation workers, the average is 20 mSv per year, for 5 years with a total dose of 100 mSv. Patients diagnosed with X-rays of the thorax receive a radiation dose of gray units, which is 1000 times greater than the general population. Still, doctors guarantee their safety and security [19,20]

In the NPP system, radiation confinement is carried out using the principle of multiple barriers. The implementation of this principle in solid fuel reactors is [21].

a) Solid RAS fission products bound to the fuel matrix in the cladding, wrapping.
b) Suppose a volatile RAS is released from the fuel matrix; the RAS will be made of strong metal in the cladding.
c) RAS leaking from the cladding is blocked by the closed flow of primary cooling water. Periodically the cooling water is demineralized, pollutants including RAS will be removed and managed,
d) Suppose there is RAS that escapes from the demineralization process and escapes from the cooling water, then there is a thick steel tube around the reactor core, preventing the RAS from being released from the reactor,
e) The air pressure inside the building is made lower than the air pressure outside. Before being discharged, the air is filtered, the RAS will stick to the filter, demineralize, and remove the RAS.
f) The exhaust chimney is high so that the RAS activity is reduced during the trip.

On the Chernobyl accident (1986), no public health impact from radiation exposure two decades after the accident. The reactor terrace and the building are buried using strong, thick concrete as a radiation shield/graveyard. Over time, RAS activity decreased. Radiation levels outside the Chernobyl graves are the same as background/natural radiation, and the dose rate is measured at 0.19 Sv per hour. Today the Chernobyl cemetery is a tourist spot without the threat of radiation hazards [12].

The Fukushima Daiichi reactor is a Generation II technology equipped with Class I enclosures. These enclosures are not strong enough to withstand enormous pressure. The current NPP design uses generations III and III+, with Mark II class enclosures. Radiation hazards can be reduced by applying distance. In the case of the Fukushima Daiichi accident, **Figure 2 (b).** shows the dose rate ten days after the accident. The highest dose rate is in the Fukushima Daiichi area at 7.47 Sv per hour, and the population has been evacuated [22].

Indonesia has 3 reactors, although not NPPs; reactors for education and research with small power still have the potential for accidents, so the radiation hazard caused is also present. The Kartini Reactor at Babarsari Yogyakarta is next to the BATAN Employee Housing, campus, hotel, and commercial centre. Likewise, the Triga 2000 Bandung reactor is closed to the ITB campus, zoo, and people's homes. The reactor with a higher power, GA Siwabessy, is located at the Center for Science and Technology Development (Puspiptek) Serpong. Although there are no complaints about radiation fear from the public, BATAN has always reported radiation safety guarantees to BAPETEN and IAEA.

3.4. Nuclear Waste Issue

NPPs directly or indirectly generate RAS. Used fuel is not called waste because there are many RAS that can be useful. For example, Uranium and Plutonium can be processed into new fuels. After all useful RAS is removed, the remainder is called waste with decreasing volume.

The fuel used by NPP is RAS, which is very toxic and has very high activity. Every 1 unit of NPP with a power of 1,000 MW produces 100 tons of fuel per year. According to the IAEA, the total amount of nuclear fuel in the world to date is around 370,000 tons [21].

Figure 3. (a) shows that the HWL group RAS waste volume is only about 3%, but it is very toxic. Its half-life varies widely, ranging from the order of less than 1 minute to billions of years. Based on the half-life, HLW is divided into 2, radioactive materials with a short half-life, less than 5 years, an

IOP Conf. Series: Earth and Environmental Science **986** (2022) 012012 doi:10.1088/1755-1315/986/1/012012

intermediate between 5-30 years, and a long half-life above 30 years. Figure 3. (b) shows the graph of the RAS decay of 1 tonne of spent PWR fuel. After being stored 40-50 years, the activity is only 1% of the original activity [23] The potential radiation hazard has been greatly reduced. Figure 4. (c) shows a photo of an example of temporary storage of used fuel and 2 people embracing a barrel.

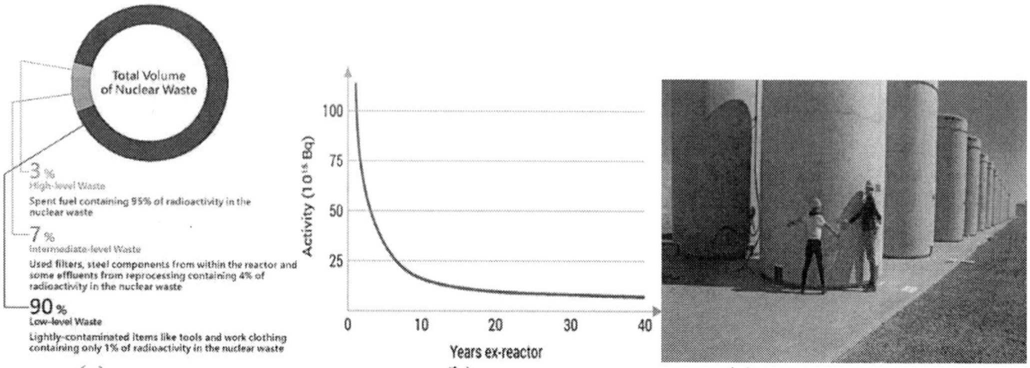

(a) (b) (c)

Figure 3. (a) Composition of radioactive material waste based on radiation levels [24]; (b) ZRA decay in 1 tonne of spent fuel [2–4]; (c) Dry cask storage at the Palo Verde Nuclear Plant, Arizona [25]

Waste treatment technology continues to evolve. The United States, since the 1940s, has mastered the technology of Plutonium Uranium Reduction Extraction (PUREX) [26]. This method is used to separate the elements uranium and plutonium from used fuel Recycling Optimization by Removing Impedimental Elements (ORIENT) technology [27] claims a permanent residual of 3.3% of the initial volume. SYLOS waste treatment technology uses lasers, first proposed by Gerard Mourou, winner of the Nobel Prize in Physics (2018). SYLOS was able to convert a million-year half-life RAS to the order of minutes [28].

Waste with a very long half-life is permanently stored deep below the soil surface has an andesitic rock structure. There are empty islands in Karimun Jawa that are eligible for permanent storage of nuclear reactor waste [29].

The nuclear industry adheres to the principle of "from the cradle to the grave" in operation. Management of NPP must be responsible for its waste comprehensively. The Government of Indonesia issued Government Regulation (PP) No. 61 of 2013 concerning Radioactive Waste Management and Government Regulation No. 58 of 2015 concerning radiation safety and security in the transportation of radioactive substances. The BATAN work unit responsible for managing radioactive waste is the Center for Radioactive Waste Technology which BAPETEN always supervises [30,31].

Handling of hazardous waste generated by any process, including NPPs, is carried out with strict protocols and requires expensive funds. However, these costs can be predicted to be included in the components of the price of electricity production. In the USA, the price of nuclear waste management is about 0.1 cents US$ per kWh [32].

The issue of radioactive waste is indeed very sensitive. Temporary and permanent storage places should be thoroughly discussed before the Indonesian government builds NPPs. It is very unwise to pass on the problem of nuclear waste to the next generation.

3.5. The issue of NPPs is Sunset Technology

Sunset technology can be interpreted as almost obsolete technology or outdated due to the absence of the latest development or development. NPP technology is not sunset technology. In fact, by the end of 2020, 5 new NPPs began to operate in China, Russia actively, and 3 newcomer countries Belarus, Turkey, and the United Arab Emirates, and there were more than 50 new nuclear power projects. In

2021, 15 NPPs will start operating, including reactors with new technologies such as fast reactors and small modular reactors [33].

Generation IV SMR NPPs have many variants, including the Molten Salt Reactor (MSR) or fused salt reactor. MSR technology uses a mixture of uranium and thorium fuel. It will be saving the use of uranium and minimizing the quantity of the half-million-year-old radioactive element actinide in the fuel used. MSR waste can thus be said to be more environmentally friendly. Japan, China, Canada, the United States of America, European Union countries are also very active in developing the MSR design. So, NPPs are not sunset technology.

3.6. NIMBY Syndrome Issues

NIMBY (Not in My Back Yard) syndrome is characterized by residents' resistance to proposed developments near them. Opposition arises for various reasons such as environmental impact, visual appearance, noise pollution, reduced property values, interference with place attachment, and other factors [34]. Communities oppose the construction of nearby facilities but will tolerate/support them if they are built far from their settlements.

The construction of an NPP will certainly cause a reaction from NIMBY, especially the community around the prospective site. However, it will also enjoy positive impacts, such as the availability of reliable and cheap electricity, better transportation, and increased welfare as an economic impact, and the availability of new jobs.

To overcome NIMBY, it is necessary to continuously socialize and learn about NPPs, their benefits, and potential dangers. Potential hazards cannot be eliminated, and they can be minimized by selecting the right technology and proven to be safe and secure. Japanese people who live near NPPs support the construction of NPPs. The existence of local economic feedback, job creation, health insurance, CSR programs must be given to the community around the NPP[37]

Bisconti Research, Inc. (2017) surveyed the acceptance of NPPs in the USA. Respondents were divided into two groups, namely neighboring groups, and the public. Neighbors of NPPs are residents who live within a 10-mile radius of 59 NPPs in the USA. It was found that 81% of NPPs' neighbors liked and supported the use of NPPs, while the support from the public was 64%. The NPP's neighboring group is more agreeable than the public. This fact shows that the issue of NIMBY in NPPs is not real. Support for NPPs should be based on awareness of the many benefits obtained from NPPs, especially reliable electricity [35].

3.7. Government Attention and Public Acceptance of Nuclear Power Plants

In Bappenas's (2016) policy brief, it has been explained that various efforts have been made to build the first NPP in Indonesia, but this has not materialized. The plan for developing NPPs in Indonesia for 3 decades has led to endless pro and con debates. The government tends to adhere to a short-sighted policy where Law Number 17 of 2007 and Presidential Regulation Number 5 of 2006 have provided a strong foundation for developing NPPs. Still, Government Regulation Number 79 of 2014 seems to annul it by placing nuclear as a last resort [31].

Surveys of Public Acceptance (PA) of NPPs have been carried out in various countries, including Indonesia. The results of the BATAN survey are shown in Figure 4. (a). Survey results 2010-2016 show that 50% of respondents accept NPPs. In 2011 there was an 11% decline, thought to be related to the Fukushima tragedy. After 2011, the level of PA increased again to reach 77.53% in 2016.

The Public Acceptance (PA) survey results in 34 provinces in Indonesia from 2011 to 2016 are shown in Figure (b). The level of public acceptance is always more than 50%.

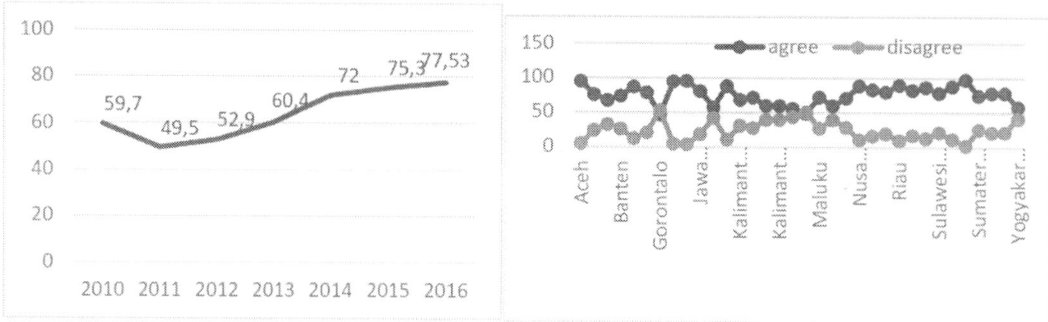

(a) (b)

Figure 4. (a) Level of Public Acceptance of NPPs [31].
(b) Public Acceptance and Rejection of NPPs in 34 Provinces in Indonesia

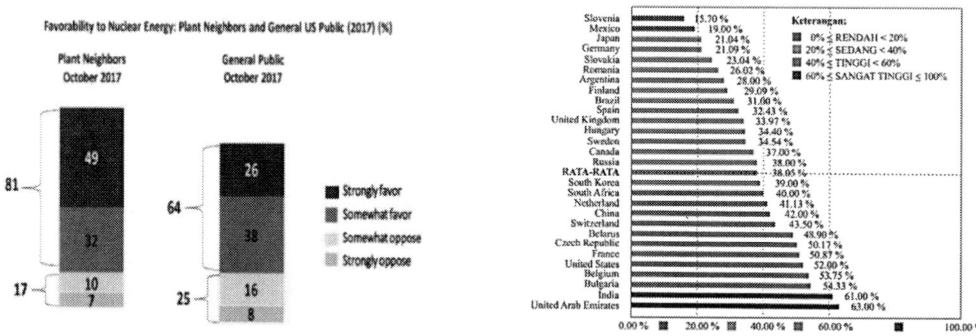

(a) (b)

Figure 5. (a) Results of the acceptance survey of NPPs in the United States (2017) [35]; (b) Grouping of countries based on the level of public acceptance of NPPs.

There are 5 reasons In **Figure 5 (a)** for the highest public acceptance, namely (1) stable electricity supply, (2) cheap electricity, (3) employment opportunities, (4) technology transfer, and (5) low pollution.

The five highest reasons for public rejection of NPPs, namely: (1) fear of NPP accidents, (2) radioactive contamination, (3) unprepared human resources, (4) nuclear weapons and, (5) power plants from other sources, are sufficient. More than 70% of people have indeed started to accept the existence of NPPs

Figure 5. (a) is the result of the analysis of the PA survey in the US, regarding the population who are "neighbors to nuclear power plants" and "far neighbors to nuclear power plants" The support of close neighbors (81%) is greater than that of distant neighbors (64%). Figure 5. (b) is the PA to NPPs in the world. The average PA is 38%. The highest PA is precisely in the United Arab Emirates, a country rich in oil. Japan, which has used NPPs, is at 20% of its revenues. The PA figure only contributes a small role to the development of NPPs. The good intentions of the government determine the development of NPPs in a country.

No regulation prohibits or hinders the use of NPPs in Indonesia. Government Regulations are legal products whose status is under the Act, and the government can issue them without having to ask for approval from the Council. If PP Number 79 of 2014 is considered one of the obstacles to developing NPPs, it is very easy for the government to perfect it. Thus, what determines the decision to build an NPP is the government's political will.

The NPP is a new technology in Indonesia. The concern and fear must be great for both the population and the government. With the description above, it would be food for reflection. The concept of hazard is usually juxtaposed with the concept of risk. The NPP hazards are minimized by site selection, technology, and strict procedures. Thus, the risk of failure and accidents of NPP can be minimized as well.

4. Conclusion

Various issues about NPPs circulating in the community have been explained. The NPPs have high risks but have high safety and security. The selection of site candidates, the design of the reactor core, construction, operation, and decommissioning are all regulated very strictly and in detail by the world body IAEA and BAPETEN in Indonesia.

The government should be given input to choose the generation 4th NPP, type of Small Modular Reactor following archipelagic countries. The development of NPPs in Indonesia should also follow the flow of research, development to proven, followed by utilization as NPPs. Still, a strategy is sought so that 5 years can achieve proper status as NPPs.

Acknowledge
This paper results from a joint study between LPPM UNS and PT ThorCon Power Indonesia, with PKS No. 001/PKS.THC/01/I/2021.

References

[1] IAEA 2020 *Climate Change and Nuclear Power 2020*
[2] JR L and Baratta AJ 2001 Introduction to nuclear engineering (NJ: Prentice hall) pp 1–5
[3] International Atomic Energy Agemcy (IAEA) 2017 *International Nuclear and Radiological Event Scale*
[4] Department of Energy Washington 1993 *DOE FUNDAMENTALS HANDBOOK NUCLEAR PHYSICS* vol 2
[5] Anon an official website of the United States Government.
[6] Abram T and Ion S 2008 Generation-IV nuclear power: A review of the state of the science *Energy Policy* **36** 4323–30
[7] IEA 2021 Data and statistics IEA
[8] International Atomic Energy Agemcy (IAEA) 2019 *Nuclear power reactors in the world* vol 36
[9] IAEA 2016 Nuclear Accident Knowledge Taxonomy
[10] Suwarto S, Indrawanto R and S A 2010 ASPEK KESALAHAN MANUSIA DALAM KECELAKAAN THREE MILE ISLAND UNIT 2 *Bul. Pengelolaan Reakt. Nukl.* **7** 66–71
[11] Anon United Nations Scientific Committee of the Effect of Atomic Radiation
[12] International Atomic Energy Agemcy (IAEA) 2006 *ENVIRONMENTAL CONSEQUENCES OF THE CHERNOBYL ACCIDENT AND THEIR REMEDIATION: TWENTY YEARS OF EXPERIENCE Report*
[13] Supriatna P 2011 Analisis Komparasi Sistem Keselamatan Reaktor Bwr Fukushima Dan Rgtt200K Akibat Gempa Bumi Dan Tsunami *Badan Tenaga Nukl. Nas.* 142–52
[14] Ryu A and Meshkati N 2014 Onagawa: The Japanese nuclear power plant that didn't melt down on 3/11
[15] Pichardo B 2011 RADIATION 101: WHAT IS IT, HOW MUCH IS DANGEROUS, AND HOW DOES FUKUSHIMA COMPARE TO CHERNOBYL?
[16] Rose T and Sweeting T 2016 How safe is nuclear power? A statistical study suggests less than expected *Bull. At. Sci.* **72** 112–5
[17] Escobar Rangel L and Lévêque F 2014 How Fukushima Dai-ichi core meltdown changed the probability of nuclear accidents? *Saf. Sci.* **64** 90–8
[18] Bukszpan D 2011 Nuclear Meltdowns and Disasters *CNBC*
[19] World Health Organization 2013 *Health risk from the nuclear accident assessment after the 2011*

Great East Japan Earthquake and Tsunami based on a preliminary dose estimation
[20] Anon https://www.radiologyinfo.org/en/info.cfm?pg=safety-xray
[21] International Atomic Energy Agemcy (IAEA) 2009 *IAEA Safety Standards for protecting people and the environment : Classification of Radioactive Waste*
[22] Sarjiati U 2018 Risiko Nuklir Dan Respon Publik Terhadap Bencana Nuclear Risk and Public Response To Fukushima Nuclear *J. Kaji. Wil.* **9** 46–61
[23] Ain Sulaiman S N, Mohamed F and Ab Rahim A N 2018 Radioactive release during nuclear accidents in Chernobyl and Fukushima *IOP Conf. Ser. Mater. Sci. Eng.* **298**
[24] World Nuclear Association What is nuclear waste, and what do we do with it?
[25] Energy For Humanity Nuclear waste: 6 reasons we changed our mind
[26] Rodríguez-Penalonga L and Yolanda Moratilla Soria B 2017 A review of the nuclear fuel cycle strategies and the spent nuclear fuel management technologies *Energies* **10**
[27] Nuclear Energy Agency 2006 *A dvanced Nuclear Fuel Cycles and Radioactive Waste Management*
[28] Overton G 2019 SYLOS ultrashort-pulse laser could solve nuclear waste problem *Laser Focus World*
[29] Wisnubroto D S 2010 Analysis of the Institutional Framework For Radioactive Waste Management in Indonesia *Atom Indones.* **35** 71–84
[30] Anon https://www.bapeten.go.id
[31] Anon https://www.batan.go.id
[32] P3Tek Balitbang ESDM 2019 *Kajian Pengembangan & Implementasi PLTT di Indonesia* (Jakarta)
[33] IAEA 2017 Instrumentation and Control Systems for Advanced Small Modular Reactors (IAEA Nuclear Energy Series) *IAEA Nucl. Energy Ser.*
[34] Uji A, Prakash A and Song J 2021 Does the "NIMBY syndrome" undermine public support for nuclear power in Japan? *Energy Policy* **148** 111944
[35] Anon http://www.bisconti.com/

Analysis of disaster literacy on student behavioral responses in efforts to reduce earthquake disaster risk at SMA Negeri 1 Klaten

M A Kesumaningtyas[1*], S H N Hafida[1*] and M Musiyam[1]

[1] Geography Education, Faculty of Teacher Training and Education, Universitas Muhammadiyah Surakarta, Indonesia

*Email: mayangarum995@gmail.com, shnh421@ums.ac.id

Abstract. Earthquakes are disasters that often occur in Indonesia. Many students in Indonesia became victims when the earthquake occurred. Therefore, the behavioral response of students when facing disasters is very important to understand. This behavioral response will not manifest spontaneously, it takes time and considerable effort to realize a good behavioral response to disasters. One ability that is currently being developed in students is disaster literacy, a good disaster literacy can help students to know the right attitude when a disaster occurs. This study aims to determine the effect of disaster literacy on student behavioral responses in an effort to reduce earthquake risk at SMA Negeri 1 Klaten. There are 2 variables, namely the independent variable with 4 indicators that is basic disaster literacy, functional disaster literacy, communicative / interactive disaster literacy and critical disaster literacy, while the dependent variable has 4 indicators, namely preparedness, awareness, action and affect. This study uses a quantitative method with a correlational design. Respondents in this study amounted to 269 students, and data collection techniques were carried out using questionnaires, interviews and documentation. The data analysis technique used the Pearson Bivariate Correlation Test. The results showed that there was a significant relationship (sig. 0.672) between disaster literacy and student behavior responses in an effort to reduce earthquake risk at SMA Negeri 1 Klaten. The existence of this significant relationship should be utilized by every school that located in disaster-prone areas to develop several programs that focus on disaster literacy for students.

1. Introduction

Indonesia is considered crucial to natural disasters. According to [1], Indonesia has several active volcanoes, and this is due to Indonesia's geographical location, which is in the path of the Pacific Ring of Fire. It has the potential to become a major natural disaster. In addition, the Indo-Australian Plate is located in the south, the Eurasian Plate is located in the north, and the Pacific Plate is located in the east. Because of the conditions, Indonesia has a high vulnerability to natural disasters. [2] stated that Indonesia is one of the countries with the second-highest disaster risk out of 15 countries in the world after Bangladesh. It was asserted by [3] that Indonesia faces high disaster risk due to the increasing vulnerability of the population and the lack of integration of sustainability principles in disaster activities.

Earthquakes are natural disasters that occur in Indonesia. According to [4] an earthquake is a natural phenomenon whose occurrence cannot be predicted and can cause material losses to mortality. Another thought is conveyed by [5] which states that an earthquake is a natural event that occurs unexpectedly due to the shaking of the earth due to the movement of rock layers located in the earth's crust that occurs due to the movement of tectonic plates. Moreover, earthquakes can also be caused by volcanic activity. Based on the geographical conditions of Indonesia and the causes of earthquakes, Indonesia is one of the most exposed countries to earthquakes.

Earthquakes can occur in various regions in Indonesia. The enormous pressure caused by three earth plates led part of the Earth's plates to break apart, dividing the islands of Indonesia into separate sections. The sections are bounded by active faults. Therefore, the Indonesian archipelago is located in an area that has the potential for a large earthquake to occur [6]. One of the islands in Indonesia that has the potential for earthquakes is Java, especially the Central Java area. The area is in the moderate to high disaster risk class. Purworejo Regency, with an index value of 218.20 in 2018, is one of the areas with a high disaster index. Then one of the districts with a moderate index is Klaten Regency in 2018 of 123.20 [7]. In addition, Klaten Regency is included in the class of high earthquake vulnerability with the second rank nationally with a score of 68 [8].

According to [9], it is necessary to integrate knowledge, understanding and skills to minimize risks or losses due to disasters. Efforts to increase knowledge, understanding and skills can be obtained through school activities. According to [10], schools are sites where a group of people join together to form a single entity, making them ideal to be used as a learning experience for disaster risk reduction. Schools can hold disaster simulations to improve preparedness for all school residents. Disaster preparedness is crucial in order to reduce the impact of disasters that cause casualties and other natural damage [11].

Empowering children from an early age can be used as the first step in helping children become more aware of the disaster risks around them. It can be known through the perception of each person. This perception can arise because of the interaction with the surrounding environment, which is influenced by social, economic, cultural and religious factors. Therefore, understanding perceptions can make students aware of disasters [12].

It is possible for students to be apathetic to the situations around them if they are unaware of and unconcerned about disaster risk reduction efforts. According to [13], understanding student responses in disaster mitigation is critical because the response can help students' self-adjustment resulting from understanding disaster risk reduction efforts that have occurred. Through understanding disasters, students are expected to be able to act quickly, actively and think critically in dealing with natural disasters. Moreover, it can build empathy in students towards disaster victims to be careful and precise when helping others.

One of the ways to improve students' behavioral responses to disaster risk reduction efforts is disaster literacy for students [14]. Disaster literacy is an effort to make students aware of the importance of dealing with disasters. Students must understand disaster literacy in order to completely comprehend the location of the Republic of Indonesia, which is located in a disaster-prone area [15]. Disaster literacy needs to be given to students in schools because students are fragile to disasters. They spend part of their time in schools separated from their parents, so the condition causes many children to become victims. The existence of disaster literacy can reduce or even eliminate losses and casualties due to disasters. Without disaster literacy in students, it can lead to a lack of information obtained about independent evacuation methods that must be carried out in the event of a disaster.

Because of disaster literacy, it can help students understand the disaster. In accordance with the statement of [16] which explains that disaster literacy can be used as an individual's capacity to read, use the information and understand disasters in disaster mitigation, preparation, response and recovery from disasters. Without disaster literacy's students, they will be easily influenced by information scattered in cyberspace whose sources are not yet verified. It creates a sense of fear and threatens them. According to [17], disaster literacy is able to encourage students to understand the efforts on what they must do regarding disasters. It is to form responsibility for students to practice independence in the event of a disaster.

Considering the condition of the Klaten Regency, which has a high vulnerability to earthquakes at the national level and supported by the statement submitted by [10] schools are expected to be able to conduct disaster simulations to improve preparedness for all school residents. SMA Negeri 1 Klaten is one of the schools in Klaten Regency with a geographical location prone to earthquakes. It is due to the location close to an active volcano. Thus, it is important to realize the importance of raising awareness and improving disaster risk reduction. The initial interview results were on February 16, 2021, with Mr Andy Rochmadi, who explained that SMA Negeri 1 Klaten carried out disaster risk reduction efforts in the form of disaster simulations, one of which was the earthquake in 2019.

The simulation was carried out because the geographical location of SMA Negeri 1 Klaten was in a disaster-prone area. Through the disaster simulation, students are trained to be alert when an actual disaster occurs. In the disaster simulation activity, students receive direction and training on how to mitigate disaster efforts to be carried out in the event of an earthquake at school. The results of the interview with Mr Andy Rochmadi on February 16, 2021, explained that the disaster risk reduction efforts carried out by SMA Negeri 1 Klaten were only carried out once a year. In addition, disaster simulations are carried out differently every year and are focused on earthquake disaster simulations due to the geographical location of the Klaten Regency, which is prone to earthquake disasters. The simulation is carried out as a lesson for students so that if a disaster occurs, they are ready to help themselves and those around them.

Disaster literacy can help SMA Negeri 1 Klaten students develop a better knowledge of disasters. The disaster-related response is an activity to understand a sign of a disaster that comes from oneself where the response occurs in a conscious state [18]. Disaster studies are required in disaster mitigation efforts for schools in areas where natural disasters, particularly earthquakes. Disaster literacy is important because it becomes a recommendation in measuring and building students' capacity for disasters.

Because Indonesia is prone to disasters, disaster literacy can be a relatively strategic study for future advances in improving student response and behaviour in creating understanding and critical power towards disasters to be used to reduce disaster risk [19]. According to [20], students' behavioural responses in disaster risk reduction efforts can contribute to student preparedness. It can change mindsets and get used to always caring about the environment, being aware of disasters, and increasing protective actions that are useful for themselves and those around them to be applied effectively in life. This study aims to determine the relationship between disaster literacy and student behavioural responses to reduce earthquake risk at SMA Negeri 1 Klaten.

2. Research Methods

This method and research is quantitative method and correlational quantitative method design. According to [21], the purpose of quantitative correlational research is conducted to determine the correlation of a variable with other variables. The correlation between one and several other variables is expressed by the correlation coefficient's significance and statistical significance. In this study, the correlation was used to determine the relationship between disaster literacy and student behavioral responses. The research was conducted at SMA Negeri 1 Klaten. The reason for choosing the location is that the school is a school that is prone to earthquakes because it is close to an active volcano, Mount Merapi. The school also organizes disaster risk reduction activities every year, focusing on earthquake disasters. Therefore, it is important to know disaster literacy to support behavioral responses to disasters.

The respondents of this study were class X and XI, with a total sample of class X 139 and class XI 130. Therefore the total sample was 269 students. Questionnaires, observation, and documentation were used to collect data in this study. Questionnaires were used to determine the relationship between disaster literacy from theory [16] and student behavioral responses from theory [22]. Likert scale questionnaire was used to measure students' opinions and attitudes towards disaster literacy and student behavioral responses. To check the validity and reliability of the questionnaire, the researcher conducted a validity and reliability test at SMA Negeri 1 Karanganom with 34 respondents. Observations were made to observe the research location. Furthermore, the last is the documentation to get secondary data, such as administrative maps. In this study, there are two variables. First, disaster literacy variable consists of four indicators: basic disaster literacy, functional disaster literacy, communicative/interactive disaster literacy, and critical disaster literacy. To measure disaster literacy indicators using descriptive statistics. Second, the behavioral responses variable also consists of 4 indicators, that is: preparedness, awareness, action, and affect. To measure behavioral response indicators using descriptive statistics. To know about correlation between disaster literacy and behavioral response, researchers using Pearson bivariate analysis. A correlation test with bivariate analysis was used, followed by a Pearson correlation test to determine the effect of disaster literacy on student behavioral responses in an effort to reduce earthquake risk at SMA Negeri 1 Klaten.

3. Results and Discussion

3.1. Results

The results of the analysis on disaster literacy for grade X and XI students at SMAN 1 Klaten can be seen in Table 1 below.

Table 1. Disaster Literacy

Indicator	Category				
	Very Good	Good	Currently	Not Good	Very Not Good
Basic Disaster Literacy	27 (10%)	222 (83%)	20 (7%)	0 (0%)	0% (0%)
Functional Disaster Literacy	23 (9%)	203 (75%)	43 (16%)	0 (0%)	0 (0%)
Communicative/Interactive Disaster Literacy	10 (4%)	126 (47%)	131 (49%)	2 (0%)	0 (0%)
Critical Disaster Litracy	44 (16%)	211 (79%)	14 (5%)	0 (0%)	0 (0%)
Total	104 (9,66%)	762 (70,82%)	208 (19,34%)	2 (0,18%)	0 (0%)

Disaster literacy of SMAN 1 Klaten students is in a good category (70,82%). It can be concluded that the disaster literacy of SMA Negeri 1 Klaten students is good, especially on fundamental disaster literacy indicators. The disaster literacy of students at SMA Negeri 1 Klaten is good because students understand the basics of disasters well, and students can participate in earthquake disaster risk reduction efforts well. The disaster literacy of SMA Negeri 1 Klaten students is good because students know and understand the basics of disaster literacy to analyze disaster-related information to find out obstacles and take steps to control themselves.

The results of the analysis of the behavioral responses of class X and XI students at SMAN 1 Klaten can be seen in Table 2.

Table 2. Behavioral Response

Indicator	Category				
	Very Good	Good	Currently	Not Good	Very Not Good
Preparedness	80 (30%)	53 (20%)	205 (76%)	11 (4%)	0% (0%)
Awareness	38 (14%)	38 (14%)	192 (72%)	38 (14%)	0 (0%)
Action	30 (11%)	202 (75%)	36 (14%)	1 (0%)	0 (0%)
Affect	14 (6%)	181 (68%)	74 (26%)	0 (0%)	0 (0%)
Total	162 (13,58%)	474 (39,73%)	507 (42,50%)	50 (4,19%)	0 (0%)

Students of SMA Negeri 1 Klaten showed a moderate response and behaviour (42,50%) to risk earthquakes, especially in the preparedness indicator (76%). Student behavioural responses are good because students are well prepared to start from knowledge, training and participation, student awareness about disasters is high to protect the environment to reduce the impact of disasters, students carry out disaster risk reduction efforts according to directions and act quickly, precisely and actively. The behavioural response of students in efforts to reduce earthquake risk is quite good because the school holds annual disaster simulations, especially earthquake disasters, to train students to prepare themselves when a disaster occurs, make students aware of the importance of disaster response, care for the environment and know the impact of the disaster itself. So that students can provide good responses and behaviour in dealing with earthquake disasters.

The results of the analysis of the relationship between disaster literacy and behavioral responses of class X and XI students at SMAN 1 Klaten can be seen in the table below.

Table 3. Relationship of Disaster Literacy to Behavioral Response

		Disaster Literacy	Behavioral Response
Disaster Literacy	Pearson correlation	1	.672**
	Sig. (2-tailed)		.000
	N	269	269
Behavioral Response	Pearson correlation	.672**	1
	Sig. (2-tailed)	.000	
	N	269	269

Table 3 is a Pearson bivariate correlation test using IBM SPSS 26. Pearson's correlation on the variable perception of disaster literacy is 0.672 and with behavioral responses 0.672. The correlation test with the 269 respondents is higher than the R table. It is because the R table with 200 samples at 5% significance is 0.138, and the R table with 300 samples at 5% significance is 0.113. According to the rules, relationship guidelines according to Sugiyono (2015). The correlation of 0.672 is included in the strong correlation, so it can be concluded that the variables X (perception of disaster literacy) and Y (response to student behavior) have a significant relationship. The value of 0.672 includes a positive correlation. Therefore it can be interpreted that the higher of disaster literacy at SMAN 1 Klaten, the higher the behavioral response of students in efforts to reduce earthquake risk.

3.2. Discussion

Based on the analysis results, this study shows a positive relationship between disaster literacy and student behavioral responses in an effort to reduce earthquake risk at SMAN 1 Klaten. The Pearson correlation test with a significance value of 0.672 was used in this analysis using R table N=269. R table with 200 samples at 5% significance, 0.138, and R table with 300 samples at 5% significance, 0.113, were the details. Thus, it can be concluded that the higher the disaster literacy rate, the higher the behavioral response of students in an effort to reduce the risk of an earthquake at SMAN 1 Klaten.

In this analysis, there is a significant relationship between disaster literacy and student behavioral responses in an effort to reduce earthquake risk at SMAN 1 Klaten. It is in accordance with what was stated by [23] that because disaster literacy can be used as a technique to minimize disaster risk, the behavioral reaction of students in disaster reduction could be increased. Even so, earthquakes are a common phenomenon in Indonesia. In addition, according to [24] student disaster literacy is often poor due to a lack of disaster prevention and aid exercises and disaster reduction efforts. The same study was also conducted by [25] explaining that students' disaster mitigation literacy was at level 3 because the problem-based learning process was running very well. According to [25], improving disaster mitigation literacy by encouraging the improvement of students' skills and awareness about disaster awareness in the surrounding environment.However, there is still a need to integrate disaster education into school learning to form disaster preparedness with a generation that is resilient to disasters.

A similar study was also conducted by [26], This explains that students have a positive effect on students' perceptions of their ability to deal with earthquake disaster risks. Students can learn on their own and have a higher knowledge of disaster risk. However, a different result from the research conducted by [27] explains that because students did not participate in disaster preparedness socialization activities conducted by schools, their preparedness behavior in dealing with disasters is poor. In line with research conducted by [28] which stated that the students' response to the earthquake was moderate; male students have less response than female students. In addition, the experience of an earthquake has a significant beneficial effect on emergency response, so it is crucial to have a disaster simulation exercise to increase students' awareness of disaster prevention and response skills. A similar study was conducted by [29] which explained that students' prevention literacy (knowledge, attitudes & skills) was highest on disaster prevention skills and students' lowest score on disaster prevention knowledge. There are similarities, namely prevention literacy is included in disaster literacy, because disaster literacy covers from disaster mitigation to disaster recovery. Despite having great disaster awareness skills, students lack action skills (behavioral response). As a result, consideration should be given to how to improve disaster preparedness in the future.

4. Conclusion

The conclusion that can be drawn based on the results of the discussion is that there is a significant relationship between perceptions of disaster literacy and student behavioral responses in an effort to reduce earthquake risk at SMA Negeri 1 Klaten in 2021. It can be proven by a significance value of 0.672, which is included in a positive correlation. Hence, if the perception of disaster literacy in students is good, it can encourage student behavioral responses to disasters. The influence of disaster literacy can increase students' knowledge about disasters and realize responsive behavioral responses in disaster risk reduction and during disasters.

Acknowledgments

Thank you to SMA Negeri 1 Karanganom for allowing us to carry out the reliability validity test at school, thank you to SMA Negeri 1 Klaten for allowing us to able to do this research and thank you to all those who have contributed to the implementation of this article.

Reference

[1] Rahma A 2018 Implementasi Program Pengurangan Risiko Bencana (PRB) melalui Pendidikan Formal *J. VARIDIKA* **30** 1–11
[2] Maplecorft 2010 Natural Disasters Risk Index 2010 **44** 2010
[3] Zulfiar M H and Jayady A 2018 Kajian Kerentanan Pada Sektor Konstruksi Dalam Pengurangan Risiko Bencana Gempa Bumi *Karkasa* **4** 1–7
[4] Subagia I W 2015 Pelatihan Mitigasi Bencana Alam Gempa Bumi Pada Siswa Sekolah Dasar Negeri 1 Pengastulan Kecamatan Seririt Kabupaten Buleleng Bali *JPI (Jurnal Pendidik. Indones.* **4** 585–98
[5] Utomo D P and Purba B 2019 Penerapan Datamining pada Data Gempa Bumi Terhadap Potensi Tsunami di Indonesia *Pros. Semin. Nas. Ris. Inf. Sci.* **1** 846
[6] Pasau G and Tanauma A 2011 Pemodelan Sumber Gempa di Wilayah Sulawesi Utara sebagai Upaya Mitigasi Bencana Gempa Bumi *J. Ilm. Sains* **11** 202–9
[7] BNPB N D M A of the R of I- 2018 IRBI (Indeks Resiko Bencana Indonesia) Tahun 2018 *Republik Indones.* 115
[8] Indonesian National Agency for Disaster Management 2013 *IRBI (Indeks Risiko Bencana Indonesia)*
[9] Maryani E 2016 Model Pembelajaran Mitigasi Bencana Dalam Ilmu Pengetahuan Sosial Di Sekolah Menengah Pertama *J. Geogr. Gea* **10**
[10] Khoirunisa N 2015 Analisa Kesiapsiagaan Siswa pada Satuan Pendidikan Muhammadiyah Terhadap Bencana Gempa Bumi di Kabupaten Sukoharjo *Konferensi Nasional Forum Perguruan Tinggi Pengurangan Risiko Bencana* pp 105–9
[11] Lee G H, Bae J W, Oh N, Hong J H and Moon I C 2015 Simulation Experiment of Disaster Response Organizational Structures With Alternative Optimization Techniques *Soc. Sci. Comput. Rev.* **33** 343–71
[12] Pahrul, Razikin; Rosalina K D A 2017 Strategi Penangulangan Bencana Banjir Berdasarkan Persepsi Masyarakat Di Kecamatan Barabai Kabupaten Hulu Sungai Tengah *J. Pendidik. Geogr.* **4** 27
[13] Desfandi M 2014 Urgensi Kurikulum Pendidikan Kebencanaan Berbasis Kearifan Lokal di Indonesia [The urgency of disaster education curriculum based on local wisdom in Indonesia} *Sosio Didakt. Sos. Sci. Educ. J.* **1** 191–8
[14] Marlyono S G, Pasya G K and Nandi 2016 Peranan Literasi Informasi Bencana Terhadap Kesiapsiagaan Bencana Masyarakat Jawa Barat *Gea. J. Pendidik. Geogr.* **16** 116–23
[15] Afrian R and Islami Z R 2019 Peningkatan Potensi Mitigasi Bencana dengan Penguatan Kemampuan Literasi Kebencanaan pada Masyarakat Kota Langsa *J. Pendidik. Geogr. Kajian, Teor. dan Prakt. dalam Bid. Pendidik. dan Ilmu Geogr.* **24** 132–44
[16] Brown L M, Haun J N and Peterson L 2014 A proposed disaster literacy model *Disaster Med. Public Health Prep.* **8** 267–75
[17] Labudasari E and Rochmah E 2020 Literasi Bencana di Sekolah: sebagai Edukasi untuk Meningkatkan Pemahaman Kebencanaan *Metod. Didakt. J. Pendidik. Ke-SD-an* **16** 41–8
[18] Pratama T K, Sarir R R, Utami N Z, Mufidah A Z, Setyadi L and Wardani P I 2020 No Respon Mitigasi Bencana Gempa Bumi di SMP M 7 Bayat dan SMP MBS 2 Prambanan *J. Geogr. Edukasi dan Lingkung.* **4** 39–49
[19] Muktaf Z M 2017 Studi Literasi Bencana dalam Perspektif Ilmu Komunikasi *Apik Ptm* 1–12
[20] Setyowati D L 2019 *Pendidikan Kebencanaan* (Semarang: Universitas Negeri Semarang)
[21] Creswell. J W 2014 *Research Design : Qualitative, Quantitative, and Mixed Methods Approaches* (California: SAGE Publications, Inc.)
[22] Luo Y, Shaw R, Lin H and Joerin J 2014 Assessing response behaviour of debris-flows affected communities in Kaohsiung, Taiwan *Nat. Hazards* **74** 1429–48
[23] Muktaf Z M 2017 Studi Literasi Bencana dalam Perspektif Ilmu Komunikasi *Konferensi APIK PTM di Prodi Ilmu Komunikasi Univeritas Muhammadiyah Ponorogo* pp 1–12

[24] Zhang D, Zhu X, Zhou Z, Xu X, Ji X and Gong A 2020 Research on Disaster Literacy and Affecting Factors of College Students in Central China *Disaster Med. Public Health Prep.* 1–7

[25] Suharini E, Meliana D, Sanjoto T and Kurniawan E 2020 The Strategy of Disaster Mitigation Literacy through Problem-Based Learning (PBL) in the School Prone to Tidal Floods

[26] Shoji M, Yoko Takafuji and Tetsuya Harada 2019 *Behavioral Impact of Disaster Education : Evidence from a Dance-Based Program in Indonesia*

[27] Widjanarko M and Minnafiah U 2018 Pengaruh Pendidikan Bencana Pada Perilaku Kesiapsiagaan Siswa Effect of Disaster Education on Student Preparedness Behavior 1–7

[28] Wei B, Su G and Li Y 2020 Evaluating the cognition and response of middle/high school students to earthquake—a case study from the 2013 Mw6.6 Lushan earthquake-hit area, China *Int. J. Disaster Risk Reduct.* **51** 101825

[29] Chung S-C and Yen C-J 2016 Disaster Prevention Literacy among School Administrators and Teachers: A Study on the Plan for Disaster Prevention and Campus Network Deployment and Experiment in Taiwan *J. Life Sci.* **10** 203–14

The Effectiveness Of Socialization Toward Students Understanding Of Tectonic Activities Disaster

S Supriyadi*, NV Saputro, M P Aji, S Linuwih

Faculty of Mathematics and Science, Universitas Negeri Semarang, Semarang, Indonesia

*Email : supriyadi@mail.unnes.ac.id

Abstract. The purpose of this research is to determine how much effectiveness rather than socialization, especially in terms of disaster which occurs because of tectonic activities . This relates to the location of Indonesia at the triple junction so it is prone to experiencing disasters due to tectonic activity. The alternative hypothesis (ha) states that there is a difference in understanding of tectonic disaster through socialization treatment. Research subjects were Indonesian students in Kuala Lumpur Malaysia. The method used is an experimental research method with a one group pretest and posttest strategy. The data obtained was analyzed in three ways which are normality test, product moment correlation test, and two party test. The normality test found that the distribution of data is normal with scale 0.007 for pretest and 0.353 for posttest based on Saphiro-Wilk test. The effectiveness is measured by t-test from SPSS. The result shows that the syg 2-tail is 0.000 which means that there is a significance in the research result. This makes the hypothesis can be accepted. The accepted hypothesis means that socialization is still a good way to make people understand about the disaster and make the chance of survival is getting higher.

1. Introduction

Indonesia is known among the geological scientist as a triple jucntion. This term relates to the position of Indonesia which is crossed by three tectonic plates. These three tectonic plates are (1) the Indo-Australian plate to the south, (2) the Eurasian plate, and (3) the Pacific plate. The consequences of the existence of tectonic plates cause Indonesia have high volcanic and seimic activity. This is marked by the occurrence of disasters that occurred in various regions in Indonesia as summarized by National Disaster Management Agency (BNPB) as proclaimed [1]. From BNPB we also notice some of natural disasters that occur in 2018-2019 like shown in Table 1.

Table 1. Natural Disasters in Indonesia 2018-2019

Kind of disaster	Amount	Victim		
		Died & Lost	Injured	Suffer & Evacuate
Eartquake	50	581	2,166	531,809
Tsunami	2	453	14,059	42,325
Earthquake and Tsunami	2	652	2,877	169,035
Volcano Eruption	63	0	56	71,157

Source: https://www.bps.go.id

Earthquake and tsunami happened a lot in Indonesia, this is once again related to the position of Indonesia. Earthquake basically is a nature phenomenon in the form of shaking on the ground that may destroy the building and also the environment on the ground [2], while tsunami is some kind of wave that triggered by vertical movement by water column in the sea [3]. Earthquake and tsunami are two disasters that can be considered as the most horrific disaster among others. This is shown in figure 1 and figure 2 bellow. From figure 1, it can be concluded that the eartquake percentage is only 0,68% from all the disasters that happened, but based on the figure 2 the percentage of victim that caused by earthquake reachs 2.5%. This fact shows that earthquake is the disaster that may caused major effect to the people. Beside the power of earthquake and tsunami, there are several factors that caused the amount of victim is so high. One of them is the less knowledge of earthquake and tsunami on Indonesian people. They usually get panic and do the wrong things so they fail to survive. This condition is getting worse because the school as the prime education place is not considered the knowledge of disaster is highly important [4], [5], [6]. That's why the role of government is still really important to make sure people have the good understanding about disasters especially earthquake and tsunami.

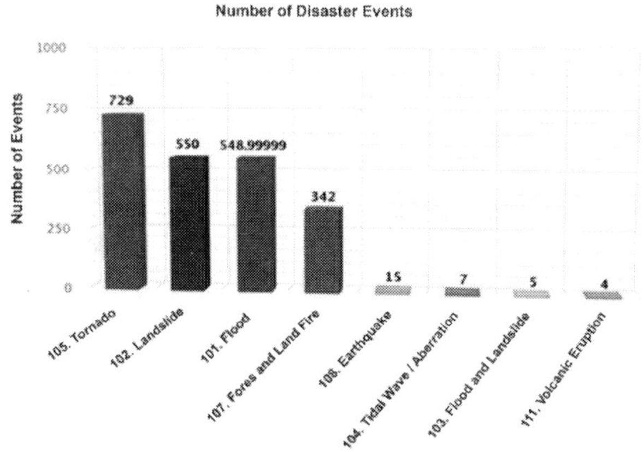

Figure 1. The Amount of Disaster Happened in 2019

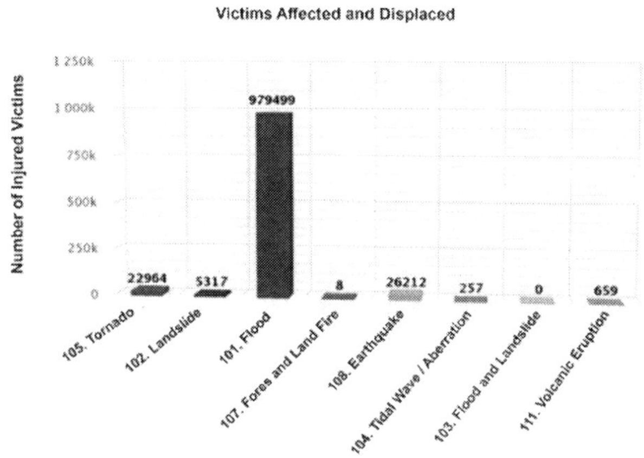

Figure 2. Victims from Several Nature Disasters in 2019

Government through BNPB is often use socialization as the main wy to educate people related to knowledge of disasters. Socialization is viewed as the easiest and cheapest way to reach the goal of disaster education. This study aims at finding out the effectiveness of socialization towards students understanding on tectonic activities disaster. To make sure that disaster education is done on the right method, the effectiveness of socialization needs to be known. This study can give students, especially in Indonesia, comprehension on the importance of socialization about natural disaster. This is in accordance with research by Yamori [7] about action research on disaster reduction education. In addition, disaster learning should be provided through a simulation model that can involve the cognitive, affective, and psychomotor domains in the context of disaster mitigation [8]. For this reason, it is necessary to develop a virtual-based disaster learning model. Disaster learning is developed with the assumption that the current focus of mitigation is not only about disaster management, but rather refers to the notion of preparedness. The purpose of disaster mitigation learning to community members is to increase citizens with the capacity to cope with disasters not only during and after disasters but also before disasters [9].

2. Methods

Hypothesis in this research is that socialization has a significant effect on students understanding of tectonic activities disaster. The research is done by using pre-experimental design method, that has been describe by Sugiyono [10] the method used is experimental, but there is still a relationship between external variables and the formation of the dependent variable. Figure 3 shows the steps in the research.

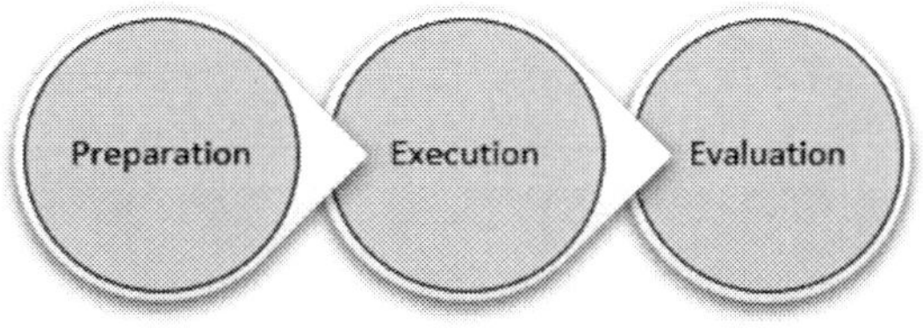

Figure 3. Steps of research

The first step of the reseacrh is preparation, in this step, the subject and also the instrument of research is being prepared. The subject is a class of student consit of student from first, second, and third grade from Sekolah Indonesia Kuala Lumpur Malaysia (SIKLM). Then, for the instrument, pre-test and post-test problem are being prepared in this step also. Those tests have the same problem set that consist of six essay questions.The execution is the day where students are given the socialization,off course before that they're also given the pretest, and after the socialization, students dothe posttest. At the evaluation step, the data (of pretest and posttest) is evaluated and calculated using statistical method which used T experiment to find out the socialization influence toward students' comprehension using SPSS program.

3. Results and Discussions

3.1 Normality Test

The normality test is used to make sure whether the data distribution is normal or not, this thing s really important because the normality will be the factor that used to decide what type of the significance test that is suitable for the data. The two most popular methods of testing the normality of data are the Saphiro-Wilk test. The Shapiro–Wilk test is a more effective method for small (n) data (n<50) [11], [10]. The normality test is done by using SPSS software, then the result is shown in Table 2. Normality test results show students are in the same position in terms of ability. It is hoped that this socialization activity will change understanding in the true form not influenced by external factors

Table 2. Test of Normality

	Shapiro-Wilk		
	Statistic	Df	Sig.
Pre test	,927	46	,007
Posr test	,973	46	,353

3.2 T-test for Significance

The significance is tested using independent variable t-testin SPSS program. The result shows in Table 3, as we can see that the sig. (2-tailed) is 0.000. It means that there is a significant difference between pretestand posttest, and socialization is proven as the effective way to increase the knowledge of disasters that caused by tectonic activities. Based on the activities that have been carried out by providing socialization, and its influence with the pre-test post-test, maximum results have not been obtained. The socialization carried out did not provide an alternative in delivering the material and did not activate students, who were more likely to be passive, active students to give responses in the form of questions, giving invisible rebuttals. The same was stated by Torani [11] in his research entitled The Importance of Disaster and Emergency Education, his research concludes that the person in charge of research and education in disaster education can be done to improve education. In this case, there should be the use of different training methods. In addition, the application of maneuvers and types of exercises should be carried out for proper planning, to increase personal knowledge and skills, and to assist as a method of evaluating individual performance. Likewise Pty stated that [12] It was found that the implementation of disaster education is related to the location of preparedness plans and the provision of information, as well as the 'Traditional' approach which may have more negligible effects, such as motivating preparedness behaviour.

Table 3. T-test result

		Levene's Test for Equality of Variances				
		F	Sig.	T	df	Sig.(2-tailed)
Disaster socialization result	Equal Variances Assumed	,157	,693	-8,047	90	,000
	Equal Variances not Assumed			-8047	87,873	,000

4. Conclusion

The research proves that socialization is the effective way to educate people about disaster especially the earthquake and tsunami disaster. This result should be followed up by some parties by increasing socialization activities in various regions. This is intended to increase community knowledge related to disasters and enhance their chances of survival when disasters strike.

The results of the socialization showed an increase in students' understanding related to the earthquake and tsunami disaster. There needs to be a socialization innovation that will be implemented. The innovation in question is related to the approach and the media used, and the evaluation applied. The approach that can be discussed is a participatory discussion with students who are more dominant than the teacher.

Acknowledgments

Acknowledgments to the Dean of the Faculty of Mathematics and Natural Sciences who has funded this research through an assignment research scheme, coordinator of the Earth Physics expertise group for the discussion

References

[1] Hasanah N 2018 Rentetan Bencana Alam di Indonesia Sepanjang 2018, retrieved from https://news.okezone.com/read/2018/12/24/337/1995339/rentetan-bencana-alam-di-indonesia-sepanjang-2018, on Saturday April 2019, 04.29 pm

[2] Fitriana F, Farid M, Mayub A and Connie C 2019 Identifikasi Koefisien Atenuasi dan Implementasinya Untuk Mengetahui Kesiapsiagaan Masyarakat Terhadap Bencana Gempa Bumi di Kota Bengkulu *PENDIPA J. Sci. Educ.* **3** 20–7

[3] Mambu B, Tamuntuan G H and Pasau G 2019 Simulasi Ketinggian dan Waktu Tiba Gelombang Tsunami di Tahuna Sebagai Upaya Mitigasi Bencana *J. MIPA* **8** 13
[4] Yusuf S, Asniar, Rachmalia and Susanti S S 2019 Disaster preparedness in the red zone schools at 13 years post tsunami 2004 *IOP Conf. Ser. Earth Environ. Sci.* **273**
[5] Kane S N, Mishra A and Dutta A K 2016 Preface: International Conference on Recent Trends in Physics (ICRTP 2016) *J. Phys. Conf. Ser.* **755**
[6] Adlim 2011 Education Reconstruction After Tsunami in Aceh (Lesson Learn from Aceh Province)
[7] YAMORI K 2008 Action Research on Disaster Reduction Education: Building a "Community of Practice" through a Gaming Approach *J. Nat. Disaster Sci.* **30** 83–96
[8] Sugiyono 2015 *Metode Penelitian Pendidikan Pendekatan Kuantitatif, Kualitatif, dan R&D* (Bandung: Alfabeta)
[9] Mishra P, Pandey C M, Singh U, Gupta A, Sahu C and Keshri A 2019 Descriptive statistics and normality tests for statistical data *Ann. Card. Anaesth.* **22** 67–72
[10] Ghasemi A and Zahediasl S 2012 Normality tests for statistical analysis: A guide for non-statisticians *Int. J. Endocrinol. Metab.* **10** 486–9
[11] Torani S, Majd P M importance of education on disasters and emergencies: A review article, Maroufi S S, Dowlati M and Sheikhi R A 2019 The Importance of Education on Disasters and Emergencies: A Review Article *J. Educ. Health Promot.* **8** 85
[12] Dufty N 2018 a New Approach To Disaster Education *Int. Emerg. Manag. Soc. Annu. Conf. Manila, Philipp. 13-16* 1–10

The importance of disaster mitigation literacy in social studies learning

D S Logayah[1,2*], E Maryani[3], M Ruhimat[3] and E Wiyanarti[3]

[1] Students Doctoral of Social Studies, Postgraduate of Universitas Pendidikan Indonesia, Indonesia.
[2] Department of Social Studies, Universitas Pendidikan Indonesia.
[3] Departemen of Social Studies, Postgraduate of Universitas Pendidikan Indonesia.

*Email: dina.logayah@upi.edu

Abstract. Education is one of the conscious and planned efforts in shaping the character of students for the cultivation of knowledge, attitudes, and skills. One of them is that social studies learning contributes to providing students' knowledge, attitudes, and skills in dealing with disasters that are objective and subjective. In sustainable development efforts, social studies learning has a crucial role in reducing disaster risk. This paper aims to reveal that Indonesia is a disaster-prone country, especially in the community, one of which is the school environment. There are approximately 126,681 education units located in moderate and high disaster risk areas. Disaster literacy seeks to minimize disasters by providing knowledge about disasters, attitudes, and skills about disaster risk reduction. This research approach uses a literature review. The research results that have been carried out on disaster literacy show that there is still a lack of knowledge and understanding of social studies teachers and students related to disasters, lack of learning tool guidelines on disaster and disaster vulnerability: physical condition, facilities, and infrastructure of schools against catastrophe.

1. Introduction

As a ring of fire region, Indonesia is one of the countries prone to natural disasters. Geological conditions and astronomical locations also influence the occurrence of disasters that often-hit Indonesia [1,2]. Government regulations regarding national disaster management policies contained in Law no. 24 of 2007 concerning disaster management and the regulation of the Minister of Home Affairs No. 33 of 2006 concerning general guidelines for disaster mitigation as a reference in knowledge about disaster studies in Indonesia [3]. There are around 126,681 academic units located in moderate and high disaster-prone areas. Disaster literacy is given to minimize disasters by providing a person with knowledge, attitudes, and skills to reduce good disaster risk. This also has an impact on the required field of disaster studies. The usefulness of the field of disaster studies is a concrete thing in dealing with disasters that often occur in Indonesia. Education is a way for someone to have quality abilities to meet their future. The competencies developed are survival skills and expertise in change, conflict, uncertainty, and complexities in life [4]. Education is also one of the conscious and planned efforts to shape the character of students through the cultivation of knowledge and skills. So that education has a goal in shaping the character of the nation's generation. Education can provide students with knowledge and skills in dealing with disasters caused by natural events or not [5]. To reduce the impact of current and future disasters caused by natural events or not, disaster mitigation education is needed by students. Disaster studies or disaster education are still not taught directly or indirectly in formal education, both basic, secondary, and higher education and disaster education is currently only delivered to the community through socialization and training that is incidental in nature and has not been structurally and massively. Lack of disaster knowledge causes the Indonesian state to have a high impact potential that causes loss of property and life. Likewise, the character of a disaster is not formed quickly, but requires a long-term process. In understanding disaster, disaster literacy is needed by students to provide a whole experience that Indonesia is in a disaster-prone area. Research conducted by Reo Kimura, et al (2017)[5] states that disaster literacy can improve disaster response for the community, because literacy provides basic knowledge, skills, and attitudes for disasters. Social studies education has a contribution in providing disaster literacy to students, because social studies is one of the subjects that has multidisciplinary/ interdisciplinary/ crossdisciplinary/ transdisciplinary approach [6] which consists of several fields of study in geography, sociology, economics, and history. It is very appropriate when disaster literacy is given to students who study social studies subjects. Social studies subjects have the task of developing students' sensitivity to social problems that occur around them. This is prepared so that students have a positive mental attitude in overcoming inequality and everyday issues. The students' self-ability can be realized if the social studies education and learning, in particular, is carried out well [4]. Disaster literacy is very important to be given to students considering that Indonesia is a disaster-prone country. However, there are very few studies that examine disaster literacy in schools, especially those contained in subjects. Therefore, the purpose of this research is to provide information and education to teachers and students regarding disaster literacy. Specifically, this paper will look at how disaster literacy is included in social studies learning.

2. Methods

This study uses a literature review method. This method can synthesize research work that has gone through the previous identification and evaluation stages [7]. The data sources are divided into two main and additional sources from scientific articles in national and international research journals, research reports, and reference books related to disaster literacy, especially in social studies learning in schools. Data collection techniques are documentation and technical data analysis in this study refers to qualitative analysis techniques. The purpose of using this library method is to foster new conceptions of studies in disaster education and social studies education and to become a consideration for future disaster research. Opportunities for developing disaster research in the future and opportunities for developing disaster research are very open considering the high potential for disasters in Indonesia.

3. Result and Discussion

3.1. Result

Disaster education has been widely recognized as having an essential role in reducing disaster risk and sustainable development [8]. Social studies education has a contribution in efforts to reduce disaster risk because social studies education as a subject and disciplinary education should have a foundation, one of which is a humanitarian, sociological, and anthropological basis which provides a system of fundamental ideas to determine the characteristics of problems that occur around humans [9]. In addition, according to [10], how important is disaster mitigation education considering that Indonesia is a disaster-prone area and to accommodate disaster education, one of which is in the junior secondary unit, namely IPS. The following are some studies and research results related to disasters as revealed by Amaral et al (2013), residents of Fukui City, Japan who have disaster education are better prepared than residents in San Francisco City who do not have disaster education. Another study conducted by Muttarak and Pothisiri (2013) stated that education can increase knowledge and skills in evaluating risk, while formal education can improve preparedness and reduce vulnerability to disasters. Another study is a Taiwanese government program that integrates educational resources to increase resilience in a disaster prevention education technology project for all students in Taiwan that include scientific and technological literacy. Education is essential to understand disasters as a whole so that they can make decisions and resolve them [8,11]. The Philippines, according to the research of Mark Anthony Catedral Mamon et al (2018) due to Typhoon Morakat, has had a disaster preparedness education strategy in project schools to promote a pilot program of natural disaster prevention education technology in 2003 to 2007-2011 for grades 1-12, students and the community. generally, in the form of e-learning, in the high school curriculum in the Philippines disaster risk reduction education is integrated in earth and life sciences to build a culture of safety and resilience to disaster risk [8]. In Pakistan, research by Shah (2018) reveals that there is an urgent need to address the lack of flood disaster preparedness in schools. In Malaysia, according to research by Akmar, Nifa, Abbas, & Lin, (2018) [12] revealed that the importance of flood disaster education programs. Related to the explanation of the research above that disaster literacy has an important role in measuring a person's ability to read and write, as well as gain further knowledge and abilities related to certain fields [13] according to Sung-Chin Chung and Cherng-Jyh Yen (2016) framework There are three conceptual dimensions of disaster prevention literacy and dimensions of disaster literacy, namely the knowledge dimension of disaster reduction, the dimension of disaster reduction attitude, and the dimension of disaster prevention skills [14].

The results of other studies reveal that it is important to raise awareness among the public and institutions about the importance of prevention before a disaster occurs through non-structural mitigation. The research results of Brown, Haun, and Peterson (2014)[15] recommend developing and evaluating disaster communication materials for vulnerable populations to improve disaster literacy and suggesting regular trials and evaluations that are used to inform the selection of media types. the use of technology in learning such as multimedia helps solve the problem of verbalism by providing videos, animated images etc. in the learning process [16]. Lee (2010) stated that disaster rescue information was not fully understood due to the lack of alignment of disaster literacy and preparedness regarding disaster recovery [17]. Another study revealed that if students are guided in the right way, they can transfer knowledge to parents through the classroom to the community [8].

3.2. Discussion

Disaster prevention literacy skills are intended to achieve optimal disaster prevention. There are three main points, namely knowledge-attitude-skills of disaster prevention. A person's knowledge to prepare for disaster prevention includes three indices: knowledge of disasters, knowledge of prevention, and knowledge of response. The attitude of disaster prevention is evaluated based on the sensitivity of disaster prevention items, values related to disaster management, and responsibility for disaster prevention. According to The National Literacy Act (Metiri Group, 2003) literacy is closely related to a person's ability to read, write, speak, and process information obtained to solve problems encountered in everyday life [16].

Another explanation was also given regarding the understanding of disaster literacy according to Brown (2014) which is the individual's capacity to read, understand, use the information to make decisions based on the context in disaster mitigation, preparedness, response, and recovery [17]. If Brown (2014)[18] says that disaster literacy is one's capacity to understand disasters, Sung-Chin Chung (2016)[14] explains that disaster mitigation literacy is a vital component of disaster prevention education to equip teachers, students, and school residents with an essential understanding of disaster prevention. And disaster literacy needs coaching first [12]. According to Sung-Chin Chung & Cherng-Jyh Yen (2016)[14], the conceptual framework of disaster prevention literacy and the dimensions of disaster literacy consist of 1. Disaster reduction knowledge dimension: disaster knowledge, preparedness knowledge, response knowledge 2. Disaster reduction attitude dimension: disaster prevention sensitivity, values related to disaster prevention, sense of responsibility for disaster prevention 3. Disaster prevention skills: preparedness actions and behavioural response. A description of disaster mitigation literacy can be seen in table 1 according to Sung-Chin Chung (2016)[14].

Table 1. Dimensions, categories, and descriptors of disaster prevention literacy

Dimension Category	Descriptor
Knowledge	
Disaster knowledge	
	Synthesize and analyze the definitions and causes of various disasters
	Explain the impacts and harms of disasters to humans and their environment
Preparedness knowledge	
	Develop the disaster mitigation procedures
	Make the action plan of disaster mitigation and preparedness
Response knowledge	
	Decide the response procedures in the event of a disaster
	Design the measures of post-disaster rescues and medical cares
Attitude	
Prevention awareness	
	Evaluate the environment comprehensively and recognize the potential hazards
	Synthesize and analyze the disaster-related information proactively
Prevention values	
	Promote the importance of disaster prevention, disaster relief, and evacuation plan
	Explain the relation between disaster prevention and social cost
Prevention sense of responsibility	
	Organize the promotion of campus and community disaster prevention
	Plan the execution of campus evacuation and shelter placement
Skills	
Preparedness action	
	Plan the escape routes and sites in the event of a disaster
	Plan and participate in the drills and training for disaster prevention relief
Response behaviors	
	Ensure self-safety and help others to escape in the event of a disaster
	Cooperate during the evacuation and shelter placement

Table 1 explains that the category of disaster mitigation literacy dimensions relates to knowledge consisting of knowledge about disasters in synthesizing and analyzing definitions and causes of various disasters; explain the impact and danger of disasters on humans and the environment. Preparedness knowledge, namely developing disaster mitigation procedures and making action plans for mitigation and disaster preparedness. Response knowledge is determining emergency response procedures when a disaster occurs and designing post-disaster rescue measures and medical care. The second category is the attitude which consists of awareness of prevention; awareness of prevention, namely evaluating the environment comprehensively and recognizing potential hazards; proactively synthesizes and analyze disaster-related information. The value of prevention is intended to emphasize the importance of disaster prevention, including strategic plans related to disaster relief, evacuation, and its relationship with social costs. Preventing a sense of responsibility is by organizing the promotion of disaster management schools and communities and planning the implementation of school evacuations and shelters. The last dimension is skilled with preparedness actions, namely planning routes and locations in the event of a disaster (evacuation) and planning and participating in exercises and training for disaster prevention and relief as well as cooperating during evacuation and shelter placement. This also agrees with the research conducted by According to Kimura, 2014 disaster literacy consists of three elements, namely: (1) Knowledge to learn about disaster management and mitigation; (2) Skills needed to learn about disaster

IOP Conf. Series: Earth and Environmental Science **986** (2022) 012015 doi:10.1088/1755-1315/986/1/012015

management and mitigation; (3) Basic competencies and attitudes for dealing with disasters [5,12]. While the disaster mitigation literacy model according to Brown (2014)[18] consists of the following images.

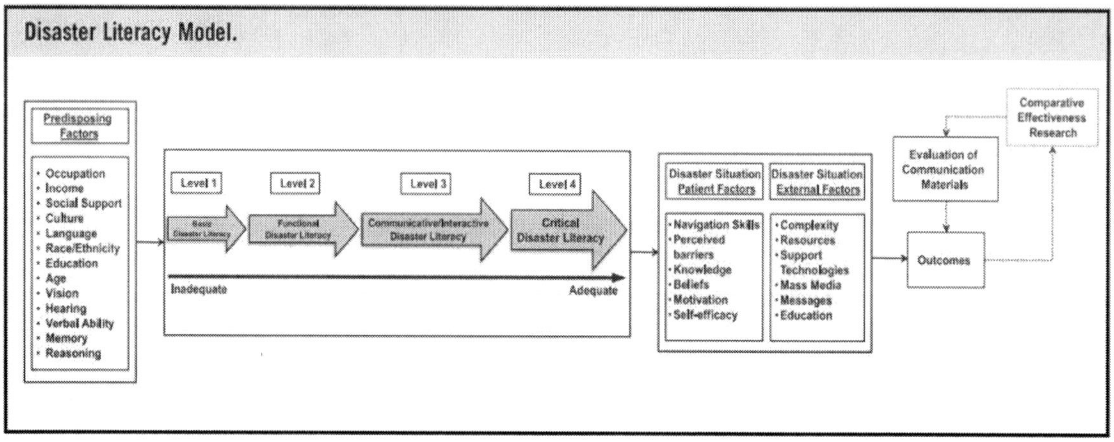

Figure 1. Disaster mitigation literacy model (Brown, 2014)[18].

The disaster mitigation literacy model is in Fig. 1 it can be explained that the stage starts from literacy so that awareness arises for read and understand the correct knowledge in the implementation of disaster mitigation and preparedness. At this stage, the tendency is that capacity is still low but is willing to follow instructions related to messages on disaster preparedness, disaster response, and recovery. Usually, messages that are simple to understand are also needed and respond to messages of information that are new in nature with messages that are quite familiar or known to the public. Phase 2 is a comprehensive capability related to disaster information, as evidenced by having followed the recommendations and instructions. This skill is very necessary for a person's provision in dealing with disasters, even though most individuals lack experience in processing information. The third stage is the motivation and self-confidence of individuals in being proactive. At this stage, the message has been able to be received well. Messages can then be modified or added according to the general terms. Messages can be different according to the needs of society and individuals. And this stage 4 is the involvement of the individual has gone further, the individual has understood broader information, understands the wider situation, especially related to environmental and social safety barriers [17]. Then how important is disaster mitigation literacy in social studies learning? Social studies education has a big challenge in its aim to keep up with current developments in this era. Increasingly advanced information technology is one of the phenomena that occur in the world today, and of course, it becomes a necessity in the 21st century.

Literacy is very much needed today, including in disaster prevention literacy. Social studies have a noble task and becomes an important foundation for the intellectual, emotional, cultural, and social development of students, namely being able to develop responsible ways of thinking, acting, and behaving as individuals and citizens of the world. In addition, social studies have the task of developing the potential of students to be sensitive to social problems that occur in society, have a positive mental attitude to improve all inequalities and be skilled in overcoming every problem that occurs daily, both those that befall themselves and those in society. This goal can be achieved when social studies education and learning in schools are well organized [6,18,19].

Social studies subjects are subjects that can integrate concepts related to disasters. Social Sciences is an integrated study material which is a simplification, adaptation, selection and modification of the concepts and skills of the disciplines of history, geography, sociology, anthropology, and economics organized scientifically and psychologically for learning purposes [9]. Described in the National Council for the Social Studies (1991) definition the social studies are concerned with human relationships. Their content is derived principally from the scholarly disciplines of economics, geography, history, political science, and sociology and includes elements from other social studies [20]. In the context of the reorganization of secondary education by a commission, namely the Commission on the Reorganization of Secondary Education, in 1916, defined social studies as the social studies are understood to be those whose subject matter related directly to the organization and development of human society, and to man. as a member of a social group [9]. Disasters have become a development issue, because the results of development that have been initiated for tens or even hundreds of years can be destroyed or damaged instantly by a disaster, the economy of the community and the state has experienced a lot of decline, many economic, social, and cultural infrastructure and facilities are damaged. Communities affected by disasters often must reorganize their lives from scratch, they must move to other places and start living in new places [4]. It is also explained in Maryani research (2016)[4] that in social studies learning about disasters can be implemented through an integrated learning model that is applied to the junior high school level, especially to teach the theme of disaster mitigation. The most appropriate learning method applied is cooperative learning and problem solving in addition to other methods [4]. In social studies learning, to improve disaster mitigation literacy in terms of knowledge, attitudes, and skills, it can be developed through a digital-based Geocaching Instructional Game (IGG) where the player's location hides a box and then shares its coordinates online [21,22]. The aim of this game is to reveal students' problem-solving skills regarding disasters and contribute to the development of students' environmental awareness and disaster awareness [5,23]. Disaster education in social studies subjects based on the results of the study above can essentially provide information about disasters embedded in social studies subjects. By equipping students with disaster literacy, students should have information and preparedness in anticipating disasters through knowledge, attitudes, and skills to be able to respond to a disaster situation effectively and efficiently.

4. Conclusion

Based on the results of the study, it can be concluded that disaster mitigation literacy is a vital component in seeing how the community's understanding and skills in disasters are. This can also be seen through understanding how critical the community is in receiving messages or information related to disasters. Disaster mitigation literacy is part of a non-structural approach that focuses on skills and understanding of disasters. The role of education can certainly affect a person's disaster mitigation literacy, especially school residents who are equipped with an understanding of disaster education in social studies subjects that can be implemented through models, methods, and learning media.

Acknowledgments

This research is one of the writings that inspired the dissertation. Of course, in the development of disaster literacy, there are still many things that need to be explored. Many thanks to the dissertation promoters who have provided many inputs related to the research being researched.

References

[1] Minanto A and Ningsih I N D K 2018 Literasi Bencana Di Sekolah Gunung Merapi Tentang Mitigasi Bencana Dan Kewarganegaraan Transformatif *WACANA, J. Ilm. Ilmu Komun.* **17** 207
[2] Desfandi M 2014 Urgensi Kurikulum Pendidikan Kebencanaan Berbasis Kearifan Lokal di Indonesia [The urgency of disaster education curriculum based on local wisdom in Indonesia} *Sosio Didakt. Sos. Sci. Educ. J.* **1**
[3] BNPB 2016 RBI Risiko Bencana Indonesia
[4] Maryani E 2016 Model Pembelajaran Mitigasi Bencana Dalam Ilmu Pengetahuan Sosial Di Sekolah Menengah Pertama *J. Geogr. Gea* **10**
[5] Kimura R, Hayashi H, Kobayashi K, Nishino T, Urabe K and Inoue S 2017 Development of a "disaster management literacy hub" for collecting, creating, and transmitting disaster management content to increase disaster management literacy *J. Disaster Res.* **12** 42–56

[6] Sapriya 2020 Implementasi-Nilai-Pancasila-PKn-REV
[7] Tomasic D 2011 Health Sciences Literature Review Made Easy: The Matrix Method, 3rd Edition *Am. J. Heal. Pharm.* **68** 2302–2302
[8] Shaw R, Shiwaku K and Takeuchi Y 2011 *Disaster Education* vol 7 (United Kingdom: Emerald Group Publishing Limited)
[9] Sapriya 2009 *Pendidikan IPS* (Bandung: Rosda Karya)
[10] Ruhimat M *Manusia, Tempat, dan Lingkungan* (Yogyakarta: Ombak)
[11] Pathirage C, Seneviratne K, Amaratunga D and Haigh R 2012 Managing disaster knowledge: Identification of knowledge factors and challenges *Int. J. Disaster Resil. Built Environ.* **3** 237–52
[12] Kagawa F and Selby D 2012 Ready for the Storm: Education for Disaster Risk Reduction and Climate Change Adaptation and Mitigation 1 *J. Educ. Sustain. Dev.* **6** 207–17
[13] Valle P V del 2016 The Structure of Health Factors among Community-dwelling Elderly People *Am. J. Orthod. Dentofac. Orthop* **2** 1–8
[14] Chung S-C and Yen C-J 2016 Disaster Prevention Literacy among School Administrators and Teachers: A Study on the Plan for Disaster Prevention and Campus Network Deployment and Experiment in Taiwan *J. Life Sci.* **10** 203–14
[15] Brown L M, Haun J N and Peterson L 2014 A Proposed Disaster Literacy Model *Disaster Med. Public Health Prep.* **8** 267–75
[16] Zavar E and Nelan M 2020 Disaster drills as experiential learning opportunities for geographic education *J. Geogr. High. Educ.* **44** 624–31
[17] Bhandari R K 2014 *Disaster Education and Management*
[18] Brown L M, Haun J N and Peterson L 2014 A proposed disaster literacy model *Disaster Med. Public Health Prep.* **8** 267–75
[19] Koike T, Hayashi H, Satake K, Tsukahara K, Kawasaki A, Amano Y, Takara K, Saya S, Nishiguchi N, Nishikawa S, Tamura K, Hiroki K, Shaw R and Ikeda T 2018 Role played by science and technology in disaster risk reduction: From framework planning to implementation *J. Disaster Res.* **13** 1222–32
[20] Maftuh B 2008 *Pendidikan Resolusi Konflik* (Bandung: UPI PRESS)
[21] Desfandi M 2014 Urgensi Kurikulum Pendidikan Kebencanaan Berbasis Kearifan Lokal di Indonesia [The urgency of disaster education curriculum based on local wisdom in Indonesia} *Sosio Didakt. Sos. Sci. Educ. J.* **1** 191–8
[22] Siriwardena M, Malalgoda C, Thayaparan M, Amaratunga D and Keraminiyage K 2013 Disaster resilient built environment: role of lifelong learning and the implications for higher education *Int. J. Strateg. Prop. Manag.* **17** 174–87
[23] Mustofa M 2020 Pendidikan Kebencanaan Berbasis Kearifan Lokal Dalam Penguatan Karakter Siapsiaga Bencana *Geodika J. Kaji. Ilmu dan Pendidik. Geogr.* **4** 200–9

IOP Conf. Series: Earth and Environmental Science **986** (2022) 012016 doi:10.1088/1755-1315/986/1/012016

Implementation of comprehensive school safety: the risk reduction and resilience education pillar in State Senior High School 1 of Pacitan and Islamic State Senior High School 1 of Pacitan, Indonesia

Deasyka Milna Syahputri[1], Yunus Aris Wibowo[1, 2, *], Siti Hadiyati Nur Hafida[1, 2], Wahyu Widiyatmoko[1, 2], Yaskinul Anwar[3], Ratih Puspita Dewi[1, 2]

[1]Geography Education Department, Universitas Muhammadiyah Surakarta, Indonesia
[2]Centre for Disaster Mitigation Study, Universitas Muhammadiyah Surakarta, Indonesia
[3]Geography Education Department, Universitas Mulawarman, Indonesia

[*]Email: yaw222@ums.ac.id

Abstract. Pacitan is vulnerable to earthquakes and tsunamis because it is located near the subduction zone between the Indo-Australian and Eurasian plates. So that it needs Disaster Risk Reduction efforts, one of the best measures is to implement the three pillars of Comprehensive School Safety. This study aims to analyze the implementation of the third pillar of Comprehensive School Safety, namely the Pillar of Risk Reduction and Resilience Education. The study was conducted at State Senior High School 1 of Pacitan and Islamic State Senior High School 1 of Pacitan because they directly face the Indian Ocean. This study used a quantitative approach with primary data obtained from filling out questionnaires by respondents. Respondents in this study were principals, teachers, and students involved in Scouting and Junior Red Cross in these two schools. The instrument was a questionnaire including teacher and education staff training, integration of disaster education into the curriculum, and extracurricular activities indicators. The results indicated that teacher and education staff and curriculum integration were not optimally carried out in these two schools. Furthermore, they have not yet established disaster mitigation extracurricular, but implementing disaster mitigation education was carried out through Scouts and Junior Red Cross.

1. Introduction

A tsunami is a natural disaster that is directly related to waves in the ocean. It is a natural disaster characterized by waves from the sea to the land with very high destructive power [1]. Rolling ocean waves occur because of a shift in the seabed, which causes seawater to move towards the mainland. The transition can be caused by earthquakes, volcanoes, volcanic eruptions, and underwater landslides [2].

The tsunami had an impact on the surrounding community [3]. The level of damage caused by a tsunami depends on the coastal area's height, period, and topography [4,5]. If the tsunami hits the coastal area, it will threaten buildings, housing, rice fields, and carry various materials [3,6,7]. The community certainly feels the impact, especially in the education sector and children [8–10]. The effect of the tsunami disaster in education, among others, was a large number of casualties from the vulnerable age group, damaged school buildings, washed away school infrastructure, and student teaching and learning activities were disrupted and delayed. Therefore, to reduce the impact of the tsunami disaster, various countermeasures were carried out, such as strengthening coastal areas with green belts, installing water breaks on the edge of coastal waters, and installing clear and easy-to-read evacuation routes. Disaster risk reduction (DRR) in the education sector is implemented in the Hyogo Framework for Action (HFA), Disaster Preparedness Schools, and 3 INEE Pillars of Disaster Safe Schools [11].

The INEE pillar is designed to interest various situations, whether crisis, disaster, or even conflict. Safe schools are learning communities that can commit to a safe culture, are aware of the risks of threats, and prepare and think about pre-and post-disaster responses to emergency response [12,13]. The Ministry of Education and Culture's Disaster Safe School Module, 2015 contains three Pillars of Disaster Safe Schools, Pillar 1 of Disaster Safe School Facilities, Pillar 2 of Disaster Risk Management in Schools, and Pillar of Disaster Risk Reduction and Prevention Education [13]. Implementing these pillars is a direction towards a Disaster Safe School, especially in the Education Pillar of Disaster Risk Reduction and Prevention [14]. By synchronizing the disaster curriculum into the formal curriculum of school education, it is easier to spread knowledge about disaster well [15,16]. There is enthusiasm in developing themselves through extracurricular activities, providing opportunities for students to get to know disaster mitigation. Schools began to introduce DRR learning to teachers/education staff and students more specifically by inserting themes that already existed in every school [17–21].

Pacitan Regency is between 110° 55" – 111° 25" east longitude and 7° 55" – 8° 17" south latitude. Located at the confluence of the Indo-Australian and Eurasian plates, Pacitan Regency is an area that is prone to disasters [22], one of which is a tsunami. Schools located in disaster-prone areas need to be equipped with knowledge and skills in responding to disasters [23,24]. State Senior High School 1 of Pacitan and Islamic State Senior High School 1 of Pacitan were the research targets. The location of the two schools is on the coast of Teleng Ria Pacitan. The schools have the same accreditation and good academic management. This type of research uses a quantitative research design which is expected to obtain accurate and natural data. The expected results in the study of schools located in disaster-prone areas have implemented the Education Pillars of Disaster Risk Prevention and Reduction even though the schools are not yet at the Disaster Preparedness School level.

2. Method

This study was conducted in State Senior High School 1 of Pacitan and Islamic State Senior High School 1 of Pacitan, Pacitan Regency, East Java Province, Indonesia. Pacitan Regency is a coastal city centered in Pacitan Bay (Pancer Door), directly adjacent to the Indian Ocean (Figure 1). Therefore, if the area is exposed to an earthquake or tsunami, the losses will be substantial.

This study consists of several stages, namely literature study, designing data collection instruments, testing the validity and reliability of the instrument, data collection, data processing, and data analysis. The literature study collected data on the tsunami hazard and the history of disaster education in Indonesia. The data collection instrument in this study was a questionnaire consisting of three leading indicators according to Module 3 of Disaster Safe School and interview guidelines (Table 1). The completed questionnaire was then tested for validity and reliability. The validity test was carried out using the Pearson product-moment test, while the questionnaire reliability test used a stability test type, namely Cronbach Alpha. The samples used to test the validity and reliability were 30 students and 30 teachers.

Furthermore, the validity of the interview guide was tested using the expert judgment technique. Interview guidelines were used to support the data from the questionnaire results. Interviews were conducted with school principals because they have the authority to determine policies and decisions in schools. Respondents in this study were teachers/educators and students involved in extracurricular Scouts and Junior Red Cross at State Senior High School 1 of Pacitan and Islamic State Senior High School 1 of Pacitan. The selection of the two extracurriculars follows the Disaster Safe Education Unit

(SPAB) module. Although these schools are not yet based on disaster preparedness schools, they have built capacity through two extracurricular activities, namely Scouts and Junior Red Cross.

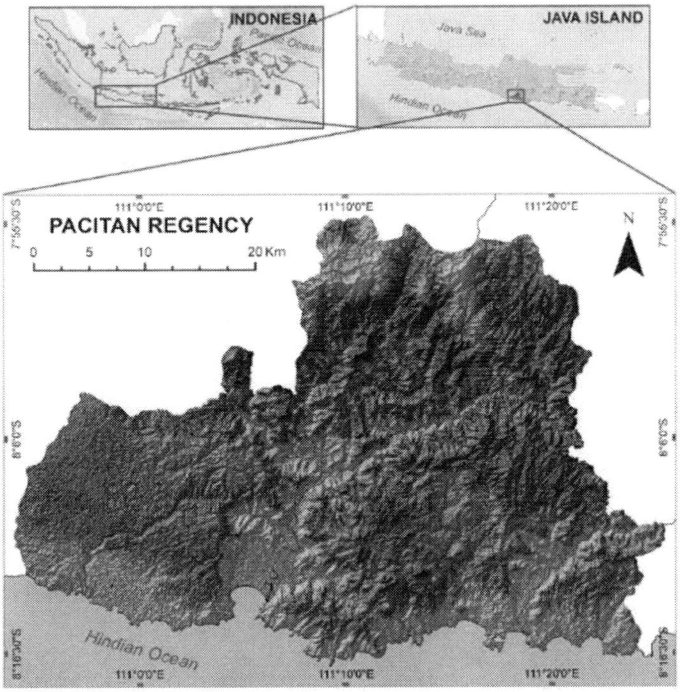

Figure 1. Research area (Source: [25,26])

Tabel 1. Research Questionnaire Indicator

Indicators	Number of question/ statement
Training for teachers and education personnel	
Attitude and Action	2
School Policy	2
Preparedness Planning	3
Resources Mobilization	3
Integrated Curriculum for disaster materials	
Prevention Education	2
Emergency Response Education	2
Rehabilitation Education	3
Reconstruction Education	3
Extracurricular	
Extracurricular activities	4
Workprogram Evaluation	4
Capacity Building Through Work Programs	2

Source: Module 3 of Disaster Safe School

Validity and Reliability tests were carried out to avoid bias. Decision-making to test the validity of the indicators is 1) If r count > r table and positive or significant value < 0.05, then the indicator items are declared valid. 2) If r count < r table and significant > 0.05, the indicator item is declared invalid. T table for N_{20} is 0.444 with the results of the validity test presented in Table 2.

Tabel 2. Questionnaire Validity Test

Variable	$r_{calculate}$	Result
Integrated Curriculum for disaster materials	0.850	Valid
	0.816	Valid
	0.816	Valid
	0.850	Valid
	0.550	Valid
	0.850	Valid
	0.923	Valid
	0.923	Valid
	0.816	Valid
	0.816	Valid
Training for teachers and education personnel	0.835	Valid
	0.841	Valid
	0.841	Valid
	0.732	Valid
	0.566	Valid
	0.835	Valid
	0.786	Valid
	0.905	Valid
	0.841	Valid
	0.827	Valid
Extracurricular	0.874	Valid
	0.773	Valid
	0.747	Valid
	0.874	Valid
	0.526	Valid
	0.874	Valid
	0.853	Valid
	0.907	Valid
	0.773	Valid
	0.773	Valid

Source: analysis result, 2021

The indicator of decision making in the reliability test, if alpha > 0.90, then the reliability is perfect. If the alpha is between 0.70 – 0.90, then the reliability is high. If the alpha is 0.50 – 0.70, then the reliability is moderate. If alpha < 0.50, then the reliability is low. The results of the reliability test are presented in Table 3.

Table 3. Questionnaire Reliability Test

Variable	Alpha	Result
Integrated Curriculum for disaster materials	0.958	Reliable
Training for teachers and education personnel	0,951	Reliabel
Extracurricular	0,950	Reliabel

Source: analysis result, 2021

The analysis technique used is descriptive statistics. Data analysis through student and teacher assessment of the implementation of the education pillars of disaster risk prevention and reduction with the type of data in the form of primary data. The collecting data through questionnaires, respondents came from students and teachers of State Senior High School 1 of Pacitan and Islamic

State Senior High School 1 of Pacitan. The data were obtained and processed using excel software to determine the amount, look for the Max and Min values to determine the class interval, and consider the category using the IF formula. The category results are converted into low (1), medium (2), and high (3). The reclaimed data were analyzed using SPSS software using Descriptive Statistics to find each school's low, medium, and high categories.

The teacher training data, integrated curriculum, and extracurricular at State Senior High School 1 of Pacitan and Islamic State Senior High School 1 of Pacitan were processed using Excel and SPSS software. The two schools are categorized into low, medium, and high grades, respectively, with the category measurement formula presented in Table 4. Each school has different category results in its application with the following calculations.

Table 4. Category Measurement Formulas

Keterangan	Rumus
Rendah	$X < M - 1SD$
Sedang	$M - 1SD \leq X < M + 1SD$
Tinggi	$M + 1SD \leq X$

M = Mean Deviation
SD = Standard Deviation

3. Results dan Discussions

3.1. Implementation of the Education Pillars of Disaster Risk Reduction and Prevention

3.1.1. Training of Teachers and Educators

Based on the data on teacher training and teaching staff indicators of attitude to action, school policies, preparedness planning, and resource mobilization, the results are presented in Figure 1. Based on Figure 3, research results from State Senior High School 1 of Pacitan and Islamic State Senior High School 1 of Pacitan show a percentage of 100% in the implementation of activities, emergency response equipment, socialization/simulation, internal school training, and training facilities. The same percentage results were also obtained by both schools, as much as 72% and 81% in disaster risk reduction through teacher training and training on disaster topics. State Senior High School 1 of Pacitan obtained higher results than Islamic State Senior High School 1 of Pacitan at the level of sensitivity and discussion with disaster insight with a percentage of 81%. In comparison, in financial training activities, the research results in Islamic State Senior High School 1 of Pacitan were higher than State Senior High School 1 of Pacitan as much as 81%.

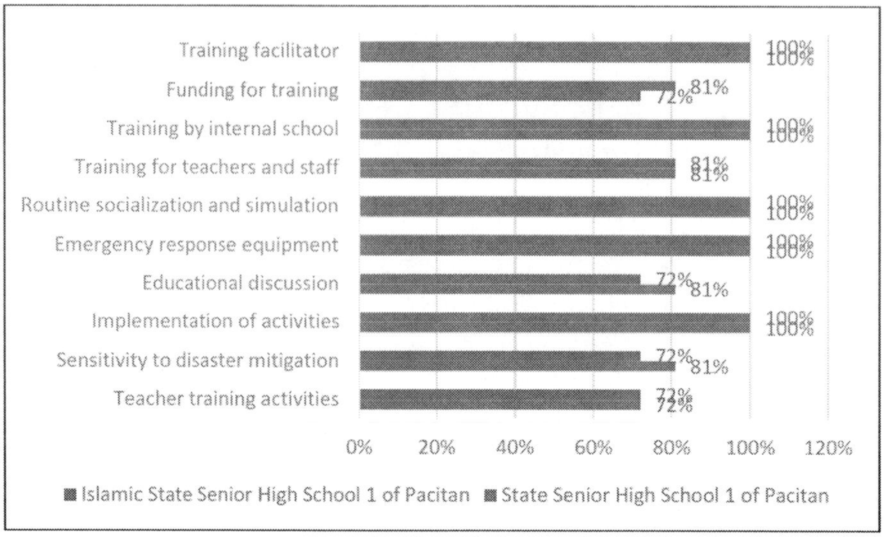

Figure 2. Training for Teachers and Education Staff

Teacher training activities related to disasters in schools can increase their insight into disaster mitigation and the sensitivity in responding to the threat of disaster. Therefore, every teacher must have a high level of sensitivity to disasters in school. Considering the pre- and post-disaster school environment's condition can also be a form of teacher sensitivity, especially in disaster-prone areas.

Table 5. Categorization of Training for Teachers and Education Personnel in State Senior High School 1 of Pacitan

		Frequency	Percent	Valid Percent	Cumulative Percent
Valid	Low	6	43.9	43.9	43.9
	Medium	3	21.4	21.4	64.3
	High	5	35.7	35.7	100.0
	Total	14	100.0	100.0	

Source: analysis result, 2021

Table 6. Categorization of Training for Teachers and Education Personnel in Islamic State Senior High School 1 of Pacitan

		Frequency	Percent	Valid Percent	Cumulative Percent
Valid	Low	2	18.2	18.2	18.2
	Medium	7	63.6	63.6	81.8
	High	2	18.2	18.2	100.0
	Total	11	100.0	100.0	

Source: analysis result, 2021

Based on the study results, the application of teacher training and teaching staff at State Senior High School 1 of Pacitan was in the low category of 42.9%. At Islamic State Senior High School 1 of Pacitan was in the medium class 63.6%. The training is mainly implemented, such as making biophore/infiltration holes in the school environment. [27] supports these results that biopore infiltration holes are critical, especially in densely populated areas, where soil absorption is minimal, and locations often have problems with rainwater. Making biopore infiltration holes aims to maximize water infiltration into the soil to increase the volume of groundwater.

Making a waste bank and processing compost properly as a form of recycling reduces the volume of waste in the school area. It can also shape the character of students in being aware of and caring for the environment. Teacher training activities also provide opportunities for discussion and dialogue as a forum to convey insights. Discussion activities were carried out between teachers by taking the theme of the environment, paying attention to post-disaster aspects starting from the impact of losses, damage to buildings and facilities. This could introduce the importance of disaster mitigation.

Judging from the supporting equipment in schools, several facilities such as telephones, buoys/inflatable boats, first aid kits are already available in both schools, State Senior High School 1 of Pacitan and Islamic State Senior High School 1 of Pacitan. Emergency response equipment is needed when a disaster occurs, so schools must always prepare, provide, and carry out proper maintenance. Moreover, logistical and medical equipment play a crucial role in disaster emergency response conditions.

3.1.2. Integrated Curriculum for disaster materials

Based on the integrated curriculum data that contains indicators of prevention education, emergency response education, rehabilitation education, and reconstruction education, the results are presented in Figure 2. Reviewing the research results in Figure 2, both schools have the same percentage of 100% on the curriculum as DRR and mitigation-based extracurricular. Integration in the subjects of the two schools yielded 90% results and risk analysis of 63%. Based on Figure 2, State Senior High School 1 of Pacitan obtained higher results than Islamic State Senior High School 1 of Pacitan in the level of school services as much as 81%, field activities in schools as much as 63%, and trauma healing and simple reconstruction as much as 100%. There are also results obtained by Islamic State Senior High School 1 of Pacitan, which are higher than State Senior High School 1 of Pacitan, namely in the learning system as much as 81%, and learning media which get results as much as 63%.

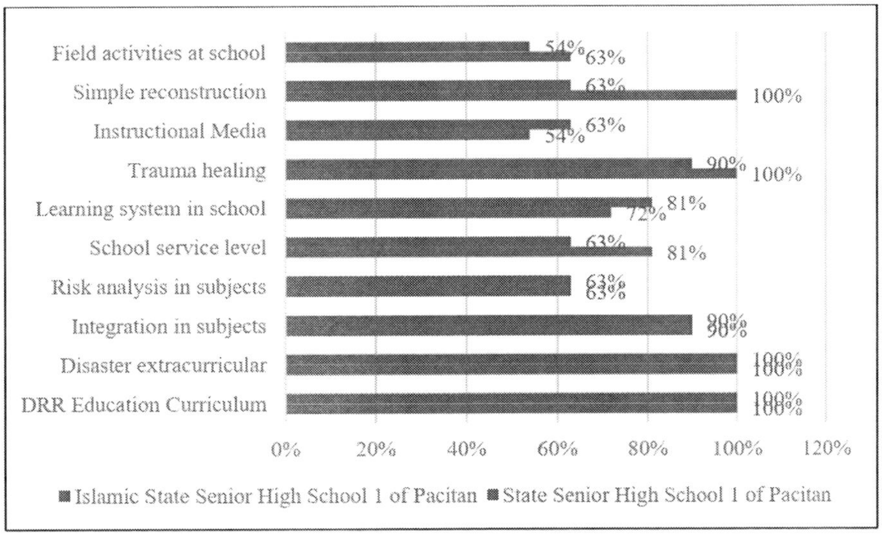

Figure 3. Integrated Curriculum for disaster materials

The integration of disaster mitigation education in schools through the curriculum is one form and model of disaster learning without adding new subjects. Supported in [28] research, an integrated curriculum in disaster education must be immediately made and implemented in schools in disaster-prone areas. These efforts provide insight to students in responding to disasters such as earthquakes, tsunamis, landslides, etc. Disaster mitigation learning is one of the materials in geography subjects.

Table 7. Categorization of Integrated Curriculum for disaster materials at State Senior High School 1 of Pacitan

		Frequency	Percent	Valid Percent	Cumulative Percent
Valid	Low	7	33.3	33.3	33.3
	Medium	9	42.9	42.9	76.2
	High	5	23.8	23.8	100.0
	Total	21	100.0	100.0	

Source: analysis result, 2021

Table 8. Categorization of Integrated Curriculum for disaster materials at Islamic State Senior High School 1 of Pacitan

		Frequency	Percent	Valid Percent	Cumulative Percent
Valid	Low	10	50.0	50.0	50.0
	Medium	1	5.0	5.0	55.0
	High	9	45.0	45.0	100.0
	Total	20	100.0	100.0	

Source: analysis result, 2021

The results of the integrated curriculum research at State Senior High School 1 of Pacitan are in the medium category of 43.9%, while at Islamic State Senior High School 1 of Pacitan, it is in the low category of 50%. The main subject is geography, and the teacher conducts lessons such as running a risk analysis by identifying disaster-prone areas in their respective regions. Disaster integration learning provides teaching and learning activities material, but teachers also try to provide the best service to students. The condition of students after the disaster made the psychological state of students vulnerable so that the teacher tried to make a recovery for the affected students. The recovery of victims is essential by instilling several principles, including (1) a sense of security, (2) providing calm, (3) instilling hope, and (3) establishing good communication.

3.1.3. Extracurricular

Scout and Junior Red Cross extracurricular data contain evaluation activities, evaluation of work programs, and capacity building through work programs (Figure 3). Scout extracurriculars at State Senior High School 1 of Pacitan and Islamic State Senior High School 1 of Pacitan have the same results in 92% of socialization, 96% of mitigation materials, and 100% of agency cooperation. Based on Figure 3, several results show State Senior High School 1 of Pacitan obtained higher results, namely in simulations as much as 84%, preparedness procedures and relations between extracurriculars as much as 96%, paying attention to disaster aspects as much as 100%, and disaster mitigation competitions as much as 92%, while for Islamic State Senior High School 1 of Pacitan get higher results in social services as much as 96%.

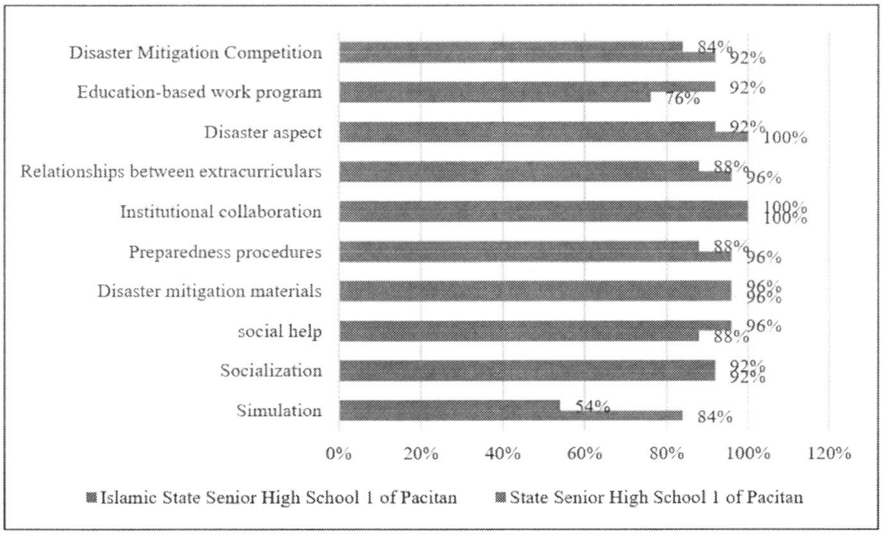

Figure 4. Scout Extracurricular

The Junior Red Cross (PMR) extracurriculars at State Senior High School 1 of Pacitan and Islamic State Senior High School 1 of Pacitan have the same results in social service activities as 100%. Based on Figure 4, several results show State Senior High School 1 of Pacitan obtained higher results, namely in the simulation as much as 80%, socialization, relations between extracurricular and preparedness procedures as much as 96%, and paying attention to disaster aspects as much as 100%. There are several results that Islamic State Senior High School 1 of Pacitan obtained higher results in mitigation material as much as 92%, cooperation with relevant agencies and mitigation-based work programs as much as 100%, and mitigation competitions as much as 88%.

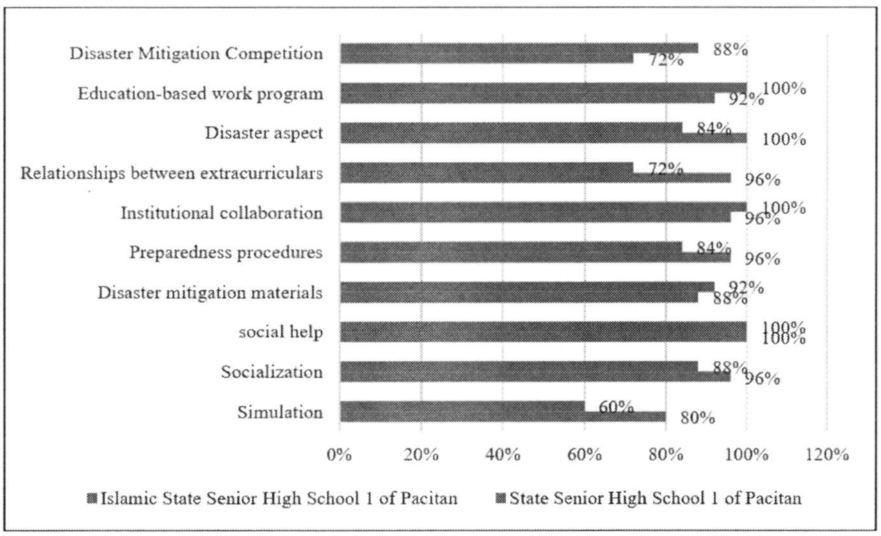

Figure 5. Junior Red Cross Extracurricular

Extracurricular activities that play an active role in disaster risk reduction are important notes in school development. Developing interactions from disaster mitigation systems into simulations aims to increase awareness. Simulations in schools held by extracurricular members in earthquake simulation scenario followed by a tsunami disaster scenario. They focused on the logistic simulation model in the distribution of aid due to the earthquake and tsunami disaster; the analysis results focus on the assistance distribution to affected victims.

Table 9. Categorization of disaster materials into extracurricular activities at State Senior High School 1 of Pacitan

		Frequency	Percent	Valid Percent	Cumulative Percent
Valid	Low	2	4.0	4.0	4.0
	Medium	7	14.0	14.0	18.0
	High	41	82.0	82.0	100.0
	Total	50	100.0	100.0	

Source: Analysis result, 2021

Table 10. Categorization of disaster materials into extracurricular activities at Islamic State Senior High School 1 of Pacitan

		Frequency	Percent	Valid Percent	Cumulative Percent
Valid	Low	2	4.0	4.0	4.0
	Medium	5	10.0	10.0	14.0
	High	43	86.0	86.0	100.0
	Total	50	100.0	100.0	

Source: Analysis result, 2021

The results of extracurricular research at State Senior High School 1 of Pacitan and Islamic State Senior High School 1 of Pacitan are in the high category, namely 82% and 86% so that the implementation of the pillars of education on prevention and disaster risk reduction in extracurricular activities is highly prioritized. Other extracurricular activities were also organized at State Senior High School 1 of Pacitan and Islamic State Senior High School 1 of Pacitan, such as social services, simulations, and competitions. Social service activities at school teach students to give each other help and care about people who certainly need help.

In extracurricular both scouts and Junior Red Cross (PMR) collaborate with agencies related to disasters such as Search And Rescue (SAR), Fire Fighter, the Indonesian National Disaster Management Agency (BNPB), and the Indonesian Red Cross (PMI). The institution has a vital role in the task force and always emphasizes students to participate in volunteer activities. As in [29] research, collaboration with specific institutions, universities, and NGOs in disaster mitigation is critical in developing and training programs.

Based on the high category research results at State Senior High School 1 of Pacitan on the application of extracurricular in terms of Table 9 valid percent as much as 82%. The application was in the medium category, namely implementing the integrated curriculum with an accurate percent of 43.9%. In contrast, teacher training and teaching staff implementation with a valid percent was 42.9% for the low category, as seen from Table 7. The study results show that extracurricular application at Islamic State Senior High School 1 of Pacitan belongs to the high category with a valid percentage of 86% (Table 10). The training of teachers and educators is classified as moderate with 63.6% (Table 9), and the implementation of the integrated curriculum is classified as low with 50% (Table 10).

4. Conclusion

Teacher training activities are carried out in the context of strengthening attitudes and disaster response actions. However, implementation policies related to teacher training in Pacitan Regency are still relatively minimal. Disaster mitigation learning in schools is contained in the main geography subject. The integration can be in student observations in demonstrations/teacher stories related to disasters (earthquakes, landslides, floods) with media used in comic media. Extracurricular work programs that are engaged in disaster mitigation can train students' skills. Some of these programs include first aid material activities, disaster rescue methods, competitions with disaster education.

Acknowledgment

This study is part of the Tri Dharma Integrated-Grant (Hibah Integrasi Tri Dharma), LPPM, Universitas Muhammadiyah Surakarta, Indonesia

References

[1] Honarmand M, Shanehsazzadeh A and Zandi S M 2020 3D numerical simulation of tsunami generation and propagation, case study: Makran tsunami generation and penetrating in Chabahar Bay *Ocean Eng.* **218** 108109
[2] Marfai M A, Cahyadi A, Fatchurohman H, Rosaji F S C and Wibowo Y A 2019 Tsunami preparedness and environmental vulnerability analysis in Kukup Beach, Gunungkidul, Indonesia *IOP Conf. Ser. Earth Environ. Sci.* **256**
[3] Martínez C, Cienfuegos R, Inzunza S, Urrutia A and Guerrero N 2020 Worst-case tsunami scenario in Cartagena Bay, central Chile: Challenges for coastal risk management *Ocean Coast. Manag.* **185**
[4] Macías J, Castro M J, Ortega S and González-Vida J M 2020 Performance assessment of Tsunami-HySEA model for NTHMP tsunami currents benchmarking. Field cases *Ocean Model.* **152** 101645
[5] Marfai M A, Sunarto, Khakim N, Cahyadi A, Rosaji F S C, Fatchurohman H and Wibowo Y A 2018 Topographic data acquisition in tsunami-prone coastal area using Unmanned Aerial Vehicle (UAV) *IOP Conference Series: Earth and Environmental Science* vol 148
[6] Ismail N, Okazaki K, Ochiai C and Fernandez G 2018 Livelihood Strategies after the 2004 Indian Ocean Tsunami in Banda Aceh, Indonesia *Procedia Eng.* **212** 551–8
[7] Marfai M A, Sunarto, Khakim N, Fatchurohman H, Cahyadi A, Wibowo Y A and Rosaji F S C 2019 Tsunami hazard mapping and loss estimation using geographic information system in Drini Beach, Gunungkidul Coastal Area, Yogyakarta, Indonesia *E3S Web of Conferences* vol 76
[8] Rush S C, Houser R and Partridge A 2015 Rebuilding Sustainable Communities for Children and Families After Disaster: Recommendations from Symposium Participants in Response to the April 27th, 2011 Tornadoes *Community Ment. Health J.* **51** 132–8
[9] Proulx K and Aboud F 2019 Disaster Risk Reduction in Early Childhood Education: Effects on Preschool Quality and Child Outcomes *Int. J. Educ. Dev.* **66** 1–7
[10] Tatebe J and Mutch C 2015 Perspectives on education, children and young people in disaster risk reduction *Int. J. Disaster Risk Reduct.* **14** 108–14
[11] Sakurai A, Sato T and Murayama Y 2020 Impact evaluation of a school-based disaster education program in a city affected by the 2011 great East Japan earthquake and tsunami disaster *Int. J. Disaster Risk Reduct.* **47** 101632
[12] Petal M, Ronan K, Ovington G and Tofa M 2020 Child-centred risk reduction and school safety: An evidence-based practice framework and roadmap *Int. J. Disaster Risk Reduct.* **49** 101633
[13] Paci-Green R, Varchetta A, McFarlane K, Iyer P and Goyeneche M 2020 Comprehensive school safety policy: A global baseline survey *Int. J. Disaster Risk Reduct.* **44** 101399

[14] Shah A A, Gong Z, Pal I, Sun R, Ullah W and Wani G F 2020 Disaster risk management insight on school emergency preparedness – A case study of Khyber Pakhtunkhwa, Pakistan *Int. J. Disaster Risk Reduct.* **51** 101805
[15] Mutasa S and Munsaka E 2019 Botswana and international policies on the inclusion of disaster risk reduction in the school curriculum: Exploring the missing link *Int. J. Disaster Risk Reduct.* **40** 101271
[16] Li T, Wang Q and Xie Z 2019 Disaster response knowledge and its social determinants: A cross-sectional study in Beijing, China *PLoS One* **14** 1–11
[17] Shah R and Lopes Cardozo M 2014 Education and social change in post-conflict and post-disaster Aceh, Indonesia *Int. J. Educ. Dev.* **38** 2–12
[18] Nakano G, Yamori K, Miyashita T, Urra L, Mas E and Koshimura S 2020 Combination of school evacuation drill with tsunami inundation simulation: Consensus-making between disaster experts and citizens on an evacuation strategy *Int. J. Disaster Risk Reduct.* **51** 101803
[19] GFDRR 2015 Towards Safer School Construction
[20] Chou J S, Yang K H and Ren T C 2015 Ex-post evaluation of preparedness education in disaster prevention, mitigation and response *Int. J. Disaster Risk Reduct.* **12** 188–201
[21] Nahayo L, Li L, Habiyaremye G, Richard M, Mukanyandwi V, Hakorimana E and Mupenzi C 2018 Extent of disaster courses delivery for the risk reduction in Rwanda *Int. J. Disaster Risk Reduct.* **27** 127–32
[22] Musiyam M 2020 Analysis of Flood-Affected Areas Due to Extreme Weather In Pacitan, Indonesia *Int. J. GEOMATE* **19**
[23] Muñoz V A, Carby B, Abella E C, Cardona O D, López-Marrero T, Marchezini V, Meyreles L, Olivato D, Trajber R and Wisner B 2019 Success, innovation and challenge: School safety and disaster education in South America and the Caribbean *Int. J. Disaster Risk Reduct.* 101395
[24] Kastolani W and Mainaki R 2018 Does Educational Disaster Mitigation Need To Be Introduced In School? *SHS Web Conf.* **42** 00063
[25] BIG 2018 Using DEMNAS Data
[26] Geospasial B I 2017 Tanahair Indonesia
[27] Yohana C 2016 Penerapan Pembuatan Teknik Lubang Biopori Resapan Sebagai Upaya Pengendali Banjir **1** 1–23
[28] Septikasari Z and Ayriza Y 2018 Strategi Integrasi Pendidikan Kebencanaan Dalam Optimalisasi Ketahanan Masyarakat Menghadapi Bencana Erupsi Gunung Merapi *J. Ketahanan Nas.* **24** 47
[29] Shiwaku K and Fernandez G 2011 Roles of School in Disaster Education *Disaster Education* vol 7 (Emerald Group Publishing Limited) pp 45–75

Disaster literacy level of teachers and students in secondary school case study in Surakarta City, Indonesia

Seno Budhi Ajar [1, 2, *] and Lintang Ronggowulan [1, 2]
[1]Geography Education, Faculty of Teacher Training and Education, Sebelas Maret University
[2]Disaster Research Center, Institute for Research and Community Service, Sebelas Maret University

[*]Email: senobudhiajar@staff.uns.ac.id

Abstract. This article contains the level of information literacy and disaster literacy skills for teachers and secondary students in Surakarta in reviewing and responding to disaster information and symptoms of disasters. This article is based on Doyle's theory that the most basic ability in information literacy is the ability to know the accuracy and completeness of information which is the basis for making sound decisions. Quantitative methods were used to test the ability of 20 teachers and 82 secondary students to examine the accuracy and completeness of disaster information and symptoms of disasters and the ability of teachers and secondary students to respond to disaster information and symptoms of disasters. the results presented in this article. First, the ability of information literacy in knowing the accuracy and completeness of information on teachers and secondary students is low because only 5.5% of teachers and 7.8% of students confirm the accuracy of disaster information, and only 5.2% of students who want to find the completeness of the information. Second, most teachers and secondary students generalize their responses to both disaster symptoms and disaster information, they will immediately evacuate if they receive disaster symptoms or information.

1. Introduction

Indonesia experiences thousands of disasters every year. In 2019, BNPB - the National Disaster Management Agency - recorded 3814 disaster events that occurred in Indonesia, both geological and hydrometrological disasters [1]. The catastrophic events include:

Table 1. Number of Disaster in Indonesia [1]

Kinds of Disasters	Number of Disasters
Tornado	1387
Flood	784
Forest and Land Fires	756
Landslide	719
Drought	123
Earthquake	30
Tidal Wave and Abrasion	18
Volcanic Eruption	7

Disaster Literacy must be applied in curriculum because Indonesia had many numbers of natural Disaster. If Knowledge and Information management of disaster literacy can be learned by students, they will more awareness of the disaster mitigation of the population [2]. In disaster management there are 3 cycles [3] consisting of:

- Pre-disaster is preparedness activities.
- Disaster occurrence is response activities.
- Post-disaster is recovery activities.

Providing knowledge about disaster mitigation is part of the pre-disaster stage in disaster management. Knowledge of disaster mitigation is instructions, directions and instructions that must be carried out not only during a disaster but also pre-disaster and post-disaster. Knowledge of disaster mitigation is part of disaster literacy. Brown et al built model of disaster literacy (figure 1) [4].

- In this model knowing and understanding knowledge of disaster mitigation is basic of disaster literacy or first level or inadequate position.
- The second level is functional disaster literacy. In this level not only know and understand but also can remember and follow the rules of instruction in knowledge of disaster mitigation.
- Communicative/interactive disaster literacy is the third level. In this level Knowledge of disaster mitigation has been represented and adapted to the needs and situation of the location and place.
- In the adequate level is critical disaster literacy. In this level is information literacy skills where the ability to understand information and process it with knowledge of disaster mitigation.

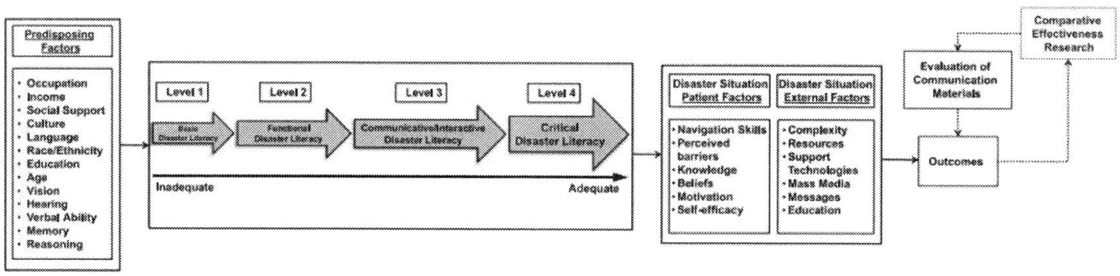

Figure 1. Disaster literacy Model [4]

The Information literacy refer to ability of manage information. ability to access Information, ability to evaluate information, and ability to use information [5]. Doyle said the measurement of information literacy are:

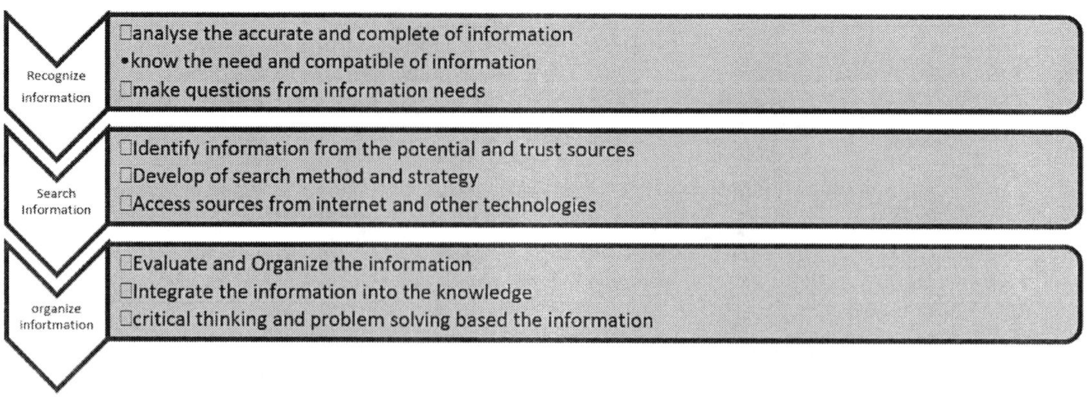

Figure 2. Measurement information literacy [5].

Knowledges are interpreted as actionable information or it may be used as to make decision and to take certain strategy but experience that we have or hear from other individual will become very valued knowledge resource [6]. In disaster mitigation knowledge, one must be able to distinguish between natural phenomena before a disaster and disaster information. natural symptoms/signs before a disaster can be used as a warning for the community to be aware of disasters such as the tsunami where natural symptoms before the tsunami are earthquakes felt in coastal areas, sudden receding water and visual observations and thunderous sounds [7]. Disaster information informs the potential, occurrence and risk of disasters and in the management system, that information still has to be collected, analyzed, managed, and disseminated. [8]. In 2019, Surakarta city experienced natural disasters as many as 243 times and only experienced 3 types of disasters, namely 3 landslides, 130 droughts and 110 hurricanes [9]. The purpose of this paper to describe the level of knowledge and information management of disaster literacy in teacher and secondary student and discuss the factors that affect teacher and secondary students' s disaster literacy level in Surakarta City.

2. Research Method

This research used mix of Doyle's theory of information literacy and Brown's disaster literacy model. The most basic ability in information literacy according to Doyle's theory is the ability to know the accuracy and completeness of information. Brown's disaster literacy model divided two level, (1) The inadequate level is knowledge of disaster literacy, (2) the adequate level is Information management in disaster literacy. The categories will be used as a grid in making a list of questions in the questionnaire, which becomes a survey tool to find out teacher and student's disaster literacy. Analysis of the survey results using the percentage technique then data would analyse with descriptive method

3. Result and Discussion

3.1. Knowledge of Disaster Literacy

Students dan teachers has pass the basic of disaster literacy or level one. 100% of teachers know and understand evacuation steps when natural disasters occur, ranging from geological disasters to meteorological disasters. On students, there are 20% of students who do not know or do not understand about evacuation during a disaster. The misunderstanding is in the salvage step and panic. Especially in the tsunami and earthquake natural disasters, there are still students who try to save property and are unable to control their fears.

The second level is ability of remember and follow the rules of instruction in knowledge of disaster mitigation. Basic evacuation knowledge instruction in accordance with BNPB pocket book (Indonesian National Disaster Management Agency) [10].

- Earthquake disaster evacuation
  The shock will be felt for a while. During this time, try to protect yourself by taking cover under a table to avoid falling objects and glass windows. Cover your head with a pillow or helmet, or stand under a door. When it feels safe, immediately run out of the house.
- Tsunami disaster evacuation
  You need to immediately get out of the beach area and find a high location. If you have reached a high area, stay there because the second and third tsunami waves are usually larger than the first wave and listen to information from the authorities via radio or other means of communication.
- Volcanic Eruption Disaster Evacuation
  Not in the recommended location to vacate. Not in a valley or watershed. Avoid open places. Protect yourself from volcanic ash
- Flood evacuation
  Unplug the tools that are still connected to electricity. Do not touch electrically charged equipment when you are standing on/in water. If you must prepare for evacuation: secure your home. If there is still time, place the furniture outside the house or in a place that is safe from flooding. The more valuable items are placed higher in the house.

- Landslide evacuation
  Evacuate immediately to stay away from the roar or the direction of the avalanche.
- Tornado Evacuation
  Avoid tall buildings, power poles, billboards, and so on. Immediately enter into a house or sturdy building.
- Forest Fire Evacuation
  Stay indoors. Seal off any smoky air that can enter the house and keep the indoor air as clean as possible. If it's too stuffy to stay indoors, seek shelter at the center Protect the respiratory tract with a mask/cloth every time you go outside

At this level, both teachers and students have the same characteristics. For the evacuation of landslides, tornadoes, forest fires, tsunamis, and volcanic eruptions, students and teachers have been able to remember and follow the appropriate evacuation procedures on the contrary for earthquake evacuation and flood evacuation, both students and teachers experienced inappropriate procedures. For earthquake disaster evacuation, as many as 50% of teachers and 51.25% of students immediately evacuated themselves out of the room when the earthquake occurred. Though this step can be done if the conditions are felt safe to go out. The same thing happened to flood evacuation, only 20% of teachers and 17.5% of students were able to remember and follow evacuation procedures correctly. The students and teachers ensure that the electricity is off as well as the electronic equipment and batteries are safe. The discrepancy that occurs in most students and teachers is that they try to immediately save property by placing it in a higher place and going to a place that is not affected by flooding without checking the condition of the electricity again.

3.2. Information managemengt of Disater literacy.

Disaster literacy is one of branch in communication science [11]. It because disaster literacy and information literacy have a same goal. Disaster literacy is an individual's ability to make decisions based on information obtained, read, understood and adapted to the context of mitigation, preparation, response, and recovery from disasters [4], especially in level 3 and level 4 model of disaster literacy explain the ability of communicative and critical disaster literacy. In this part according to the purpose of information literacy. Information Literacy is designed to develop of critical questions and approaches related to the design, implementation and evaluation of Information [12].

Level 3 and 4 at model of disaster literacy are the communicate and critical in disaster literacy. in particular, it can distinguish between natural symptoms/signs and information. First basic of information literacy, that is the ability to know the accuracy and completeness of information which is the basis for making sound decisions, it is the most important in adequate level disaster literacy because this ability can distinguish between the true news and fake news. the way to find out the authenticity of the news is the way we are able to find the source of the information, find the source be verified, find reliable news and find the trusted media platform or information source [13]. but public must Increasing awareness recognizes the signs and symptoms of the occurrence of disasters caused by the dynamics of the lithosphere because it is one of major disaster mitigation measures [2].

The adequate level of disaster literacy of teachers and secondary students in Surakarta, Indonesia is low, because the ability of information literacy in knowing the accuracy and completeness of information on teachers and secondary students is low. Just only 5.5% of teachers and 7.8% of students confirm the accuracy of disaster information. The majority of other teachers and others immediately believed the news which they got, so they immediately evacuated themselves. This is also exacerbated by the fact that most teachers and secondary students are reluctant to seek complete information. they immediately evacuate if they receive disaster news even if it is a piece of information. only 5.2% of students who want to find the completeness of the information.

In previous research, fadilah [14] said that the college students perform best in disaster effect. Suffering from the disaster personally in their past time has left awareness and preparedness to cope with disaster hazard however, the students do not really understand about knowledge of potential factors to predict disaster events. In this result show most teachers and secondary students know the natural symptom before the disaster and the know how to evacuate themselves, but most of teachers and students too trusting of disaster information. This makes most teachers and secondary students not ensure the accuracy and completeness of information so most of teacher and secondary students generalize their responses to both disaster symptoms and disaster information, their respond is same, they will immediately evacuate themselves.

4. Conclution

Teachers and secondary students have good quality in knowledge of disaster literacy but they have low quality of information management in disaster literacy. It made teachers and secondary students can not to tell the differences between natural symptom before disaster and disaster information, they directly evacuate themselves when they got the natural symptom before disaster and disaster information. in fact, if you get a natural sign before a disaster, it is correct to immediately evacuate, but if you get disaster information, the material and completeness must be confirmed. This is what makes the public may easily exposed to hoax news because most Indonesians admit that sometimes they can tell the difference, sometimes they can't tell whether the information they receive is a hoax or not [15].

References

[1] Badan Nasional Penanggulangan Bencana 2021 Kejadian Bencana 2019
[2] Kastolani W and Mainaki R 2018 Does Educational Disaster Mitigation Need To Be Introduced In School? *SHS Web Conf.* **42** 00063
[3] KHAN H, VASILESCU L G and KHAN A 2008 Disaster Management Cycle – a Theoretical Approach *Manag. Mark.* **6** 43–50
[4] Brown L M, Haun J N and Peterson L 2014 A Proposed Disaster Literacy Model *Disaster Med. Public Health Prep.* **8** 267–75
[5] Doyle C S 1992 Outcome Measures for Information Literacy within the National Education Goals of 1990. Final Report to National Forum on Information Literacy. Summary of Findings. *Natl. Forum Inf. Lit.* 1–18
[6] Nadeak B 2015 Correlation between Knowledge, Experience and Common Sense, with Critical Thinking Capability of Medical Faculty's Students at Indonesia Christian University *J. Educ. Pract.* **6** 45–55
[7] Marwanta B 2005 Tsunami Di Indonesia Dan Upaya Mitigasinya *Alami J. Teknol. Reduksi Risiko Bencana* **10**
[8] Mutasa M 2013 Invesigaing the signiicance of disaster informaion management *Jamba J. Disaster Risk Stud.* **5** 1–6
[9] Badan Pusat Statistik Kota Surakarta 2020 Jumlah Kejadian Bencana Alam Menurut Kecamatan di Kota Surakarta 2019
[10] BNPB 2019 *Buku Saku Tanggap Tangkas Tangguh Menghadapi Bencana, Cetakan Keempat*
[11] Muktaf Z M 2017 Disaster Literacy in Communication Perspective *Apik PtmInternational Conf. Call Pap. SILAT APIK PTM* 1–12
[12] Wilson C 2012 Media and information literacy: Pedagogy and possibilities *Comunicar* **20** 15–22
[13] Wilson C 2019 Media and information literacy : Challenges and opportunities for the world of education 1–17
[14] Fadilah M, Permanasari A, Riandi and Maryani E 2020 The level of disaster literacy of earthquake-experienced students in Mathematics and Science Faculty of State University in Indonesia *J. Eng. Sci. Technol.* **2020** 30–8
[15] Juditha C 2020 People Behavior Related To The Spread Of Covid-19's Hoax *J. Pekommas* **5** 105

IOP Conf. Series: Earth and Environmental Science **986** (2022) 012018 doi:10.1088/1755-1315/986/1/012018

Disaster management effort in peatland and coastal areas through cross sectoral collaboration, case study in Bukit Batu District, Bengkalis Regency, Riau Province

W Purwanto[1], R Hidayat[1] and S B Ajar[2]

[1] CSR Department PT Kilang Pertamina Internasional RU II Sungai Pakning,
[2] Geography Department Sebelas Maret University

*Email: wahyup.geo09@gmail.com

Abstract. This paper describes the disasters that occur in peatland and coastal areas, especially in Bukit Batu District, Bengkalis Regency, Riau Province. In addition, it is also explained about disaster management with cross – sectoral cooperation in accordance with the role of each sector so as to reduce the impact and even overcome the disaster. In this research paper using qualitative research methods and conducted with open interviews and filed surveys. As a result, with this cross-sectoral collaboration, land and forest has stoppes and the abrasion on the coast of Pangkalan Jambi Village has stopped and has even begun to form new land.

1. Introduction

Peatlands have a very important role for the ecosystem on them. Not only serves as a support for the life of the living creatures on it but also has ecological functions such as flood control and global climate control. The characteristic of peat that has many fine roots causes it to be able to hold water and become a place for storing carbon which is quite high. However, various problems often arise in peatlands due to mismanagement of land, causing peatlands to dry out which has the potential for land and forest fires. In addition to the high potential for fires due to dry peatlands, there are also frequent flooding problems in peatlands due to reduced land cover so that peatlands are no longer able to accommodate rainwater in their areas. So that the balance of the peat ecosystem becomes very important in the sustainability of life on it.

Indonesia is one of the countries with a fairly large peatland area in the world. The total area of peatland in Indonesia in 2019 was 13.43 million ha, according to data from the Agricultural Research and Development Agency (Balitbangtan) spread across several provinces in Indonesia. The extent of this peatland can be a potential as well as a big problem for the local community, including Riau Province.

In Sungai Pakning, Bengkalis Regency, Riau Province, peatlands often become a problem for the surrounding community, because land and forest fires often occur during the dry season and floods during the rainy season. Even disasters like this keep repeating themselves every year, causing various kinds of losses. This condition is exacerbated by the increasingly damaged peat ecosystem due to deforestation.

On the other hand, there are still many people living on peatlands who don't understand the condition of the land they live in, or even don't know what peat is at all. From these problems that should be a trigger for the community to survive with ecological and social resilience, so that it will greatly help the community to survive on peatlands which are known as a source of disaster for the surrounding community.

Not only land and forest fires and floods, the location of the Sungai Pakning and its surroundings on the coast is also prone to abrasion, so this is also a threat to the surrounding community. In addition, land subsidence is also a long-term threat for them. When viewed in the field, there has been a land subsidence of approximately 10-50 cm.

Then on the coastal side there is also have abrasion disaster, in Pangkalan Jambi Village in the last 20 years it has abrasion of approximately 115m, which has caused the displacement of local people's settlements to higher locations. Disaster management on the peatland and coast needs collaboration by all parties, both from the community, government and private sector.

In Bukit Batu District, Bengkalis Regency, Riau Province, the results of this cross-sectoral collaboration are starting to show results. Several disaster management activities in this peatland coastal area are slowly starting to be overcome, ranging from land and forest fires to abrasion, but have not been fully resolved because the disaster-affected areas are not only in Bukit Batu District but also almost all coastal areas of Riau Province have a same disaster.

2. Literature

Disaster according to the Law of the Republic of Indonesia number 24 Article 1 of 2007 is an event or series of events that threaten and disrupt the lives and livelihoods of the community caused, both by natural factors and/or non-natural factors as well as human factors, resulting in human casualties, damage environment, loss. property and psychological impact [1].

In disaster management, it is necessary to have good disaster management so that disaster management becomes effective. Disaster management is a science that studies disasters and all aspects related to disasters, especially disaster risk and how to avoid disaster risk [2]. Disaster management is a dynamic process about the functioning of management functions that we know so far, for example the functions of planning, organizing, actuating, and controlling. Effective disaster risk management requires a combination of four concepts, namely over all hazards, comprehensive, integrated and community preparedness.

An integrated approach to effective disaster management requires the active collaboration of various stakeholders. So it can be said that in disaster management, one of the factors for successful handling is integrated handling or by means of cross-sectoral cooperation. Cristian, Y., et al. (2021) [3] explains that cross-sectoral cooperation can be a solution in solving problems or disasters on the coast. Greenwood, M. (2007)[4] argues that stakeholder involvement can be understood as a practice carried out by an organization to positively involve various parties in an activity within it. Involving stakeholders from various types of actors can provide a positive impetus for the achievement of a common goal, namely disaster management in Bukit Batu District. In this case, cross-sectoral cooperation in disaster management in Bukit Batu District involves the Government, the Community, and the surrounding companies.

Purwanto, W (2021) [5], explained that a company's CSR if it is well structured can be one of the solutions in disaster management, of course, in collaboration with various parties. Widhagdha, 2018 [6] , explained that disaster-prone land can be converted into locations that provide many benefits if managed properly, one of which is on peatlands.

3. Method

This research was conducted using qualitative methods to produce descriptive data from the observed research objects. While the approach is carried out by case studies by looking at the types of disasters and how to handle them in the field. Data was collected through participation observation and focus group discussions (FGD). Then the sample is carried out by purposive sampling on groups involved in disaster management

4. Result and Discussion

Living in peat coastal areas can be both a disaster and a blessing for the community, depending on how to address the potential and problems that exist in this peatland coastal area. There are not only many potential disasters that exist on the peatland coast but also benefits that nature provides for the people who live there.

4.1. Disasters in Peatland and Coastal Area

4.1.1. Abrasion in Coastal Area

The location of Sungai Pakning which are located on the east coast of Sumatra Island causes a large potential threat of abrasion. Located in front of the Bengkalis Strait as a route for large ships to make the big wave. In addition, the presence of the north wind also contributed to the magnitude of the wave force that hit the coastal area of the Sungai Pakning area.

One of the areas affected by large abrasion is Pangkalan Jambi Village. The abrasion that hit Pangkalan Jambi Village has eroded the land area by approximately 115m. This causes the displacement of the location of community settlements to higher ground, because the area they originally occupied has become prone to abrasion or even has been abraded.

Abrasion is also getting worse with the damage to mangroves due to illegal logging, thus making abrasion faster. People still use mangrove wood for charcoal and "trocok" or foundations for houses, so many mangroves are being cut down illegally. These are some of the factors that cause abrasion to get worse in coastal areas.

Figure 1. Abrasion Conditions in Pangkalan Jambi Village

4.1.2. Land and Forest Fire

The next problem when in peatland areas is the haze disaster due to land and forest fires. In fact, it often has an impact on neighboring countries such as Malaysia and Singapore. This land and forest fire has also become a routine disaster that occurs every year during the summer season.

The losses suffered were also a lot for the community, government and companies operating in near of the fire location. Although the number of land and forest fires continues to decline every year, these fire disasters keep repeating themselves.

In Sungai Pakning, land and forest fires are also a regular occurrence every year, the largest being around the period from 2013 to 2016. The causes of these land and forest fires occur due to several factors, including natural factors and intentional acts committed by irresponsible people.

On the other hand, the community has not been educated much about the potential and problems in peatlands, so they are not fully aware of how to manage peat properly and correctly. There are still many people who do agriculture by burning peat land (merun) before planting, even though this is very dangerous and increases the potential for land and forest fires around it.

In addition to agricultural land, land clearing also often occurs for plantation lands that have a wider scope. Burning for plantations is more dangerous because it covers a wider area and takes a long time to be extinguished in the event of a fire.

Figure 2. Land and forest fire in Kampung Jawa, Kelurahan Sungai Pakning

4.1.3. Land Subsidence

Land subsidence is one of the most common disasters in peat areas. But many are not aware of it. This is because many people still think that disasters on peatlands are just a matter of land and forest fires. Even though the process of land subsidence is unconsciously also having a negative impact on the community, such as the widespread flooding, the entry of tidal water that is getting farther inland, to affect the quality of ground water, especially during the dry season which causes groundwater to become salty due to intrusion sea water to land area.

In Kampung Jawa, Sungai Pakning, Bukit Batu District, Bengkalis Regency, land subsidence also occurred in several locations with various land subsidence. From the results of measurements in the field, locations with buildings and settlements have a higher decrease in the number of land compared to land that uses land as gardens. This is a measurement of land subsidence in Kampung Jawa, Sungai Pakning, Bukit Batu District, Bengkalis Regency:

Table 1. Land subsidence in Kampung Jawa, Kelurahan Sungai Pakning

No	Coordinate	Land subsidence	Land use
1	Lat 1^0 20'52"N, Long 102^0 6'54"E	7 cm	Pineapple plantation
2	Lat 1^0 20'52"N, Long 102^0 7'0"E	14 cm	Palm oil plantation
3	Lat 1^0 20'53"N, Long 102^0 7'8"E	10 cm	Pineapple plantation
4	Lat 1^0 20'55"N, Long 102^0 7'17"E	30 cm	Settlement
5	Lat 1^0 20'55"N, Long 102^{0x}7'17"E	15 cm	Settlement
6	Lat 1^0 20'59"N, Long 102^0 7'32"E	35 cm	Settlement and building
7	Lat 1^0 21'3"N, Long 102^0 7'51"E	27 cm	Settlement and public health centre
8	Lat 1^0 21'1'N, Long 102^0 8'40"E	22 cm	Settlement and rubber plantation

The land subsidence data is the value of land subsidence from 2013-2021, based on the results of calculations in the field on locations around the bridge that experienced land subsidence and interviews with the surrounding community regarding the construction of the bridge. Based on these data, in Kampung Jawa there was a decrease between 7 cm to 35 cm in that period.

4.2. Cross-Sectoral Participation in disaster management

In disaster management, it cannot be carried out alone, cross-sectoral collaboration is needed so that the disaster management process can be successful. Collaboration between the community, government and the private sector has their respective roles in disaster management, these is:

4.2.1. Community

The role of the community in handling disasters, both land and forest fires and abrasion is very vital, because the community is a group directly dealing with disasters that occur, so that success in disaster management depends also on the alertness of the local community. In Bukit Batu Subdistrict, there are several community groups that take part in handling land and forest fires and abrasion, including the "Masyarakat Peduli (MPA)", "Kelompok Pengawas Mangrove", farmers and fisherman groups around the location.

The Masyarakat Peduli Api have a very important role in handling land and forest fires. Groups that voluntarily carry out daily patrols and are always ready to put out fires when there is a fire. In addition, in Bukit Batu District, MPA members are encouraged to provide lessons for elementary schools in their location regarding the dangers of land and forest fires, so that many school children understand about land and forest fires.

Figure 3. Masyarakat Peduli Api (MPA)

Not only the Masyarakat Peduli Api, farmer groups and forest honey hunter groups also have a role in handling land and forest fires. Farmers go to their plantation as well as patrol around their respective gardens, so that apart from carrying out their daily activities as farmers, they also indirectly help protect their area from land and forest fires. Then the forest honey hunter groups have also been educated to conduct patrols in the forest when they are looking for forest honey, so that apart from getting honey, they also take part in preventing land and forest fires in their area.

In handling abrasion, there is also a Kelompok Pengawas Mangrove which plays a very important role in maintaining the mangrove ecosystem in Pangkalan Jambi Village. They are also assisted by local fishermen to carry out patrols in the mangrove forest area. Because these voluntary groups in protecting forests and land in peatland and mangrove areas in coastal areas, finally in the Bukit Batu District area there were no very large land and forest fires like in previous years, even up to zero fires and in coastal areas, abrasion has stopped and has started to form a new land in Pangkalan Jambi Village.

4.2.2. Government
The role of the government in handling land and forest fires and abrasion in Bukit Batu District can be seen from the active role of the Village and District governments which are very active in carrying out activities in disaster management. In Bukit Batu, the government has prepared village funds budget for the construction of infrastructure for disaster management, such as water reservoirs, canals, and canal blocking. In addition, the village government has also formed a task force called "Masyarakat Peduli Bencana (MPB)" which aims to handle disasters in each region.

The village and sub-district governments also collaborate with several elements such as the Indonesian National Army (TNI) and the Indonesian National Police (Polri) in disaster management, while also involving the private sector in their area. There is also the role of "Manggala Agni" from the Ministry of Forestry and Environment (KLHK) who always patrols and is ready to help when land and forest fires occur in the Bukit Batu area.

So it can be said that the government's role here is also very important in handling disasters in the Bukit Batu area. In addition, as a regulator, the government also makes regulations for the community regarding disaster prevention in its area.

Figure 4. Land and Forest Fire Management Training

4.2.3. Private Sector
The role of the private sector in handling land and forest fires and abrasion in Bukit Batu District is also very crucial through the Corporate Social Responsibility (CSR) program. PT Pertamina RU II Sungai Pakning is one that is engaged in disaster management in Bukit Batu District. Through the Pertamina RU II Sungai Pakning CSR program, mobilizes the Masyarakat Peduli Api (MPA), conducts routine firefighting training, and help with extinguishing equipment. Pertamina RU II Sungai Pakning also carried out fireman certification for the MPA. There are also entrepreneurial activities for MPA that support MPA's economy so as to have an economic impact on MPA.

In addition, there is also the Sekolah Cinta Gambut program and the Arboretum Gambut which are educational programs for school students or the general public regarding the management, utilization and prevention of disasters in peatlands. This program is able to increase community knowledge in good peatland management. There is also a program to use fire-prone lands into productive land, such as pineapple, lemongrass and wood plantations, so as to prevent land and forest fires on peatlands.

Figure 5. Sekolah Cinta Gambut Program in Arboretu, Gambut

Then in handling the abrasion disaster, PT Pertamina RU II Sungai Pakning made several CSR programs, including build the Triangle Mangrove Barrier (Trimba) which is a breakwater with natural materials with the aim of breaking waves, catching sedimentation and protecting mangrove plantings so as to increase success in planting mangroves. With this program, the existing abrasion in Pangkalan Jambi Village has stopped, and is able to add land through the sediment formed from the installation of TRIMBA.

Figure 6. Before and After Program

Often the constraint in conservation is in conflict with the economy. People who used to cut down mangroves to sell their wood now have to take care of the mangroves, of course there will be a decrease in income for the community. Based on this, PT Pertamina RU II Sungai Pakning created Mangrove Ecotourism in Pangkalan Jambi Village, the goal is to maintain mangrove conservation but the community still gets economic benefits from the conservation. Not only tourism but also a program for the utilization of mangrove products which are processed into snacks as a product of the mangrove ecotourism. With the ongoing conservation and economic benefits, this program is able to run sustainably.

5. Conclusion

The success in handling land and forest fire disasters and abrasion in Bukit Batu District is due to good cross-sectoral collaboration from the community, government to the private sector. Each of these elements has a role and function of each. The impact is that in the last few years the Bukit Batu District has been free from severe fires or zero fires and the abrasion disaster, especially in Pangkalan Jambi Village, has stopped.

References

[1] Badan Nasional Penanggulangan Bencana 2007 *UNDANG-UNDANG REPUBLIK INDONESIA NOMOR 24 TAHUN 2007 TENTANG PENANGGULANGAN BENCANA* (Jakarta)
[2] Nurjanah 2012 *Manajemen Bencana* (Bandung: Alfabeta)
[3] Christian Y, Budiman M K, Purwanto W and Damar A 2021 Supporting community-based mangrove forest management as Essential Ecosystem Area in Sungai Pakning, Riau *IOP Conf. Ser. Earth Environ. Sci.* **744**
[4] Greenwood M 2007 Stakeholder engagement: Beyond the myth of corporate responsibility *J. Bus. Ethics* **74** 315–27
[5] Purwanto W, Hidayat R and Utami A 2020 Disaster Management of Forest and Peatland Fire Through The Berdikari Gambut Village Program in Pakning River, Bengkalis District *Soc. Humanit. Educ. Stud. Conf. Ser.* **3** 124–32
[6] Widhagdha M et al 2018 *CSR: Pemberdayaan Masyarakat di Lahan Gambut* (Surakarta: Arjuna Wijaya Karya)

Ecosystem conservation of Mangrove Education Center (MEC) in handling abration disaster at Pangkalan Jambi village based on stakeholder engagement perspective

T A Seva[1*], W Purwanto[1] and A R Latuconsina[1]

[1] CSR Department PT Kilang Pertamina Internasional RU II Production Sungai Pakning

*Email: tahsaseva@gmail.com

Abstract. The coastal communities of Pangkalan Jambi Village lives in an abrasion-prone situation. This vulnerability can't be separated from natural factor namely hydrodynamic phenomena worsened by lack of mangrove plants, which also caused by some destructive human activities that ruined the ecosystem. This phenomena, simultaneously creates an abrasion-prone situation on the Pangkalan Jambi Village's coastal area, which is marked by the coastline decline on the village coast up to 115 meters. Hence, this study aims to examine an innovative idea to restore and conserve the mangrove ecosystem of village coastal area. Using descriptive qualitative approach with conceptual framework of stakeholder engagement, it focused on analyzing the idea of conservation of mangrove areas initiated by CSR PT Kilang Pertamina Internasional RU II Production Sungai Pakning in partnership with various stakeholders such as local government and also civil society organization as well. Activities that performed to conserve coastal mangrove areas in Pangkalan Jambi Village started from mangrove planting practice using Hybrid Engineering and Triangle Mangrove Barriers (Trimba), and developing Mangrove Education Center (MEC) area as a tourism spot as well as a learning laboratory for mangrove ecosystems.

1. Introduction

Based on data from the Pangkalan Jambi Village Government in 2021, Pangkalan Jambi Village is located in Bukit Batu District, Bengkalis Regency, Riau Province. The village area covering an area of 10 km2 has a population of 1.259 people spread across 12 Neighborhood Units (RT) and 6 Community Units (RW). The eastern region of Pangkalan Jambi Village is a coast directly adjacent to the waters of the Bengkalis Strait which is prone to abrasion phenomena. The lack of mangroves and coastal hydrodynamic phenomena such as waves, currents, and tides are the driving factors for abrasion. Technically, the abrasion condition is not only influenced by the hydrooceanographic process. However, there are natural factors from land that can change the coastline, such as the result of sedimentation and erosion from the impact of tidal currents (floods) and changes in river flow.

In 2004, a group of fishermen from Pangkalan Jambi Village who live in coastal areas formed a group called "Harapan Bersama". The group is a manifestation of the fishermen's concern about the condition of the coastal area affected by abrasion. The rate of abrasion phenomenon is caused by the hydrodynamic phenomenon of the coast which is driven by irresponsible human actions in cutting down mangrove trees illegally in the coastal area of Pangkalan Jambi Village. According to (Purwanto et al., 2020) the coastal land of Pangkalan Jambi Village has regressed by 115 meters or in the last 20 years each year it has decreased by 5 meters. These ecological problems require solutions and interventions from various stakeholders. One of the welfare actors who played a role in overcoming the ecological crisis was PT Kilang Pertamina Internasional RU II Production Sungai Pakning.

Pangkalan Jambi Village is located in the Ring 1 area of the operational area of PT Kilang Pertamina Internasional RU II Production Sungai Pakning. This means that the Pangkalan Jambi Village area will be directly affected by the company's operational activities both in terms of asset placement and distance. So that in business ethics, the company is responsible for providing compensation both socially and environmentally to the people of Pangkalan Jambi Village. This form of compensation is realized through Corporate Social Responsibility (CSR) and biodiversity programs, as a form of intervention and the role of the company in overcoming environmental crisis problems in the coastal area of Pangkalan Jambi Village. The company provides assistance to the "Harapan Bersama" group, and encourages synergy among other welfare actors in order to support the success of mangrove planting in the coastal area of Pangkalan Jambi Village. The mangrove planting effort is targeted to protect coastal areas from the impact of abrasion disasters.

Furthermore, in 2018 the company also developed the potential of the area on the coast of Pangkalan Jambi Village by creating a mangrove ecotourism area. This is marked by fostering the empowerment of coastal women's groups by utilizing the potential of marine products and mangrove plants to then be processed into food and drink products. This mangrove planting and community development zone on the coast of Pangkalan Jambi Village was later known as the Mangrove Education Center (MEC). The direction of development of the Mangrove Education Center (MEC) emphasizes the synergy between stakeholders in efforts to restore and preserve mangrove ecosystems, as well as to develop groups from a social and economic perspective. Departing from the issues above, it is interesting to study from the perspective of stakeholder engagement regarding how the practice of developing the MEC ecosystem conservation area on the coast of Pangkalan Jambi Village is? and how is the analysis of stakeholder involvement in the development of the MEC ecosystem conservation area on the coast of Pangkalan Jambi Village?

2. Theoretical Framework

2.1. Stakeholder Engagement

[1] Argues that stakeholder involvement can be understood as a practice carried out by an organization to positively involve various parties in an activity within it. Involving stakeholders from various types of actors can provide a positive impetus for the achievement of common goals in conserving the mangrove ecosystem in the coastal area of Pangkalan Jambi Village. However, it can also reduce social conflicts that may occur if other actors such as local government, other private parties and community groups at large are not involved. When many actors are involved in a program or activity, it will be seen how big the level of participation and decision making will be. There are ten levels of stakeholder involvement which are depicted through the spectrum in the form of remain passive, monitor, advocate, inform, transact, consult, negotiate, involve, collaborate and empower in the AA1000 Stakeholder Engagement Standard guide [2].

Stakeholder involvement is then grouped to find categories of involvement through stakeholder mapping activities. The results of the mapping will later be used to support the company's needs to see the order of stakeholders who have a main role to those who have a minority role. Thus, a coherent and effective mapping plan is needed in obtaining the desired data results. Morris & Baddache [3] suggest that there are four stages that must be carried out in conducting stakeholder engagement mapping activities, namely by identifying, analyzing, mapping and prioritizing. The identification stage is aimed at knowing and determining who are and are not stakeholders of the company [3]. After we identify, the next step is to analyze the relationship between the company and its stakeholders using certain variables to arrange the priority level of the stakeholders. Next, we describe or describe the results of the identification and analysis into a visualization in the form of modeling from stakeholder mapping. From there, we will know the strengths and interests of each stakeholder which is then used to determine the priority scale of each of these stakeholders. Only then will we determine the appropriateness of attitudes and actions that will later be used to manage relationships with stakeholders according to the order of priorities that have been determined.

After we get the data from the mapping of stakeholder involvement, the next step is to develop a model of the form of engagement. [4] suggest that there are four categories or groups of stakeholders by presenting a model matrix with four quadrants divided by 2 axes of power/influence and interest. Through this model, it will be easier for companies to determine the appropriate method of engagement with each stakeholder and at the same time to know which stakeholders should be prioritized [3].

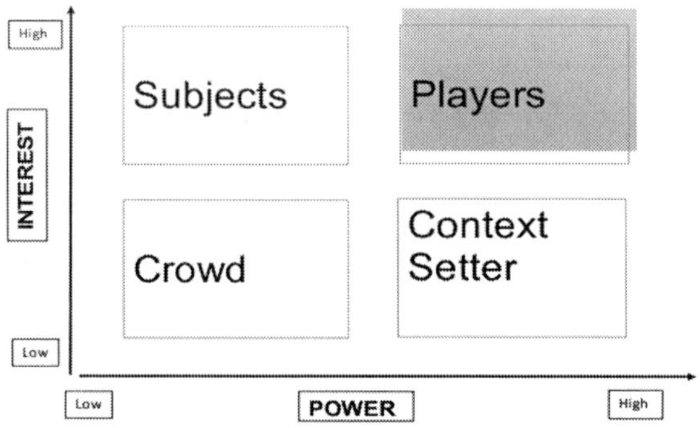

Source : [4]

Figure 1. Stakeholder Influence and Stakeholder Matrix Modeling

Quadrant 1 is a "key player" which is a stakeholder who has high power and interest and is able to support or sabotage the company's strategy [4]. While quadrant 2 "subjects" are stakeholders who have high interest but low influence. Furthermore, quadrant 3 is "crowd" which shows low interest and influence [5]. The last quadrant is "context setters" which shows stakeholders who have high influence but low in terms of interest or importance.

3. Method

This research was conducted using qualitative methods to present descriptive data findings with a case study approach. Data collection was done by identifying through observation, Forum Group Discussion (FGD), interviews and documentation. Informants in this study were determined by non-probability sampling, namely by purposive technique. The object of research to be studied is the MEC coastal area in Pangkalan Jambi Village which is the CSR target area of PT Kilang Pertamina Internasional RU II Production Sungai Pakning. Thus, the results of the stages of identification, analysis, mapping and prioritizing stakeholders are aimed at supporting the development of CSR at PT Kilang Pertamina Internasional RU II Production Sungai Pakning in synergy in implementing conservation and group empowerment in MEC Pangkalan Jambi Village.

The determination of the category of informants which is a collection of stakeholders resulted from scoring using a list of questions to measure how much influence and level of interest they have on the implementation of the Mangrove Education Center (MEC) ecosystem conservation program and activities for handling abrasion disasters in Pangkalan Jambi Village. The question instrument in scoring was adopted from Roslinda et al. [6] by using five gradations of data measurement, then the score results are accumulated and mapped into the stakeholder analysis model matrix grid.

4. Results and Discussion

The practice of developing the MEC ecosystem conservation area on the coast of Pangkalan Jambi Village, started in 2017 through the CSR programs of PT Kilang Pertamina Internasional RU II

Production Sungai Pakning which focused on mangrove restoration efforts and continued in the following years with the aim of increasing biodiversity and empowerment of the Harapan Bersama Group. Members of the Harapan Bersama Group are a mixture of fishing fishermen groups and community groups outside the fishing fishing groups. The restoration of the coastal area of MEC in Pangkalan Jambi Village was carried out by improving the quality of the 7 km long coastal environment through the selection of mangrove seedlings of the api-api and mangrove species as well as planting 2.500 mangrove seedlings. Until the provision of good and correct mangrove cultivation training by continuing to provide regular assistance to the Harapan Bersama Group. The Pangkalan Jambi Village Government also contributed by realizing village funds for the construction of boat moorings for fishing groups. The construction of this facility is intended to provide infrastructure support for fishing groups in facilitating their access to sea. Thus, it is expected to increase the marine catch of the group.

The development of the mangrove conservation CSR program continues, marked by the creation of a mangrove nursery, replanting 10.000 new mangrove seedlings, developing the Mangrove Education Center (MEC) area as an ecotourism location and applying the installation of Wave Breakers (APO) in stages in the form of hybrid engineering along 1.000 meters in 2018. However, this hybrid engineering type APO did not last long when the waves hit, which was only able to last for 1 week. Furthermore, the hybrid engineering type was modified by adding components in the form of crossed wood which were arranged on top of the previous hybrid engineering. As a result, this type of APO can withstand the waves for a period of 1 year and can be useful as a means of catching mud sediments, where the sediment that has formed on the MEC coast of Pangkalan Jambi Village until 2020 is 21 meters wide with a depth of 30-60 cm/year.

Then in 2019, many actors began to be involved in the development of the MEC area, both in terms of tourism and in terms of conservation. The actors involved are PT Kilang Pertamina Internasional RU II Production Sungai Pakning through the implementation of CSR programs which are marked by the addition of planting 5.000 mangrove seedlings which are attended by elementary school students from the Cinta Gambut School Network, ecotourism development by building a wooden track around the MEC area, training diversification of marine catches and mangrove plants into processed food and drink products that have standardized quality through a PIRT (Home Industry Products) permit and halal certification from LPPOM MUI Riau. In addition, the development of APO is also carried out with the innovation of the Triangle Mangrove Barrier (TRIMBA) which uses *nibung* wood as its raw material. TRIMBA is the result of a combination of hybrid engineering modifications, straight wood type APO and additional mounding techniques placed in front of the APO as a breakwater [7]. The next actor is the Pangkalan Jambi Village Government which allocates village funds for the continued construction of boat moorings for fishing groups that have been started since 2017 and training assistance for mangrove tourism development. The Department of Marine Affairs and Fisheries of Bengkalis Regency also took part in providing assistance in the form of the construction of a 36 meter long boat mooring, the construction of 1 saung infrastructure and 1 toilet. The construction of this infrastructure is intended so that the Harapan Bersama Group can have public facilities in the form of a place to dock for ships, a place for deliberation activities between members to take place, and a place for groups to rest when they want to go or return from fishing. In addition, the Bengkalis Regency Cooperatives and Micro, Small and Medium Enterprises (UMKM) Office also frequently visits through the UMKM exhibition to promote the processed products of the Pokja Jaya Bersama. Pokja Jaya Bersama is one of the working groups of the Harapan Bersama Group in charge of processing marine catches and mangrove plants to be processed into food and beverage products. In its development, the Harapan Bersama Group has 3 working groups including the Fish Cultivation Working Group, the Mangrove Working Group and the Product Management Working Group.

Source: company data

Figure 2. Implementation of TRIMBA at MEC Pangkalan Jambi Village

Then in 2020, the Pangkalan Jambi Village Government also re-realized village funds to support the maintenance of fishing boat bridges/moorings, construction of boat moorings and the manufacture of coastal protection sheeting on fishing roads. The Bengkalis Regency Library and Archives Service also took part in assisting the promotion of MEC ecotourism in Pangkalan Jambi Village through literacy enrichment and social media. The Department of Environment (DLH) of Bengkalis Regency and the Department of Environment and Forestry (DLHK) of Riau Province also have a role in monitoring and supporting conservation programs in the MEC coastal area of Pangkalan Jambi Village through the Setia Lestari Bumi program and the Climate Village Program (PROKLIM). still running today. Through the support of the Bengkalis Regency DLH and the Riau Province DLHK, Rukun Hamlet, which is where the MEC is located, is a candidate for the Main PROKLIM candidate in 2021. PT Kilang Pertamina Internasional RU II Production Sungai Pakning is also still involved in efforts to conserve mangrove ecosystems and improve the group's economic welfare. in the MEC area of Pangkalan Jambi Village. This is done by providing capacity building training for tour guides in the form of outbound activities with resource persons from Pertamina Pecinta Alam (Patrapala). Aims to encourage the skills of group members as tour guides in serving tourists at MEC Pangkalan Jambi Village. In addition, the company also carries out monitoring and evaluation of biodiversity in the development of MEC Pangkalan Jambi Village in 2020. As a result, the findings from the monev report show a positive impact from the application of TRIMBA, where with this type of APO modification it is able to reduce the rate of waves hitting the mainland and capable of forming sediments along 300 meters with a width of 50 meters and a depth of ± 60-70 cm. The new sediments formed in fact also had a positive impact on the addition of 8 new benthic animal species in the seedling area that had never existed before. The Harapan Bersama group has also been able to cultivate mangroves with the results of their captive breeding being sold to various areas such as Bengkalis and Dumai. Therefore, the average mangrove seedling that has been planted from 2017 to 2020 is ± 10.000 seedlings, followed by cross-sectoral institutions/organizations. Where this year, the Forum Gambut Muda (FORGADA) is also involved in the replanting of mangrove seedlings.

Source: company data

Figure 3. Mangrove seedlings reach cross-sectoral

Furthermore, this year, PT Kilang Pertamina Internasional RU II Production Sungai Pakning through the CSR annual work plan (Renja), the company has realized several activities that are integrated into the Permata Hijau Program. The purpose of the program is the company's efforts to develop an innovative ecotourism concept based on mangrove conservation and group empowerment. The activities carried out include group institutional administration training, development and improvement of processed food products typical of mangroves, development of the mangrove area fisheries sector, reforestation of mangrove areas with stakeholders, improvement of mangrove tourism tracks and construction of a hybrid solar dryer system. Then through the Program Kampung Gambut Berdikari, the company carried out teacher training activities for the Sekolah Cinta Gambut Network with resource persons from the Geography Education Study Program, Faculty of Teacher Training and Education, Universitas Sebelas Maret. This training brought a group of Sekolah Cinta Gambut Network teachers to the MEC location in Pangkalan Jambi Village to stimulate them to create a collection of questions with the theme of mangrove ecosystem conservation.

4.1. Identification of Stakeholders in the Implementation of MEC Ecosystem Conservation

The actors involved have different portions and roles of involvement. Therefore, the results of the stages of identification, analysis, mapping and prioritizing stakeholders are directed to support the role of PT Kilang Pertamina Internasional RU II Production Sungai Pakning in improving the quality and quantity of conservation and group empowerment. The following is a table of identification results of a list of stakeholders involved in MEC ecosystem conservation activities in Pangkalan Jambi Village:

Tabel 1. Stakeholders Mapping

No.	Institution	Role in the implementation of MEC ecosystem conservation in Pangkalan Jambi Village
1	PT Kilang Pertamina Internasional RU II Production Sungai Pakning	1.As a corporation that carries out social and environmental responsibilities as a form of commitment to sustainability; 2.Determine strategies and targets for achieving sustainable conservation; 3.Planning activities in the implementation of the conservation and revitalization program for the mangrove area at MEC, Pangkalan Jambi Village; 4.Monitor and evaluate the implementation of programs and activities.
2	Pangkalan Jambi Village Government	1.Get involved in providing MEC conservation development assistance through the realization of village funds; 2.Supporting mangrove conservation in MEC by issuing village regulations; 3.Support the prohibition of illegal logging of mangroves through the establishment of notice boards; 4.Support the formation of the Harapan Bersama Mangrove Cultivation Working Group which was ratified through a Village Decree
3	Harapan Bersama Group	Carry out CSR programs and activities that have been planned with PT Kilang Pertamina Internasional RU II Production Sungai Pakning which aims to maintain the MEC ecosystem through conservation activities and improving the group's economy.
4	Marine and Fisheries Services	The role of the Bengkalis Regency Marine and Fisheries Service in supporting the success of the Mangrove Education Center in Pangkalan Jambi Village is to build boat moorings around the mangrove location, thus supporting the economy of fishermen as mangrove conservation activists. Because often fishermen or other conservation groups are reluctant to carry out conservation because it clashes with their economic needs. With this support, fishermen still get economic benefits when carrying out mangrove conservation.
5	Environmental services	The Environmental Service at the Bengkalis District Environmental Service (DLH) and the Riau Province Environmental and Forestry Service (DLHK) strongly supports conservation programs through its programs such as PROKLIM, Setia Lestari Bumi, and others. The Environmental Service also provides assistance and monitoring of The success of mangrove planting in Pangkalan Jambi Village.
6	Other Offices and Ministries	Several other agencies and ministries are also heavily involved, especially in local and national promotions. The Department of Cooperatives and Micro, Small and Medium Enterprises (UMKM) of Bengkalis Regency often visits the UMKM exhibition. Then the Bengkalis Regency Library and Archives Service helped promote through literacy enrichment and social media, until the Ministry of Home Affairs made Pangkalan Jambi Village an example of a village that was successful in managing mangrove conservation through cross-sectoral collaboration.
7	Pertamina Pencinta Alam (PATRAPALA)	Provide capacity building for group members in the form of outbound training so that it can be applied in the development of tourism based on mangrove conservation.
8	Public Administration Study Program, Universitas Riau	Explore knowledge through research such as theses, theses and dissertations at the MEC location in Pangkalan Jambi Village.
9	Forum Gambut Muda (Forgada)	One of the assisted groups of PT Kilang Pertamina Internasional RU II Production Sungai Pakning which also has a role in helping mangrove conservation efforts such as planting mangroves.
10	Sekolah Cinta Gambut network	One of the assisted groups of PT Kilang Pertamina Internasional RU II Production Sungai Pakning which makes MEC a learning laboratory for elementary school students in educating them to love the peat environment and the coastal environment is no exception.
11	Geography Education Study Program, Faculty of Teacher Training and Education, Universitas Sebelas Maret	An educational institution that provides knowledge enrichment training for elementary school (SD) teachers who are members of the Cinta Gambut School network to create a collection of questions at the MEC mangrove conservation location.

Source: Processed from primary data

4.2. Analysis of Stakeholder Involvement in the Implementation of MEC Ecosystem Conservation
After we identify the actors involved in the MEC conservation activities in Pangkalan Jambi Village, it is necessary to analyze the influence and interests of each of these actors and their relationship with PT Kilang Pertamina Internasional RU II Production Sungai Pakning. Thus, the results of the analysis are as follows:

a. PT Kilang Pertamina Internasional RU II Production Sungai Pakning
The company's intervention is the dominant factor in the development of the MEC conservation area in Pangkalan Jambi Village which until now has developed into an ecotourism area. This is evidenced by the company's guidance to coastal community groups in Pangkalan Jambi Village through the implementation of CSR programs since 2017. Community development funds have been disbursed by the company to assist groups in overcoming abrasion problems and overcoming the failure of mangrove planting due to lack of group knowledge. There were innovations which were later developed by the company together with the group in the form of breakwaters in the form of hybrid engineering, hybrid engineering with modifications and the Triangle Mangrove Barrier (TRIMBA). The company has a very high influence, due to the existence of an intense mentoring relationship with community groups. Making companies trusted by groups to help them deal with abrasion disasters while helping to improve their economy through the development of ecotourism. The level of interest owned by the company is classified as very high to maintain its existence in carrying out sustainable community development through the implementation of CSR programs in the MEC area of Pangkalan Jambi Village. During a period of four years from the start of the company's mentoring in 2017 to 2020, it was able to create a benefit value of 3,5 times the program costs incurred. This value is generated through the calculation of SROI (Social Return On Investment), where the investment issued by the company starting from 2017 to 2020 amounting to Rp. 690.178.439,04 managed to create benefits felt by the group equivalent to Rp. 2.417.149.336,48. The achievement of perceived benefits was largely contributed by the success in mangrove nurseries, the reduction of abraded shorelines and media exposure.

b. Pangkalan Jambi Village Government
The influence that the Pangkalan Jambi Village Government institution has is very high, this is because this institution has the responsibility to grant partnership permits between the Harapan Bersama Group and companies or with other institutions and organizations in an effort to deal with abrasion disasters and empower groups. The Pangkalan Jambi Village Government also plays a role in providing input on the strategies and policies that will be taken by PT Kilang Pertamina Internasional RU II Production Sungai Pakning in implementing CSR programs in the MEC area. In addition, the Pangkalan Jambi Village Government also has the authority to issue policies that encourage mangrove conservation activities as well as allocate village funds to support these activities. From 2017 to 2020, the Pangkalan Jambi Village Government has contributed to allocating village funds of Rp. 305.153.300 to support the development of ecotourism based on mangrove conservation and improvement of the fishermen's economy. On the other hand, the interest of the Pagkalan Jambi Village Government is considered very high seen from the gradual contribution it has made through the allocation of village funds aimed at supporting mangrove planting and ecotourism development in MEC. Another reason is to advance the village by making the reputation of the village name better.

c. Harapan Bersama Groups
The Harapan Bersama Group has a fairly high influence in expressing its aspirations to provide input to the Community Development Officer team and management regarding the planning and strategy for implementing CSR programs in the MEC area of Pangkalan Jambi Village. The group also has a high interest in managing the MEC area of Pangkalan Jambi Village, where there is a periodic performance and success assessment of the implementation of the CSR program conducted by PT Kilang Pertamina Internasional RU II Production Sungai Pakning which is

contained in the CSR monitoring and evaluation report. Monitoring and Evaluation activities are intended to see the enthusiastic development and progress of the group in developing ecotourism based on mangrove conservation at MEC Pangkalan Jambi Village.

d. Bengkalis Regency Marine and Fisheries Service
The Department of Marine Affairs and Fisheries of Bengkalis Regency has a fairly high influence on the implementation of the conservation of the MEC ecosystem in Pangkalan Jambi Village, this can be seen from the absence of regular guidance by this institution. So that the policies and strategies of PT Kilang Pertamina Internasional RU II Production Sungai Pakning in the implementation of the CSR program do not have an inherent impact on this institution. On the other hand, the importance of this institution is considered quite high, seen from its form of concern in 2019 which provided infrastructure assistance for fishing groups to help improve infrastructure facilities when they wanted to go and return from fishing.

e. DLH Bengkalis Regency and DLHK Riau Province
This institution has a high enough influence on the implementation of conservation and ecotourism activities at MEC Pangkalan Jambi Village. This is because his presence is able to change the CSR implementation strategy of PT Kilang Pertamina Internasional RU II Production Sungai Pakning with the assessment criteria in the Proklim and Setia Lestari Bumi assessment categories. The interest of this institution is considered high because the policies and strategies for implementing CSR PT Kilang Pertamina Internasional RU II Production Sungai Pakning do not have a major impact on the institution. However, it has a big impact on the encouragement of changes in the actions of the Harapan Bersama Group and the whole community in Pangkalan Jambi Village to continue to improve the quality of the MEC mangrove environment with conservation efforts.

f. Other Offices and Ministries
This group of institutions has a fairly high influence and interest in the implementation of conservation and or ecotourism in the MEC area of Pangkalan Jambi Village. This is because the role and involvement of this institution is deemed sufficient to support and support mangrove conservation activities at MEC Pangkalan Jambi Village. However, the presence of these institutions has no influence on the formulation of CSR policies and strategies carried out by PT Kilang Pertamina Internasional RU II Production Sungai Pakning in the MEC area of Pangkalan Jambi Village.

g. Pertamina Pencinta Alam (Patrapala), Public Administration Study Program Universitas Riau, Forum Gambut Muda (Forgada), Sekolah Cinta Gambut Network, Geography Education Study Program Faculty of Teacher Training and Education Universitas Sebelas Maret
This stakeholder group belongs to the category that has less high influence. This is because they are passive actors who were assigned by PT Kilang Pertamina Internasional RU II Production Sungai Pakning to help enrich the knowledge of the Harapan Bersama Group and the assisted groups outside of mangrove conservation and to help in the planting of mangroves. As for the level of importance, Patrapala has a low criteria because all conservation activities carried out by PT Kilang Pertamina Internasional RU II Production Sungai Pakning through CSR development and biodiversity do not have much impact on the continuity of activities in the Patrapala group. On the other hand, the Public Administration Study Program at the Universitas Riau, the Forum Gambut Muda (Forgada), the Sekolah Cinta Gambut Network and the Geography Education Study Program at the Faculty of Teacher Training and Education at Universitas Sebelas Maret have criteria of a fairly high level of importance because all development and implementation activities of conservation are deemed sufficient. important to them.

In more detail the results of the analysis are presented in a scoring calculation table using five questions adopted from Roslinda et al.[6], the following are the results of scoring the level of influence and interest of each stakeholder:

Tabel 2. The level of influence and interest of stakeholders in the implementation of MEC conservation in Pangkalan Jambi Village

No	Stakeholders	Influence					Total Value	Interest					Total Value
		U 1	U 2	U 3	U 4	U 5		U 6	U 7	U 8	U 9	U 10	
1	PT Kilang Pertamina Internasional RU II Production Sungai Pakning	5	5	5	5	5	**25**	5	5	5	5	5	**25**
2	Pangkalan Jambi Village Government	5	5	5	4	5	**24**	5	5	5	5	5	**25**
3	Harapan Bersama Group	3	2	5	1	5	**16**	5	5	5	5	5	**25**
4	Marine and Fisheries Services	4	2	1	2	5	**14**	3	3	1	3	2	**12**
5	Environmental services	5	1	1	1	5	**13**	5	2	3	5	5	**20**
6	Other Offices and Ministries	5	1	1	1	3	**11**	3	2	3	3	3	**14**
7	Pertamina Pencinta Alam (PATRAPALA)	2	1	1	1	2	**7**	2	2	2	2	2	**10**
8	Public Administration Study Program, Universitas Riau	2	1	1	1	4	**9**	3	2	3	3	3	**14**
9	Forum Gambut Muda (Forgada)	2	1	1	1	3	**8**	5	2	3	4	2	**13**
10	Sekolah Cinta Gambut network	2	1	1	1	3	**8**	5	2	3	4	2	**14**
11	Geography Education Study Program, Faculty of Teacher Training and Education, Universitas Sebelas Maret	2	1	1	1	3	**8**	3	1	3	3	2	**12**

Information:
U1= The ability of stakeholders to fight for his aspirations related to the implementation of conservation in MEC, U2= Contribution of facilities provided for implementation/development of MEC conservation areas, U3= Institutional/HR/Facilitator capacity assigned by stakeholders related to MEC conservation area implementation/development, U4= Budget support from stakeholders used for the implementation/development of MEC conservation area,U5= The need for cooperation in the implementation/development of MEC conservation area, U6= Perception of the importance of implementation/development of MEC conservation area, U7= Stakeholder engagement in the implementation/development of the area MEC conservation, U8= Stakeholder programs/activities related to the implementation/development of MEC conservation area, U9= Implementation/development benefits MEC conservation area for stakeholders, U10= Stakeholder dependence on the implementation/development of MEC conservation areas.

Source: Processed from primary data

The next step is to determine the quadrant position of each stakeholder by mapping their influence and interests. This is supported by in-depth interviews and Focus Group Discussions (FGD) to obtain accurate data using the scoring assessment method through questions given to stakeholders. The following is a matrix of stakeholder quadrants based on the analysis of the level of influence and level of importance discussed previously:

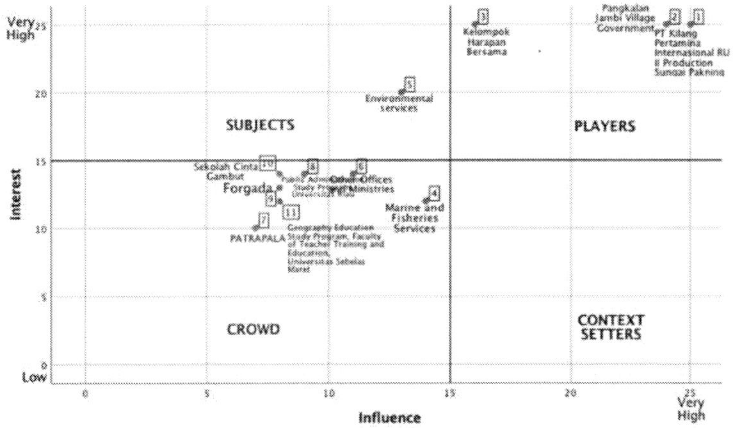

Source: Processed from primary data

Figure 4. Stakeholder classification quadrant in the implementation of conservation in the MEC area of Pangkalan Jambi Village

IOP Conf. Series: Earth and Environmental Science **986** (2022) 012019 doi:10.1088/1755-1315/986/1/012019

Based on the table above, the stakeholders who play a role in the implementation of conservation in the MEC area of Pangkalan Jambi Village are classified into 4 quadrants, namely: subjects, crowd, context setters, and players. Stakeholders who fall into the subject category are Environmental Services (DLH Bengkalis Regency and DLHK Riau Province). Then the stakeholders who fall into the crowd category are Marine and Fisheries Services, Other Offices and Ministries, Pertamina Pecinta Alam (Patrapala), Public Administration Study Program at Universitas Riau, Forum Gambut Muda (Forgada), Sekolah Cinta Gambut Network and Geography Education Study Program Faculty of Teacher Training and Education at Universitas Sebelas Maret. Meanwhile, in the context setters category, there are no stakeholders in this quadrant of the category. Finally, the players category, the stakeholders who fall into this quadrant category are Pangkalan Jambi Village Government, Harapan Bersama Group and PT Kilang Pertamina Internasional RU II Production Sungai Pakning.

Tabel 3. Forms of Stakeholder Engagement

No	Stakeholders PT Kilang Pertamina Internasional RU II Production Sungai Pakning	Engagement level	Engagement Strategy	Engagement Method
1	Pangkalan Jambi Village Government	Collaborate	Manage closely and influence actively	Coordination periodically using two-way communication, patnerships, joint ventures & projects, collaborative platforms.
2	Harapan Bersama Groups	Empower	Manage closely and influence actively	Training, conducting regular meetings, two-way communication, provision of conservation support facilities and infrastructure and capacity building for the Harapan Bersama Group. Integration of stakeholders into strategy and operations of the organisation.
3	Marine and Fisheries Services	Consult/Involve	Keep satisfied	Intense discussion, involve in visiting & survey conservation areas , focus group discussions, online engagement tools.
4	Environmental services (DLH Bengkalis Regency dan DLHK Riau Province)	Consult/Involve	Keep satisfied	Focus group discussions, involve in monitoring conservation areas, limited meetings.
5	Other Offices and Ministries	Consult	Keep satisfied	Workshop, surveys and public meetings.
6	Pertamina Pencinta Alam (PATRAPALA)	Inform	Keep informed	Whatsapp, Telephone, and letter.
7	Public Administration Study Program, Universitas Riau	Consult	Keep satisfied	Surveys and focus group discussion.
8	Forum Gambut Muda (Forgada)	Collaborate	Manage closely and influence actively	Live dialogue and maintain intensive two-way communication and training.
9	Sekolah Cinta Gambut network	Collaborate	Manage closely and influence actively	Live dialogue, maintain intensive two-way communication, training and online collaborative platforms.
10	Geography Education Study Program, Faculty of Teacher Training and Education, Universitas Sebelas Maret	Consult	Keep satisfied	Surveys and focus group discussion.

Source: Processed from primary data

The results of the matrix modeling require supporting strategies to increase stakeholder involvement. This is done by increasing the level of involvement of stakeholders who have a low level of interest/influence and interest. Meanwhile, for stakeholders who have higher levels, it is necessary to maintain relationships by maintaining and increasing the level of more comprehensive engagement communication. The Harapan Bersama Group and Pangkalan Jambi Village are "Players" who have a very high interest in handling abrasion disasters through conservation in the MEC area and are supported with a very high influence to influence the planning, strategy, implementation, monitoring and evaluation as well as publication of conservation activities. Thus, it is necessary to maintain the relationship between the two parties by using an active influence strategy and carefully managing the cooperative relationship. The method that can be used to build engagement is by two-way participatory communication followed by routine coordination, common purpose, forming a joint collaboration forum and carrying out activities together according to their roles and capacities supported by the creation of mutually beneficial cooperative relationships. The level that will be used to encourage the involvement of Pangkalan Jambi Village is to collaborate, while for the Harapan Bersama Group is to empower.

DLH Bengkalis Regency and DLHK Riau Province as well as the Department of Marine Affairs and Fisheries of Bengkalis Regency are representatives of stakeholders at the level of government institutions, which in the case of this study have an important role in the implementation of conservation activities and handling abrasion. Therefore, this institution is important to be involved and useful as a means of consultation in providing advice and input on all actions that have been taken by PT Kilang Pertamina Internasional RU II Production Sungai Pakning in collaboration with the Harapan Bersama Group and the Pangkalan Jambi Village Government in developing mangrove conservation methods. and ecotourism areas. The method of engagement that can be used by companies is limited two-way communication according to the selected subject matter. Furthermore, assessing the capacity of the institution of the Public Administration Study Program at Universitas Riau, Geography Education Study Program Faculty of Teacher Training and Education at Universitas Sebelas Maret as well as other offices and ministries provide opportunities for companies to be able to manage and improve cooperative relationships through consultation actions with the intent and purpose to the three institutions can provide more voluntary support to increase the capacity and quality of mangrove conservation implementation and provide input on existing local ecotourism management. The method of engagement that should be carried out is to establish two-way limited communication, followed by conveying the company's constructive interest in developing the MEC conservation area in Pangkalan Jambi Village. This can be done by establishing special group discussions, workshops, surveys and public meetings. Meanwhile, Pertamina Pecinta Alam (PATRAPALA) is a stakeholder whose involvement is passive because its duties and responsibilities come from the company's command to make it a resource for training guides with less strong institutions than stakeholders from educational and government institutions. Thus, the level of involvement that will be carried out by the company is still limited to informing the publication of conservation-based ecotourism development in the MEC area in Pangkalan Jambi Village as an effort to offer more medium-term collaboration.

Meanwhile, the Sekolah Cinta Gambut Network and Forum Gambut Muda (Forgada) are groups that are the result of the community development development program conducted by PT Kilang Pertamina Internasional RU II Production Sungai Pakning in the field of education. The portions and roles of these two groups that were previously passive are now managed by increasing the intensity of collaboration to foster aspirations in the form of bright ideas to help companies in the development of conservation-based ecotourism in the MEC area of Pangkalan Jambi Village. Therefore, companies need to carefully manage the communication that is built through training activities, direct discussions, and creating online collaboration platforms in the form of digital learning applications and so on.

4.3. Stakeholder Engagement Program Results

After analyzing the involvement of stakeholders in the implementation of the conservation of the MEC ecosystem in Pangkalan Jambi Village which is now also developing into an ecotourism area. There are several programs that are the result of multi-stakeholder collaboration to support the handling of abrasion in the MEC area. The programs are:

a. Mangrove tree planting involving cross-sectoral stakeholders such as the Governor of Riau, temporary person in charge of the Regent of Bengkalis Regency, GM of PT Kilang Pertamina Internasional RU II Production Sungai Pakning, Office of Library and Archives of Bengkalis Regency, Department of Environment of Bengkalis Regency, Department of Environment and Forestry of Riau Province, BAPPEDA of Bengkalis Regency, Sub-district of Bukit Batu, Sub-district of Bandar Laksamana, Sub-district of Siak Kecil, Department ofTourism, Culture, Youth and Sports (Disparbudpora) of Bengkalis Regency, the Department of Trade and Industry of Bengkalis Regency and many other stakeholders are involved. This involvement is integrated in 1 ceremonial activity for the inauguration of MEC and its ecotourism in 2020 by the Governor of Riau. However, periodically planting mangroves every year starting from 2017 until now involves at least 3 stakeholder actors outside PT Kilang Pertamina Internasional RU II Production Sungai Pakning, namely the Harapan Bersama Group and the Pangkalan Jambi Village Government.
b. Fisherman Wharf Construction Program
It is a program initiated by the Bengkalis Regency Maritime Affairs and Fisheries Service which collaborates with the Harapan Bersama Group which is no exception also coordinating with the company and the Pangkalan Jambi Village Government to build a fishing pier in the form of building a boat mooring on a river that juts into the estuary, construction of 1 MCK infrastructure and 1 saung infrastructure. In addition, the Pangkalan Jambi Village Government previously in 2017 had also contributed to the stage of building a boat mooring that jutted into the sea.
c. Ecotourism Development Program
It is a collaborative program which reflects the company's intense involvement with the Harapan Bersama Group and the Pangkalan Jambi Village Government in building mangrove wood tracks, tourist shelters, boat moorings, making coastal protection sheeting on fishing roads, building mangrove tourism towers and many more. At the same time, increasing preparedness, mitigation, recovery and development efforts in the MEC coastal area as a result of the encouragement of the criteria for the assessment of the DLH Bengkalis Regency and DLHK Riau Province in the Climate Village Program (Proklim) and Setia Lestari Bumi.
d. Breaker Equipment Development Program (APO)
This program is implemented by collaborating between the company and the Harapan Bersama Group in the practice of installing APO, starting from the application of ordinary type hybrid engineering to modification to TRIMBA.

5. Conclusion

Programs and activities to support the conservation efforts of the MEC ecosystem in Pangkalan Jambi Village have involved the contributions of various cross-sectoral stakeholders. However, this involvement is still limited to stakeholders who are representatives of actors from the government and civil society organizations. As for the involvement of the private sector, it can be said that it is less heterogeneous, where only one actor plays a role and is actively involved in the conservation activities. This private actor is PT Kilang Pertamina Internasional RU II Production Sungai Pakning. Although many actors are involved, this is only able to contribute to the stages of disaster management at the level of prevention, mitigation, recovery and development. While the stages of emergency preparedness and response have not been provided by the company, either in the form of education and training. Where the company is only limited to providing education related to the readiness and feasibility of ecotourism opening infrastructure that is safe for tourists, as well as providing tricks for groups to save tourists in the event of an accident such as drowning.

References

[1] Greenwood M 2007 Stakeholder engagement: Beyond the myth of corporate responsibility *J. Bus. Ethics* **74** 315–27
[2] ACCOUNTABILITY 2015 *Stakeholder Engagement Standard 2015*
[3] Dachi C S, Djakman C D, Akuntansi M, Ekonomi F and Indonesia U 2020 Penerapan Stakeholder Engagement dalam Corporate Social Responsibility: Studi Kasus Pada Rumah Sakit Mata X *J. Ris. Akunt. dan Keuang.* **8** 291–306
[4] Ackermann F and Eden C 2011 Strategic Management of Stakeholders: Theory and Practice *Long Range Plann.* **44** 179–96
[5] Ritabulan R, Basuni S, Santoso N and Bismark M 2015 Stakeholders Mapping of Utilization of Mangrove Forest : Case of Mangrove Utilization as Raw Material for Charcoal by Community in Batu Ampar , Kubu Raya District , West Kalimantan Province *Int. J. Sci. Basic Appl. Res.* **24** 348–59
[6] Chairil Ichsan A, Soekmadi R, Adiwibowo S and Kusmana C 2017 Peran Pemangku Kepentingan Dalam Pelaksanaan Model Desa Konservasi Di Taman Nasional Gunung Rinjani *J. Anal. Kebijak. Kehutan.* **14** 47–59
[7] Purwanto W, Hidayat R and Utami A 2020 Disaster Management of Forest and Peatland Fire Through The Berdikari Gambut Village Program in Pakning River, Bengkalis District *Soc. Humanit. Educ. Stud. Conf. Ser.* **3** 124–32

IOP Conf. Series: Earth and Environmental Science **986** (2022) 012020 doi:10.1088/1755-1315/986/1/012020

Disaster mitigation-based environmental management model: a study on ten thousand hills, Tasikmalaya City, West Java

S Fadjarajani[1*] and R As'ari[1]

[1] Departement of Geography Education, Faculty of Teacher Training and Education, Siliwangi University Tasikmalaya, Indonesia

*Email: sitifadjarajani@unsil.ac.id

Abstract. Tasikmalaya seen from the physiographical aspect has its own uniqueness because it has many hills known as the ten thousand hills. The Ten thousand has functions, namely: geological, ecological, hydrological, aesthetic, economic, micro-climatological, natural defense/fortress (buffer zone), as well as education and tourism functions. The method used in this research is descriptive with field observation data collection techniques based on Satellite Imagery analysis.. The method used in this research is descriptive with field observation data collection techniques based on Satellite Imagery analysis. Data analysis is based on satellite imagery data processing with ArcGIS which was developed. The facts on the ground show that the extinction rate of the hill reaches 70%. One of the driving factors for the extinction of the hill was the mining of mineral. The distribution pattern and factual conditions in the field showed that 41.67% of the hills experienced extinction and 58.33% of the hills were heavily damaged by sand mining. This disaster mitigation-based environmental management model includes: 1) Zoning of hill areas based on conservation functions, 2) community participation based on economic improvement. Ten thousand hill zoning and community participation based on economic improvement can be used as role models for environmental management based on disaster mitigation in Tasikmalaya which is located in the ten thousand hill area.

1. Introduction

Tasikmalaya seen from the physiographical aspect has its own uniqueness because it has many hills. The hills are spread to the southeast from the mouth of the depression of Mount Galunggung which is north of Tasikmalaya City with varying heights. These hills in the community in Sundanese are known as "Gunung Sarewu" or Bukit Ten Thousand Tasikmalaya. According to the Geological Dictionary, hills are natural heights generally up to <300 m; hills are areas of a collection of hills. Quoted the Geologist Escher in 1925 who argued that the hills found in Tasikmalaya were the result of a major landslide in prehistoric times to the southeast of Galunggung Volcano. The hills were first called the Ten Thousand Hills (The Ten Thousand Hills) by a Dutch geologist Van Bemmelen (1949) in [1], [2] , because there were quite a number of approximately 3,684 hills at that time. Furthermore, the existence of the hill attracted attention and was popularized by several geologists such as Kusumadinata (1979), Bronto (1982), and Ahman Sya (1996) in [2].

The hill of ten thousand has provided such great benefits to the lives of the people around it. From a hydrological point of view, the existence of Bukit Ten Thousand serves as a water catchment area that will be able to maintain the stability of the source and depth of groundwater. However, with the decreasing number of the Ten Thousand Hills, groundwater sources are felt to be decreasing and their availability is getting deeper. The longer the Tasikmalaya area will become dry, arid, and water shortages, so that it no longer has an aesthetic or environmental beauty that is adequate for life. In a study in 1994 – 1995 it was found that one of the hills as a sample had a plant species richness of not less than 20 species.

The study of the Tasikmalaya area which studies the environment has been widely studied in [3]–[6]. Tasikmalaya as a location that has a fairly high level of disaster risk, it needs attention from various stakeholders to study in depth as an effort to reduce the negative impact if one day a disaster occurs. [7], [8]. However, the existence of the hills in Tasikmalaya is poorly understood about its function for human survival. The community only views the function of the hill from an economic perspective without looking at the function from the other side. Seeing mining activities and the decline in the number of hills, it is necessary to have a hill protection zoning to be used as an effort to manage the environment in a sustainable manner. The problem raised in this research is how the pattern of utilization of Bukit Ten Thousand as a function of conservation and protection zones for the use of environmental conservation in Tasikmalaya. The purpose of this study was to determine the pattern of utilization of Bukit Ten Thousand as a conservation function and a protection zone for the use of environmental conservation. The application of the results of this study is expected to be accepted by all parties as a reference for monitoring the level of damage to Bukit Ten Thousand in the context of environmental conservation.

2. Methods

The research location is in Indihiang Subdistrict and Bungursari Subdistrict, Tasikmalaya City. The method used in this research is a survey. The data collection techniques used are observation, interviews, documentation or literature studies. Research activities include collecting data, compiling and classifying data, then data analysis. The analysis was carried out descriptively based on the results of observations, interviews, and literature reviews related to environmental functions. To zoning the distribution of Bukit Ten Thousand, all hills are identified to be classified according to the criteria. Satellite imagery assistance is required for hill inventory. In addition, qualitative and quantitative data from the field are entered into the database and combined with graphical data using ArcGIS software.

3. Result and Discussion

3.1 Damage Rate on Ten Thousand Hills in Tasikmalaya

Tasikmalaya physiography is closely related to the activity of Mount Galunggung. The formation of hills scattered in Tasikmalaya is caused by the efflata of Mount Galunggung to the southeast. As a result, the research area, namely Bungursari and Indihiang Subdistricts, physiographically has the characteristics of many hills with vegetation cover tending to forest and mixed gardens. Figure 1 shows the location of Mount Galunggung and Bukit Ten Thousand. The geological structure of the research area is closely related to the presence and activity of Mount Galunggung. Most of the research area contains rocks resulting from pyroclastic deposits from the eruption of Mount Galunggung. According to experts, the formation of tens of thousands of hills scattered in Tasikmalaya is the result of the activity of Mount Galunggung which erupted in 1822. The rock content in the hills in the study area contains rocks resulting from pyroclastic deposits and belongs to the C excavation material [9]. can be used by the community to be used as land for livelihoods as a mining area of C-Gahan, which can be profitable from an economic point of view.

The Ten Thousand hills has provided great benefits to the lives of the surrounding community. The condition of the remaining hills at this time is still mostly overgrown with vegetation that is closer to forest vegetation. Ecologically, this situation provides protection against the hydrological and weather systems in Tasikmalaya. The distribution pattern of Bukit Ten Thousand is included in the pattern of uneven distribution (random pattern). In general, the farther away from Mount Galunggung, the relatively smaller size. Ten Thousand Hills is currently in poor condition, every year the ten thousand hills extinction rate is increasing. Figure 1 presents the damage level of a hill part of Bukit Ten Thousand.

The year 2009 | The year 2010 | Present condiition

Figure 1. Development of Ten Thousand Hills Damage Level

The results showed that the number of hills in Indihiang and Bungursari sub-districts was not evenly distributed. Of the six sub-districts in Indihiang Sub-district, there are three sub-districts that no longer have hills because they have disappeared, namely Parakansag Village, Sirnagalaih Village, and Panyingkiran Village with one hill remaining in a damaged condition. The extinction of the hills in the three urban villages is due to the expansion of urban areas [10]. The condition of the hills in the two sub-districts is very worrying with the number of hills being damaged and in the process of mining minerals C. The number of hills in each sub-district in Indihiang and Bungursari sub-districts can be seen in Figure 1. Facts on the ground show that there are many hills that have become extinct which are now only in name.

3.2. Ten Thousand Hills Zone Zoning For Environmental Management

Ten Thousand Hills in Indihiang and Bungursari sub-districts are in a state of extinction, as can be seen from the many abandoned excavations. This allows the degradation of environmental quality in Tasikmalaya. The management of ten thousand hills will help preserve the environment, this is in accordance with Law No. 32 of 2009 [11] concerning Environmental Protection and Management which states that the environment is a unitary space with all objects, power, circumstances, and living things, including humans and their behavior. that affect nature itself, the continuity of life, and the well-being of humans and other living things.

Zones are areas, lane areas, or zones. Zoning is a grouping of an area that has one thing in common or uniformity. Zoning in landscape science is the formation of zones or areas that have certain uniformity in an area with regional restrictions or topographical formations. This zoning is used to preserve the Ten Thousand Hills in Tasikmalaya, especially in Indihiang and Bungursari sub-districts. Zone I is a zone where the condition of the ten thousand hills that are still intact (in the form of a hill) has no mining activity. Zone II is a zone where ten thousand hills are damaged and in the process of mining minerals C. Zone III is a zone where ten thousand hills are extinct and have been converted. The zoning criteria that have been determined above, the Ten Thousand Hills Zoning Map in Indihiang and Bungursari Districts can be presented in Fig.2.

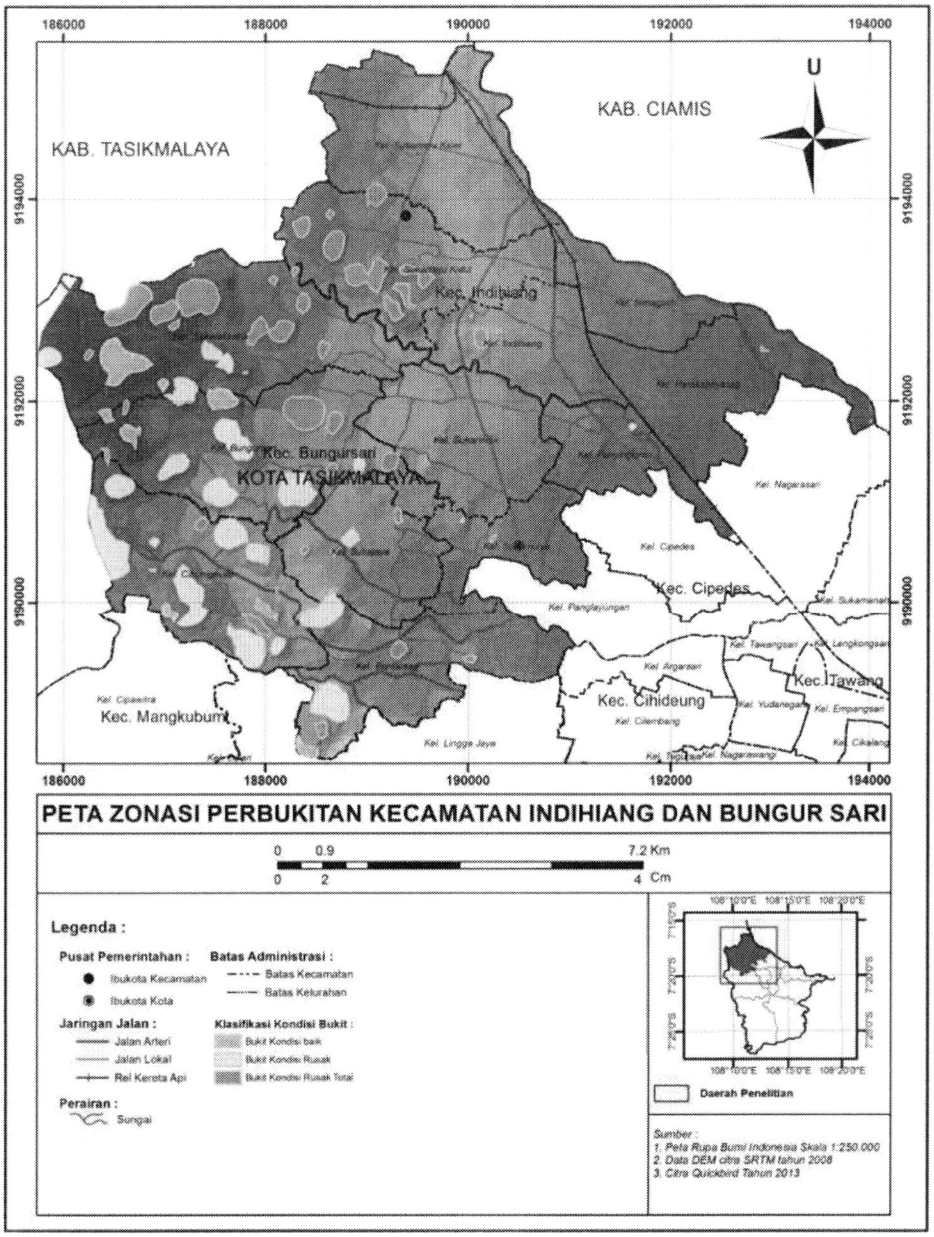

Figure 2. Ten Thousand Hills Zoning Map In Indihiang and Bungursari Districts, Tasikmalaya City

The rapid rate of hill extinction has an impact on the environmental conditions of the people in Indihiang District and Bungursari District, Tasikmalaya City. Society in general has felt the negative impact of the destruction and extinction of Bukit Ten Thousand. Some of the problems faced by the community include the depth of the dug wells has increased in depth, farming communities who work on rainfed rice fields feel that the dry season their cultivated fields experience drought, and if the rainy season the quantity of water in the rice fields recedes quickly and even landslides often occur [12] [13]. The hills, which are quite numerous, are a source of life and prosperity. This can be observed from several functions of the existence of these hills, including geological functions, ecological functions, hydrological functions, aesthetic functions, economic functions, defense functions, as well as education and tourism functions.

1. Geological Function Geologically, these hills are specific natural formations. Its existence can serve as a natural fortress from possible Galunggung lava floods. On the other hand, the rocks and sand that exist on each hill are used by the community for road paving, foundations, dams, concrete construction, and others. The sheet structure is widely used as an outboard stone. This makes it necessary to arrange utilization for mining.
2. Ecological Function From an ecological point of view, Bukit Ten Thousand has a role as an open green area that is useful for maintaining the balance of the micro-ecosystem in Tasikmalaya. Ecosystem is an arrangement of elements of the environment which is a unified whole and influences each other in shaping the balance, stability, and productivity of the environment [11].
3. Hydrological Function From the hydrological point of view, the existence of Bukit Ten Thousand serves as a water catchment area that will be able to maintain the stability of the source and depth of groundwater. Groundwater is water that is in a saturated area below the ground surface [14]. With the ten thousand hills, an aquifer will be formed, namely a pocket of water that is in the ground. Thus, from a hydrological point of view, Bukit Ten Thousand Hills is very useful for meeting water for domestic and agricultural needs, because on every hill that is still intact there are several locations of springs that can be utilized by the community. Water conservation in principle is the use of rainwater that falls to the ground for agriculture as efficiently as possible, and regulating the flow time so that there is no destructive flood and there is enough water in the dry season [15]. So with the existence of ten thousand hills, it becomes an indirect form of water conservation [16].
4. Aesthetic Function Vegetation is a plant that is found on the surface of the earth or somewhere [17]. In the mechanism of shared life, there is a close interaction, both between the individual constituents of the vegetation itself and with other organisms so that it is a system that lives and grows and is dynamic. The existence of an intact hill provides its own benefits in terms of aesthetics, considering the function of land cover vegetation (plants) that live on hill land will look green, so Tasikmalaya will look beautiful and shady. In addition, the protection of the hills shows that environmental ethics in Tasikmalaya are maintained. Considering that Environmental Ethics does not only talk about human behavior towards nature, but also talks about the relationship between all life in the universe.
5. Economical Function Economic benefits from the exploitation of Bukit Ten Thousand Most of the content of Ten Thousand Hills in Indihiang and Bungursari sub-districts is mineral C. Mineral C is non-strategic and non-vital minerals [18]. Ownership of existing hills is owned by individuals so that people who are generally farmers can easily earn income quickly by selling sand and rocks found on hill land. Various ways are done, including selling the hill, or mining it yourself and selling only the sand and rock content.
6. Education Function The educational function in Ten Thousand hills can be used as a means to communicate knowledge of earth and environmental concepts to the community or in schools. This is done so that the community understands the nature of Bukit Ten Thousand so that they behave wisely with the conditions of the surrounding environment. Bukit Ten Thousand is a natural laboratory that can be used as a place for research activities in collaboration with universities, and contacts with local experts. The collaboration between the parties is intended so that there is a complete synergy so that the research outputs produced will be more useful for people's lives. Then it can be used as a learning object for excursion studies of school children and teachers, seminars and scientific lectures.
7. Tourism Function Tourism is everything related to tourism, including the exploitation of tourist objects and attractions and related businesses in that field [19]. The tourism potential of Ten Thousand Hills cannot be separated from the condition of geodiversity which has its own

uniqueness. Existing geodiversity will be related to biodiversity and cultural diversity because the existing elements will produce a reciprocal relationship as a unique phenomenon. Geopark is an area that has outstanding geological elements including archaeological, ecological, and cultural values in it where local people are invited to participate in protecting and enhancing the function of natural heritage. Geopark is a tourism management concept in accordance with the current condition of Ten Thousand hills.

8. Defense Function Mount Galunggung is an active volcano type A, namely a volcano that has experienced a magmatic eruption at least once after 1600. Mount Galunggung has erupted after 1600. Its eruption history was recorded in 1822, 1894, 1918 and finally 1982 [20]. The Galunggung eruption belongs to the VEI (Volcano Explosion Index) 4-5 which is characterized by a plume of 10 - 25 km and a volume of > 0.1 km3. Eruptions that occur at Galunggung are usually accompanied by vertical air columns, lava flows, lava flows and pyroclastic flows or hot clouds.

4. Conclusion

From the results of the research, it is known that in the Indihiang sub-district there are 22 hills in intact condition and 9 hills that are damaged and are being exploited. For Bungursari District, there are 56 hills that are still intact and 40 hills in damaged condition and are being exploited. The factual conditions in the field show that 41.67% of the hills are damaged, while 58.33% of the hills are in the intact category but experience a decrease in number every time due to the sand mining process. The distribution pattern of Bukit Ten Thousand is included in the pattern of uneven distribution (random pattern). In general, the farther away from Mount Galunggung, the relatively smaller size. Efforts to protect hills by forming hill zoning for groundwater resource protection areas are expected to prevent environmental crises, especially water scarcity in Indihiang District and Bungursari District, Tasikmalaya City. The existence of an understanding of the importance of the preservation of the hills and the protection zone of Bukit Ten Thousand which are still intact can form global citizenship with the principle of eco-efficiency for sustainable development, so that the community will not experience the negative impact of the extinction of the Bukit Ten Thousand.

References

[1] Fadjarajani S and As'ari R 2016 "Pemanfaatan Bukit Sepuluh Ribu untuk Pengembangan Model Pembelajaran Berbasis Lingkungan," *Pros. Semin. Nas. Geogr. UMS 2016 "Upaya Pengurangan Resiko Bencana Terkait Perubahan Iklim,"* pp. 487–494

[2] As'ari R 2018 "Pengetahuan dan Sikap Masyarakat dalam Melestarikan Lingkungan Hubungannya dengan Perilaku Menjaga Kelestarian Kawasan Bukit Sepuluh Ribu di Kota Tasikmalaya," **4** 1

[3] Ramadhan Q S, Sianipar J Y, and Praptopo A K 2016 "Volcanostratigraphic Approach for Evaluation of Geothermal Potential in Galunggung Volcano Volcanostratigraphic Approach for Evaluation of Geothermal Potential in Galunggung Volcano,"

[4] Gufroni A I, Rachman A N, Hiron N, and Malik Y A 2013 "Implementasi Google Maps API dalam aplikasi mobile penghitung jarak aman dari dampak kemungkinan letusan Gunung Galunggung," *Semin. Nas. Apl. Teknol. Inf.*, **1** pp. 12–16

[5] Suryana, Parikesit, and Iskandar J 2018 "Struktur Vegetasi Kawasan Hutan Pada Zona Ketinggian Berbeda di Kawasan Gunung Galunggung Kabupaten Tasikmalaya Jawa Barat," *J. Ilmu Lingkung.* **16,** 2 p. 130

[6] As'Ari R, Rohmat D, Maryani E, and Ningrum E 2019 "Identification of Galunggung Volcano Potential Area for Geographic Education Field Laboratory Development," *IOP Conf. Ser. Earth Environ. Sci.* **286,** 1

[7] Suryo I and Clarke M C G 1985 "The occurrence and mitigation of volcanic hazards in Indonesia as exemplified at the Mount Merapi, Mount Kelut and Mount Galunggung volcanoes," *Q. J. Eng. Geol. Hydrogeol.* **18,** 1 pp. 79–98

[8] Siswanto T and Hartini 2015 "Penentuan Desain Kriteria Pada Aplikasi Mitigasi Bencana," in *KOnferensi Nasional Sistem & Informatika 2015*, pp. 9–10
[9] Swilling M 2014 "Social Dynamics : A journal of African studies Reconceptualising urbanism , ecology and networked infrastructures," no. October 2014, pp. 37–41
[10] Fadaee S "Post-contentious politics and Iran ' s first ecovillage," *Int. J. Justice Sustain.*, **9839**, no. March, 2016
[11] Indonesia 2009 *Law of the Republic of Indonesia Number 32 of 2009 concerning Environmental Protection and Management.*
[12] Budiharjo E 1991 *Arsitektur dan Kota di Indonesia.* Yogyakarta: Gajah Mada University Press
[13] Hawasly M, Corne D, and Roaf S 2014 "Social networks save energy : optimizing energy consumption in an ecovillage via agent-based simulation," no. October 2014, pp. 37–41
[14] Asdak C 2002 *Hidrologi dan pengelolaan daerah aliran sungai.* Yogyakarta: Gadjah Mada University Press
[15] Arsyad S 2012 *Konservasi Tanah dan Air*, III. Bogor: IPB Press
[16] Keraf A S 2005 *Etika Lingkungan HIdup*. Jakarta: Kompas
[17] Suharini E and Palangan A 2014 *Biogeografi*. Yogyakarta: Penerbit Ombak
[18] Sukandarrumidi 2018 *Bahan Galian Industri*. Yogyakarta: Gadjah Mada University Press
[19] Fadjarajani S, Hendriawan N, and Asari R 2019 "Modeling of Ecotourism Development in Tasikmalaya District , West Java,", **306** no. Isseh 2018 pp. 112–115.
[20] As'ari R, Rohmat D, Maryani E, and Ningrum E 2019 "Identification of Galunggung Volcano Potential Area for Geographic Education Field Laboratory Development Identification of Galunggung Volcano Potential Area for Geographic Education Field Laboratory Development,"

Identification of flood prone areas based on Geographic Information System (GIS): a case study of Buntulia District, Pohuwato Regency, Gorontalo Province

Wandi Asari , Sri Maryati *, Syahrizal Koem
Department of Earth Science and Technology, Universitas Negeri Gorontalo

*Email : sri.maryati@ung.ac.id

Abstract. This study aims to identify flood-prone areas in the Buntulia district. The types of data used in this study are primary data and secondary data. Data analysis techniques used to determine flood-prone areas are scoring and overlaying of three parameters, namely, slope maps, rainfall maps, and land use maps. The results of this study are to determine flood-prone areas based on Geographic information systems (GIS) in Buntulia District, Pohuwato Regency, Gorontalo Provinsi. To find out the number of flood-affected areas, it is carried out using a mapping of flood-prone areas, so that it is divided into four levels of flood vulnerability, namely the very vulnerable level of 1275 hectares (2.93%), the vulnerable level of 4493 hectares (10.35%), moderate level vulnerable area of 21406 hectares (49.31%), and the level of the non-vulnerable area of 16221 hectares (37.37%).

1. Introduction

Natural disasters in Indonesia have a high potential impact on infrastructure damage, economic losses, and casualties. The disaster that occurred is a serious concern. The increasing risk of disasters causes concern. Geographically, Indonesia is located in the ring of fire, increasing the risk of volcanic eruptions, storms, floods, hurricanes, earthquakes, and tsunamis. The increase in geological and hydrometeorological disasters impacts the environment, life, and economy [1,2].

Disaster events in Indonesia show an increase in its symptoms, such as extreme rain or rain with high intensity. A flood is an event where the water in the channel increases and exceeds its carrying capacity. Floods that occur have distinctive characteristics. There are various floods, namely upstream, extreme rain, postal, and flash floods. Flash floods occur due to high rainfall, with flooding events in about 5-6 hours. In addition, flash floods and heavy rains can trigger landslides [1].

One of the impacts of the flood is the inundation of residential areas and agricultural land. Other impacts of floods include economic losses and infrastructure damage. Floods are natural disasters that are usually caused by both natural and human factors. One of the factors causing the flood disaster in the Buntulia Sub-district comes from careless management of waste by the community. Land use that is not following conservation regulations has resulted in reduced vegetation in the Buntulia Sub-district that triggers the incidence of the flood.

Natural disasters hit almost all regions in Indonesia, one of which is Gorontalo Province. Gorontalo Province has five districts and one municipality, Boalemo District, Bone Bolango District, Gorontalo District, Gorontalo Utara District, Pohuwato District, and Gorontalo Municipality. Every district and municipality in Gorontalo Province has an area often hit by floods during the rainy season.

Based on Statistics Indonesia in Pohuwato District (2018), Buntulia District is a flood susceptible area. Buntulia District consists of seven villages, namely Buntulia Utara, Taluduyunu, Hulawa, Buntulia Tengah, Karya Indah, Sipatana, and Taluduyunu Utara. Three villages in Buntulia Sub-district that often experience floods are Taluduyunu, Hulawa, and Karya Indah. These areas are very susceptible to flood during the rainy season. Buntulia Sub-district is traversed by two major rivers, the Taluduyunu River and the Botudulanga River, which flow into Tomini Bay. Most of the economic activities of the Buntulia Sub-district community are in the form of plantations and mining. The expansion of rural activities, including industry, economy, settlements, and the hydrological cycle, has resulted in several areas becoming susceptible to floods.

According to the community, the flood in Buntulia District overflowed up to 50 centimeters when high rainfall occurs within 1-2 hours, inundating several villages around the riverbanks. One of the efforts to overcome the impacts of the flood is understanding the causes of flood and flood susceptible areas that depend on the area's climatological, hydrological, and physical characteristics.

An area is classified as susceptible to flood if it has high rainfall intensity, low soil capacity or water-saturated soil, impermeable surface, and damaged forest conditions [3]. One of the disciplines that enable to map in flood studies is Geographic Information Systems (GIS), which is applied to identify and map flood susceptible areas.

2. Methods

2.1 Study Location

This study was conducted in Buntulia Sub-district, Pohuwato District, Gorontalo Province, with 434.08 Km^2. Buntulia Sub-district administratively consists of seven villages namely, Buntulia Tengah, Buntulia Utara, Hulawa, Karya Indah, Sipatana, Taludyunu, and Taluduyun Utara, which are shown in Figure 1.

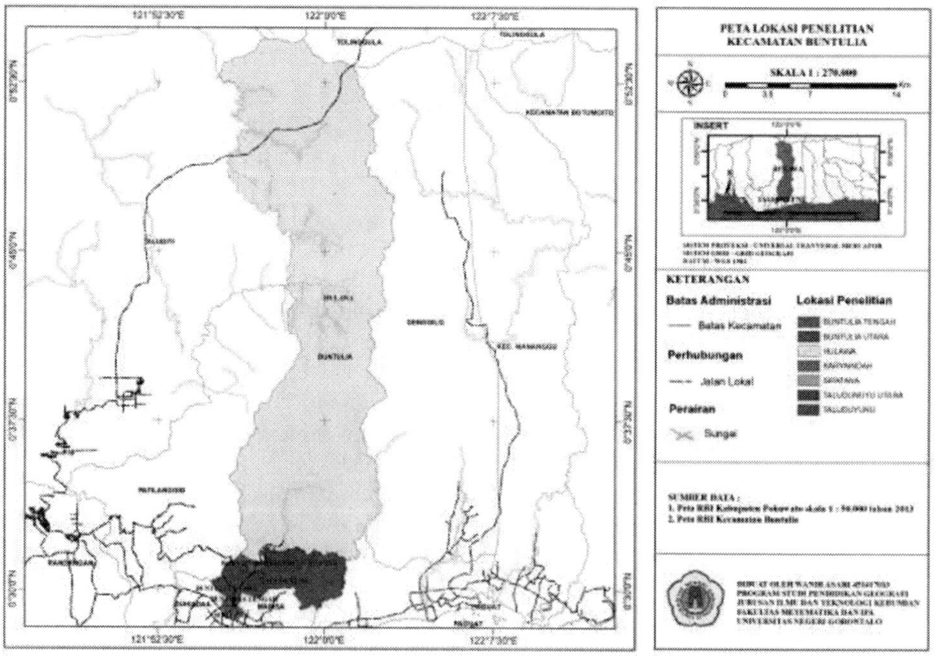

Figure 1. Study Location Map

2.2 Types of Data
This study used primary and secondary data

2.2.1 Primary Data
a. Community perception of flood
b. Community interviews about flood susceptible issues
c. Community interviews about flood characteristics that include return period (frequency of flooding), inundation depth, and duration of inundation.

2.2.2. Secondary Data
a. Slope map to classify slope classes in each area in the study location.
b. Rainfall map to classify rainfall in the study location.
c. Land use map to classify land in the study location.

Table 1. Types and Sources of Data

No	Types	Sources	Purpose
1	Slope	DEMNAS Imagery of Pohuwato District	classify slope classes in each area in the study location
2	Rainfall	Meteorological, Climatological, and Geophysical Agency of Bonebolango	classify rainfall in the study location
3	Land Use	Geospatial Information Agency	classify land in the study location

(Source: Author)

2.3 Data analysis method
The method used to determine flood susceptible areas was scoring on each variable and direct interviews with the community in the study location. The scoring method for each variable resulted from multiplying and adding the factors and variables used to determine the flood susceptible area by subtracting the highest and lowest values with the previously determined hazard class. Meanwhile, observation is a method of collecting data systematically through observation and recording the phenomenon in the study. Researchers conduct direct observations on the study location and interacted intensely with the community during data collection

2.3.1. Score and Weighting
The distribution of flood susceptible areas was determined by scoring the value of each parameter. These parameters were slope, rainfall, and land use that were given a different score according to the influence of the parameter on the effect of a flood. The score value was adjusted to the conditions and effects of flooding. The scoring was defined as giving a score to each class. According to Suherlan (2001)[4], the scoring was based on the impact of the class on the magnitude of the flood. Table 2 shows the weighting factors for the classification of flood susceptible areas in the Buntulia Sub-district.

Table 2. Rating of Flood Susceptible Classification

No	Parameter	Skor	Bobot
1	Slope	1-5	30
2	Rainfall	1-5	30
3	Land Use	1-5	30

Based on Table 2 each parameter was given a score of 1-5, and each parameter was given a weight of 30 because each parameter had an equal effect. In determining flood susceptible areas, each score and weight were summed to obtain the highest and lowest values used as the reference in dividing the number of classes and class intervals of flood susceptibility.

2.3.2. Total Score of the Summation

After all the parameters in determining flood susceptibility are overlaid, the next step is the summation of the total scores to obtain the total value of all overlaid parameters. The equation used to calculate the total score in determining flood susceptibility [5] was:

Flood susceptibility = 30(CH) + 30(KL) + 30(PL)

where: CH = Rainfall

KL = Slope Class

PL = Land Use

30,30,30, = Weighting Factor/Weight

The weighting factor for each parameter was given the same value, namely 30, because each parameter had an equal effect. In rainfall, a weighting factor of 30 was given with an influence on flood susceptibility; the higher the rainfall, the greater the incidence of flooding. The slope was given a weighting factor of 30. This physical land parameter has a significant effect on flood susceptibility. The steeper the slope in heavy rain, the rain always flows to low areas until it reaches plain areas. Land use is an influential factor in flood susceptibility. It has a weighting factor of 30 since excessive land use can cause the surrounding vegetation to decrease, harming the environment, and causing flood susceptibility.

2.3.3. Class Interval Determination

The number of classes used in this study was 4 (four). The score of each physical parameter had been multiplied by the weighting factor. The level of flood susceptibility is obtained if the value of flood susceptibility is high. The score is the result of the sum of the highest scores of each parameter—the lower the flood susceptibility score, the lower the flood susceptibility level. The flood susceptibility score for each parameter can be seen in Table 3.

Table 3. Flood Susceptibility Criteria

No	Class Interval	Class Interval
1.	90-157	Non-susceptible
2.	158-224	Moderately Susceptible
3.	225-291	Susceptible
4.	292-358	Very Susceptible

(Sumber: Author's Calculation Result)

The formula used to determine the class interval class is: :

$$\text{Class Interval} = \frac{\text{Highest Score} - \text{Lowest Score}}{\text{Number of Classes}}$$

$$\text{Class Interval} = \frac{358 - 90}{4} = 67$$

After the overlay, the determination of the total score was beneficial in determining the number of flood susceptibility classes and the flood susceptibility class intervals. The class interval was used to classify the overlay results into the level of flood susceptibility. The number of classes and class intervals can be calculated using the Sturges equation, as follows:

$$k = 1 + 3.3 \log n$$

Source : [6]
where:
k = number of classes n = number observation

$$R = \text{Maximum Value} - \text{Minimum Value}$$

where: R = the difference between the largest and the smallest data

$$i = \text{range : class}$$

The value of an area's susceptibility to flooding is determined from the total score of each flood parameter. Areas that are very susceptible to flooding will have a high total score and vice versa. After obtaining the flood susceptibility value, the map is analyzed using overlay analysis. The results of the analysis are presented in the form of a flood susceptibility map.

3. Results And Discussions

3.1 Results

Analysis of the level of susceptibility to flooding was the output of this study. Analysis of the level of susceptibility to flooding was carried out to provide information to the surrounding community to be more careful and alerted to the flood potential. Analysis of the level of flood susceptibility was carried out by combining the three parameters used in this study, namely the slope class, rainfall, and land use in Buntulia Sub-district. The rainfall data were obtained from the Meteorological, Climatological, and Geophysical Agency of Bonebolango District in the form of rainfall data from 2016-2020. The slope data were obtained from the DEMNAS Imagery of Pohuwato District, and the land use data were obtained from the Geospatial Information Agency. Each data was georeferenced and then digitized to obtain new information in the form of vector data.

After the data for each parameter became vector data, it was overlaid using union overlay analysis before being scored and weighted. The score given for each parameter ranged from 1,2,3,4,5. The higher the score, the higher the level of susceptibility. The weighting was based on the effect of the parameters on the flood potential.

3.1.1 Slope

Table 4. Score and Weight of Slope

Slope Class	Slope%	Score	Weight	Weight x Score	Area (Ha)	Area %
Flat	0-8%	5	30	150	1521	3.50%
Plain	8-15%	4	30	120	2283	5.26%
Moderately Steep	15-25%	3	30	90	4410	10.16%
Steep	25-45%	2	30	60	14031	32.32%
Very Steep	>45%	1	30	30	21163	48.75%
Total					**43408**	**100.00%**

Slope class is an essential parameter in analyzing the level of susceptibility to flooding with a weight of 30%. The very steep slope class with a slope of >45% dominates the Buntulia Sub-district area with an area of 21163 Ha or about 48.75% of the total area of the Buntulia Sub-district. Areas with a very steep slope have a high flood potential.

In general, the morphological description of the Buntulia Sub-district administrative area consists of mountainous land, hills, and lowlands. The description of the morphology of the Buntulia Sub-district is based on the results of field observations. This description is also supported by the results of slope class mapping using the DEMNAS Imagery of Gorontalo in 2020. Class parameters are the most critical parameters in analyzing the flood potential. The slope class classification in this study used DEMNAS Imagery of Gorontalo in 2020.

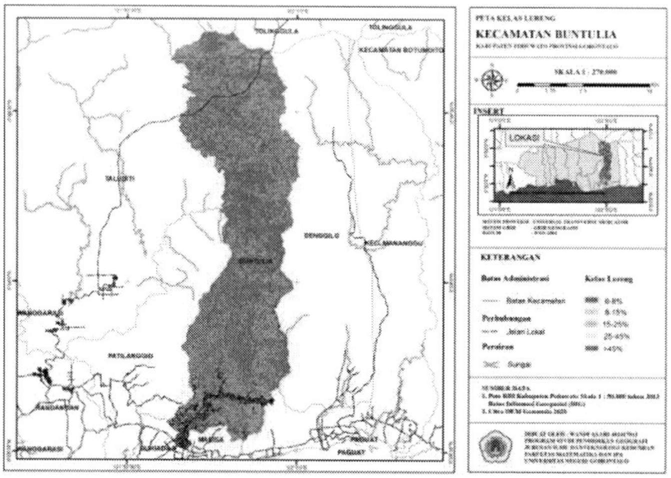

Figure 2. Slope Map of Buntulia Sub-district

3.1.2 Rainfall

Table 5. Rainfall Score and Weight

Description	Rainfall (mm/year)	Score	Weight	Weight x Score	Area (ha)	Area %
Very wet	>3000mm	0	0	0	0	0
Wet	2501mm-3000mm	3	30	90	4350	10.02%
Moderate/humid	2001-2500mm	3	30	90	7451	17.16%
Dry	1501mm-2000mm	2	30	60	16452	37.90%
Very dry	<1500mm	1	30	30	15155	34.91%
	Total				**43408**	**100.00%**

Rainfall is the most critical parameter in the level of susceptibility to flooding, with a weight of 30%. The prediction is that if the rainfall is high, the potential for flooding is higher. The rainfall in Buntulia Sub-district was in 2001-2500 mm/year, which is included in the moderately humid category. Rainfall in Buntulia Sub-district can be classified in the range of 2001-2500 mm/year that makes it categorized as moderate/humid with an area of 10.02%, with the prediction that if high rainfall occurs, it can cause flooding.

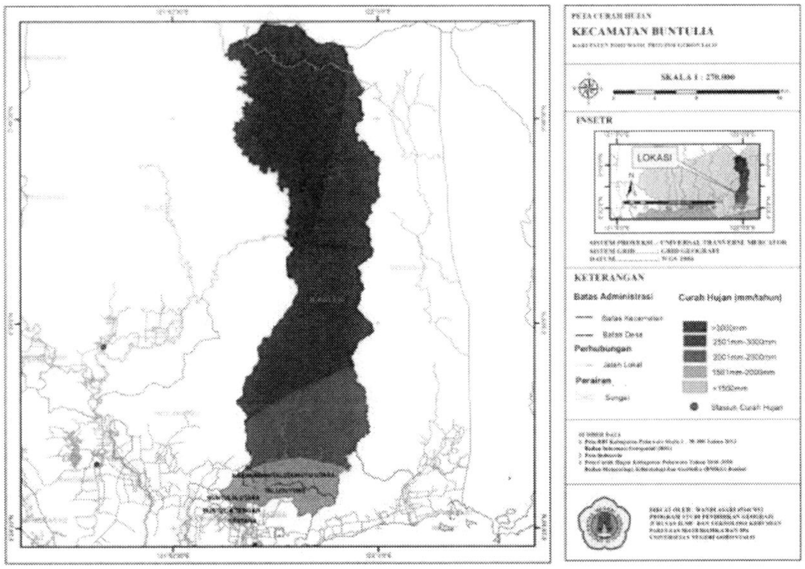

Figure 3. Rainfall Map of Buntulia Sub-district

3.1.3 Land Use

Table 6. Land Use Score and Weight

Land Use	Score	Weight	Weight x Score	Area (ha)	Area %
Forest	1	30	30	39498	90.99%
Sand	1	30	30	2	0.004%
Plantation	4	30	120	803	1.85%
Settlement	3	30	90	96	0.22%
Ricefield	3	30	90	205	0.47%
Shrub	3	30	90	7	0.02%
Mine	3	30	90	130	0.30%
Bare Land	2	30	60	20	0.05%
Moor	4	30	120	2647	6.09%
		Total		**4340**	**100.00**

Land use parameters are essential parameters in determining the level of susceptibility to flooding with a weight of 30% with a prediction that massive land management by cutting down trees, and excessive land exploration has a high flood potential because the trees that hold water and prevent erosion are reduced. The table illustrates that moor and plantation dominate the land use with a percentage of 6.09% and 1.58% respectively of the total area of Buntulia Sub-district.

In general, the morphology of the Buntulia Sub-district consists of mountainous land, hills, and lowlands. This description is also supported by the results of land use mapping using the data obtained from the Geospatial Information Agency. Land use in Buntulia Sub-district was classified into nine land use classes: forest, plantation, settlement, rice field, shrubs, mine, bare land, dry land, and sand. Settlements, ricefields, plantations, and moor were scattered in Hulawa Village, Karaya Indah Village, Taluduyunu Utara Village, Taluduyunu Village, Buntulia Utara Village, Buntulia Tengah Village, and Sipatana Village. Meanwhile, mines, sand, shrubs, bare land, and forests were scattered in Hulawa Village, Taluduyunu Utara Village, and Taluduyunu Village. Land use located in rural areas and rivers is more susceptible to flooding.

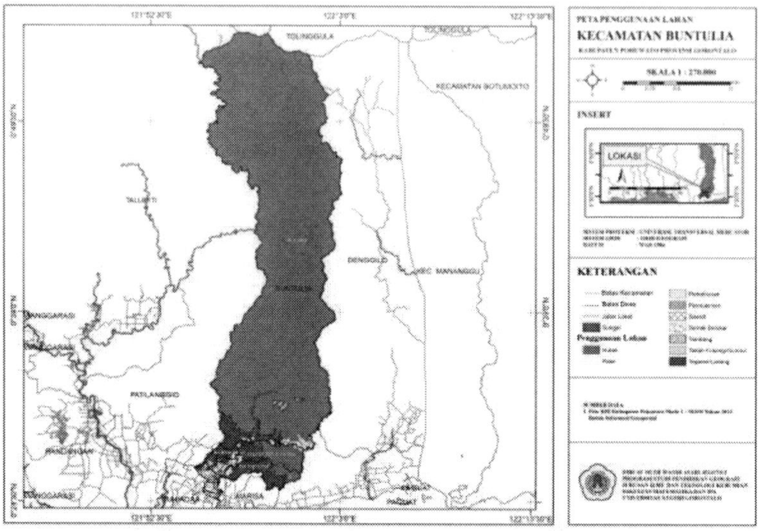

Figure 4. Land Use Map of Buntulia Sub-district

3.1.4 Total Value and Map of Flood Susceptibility Level

The total value was obtained from the sum of the scores and weights of each parameter after overlaying all the parameters used in the classification of the flood susceptibility map in the Buntulia Sub-district.

Table 7. Parameter Total Value

No	Parameter	Score Highest	Score Lowest
1	Slope Class	150	30
2	Rainfall	90	30
3	Land Use	120	30
Total		**360**	**90**

Based on observations in seven villages in Buntulia Sub-district, three villages were identified as the place where floods occurred. Thus, the class was divided into several classes of susceptibility levels using the Sturges Equation.

$k = 1 + 3.3 \log n$

$= 1 + 3.3 \log 7$

$= 4$

where:

k = number of classes

n = number of observations

R = Maximum Value – Minimum Value

$= 360 - 90$

= 270
where : R = the difference between the largest and the smallest data
i = range : class
= 270 : 4
= 67
where: i = class interval

The susceptibility level analysis showed four categories: non-susceptible, moderately susceptible, susceptible, and very susceptible, with class intervals of 67.

Table 8. Flood Susceptibility Class

No	Class Interval	Class Category	Area (Ha)	Area %
1.	90-157	Non-susceptible	7272	16.75%
2.	158-224	Moderately Susceptible	26430	60.89%
3.	225-291	Susceptible	8326	19.18%
4.	292-358	Very Susceptible	1380	3.18%
		Total	**43408**	**100.00%**

Table 9. Flood Susceptibility in the Buntulia Sub-district

Village	Flood Susceptibility Area (Ha)			
	Non-susceptble	**Moderately Susceptible**	**Susceptible**	**Very Susceptible**
Hulawa	15075	20296	3423	773
Karya Indah	7	62	75	83
Taluduyunu Utara	786	458	235	191
Taluduyunu	342	541	481	79
Buntulia Utara	3	30	188	66
Buntulia Tengah	2	8	58	53
Sipatana	6	11	30	30
Total	16221	21406	4493	1275
Area of the Buntulia Sub-district	43408			
Percentage (%)	37.37%	49.31%	10.35%	2.93%

The flood susceptibility class in Buntulia Sub-district was divided into 4, namely non-susceptible, moderately susceptible, susceptible, and very susceptible. All seven villages had different flood susceptibility classes. The very susceptible class was found in Hulawa Village with an area of 773 ha, Karya Indah Village with an area of 83 ha, Taluduyunu Utara Village with an area of 191 ha,

IOP Conf. Series: Earth and Environmental Science **986** (2022) 012021 doi:10.1088/1755-1315/986/1/012021

Table 10. Flood Characteristics in the Buntulia Sub-district

No	Village	Flood Characteristics	
		Duration	**Depth (cm)**
1.	**Hulawa**	1-2 Jam	<50
2.	**Karya Indah**	2-6 Jam	<50
3.	**Taluduyunu Utara**	1-2 Jam	<50
4.	**Taluduyunu**	-	-
5.	**Buntulia Utara**	-	-
6.	**Buntulia Tengah**	-	-
7.	**Sipatana**	-	-

The flood characteristics in the Buntulia Sub-district had different durations and equal depth. Three villages were affected by the flood, and four villages were not affected by the flood. The villages affected by the flood were Hulawa, Karya Indah, and Taluduyunu Utara, with different durations and equal depth. Hulawa Village had a duration of 1-2 hours with a depth of <50 cm, Karya Indah Village had a duration of 2-6 hours with a depth of <50 cm, and Taluduyunu Utara Village had a duration of 1-2 hours with a depth of <50 cm. The villages not affected by the flood were Taluduyunu, Buntulia Utara, Buntulia Tengah, and Sipatana.

The overlay was conducted using union overlay analysis by involving spatial elements in the form of each parameter map and attribute. After the class and class interval were determined, the susceptibility area was calculated and mapped using three parameters, namely slope class, rainfall, and land use.

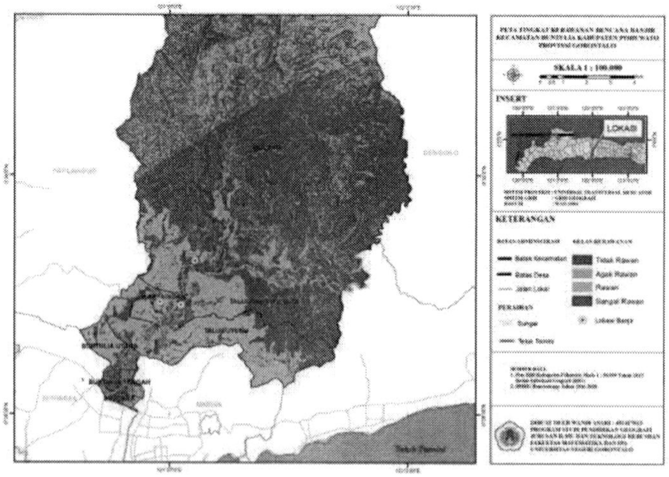

Figure 5. Flood Susceptibility Map

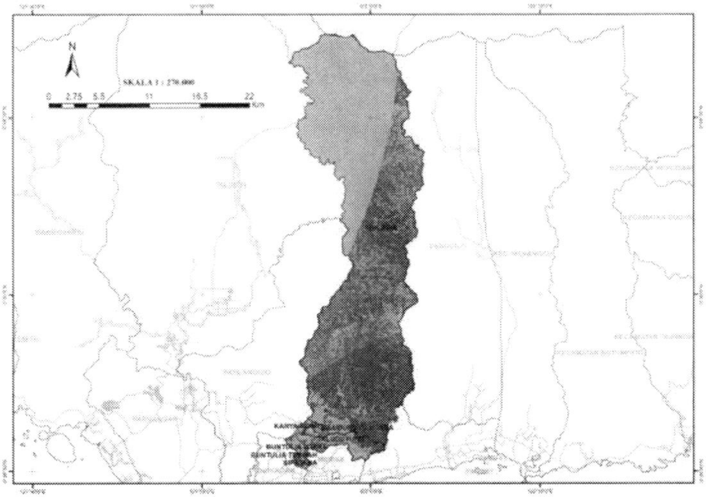

Figure 6. Flood Susceptibility Map

Thee Buntulia Sub-district is an area that is susceptible to flooding. The slope class is very steep with >45%, the rainfall was in the moderate/humid category with 2001-2500 mm/year, and the land use is dominated by moor with 6.09%, plantations with 1.85%, and mines with 130%. It is traversed by two major rivers, the Taluduyunu River and the Botudulanga River, flowing into Tomini Bay.

The results of the flood susceptibility analysis, the Buntulia Sub-district was dominated by the very susceptible category with 2.93 % of the total area symbolized by red on the map, followed by the susceptible category with an area of 10.35% symbolized by brown on the map, moderately susceptible category with an area of 49.31% symbolized by light green, and the non-susceptible category with an area of 37.37% symbolized by dark green.

3.2 Discussion

3.2.1 Flood Susceptibility Analysis

Analysis of susceptibility to flooding is a way of determining the flood disaster suscep ibility of an area. The level of flood susceptibility in the Buntulia Sub-district is described through a map of the flood susceptibility level. In this study, analyzing the level of susceptibility to flooding used descriptive qualitative analysis methods and scoring. The use of this method was to make it easier for researchers to describe flood susceptible areas in Buntulia Sub-district.

The scoring method was used to give a value for each parameter for further processing to produce data on the level of susceptibility to flooding in the Buntulia Sub-district. The parameters used in determining the level of susceptibility to flooding were slope class, rainfall, and land use obtained from the Meteorology, Climatology and Geophysics Agency (BMKG), DEMNAS Imagery in 2020, and the Geospatial Information Agency (BIG). The attribute table of those data was georeferenced for further processing.

A field survey was conducted to obtain coordinate data of flood locations spread in three villages, namely Hulawa, Karya Indah, and Taluduyunu Utara. The field survey aimed to obtain direct information through observations and interviews with the local community. Field conditions can also determine the duration and the depth of the inundation and observe the potential losses experienced by the local community.

The parameters used in this study were slope class, rainfall, and land use. In terms of slope class, the Buntulia Sub-district is a very steep area with 48.75% of its area has a slope of > 45%. Water flowing from upstream will quickly flow downwards so that residential areas located in lowlands will be affected by floods. Then the rainfall in Buntulia Sub-district is in the range of 2001-2500 mm/year with moderate/humid rainfall. Thus, a fairly high amount of rain can cause flooding since the area is traversed by two large rivers that flow to Tomini Bay. The land use in Buntulia Sub-district is dominated by dry land, plantations, and mining sites. The use of dry land, plantations, and mining on a large scale can damage the vegetation that functions as a water reservoir and erosion barrier. The plantation covers 1.85%, the dry land covers 6.09%, and the mining sites cover 0.30% of the total area of the Buntulia Sub-district.

After each parameter was georeferenced, the attribute table was converted into vector data to assess each parameter. The assessment was given with intervals of 1,2,3,4, and 5. The higher the rating assigned, the higher the potential for susceptibility. The weights were given based on the effects of parameters on the potential for flood disasters. Each parameter was given a weight of 30% since all parameters have an equal effect.

After assigning values and weights to each parameter, the union overlay analysis was conducted to obtain data. Then, the values were summed to obtain the total value used as a class division for the level of flood susceptibility. Thus, four classes of flood susceptibility were obtained, namely non-susceptible, moderately susceptible, susceptible, and very susceptible. After obtaining the number of classes, the dissolve analysis was conducted to divide the susceptibility area by a total of 4 classes in the category of flood susceptibility.

The final result showed that there were four classes of flood susceptibility levels. Very susceptible with an area of 2.93% symbolized with red, susceptible with an area of 10.35% symbolized with brown, moderately susceptible with an area of 49.31% symbolized with light green, and non-susceptible with an area of 37.37% symbolized with dark green. The very susceptible category has a small area and only affected three villages, namely Hulawa, Karya Indah, and Taluduyunu Utara.

4 Conclusions

The level of flood susceptibility in the Buntulia Sub-district was classified into four categories, namely very susceptible with an area of 1275 ha (2.93%) which spread in Hulawa Village, Karya Indah Village, and Taluduyunu Utara Village, susceptible with an area of 4493 ha (10.35%) which extend in all villages in Buntulia Sub-district, moderately susceptible with an area of 21406 ha (49.31%) which spread in all villages in Buntulia Sub-district, and non-susceptible with an area of 16221 ha (37.37%) which extend in all villages in Buntulia Sub-district. The results showed that the areas affected by the flood disaster were Hulawa Village, Karya Indah Village, and Taluduyunu Utara Village.

References

[1] Adi S 2013 Characterization of Flash Flood Disaster in Indonesia (Karakterisasi Bencana Banjir Bandang di Indonesia) *J. Sains dan Teknol. Indones.* **15** 42–51
[2] BNPB 2016 RBI Risiko Bencana Indonesia
[3] P A R 2017 *Pemetaan kawasan rawan banjir berbasis sistem informasi geografis (SIG) untuk menentukan titik dan rute evakuasi* (Makasar)
[4] Suherlan 2001 *Zonasi tingkat kerentanan banjir kabupaten bandung menggunakan Sistem informasi geografis* (Bogor)
[5] B P A 2009 *Pemetaan Lokasi Rawan dan Resiko Bencana Banjir di Kota Surakarta Tahun 2007* (Universitas Sebelas Maret)
[6] Sugiyono 2016 *Metode Penelitian Kuantitatif, Kualitatif dan R&D* (Alfabeta)

IOP Conf. Series: Earth and Environmental Science **986** (2022) 012022 doi:10.1088/1755-1315/986/1/012022

Land surface temperature (LST) and soil moisture index (SMI) to identify slope stability

S T J Putro[1*], N Arif[1] and T Sarastika[1]

[1] Department of Geography Education, Faculty of Social Science. Universitas Negeri Yogyakarta, Indonesia.

*Email: sutanto.tp@uny.ac.id

Abstract. Scientists widely use satellite images for scientific purposes, including investigation on earth science and environmental issues. Developing of many environmental models is due to replicating the natural process. Landslide is a known natural process controlled by slope stability which incorporates many parameters such as soil water content, morphology, and meteorological factor. Both LST and SMI were derived from satellite images, while SMI was the derivation of LST, meanwhile the use of both parameters in determining slope stability was rarely done. This research explores the use of LST and SMI in slope stability modeling. The LST analysis was calculated based on SEBAL (Surface Energy Balance Algorithms) using Landsat 8 imagery. The LST was then used to construct the SMI. Slope stability (FS) was calculated using the Selby model. All those variables were then cross-plotted in a regression to find the R2 value. The result shows a weak connection between FS-LST and FS-SMI with the R2 value of 9,09% and 8,16%. A stronger connection is only demonstrated in FS-Slope regression with a value of 70,98%. The weak R2 indicates that the model is not fit to calculate the FS of the Selby model. The LST and SMI were derived from satellite images and did not directly correspond to the soil characteristic as SMI was derived from LST and vegetation indices. Further empirical data collection needs to be used to build a better model on FS.

1. Introduction

1.1. Background

Methods and technologies for environmental monitoring have been developing rapidly since the 19 century. Those changes bring a significant role in how we assess the data for environmental modeling. Landslide is a phenomenon often set using modern technology such as remote sensing that coexist with GIS [1]. Many researchers have developed ways to mitigate landslides using various remote sensing data and analysis resulting from different results and advantages [2,3]. Several studies suggested using artificial intelligence and machine learning in landslide studies and monitoring [4–7]. Nonetheless, remote sensing technology may not represent the actual condition of landslides.

Landslide is an intricate natural phenomenon [8] as a part of earth`s geological cycle [9], exist as a combination as many factors [8,10] such as geology [11], geomorphology [12][13], hydrology [12], soil [12], landuse [12], and anthropogenic influence [14][15]. The complex influence of many factors contributes to the complexity of the models to determine the landslide occurrence. Multiple factors and classification on landslide triggering factors had been explained by Froude and Petley [16]. Landslide modeling and mitigation are important to reduce risk as part of landslide disaster management [17–19].

Globally, landslides have affected many aspects of the environmental process. Landslides have led to assets damage and loss, fatalities, economic failure [16], some lead to mental deterioration of the affected population [20,21]. From 1998 to 2017, 1.3 million people died, 4.4 billion victims were affected by climate and geophysical-related disasters. From those, a landslide occurred 378 times (5.2%), with 4.8 million affected (0.1%) and 18,414 deaths (1%) [22].

As a natural process of the earth, its occurrence is necessary, but on the other side, it causes some damage to life and the environment. So many researchers attempt to develop methods to anticipate the landslide occurrence and early warning system (EWS) on landslides. Those attempts depend on the parameters used to run the models [23]. Too many parameters will lead to the high resources needed to run the models. Even after running sensitivity analysis to filter the most related parameter to the study, there are possibilities where data is not available, typical of developing countries [24,25]. Those countries must decide the most available, critical, and related parameters to build environmental models.

Environmental modeling adopts information and technological advances, but the development depends on the ability of the developer to obtain them. The development of landslide modeling had been discussed elaborately by Brunsden in 1999 [26]. He construes briefly on temporal progress of landslide modeling encompassed static, dynamic, rheologic, and hydrological approaches for landslide modeling. New models have already updated the model he described in his work. The new models are getting more complicated by engaging artificial intelligence and machine learning [4–7]. Nowadays, some landslide mitigation systems apply those models in risk reduction efforts, especially in an early warning system (EWS) [27–30].

Indonesia has diverse landforms, geology, and environmental conditions that lead to various disasters in Indonesia. Thus, Indonesia is known as "the supermarket of disaster" [31]. Based on BNPB (National Disaster Management Authority/NDMA) data, the landslide has become a significant disaster in Indonesia and occupied 19.16% of total disasters from 2019 through 2020 [32]. In Yogyakarta Province, landslides occupy 54.98% (436 out of a total of 793 disaster events). The study area is prone to landslides [33,34], triggered by many factors such as geomorphology, soil, hydroclimatology, anthropology, and seismic activities [35].

Complete and detailed environmental data and resources needed for landslide modeling are scarce. Therefore, landslide potential hazard mapping needs the best and most straightforward approach. The approach calculates satellites images with an environmental parameter that plays an essential role in the landslide events. Landslide is closely related to the physical and anthropologic condition [15]. The state represents factors that affected landslides are land surface temperature (LST) and soil moisture index (SMI) [36,37]. Both parameters are quickly drawn from satellite image analysis spatially and temporally, depending on the image availability of various providers.

While LST and SMI were quickly achieved, we understand that landslide is complicated and depends on many factors. For users with limited resources, uncertainty arose whether both parameters were sufficient or not to understand the landslide occurrence. Thus in this study, we tried to determine the significance among LST and SMI towards slope stability (safety factors/fs) as one of the simplest landslide models developed by Selby [38] based on the infinite slope model.

1.2. Study Area

The study area is a trans administrative boundary watershed called Ngijo Watershed, and it has a drainage area of 61.5 Km^2 stretchings across three regencies. Ngijo watershed is the sub-watershed of Opak river and contributes to 4.4% of its total area. The main channel drainage the water has a total length of approximately 19 Km's. The site consists of rough terrain in the upper area and relatively flat in the lower area. Perrenial and intermittent streams supply the river. Thus, there will be water along the main river across the year. The study area lies at the border of Yogyakarta and Central Java Province (see **Figure 1D**). Tertiary and Quarternary materials build up this area. The tertiary materials consist of Semilir Formation (old pyroclastic materials such as tuffs and breccia), Nglanggran Formation (volcanic breccia, lava, and tuff), and Kebobutak Formation (andesitic breccia, tuff, and andesitic lava flow).

Quarternary materials called young volcanic deposits of Merapi Volcano is inhabits the flat area, consists of tuff, ash, breccia, and lava flow (see **Figure 1B**).

The geologic and tectonic forces, combined with exogenic processes, shape the landscape of this watershed into several parts, in general as a hilly and flat area. According to slope classification by van Zuidam [39], there are six out of seven class presents, only the class of more than 140 % slope is absent, with the steepest slope reaching 93%. Flat to sloping (0-15% slope) occupies the eastern part of the flat area, while the moderate to very steep slope occupies the hilly area (see slope map in **Figure 1A**).

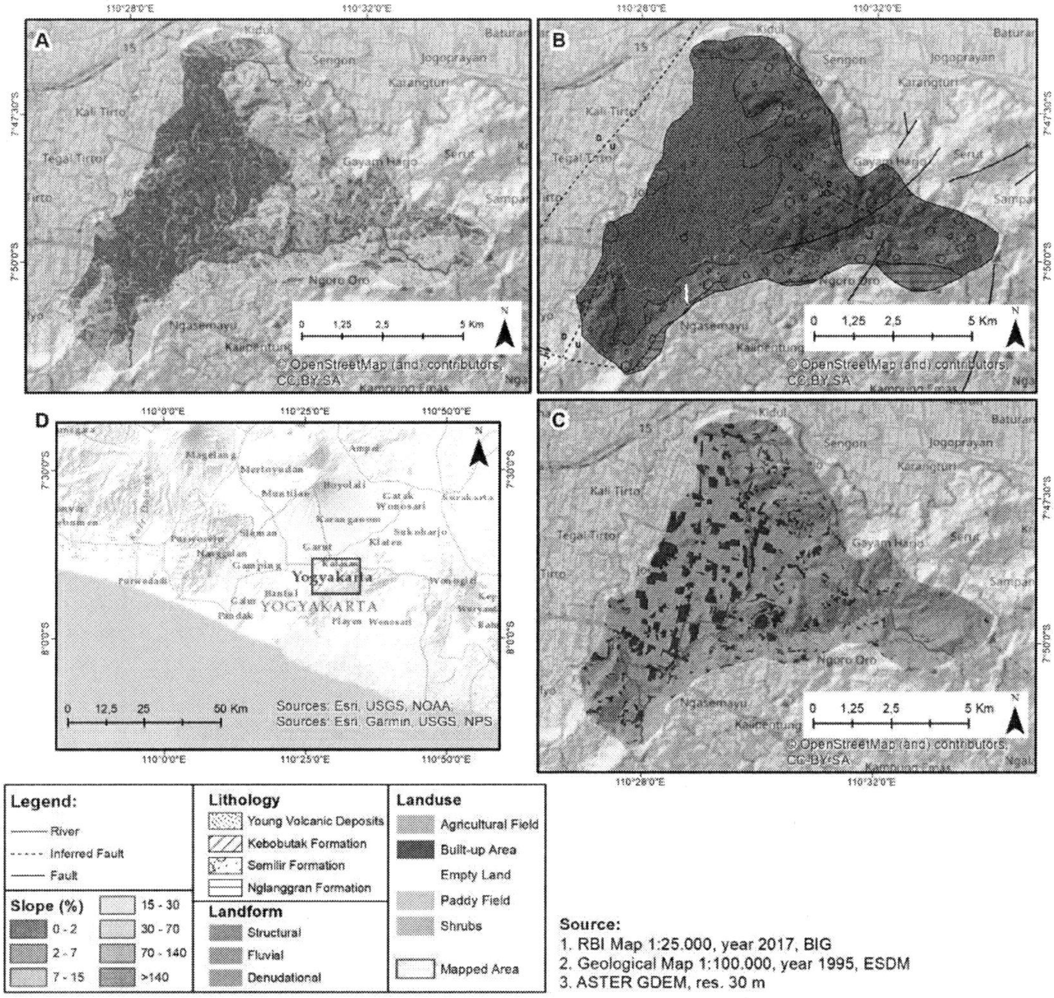

Figure 1. A) Slope map of the study area; B) The general geomorphology map of Ngijo Watershed; C) The Landuse area over the study area; D) Inset map shows the study area.

The landscape configuration is mainly built from structural, denudational, and fluvial processes (**Figure 1B**). Fault scarp is developed along the main fault with NW-SW direction. The watershed is mainly located on the footwall of the fault. Several minor faults are found in the hilly area, which controls the surface topography and the river following its fault line. The denudational process is characterized by the dissected surface controlled by the exogenic process. Some of the exogenic processes is landslide triggered by unstable slope and many other factors.

IOP Conf. Series: Earth and Environmental Science **986** (2022) 012022 doi:10.1088/1755-1315/986/1/012022

Human activities had altered the landscape of the study area. Some areas are still left intact with their origin vegetation, and the others develop into agricultural areas and settlements. Those artificial landuses (see **Figure 1C**) occupy mainly in the flat fluvial area. Nevertheless, a small number of them reside in the very steep area of Ngijo Watershed, which is prone to landslide occurrence.

2. Methods

The data used in this study were LST, SMI, and FS. Those data then crossed each other to find the significance of each value compared to the safety factor (fs). LST and SMI were derived from Landsat 8 imagery, recorded on June 25th, 2019, processed using methods developed by Sebal [40,41]. SMI was calculated based on LST data as a method proposed by Potić in 2017 [42]. There are ten steps in LST calculation as follows:

Step 1. DN value conversion to radian for thermal band (band 10 and 11) using equation 1.

$$L\lambda = MLQcal + AL \quad (1)$$

Where:

$L\lambda$ = spectral radiance (W/(m^{2}sr.µm)
$Qcal$ = pixel value (DN)
ML = rescaling constant (RADIANCE_MULT_BAND_x, where x is the band used)
AL = incrementing constant (RADIANCE_ADD_BAND_x, where x is the band used)

Step 2. DN value conversion for the non-thermal band using equation 2.

$$\rho\lambda' = MpQcal + Ap \quad (2)$$

Where:

$\rho\lambda$` = image reflectance value
Qcal = pixel value (DN)
M_P = rescaling constant (RADIANCE_MULT_BAND_x, where x is the band used)
A_P = incrementing constant (RADIANCE_ADD_BAND_x, where x is the band used)

Correction of $\rho\lambda$ using sun angle correction by having equation 3.

$$\rho\lambda = \frac{MpQcal + Ap}{cos(\theta sz)} = \frac{MpQcal + Ap}{sin(\theta sz)} \quad (3)$$

Where:

$\rho\lambda$ = TOA planetary reflectance
$\rho\lambda'$ = image reflectance value
θSE = sun elevation angle
θSZ = aenith angle; θSZ = 90° - θSE

Step 3. Calculate albedo value using formula 4–7.

$$\alpha = \frac{\alpha_{toa} - \alpha_{path-radiance}}{\tau_{SW}{}^2} \quad (4)$$

$$\alpha_{toa} = \sum(\omega\lambda \times \rho\lambda) \quad (5)$$

$$\omega\lambda = \frac{ESUN\lambda}{\sum ESUN\lambda} \quad (6)$$

$$\tau_{SW}{}^2 = 0.75 + 2 \times 10^{-5} \quad (7)$$

Where:

α	= albedo
αtoa	= albedo in atmosphere surface
$\alpha_{path\text{-}radiance}$	= albedo path radiance
$\rho\,\lambda$	= reflectivity for band λ
τsw	= shortwave transmissivity of air
$\omega\lambda$	= weighting coefficient for band λ

Step 4. NDVI (Normalized Difference Vegetation Index) calculation by formula 8.

$$NDVI = \frac{\alpha_{nir} - \alpha_{red}}{\alpha_{nir} + \alpha_{red}} \tag{8}$$

Where:

α_{nir} = reflectance of near-infrared band
α_{red} = reflectance of the red band

Step 5. WDVI (Weighted Difference Vegetation Index) calculation using equation 9.

$$WDVI = NIR - \alpha RED \tag{9}$$

Where:

NIR = value of band 5 (near-infrared)
α = coefficient of the relationship between the near-infrared band and the infrared band
RED = value of band 4 (red)

Step 6. SAVI (Soil Adjusted Vegetation Index) calculation using equation 10.

$$SAVI = \frac{(\alpha_{nir} - \alpha_{red})(1+L)}{(\alpha_{nir} + \alpha_{red}+L)} \tag{10}$$

Where:

α_{nir} = reflectance of near-infrared band
α_{red} = reflectance of the red band
L = Monin-Obukhuv length
= $1 - 2 X a X NDVI X WDVI$
a = 1.6

Step 7. Calculate LAI (Leaf Area Index) using equation 11.

$$LAI = (ln(\frac{0.69-SAVI}{0.59})) \times 0.91^{-1} \tag{11}$$

Where:

LAI = 11 x $SAVI^3$, if $SAVI \leq 0.817$, or
LAI = 6, if $SAVI > 0.817$

Step 8. Calculate object emissivity using formula number 12-14.

$$\varepsilon NB = 0.97 + 0.0033 \times LAI;\ \text{for } LAI<3 \tag{12}$$

$$\varepsilon 0 = 0.97 + 0.0033 \times LAI;\ \text{for } LAI<3 \tag{13}$$

$$\varepsilon NB = 0.98\ and\ \varepsilon 0 = 0.98\ when\ LAI \geq 3 \tag{14}$$

Note:
εNB value is used in surface temperature calculation
ε0 value used for calculating the total surface longwave radiation emission

Step 9. Calculate corrected thermal radiance (R_c) using equation 15.

$$R_c = \frac{L6-Rp}{\tau NB} - (1-\varepsilon NB)Rsky \qquad (15)$$

Where:

L6 = Spectral radiance of corrected band 6
Rp = path radiance
Rsky = thermal radiance in the narrow band when the sky is clear
τNB = air transmission in the narrow band

Step 10. Calculate surface temperature (T_s) using equation 16.

$$T_s = K2 \times (ln(\frac{\varepsilon NB\, K1}{Rc} + 1))^{-1} \qquad (16)$$

Where:
T_s = surface temperature (kelvin)
K1,, K2 = constant from landsat (see **Table 1**)
Rc = corrected thermal radiance value

Table 1. K1 and K2 value

	Band 10	**Band 11**
K1	774.89	480.89
K2	1,321.08	1,201.14

We calculate the surface temperature for bands 10 and 11 separately, then have the actual surface temperature value by extracting the average values from both bands. The last step was converting the kelvin value to the centigrade unit.

Many methods are used to measure soil moisture. One of the reliable methods is remote sensing and GIS because it can cover a large area and requires a short time. Several previous studies have used remote sensing to map soil moisture with various approaches, including land surface temperature [43,44], and vegetation indices transformation [45,46]. In this study, soil moisture estimation uses the LST approach to reduce the soil moisture index and NDVI to extracted vegetation density information. NDVI and soil surface moisture have a solid positive correlation [44], and vegetation becomes the most influential soil moisture parameter. The vegetation cover increase will deplete soil moisture content (SWC) [47]. LST is a state, controlled by the energy balance of the surface, atmosphere, thermal properties of the surface, and subsurface media [48–50]. If the surface temperature is low, the surface has low infiltration indicating high soil moisture [51]. Equation 17 describe the SWI calculation [42,51,52].

$$SMI = \frac{(LST_{max}-LST)}{(LST_{max}-LST_{min})} \qquad (17)$$

Where:
LST_{max} = maximum surface temperature (°C)
= a_1 x NDVI + b_1
LST_{min} = minimum surface temperature (°C)
= a_2 x NDVI + b_2
a_1, a_2, b_1 and b_2 derived from linear regression which define warm and cold edge of the data.

$$fs = \frac{Slope\ strength}{Stress\ acting\ on\ the\ slope} \qquad (18)$$

The calculation of slope stability, known as the factor of safety, is based on the infinite slope model developed by Brundsen and Prior in 1984 [53]. The method was also perfected by Selby in 1993 [38,54].

Both ways possess similarities applicable for shallow landslides using soil parameters resulted in the ratio between available shear strength and mobilized shear strength. FS value is lower than 1, known as unstable slope where landslide occurs. Simplified fs calculation represented in equation 18 [55]. Saga GIS was used to process the data to obtain the fs value as a raster map. It needs input from the slope grid and soil parameters representing its mechanics such as soil thickness, saturation, friction, density, and cohesion [54].

We used a scatter plot to find the relation among each parameter by plotting two parameters respectively. We used the scatter plots for slope vs fs, slope vs LST, slope vs SMI, fs vs LST, and fs vs SMI (see **Figure 4** for the results). The R^2 resulted from the regression express the strength between the parameters. It shows how well it represents the changes of dependant variables.

3. Result and Discussion

3.1. Result

3.1.1. LST and SMI.

According to Landsat 8 image processing, LST was obtained for July 25th, 2019 (see **Figure 2A**). We can infer from the figure that the high temperature is located mainly in the valley, where the built-up area emerged. It is in accordance that high temperature is relevant to the changes of natural land uses by anthropogenic force [56,57]. The second place where the high temperature is located in the low vegetation-covered area, occupied primarily by shrubs and grass. The opposite is that low temperature can be found in the vegetated area with minimum human interference and primarily in the hilly area.

The SMI (**Figure 2B**) represents the same pattern as the LST derived from LST. The area with a high temperature has low soil moisture as the higher evaporation indicates that soil moisture is sensitive to evapotranspiration [45]. Thus, as settlement and human-influenced area (built-up) have a higher temperature, the soil moisture in these areas are also lower. Meanwhile, vegetated areas, primarily occupied hills and slopes, tend to have a higher SMI.

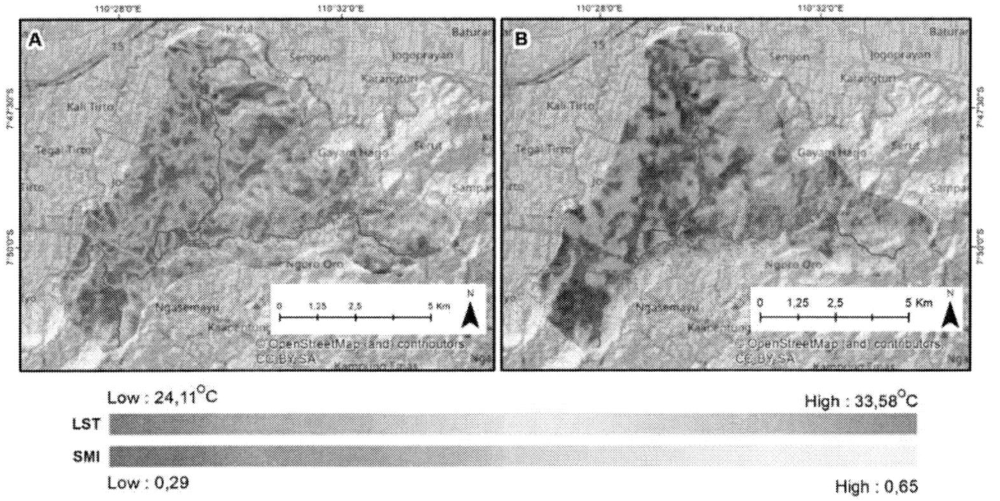

Figure 2. Land Surface Temperature (A); and Soil Moisture Index (B) in Ngijo Watershed

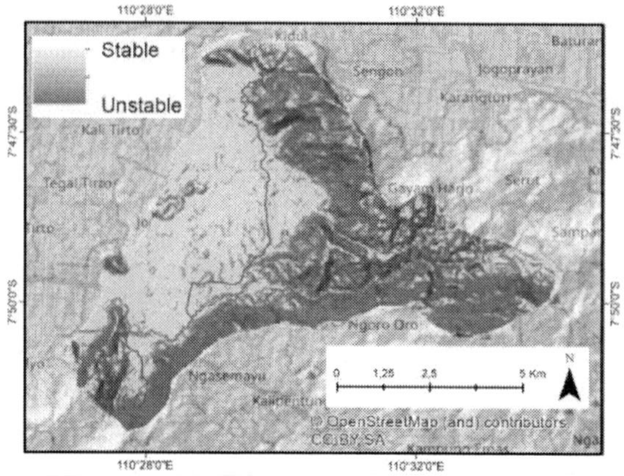

Figure 3. Slope stability (fs) map of Ngijo watershed

3.1.2. Slope Stability, Safety Factor (fs).

Based on the calculation made using SAGA GIS, we obtained a raster of the safety factor of the Ngijo watershed (see **Figure 3**). Mainly, the stable area means safe from the landslide, occupy the relatively flat areas. The unstable area, marked red, is clustered in the hilly area, occupying the denudational and structural area. We can infer that around 50% of the regions are susceptible to unstable slope leads to landslides.

3.2. Discussion

Spatially, the comparison of the pattern of each map shown in **Figures 2** and **3** displays a similar pattern as a landslide or slope stability is closely related to the topographic condition, which is hilly with steep up to very steep slopes [17]. The visual pattern of the resulted map suggests that the stability of the slope matches the landuse pattern. The unstable slope occupies the vegetated agricultural area, which represents low surface temperature and high soil moisture. Vegetation cover is closely related to soil moisture. Vegetation tends to lower the surface temperature and block wind movement above the ground, thus reducing evaporation, but the maturity of the vegetation controls the rate of transpiration. As a result, the moisture in the vegetated area tends to be high [58]. Soil moisture will trigger the slope movement by increasing shear strength due to the pore's water pressure increase [59].

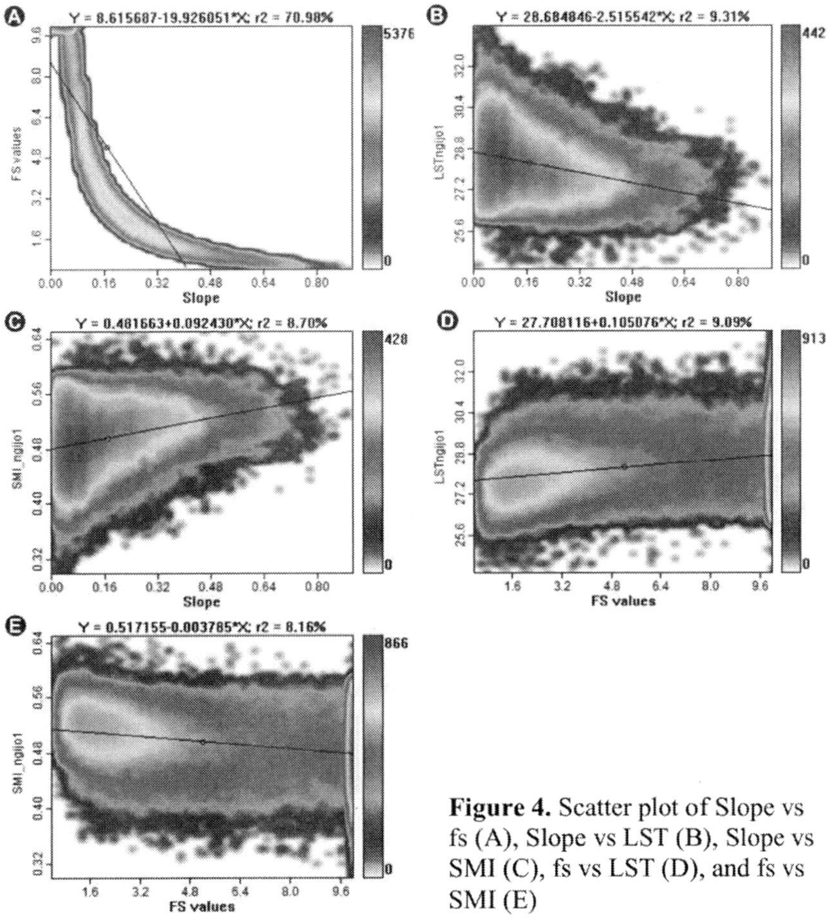

Figure 4. Scatter plot of Slope vs fs (A), Slope vs LST (B), Slope vs SMI (C), fs vs LST (D), and fs vs SMI (E)

Slope stability is heavily related to the morphology of the terrain, mainly controlled by the slope steepness [3,12,14,15,29,60]. That condition is clearly shown by the plotting of fs and slope result of 70.98% of R^2 (**Figure 4**). The value is distributed evenly and closed among them, forming a distinguishable and regular curve. It indicates that the changes will follow the changes in slope values in the slope stability calculation. By the formula Brunsden and Prior (1984) [53], it is clear that fs has a linear relation with slope steepness. It is implied that the fs value will be worse with the increasing slope steepness.

Based on geomorphological basic, morphology is one of the central controllers of the surface process. It controls the hydrological condition such as moisture, infiltration, and runoff [55] and vice versa. Its morphological signatures can provide a glimpse of its hydrological properties, as an example of the morphology of the karst region [61].

The fs pattern in **Figure 3** matches the distribution of vegetation cover over the study area, where an unstable slope is recognized. Partially the phenomenon described as vegetation cover altered the rainfall convert into an overland flow, and it regulates the water flow so the water can infiltrate. The infiltrated water will fill the pores, and at some point, when it is saturated enough and powered by the steep slope angle, it triggers the mass movement. The situation will be worse if the steep slope is bare or with minimum vegetation covers. Another morphological aspect responsible for defining the moisture is the aspect. Aspect, known as the facing direction of the slope, will control the humidity by its proportion and duration in receiving radiation/sunlight, which enables vegetation to settle and grow.

We thought that LST and SMI would represent the actual condition of the physical variables as the result of several parameters behind, such as the mixture of land uses, soil, hydrometeorology, and geology-geomorphological condition. Those incorporating parameters seen as the process resulted in the fs value. However, **Figure 4** respectively provides no distinct relation among them. It represents by the low R^2 values and their scattered pattern. The low value does not mean that it has no relation at all. The use of LST and SMI directly to analyze the landslide occurrence is not entirely a misstep.

Nevertheless, it needs the fundamental parameter to have an adequate analysis of the model. We can fulfill those data by having the fieldwork assess the data directly as a sample to build the model. It does not mean that remote sensing application in a landslide modeling is a mistake, but it needs to be validated for a better and reliable result. The result is according to the theory, but the low R2 suggests the non-direct relation among the analyzed parameters.

4. Conclusion

The use of models in detecting slope stability as part of landslide modeling needs considerable precautions. LST and SMI are not showing a high relation to FS in this trial. Slope vs fs is the only parameter that indicates high ties among the others. It is better to have adequate field data as the input and validator for building the models, as it represents the actual condition of the physical properties onsite.

Acknowledgment

We had not received any funding for this paper and the publication. The first author acted as the corresponding author and drafted the manuscript, layout, and collaborating ideas, while LST and SMI data processing and their methodology are the second author's works. The third author provided data and supported the ideas, especially concerning landslides and their basic factors.

References

[1] Shahabi H and Hashim M 2015 Landslide susceptibility mapping using GIS-based statistical models and Remote sensing data in tropical environment *Sci. Rep.* **5** 1–15

[2] Zhao C and Lu Z 2018 Remote sensing of landslides-A review *Remote Sens.* **10** 8–13

[3] Golovko D, Roessner S, Behling R, Wetzel H U and Kleinschmit B 2017 Evaluation of remote-sensing-based landslide inventories for hazard assessment in southern Kyrgyzstan *Remote Sens.* **9** 1–22

[4] Ghorbanzadeh O, Blaschke T, Gholamnia K, Meena S R, Tiede D and Aryal J 2019 Evaluation of different machine learning methods and deep-learning convolutional neural networks for landslide detection *Remote Sens.* **11**

[5] Arabameri A, Pradhan B, Rezaei K and Lee C W 2019 Assessment of landslide susceptibility using statistical- and artificial intelligence-based FR-RF integrated model and multiresolution DEMs *Remote Sens.* **11**

[6] Mohan A, Singh A K, Kumar B and Dwivedi R 2020 Review on remote sensing methods for landslide detection using machine and deep learning *Trans. Emerg. Telecommun. Technol.* 1–23

[7] Nhu V H, Mohammadi A, Shahabi H, Ahmad B Bin, Al-Ansari N, Shirzadi A, Clague J J, Jaafari A, Chen W and Nguyen H 2020 Landslide susceptibility mapping using machine learning algorithms and remote sensing data in a tropical environment *Int. J. Environ. Res. Public Health* **17** 1–23

[8] Merritt A J, Chambers J E, Murphy W, Wilkinson P B, West L J, Gunn D A, Meldrum P I, Kirkham M and Dixon N 2014 3D ground model development for an active landslide in Lias mudrocks using geophysical, remote sensing and geotechnical methods *Landslides* **11** 537–50

[9] Davies T 2015 Landslide Hazards, Risks, and Disasters: Introduction *Hazards and Disasters Series Landslide Hazards, Risks, and Disasters* ed J F Shroder and T Davies (Amsterdam: Elsevier) pp 1–16

[10] Mccoll S T 2015 Landslide Causes and Triggers *Hazards and Disasters Series Landslide Hazards, Risks, and Disasters* ed J F Shroder and T Davies (Amsterdam: Elsevier) pp 17–42
[11] Fan W, Wei X sheng, Cao Y bo and Zheng B 2017 Landslide susceptibility assessment using the certainty factor and analytic hierarchy process *J. Mt. Sci.* **14** 906–25
[12] Aditian A, Kubota T and Shinohara Y 2018 Comparison of GIS-based landslide susceptibility models using frequency ratio, logistic regression, and artificial neural network in a tertiary region of Ambon, Indonesia *Geomorphology* **318** 101–11
[13] Sutanto Trijuni Putro; Fitria Nucifera 2017 Klasifikasi Bentuklahan Secara Otomatis Menggunakan Topographic Position Index (TPI) *J. Geogr.* **14** 75–83
[14] Skilodimou H D, Bathrellos G D, Koskeridou E, Soukis K and Rozos D 2018 Physical and anthropogenic factors related to landslide activity in the northern Peloponnese, Greece *Land* **7** 1–18
[15] Hadmoko D S, Lavigne F, Sartohadi J, Gomez C and Daryono D 2017 Spatio-Temporal Distribution of Landslides in Java and the Triggering Factors *Forum Geogr.* **31** 1–15
[16] Froude M J and Petley D N 2018 Global fatal landslide occurrence from 2004 to 2016 *Nat. Hazards Earth Syst. Sci.* **18** 2161–81
[17] Nsengiyumva J B, Luo G, Nahayo L, Huang X and Cai P 2018 Landslide Susceptibility Assessment Using Spatial Multi-Criteria Evaluation Model in Rwanda *Int. J. Environ. Res. Public Health* **15** 243
[18] Almeida S, Holcombe E A, Pianosi F and Wagener T 2017 Dealing with deep uncertainties in landslide modelling for disaster risk reduction under climate change *Nat. Hazards Earth Syst. Sci.* 225–41
[19] Jaiswal P and Westen C J Van 2013 Use of quantitative landslide hazard and risk information for local disaster risk reduction along a transportation corridor: a case study from Nilgiri district , India *Nat Hazards* 887–913
[20] Asnakew S, Shumet S, Ginbare W, Legas G and Haile K 2019 Prevalence of post-traumatic stress disorder and associated factors among Koshe landslide survivors, Addis Ababa, Ethiopia: A community-based, cross-sectional study *BMJ Open* **9** 1–8
[21] Sonpaveerawong J, Limmun W and Chuwichian N 2019 Prevalence of psychological distress and mental health problems among the survivors in the flash floods and landslide in Southern Thailand *Walailak J. Sci. Technol.* **16** 255–64
[22] CRED and UNISDR 2017 *Economic Losses, Poverty & DISASTERS 1998-2017*
[23] Huo X, Gupta H, Niu G Y, Gong W and Duan Q 2019 Parameter Sensitivity Analysis for Computationally Intensive Spatially Distributed Dynamical Environmental Systems Models *J. Adv. Model. Earth Syst.* **11** 2896–909
[24] Gorji T, Sertel E and Tanik A 2017 Monitoring soil salinity via remote sensing technology under data scarce conditions: A case study from Turkey *Ecol. Indic.* **74** 384–91
[25] Putro S T and Prasetiyowati S H 2019 Challenges in collecting primary data for environmental research purposes: a case study in Parangtritis sand dune, Yogyakarta *IOP Conf. Ser. Earth Environ. Sci.* **243** 012004
[26] Brunsden D 1999 Some geomorphological considerations for the future development of landslide models *Geomorphology* **30** 13–24
[27] Merghadi A, Yunus A P, Dou J, Whiteley J, ThaiPham B, Bui D T, Avtar R and Abderrahmane B 2020 Machine learning methods for landslide susceptibility studies: A comparative overview of algorithm performance *Earth-Science Rev.* **207** 103225
[28] van Natijne A L, Lindenbergh R C and Bogaard T A 2020 Machine learning: New potential for local and regional deep-seated landslide nowcasting *Sensors (Switzerland)* **20** 1–18
[29] Thirugnanam H, Ramesh M V and Rangan V P 2020 Enhancing the reliability of landslide early warning systems by machine learning *Landslides* **17** 2231–46
[30] Ma Z, Mei G and Piccialli F 2020 Machine learning for landslides prevention: a survey *Neural Comput. Appl.* **8**

[31] Vun J, Stantot-Geddes Z, Sudarmo S P and Kryspin-Watson J 2018 Safeguarding Indonesia's development from increasing disaster risks *Wold Bank Blogs*
[32] BNPB 2021 Infografis *Badan Nas. Penanggulangan Bencana*
[33] Zakaria M F and Maisarah S M 2019 Identifikasi Bidang Gelincir Pada Daerah Rawan *J. Geofis. Eksplor.* **5** 55–63
[34] Santoso D H, Muryani E and Permatasari A Z 2019 Pengendalian Longsor di Daerah Desa Sumberharjo, Kecamatan Prambanan, Kabupaten Sleman, Daerah Istimewa Yogyakarta *J. Sains Teknol. Lingkung.* **5** 61
[35] Saputra A, Sartohadi J, Hadmoko D S and Gomez C 2016 Geospatial Assessment of Coseismic Landslides in Baturagung Area *Forum Geogr.* **29** 99
[36] Tajudin N, Ya'acob N, Ali D M and Adnan N A 2021 Soil moisture index estimation from Landsat 8 images for prediction and monitoring landslide occurrences in Ulu Kelang, Selangor, Malaysia *Int. J. Electr. Comput. Eng.* **11** 2101–8
[37] Yaa'Cob N, Abd Rashid Z N A, Tajudin N and Kassim M 2020 Landslide Possibilities using Remote Sensing and Geographical Information System (GIS) *IOP Conf. Ser. Earth Environ. Sci.* **540**
[38] Selby M J 1993 *Hillslope Materials and Processes* (Oxford: Oxford University Press)
[39] van Zuidam R A 1983 *Guide to Geomorphologic Aerial photographic interpretation and mapping* (Enschede, Netherland: ITC)
[40] Bastiaanssen W 1995 *Regionalization of surface flux densities and moisture indicators in composite terrain.* (Wageningen: Agricultural University of Wagieningen)
[41] Bastiaanssen W.G.M., M. Meneti, R.A. Feddes and Holtslag A A . 1998 A remote sensing surface energy balance algorithm for land (SEBAL). 1. Formulation *J. Hydrol.* **212–213** 198–212
[42] Potić I, Bugarski M and Matić-Varenica J 2017 Soil Misture Determination Using Remote Sensing Data For The Poverty Protection and Increase of Agriculture Production *Proceeding of 2017 World Bank Conference in Land and Poverty* (Washintin DC, USA: World Bank)
[43] Amani M, Salehi B, Mahdavi S, Masjedi A and Dehnavi S 2017 Temperature-Vegetation-soil Moisture Dryness Index (TVMDI) *Remote Sens. Environ.* **197** 1–14
[44] Mohamed E S S, Ali A, El-Shirbeny M, Abutaleb K and Shaddad S M M 2020 Mapping soil moisture and their correlation with crop pattern using remotely sensed data in arid region *Egypt. J. Remote Sens. Sp. Sci.* **23** 347–53
[45] Joiner J, Yoshida Y, Anderson M, Holmes T, Hain C, Reichle R, Koster R, Middleton E and Zeng F W 2018 Global relationships among traditional reflectance vegetation indices (NDVI and NDII), evapotranspiration (ET), and soil moisture variability on weekly timescales *Remote Sens. Environ.* **219** 339–52
[46] Schnur M T, Xie H and Wang X 2010 Estimating root zone soil moisture at distant sites using MODIS NDVI and EVI in a semi-arid region of southwestern USA *Ecol. Inform.* **5** 400–9
[47] Yang T, Ala M, Zhang Y, Wu J, Wang A and Guan D 2018 Characteristics of soil moisture under different vegetation coverage in Horqin Sandy Land, northern China *PLoS One* **13** 3–6
[48] Weng Q 2009 Thermal infrared remote sensing for urban climate and environmental studies: Methods, applications, and trends *ISPRS J. Photogramm. Remote Sens.* **64** 335–44
[49] Guha S and Govil H 2021 An assessment on the relationship between land surface temperature and normalized difference vegetation index *Environ. Dev. Sustain.* **23** 1944–63
[50] Nwaerema P, Vincent O N, Amadou C and Morrison A I 2019 Spatial Assessment of Land Surface Temperature and Emissivity in the Tropical Littoral City of Port Harcourt, Nigeria *Int. J. Environ. Clim. Chang.* **9** 88–103
[51] Saha A, Patil M, Goyal V C and Rathore D S 2018 Assessment and Impact of Soil Moisture Index in Agricultural Drought Estimation Using Remote Sensing and GIS Techniques *Proceedings* **7** 2
[52] Vani V, Pavan Kumar K and Ravibabu M V 2019 Temperature and vegetation indices based surface soil moisture estimation: A remote sensing data approach *Proceeding of International*

IOP Conf. Series: Earth and Environmental Science **986** (2022) 012022 doi:10.1088/1755-1315/986/1/012022

Conference on Remote Sensing for Disaster Management Springer Series in Geomechanics and Geoengineering ed P Rao, K Rao and S Kubo (Springer, Cham) pp 281–9

[53] Cristina Magdalena C, Sanda M, Ernest O and Ioan B 2017 Comparative analysis of large scale scenario-based landslides hazard zonation maps. A case study in Iasi City, Romania *Rev. Geomorfol.* **19** 145–54

[54] Günther A 2012 Module SAFETYFACTOR / SAGA-GIS Module Library Documentation (v2.2.1) *Modul. SAFETYFACTOR*

[55] Huddart D and Stott T 2010 *Earth Environments. Past Present and Future* (West Sussex, England: A John Wiley & Sons, Ltd)

[56] Das S and Angadi D P 2020 Land use-land cover (LULC) transformation and its relation with land surface temperature changes: A case study of Barrackpore Subdivision, West Bengal, India *Remote Sens. Appl. Soc. Environ.* **19** 100322

[57] Tran D X, Pla F, Latorre-Carmona P, Myint S W, Caetano M and Kieu H V. 2017 Characterizing the relationship between land use land cover change and land surface temperature *ISPRS J. Photogramm. Remote Sens.* **124** 119–32

[58] Ranney K J, Niemann J D, Lehman B M, Green T R and Jones A S 2015 A method to downscale soil moisture to fine resolutions using topographic, vegetation, and soil data *Adv. Water Resour.* **76** 81–96

[59] Ray R L and Jacobs J M 2007 Relationships among remotely sensed soil moisture, precipitation and landslide events *Nat. Hazards* **43** 211–22

[60] Kundu S, Saha A K, Sharma D C and Pant C C 2013 Remote Sensing and GIS Based Landslide Susceptibility Assessment using Binary Logistic Regression Model: A Case Study in the Ganeshganga Watershed, Himalayas *J. Indian Soc. Remote Sens.* **41** 697–709

[61] Haryono E, Trijuni Putro S, Suratman and Sutikno 2017 Polygonal karst morphology of Karangbolong area, Java-Indonesia *Acta Carsologica* **46**

Analysis of landforms, early warning system, locations and evacuation paths for reducing flood disaster risk in Sragen, Central Java, Indonesia

Sutarno*

Disaster Research Center, Universitas Sebelas Maret, Jl. Ir. Sutami No.36, Kentingan, Jebres, Surakarta

*Email: sutarno@staff.uns.ac.id

Abstract. One of the areas in Central Java Province that is prone to flooding is Sragen Regency. The distribution of flood disasters occurs in commensurate rivers and river confluences in lowlands which are dominated by settlements and agricultural land. The purpose of this study was to analyze the distribution of landforms along riverbanks in residential areas and rice fields, refugee shelter locations and evacuation routes for handling disaster risk reduction. The research method is grounded research by recording the population census and public facilities at the affected residential locations. Determination of evacuation locations and routes by using evacuation criteria parameters and safe distance parameters to determine the classification of refugee evacuation locations. The results of the study found that there was a lack of understanding of the landforms of river borders for agricultural and residential land, the utilization and loss of embankments on comparable rivers, especially in areas of interconnection between rivers in the lowlands. To minimize the risk of flooding, it is necessary to add EWS, river border land that is already owned by the community, the use of right of use must be gradually returned as a river border and the selection of the right location and route evacuation.

1. Introduction

Natural disasters are a series of events that threaten and disrupt people's lives, both caused by natural/non-natural factors and human factors, resulting in fatalities, environmental damage, property and psychological losses [1]. Indonesia is considered as one of the countries in the world that is prone to disasters [2]. Indonesia is located in one of the most active disaster hot spots in the world so it has the potential to be exposed to natural disasters [3]. Natural disasters that often occur are earthquakes, tsunamis, volcanic eruptions, floods, landslides, droughts and forest fires. Hydrometeorological disasters such as floods, hurricanes and landslides still dominate in Indonesia [4].

Based on the BNPB database, the incidence of floods has the highest percentage of 28% compared to other natural disasters [5]. Floods are the most common disaster event in Java caused by overflowing rivers, coasts, and flash floods [6]. Floods are not only caused by extreme rainfall and sea level rise, but can also be caused by uncontrolled development [7]. Anticipation is needed during the rainy season because flooding becomes an annual disaster that is closely related to climate change [8]. The impact of the flood disaster caused economic, social and environmental losses [9]. Flood victims rank first in the world, reaching 55%. The percentage of flood events in Indonesia reached 38% of all disasters, of which 63.7% occurred in Java and 37.4% occurred in Central Java [10].

Sragen is one of the regencies in Central Java Province that subscribes to flood disasters. The area of Sragen Regency is 941.55 km2 which flows through the Bengawan Solo River and divides the Sragen Regency into 2 (two) parts. The southern part of the Bengawan Solo river is influenced by Lawu volcanic rocks and the northern part is influenced by the South Kendeng limestone mountains [11]. According to BPBD Sragen [12], there are 14 sub-districts out of 20 sub-districts which are flood-prone areas spread on the right and left of the Bengawan Solo river and its confluence with the Grompol, Mungkung, Sragen and Natan sub-watersheds as well as the Sawur sub-watershed with basin and lowland landforms.

Natural disaster events not only destroy and cause loss of life, but also the livelihoods of the victims [13]. The biggest economic loss in Sragen due to flood disaster is Rp. 240 billion, which damaged more than 100 hectares of agricultural land, flooded 14,437 houses and displaced 51,946 residents [14]. The flood disaster is still ongoing in Sragen, although the local government and private parties (NGOs) have made efforts to repair, control and mitigate flood disasters. Flood control efforts carried out such as installing early warning system (EWS), evacuation route, command post, disaster education in order to increase awareness and psychological readiness of the community [15,16].

Resilience is not a term commonly applied in urban and disaster-related contexts, although it is clear that disasters such as floods are a major problem that requires attention [17]. Public perception of flood risk can improve self-protection [18]. Public awareness of flood risk is a key strategy to improve disaster risk preparedness [19]. In order to reduce flood disaster losses in Sragen, a research was conducted by studying the landform form of the flood location, the shape of the river border and the causes of the disaster. Furthermore, to find out the number and function of EWS as well as evacuation routes for flood-affected communities as an effort to reduce disaster risk.

2. Methods

The total area of Sragen Regency is 941.55 sq km (9415500 divided into: Wetland (rice fields) covering 40,037.93 Ha (42.52%) and Dry Land 54,117.88 Ha (57.48%). The area of the research area uses the Sragen Flood Prone Map, flood data from the Sragen Central Statistics Agency (BPS) 2018, 2019, 2010. As for the sources of flood disaster information obtained from BPBD, BPS and news from newspapers, questionnaires, field data and other supports such as the use of rollmeters, cameras and clinometers. The research was carried out in the Sragen area from February to May 2021.

The research method used is descriptive qualitative with a field survey/interview approach using questionnaires and secondary data analysis. The observation locations were selected in flood-prone affected areas and the confluence of the main river Bengawan Solo with river branches. in the Sragen area and around the river that crosses the highway and village connecting roads. The parameters studied are the shape and location of the landform, the number and function of the EWS, the location of the Command Post affected since the early 2018-2021.

3. Results and Discussions

The disaster hazard map of Sragen district (Revised 2008) and data from the 2018 Central Bureau Statistics, 2019 and 2020 and data from the BPBD of Sragen Regency (2018) were obtained, which were used as a reference for field survey guidelines. There are 14 sub-districts in Sragen Regency, of which 20 sub-districts are flood-prone areas, namely to the south of the Solo River and to the north of the Solo River. There are 8 sub-districts. The flood area to the south of Bengawan Solo includes the Districts of Masaran, Sidoharjo, Sragen, Karangmalang, Ngrampal, Kedawung, Gondang, and Connectmacan Districts. Meanwhile, to the north of the Solo River, there are 6 sub-districts including Plupuh, Tanon, Sukodono, Gesi, Tangen, and Jenar sub-districts (Figure 1).

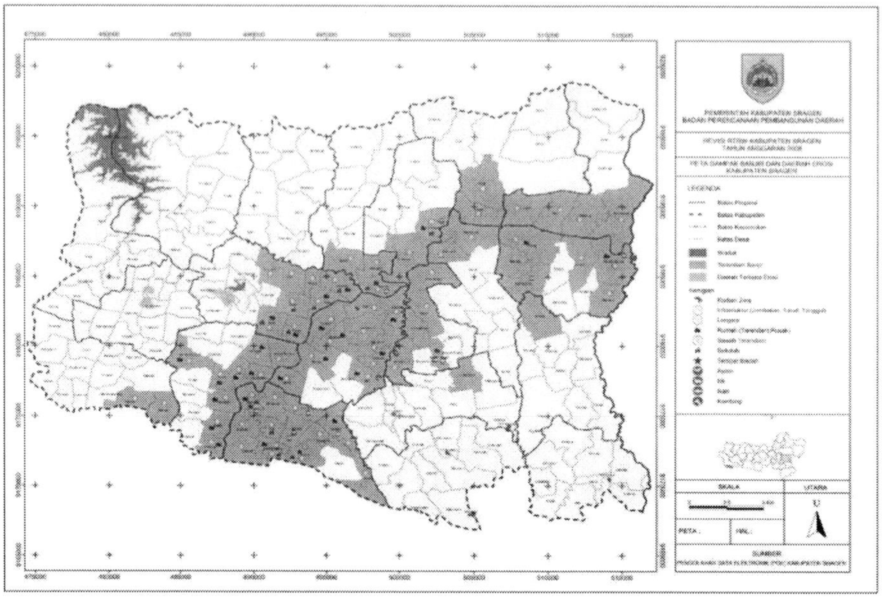

Figure 1. Map of Flood Prone in Sragen Regency [20]

From the Central Bureau Statistics (BPS) in 2018, data is obtained as shown in the table 1.

Table 1. Land Area Affected by Flood Disaster in Sragen in 2018

Districts	Total Districts	Flood Affected Village	Total Villages Affected by Flood	Land Area Affected by Flood (hectare)
Sidoharjo	1	8	8 village	4.837,41
Tanon	1	6	6 village	5.259,59
Plupuh	1	5	5 village	5.040,92
Masaran, Ngrampal	2	4	8 village	8.603,20
Sragen, Gondang	2	2	4 village	7.594,33
Mondokan	1	1	1 village	5.070,90
Total	8		32 village	36.415,35

Source: [21]

From table 1 above, it can be seen that there are 8 sub-districts covering .32 villages affected by the flood disaster with an area of 36,415.35 ha and locations that are not affected by floods covering an area of 61,284.47 hectares. The data table above shows that almost one village in the Mondokan sub-district was inundated and the worst with an area of 5,070,90 hectares, followed by 8 areas in Masaran and Ngrampal with an area of 8,603.20 hectares and 4 areas in Sragen and Gondang. In mid-2018, 4 EWS installations were added, bringing the total to 7 EWS and all functioning well. In 2018, the flood disaster did not result in any casualties, while accurate data were not obtained for materials.

From the data from the Central Statistics Agency for 2019, data is obtained as in table 2.

Table 2. Area of Land Affected by Flood Disaster in Sragen in 2019

Districts	Total Districts	Flood Affected Village	Total Villages Affected by Flood	Land Area Affected by Flood (hectare)
Tanon	1	2 village	2 village	5.259,59
Plupuh, Masaran, Sidoharjo, Karangmalang, Miri	5	1 village	5 village	23.891,01
Total	6		7 village	29.150,60

Source: [22]

From the 2019 table, it can be seen that there are 6 sub-districts covering 7 flood-affected villages with an area of 29,150.60 hectares, of which one village is in Miri District. The number of EWS build is fixed at 7 EWS and all of them function properly so that they can provide information to flood-affected communities, there are no casualties and material losses are not recorded [22].

Table 3. Area of Land Affected by Flood Disaster in Sragen in 2020

Districts	Total Districts	Flood Affected Village	Total Villages Affected by Flood	Land Area Affected by Flood (hectare)
Kalijambe, Plupuh	2	2 village	4 village	9.779,75
Masaran, Karangmalang, Sambirejo, Ngrampal, Kedawung, Sragen	6	1 village	6 village	26.025,62
Total	8		10 village	35.805,37

Source: [23]

From table 3 above, there are 8 sub-districts covering 10 flood-affected villages with an area of 35,805.37 ha and locations not affected by flooding 61,894.47 hectares. In 2020, it was found that 2 EWS were damaged or not functioning, namely in Sapen and in Sragen rivers. This year there were no fatalities and no nominal figures for losses in rupiah were found.

Table 4. Number of Districts and Villages Affected by Flood

Districts	Flood Affected Village
Sragen	2 Village
Masaran,	4 Village
Sidoarjo	8 Village
Karangmalang,	1 Village
Sambirejo,	1 Village
Gondang	2 Village
Kedawung,	1 Village
Ngrampal	4 Village
Kalijambe,	2 Village
Plupuh	5 Village
Tanon	6 Village
Mondokan	1 Village
Sukodono	1 Village
Miri	1 Village
Total	39 village

Table 5. Areas Affected by Flood Disasters

Year	Number of districts affected by flooding	Total Villages Affected by Flood	Land Area Affected by Flood (hectare)
2008*	14	-	-
2018	8	32 village	36.415,35
2019	6	7 village	29.150,60
2020	8	10 village	35.805,37

Sources: [21–23] (*= RTRW Data 2008)

The results of the distribution of flood disasters in 2018, 2019, 2020 and combined with the revised map from BAPPEDA in 2008 are presented in Figure 2.

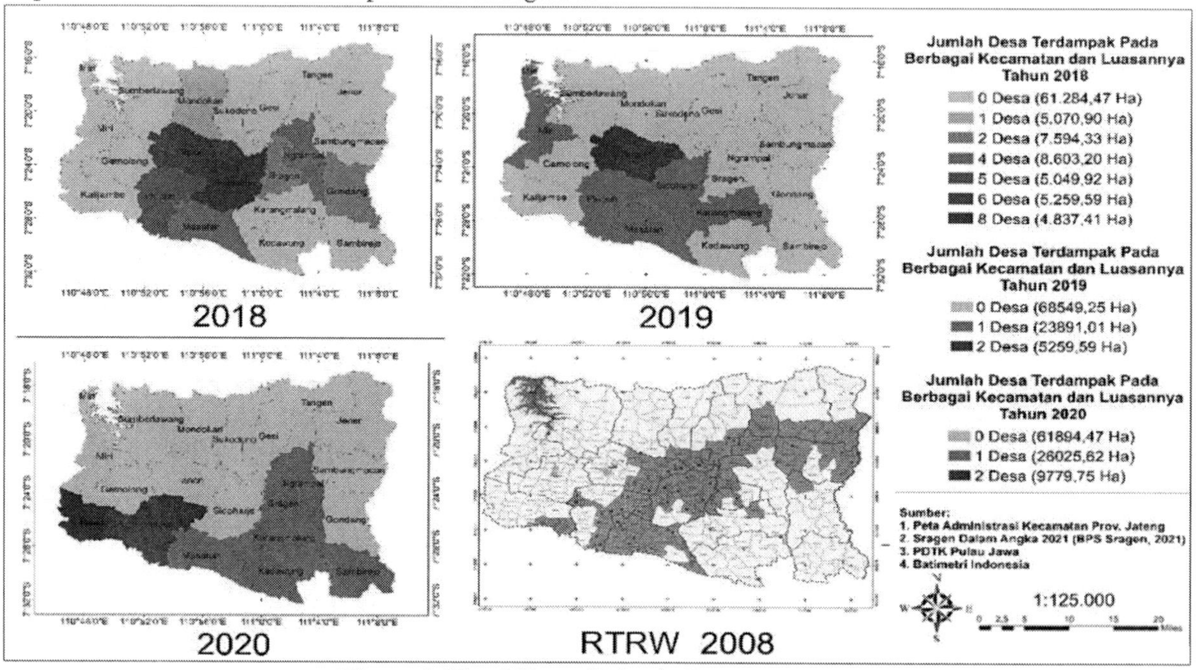

Figure 2. Flood Hazard Map in Sragen

The number of sub-districts affected by floods in 2008 was 14 sub-districts (blue), in 2018 it was reduced to 8 sub-districts covering 32 affected villages, in 2019 it was reduced again to 6 sub-districts covering 7 villages. From 2018 there were 32 flood-affected villages to 7 villages with a reduction in the affected area from 36,415.35 hectares in 2018 to 29,150.60 hectares in 2019. This happens as a result. continue to carry out disaster socialization and physical development in some areas of the Bengawan Solo river and tributaries in the Sragen area as well as the installation of EWS in the affected areas and provide good information. Meanwhile, the flood disaster in one village of Miri sub-district was caused by a broken river embankment.

In Figure 2, it appears that there has been a shift in flood locations and a reduction in flood locations. Based on the results of the interview and data BPBD Sragen (2021), it was explained that the socialization of flood disasters has been and is often carried out and a lot of physical development has been carried out starting from river alignment, construction of embankments on river borders and the alertness of the community in receiving information about a flood disaster either from the EWS sound source, from community leaders and community leaders. from neighbors.

In 2019 to 2020 the affected area increased from 6 sub-districts covering 7 villages with an area of 29,150.60 hectares, increasing to 8 sub-districts covering 10 villages with an area of 35,805.37 hectares. This is also caused by the collapse of river embankments in the Kedawung and Sidoharjo sub-districts due to high rainfall and damage to 2 EWS that are not functioning and most of the river borders are still widely used for agricultural land. While the river border around the protocol road bridge, both State roads, Provincial roads and Regency roads are used for residential/business buildings and the land has been certified with ownership rights (SHM). It should be returned to its original function as soon as possible.

Below is a picture of the riparian zone that should be obeyed in order to reduce the occurrence of floods [24].

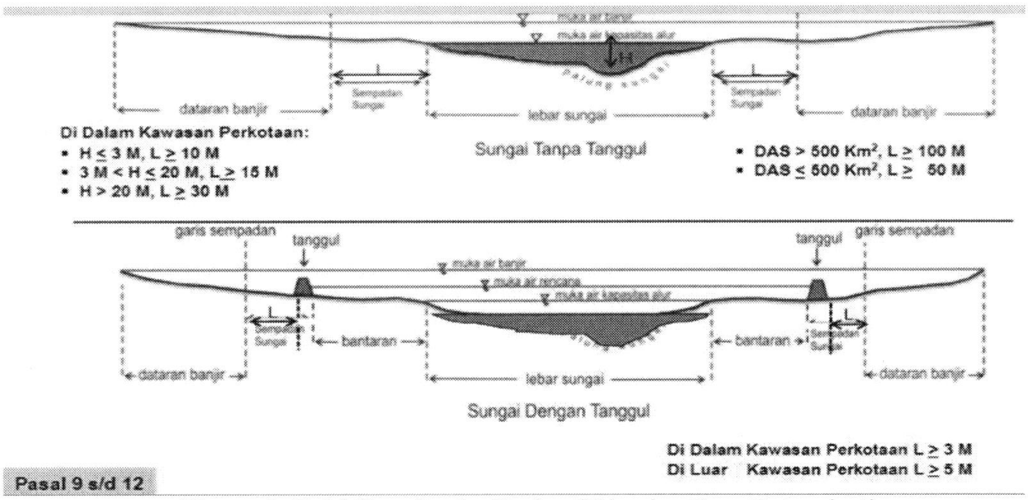

Figure 3. Guidelines for the Study of Riparian Zone Determination

While the conditions and conditions at the flood disaster location are on average not dammed as shown figure 4.

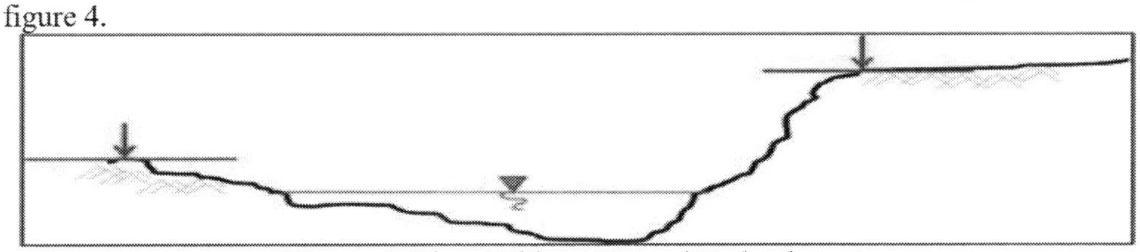

Figure 4. River Without Natural Embankments

The two arrow lines in the image above indicate the asymmetrical edge of the riverbed and no natural embankments or river embankments. that river borders are not symmetrical and without embankments, even river border land is mostly used for agricultural land so that it is the beginning of river overflow or flooding.

while according to [25], based on terrain characteristics, relief morphology is flat until flat terrain irregular surface may exist, natural levees, abandoned channel possible, process: river actions, ie vertical and lateral erosion or deposition of clastic material. In general, poorly drained terrain.

Riparian zone function to protect rivers and the environment in a sustainable manner, a river embankment is needed [24].

IOP Conf. Series: Earth and Environmental Science **986** (2022) 012023 doi:10.1088/1755-1315/986/1/012023

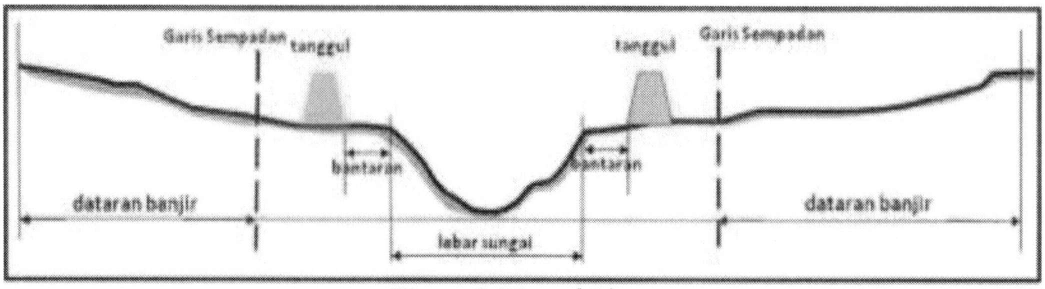

Figure 5. Normal River

Judging from the distribution of thematic maps in dark blue to blue, it appears that the 2018 2019 and 2020 Flood Maps have narrowed the affected area due to improvements in the construction of physical infrastructure and community alertness. However, it also experienced a shift in the location of the flood disaster. This is due to the occurrence of broken embankments especially in Pengkok, Kedawung and in Miri District due to high rainfall and overflowing of river water at the confluence of the Bengawan Solo river with rivers as well as in the shape of a meander river with slow flow velocity in Sragen.

The flood disaster also affected the psychology of the community, from 50 respondents consisting of 25 male respondents and 25 female respondents aged 30-60 years. Of the 50 respondents who stated that 37 respondents were very prepared to face the flood disaster with all its consequences, while 9 respondents were somewhat prepared and the other 4 respondents were old women had severe psychological consequences, considering that 4 respondents had been traumatized by the flood disaster.

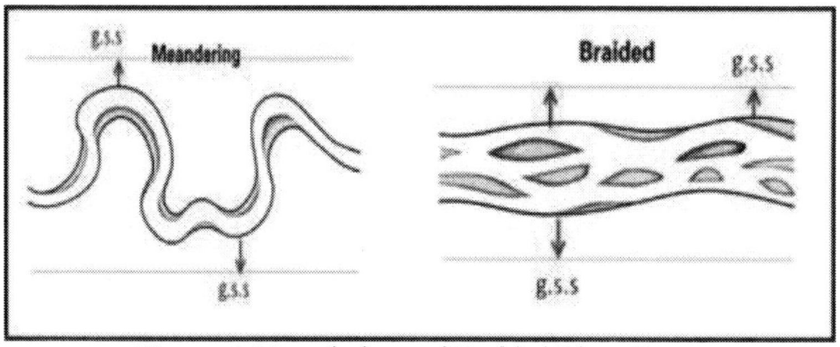

Figure 6. Meandering and Braided River Types

Meanders like the picture above indicate that the area is a relatively flat area so that the water flows slowly/less smoothly. With high rainfall, it is easy to overflow to the right and left of the river [26]. The location of evacuation posts from 39 villages has met the requirements in terms of room capacity, concrete and asphalt evacuation routes, preparation of infrastructure, public kitchens and a lack of toilets.

4. Conclusions

The socialization of flood preparedness education is still being carried out even though the sound quality of the EWS siren is weak. Command posts and evacuation site infrastructure continue to be improved, especially infrastructure for evacuation routes and toilets. It was found that the condition of the violated river border was used for the purpose of increasing the area of agricultural land. Risk management in high-risk areas can save costs in preventing disaster losses and improve disaster preparedness, leading to a faster and more planned recovery [27]. People living in flood-prone areas of Sragen Regency need to increase their awareness regarding the importance of disaster risk reduction.

References

[1] RI 2007 *Undang-Undang Republik Indonesia No. 24 Tahun 2007 Tentang Penanggulangan Bencana* (Indonesia)
[2] Kusumastuti R D, Viverita, Husodo Z A, Suardi L and Danarsari D N 2014 Developing a resilience index towards natural disasters in Indonesia *Int. J. Disaster Risk Reduct.* **10** 327–40
[3] Skoufias E, Strobl E and Tveit T 2017 Natural Disaster Damage Indices Based on Remotely Sensed Data: An Application to Indonesia *Nat. Disaster Damage Indices Based Remote. Sensed Data An Appl. to Indones.*
[4] Muhari A 2021 Review Bencana Juli 2021: Bertemunya Bencana Hidrometeorologi Basah dan Kering *bnpb.go.id*
[5] BNPB 2021 Data Informasi Bencana Indonesia *dibi.bnpb.go.id*
[6] Handayani W, Rudiarto I, Setyono J S, Chigbu U E and Sukmawati A M awanah 2017 Vulnerability assessment: A comparison of three different city sizes in the coastal area of Central Java, Indonesia *Adv. Clim. Chang. Res.* **8** 286–96
[7] Malalgoda C, Amaratunga D and Haigh R 2016 Overcoming challenges faced by local governments in creating a resilient built environment in cities *Disaster Prev. Manag.* **25** 628–48
[8] Al-Amin A Q, Nagy G J, Masud M M, Filho W L and Doberstein B 2019 Evaluating the impacts of climate disasters and the integration of adaptive flood risk management *Int. J. Disaster Risk Reduct.* **39** 101241
[9] Palinkas L A 2020 Fleeing flooding: Asia and the Pacific *Global Climate Change, Population Displacement, and Public Health: The Next Wave of Migration* (Los Angeles: Spinger)
[10] Bakornas PB 2007 *Pengenalan Karakteristik Bancana dan Upaya Mitigasinya Di Indonesia* (BNPB)
[11] Pannekoek A 1949 *Outline of the geomorphology of Java* (Leiden: S.I Publisher)
[12] Rahayu T 2020 Hati-Hati! 14 Kecamatan di Sragen Rawan Banjir *Solopos*
[13] Sawada Y and Takasaki Y 2017 Natural Disaster, Poverty, and Development: An Introduction *World Dev.* **94** 2–15
[14] BPBD 2021 *Data Banjir Kabupaten Sragen* (Sragen)
[15] Tanaka K 2005 The impact of disaster education on public preparation and mitigation for earthquakes: A cross-country comparison between Fukui, Japan and the San Francisco Bay Area, California, USA *Appl. Geogr.* **25** 201–25
[16] Finnis K K, Johnston D M, Ronan K R and White J D 2010 Hazard perceptions and preparedness of Taranaki youth *Disaster Prev. Manag. An Int. J.* **19** 175–84
[17] Handayani W, Fisher M R, Rudiarto I, Sih Setyono J and Foley D 2019 Operationalizing resilience: A content analysis of flood disaster planning in two coastal cities in Central Java, Indonesia *Int. J. Disaster Risk Reduct.* **35** 101073
[18] Wang Z, Wang H, Huang J, Kang J and Han D 2018 Analysis of the public flood risk perception in a flood-prone city: The case of Jingdezhen city in China *Water (Switzerland)* **10**
[19] Onwuemele A 2018 Public perception of flood risks and disaster preparedness in lagos megacity, Nigeria *Acad. J. Interdiscip. Stud.* **7** 179–85
[20] BAPPEDA 2008 *Sragen in Figure* (Sragen: Badan Pusat Statistik Kabupaten Sragen)
[21] BPS 2018 *Kabupaten Sragen dalam Angka Tahun 2018* (Sragen: Badan Pusat Statistik Kabupaten Sragen)
[22] BPS 2019 *Kabupaten Sragen dalam Angka Tahun 2019* (Sragen: Badan Pusat Statistik Kabupaten Sragen)
[23] BPS 2020 *Kabupaten Sragen dalam Angka Tahun 2020* (Sragen: Badan Pusat Statistik Kabupaten Sragen)
[24] RI 2015 *Permen PUPR RI NOMOR 28/PRT/M/2015 TENTANG PENETAPAN GARIS SEMPADAN SUNGAI DAN GARIS SEMPADAN DANAU DENGAN* (Indonesia)

[25] Van Zuidam R . and Van Zuidam C 1979 *Terrain Analisys amd Clasification Using Photographs. A Geomorphological Approach* (Netherland: International Institude for Aerial Survey and Earth Science (ITC))
[26] Kodoatie R J and Sjarief R 2005 *Pengelolaan Sumber Daya Air Terpadu* (Yogyakarta: Andi)
[27] Dilley M, Chen R S, Deichmann U, Lerner-Lam A L, Arnold M, Agwe J, Buys P, Kjekstad O, Lyon B and Yetman G 2005 *Natural Disaster Hotspots A Global Risk Analysis* (Washington D C: The World Bank and Columbia University)

Identification of coastal typology: potential resources and hazards

Y A Wibowo[1,8*], L Ronggowulan [2], H Fatchurohman [3], M Nursaputra [4], D A Arief [5], L Permonojati [6], Suriadi [6], D Kurniawan [6] and R Afrizal [7]

[1] Geography Education Department, Universitas Muhammadiyah Surakarta, Indonesia
[2] Geography Education Department, Universitas Sebelas Maret Surakarta, Indonesia
[3] Department of Earth Technology Vocational College, Universitas Gadjah Mada, Indonesia
[4] Faculty of Forestry, Universitas Hasanudin, Indonesia
[5] Department of Geography, Universitas Negeri Padang, Indonesia
[6] Master Program in Planning and Management of Coastal Area and Watershed, Universitas Gadjah Mada, Indonesia
[7] Urban and Regional Planning Department, Kuantan Singingi Islamic University, Indonesia
[8] Center for Disaster Mitigation Study, Universitas Muhammadiyah Surakarta, Indonesia

*Email: yaw222@ums.ac.id

Abstract. Coastal areas are unique ecosystems because of the potential for abundant natural resources and hazards that can cause losses. Therefore, this study aims to identify the typology and coastal dynamics of Kukup, Ngrawe, Sepanjang, Drini, and Krakal, Indonesia, for analyzing the potential of natural resources and hazards. A rapid-integrated assessment was carried out to identify typology, genesis, and dynamics of coastal areas by conducting field surveys at locations that have been determined by purposive sampling. The results showed that wave erosion coast was formed in Ngrawe, Kukup, Drini, and Krakal. The marine deposition coast was formed in Ngrawe, Kukup, Sepanjang, Drini, and Krakal. Moreover, the coast built by organism can be found in Sepanjang, Drini, and Krakal. In contrast, the structurally shaped coast was formed in the western part of the Krakal. The coastal dynamics begin with geodynamic processes that form structurally shaped coast, hydrodynamics that form marine deposition coast, wave erosion coast, and ecodynamics that form coast built by organism typology. The potential of natural resources consists of fishery potential, agriculture, and natural beauty for tourism objects. In contrast, the potential hazards consist of earthquakes, tsunamis, tidal waves, rockfall, rip currents, topographic changes, and land-use changes. The coastal area management planning recommendations were Spatial Planning Regulation, Early Warning System, Evacuation Route and assembly point (shelter), Digital-based Information Dissemination, and building retrofitting.

1. Introduction

Coastal areas are unique ecosystems because of the potential for abundant natural resources and hazards that can cause losses. The potency of plentiful natural resources with high economic value promises prosperity for humans if appropriately directed [1–5]. The potential can be fisheries, mining minerals, geothermal energy, wind energy, and natural scenery for tourist objects [6]. However, even though it has abundant natural resource potential, coastal areas also have potential hazards/threats that can cause harm to humans [7–9]. The hazards/threats are geological, hydrometeorological, extraterrestrial hazards, and hazards triggered by human activities. Therefore, a comprehensive coastal area management plan is needed, especially for countries with long coastlines.

Indonesia is rich in coastal and marine resources. However, it also causes several areas of Indonesia to be potentially hit by disasters such as tsunamis, tidal waves, intrusions, sea-level rise, and high waves [10–12]. Disasters that potentially occur in coastal areas of Indonesia can be identified based on their physical and socio-cultural characteristics. The northern coastal area of Java Island is identical to intrusion, land subsidence, and coastal flooding [13–15]. The western coastal regions of Sumatra Island, South Java Island, Bali, Lombok, Sumba, and Sumbawa are potentially exposed to earthquakes, tidal waves, and tsunamis. In addition, human activities in coastal areas such as large amounts of groundwater abatement and unplanned coastal spatial planning can accelerate environmental pollution, intrusion, abrasion, and land subsidence [16]. Therefore, a holistic management plan is needed by considering the physical and social characteristics of the area, natural resources, disaster risk, and the sustainability of the coastal ecosystem [17–19].

The coastal area of Gunungkidul has abundant natural resources that the community can utilize to improve their welfare. The potential for fisheries, agriculture, and tourism are the three most prominent potentials in the coastal area of Gunungkidul. The fisheries that develop in Gunungkidul are the potential for marine fishing. In addition, Gunungkidul is also has a high potential for seaweed and ornamental fish. Furthermore, the agricultural potential in the coastal area of Gunungkidul is dominated by dry land crops or fields. There is potential for farming by field system, although there is a minimum water supply for agriculture. The potential that makes Gunungkidul widely known is tourism potential. The Gunungkidul coastal area consists of rocky beaches with hilly reliefs and sandy beaches with white sand material in small bays (pocket beaches) that are interesting to visit. The potential of natural resources in an area can improve the community's welfare economically [20].

In addition to the potential of abundant natural resources, the coastal area of Gunungkidul also has the potential for disasters that threaten people's lives. The coastal area of Gunungkidul is part of the southern zone of Java which is directly opposite the Indian Ocean. Therefore, the Gunungkidul coastal area can be affected by geological disasters such as earthquakes and tsunamis. Moreover, other hazards such as tidal waves, rip currents, and rockfalls also potentially cause losses. Most of the coastal areas of Gunungkidul, namely Kukup, Ngrawe, Sepanjang, Drini, and Krakal, have developed into tourist attractions and many community activities. The five areas are part of the Gunungsewu karst area, designed as a tourist attraction. Therefore, efforts are needed to analyze coastal characteristics, identify natural resources, identify potential hazards, and disaster-resilient coastal area management planning. They are conducted to optimize the utilization of natural resources while still paying attention to preserving nature and reducing disaster risk. This study aims to identify the coastal characteristics and coastal dynamics of Kukup, Ngrawe, Sepanjang, Drini, and Krakal, Indonesia, for analyzing the potential of natural resources and hazards.

The area was chosen as the study location because it has unique characteristics and is a potential tourist attraction in Gunungkidul. Data collection techniques were carried out by rapid-integrated assessment and field surveys. The data analysis technique was carried out qualitatively and based on a literature study. The results of the analysis are recommendations for disaster risk-based coastal area management planning.

2. Study area

The study area covers Kukup, Ngrawe, Sepanjang, Drini, and Krakal coastal areas in Gunungkidul Regency, Yogyakarta Special Region, Indonesia (Fig. 1). Those locations are part of the karst landscape area with limestone material. The karst landscape was due to the influence of tectonic activity in the past. Furthermore, the karst area in Gunungkidul Regency is part of the Gunungsewu karst area in the

Southern Mountain Zone of western East Java. This mountain range is west-east with karst hills by 25 meters to 150 meters [21]. Geomorphologically, the Gunungkidul coastal area consists of solutional, marin, and structural landforms. The primary constituent material of the western part of the coastal area of Gunungkidul is geologically composed of the Wonosari Formation (Tmwl), which is composed of reef limestone, layered limestone, chipped limestone and sandy, and marl limestones [22].

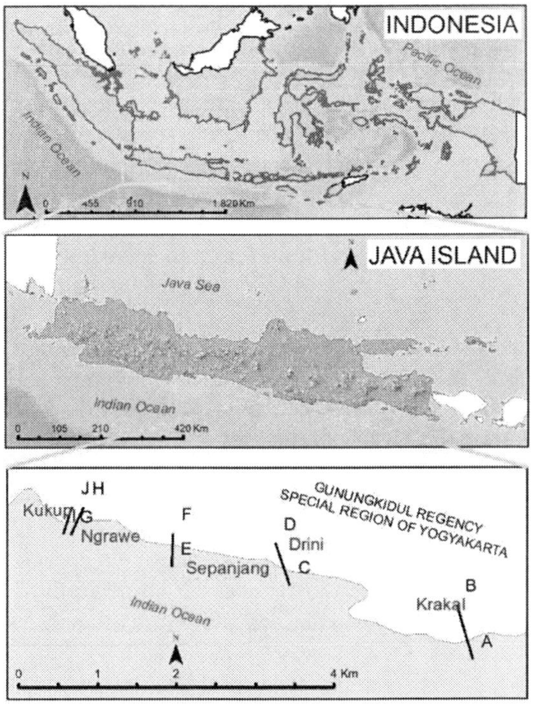

Figure 1. Study area
(Sources: http://tanahair.indonesia.go.id/portal-web/download/perwilayah, [23], and field surveys, 2021)

3. Research method

A rapid-integrated assessment was carried out to identify typology, genesis, and dynamics of coastal areas. The Kukup, Ngrawe, Sepanjang, Drini, and Krakal coastal areas have been determined by purposive sampling because of the relief variety based on identification from DEMNAS data [23]. Variations in relief can be used as a basis for the identification of coastal typology. Furthermore, the area has also developed into a center for community activities.

3.1. Coastal zoning

The coastal zoning indicator used was zoning based on [24–26], which includes coastal, coast, shore, breaker zone, and offshore (sea) areas. The determination of the boundaries between coastal zones was carried out through a tentative map compiled from DEMNAS. Then, it was validated directly through a field survey.

3.2. Typology of coastal areas

Identification of coastal areas was carried out by qualitative observation, including topography, material, land use, and landform. The Shepard classification was used to analyze the typologies of coastal areas. [27] divides coastal areas into two main typologies based on their genesis, namely primary and secondary coastal typologies. Primary coastal typology was formed not as a result of marine activity but by the process of volcanism, diastrophism, and the influence of erosion and sedimentation from the mainland. Primary coastal typology consists of land erosion, sub-aerial deposition, volcanic, and structurally shaped coast. Secondary coastal typology was formed due to marine activities such as waves and the activities of the organism. This typology can be developed starting with a primary coast or directly formed due to marine activity. Secondary coastal typology consists of wave erosion coast, marine deposition, and coast built by organisms.

3.3. Coastal area potential resources

Analysis of coastal area potential resources was carried out based on field surveys and interpretation of Small-Format Aerial Photograph taken by UAV-Drone. That was conducted on several spatial aspects in coastal areas. Potential resources in coastal areas include elements of air, water, and land space. That space contains four main categories of resources, namely space, abiotic minerals, and biotic resources. Space includes space for users' activities like agriculture, fishing, mariculture, industry, housing, recreation, etc. Abiotic minerals include building materials (e.g., sand and rock), oil and gas, water; not to mention air in good qualities can be included in this category. Furthermore, biotic resources include flora and fauna that are utilized for the provision of protein and carbohydrates.

3.4. Potential hazards/threats to coastal areas

Potential hazards/threats to coastal areas were identified based on geological settings, geomorphological characteristics, and community activities in coastal areas. Geological and geomorphological data were obtained from Geological Maps, DEMNAS, and field observations. In addition, the analysis was also conducting a literature study on the history of disaster events and literature studies.

3.5. Disaster-resilient coastal area management planning

Disaster-resilient coastal area management planning was a management concept based on identifying potential resources and hazards of coastal areas. Disaster-resilient coastal area management planning in this study included attributes or facilities to reduce the vulnerability of the coastal regions. These attributes or facilities included statutory regulations and physical facilities to reduce disaster risk.

4. Result and discussion

4.1. Coastal zoning

The coastal zone is a transitional area between land and sea; the seaward boundary includes the breaker zone to the continental shelf. The landward boundary is the final boundary for marine influences such as topography, cliffs, or other marine activities (Fig. 2). Slightly different, the coast in some articles is mentioned as the same zone as the coast, but they have differences. The coast is part of the coastal zone, emphasizing the geomorphological processes that work in coastal areas. The coast boundary towards the land is the final boundary of the sea-influenced features. The shore zone borders the coast towards the sea. The boundary between the two is the highest tide line, commonly referred to as the coastline, while the shore zone boundary towards the sea is the lowest low tide line or the so-called shoreline. Tidal conditions strongly influence a shore, so it is referred to as a tidal zone under normal conditions. A shore physically can consist of material in the form of sand (beach) resulting from material deposition by waves or shore without a beach on a coastal area in the state of a cliff.

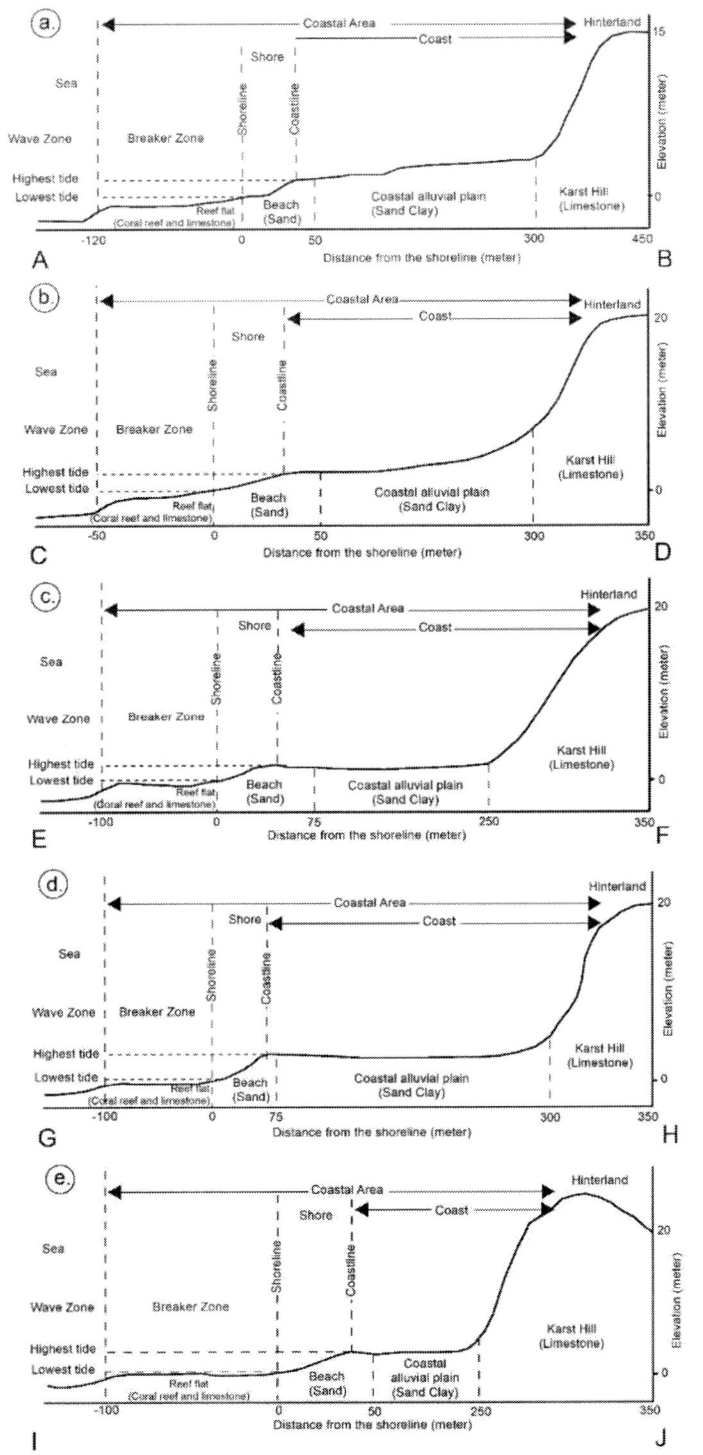

Figure 2. Cross-sectional profiles of the Krakal, Sepanjang, Kukup, Ngrawe, and Drini coastal area (Sources: http://tanahair.indonesia.go.id/portal-web/download/perwilayah, [23], and field surveys, 2021)

Furthermore, the lowest low tide line and the highest high tide line are also the boundaries of the breaker zone, where the breaker zone line at the lowest ebb is the boundary of the offshore area towards the land. The offshore area is a sea area that is directly adjacent to the coastal zone. The offshore area has no longer experienced material deposition from the land area, so the dominant processes are marine.

The analysis showed that Kukup, Ngrawe, Sepanjang, Drini, and Krakal consist of almost similar coastal zones because these areas were still in the same landscape and driving force, not to mention forming material. The difference was the dynamics of the coastal zone, for example, the development of the management of the coastal area to become a tourist center. The growth of Kukup, Ngrawe, Sepanjang, Drini, and Krakal to become tourist attractions also encourages the development of restaurants, hotels, attraction infrastructures, and other means of fulfilling needs.

4.2. Typology of coastal areas

Qualitative observation carried out the typology identification, including topography, materials, land use, and landforms (Fig. 3). The typological classification of coastal areas used was according to Shepard [27].

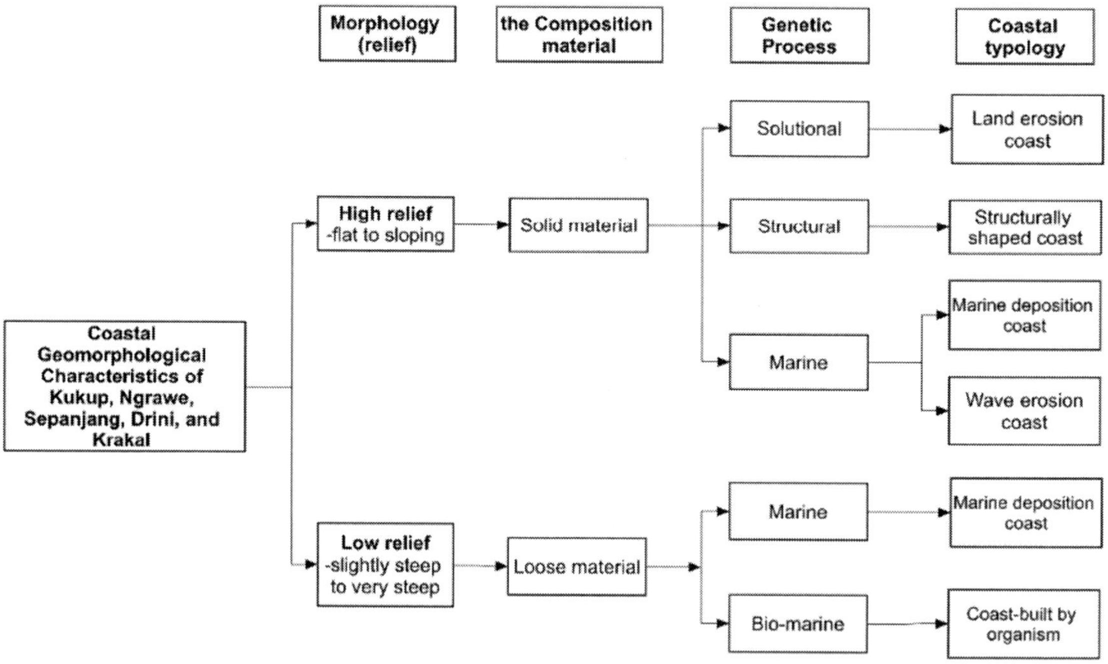

Figure 3. Coastal typology analysis of Kukup, Ngrawe, Sepanjang, Drini, and Krakal

The primary coastal typology in the Kukup, Ngrawe, Sepanjang, Drini, and Krakal area consists of a land erosion coast and a structurally shaped coast (fault coast). It was formed due to the lifting of the Wonosari Formation at the end of the Pliocene to the beginning of the Pleistocene. Secondary coastal typology was constructed due to the influence of marine activity. This typology can be developed by starting with primary coasts or directly formed due to marine activity. The secondary coastal typology in the Kukup, Ngrawe, Sepanjang, Drini and Krakal area consists of a marine deposition coast, a coast built by the organism, and a wave erosion coast.

The formation of typology begins with a structurally shaped coast (fault coast). This typology was strongly influenced by tectonic activity, namely the uplift of southern Java. This typology was characterized by the appearance of cliffs directly adjacent to the sea (cliff). The appearance of this typology is presented in Fig. 4, where shadows in Small-Format Aerial Photograph characterize the cliff.

Figure 4. Structurally Shaped Coast at Krakal coastal area
(Source: Small-Format Aerial Photograph)

After forming a structurally shaped coast (fault coast) typology, the other typology created was the wave erosion coast. It was developed because of the interface zone between the cliff (fault coast) and the waves from the ocean. The energy from the strong waves intensively hit the cliff with limestone as the primary material to produce a recess called a notch. Notch in a relatively long time will get deeper and collapse. The material from the collapse will stand as a separate object around the cliff and was called a stack (rock collapse at the bottom of the cliff). Apart from notches and stacks, this interaction also produced distinctive formations such as marine terraces, beaches, sea arcs, and separate islands. These typical formations can be observed through aerial photography or direct observation in the field (Figure 5).

Figure 5. Wave Erosion Coast at Kukup coastal area
(Source: Small-Format Aerial Photograph)

Furthermore, the typology formed was the marine deposition coast. This typology was depicted by the formation of coastal sandstone with marine sedimentary material. This typology was developed by wave action that deposited material from coastal erosion and underground rivers along the shore area. Moreover, this typology was also dominantly shaped in the bays, which were known as pocket beaches. The appearance of white sand in the bays produces a beautiful panorama. Most coastal areas with pocket beaches develop into tourist attractions such as Baron, Kukup, and Krakal.

Figure 6. Marine Deposition Coast at Sepanjang coastal area
(Sources: Small-Format Aerial Photograph)

The land erosion coast typology was identified from erosion on a cliff with hilly morphology. In addition, the dominant process was the chemical erosion process in the karst landscape or well-known as the solutional process. The coast built by organism typology was shaped due to the activities of animals or plants, including coral reefs formed by algae and oysters or plants such as mangroves or marsh grasses. This typology was generally shaped in areas with tropical climates such as Indonesia. This typology was not formed throughout the year but can be found in certain seasons and was characterized by seagrass and seaweed fields. The coastal areas of Kukup, Ngrawe, Sepanjang, Drini, and Krakal were overgrown with seagrass species of *Thalassia hemprichii* and also various types of seaweed, including *G. salicornia, Padina gymnospora, Gracillaria lichenoides, Ulva lactuca, G. reticulata, Chaetomorpha crassa, Valonia aegagropila, Acanthopora muscoides,* dan *Caulerpa racemose* [28].

4.3. Coastal area potential resources

Potential resources that can be utilized include fishery resources, natural beauty for tourism objects, agriculture. The fishery potential developing in Gunungkidul was marine fisheries, and most fishermen concentrate their post-catch activities on the Baron Coast, west of Ngrawe. The types of fish caught in Gunungkidul are lobster, stingray, tuna, pomfret, anchovies, tuna, and skipjack tuna [29]. The fishery potential developing in Gunungkidul was marine fisheries and most fishermen concentrate their post-catch activities on the Baron Coast, west of Ngrawe. The types of fish caught in Gunungkidul were lobster, stingray, tuna, pomfret, anchovies, tuna, and skipjack tuna. In addition, ornamental fisheries and grass were also increasing due to the physical characteristics of the Gunungkidul coastal area, which strongly support the requirements for growing seaweed and ornamental fish. Types of ornamental fish caught in Gunungkidul include Butterfly, Blue Stone, Tragger, and Scorpion [29]. Furthermore, the coastal areas of Ngrawe, Kukup, Drini, Sepanjang, and Krakal have shallow terraces or areas that can be used for seaweed cultivation because the sun can still penetrate the water.

The agricultural potential that can be developed in the coastal areas of Ngrawe, Drini, Kukup, Sepanjang, and Krakal was corn, soybean, peanut, and cassava farming [29]. These types of plants can grow and develop even though there was no abundant water supply. These types of plants were Gunungkidul's primary commodities and provided a significant contribution to regional income.

The most promising potential was tourism because it was supported by the physical characteristics of the Gunungkidul coastal area with small bays and white sand beaches (Fig. 7, 8, 9, 10). Sand material in Ngrawe, Kukup, Drini, Sepanjang, and Krakal comes from limestone, dissolved by water, and deposited back into the mainland to form the beach. The panorama of the coastal area of Gunungkidul was as beautiful as Bali and Lombok. However, Gunungkidul's local culture was not as attractive as Bali and Lombok, closely related to local wisdom. Therefore, the culinary potential of Gunungkidul has to be developed simultaneously with tourism potential to increase for tourists.

Figure 7. The landscape of Ngrawe and Kukup coastal area
(Sources: Small-Format Aerial Photograph)

Figure 8. The landscape of Sepanjang coastal area
(Sources: Small-Format Aerial Photograph)

Typical culinary potentials such as *tiwul*, *fried locust*, and *sayur lombok ijo* were exciting to develop. *Tiwul* is a staple food substitute for rice made from cassava. It is often used as a substitute for rice by the Javanese because of its soft texture and can be processed in various ways. Locusts are not commonly consumed, but Gunungkidul can be processed into delicious food with a savory, sweet, and spicy taste. Locusts are not dangerous for human consumption because they contain high protein.

Furthermore, *sayur lombok ijo* is also very delicious. *Lombok* means chili, and *Ijo* means green (color). Based on these names, it can be identified that this food consists of red chilies, green chilies, coconut milk, *tempe*, tofu, eggplant, onions cooked in a savory sauce. The development of this extraordinary culinary can increase tourist attraction and move the rural economy because these foods are the daily menu of the people of Gunungkidul.

Figure 9. The landscape of Drini coastal area
(Sources: Small-Format Aerial Photograph)

Figure 10. The landscape of Krakal coastal area
(Sources: Small-Format Aerial Photograph)

4.4. Potential hazards/threats to coastal areas

4.4.1. Earthquake
The Krakal, Drini, Sepanjang, Ngrawe, and Kukup coastal areas are directly adjacent to the Hindian Ocean, where a megathrust zone potentially triggers earthquakes. Earthquakes cannot be predicted when they occur, their magnitude, epicenter, and return period. Earthquakes can be a single hazard and can cause considerable losses to the community, both property, psychological, and fatalities. Furthermore, an earthquake can also be followed by other catastrophic events, namely a tsunami if its magnitude, epicenter, and depth meet the requirements for triggering a tsunami.

4.4.2. Tsunami
A tsunami is a disaster that is very destructive and potentially causes considerable losses to humans. Earthquake events can be triggered by tectonic activity potential to cause a tsunami that can cause harm

to people in the coastal areas of Krakal, Drini, Sepanjang, Ngrawe, and Kukup [30,31]. Tsunamis are generally triggered by earthquakes in subduction zones, seafloor landslides, seabed volcanic eruptions, and extraterrestrial forces [32]. Based on the geological setting of Indonesia, earthquakes in subduction zones and seafloor volcanic eruptions are the dominant triggers for tsunamis that occur in Indonesia [30]. The coastal area of Gunungkidul is directly opposite the Indian Ocean that is prone to be affected by a tsunami. Historical data on tsunami events shows that Indonesia has been frequently affected by tsunamis since the 1800s [33]. Tsunami events in the last 30 years that caused casualties occurred in 1994 and 2006. In June 1994, the southeastern region of East Java and July 2006 hit the Widarapayung, Cikembulan, Parang Kusumo, Pangandaran areas with wave heights of 1.8 - 4.6 meters [34].

4.4.3. Rip Current

Rip currents are a channel of powerful, narrow, fast-moving water that can carry floating objects away from the shore, presenting one of the most common coastal hazards to swimmers [35]. Unfortunately, most tourists are unaware of rip currents and their threats and how to avoid them. Rip currents are narrow channels of fast-moving water that are widely distributed along the southern coast of Java. Rip Current moves up to eight feet per second. They occur on almost any beach where waves break and act as "sea rivers", marine organisms, moving sand, and other offshore material. In terms of their location, the coastal areas of Krakal, Drini, Sepanjang, Ngrawe, and Kukup can form rip currents. Therefore, it needs warning signs and strict supervision of tourist activities in areas prone to rip currents because most of the victims dragged by rip currents were found dead.

4.4.4. Tidal wave

Tidal waves are waves of sea water due to the gravitational force between the earth, sun and moon. Tidal waves are different from tsunamis and are not related to tsunamis. Based on the energy generation, the tsunami and tidal waves are very different. Tsunamis are caused by large earthquakes, volcanic eruptions, underwater erosion, or land erosion so that large amounts of debris fall into the water. Tidal periodically hits the coastal areas of South Java and causes coastal dynamics to cause losses to the community. This is because the tidal wave causes damage to buildings located in coastal regions in some areas of Yogyakarta [36]. Therefore, mitigation measures such as planning for the development of tourist areas must be carried out, especially for the arrangement of tourist spaces that are directly adjacent to the sea

4.4.5. Rockfall (in the structurally shaped coast area)

The coastal areas of Krakal, Sepanjang, Drini, Ngrawe, and Kukup consist of beaches with sandy and rocky materials. In the regions with rocky materials, there is a high potential for a rockfall to occur and cause casualties because the area has developed into a tourist attraction. The Krakal, Drini, Sepanjang, Ngrawe, and Kukup material is dominated by limestone, which is very massive but easily soluble in water. This results in the emergence of cracks in the existing rock. As time flies, the fracture will enlarge, causing the rock to separate into separate parts. The isolated rocks will likely become rockfalls, especially in areas with wave erosion coast typologies and fault coasts. If the incident is not identified earlier, it can endanger visitors to the tourist attraction.

4.4.6. Topographic changes

Topographic changes are likely to occur in the coastal areas of Drini, Krakal, Sepanjang, Ngrawe, and Kukup due to the influence of landform factors and land resource potential. Those were dominated by landforms from solutional processes with materials in the form of limestone and various surface reliefs so that they have the potential to be mined. If mining is carried out, the flow system in a typical karst area can be disturbed or even changed. Topographical changes associated with land use change and mining significantly alter drainage networks and the direction of runoff. This will cause changes in connectivity that can affect the hydrological system. This can affect the quality and quantity of water resources in the region. In addition, the potential of land resources, especially for the development of tourism objects, will trigger changes in land use from open land/agricultural land to built-up land such as settlements, inns, restaurants, shops, etc. The process of land-use change must carry out engineering or relief changes. Changes in relief will affect topographic conditions to increase the area's vulnerability to rockfall hazards.

4.4.7. Landuse changes
The coastal areas of Krakal, Drini, Sepanjang, Ngrawe, and Kukup were experiencing land-use changes that potentially increased disasters' vulnerability. The dominant change that occurs was the growth of tourist facilities. Changes in land use in the area can improve the welfare of the community. However, on the other hand, it creates pressure on coastal ecosystems and likely causes damage if it is not accompanied by strict and appropriate coastal area management regulations [37,38].

The Krakal, Drini, Sepanjang, Ngrawe, and Kukup areas have developed into attractive tourist centers so that many local and foreign tourists visit. This increase is creating many opportunities to do business in the tourism sector. Therefore, many people, especially developers or entrepreneurs, have established tourism infrastructure facilities such as lodging, hotels, restaurants, souvenir shops, etc. [39]. These changes have the potential to be uncontrolled, so that the potential for damage to the coastal environment can occur. Moreover, the vulnerability to disasters will increase, the potential for losses also increases.

4.5. Disaster-resilient coastal area management planning

4.5.1. Spatial Planning Regulation
The coastal areas of Krakal, Drini, Sepanjang, Ngrawe, and Kukup are part of the Gunungsewu karst area, which has been designated as a UNESCO Global Geopark (UGG) since 2019. Therefore, sustainable ecosystem management and disaster risk reduction must be critical in determining coastal spatial planning policies [40]. In addition, spatial planning in the area must also be based on disaster risk reduction, as stated in [41–43]. To achieve an optimal form of planning, monitoring, and evaluation of spatial planning, policies must be synchronized from the central to district/city levels. Therefore, it can be concluded that a strict spatial policy and considering disaster risk is needed for the Krakal, Drini, Sepanjang, Ngrawe, and Kukup areas because these areas have developed into new Center Business District, namely tourist attractions.

4.5.2. Early Warning System (EWS)
EWS is a series of warning systems for signs or events of a disaster. An early warning system is needed in the Krakal, Drini, Sepanjang, Ngrawe, and Kukup areas because of the many potential threats in these areas. However, the EWS must be made with language and signs that are readily accepted by the community. Along with the development of the digital era, an accurate EWS with various media is needed in disaster-prone areas. In addition, dissemination of the existence of EWS in the Krakal, Drini, Sepanjang, Ngrawe, and Kukup areas to the public or visitors must be carried out. Indeed, EWS does not guarantee any victims if a disaster strikes, but it helps reduce the potential impact of disasters on humans. The earlier the warning through the EWS is given, the faster the community will receive and respond so that there is enough time to save themselves.

4.5.3. Evacuation Route and Assembly Point (Evacuation Place)
An evacuation route is necessary, especially in areas that have developed into tourist attractions. It is intended so that visitors have the opportunity to save themselves through the path that has been provided. Evacuation routes from crowded centers or tourist attractions are essential and clear directions must be given to reduce disaster risk. It must be equipped with clear advice and visible to visitors. The final destination of the evacuation route is an assembly point (evacuation place) as a temporary shelter until the rescue teams arrive. The evacuation place must be safe from disaster; for example, in the case of a

tsunami hazard, the evacuation place must be in a high place (the highest place from the surrounding area), assuming that the incoming tsunami waves will not destroy the site.

4.5.4. Diseminasi Informasi berbasis Digital
Dissemination of information is essential, especially about potential disasters. Nowadays, the dissemination of information can be done more quickly and widely by utilizing digital technology. One of the digital technologies that are easily accessible and fast in disseminating information is social media. Social media contains information about tourist attractions in Krakal, Drini, Sepanjang, Ngrawe, and Kukup. Information sharing is not only about the potential for attracting tourism but also the potential hazards. This is done so that visitors to the tourist attraction have prepared themselves both in terms of knowledge and technically before visiting it. Information disseminated can include tourism potential, potential hazards, EWS, evacuation routes, assembly points, and how to evacuate yourself. The hope is that if tourists before visiting have received this information when they arrive at the location, they can immediately find instructions for evacuation.

4.5.5. Building retrofitting
Building retrofitting is significant for residential and tourism infrastructure in the Krakal, Drini, Sepanjang, Ngrawe, and Kukup areas. Building retrofitting carried out in the area is planned and implemented based on earthquake risk. This was done because of the potential for earthquakes and the condition of existing buildings that did not meet earthquake-resistant building standards. Buildings are reinforced to improve their systems or structures after initial construction. This work can improve the amenities for occupants and the performance of the building in the event of an earthquake.

5. Conclusion
The Krakal, Drini, Sepanjang, Ngrawe, and Kukup were composed of wave erosion coast, structurally shaped coast, marine deposition coast, land erosion coast, and coast-built by an organism. Potential resources that can be utilized include fishery resources, natural beauty for tourism objects, agriculture, and culinary. Furthermore, the potential disasters that threaten the area were earthquake, tsunami, Rip current, tidal wave, rockfall (in the structurally shaped coast area), topographic changes, and land-use changes. Therefore, disaster-resilient coastal area management planning was needed to reduce disaster risk, such as Spatial Planning Regulation, Early Warning System, Evacuation Route and assembly point (shelter), Digital-based Information Dissemination, and building retrofitting. Studies on the characteristics of coastal areas, which include potential resources and potential hazards/threats, need to be carried out in-depth in Indonesia. The utilization of natural resources in coastal areas must be based on sustainable development and disaster risk reduction. This study has only analyzed several coastal regions of Gunungkidul and has not carried out quantitative or technical analysis, so it is necessary to deepen the research regionally, data collection techniques, and analytical techniques to formulate appropriate policy recommendations.

References
[1] Dávila O G, Stithou M, Pescaroli G, Pietrantoni L, Koundouri P, Díaz-Simal P, Rulleau B, Touili N, Hissel F and Penning-Rowsell E 2014 Promoting resilient economies by exploring insurance potential for facing coastal flooding and erosion: Evidence from Italy, Spain, France and United Kingdom *Coast. Eng.* **87** 183–92
[2] Landry C E, Shonkwiler J S and Whitehead J C 2020 Economic Values of Coastal Erosion Management: Joint Estimation of Use and Existence Values with recreation demand and contingent valuation data *J. Environ. Econ. Manage.* **103** 102364
[3] Venancio K K, Garcia P D, Gireli T Z and Corrêa T B 2020 Hydrodynamic modeling with scenario approach in the evaluation of dredging impacts on coastal erosion in Santos (Brazil) *Ocean Coast. Manag.* **195**
[4] Ardeshiri A, Swait J, Heagney E C and Kovac M 2019 Willingness-to-pay for coastline protection

in New South Wales: Beach preservation management and decision making *Ocean Coast. Manag.* **178** 104805

[5] Yin P, Duan X, Gao F, Li M, Lü S, Qiu J and Zhou L 2018 Coastal erosion in Shandong of China: status and protection challenges *China Geol.* **1** 512–21

[6] Sangha K K, Stoeckl N, Crossman N and Costanza R 2019 A state-wide economic assessment of coastal and marine ecosystem services to inform sustainable development policies in the Northern Territory, Australia *Mar. Policy* **107** 103595

[7] Dunning K H 2020 Building resilience to natural hazards through coastal governance: a case study of Hurricane Harvey recovery in Gulf of Mexico communities *Ecol. Econ.* **176** 106759

[8] Domingues R B, Santos M C, de Jesus S N and Ferreira Ó 2018 How a coastal community looks at coastal hazards and risks in a vulnerable barrier island system (Faro Beach, southern Portugal) *Ocean Coast. Manag.* **157** 248–56

[9] Mukhopadhyay A, Ghosh P, Chanda A, Ghosh A, Ghosh S, Das S, Ghosh T and Hazra S 2018 Threats to coastal communities of Mahanadi delta due to imminent consequences of erosion – Present and near future *Sci. Total Environ.* **637–638** 717–29

[10] Kamil P A, Utaya S, Sumarmi and Utomo D H 2020 Improving disaster knowledge within high school students through geographic literacy *Int. J. Disaster Risk Reduct.* **43** 101411

[11] Marfai M A, Sunarto, Khakim N, Cahyadi A, Rosaji F S C, Fatchurohman H and Wibowo Y A 2018 Topographic data acquisition in tsunami-prone coastal area using Unmanned Aerial Vehicle (UAV) *IOP Conference Series: Earth and Environmental Science* vol 148

[12] Nurhidayah L and McIlgorm A 2019 Coastal adaptation laws and the social justice of policies to address sea level rise: An Indonesian insight *Ocean Coast. Manag.* **171** 11–8

[13] Marfai M A 2014 Impact of sea level rise to coastal ecology: A case study on the northern part of java island, indonesia *Quaest. Geogr.* **33** 107–14

[14] Wibowo P L A, Hartoko A and Ambariyanto A 2015 Land Subsidence Affects Coastal Zone Vulnerability (Pengaruh Penurunan Tanah Terhadap Kerentanan Wilayah Pesisir) *Ilmu Kelaut. Indones. J. Mar. Sci.* **20** 127

[15] Dewi R S and Bijker W 2020 Dynamics of shoreline changes in the coastal region of Sayung, Indonesia *Egypt. J. Remote Sens. Sp. Sci.* **23** 181–93

[16] Rumson A G, Hallett S H and Brewer T R 2017 Coastal risk adaptation: the potential role of accessible geospatial Big Data *Mar. Policy* **83** 100–10

[17] Sánchez-García E, Balaguer-Beser A and Pardo-Pascual J E 2017 C-Pro: A coastal projector monitoring system using terrestrial photogrammetry with a geometric horizon constraint *ISPRS J. Photogramm. Remote Sens.* **128** 255–73

[18] Kerguillec R, Audère M, Baltzer A, Debaine F, Fattal P, Juigner M, Launeau P, Le Mauff B, Luquet F, Maanan M, Pouzet P, Robin M and Rollo N 2019 Monitoring and management of coastal hazards: Creation of a regional observatory of coastal erosion and storm surges in the pays de la Loire region (Atlantic coast, France) *Ocean Coast. Manag.* **181**

[19] Triyanti A, Bavinck M, Gupta J and Marfai M A 2017 Social capital, interactive governance and coastal protection: The effectiveness of mangrove ecosystem-based strategies in promoting inclusive development in Demak, Indonesia *Ocean Coast. Manag.* **150** 3–11

[20] Wibowo Y A and Ronggowulan L 2021 Potential analysis and community-based sustainable tourism development strategy (a case of kampung menjing, sukoharjo regency, central java, indonesia) *IOP Conf. Ser. Earth Environ. Sci.* **683**

[21] Marfai M A, Cahyadi A and Anggraini D F 2013 Tipologi, dinamika, dan potensi bencana di pesisir kawasan karst kabupaten gunungkidul *Forum Geogr.* **27** 147–58

[22] The Indonesian Geological Research and Development Centre 1977 Peta Geologi Lembar Yogyakarta (Geological Map of Yogyakarta)

[23] BIG 2018 Using DEMNAS Data

[24] Beatley T, Brower D and Anna, K S 1994 *An Introduction to Coastal Zone Management* (Washington DC: Island Press)

[25] Bird E 2008 *Coastal Geomorphology: an Introduction (Second Edition)* (Chichester: John Wiley & Sons Ltd)
[26] Davidson-Arnott R 2010 *Introduction to Coastal Processes and Geomorphology* (New York: Cambridge University Press)
[27] Shepard F P 1973 *Submarine Geology* (New York: Harper and Row)
[28] Yusron E 2015 Di Pantai Gunung Kidul , Yogyakarta Community Structure of Echinoderms (Holothuroidea , Echinoidea , and Ophiuroidea) in Seagrass Beds At the Beach Area of Gunung Kidul , Yogyakarta **24** 73–82
[29] Badan Pusat Statistik (Central Bureau for Statistic) 2021 *Kabupaten Gunungkidul dalam Angka 2021* (Gunungkidul: Badan Pusat Statistik (Central Bureau for Statistic))
[30] Marfai M A, Cahyadi A, Fatchurohman H, Rosaji F S C and Wibowo Y A 2019 Tsunami preparedness and environmental vulnerability analysis in Kukup Beach, Gunungkidul, Indonesia *IOP Conf. Ser. Earth Environ. Sci.* **256**
[31] Marfai M A, Sunarto, Khakim N, Fatchurohman H, Cahyadi A, Wibowo Y A and Rosaji F S C 2019 Tsunami hazard mapping and loss estimation using geographic information system in Drini Beach, Gunungkidul Coastal Area, Yogyakarta, Indonesia *E3S Web of Conferences* vol 76
[32] Honarmand M, Shanehsazzadeh A and Zandi S M 2020 3D numerical simulation of tsunami generation and propagation, case study: Makran tsunami generation and penetrating in Chabahar Bay *Ocean Eng.* **218** 108109
[33] Triyono R, Prasetya T, Daryono, Anugrah S D, Sudrajat A, Setiyono U and Gunawan I 2018 *Katalog Tsunami Indonesia Tahun 416-2018*
[34] Setiyono U, Gunawan I, Priyobudi, Yatimantoro T, Imananto R T, Ramdhan M, Hidayanti, Anggraini S, Rahayu R H, Hawati P, Yogaswara D S, Julius A M, Apriani M, Harvan M, Simangunsong G and Kriswinarso T 2019 *Katalog Gempabumi Signifikan dan Merusak 1821-2018* vol 1, ed T Prasetya and Daryono (Jakarta: Pusat Gempabumi dan Tsunami Kedeputian Bidang Geofisika Badan Meteorologi, Klimatologi dan Geofisika)
[35] Fatchurohman H, Khasanah A N and Cahyadi A 2021 Identification of Rip Current Hazards Using Fluorescent Dye And Unmanned Aerial Vehicle (A Case Study Of Drini Beach, Gunungkidul, Indonesia) *Nat. Hazards Earth Syst. Sci. Discuss.* **2021** 1–20
[36] Indratmoko S, Al Hakim I B and Satrio Guntoro W 2019 Analysis of People Response on Twitter Towards Tidal Wave Disaster in the Southern Coast of Yogyakarta Special Province (Case Studies: Parangtritis Beach, Bantul Regency) *KnE Soc. Sci.*
[37] Aswani S 2019 Perspectives in coastal human ecology (CHE) for marine conservation *Biol. Conserv.* **236** 223–35
[38] Chiang Y C and Chang H P 2018 Cultural dimensions of risk perceptions: A case study on cross-strait driftage pollution in a coastal area of Taiwan *J. Environ. Manage.* **206** 123–33
[39] Bañas J C C, Subade R F, Salaum D N and Posa C T 2020 Valuing vanishing coasts: The case of Miagao coastline in Southern Iloilo, Philippines *Ocean Coast. Manag.* **184** 105008
[40] Rangel-Buitrago N, Neal W J and de Jonge V N 2020 Risk assessment as tool for coastal erosion management *Ocean Coast. Manag.* **186** 105099
[41] Law of the Republic of Indonesia Number 24 of 2007 concerning Disaster Management 2007 *Law of the Republic of Indonesia Number 24 of 2007 concerning Disaster Management*
[42] Law of the Republic of Indonesia Number 1 of 2014 concerning Amendments to Law Number 27 of 2007 concerning Management of Coastal Areas and Small Islands 2014 *Law of the Republic of Indonesia Number 1 of 2014 concerning Amendments to Law Number 27 of 2007 concerning Management of Coastal Areas and Small Islands*
[43] Law of the Republic of Indonesia number 26 of 2007 concerning Spatial Planning 2007 *Law of the Republic of Indonesia number 26 of 2007 concerning Spatial Planning*

IOP Conf. Series: Earth and Environmental Science **986** (2022) 012025 doi:10.1088/1755-1315/986/1/012025

Modelling of Volcano Lawu Fault Structure Using Gravity Anomaly to Determine Landslides Potential

Budi Legowo*, **Rahmi Nailatunisrina, Hery Purwanto, Budi Purnama and Wiwit Suryanto**
Laboratory of Geophysics, Department of Physics, Sebelas Maret University
Jl. Ir. Sutami 36A Surakarta 57126, INDONESIA

*Email: pakbeel@staff.uns.ac.id

Abstract. Gravity anomaly data is used in modelling the Lawu volcano fault. The fault modelling uses a total of 10,126 data points for free air anomaly GGMplus 2013 from http://ddfe.curtin.edu.au/ with a north-south direction with a grid of ± 200 meters. Separation of regional and residual anomalies using a moving average filter and Modelling using the forward modelling method. The modelling results show the distribution of rock densities of 2.52 gr/cm3 and 2.25 gr/cm3, which are composed of tuff and breccia with lava insertion, andesite lava with a density of 2.48 gr/cm3 and 2.9 gr/cm3, tuff lapilli, and pumice stone with a density of 2.35 g/cm3 and the last is andesite and basalt rocks with a density of 2.94 g/cm3. The modelling results also show two faults, namely the Sidoramping-Lawu normal fault and the Cemorosewu fault. From the type of rock and the location of the fault, it can be seen that Mount Lawu is an area highly vulnerable to landslides.

1. Introduction

Landslide is a natural phenomenon that frequently happens. In many cases leading to significant economic losses and even fatalities. According to Pusat Vulkanologi dan Mitigasi Bencana Geologi, Indonesia has around 918 vulnerable areas of landslides. On average, more than one hundred people died due to the disaster of this landslide [1]. Knowing landslide-prone areas is very important as an early warning to reduce the damaging effect as little as possible. Landslide happen especially in hilly and mountainous areas such as Mount Lawu.

Lawu is a volcano located near the border between the provinces of Central Java and East Java. Mount Lawu was active during the quarter, and the youngest eruption probably occurred during the Holocene [2]. The geological structure of Mount Lawu can be seen from the regional geological map in the form of faults trending west-east and northwest-southeast [3]. Fault identification can be done using the gravity method, which is used to measure the density of rocks on the Earth [4]. This method is used to identify rock lithology and subsurface structures based on variations in Earth's gravity caused by variations in rock density [5].

Several studies using the gravity method have been carried out to examine the subsurface structure of the Mount Lawu area. However, research reporting the use of GGMplus data (Global Gravity Model) is still scarce. GGMplus data refers to a combination of GRACE and GOCE data with EGM2008, combined with the results of forwarding modelling of topographic gravity based on spherical harmonic expansion to increase the resolution [6]. Research using GGMplus data has been successfully applied to the areas of South Serayu [7] and Nabire [8]. The results obtained are modeled according to each geological map. This study uses GGMplus 2013 data with the modelling slices adjusted to the cross-sections on the geological map. This study focuses on identifying the distribution of gravity anomalies and identifying the fault types in the area of Mount Lawu. Gravitational anomaly value can be used as a reference to determine the type and structure of the rock beneath the surface. Rock types and structures, mutually with knowledge of fault locations, can determine the vulnerability to landslides in the Mount Lawu area.

2. Description of the Study Area

The research location is Mount Lawu with coordinates 111o 18' 32" to 111o 02' 27" South Latitude and -7o 41' 20" to -7o 41' 20" East Longitude with an area of ±506 km2 as shown in figure 1. Based on the tectonic analysis, Lawu Tua volcano initially took place during the Pleistocene. Then there was volcanic activity in the middle of the Pleistocene, which caused the collapse of the northern slopes of Lawu Tua. A graben-like basin forms a crescent moon pass, convex to the north (the Cemorosewu fissure). In the south, there is another graben (Jobolarangan Graben) which is more or less parallel to the first graben. This graben extends from Mount Jobolarangan (Selatan Cemorosewu) in the east to the west and ends in the south of Tawangmangu. Almost the entire north side of the research area is occupied by Lawu Muda products, while the southern complex is filled with Lawu Tua products (Jobolarangan) [9].

Figure 1. Map of Complete Bouguer Anomaly.

3. Material and Method

This study uses GGMplus 2013, a secondary data model of global gravity (Global Gravity Model) with the highest resolution currently, with a grid data distance of approximately 200 m in the North-South direction [7]. The data can be downloaded from the website http://ddfe.curtin.edu.au/gravitymodels/GGMplus. The data obtained from the website is a free air anomaly processed to produce a simple Bouguer anomaly through calculations using the Nettleton method. The simple Bouguer anomaly value is then corrected again through terrain correction with the help of a DEM (Digital Elevation Model) map to determine the elevation value of each data point.

The subsurface model is obtained from the forward modelling of the residual gravity anomaly. This anomaly is obtained from reducing the complete Bouguer anomaly with the residual anomaly generated from the moving average filter. This filter is obtained from spectrum analysis which converts the gravitational anomaly in one slice into the wavenumber domain (k) using Fast Fourier Transform (FFT).

The modelling is done by making slices on the residual anomaly. The slices are adjusted to the slices on the geological map of the research area. The depth obtained from the spectrum analysis is used to carry out this modelling process. The modelling process was carried out using Geosoft Oasis Montaj software. The model that has been obtained is then used as material for quantitative interpretation.

4. Result and Discussion

The description of the subsurface model and the value of the Lawu density distribution in this study is based on the analysis of gravity data which is processed to interpret the results. In addition, it is also correlated with geological information obtained from the geological map of the area. Meanwhile, it is necessary to use seismic methods to see the subsurface structure more profoundly and in more detail [3].

4.1. Bouguer Anomaly

The Nettleton method's density value, which has a correlation of 0 (uncorrelated) and is considered to affect the anomaly change, is 2.34 gr/cm3. The density value is used to perform the Bouguer correction. The result of the Bouguer correction is a simple Bouguer anomaly which is then corrected again using terrain correction with DEM maps as elevation data. The results of the field correction are plotted into a complete Bouguer anomaly map, as shown in figure 2.

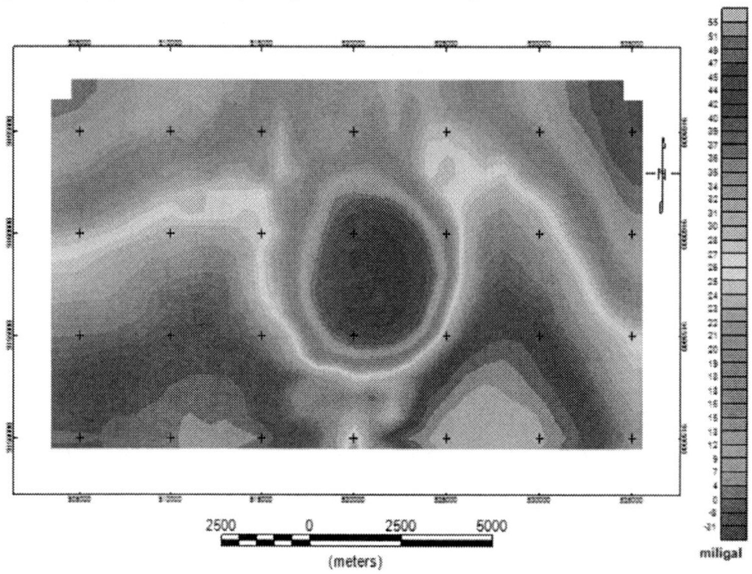

Figure 2. Map of Complete Bouguer Anomaly.

In figure 2, the lowest anomaly value is -31 mGal and the highest is 55 mGal. The difference in rock density at the survey site causes differences in the color of the CBA. (Complete Bouguer Anomaly) map. The high anomaly is marked by orange to purple color with a Bouguer anomaly value range of 34 to 55 mGal which is spread in the southern area of the research area towards the center of the research site. Based on the regional geological map, the area with the high anomaly is the complex of Mount Lawu Tua (Jobolarangan), whose rocks are Pleistocene in age.

Moderate anomalies are indicated by a light green to yellowish-orange color with anomaly values ranging from 20 to 32 mGal, located in the north-central to the southwest and south-eastern areas of the research site. The anomaly is in the Lawu Muda Volcano complex, which rocks are Holocene in age. In contrast, the low anomaly is marked with dark blue to green color with an anomaly value range of -31 to 19 mGal and was found in the middle and also the northwest and north-eastern part of the research site. A filtering process is carried out to analyze further and see the depth limit of the anomaly and the location of the fault.

4.2. Moving Average

The spectrum analysis process was carried out to determine the window width value used to separate regional anomalies from residuals using a moving average filter. For spectrum analysis, it is necessary to input distance with gravity anomaly value resulting from slicing on a complete Bouguer anomaly map to be further processed with Fast Fourier Transform.

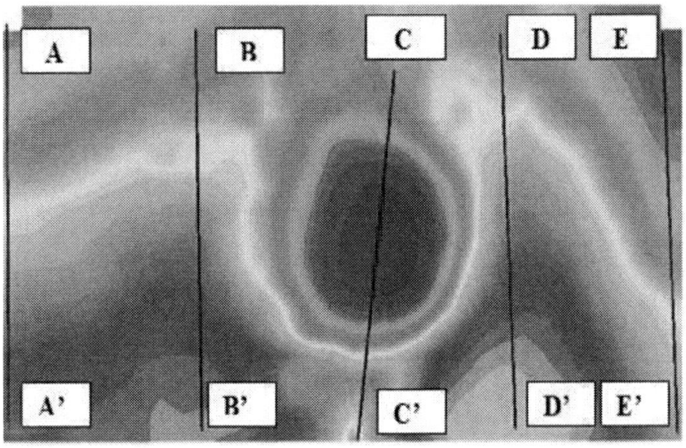

Figure 3. Slices for spectrum analysis on Complete Bouguer Anomaly.

Five paths were used on the complete Bouguer anomaly map. Spectrum analysis was performed for each path and resulted in the estimated depth of the regional and residual anomalies, as shown in table 1. The average value for the depth of the residual anomaly was used as the depth limit when conducting subsurface modelling. In this study, the estimated depth of the residual anomaly was 419.41 m. The average window width (N) of 33.2012 (rounded to 33) was used to separate regional and residual anomalies using a moving average filter.

Table 1. Power spectrum analysis to get windowing value.

Slice	Regional depth (m)	Residual depth (m)	C1	C2	k cut-off	λ	N
A-A'	-4252,2	-442,14	8,1743	4,9437	0,000848	7410,176	37,05088
B-B'	-5270,8	-461,65	8,6573	4,8334	0,000795	7902,084	39,51042
C-C'	-2319,9	-401,32	7,0302	4,7591	0,001184	5307,91	26,53955
D-D'	-5225,9	-442,56	8,6684	4,8327	0,000802	7835,496	39,17748
E-E'	-2106	-349,38	6,9269	4,6011	0,001324	4745,537	23,72768
Average	-3834,96	-419,41				Window length	33,2012

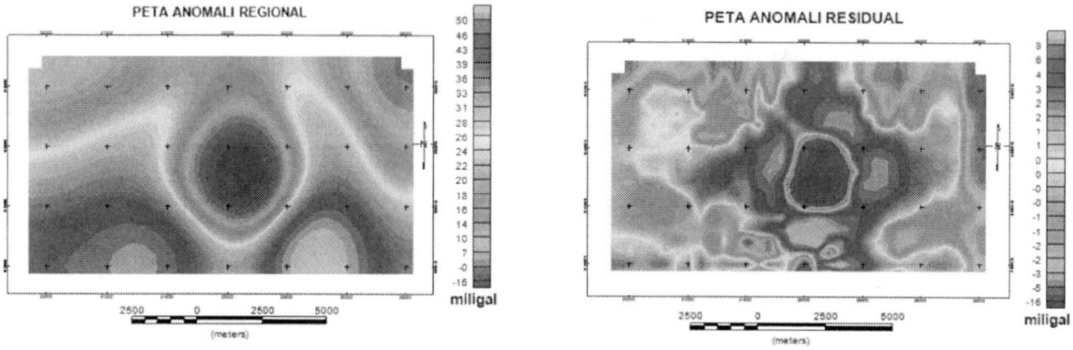

a b

Figure 4. (a) Map of the regional anomaly **(b)** Map of the residual anomaly.

As regional anomaly does not allow to describe of the subsurface structure in a detailed manner, a shallower structure is needed in residual anomalies. A moving average filter is used to separate the two anomalies, and a regional anomaly map and a residual anomaly map are generated, as shown in figure 4. Regional anomalies are deep anomalies which source comes from the inner Earth, such as the Earth's crust. The characteristic of this anomaly is that it has a low frequency. In figure 4(a), the anomaly value is -15 to 50 mGal. In contrast, the residual anomaly is a shallow anomaly characterized by high frequency. Comparing the results of the two maps, the residual anomaly exhibited more complex pattern because it describes an anomaly pattern with a shorter wavelength. The range of anomaly values on the residual anomaly map was -16 to 8 mGal, as displayed in in figure 4(b) with a low anomaly located in the middle of the research site.

4.3. 2D Modelling

The results of the forward Modelling show the cross-section of each track. Line A-B crosses from the northwest to the southeast of the research site. This track passes through the southern slopes of Mount Lawu which has a tight alignment. The lineaments on the southern slope contain the Sidoramping-Lawu Fault and the Cemorosewu Fault. The Modelling is carried out by taking into account the results of residual anomalies and their correlation with the regional geological map of the research site.

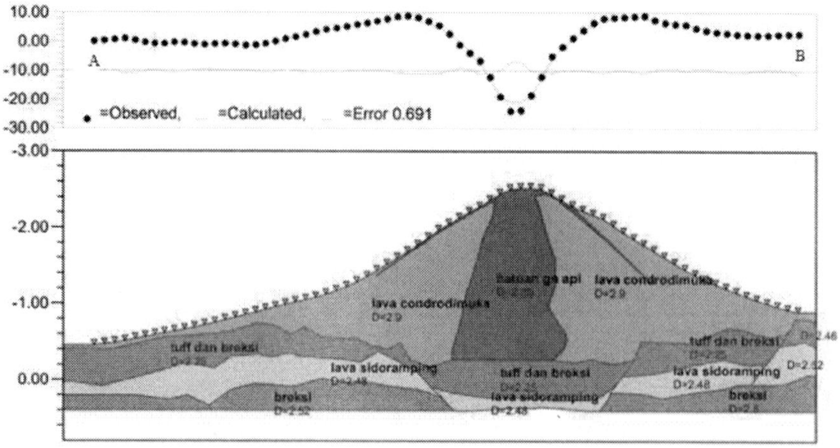

Figure 5. Structure model slice A-B based on forward modelling of the gravity data.

IOP Conf. Series: Earth and Environmental Science **986** (2022) 012025 doi:10.1088/1755-1315/986/1/012025

The subsurface structure on the A-B path is modeled, consisting of several layers of rock. The northwest area of the research site is Lawu Volcanic Rock (Qvl), with a density of 2.25 gr/cm3 which is composed of tuff and breccia with lava insertion. The rocks of the Lawu Volcano are Holocene in age. The next layer is the Condrodimuka Lava (Qvcl), with a density of 2.9 g/cm3 which is bounded by two faults, namely the Sidoramping-Lawu fault and the Cemorosewu fault. The Sidoramping-Lawu Fault limits this lava that flows to the northwest, while the Cemorosewu Fault limits the lava flow to the south. The Condrodimuka Lava is Holocene in age. Next is the Sidoramping Lava layer (Qvsl) which has a density of 2.48 gr/cm3. This layer is composed of andesite and profinite. Sidoramping Lava has a flow structure originating from the Mount Sidoramping complex and the flow direction is westward. The Sidoramping Lava is Pleistocene in age. The last layer in the research site is the Jobolarangan Breccia (Qvjb), which density of 2.52 g/cm3. This layer is found on the south-eastern slope of the research site and the eastern region of Mount Jobolarangan.

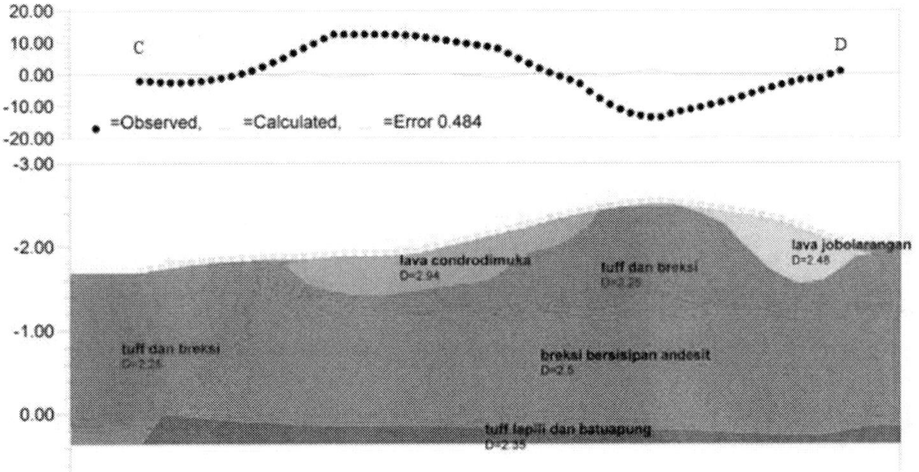

Figure 6. Structure model slice C-D based on forward modelling of the gravity data.

The second trajectory made is the C-D trajectory. The results of the C-D trajectory modelling is displayed in figure 6. The C-D trajectory crosses from the southwest to the northeast of the research site and crosses one fault, namely the Cemorosewu Fault. The first layer is the Condrodimuka Lava layer (Qvcl), with a density of 2.94 gr/cm3. Unlike in the previous trajectory, the Condrodimuka Lava layer seen in this trajectory is only one piece and leads to the south which the Cemorosewu Fault bounds. The second layer is Jobolarangan Lava (Qvjl), with a density of 2.48 gr/cm3. This layer is Pleistocene in age and is found in the Mount Jobolarangan complex.

The third layer is the Lawu Volcano Rock layer (Qvl), with a density of 2.25 gr/cm3 which consists of tuff and breccia. The next layer is the Jobolarangan Breccia (Qvjb) which has a density of 2.5 gr/cm3. This layer is a volcanic breccia with andesitic lava insertion. In the research area, it is spread to the southeast of the Lawu Tua Volcano complex. The last layer is Jobolarangan Tuff (Qvjt), with a density of 2.35 gr/cm3.

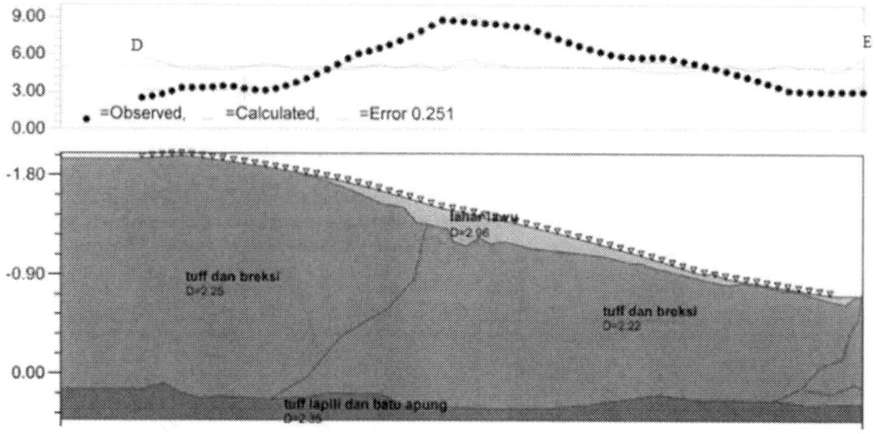

Figure 7. Structure model slice D-E based on forward modelling of the gravity data.

The last trajectory made is the D-E trajectory. The results of the D-E trajectory modelling is displayed in figure 7. The D-E trajectory crosses from the southeast of the center of the research site to the north. On the D-E trajectory, it does not pass through any faults. The first layer is Lahar Lawu (Qlla), with a density of 2.94 gr/cm3. Its distribution fills the plains at the foot of the volcano. The second layer is Lawu Volcano Rock (Qvl) which consists of tuff and breccia with a density of 2.22 gr/cm3 to 2.25 gr/cm3. The last layer is Jobolarangan Tuff (Qvjt), with a density of 2.35 gr/cm3.

Based on 2D forward modelling, the density values and types of rocks can be seen that fill the subsurface of Mount Lawu. Most of the component rocks are breccia, andesite, and lava which are igneous rocks. Volcanic rock or igneous rock is formed directly from volcanic activity, both active and inactive [10]. Igneous rock has a less stable binding capacity because plate shifting activity can still occur. This activity is triggered by magma activity beneath the volcano. Faults occur when rock is subjected to low pressure and temperature so that it becomes brittle, resulting in boundaries between rock structures. The existence of faults also needs to be considered as a factor of vulnerability to landslides [11]. The component rocks in the form of igneous rocks and several faults in the study area can be concluded that the Mount Lawu area is highly vulnerable to landslides.

Previous research in 2017 analyzing the dangers in the slopes of Mount Lawu using the weighted overlay sum technique can be concluded that the level of hazard in the research area is classified into three classes, namely low, medium and high hazard. From this classification, 17 villages have a high hazard. The resulting hazard level can be used as a continuation to formulate a landslide risk zone in the slopes of Mount Lawu, Karanganyar Regency, Central Java [12].

5. Conclusion

In the center of the study area, composed of tuff and breccias with lava insertion, then andesite lava is found in the southwest direction of the study area near the Cemorosewu Fault. Lapilli tuff and pumice scattered southeast of the study area near the Cemorosewu Fault. From the rock density, it is concluded that the rock structure that dominates Mount Lawu is volcanic rock or commonly called igneous rock. The characteristics of igneous rocks are less varied and unstable structures. The modelling trajectory applied to the research area can identify two faults, namely the Sidoramping-Lawu Fault and the Cemorosewu Fault. The Sidoramping-Lawu Fault is a normal fault that extends across the center of the study area and has a north-south direction. While the Cemorosewu Fault is in the south of the study area and has a west-east direction. The unstable rock structure and faults identified are related to the high potential for landslides in the Mount Lawu area.

References

[1] Badan Geolgi-DESDM 2008 *Kajian Bahaya Gerakan Tanah dan Perencanaan* (Bandung: Badan Geologi Departemen Energi dan Sumberdaya Mineral)

[2] Bemmelen V 1949 The Geology of Indonesia *Gov. Print. Off. Hague* **IA** 732 pp

[3] Helmi H, Kurniawan H and Adam W 2020 GUNUNG LAWU Kontrol geologi terhadap permunculan manifestasi panas bumi di kawasan gunung lawu *Kurvatek* **5** 1–9

[4] Nabighian M N, Grauch V J S, Hansen R O, LaFehr T R, Li Y, Peirce J W, Phillips J D and Ruder M E 2005 The historical development of the magnetic method in exploration *Geophysics* **70**

[5] Hasanah N 2018 Rentetan Bencana Alam di Indonesia Sepanjang 2018, retrieved from https://news.okezone.com/read/2018/12/24/337/1995339/rentetan-bencana-alam-di-indonesia-sepanjang-2018, on Saturday April 2019, 04.29 pm

[6] Hirt C, Claessens S, Fecher T, Kuhn M, Pail R and Rexer M 2013 New ultrahigh-resolution picture of Earth's gravity field *Geophys. Res. Lett.* **40** 4279–83

[7] Handayani N T 2018 *Interpretasi Struktur Bawah Permukaan Cekungan Serayu Selatan Bagian Timur Menggunakan Analisis Data Gravitasi Ggmplus (V.2013)* (Universitas Gadjah Mada)

[8] Sudrajad B 2018 *Subsurface Structure Modelling Of Nabire Regency Area In Northern Bird's Neck Papua Using Threedimensional (3d) Inverse Modelling And Horizontal Derivative Analysis Based On Gravity Anomaly Data Ggmplus* (Universitas Gadjah Mada)

[9] Hartono U 1994 *The petrology and geochemistry of the Wilis and Lawu volcanoes, East Java, Indonesia* (University of Tasmania)

[10] Hartono M A 2017 *Permodelan risiko bencana tanah longsor Kecamatan Ngargoyoso, Kabupaten Karanganyar* (Universitas Sebelas Maret)

[11] Akhadi M, Harahap D, Subakti H and Wandono 2021 The Utilization of GGMplus Data Using Derivative Method and 3D Modeling of Subsurface Structure for Trienggadeng Fault Identification in Pidie Jaya, Aceh Region *J. Phys. Conf. Ser.* **1805**

[12] Ramadhani N I and Idajati H 2017 Identification of the Landslide Hazard Level, Case Study: Mount Lawu Slope Area, Karanganyar Regency, Central Java *J. Tek. ITS* **6** 87–90

Community and infrastructure preparedness towards the implementation of sister village disaster mitigation program

Rivani[1] **and E T W Mei**[2*]

[1,2] Department of Development Geography, Faculty of Geography, Universitas Gadjah Mada, Indonesia

*Email: rivani@mail.ugm.ac.id

Abstract. As one of the most active volcanoes in Indonesia, Merapi Volcano has potential risk towards the inhabitants living nearby. After the 2010 paroxysmal eruption, Sister Village program was established in Magelang Regency as an alternative strategy to a better disaster management by evacuating people living in the highest volcanic disaster-prone area (affected villages) to safer zones (buffer villages). By taking location in Tamanagung Village, Muntilan Sub-district, Magelang Regency as buffer village, this research tries to understand how the Sister Village program is implemented in this area, specifically to understand community and infrastructure preparedness. The study used both quantitative and qualitative data taken directly in the field by using questionnaires towards 40 respondents, checklists, and interview with keypersons. Community preparedness were analyzed using four parameters: knowledge, policy and guideline, emergency response plan and resource mobilization. The results showed that among four parameters, only knowledge parameter showing a good preparedness, while the four others still need to be improved. The use of information technology by the public, such as Whatsapp application, is widely growing. In terms of infrastructure preparedness in the buffer village, some enhancements in terms of capacity, infrastructure safety, health aspects, and comfort are still necessary.

1. Introduction

Indonesia is a country located above the 'Ring of Fire' or the Pacific ring of fire. On this path, lies a row of the most active volcanoes in the world. Indonesia has the most mountains in the world. There are 127 active volcanoes and only 69 mountains that can be monitored with tools, especially seismic equipment which is the minimum standard [1]

Disasters cause material and non-material losses. According to Law no. 24 of 2007 [2], disasters cause human casualties, environmental damage, property losses and psychological impacts. The increasing frequency of disaster events that cause damage and loss requires anticipatory efforts to reduce or minimize the impact of economic and non-economic losses. According to the International Strategy for Disaster Reduction (ISDR), 2009 [3] disaster risk reduction efforts are a long-term activity and are part of sustainable development, using knowledge, innovation, and knowledge to build survival and resilience in society.

Merapi Volcano is one of the volcanoes in Indonesia which shows the symptoms of the most active volcanism in the world [4]. According to the Directorate of Volcanology in 1979 in Asriningrum, 2004, Mount Merapi is one of the 129 active volcanoes and is a critical volcano or has the potential to erupt (www.pu.go.id). The catastrophic eruption of Mount Merapi which occurred in 2010 has caused a lot of losses. Which when the eruption occurred, there was panic and chaos in the process of evacuating residents, unclear destination of the place to evacuate, uneven fulfillment of logistics for refugees, and so on.

The experience of the Merapi Volcano eruption that occurred in 2010 has provided valuable lessons for all parties regarding the importance of better disaster management policies in the future, especially in terms of handling refugees. This is the background for the Magelang Regency Government in developing the Sister Village Program (Desa Bersaudara). The Sister Village program can help organize the evacuation process of residents in affected villages, in addition to the facilities and distribution of logistics to the refugees can also be divided equally because with the program, data collection in both villages is more accurate [5]. So that the Sister Village Program is one of the effective solutions to use if an eruption occurs at any time.

Ngargomulyo Village and Tamanagung Village are one of the sister villages in Magelang Regency. Both have their respective roles and functions in its implementation. Ngargomulyo Village is an affected village in this program because of its location in the Disaster-Prone Area (KRB) III. Meanwhile, Tamanagung Village which is in Muntilan District was chosen as a buffer village or partner village because this village is one of the villages that has a location far from disaster-prone areas and is relatively safe from the dangers of the eruption of Mount Merapi Volcano.

2. Method

This type of research used in this study using mixed methods. This study aims to identify how the condition of community and infrastructure preparedness after the launch of the Sister Village Program. This research begins by examining the existing theories of knowledge so that the causes of the problem arise. These problems were then tested to determine the level of community preparedness based on data obtained from the field. The data obtained from the field are in the form of preparedness level scores in quantitative form as well as data and information from in-depth interviews.

3. Result and Discussion

The research was conducted using a questionnaire distributed to respondents, namely the community of Tamanagung village. The questionnaire consists of nine questions aimed at determining the level of community knowledge regarding the Merapi Volcano Eruption, Disaster-Prone Areas, the Sister Village Program and the attitude of accepting refugees. The first question contains a sequence of levels of volcanic activity status. The second and third questions cover Disaster Prone Areas (KRB). In questions four, five, six, and seven, questions related to knowledge about the Sister Village Program are presented, with details in the fourth question containing the definition of the Sister Village Program. The fifth question contains the role of the Buffer Village. The sixth question is about the year the Sister Village Program was founded and the seventh question is about the couple from Tamanagung Village in the Sister Village Program. For the last two questions, namely the eighth and ninth questions, knowledge questions related to the attitude that should be carried out in accepting refugees are presented.

Based on the results of the study, it is known that the level of knowledge of the people of Tamanagung Village related to disasters and Sister Village is quite high. It can be proven that 70% of respondents who have a high score of 7-9 which means the respondent answered more than 7 questions correctly. Only 25% of respondents are classified as having moderate knowledge, while the rest (5%) have low knowledge regarding the Merapi Volcano Eruption, Disaster Prone Areas, Sister Village Program and the attitude of accepting refugees.

Based on the results of knowledge, it is also known that questions related to the attitude of accepting refugees, the definition of sister village, and the village of the couple from Taman Agung village are common knowledge that is known to most people. This can happen because of the socialization efforts to the community that have been carried out by the government. It is explained that the Village Government and Tamanagung Volunteers actively participate in socializing their village residents as an effort to build social bonds and mobilize the Tamanagung Village community to participate and participate in Sister Village [6]. In addition, the results of the study indicate that questions related to the year the Sister Village Program was established, the order of the status levels of Volcano activity, and the position of Tamanagung Village in a Disaster-Prone Area are knowledge that most of the community does not know. It is known that basic information related to the sister village program is not widely known and understood by most of the community. In addition, disaster information is also not widely understood by the public, such as knowledge about Catur Gatra Merapi, although this has been widely discussed considering the location of Tamanagung Village is a Disaster-Prone Area. These two things are important things that need to be prioritized in the next disaster socialization.

Based on the results of research that refers to questions related to the level of family preparedness for disasters, it is known that there are still many shortcomings in the preparation of anticipatory efforts made by the community to deal with conditions if a disaster occurs. In addition, the community is still lacking in preparing to accept refugees due to disasters. However, based on survey data, it is shown that the number of respondents who show high preparedness (9 people) for disasters is comparable to respondents who lack disaster preparedness (11 people). This shows that the knowledge related to disasters has not been evenly distributed by the community of Tamanagung village and there are some people who have not been actively involved in the Sister Village Program due to a lack of public awareness about the importance of reducing disaster risk.

Based on the data it is also known that the number of low scores obtained by respondents amounted to 17 or reached 42.5% of all respondents. This shows that there are still many citizens who have low knowledge in the field of policies and guidelines. These results can occur because socialization activities regarding disaster preparedness have not been followed by some people, which causes inequality in the level of knowledge between residents regarding community preparedness in dealing with disasters.

Based on the survey results, it is known that the third question which contains about the Disaster-Prone Areas (KRB) is the most answered correctly by the respondents. This shows that most people already have the awareness and guidance to save themselves when the eruption occurs. However, as previously explained, there are some people who have not participated in the related socialization, which can be seen through the respondents' answers to the first question regarding the presence of family members who have attended disaster preparedness seminars. The absence of some community members was due to the lack of community willingness to participate in the socialization and the busyness of the community in earning a living which made it possible for them not to attend the socialization related to the disaster that was carried out.

In the emergency response aspect, which refers to the level of community preparedness in carrying out first aid and the community's readiness to accept refugees in an emergency, it is known that some of the respondents have prepared first aid kits and know and store emergency telephone numbers which are very important aspects in an emergency such as when disaster. In addition, the survey results also show that some respondents have been able to provide locations for the evacuation of residents affected by the disaster and their means of transportation. In addition, some other respondents indicated that there was no attempt to prepare a first aid kit, storage of emergency numbers, or availability to provide evacuation locations. This can happen due to the lack of space owned by the houses/areas belonging to some communities which makes it impossible for them to provide a location for refugees. In addition, some of the people who became respondents could not be used as a benchmark in terms of providing space or rooms for refugees, especially during an emergency.

4. Conclusion

This study looks at the implementation of the Sister Village Program on community preparedness and infrastructure in disaster mitigation efforts in Tamanagung Village, which is a Mitra Village. From the research results obtained, it can be concluded several things on each aspect in the context of community preparedness and infrastructure.

Based on the results of research on community preparedness, it was found that:

a. Public knowledge about volcanic disasters and the Sister Village program is quite good.
b. Preparedness for aspects of policies and guidelines needs to be improved through socialization, training, and decision-making contexts at the level of family agreement.

c. Aspects of the emergency response plan need to be improved, for example through the readiness of shelters to accept refugees from partner villages or vehicles for refugees.
d. The use of mobile phones is the most important thing for a disaster warning system, so that its use can be continued, for example through the WhatsApp messaging application.
e. In the aspect of resource mobilization, community participation needs to be increased through disaster mitigation socialization activities that are more targeted (youth, mothers, children, the elderly and not only addressed to the head of the family).

The location with the best level of preparedness is the Tamanggung Village Multipurpose Building, while the location with the lowest level of preparedness is the PPP Sidoharjo Building. The preparedness of the refugee barracks infrastructure certainly needs to be improved both in terms of capacity, infrastructure safety, health, and comfort aspects so that evacuation in the context of Sister Village is easier to implement.

References

[1] LIPI-UNESCO/ISDR 2006 *Study of Community Preparedness in Anticipating Earthquake and Tsunami Disasters (Kajian Kesiapsiagaan Masyarakat dalam Mengantisipasi Bencana Gempa Bumi dan Tsunami)* (Deputy for Earth Sciences, Indonesian Institute of Sciences)
[2] Badan Nasional Penanggulangan Bencana 2007 *UNDANG-UNDANG REPUBLIK INDONESIA NOMOR 24 TAHUN 2007 TENTANG PENANGGULANGAN BENCANA* (Jakarta)
[3] UNISDR 2009 *Terminology on Disaster Risk Reduction. United Nations International Strategy for Disaster Reduction, Geneva, Switzerland* (Geneva)
[4] Asriningrum W, Noviar H, Bidang Analisa Sistem P and Bidang Pemanfaatan P 2004 Pengembangan Metode Zonasi Daerah Bahaya Letusan Gunung Api Studi Kasus Gunung Merapi *Jauh dan Pengolah. Data Citra Digit.* **1** 66–75
[5] Mei E T W, Fajarwati A, Sudibyo J, Sari I M, Safitri D, Rusmawati F, Wiwaha A A and Nasution N A R 2018 *Sister Village (Strategi Alternatif Mitigasi Bencana Gunung Api)* (Yogyakarta: UGM Press)
[6] Nurwidianingrum A 2015 *Modal Sosial di dalam Upaya Pengurangan Risiko Bencana Merapi "Sister Village" (Studi Kasus Desa Ngargomulyo Kecamatan Dukun dan Desa Tamanagung Kecamatan Muntilan)* (Universitas Gadjah Mada)

Worst-Case tsunami inundation modeling using high-resolution UAV-DEM in various coastal typologies, case study Gunungkidul coastal area

Hendy Fatchurohman [1,*], Ahmad Cahyadi [2], Taufik Hery Purwanto [1]
[1]Department of Earth Technology, Vocational College, Universitas Gadjah Mada
[2]Department of Environmental Geography, Faculty of Geography, Universitas Gadjah Mada

[*]Email: hendy.fatchurohman@ugm.ac.id

Abstract. A tsunami is a significant threat to the coastal community. The southern coast of Java is considered a tsunami-prone area due to its location facing the Indian Ocean. Gunungkidul Regency in Yogyakarta Special Region is well known for its attractive coastal tourism. Numerous visitors are exposed to tsunami hazards that can be occurred unpredictably. The unavailability of a reliable tsunami hazard map leads to an ineffective evacuation plan. This study aims to implement UAV data for tsunami hazard mapping in different coastal typologies. This study was conducted at Baron Beach, Drini, and Wediombo. The tsunami inundation model was generated from high-resolution UAV-DEM. The worst inundation scenario was set at 12 meters. Based on the results, Baron and Drini are considered highly susceptible to tsunami hazards due to their topographic conditions, mostly located on the flat karst valley. Wediombo is categorized as the primary coast that is formed by the volcanic process. The inundation model at Wediombo only covers a small extent of the area. Volcanic origin morphology formed an undulating to hilly topography that plays an essential role as a natural barrier for the tsunami. However, minimum access and narrow pedestrian street become serious disadvantages for tsunami evacuation route planning at Wediombo Beach.

1. Introduction

The southern coast of Java is highly susceptible to tsunamis due to its location facing the Indian Ocean, where the earthquake continuously occurs [1–3]. Tsunamis might be one among the foremost destructive events that threaten coastal communities. Ineffective evacuation planning and preparedness increase the risk of tsunami disaster [4,5]. Gunungkidul regency has the longest coastline in Yogyakarta Special Region, with various coastal typologies [6]. Gunungkidul is well-known for its beautiful beaches along its coastal area. A large number of tourists visit annually raise the vulnerability level as more people are exposed to tsunami hazards [7]

Located adjacent to the Indian Ocean, the southern coast of Java is exposed to near-field tsunami hazards. A near-field tsunami can be catastrophic because of its potential damage [8]. Near-field tsunami occurrences require immediate evacuation due to short wave arrival on the coastline. Pedestrian evacuations to the higher place are considered the most effective scenario compared to other schemes, anticipating road damage and heavy traffic caused by the previous earthquake [9]. The physical condition of the coastal area will affect the tsunami susceptibility. The coastal area is a transition zone between land and sea that forms various typologies[10]. Each coastal typology has its susceptibility to tsunami hazards. Gunungkidul has various coastal typologies, despite most of the area formed in a massive limestone formation and karst areas.

A tsunami hazard map plays an essential role as the primary source for evacuation planning and disaster management. Accurate tsunami hazard modeling requires high-resolution topographic data. However, the Indonesian government only provides medium-scale topographic data generated from the 12.5 m interval contour map, which is unreliable for tsunami hazard modeling. Inaccurate tsunami modeling will result in poor evacuation planning and disaster management [11]

Utilizing national topographic data with a medium-scale resolution for tsunami hazard modeling in a small area is pointless. However, this problem is solved by obtaining remote sensing images using an Unmanned Aerial Vehicle (UAV). UAVs provide up-to-date and low-cost information and are suitable for creating topographic maps of small areas [11]. UAV photogrammetry produces high spatial resolution aerial images, while Digital Elevation Models (DEMs) for tsunami hazard modeling [12]. The use of Unmanned Aerial Vehicle (UAV) for topographic mapping offers the solution for tsunami hazard modeling in Gunungkidul. UAV-based topographic mapping is considered a feasible, low-cost, and accurate tool to provide high-resolution aerial imagery and digital elevation model data [13]. UAVs also have many advantages in coastal monitoring and studies, including short preparation, feasibility, high repeatability and temporal resolution, and low-security risks [14]. This study aims to implement UAV technology for tsunami hazard mapping in various beach typologies.

2. Study Area

Massive tourism activities in Gunungkidul Regency multiply the number of people exposed to the tsunami threat. This study was conducted in Baron Beach, Drini Beach, and Wediombo Beach (Fig.1). These three beaches are the most popular among others with high visitors number. Baron Beach is categorized as a land-erosion beach with the accumulation of loose materials derived from Bribin Underground River, as the estuary located there. Baron Beach formed a narrow karst valley as the result of the dissolution process. Baron Beach Located at UTM 49S 450278 East and 9101408 North, 50 km to the South-East of Yogyakarta City. Drini Beach is approximately three kilometres to the east from Baron. Located at UTM 49S 453471 East and 9100378 North. Drini Beach is mainly controlled by the wave erosion process, resulting in broad reef flat and wave erosion features such as stacks and notch [6]. Wediombio beach is located at UTM 49S 468188 East and 9095026 North. Wediombo Beach is located 75 km in the South-East of Yogyakarta City, with an approximately of two hours drive. Wediombo formed as a volcanic coast, located in the proximal facies of an ancient volcano [15]. Rainfed agriculture and agroforestry are the primary land use in the Wediombo area.

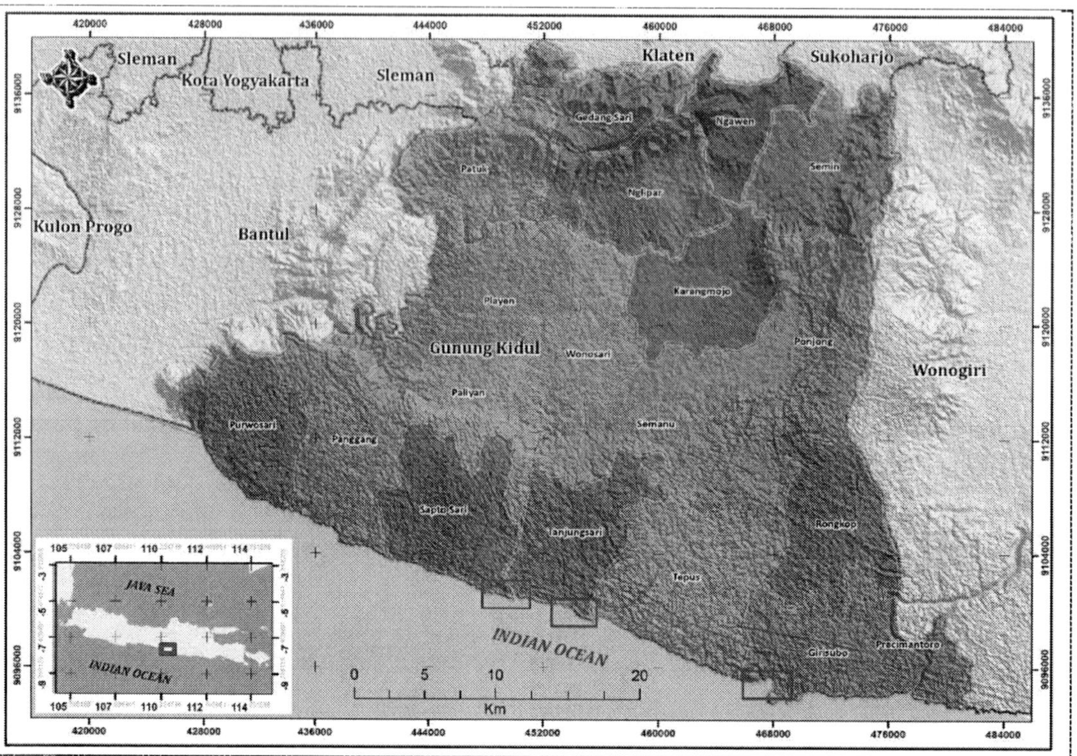

Figure 1. Research area location (inside red squares), from left to right: Baron, Drini, Wediombo

3. Methods

3.1. Data Acquisition

The aerial photographs were acquired using a multirotor quadcopter-type UAV (DJI Mavic 2 Enterprise). The DJI Mavic 2 Enterprise is equipped with 1/2.3" (CMOS) sensor, with 12.71 MP camera resolution and 35 mm focal length. This vehicle also includes GPS for positioning and the ability to fly autonomously. The flight route was set automatically to capture images with 250 m altitude using *Drone deploy*. At least 75 % overlap and 65 % side-lap arranged during the flight for sufficient Structure from Motion (SfM) photogrammetry analysis. The aerial photograph data acquisition was conducted in July 2020. The flight mission at Baron Beach area covers 224 hectares. The number of photos taken was 626 photos with a 55 minutes flight time. The Drini Beach flight route covers 110 hectares, with the number of photos taken was 315 with an estimated data collection time of about 21 minutes. The Wediombo Beach area covers 284 hectares with 489 photos. As an example, the flight mission planning using DroneDeploy for each location is depicted in Fig. 2.

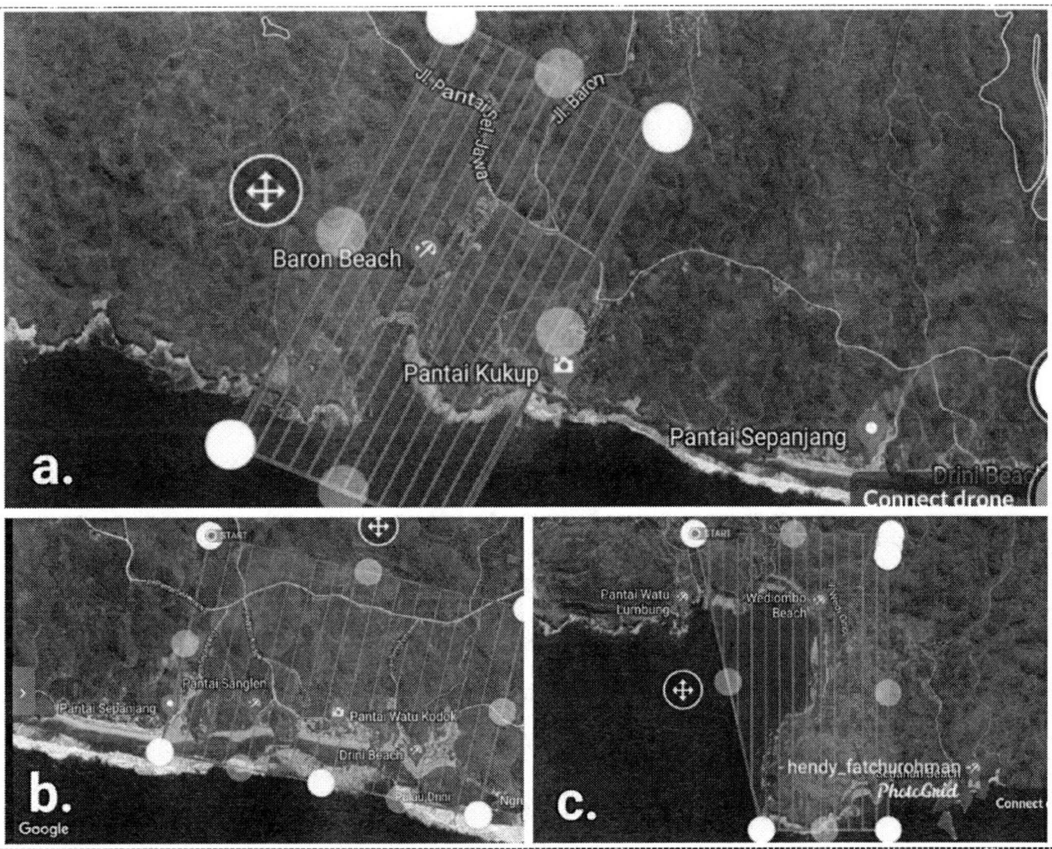

Figure 2. Flight route-planning using Drone Deploy (a. Baron, b. Drini, c. Wediombo)

Photogrammetry analysis was done using Agisoft Metashape to generate an Orthomosaic image and DEM. One Photo with orthogonal projection is formed by combining several images with central projection. Structure from Motion (SfM) processing was performed from the overlapping images of the study area and detected stationary objects on the ground from the moving camera platform [16]. All the photographs from each location were aligned based on the camera coordinates and automatic detection of points. Through the SfM process, the set of images from each site produced scattered point clouds, which then form dense point clouds and 3D structures, respectively [17].

Plots of important georeferenced data are used to provide accurate digital elevation. This plot is carried out in two ways, namely directly or indirectly. Several ground markers are evenly placed in the research area for Ground Control Point (GCP) measurements. The markers are made from a printed flexy sheet with visible color and identifiable patterns (Fig.3). The number of markers for each location depends on the topographical characteristics and the extension of the area of interest [18]. The markers are used as a terrestrial reference for GCP measurements. Therefore identifiable shape and visible colors are necessary for marker identification during the georeferencing process [19]. GCP coordinates measured using Real-Time Kinematic (RTK) method and processed during point cloud generation to improve the accuracy during orthophoto and DEM generation.

Figure 3. Ground Control Point site measurement on the marker (a) and marker appearance from the aerial photograph (b)

3.2. Tsunami inundation model

The tsunami inundation was modeled with a raster calculation that saw the worst tsunami scenario on the South Coast of Java. DEM is used to determine the inundation area in the tsunami height scenario. According to recent research, a seismic gap located on the southern coast of Java has the potential to trigger an 8.8 Mw earthquake, compensating for a 400 year return period and causing waves as high as 12 meters. [20]. Based on this research, the 12 meter wave height scenario is used to determine the coverage of the inundated area, thus reflecting the vulnerability of the area. Raster filtering to find out pixels whose value is less than/equal to 12 meters. The filtering for inundated terrain was done using the spatial analyst tool in ArcGIS 10.5. A raster calculator from the map algebra tool was used to perform a binary selection of pixels with 12 meters or less value. The selected raster was then reclassified to form an independent dataset classified as inundated pixels. The raster is converted to a polygon after the selection of the flooded pixels with the specified value. The filtering process further uses the 'select by location' command to remove polygons that are not connected to the shoreline [6]. From the inundated polygon, the area affected by the worst tsunami scenario was determined.

4. Results and discussion

The UAV produced image sequences with approximately 7-9cm/pixel Ground Sample Distance (GSD) from 200-250 m flight altitude in this study. This accuracy certainly provided highly detailed elevation data for DEM extraction. The topographic data of the existing RBI maps were corrected using high-resolution DEM. DEM reconstruction resulting from photo orthorectification can support high precision measurements [21,22]. SfM processes for orthomosaic and DEM in the Agisoft Metashape software depicted in Fig. 4.

Figure 4. Structure from Motion (SfM) processes example of Baron Beach. From left to right: point clouds; dense clouds; orthomosaic image; and Digital Surface Model (DSM).

A high-resolution Digital Elevation Model was required to perform an accurate and reliable tsunami inundation model. The maximum horizontal and vertical extent of tsunami inundation have been studied for years but remain challenging since the high resolution of bathymetric and topographic data are required for tsunami hazard modeling [23]. In this study, the digital elevation model generated from UAV photogrammetry has an average of 15 cm spatial resolution and 39.3 points/m² point density. UAV-derived DEMs can record rough topography features that cannot be measured in the terrestrial survey [22]. The DEM results clearly show a high level of detail on the terrain morphology and land covers of each location (Fig.5)

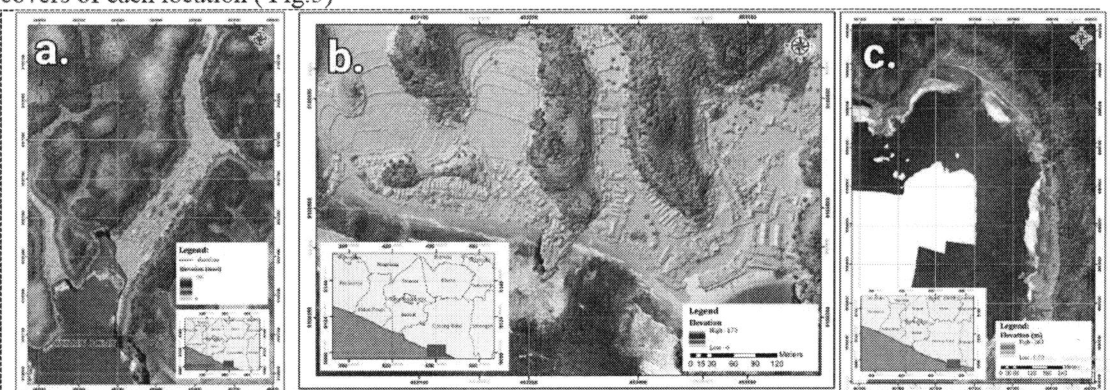

Figure 5. Digital Surface Model results from Structure from Motion processing (a) Drini Beach, (b) Baron Beach, and (c) Wediombo Beach

According to the models, each site has different characteristics in terms of tsunami susceptibility. Morphological characteristics of each beach typology play a significant role in tsunami susceptibility. The direction of tsunami run-up has always flown almost perpendicular to the coastline, and the topography controlled the backwash flow [24]. The topographic elevation for each site is the main factor affecting tsunami susceptibility. The lower the elevation, the higher the exposure is [25].

From the coastal typology perspective, Baron Beach is categorized as a land-erosion beach with the accumulation of loose materials derived from the Bribin Underground River. Baron Beach formed a narrow karst valley as the result of the karst dissolution process [6]. The whole area of Baron valley is relatively flat, with less than 3^0 slopes extended more than 800 meters landward (Fig.6 right). The elevation difference from the shoreline as the starting point of the tsunami inundation model to the farthest is only 14 meters. Due to its low-lying elevation, the whole area is completely inundated by the inundation model. All the buildings and infrastructures on the valley will be highly impacted if the tsunami occurs, as its located in the hazard zone of the worst-case inundation scenario. Baron Beach is the most susceptible site among the other sites because of its topographic condition. Most of the buildings, including shops, restaurants, hotels, and food stalls, are located in the narrow karst valley, potentially increasing the tsunami risk in this area.

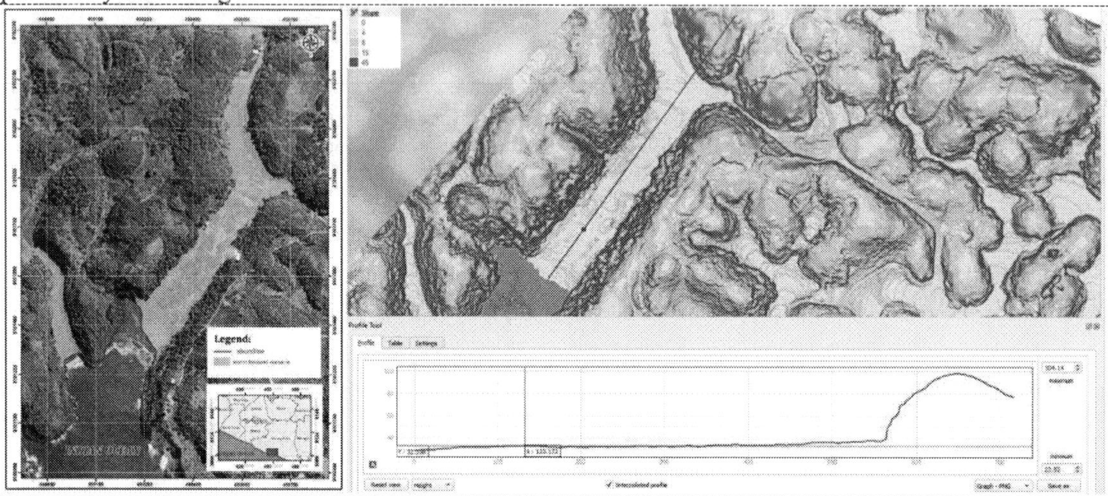

Figure 6. Worst-case tsunami inundation model (left); Slope and terrain profile of Baron Beach (right)

Drini beach morphology is more complex than Baron Beach. The typology's initial state was primary coast dominated by tectonic uplift and land erosion process. However, the secondary process also plays a significant role in forming complex coastal typology. There are several strong evidence of secondary coast characteristics such as marine deposition, wave erosion, and coastal organism [6].

Figure 7 shows that he model recommends vertical tsunami assembly points for evacuation scenarios as the whole valley is completely inundated. All the buildings and infrastructures in the valley will be highly impacted if the tsunami struck the coast. The configuration of karst hills and valleys at Drini Beach makes the model inundate a broader area than Baron. However, there are many options for vertical and horizontal evacuation. The spaces among the karst hills are wide enough to perform horizontal evacuation to the assembly points, even though vertical evacuation is the best scenario for near-field tsunami [26]. Some of the hills surrounding Drini Beach are high enough for a tsunami vertical evacuation shelter. The main disadvantage of performing vertical evacuation is that the slopes are too steep to climb without any artificial structure.

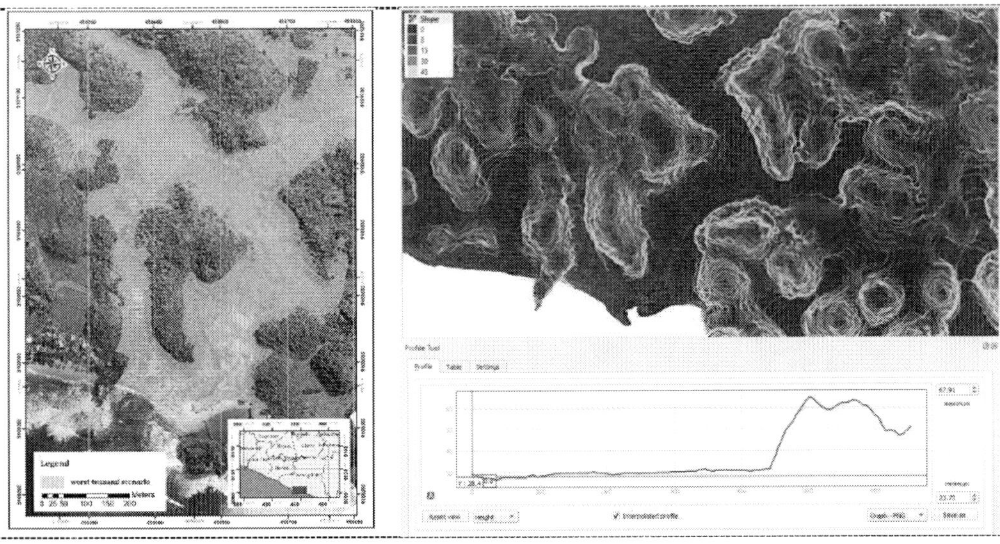

Figure 7. Worst-case tsunami inundation model (left); Slope and terrain profile of Drini Beach (right)

Wediombo beach is considered the primary coast with volcanic origin. Wediombo is situated in the ancient volcano complex with plenty of volcanic products. Lava flows and volcanic breccias dominated the bay, with enormous intrusive rocks on the west that forms Gunung Batur [15]. As the ancient volcanic complex indicated to be the central-medial facies, the topography is dominated by steep hills perpendicular to the Wediombo bay. The hills composed of volcanic breccia formed steep slopes, play an essential role as a natural barrier from the tsunami. However, the evacuation scenario must be carefully planned since there is only one way out from the beach to the safe area. Figure 8. depicts the tsunami inundation model that only covers the bay without significant threat to the community.

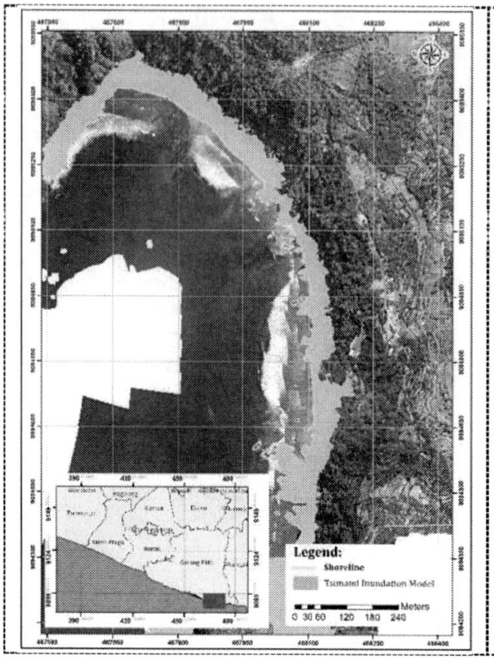

Figure 8. Worst-case tsunami inundation model of Wediombo Beach.

5. Conclusion

The tsunami inundation model was generated from high-resolution UAV-DEM. The worst inundation scenario was set at 12 meters. Based on the models, Baron and Drini are considered highly susceptible to tsunami hazards due to their topographic conditions, mostly located on the flat karst valley. Wediombo is categorized as the primary coast that is formed by the volcanic process. The inundation model at Wediombo only covers a small extent of the area. Volcanic origin morphology formed an undulating to hilly topography that plays an essential role as a natural barrier for the tsunami. However, minimum access and narrow pedestrian street become serious disadvantages for tsunami evacuation route planning at Wediombo Beach. Utilizing the hills surrounding the area as tsunami vertical evacuation points is recommended to minimize fatalities.

Acknowledgment

This study is part of the 2020 Junior Lecturer Research Grant of Universitas Gadjah Mada (Hibah Peningkatan Kapasitas Dosen Muda Universitas Gadjah Mada Tahun 2020) contract number: 646/UN1.P.III/KPT/HUKOR/2020, titled " Pemetaan Dampak Near-Field Tsunami pada Berbagai Jenis Tipologi Pesisir menggunakan Teknologi Unmanned Aerial Vehicle (UAV) (Studi Kasus Kawasan Pesisir Kabupaten Gunungkidul, D.I. Yogyakarta". The authors gratefully acknowledge the research funding and support from the Universitas Gadjah Mada. We also thank all the students from the Department of Earth Science, Vocational College, Universitas Gadjah Mada, who participated in data collection, data processing, and paperwork related to this study.

References

[1] Lavigne F, Gomez C, Giffo M, Wassmer P, Hoebreck C, Mardiatno D, et al 2007 Field Observations of The 17 July 2006 Tsunami in Java. Nat Hazards Earth Syst Sci **7** 177–83.

[2] Marfai MA, King L, Singh LP, Mardiatno D, Sartohadi J, Hadmoko DS, et al 2008 Natural hazards in Central Java Province, Indonesia: An overview. Environ Geol **56** 335–51. https://doi.org/10.1007/s00254-007-1169-9.

[3] Okamoto T, Takenaka H 2009 Waveform inversion for slip distribution of the 2006 Java tsunami earthquake by using 2.5D finite-difference Green's function. Earth, Planets Sp **61** 17–20. https://doi.org/https://doi.org/10.1186/BF03352919.

[4] Mas E, Koshimura S, Imamura F, Suppasri A, Muhari A, Adriano B 2015 Recent Advances in Agent-Based Tsunami Evacuation Simulations: Case Studies in Indonesia, Thailand, Japan and Peru. Pure Appl Geophys **172** 3409–24. https://doi.org/10.1007/s00024-015-1105-y.

[5] Henry KD, Wood NJ, Frazier TG 2017 Influence of road network and population demand assumptions in evacuation modeling for distant tsunamis. Nat Hazards **85** 1665–87. https://doi.org/10.1007/s11069-016-2655-8.

[6] Marfai M, Fatchurohman H, Cahyadi A 2020 Pesisir Gunungkidul. Gadjah Mada University Press.

[7] Marfai MA, Sunarto S, Khakim N, Fatchurohman H, Cahyadi A, Wibowo YA, et al 2019 Tsunami hazard mapping and loss estimation using geographic information system in Drini Beach, Gunungkidul Coastal Area, Yogyakarta, Indonesia. E3S Web Conf 76. https://doi.org/10.1051/e3sconf/20197603010.

[8] Wood N, Schmidtlein M 2012 Anisotropic Path Modeling to Assess Pedestrian-Evacuation Potential from Cascadia-Related Tsunamis in The US Pacific Northwest. Nat Hazards **62** 275–300. https://doi.org/10.1007/s11069-011-9994-2.

[9] Jones JM, Ng P, Wood NJ 2019 The Pedestrian Evacuation Analyst—Geographic Information Systems Software for Modeling Hazard Evacuation Potential. Geol. Surv. Tech. Methods, B. 11, chap. C9 25.

[10] Shepard F, States U, By CC. Chapter 3 Coastal Classification and Morphology 1992.

[11] Marfai MA, Sunarto S, Khakim N, Cahyadi A, Rosaji FS, Fatchurohman H, et al 2018 Topographic data acquisition in tsunami-prone coastal area using Unmanned Aerial Vehicle (UAV). IOP Conf Ser Earth Environ Sci **148** 0–7. https://doi.org/10.1088/1755-1315/148/1/012004.

[12] Marfai MA, Fatchurohman H, Cahyadi A 2019 An Evaluation of Tsunami Hazard Modeling in Gunungkidul Coastal Area using UAV Photogrammetry and GIS. Case Study: Drini Coastal Area. E3S Web Conf 125. https://doi.org/10.1051/e3sconf/201912509005.

[13] Chen B, Yang Y, Wen H, Ruan H, Zhou Z, Luo K, et al 2018 High-resolution monitoring of - Beach topography and its change using unmanned aerial vehicle imagery. Ocean Coast Manag;160:103–16. https://doi.org/10.1016/j.ocecoaman.2018.04.007.

[14] Gonçalves JA, Henriques R 2015 UAV photogrammetry for topographic monitoring of coastal areas. ISPRS J Photogramm Remote Sens **104** 101–11. https://doi.org/10.1016/j.isprsjprs.2015.02.009.

[15] Hartono G 2006 Asal-usul pembentukan Gunung Batur di daerah Wediombo, Gunungkidul, Yogyakarta. Indones J Geosci **2** 143–58. https://doi.org/10.17014/ijog.vol2no3.20073.

[16] Turner IL, Harley MD, Drummond CD 2016 UAVs for coastal surveying. Coast Eng **114** 19–24. https://doi.org/10.1016/j.coastaleng.2016.03.011.

[17] Cook KL 2017 An evaluation of the effectiveness of low-cost UAVs and structure from motion for geomorphic change detection. Geomorphology **278** 195–208. https://doi.org/10.1016/j.geomorph.2016.11.009.

[18] Nex F, Remondino F 2014 UAV for 3D mapping applications: A review. Appl Geomatics **6** 1–15. https://doi.org/10.1007/s12518-013-0120-x.

[19] Bemis SP, Micklethwaite S, Turner D, James MR, Akciz S, T 2014 Thiele S, et al. Ground-based and UAV-Based photogrammetry: A multi-scale, high-resolution mapping tool for structural geology and paleoseismology. J Struct Geol **69** 163–78. https://doi.org/10.1016/j.jsg.2014.10.007.

[20] Widiyantoro S, Gunawan E, Muhari A, Rawlinson N, Mori J, Hanifa NR, et al 2020 Implications for megathrust earthquakes and tsunamis from seismic gaps south of Java Indonesia. Sci Rep **10** 1–11. https://doi.org/10.1038/s41598-020-72142-z.

[21] Uysal M, Toprak AS, Polat N 2015 DEM generation with UAV Photogrammetry and accuracy analysis in Sahitler hill. Meas J Int Meas Confed **73** 539–43. https://doi.org/10.1016/j.measurement.2015.06.010.

[22] Kršák B, Blišťan P, Pauliková A, Puškárová P, Kovanič L, Palková J, et al 2016 Use of low-cost UAV photogrammetry to analyze the accuracy of a digital elevation model in a case study. Meas J Int Meas Confed **91** 276–87. https://doi.org/10.1016/j.measurement.2016.05.028.

[23] Dias F, Dutykh D, O'Brien L, Renzi E, Stefanakis T 2014 On the modelling of tsunami generation and tsunami inundation. Procedia IUTAM **10** 338–55. https://doi.org/10.1016/j.piutam.2014.01.029.

[24] Umitsu M, Tanavud C, Patanakanog B 2007 Effects of landforms on tsunami flow in the plains of Banda Aceh, Indonesia, and Nam Khem, Thailand. Mar Geol **242** 141–53. https://doi.org/10.1016/j.margeo.2006.10.030.

[25] Sinaga TPT, Nugroho A, Lee YW, Suh Y 2011 GIS mapping of tsunami vulnerability: Case study of the Jembrana regency in Bali, Indonesia. KSCE J Civ Eng **15** 537–43. https://doi.org/10.1007/s12205-011-0741-8.

[26] Wang H, Mostafizi A, Cramer LA, Cox D, Park H 2016 An agent-based model of a multimodal near-field tsunami evacuation: Decision-making and life safety. Transp Res Part C Emerg Technol **64** 86–100. https://doi.org/10.1016/j.trc.2015.11.010.

IOP Conf. Series: Earth and Environmental Science **986** (2022) 012028 doi:10.1088/1755-1315/986/1/012028

Stakeholders engagement in risk communication for peatland fire management

M F Widhagdha[1], A Dewi[1]

[1] PT Kilang Pertamina Internasional Unit Sungai Pakning

*Email: arjuna.miftah@gmail.com

Abstract. Bengkalis Regency, Riau Province is one of the areas prone to fires, especially peat areas since 2014 until now. One of the area in Bengkalis Regency which also the center of peatland fires is Bukit Batu District. To tackle the occurrence of peatland fires in the Bukit Batu District, a number of stakeholders have collaborated and are actively involved in the process of overcoming peatland fires starting from prevention, mitigation, preparedness, emergency response and recovery. This study was conducted using qualitative approach by collecting data through field observations and interviews with in actors of peatland fires management in Bukit Batu District consisting of the Government, Corporations, Academics, Mass Media, and Community Groups. The result shows that stakeholder involvement have a role in increasing awareness of peatland fires and stakeholders are actively involved in risk communication through the use of communication media such as dialogue, counseling and coordination between stakeholders both face-to-face and online by using Whatsapp as media communication. The stakeholders engagement also plays a role in mobilizing the community to activate community self-reliance in extinguishing land fires and creating added value in community empowerment activities so as to prevent land fires.

1. Introduction

Indonesia is an archipelagic country that has a lot of uniqueness as the meeting place of three tectonic plates namely Eurasia, Indo-Australia, and the Pacific which causes Indonesia to have many risks of natural disasters. According to Regulation Number 24 of 2007[1], Disaster is an event or series of events that threaten and disrupt people's lives and livelihoods caused by natural factors, non-natural factors, and social or human factors that can result in human casualties, environmental damage, property losses, and psychological impact. There are three types of disaster such Natural Disasters, Non-Natural Disasters, and Social Disasters.

According to the National Disaster Management Agency (BNPB), there are twelve types of disasters in Indonesia. Those are Forest and Land Fires, Earthquakes, Tsunamis, Volcanic Eruptions, Landslides, Floods, Flash Floods, Droughts, Tornados, Extreme Waves and Abrasion, Epidemics and Disease Outbreaks, and Technological Failures.

Forest and Land Fires are a national disaster that has hit Indonesia for a long time. Especially the province of Riau, as an area with extensive peatlands where the case of forest and land fires has been a disaster that has haunted the community for a long time. Peat soil has a characteristic depth of up to several meters in the ground, so that when forest and land fires occur, the extinguishing process is difficult and takes a long time. Based on data from the Ministry of Environment and Forestry, Riau

Province [2] has a large area of forest and peatland. The biggest fire in Riau occurred in 2015 with a burned area of 183,808 ha [3].

Land and forest fires occur due to two factors, intentional and unintentional. Intentional cases of forest and land fires are caused by human error, which oftenly occur due to industrial activities or human negligence. Generally, intentional fires aim to clear peatlands in preparation for seasonal plantations and development. Accidental fires, on the other hand, are caused by hot weather, open burning of trash, random throwing of lit cigarette butts, and other acts of neglect.

The forest and land fires disaster has an impact on various aspects of life, both in terms of health, the environment, and the economy. Karhutla disaster is the cause of the high number of ARI (acute respiratory infection) cases in Riau province. Riau Provincial Health Office in 2015 cases of ARI reached 37,396 cases, and Pekanbaru was in the top rank, namely 8,661 cases, followed by Siak 4,539 cases, then Dumai 3,467 cases and the rest were spread in various other districts in Riau province [2].

In handling forest and peatland fires, cross-sectoral coordination and participation is needed to expedite the activities. The parties involved in handling forest and peatland fires based on their main tasks and functions include the Regional Government, Manggala Agni, the Fire Service, the Regional Disaster Management Agency, the Environment Service, and the activities of handling forest and land fires are also supported by volunteers who are currently is in one community organization, called Masyarakat Peduli Api.

Table 1. List of stakeholders in handling forest and peatland fires in the Sungai Pakning

No	Stakeholders	Competency	Working Area
1.	Bukit Batu Districts Government	Governance	Bukit Batu District
2.	Manggala Agni Of Riau Province	Forest Fire Management	Riau Province
3.	Sungai Pakning Ward Government	Governance	Sungai Pakning Ward
4.	Police Departement Of Bukit Batu District	Law enforcement	Bukit Batu District
5.	Military Of Bukit Batu District	Safety and Security	Bukit Batu District
6.	Masyarakat Peduli Api	Forest Fire Management	Bukit Batu District
7.	PT Kilang Pertamina Internasional Unit Sungai Pakning	Oil Processing Company	K Bukit Batu District
8.	Universitas Riau	Academics	Riau Province
9.	Journalist / Media	Publication	Bengkalis Region

2. Research Method

This investigation uses a qualitative approach, to understand the meaning of individuals or groups in social problems or human problems [4]. The process of investigation includes questions, and procedures. This form of research uses a case study. According to Idrus [5] case studies are an intensive investigation of an individual, but sometimes case studies can also be used to investigate small social units such as families, schools, and groups. Data collection techniques were carried out through interviews with actors involved in mitigating forest and land fires and field observations, namely:

2.1. Government

Bukit Batu District, Pakning River Exit, Manggala Agni Korwil Riau

2.2. Corporation

Pertamina RU II Sungai Pakning

2.3. Academics
UNRI Researcher

2.4. Mass Media
Antara News Agency reporter

2.5. Community Group
MPA Sungai Pakning, MPA Sungai Selari, MPA Desa Pangkalan Jambi

3. Result and Discussion

3.1. Stakeholder Engagement
Stakeholders are parties from internal or external organizations who have an interest and influence on the performance of the organization concerned. Stakeholders can be organizations, communities, socio-economic groups, governments, or institutions that come from various dimensions at every level of society [6]. In short, stakeholders can come from local government groups, non-government institutions, and the community. According to Crosby (1992)[7], stakeholders can be classified into three groups, namely Main Stakeholders, Supporting Stakeholders, and Key Stakeholders.

Main Stakeholders are parties who get positive or negative impacts from an activity. Supporting Stakeholders are parties who assist in the process of delivering activities. Key Stakeholders are parties who have a strong and important influence related to problems, needs, and concerns for the smooth running of activities [6]. In handling forest and land fires, the synergy of stakeholders from all levels is needed in order to achieve the goals optimally. The classification of Stakeholders in handling forest and land fires can be seen as follows:

Table 2. List of stakeholders in handling forest and peatland fires in the Sungai Pakning

No	Stakeholders	Competency	Working Area	Contribution in Handling Forest and Land Fires
1.	Bukit Batu Districts Government	Governance	Bukit Batu District	▪ Formulation and determination of disaster management policies. ▪ Preparation of disaster management plan. ▪ Determination of technical guidelines for disaster management. ▪ Development of Human Resources (Training and Education) related to disaster management. ▪ Establishment of institutions for disaster management. ▪ Development of disaster management institutions. ▪ Development of infrastructure related to disaster management activities.
2.	Manggala Agni Of Riau Province	Forest Fire Management	Riau Province	▪ Carry out prevention, extinguishing, and post-fire handling activities.

				▪ Conducting socialization related to forest and land fire prevention. ▪ Conducting training related to fire and forest fire suppression. ▪ Monitoring and patrolling hot spots
3.	Sungai Pakning Ward Government	Governance	Sungai Pakning Ward	▪ Provide support in the implementation of disaster management. ▪ Coordinate overall Karhutla disaster management activities to all stakeholders in Sungai Pakning Village.
4.	Police Departement Of Bukit Batu District	Law enforcement	Bukit Batu District	▪ Monitoring and patrolling hot spots ▪ Securing the surrounding location of a disaster ▪ Inform and assist community regarding disaster mitigation and evacuation activities.
5.	Military Of Bukit Batu District	Safety and Security	Bukit Batu District	▪ Monitoring and patrolling hot spots ▪ Assist in the extinguishing process ▪ Securing the surrounding location in the event of a disaster ▪ Inform and assist the community regarding mitigation and evacuation activities.
6.	Masyarakat Peduli Api	Forest Fire Management	Bukit Batu District	▪ Doing fire fighting ▪ Monitoring and patrolling hot spots ▪ Informing the local community regarding information on forest and land fire disasters
7.	PT Kilang Pertamina Internasional Unit Sungai Pakning	Oil Processing Company	Bukit Batu District	▪ Support hot spot monitoring activities. ▪ Support extinguishing activities (HR, and logistics). ▪ Conducting firefighting and training to increase MPA capacity. ▪ Facilitate firefighting infrastructure and facilities.
8.	Universitas Riau	Academics	Riau Province	▪ Conducting research related to forest and land fire disaster management

9.	Journalist / Media	Publication	Bengkalis Region	▪ Publish articles related to the latest information on forest and land fire disasters

3.2. Risk Communication

Disaster risk communication is part of environmental communication which focuses on how many people communicate the effects to the public regarding environmental crises and human relations with nature. Therefore, the media has an important role in providing education and early warning related to natural disasters, by gathering information about the needs of victims of natural disasters for government organizations and volunteers as well as being a potential channel for discussion and feedback from the public about natural disasters.

Then, follow-up plans related to handling activities are needed through cross-sectoral collaboration involving various stakeholders from various fields of competence.

Disaster risk communication is a dynamic and interactive process involving exchanges between various groups. Covello & Sandman [8] explain that the principle of involving the community in terms of risk and communication, marks one of the important differences from risk communication. Because disasters contain dramatic elements, not artificial, problematic, solutions, and action-reactions that arise from various human characters.

Therefore, in delivering messages and information in disaster risk communication, it is necessary to pay attention to the context in activities in disaster management based on Permenlhk No 1 of 2021[9] which includes the following stages:

- Prevention
- Mitigation
- Preparedness
- Emergency Response
- Recovery

The following management stages can be achieved effectively through cross-sectoral collaboration with various parties as follows.

Table 3. Media Stakeholder Engagement Forum

No	Disaster Management Stages	Stakeholder Engagement Forum	Media Forum	Pemangku Kepentingan Yang Terlibat
1.	Prevention	Stakeholder Meeting	Discussion Forum	Bukit Batu Districts Government Sungai Pakning Ward Government, Police Departement Of Bukit Batu District, Military Of Bukit Batu District, Masyarakat Peduli Api, Journalist, PT Kilang Pertamina Internasional Unit Sungai Pakning
		Scientific Information Dissemination	Discussion Forums, Scientific Articles	Universitas Riau, PT Kilang Pertamina Internasional Unit Sungai Pakning, Masyarakat Peduli Api
2.	Mitigation	Daily Coordination	Grup Whatsapp, Phone Call, HT	Manggala Agni, Bukit Batu Districts Government Sungai Pakning Ward Government, Police Departement Of Bukit Batu District, Military Of Bukit Batu District, Masyarakat Peduli Api, Journalist,

				PT Kilang Pertamina Internasional Unit Sungai Pakning.
3.	Preparedness	Disaster Response Training	Discussion Forum, Whatsapp, HT	Manggala Agni, Bukit Batu Districts Government Sungai Pakning Ward Government, Masyarakat Peduli Api, Journalist, PT Kilang Pertamina Internasional Unit Sungai Pakning
4.	Emergency Response	Forest And Fire Extinguishing	Discussion Forum, Whatsapp, HT	Manggala Agni, Bukit Batu Districts Government Sungai Pakning Ward Government, Masyarakat Peduli Api, Journalist, PT Kilang Pertamina Internasional Unit Sungai Pakning
		Ministration	Discussion Forum	PT Kilang Pertamina Internasional Unit Sungai Pakning.
5.	Recovery	Processing Burned Peatlands Into Agriculture	Discussion Forum	Farmers Community, PT Kilang Pertamina Internasional Unit Sungai Pakning.

3.3. Collaboration in Community Empowerment

The management of forest and land fires is carried out through the development of community potential. Through various discussions by the stakeholders involved, the community empowerment program becomes a solution and a long-term plan in handling forest and land fires. The empowerment programs implemented are:

3.3.1. *Fire Mitigation Based on Masyarakat Peduli Api.* This activity is a capacity building for members of the Masyarakat Peduli Api group in handling forest and peatland fires. This program is implemented in collaboration with Manggala Agni, Bukit Batu Districts Government Sungai Pakning Ward Government, Masyarakat Peduli Api, and PT Kilang Pertamina Internasional Unit Sungai Pakning. Training to improve the ability to handle forest and land fires was given to members of the Masyarakat Peduli Api by providing Patrol and Fire Extinguishing Materials and Practices in collaboration with Manggala Agni, Riau province. In addition, MPA members are also facilitated with personal extinguishing equipment (coveralls, safety shoes, safety helmets) and equipment for meeting water needs (hydrant pumps, extinguishing hoses, and nozzles). Community empowerment activities are also carried out in terms of improving the community's economy through entrepreneurship training and professional certification. Entrepreneurship training is carried out in several fields, namely catfish cultivation, screen printing and peatland farming. Professional Certification is carried out to support MPA members to be able to work in companies according to their abilities. The certifications carried out to date are the professional certifications of Safetyman, Fireman, and Welder. This certification program has created 20 competent human resources and have worked in companies according to their respective competencies.

Through community empowerment programs, the capacity for handling forest and land disasters can be improved through training and at the same time improving the economic welfare of the community.

3.3.2. Integrated Pineapple Farming Development. Stakeholders see the potential for handling forest and forest fires through pineapple farming. Based on related researches, peatlands have characteristics that absorb a lot of water and have a high acidity level. Therefore, suitable plants are needed to survive on peatlands. The results of stakeholder discussions lead to the suitability of pineapple farming as an activity to prevent forest and land fires. Pineapple plants have strong characteristics, and do not require a lot of water and care so they are suitable for planting on peatlands. Pineapple planting was carried out

IOP Conf. Series: Earth and Environmental Science **986** (2022) 012028 doi:10.1088/1755-1315/986/1/012028

as much as 14.5 ha and it has grown to 25 ha at this time. Besides being able to increase the productivity of abandoned peatlands, this plant can be a barrier to prevent forest fires from spreading further.

The Integrated Pineapple Farming Program also improves the economic welfare of the community through the sale of fresh pineapples and processed products such as Sweets, Pineapple Chips, Pineapple Sticks, Pineapple "Dodol", Pineapple Jam, and Pineapple Syrup.

4. Conclusion

Activities for handling forest and land fires in the Bukit Batu sub-district need to involve stakeholders in order to achieve maximum results. Stakeholder involvement is implemented in all aspects of disaster management stages from prevention, mitigation, preparedness, emergency response, to recovery. The use of media in stakeholder engagement activities also supports the success of communication and the running of forest and land fire disaster management activities. Based on the results and previous discussions, it can be concluded that the contribution of stakeholders and the results of forest and land fire management activities are in the following table:

Table 4. Stakeholder Engagement Contribution

Program	Stakeholders involved	Contribution of Stakeholders in the program	Result
Fire Mitigation Based on Masyarakat Peduli Api	PT Kilang Pertamina Internasional Unit Sungai Pakning.	MPA Coaching MPA Training MPA Community Empowerment MPA Professional Certification MPA Tool Help	Reduced forest and peatland fires since 2015-2020 Increased community participation in the Fire Care Community Group from 20 people (2015) to 100 people (2020)
	Mangga Agni of Riau Province	MPA Training	Improve MPA skills and professions to become: Fireman, Safetyman, Welder
	Military of Bukit Batu District	MPA Coaching	The Creation of Peatland Fire Management Innovations: 1. Peat Nozzle 2. Portable Hydrant Well
	Police Departement of Bukit Batu District	MPA Coaching	
	Bukit Batu District Government	MPA Coaching	
Integrated Pineapple Farming Development	PT Kilang Pertamina Internasional Unit Sungai Pakning.	Farmer Group Development, Assistance with land clearing, seeds and fertilizers	Reduced forest and peatland fires since 2015-2020 Increasing abandoned land is processed into pineapple land up to 25 ha
	Bukit Batu District Government	Farmer Group Development.	Increasing the economic welfare of the community The creation of SMEs processing pineapple products

References

[1] Badan Nasional Penanggulangan Bencana 2007 *UNDANG-UNDANG REPUBLIK INDONESIA NOMOR 24 TAHUN 2007 TENTANG PENANGGULANGAN BENCANA* (Jakarta)

[2] Dinas Kehutanan Provinsi Riau 2015 *Profil Penanganan Kebakaran Hutan di Provinsi Riau Tahun 2015* (Pekanbaru)

[3] Anon http://sipongi.menlhk.go.id

[4] Creswell. J W 2014 *Research Design : Qualitative, Quantitative, and Mixed Methods Approaches* (California: SAGE Publications, Inc.)

[5] M I 2009 *Metode penelitian Ilmu Sosial* (Yogyakarta: PT. Gelora Akasara)

[6] Iqbal M 2007 Analisis Peran Pemangku Kepentingan dan Implementasinya Dalam Pembangunan Pertanian *J. Litbang Pertan.* **26** 89–99

[7] Crosby B 1991 Stakeholder Analysis : A Vital Tool for Strategic Managers *U.S Agency Int. Dev.* 1–6

[8] V C and P S 2001 *Risk communication : evolution and revolution in solution in an environmental in Peril* (John Hopkins University Press)

[9] Menteri Lingkungan Hidup dan Kehutanan 2021 *Peraturan Mentri Lingkungan Hidup dan Kelautan Republik Indonesia No. 1 tahun 2021*

Socioeconomic influence on community preparedness in the city of land disaster in Sukanagara Subdistrict, Cianjur District

E Maryani*, R Erliyandi, H Murtianto

Departemen of Geography Education, Faculty Social Science Education, Indonesia University of Education

*Email: enokmaryani@upi.edu

Abstract. This research is based on Sukanagara District which has a high level of landslide susceptibility. This study aims to analyze the effect of education level, age, employment status and income, and housing conditions on community preparedness in dealing with landslides in Sukanagara District, Cianjur Regency. This research is a descriptive research with a quantitative approach. Based on the results of the study, the variables that affect the level of readiness in education. Meanwhile, age, employment status, and income, as well as living conditions had no effect. Suggestions for research related to efforts to increase community preparedness are to increase the level of public education and provide supporting facilities for preparedness by the government.

1. Introduction

Sukanagara Subdistrict is located in Cianjur Regency and has a high landslide hazard [1] In 2010-2020 there have been seven disasters and caused losses and casualties to the community.

Table.1 Land Landslide Incident in Sukanagara District, Cianjur Regency

No	Time of occurance	Village Scene
1.	3/12th/2010	Sukamekar
2.	3/11th/2010	Sukamekar
3.	9/29th/2014	Sukanagara
4.	12/25th/2014	Sukajembar
5.	9/26th/2016	Ciguha
6.	3/20th/2019	Gunungsari
7.	4/9th/2020	Sukanagara

Disaster events are closely related to preparedness in the face of disasters. Preparedness measures are required in anticipation to minimize the number of fatalities and losses [2]. Preparedness is an effort focused on developing a plan in the face of disasters [3]. The level of disaster preparedness of a community is closely related to other factors in society. demographic factors and socioeconomic conditions have an influence on preparedness in the face of disasters [4]. Some factors of socioeconomic conditions of the community that are closely related to the level of preparedness include the level of public education, age level, employment and income status and the condition of people's residence. The level of education factor in research [5] has an influence on preparedness, people who have education at the final level (SMA) and universities tend to have a higher preparedness index value compared to people who have lower education, namely at the elementary school level (SD) and junior high school (SMP). The results of the study were supported by [6] who explained that the higher the level of education that is woven by a person, the higher the level of preparedness in the face of disasters. Age factors related to response and attitude in using knowledge to preparedness disasters [7]. The factors of employment status and income also have a powerful influence on the level of community preparedness in dealing with disasters [4]. Then, factors of residential conditions related to the utilization of appropriate types of buildings and can reduce the risk of disasters [8].

Society is the most important actor who experienced the disaster while feeling the direct impact of the disaster [9]. This explanation makes researchers want to examine the influence of socio-economic conditions on community preparedness in dealing with landslides in Sukanagara District, Cianjur. The purpose of this study was to analyze the effect of education level, age, employment status and income, and housing conditions on the level of community preparedness in dealing with landslides in Sukanagara District, Cianjur Regency.

2. Method

The research uses a quantitative approach and a descriptive type of research. The population in this study is the people of Sukanagara subdistrict cianjur district. Sampling technique used is probability random sampling using cluster random sampling. The technique of calculating samples using slovin technique so that it is known the number of research samples amounting to 100 respondents scattered throughout the village in sukanagara subdistrict, namely jayagiri village, Ciguha, Sukakarya, Sukarame, Sukalaksana, Sukanagara, Gunungsari, Sindangsari, Sukajembar and Sukamekar which is also based on the distribution of residents who are in the area of landslide disasters medium to high. The research was conducted for ± 6 months from September 2020 to February 2021. Analysis techniques used in this study

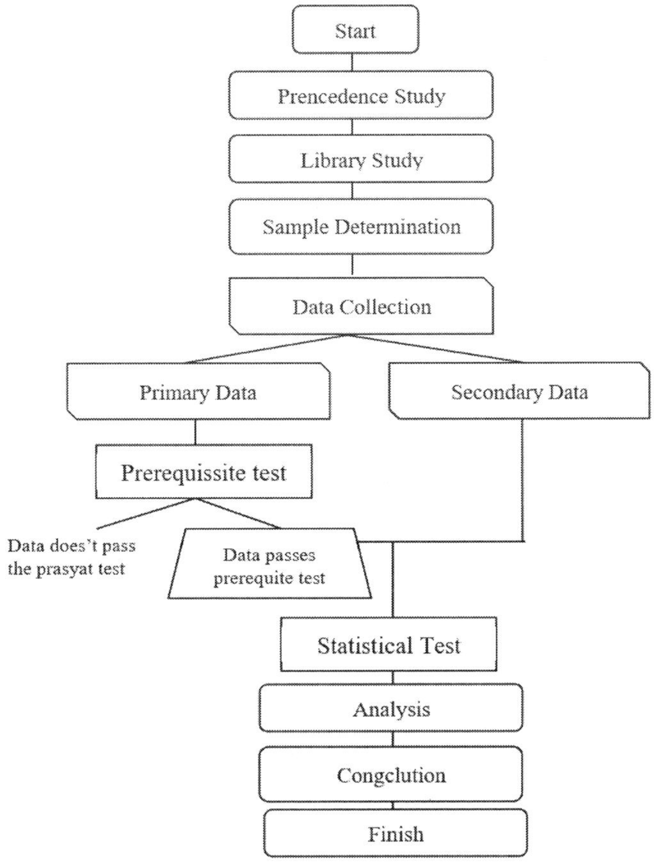

Figure 1. Research Flowchart

3. Result and Discussion

3.1 Prerequisite Assumption Test Results (Classic Assumptions)

3.1.1 Normality Test
The normality test used in this study was the Kosmogorov-Smirnov normality test. This normality test aims to detect whether the residual value of a distributed research variable is normal or not.

Table 2. Normality Test Results

		Unstandardized Residual
N		100
Normal Parameters[a,b]	Mean	,0000000
	Std. Deviation	236,99613889
Most Extreme Differences	Absolute	,066
	Positive	,056
	Negative	-,066
Test Statistic		,066
Asymp. Sig. (2-tailed)		,200[c,d]

a. Test distribution is Normal.

b. Calculated from data.

c. Lilliefors Significance Correction.

d. This is a lower bound of the true significance.

In the table, it can be seen that the significance value is 0.22. This means that the data is normally distributed because the significance value is > 0.05.

3.1.2 Linearity Test

The linearity test aims to determine the linear relationship between the independent variable and the dependent variable. As for the assumptions used in the linearity test, if the linearity significance value is > 0.05, then the independent and dependent variable data have a linear relationship. Meanwhile, if the linearity significance value is <0.05, then the independent variable and the dependent variable are declared to have no linear relationship.

Table 3. Linearity test results for Education Level variables

ANOVA Table

			Sum of Squares	df	Mean Square	F	Sig.
Preparedness * Education	Between Groups	(Combined)	3459943,376	3	1153314,459	20,152	,000
		Linearity	3214523,834	1	3214523,834	56,169	,000
		Deviation from Linearity	245419,543	2	122709,771	2,144	,123
	Within Groups		5494067,624	96	57229,871		
	Total		8954011,000	99			

Table 4. Linearity test results for Age variables

ANOVA Table

			Sum of Squares	df	Mean Square	F	Sig.
Preparedness * Age	Between Groups	(Combined)	357527,264	2	178763,632	2,017	,139
		Linearity	67631,718	1	67631,718	,763	,385
		Deviation from Linearity	289895,546	1	289895,546	3,271	,074
	Within Groups		8596483,736	97	88623,544		
	Total		8954011,000	99			

ANOVA Table

			Sum of Squares	df	Mean Square	F	Sig.
Preparedness * Age	Between Groups	(Combined)	357527,264	2	178763,632	2,017	,139
		Linearity	67631,718	1	67631,718	,763	,385
		Deviation from Linearity	289895,546	1	289895,546	3,271	,074
	Within Groups		8596483,736	97	88623,544		
	Total		8954011,000	99			

Table 5. Linearity test results for job status and income variables

ANOVA Table

			Sum of Squares	df	Mean Square	F	Sig.
Preparedness * Employment and Income	Between Groups	(Combined)	891573,542	3	297191,181	3,539	,018
		Linearity	650876,018	1	650876,018	7,750	,006
		Deviation from Linearity	240697,523	2	120348,762	1,433	,244
	Within Groups		8062437,458	96	83983,724		
	Total		8954011,000	99			

Table 6. Linearity test results for variable living conditions

ANOVA Table

			Sum of Squares	df	Mean Square	F	Sig.
Preparedness * Living Conditions	Between Groups	(Combined)	409702,238	3	136567,413	1,534	,211
		Linearity	300074,131	1	300074,131	3,371	,069
		Deviation from Linearity	109628,107	2	54814,054	,616	,542
	Within Groups		8544308,762	96	89003,216		
	Total		8954011,000	99			

Based on the results of linearity tests on all research variables, can be seen in tables 3, 4, 5, 6 then it can be known that the entire value of the significance of linearity for all variables > 0.05 means that all variables are declared to have a linear relationship with the variable preparedness.

3.1.3 Multicolinearity Test

This multicolinearity test aims to test the symptoms of multicolinearity. Symptoms of multiconinearity are symptoms of interrelationship or strong relationships between independent variables.

Table 7. Multicolinearity Test Results

Coefficients[a]

Model	Unstandardized Coefficients		Standardized Coefficients			Collinearity Statistics	
	B	Std. Error	Beta	t	Sig.	Tolerance	VIF
1 (Constant)	322,306	200,221		1,610	,111		
Education	7,360	1,100	,579	6,690	,000	,872	1,147
Age	-,780	1,769	-,037	-,441	,660	,942	1,061
Employment and Income	,572	1,130	,045	,506	,614	,820	1,219
Living Conditions	1,957	1,343	,125	1,458	,148	,886	1,129

a. Dependent Variable: Preparedness

3.1.4 Heteroskedastisitas Test

The purpose of using this test is to detect the inequality of variance from one observation to another in the regression model. The heteroscedasticity test used in this study is the scatterplot heteroscedasticity test, and the output of this test is shown in **Figure 2**:

IOP Conf. Series: Earth and Environmental Science **986** (2022) 012029 doi:10.1088/1755-1315/986/1/012029

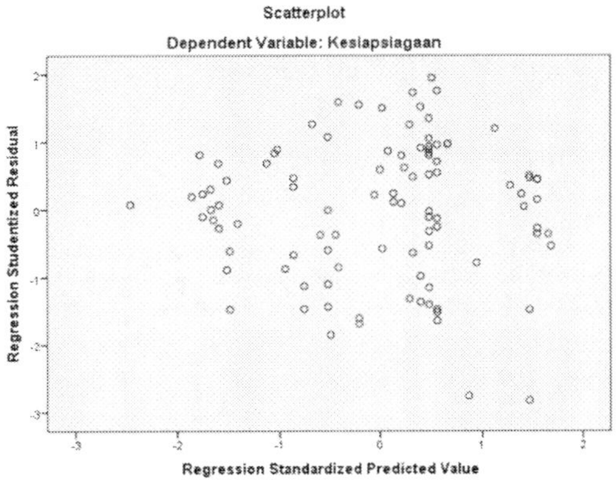

Figure 2. Scatterplot Test Results

To be able to determine the decision on the test of heteroskedastisitas scatterplot then the following considerations are used:
• If the dots spread above, below or between the numbers 0
• Point not gathered above or below only
• Dots do not form wavy patterns
• Point spread is not patterned

If the data and output results include the above criteria, then the data is declared to have no symptoms of heteroskedastisitas. So the data in the picture above is free from symptoms of heteroskedastisitas.

3.2 Multiple Linear Regression Tests
Analysis of multiple liear regression tests in this study using T test, F test as well as coefesien determination analysis

3.2.1 T Test
The t test is used in analyzing the effect of the independent variable on the y variable partially.
The results of multiple linear regression analysis using the f test are shown in **table 8**.

Table 8. Test t result

Coefficients[a]

Model		Unstandardized Coefficients		Standardized Coefficients		
		B	Std. Error	Beta	t	Sig.
1	(Constant)	322,306	200,221		1,610	,111
	Education	7,360	1,100	,579	6,690	,000
	Age	-,780	1,769	-,037	-,441	,660
	Employment and Income	,572	1,130	,045	,506	,614
	Living Conditions	1,957	1,343	,125	1,458	,148

a. Dependent Variable: Preparedness

IOP Conf. Series: Earth and Environmental Science **986** (2022) 012029 doi:10.1088/1755-1315/986/1/012029

To be able to make decisions on hypotheses in the t test, it can be the value of output significance in the t test. If the significance value < 0.05 or the calculated t value > t table then the hypothesis is accepted or there is a meaningful relationship between the free and bound variables. Conversely if the significance value > 0.05 or the calculated t value < t table then the hypothesis is rejected or does not exist meaningful relationship between a free variable and a bound variable.

3.2.2 F Test

The f test analysis was conducted to analyze the influence of free variables on simultaneously bound variables. This study found that socio-economic conditions affect the people's preparedness of Sukanagara District in dealing with landslides. The results of the analysis using SPSS results in **table 9**.

Table 9. Test f result

ANOVA[a]

Model		Sum of Squares	df	Mean Square	F	Sig.
1	Regression	3393461,185	4	848365,296	14,494	,000[b]
	Residual	5560549,815	95	58532,103		
	Total	8954011,000	99			

a. Dependent Variable: Preparedness

b. Predictors: (Constant), Living Conditions, Education, Age, Employment and Income

The conclusion from the results means that socio-economic conditions simultaneously affect the community's preparedness in the face of landslide disasters in Sukanagara District, Cianjur Regency.

3.2.3 Coefesien Determination

A determining coefesien value is a value that indicates how much influence a free variable has on a bound variable. The large influence of variable socioeconomic conditions on preparedness is shown by R square in the table. The output table of the results is as follows

Table 10. Coefesien Determination Results

Model Summary

Model	R	R Square	Adjusted R Square	Std. Error of the Estimate
1	,616[a]	,379	,353	241,93409

a. Predictors: (Constant), Living Conditions, Education, Age, Employment and Income

Table 10. shows the magnitude of R square of 0.397, meaning that socio-economic conditions affect community preparedness in dealing with landslides in Sukanagara District, Cianjur Regency by 37.9%.

3.3 The Influence of Education Level on Community Preparedness

The level of education in Sukanegara District is complete, starting from kindergarten, elementary school, junior high school, high school, to higher education. The results of interviews with respondents in ten villages of the research site are as follows

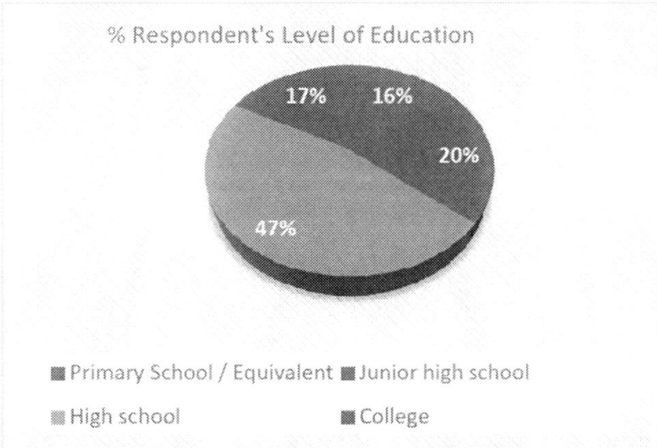

Figure 3. Percentage of Respondents' Education Level

The results of the t-test can be concluded that the level of education of the people in Sukanagara District affects their preparedness in dealing with landslides. The level of education directly affects the level of knowledge and attitudes. So it can be drawn the relationship of the influence of education level on community preparedness, the higher a person's level of education, the higher the level of readiness of a person in dealing with landslides. This is supported by previous research by [6] concluding in his research that the level of education factors are very influential on preparedness in the face of disasters.

3.4 The Influence of Age on Community Preparedness

Age variables are categorized into three categories, namely respondents with the age of 20-40 years, 40-60 years and respondents with the age of >60 years. The percentage of respondents' age based on the results interviews in ten villages at the research site are as follows

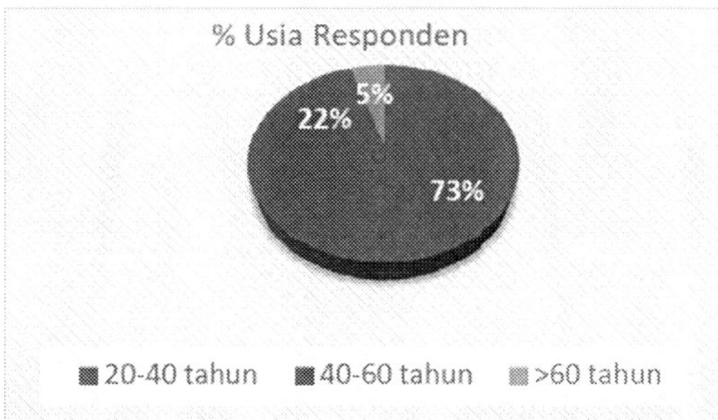

Figure 4. Respondent Age Percentage

The results of statistical analysis of education variables using the t-test showed an effect of education level on community preparedness in dealing with landslides. Educational stories directly affect the level of knowledge and attitudes. So that it can be drawn the relationship between the influence of education level on community preparedness, the higher a person's education level, the higher the level of readiness of a person in dealing with landslides in Sukanagara sub-district, Cianjur district. A number of studies in line with this, one of which is research from [4] which mentions that age has no effect on preparedness.

3.5 Effect of Employment and Income Status on Community Preparedness

The employment status of respondents is grouped into two, namely respondents who have a permanent job and respondents who do not have a permanent job. While the respondent's income is categorized into a society with income of Rp.0-1 million/month, Rp. 1-3 million/month, Rp. 3-5 million/month, Rp. 5-7 million/month, and Rp. >7 million/month. The results of the percentage of employment status and income based on the interview results of respondents as follows

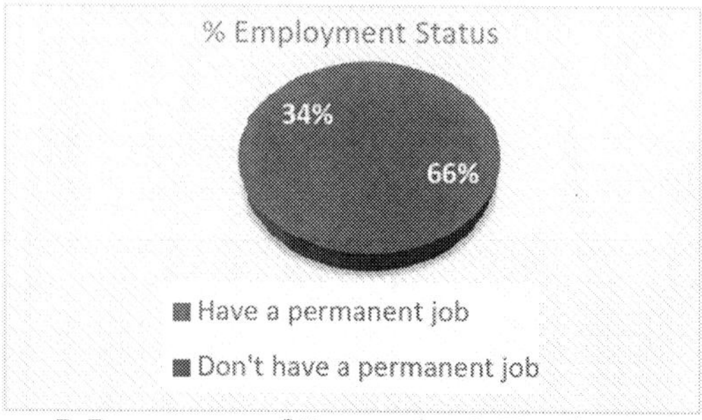

Figure 5. Percentage of Respondent's Employment Status

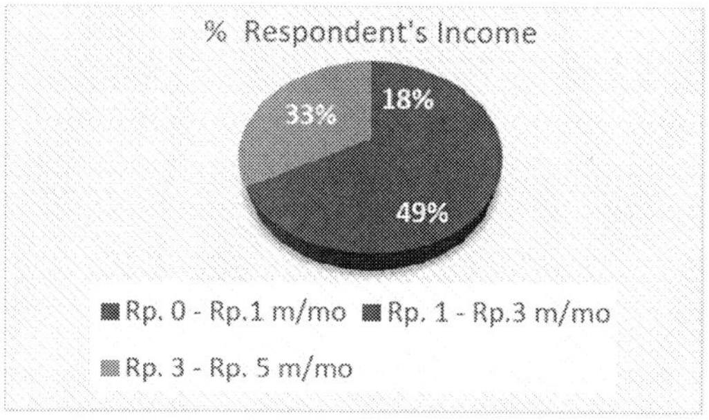

Figure 6. Percentage of Respondent's Income

The results of the t-test of the employment status variable indicate that there is no significant effect between employment status and income on community readiness. In this case, the level of work and payment of respondents does not provide a positive direction on the level of community preparedness in dealing with landslides in Sukanagara District, Cianjur Regency. This is not in line with the results of the study [4] which mentions the work has an effect on community preparedness.

3.6 Effect of Housing Conditions on Community Preparedness

Measurement of people's living conditions is categorized into home ownership and the type of home material used. Community home ownership is classified into three namely private ownership, renting and hitchhiking. While the type of community home material is categorized into a permanent home, a semi-permanent house and a not permanent home. The results of interviews to respondents are obtained data as follows.

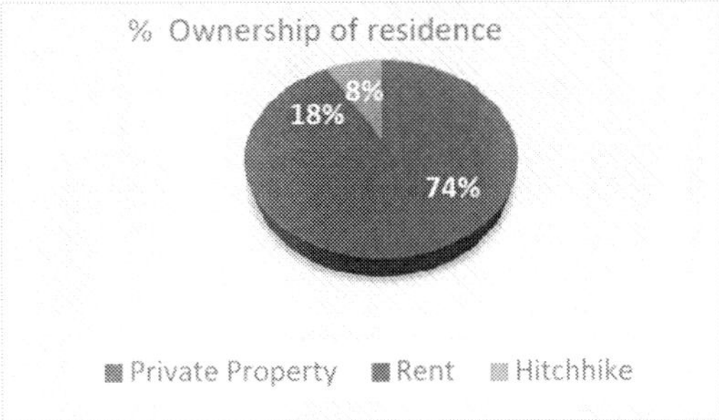

Figure 7. Percentage of Respondents' Residence Ownership

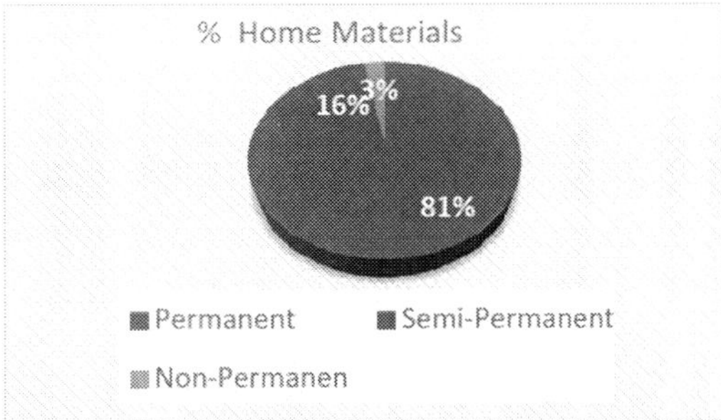

Figure 8. Percentage of Respondent's Home Materials

Tests on the variables of employment status and income show no effect of housing conditions on community readiness in dealing with landslides.

The residence of the people of Sukanagara District does not give a tendency to vary the level of community preparedness in dealing with landslides in Sukanagara District, Cianjur. This is in line with research [10] which states that there is no influence between the type of residence and the level of preparedness of the community in the face of disasters

4. Conclusion

The results of this study are (1) the education variable affects the level of community preparedness in dealing with landslides, (2) the age variable does not affect the level of community preparedness in dealing with landslides, (3) employment status and income. variable does not affect the level of community preparedness. in dealing with landslides and (4) housing condition variables do not affect the level of community preparedness in dealing with landslides in Sukanagara District, Cianjur Regency.

References

[1] BPBD Kabupaten Cianjur 2015 *Renstra Perubahan 2016-2021* vol 3

[2] Y H R, E H and Romadhoni 2020 Analisis Faktor-Faktor Kesiapsiagaan Masyarakat Rawan Bencana *J. Ilmu Keperawatan Jiwa* **3** 131 – 142

[3] Rahmayani 2018 Faktor-Faktor Yang Mempengaruhi Kesiapsiagaan Kepala Keluarga Dalam Menghadapi Bencana Gempa Bumi **1** 64–70

[4] Deviani E and Indra S 2013 Analisis Kesiapsiagaan Dan Kerentanan Perempuan Di Wilayah Pesisir Dalam Menghadapi Bencana Gempa Dan Tsunami Di Banda Aceh *Idea Nurs. J.* **4**

[5] Fitriningtiyas K 2014 *Hubungan Tingkat Pendidikan Formal Dengan Kesiapsiagaan Dalam Menghadapi Bencana Gempa Bumi Masyarakat Desa Jabung Kecamatan Gantiwarno Kabupaten Klaten* (Universitas Muhammadiyah Surakarta)

[6] Maryanti S, Lestari E, Putri W, Wardani A R and Haris F 2017 Hubungan Tingkat Pendidikan Masyarakat terhadap Kesiapsiagaan Bencana Tanah Longsor di Kelurahan Giritirto Kecamatan Wonogiri *Pros. Semin. Nas. Geogr. UMS 2017* 319–30

[7] Purwoko A, Sunarko and Putro S 2015 Pengaruh Pengetahuan Dan Sikap Tentang Resiko Bencana Banjir Terhadap Kesiapsiagaan Remaja Usia 15 18 Tahun Dalam Menghadapi Bencana Banjir Di Kelurahan Pedurungan Kidul Kota Semarang *J. Geogr. Media Inf. Pengemb. dan Profesi Kegeografian* **12** 214–21

[8] Sinaga R and Lesmana C 2019 Kajian Evaluasi Kinerja Bangunan terhadap Siaga Bencana Gempa *J. Tek. Sipil* **14** 77–103

[9] Wiratri A 2018 Menilik ulang arti keluarga pada masyarakat Indonesia *J. Kependud. Indones.* **13** 15–26

[10] Nuzuar A 2017 Analisis Kesiapsiagaan Masyarakat dalam Menghadapi Ancaman Bencana Gempa Bumi dan Tsunami di Kecamatan Padang Barat Kota Padang Tahun 2017 *J. Kesehat. Masy. Andalas* **4(21)** 8

Analysis of slip surface by using geoelectrical array at Wonoreja Karanganyar Regency as mitigation of landslide disaster

Budi Legowo*, Sorja Koesuma, Alfitra Adam Primastawan, Darsono, Iwan Yahya, Harjana
Department of Physics, Sebelas Maret University

*Email: pakbeel@gmail.com

Abstract. Landslide disaster often occur in areas with hilly topography such as Wonoreja Village. The occurrence of landslides can be predicted by identifying the slip surface. Slip surface is one of the factors that can cause landslides disaster. The aims of the research is to figure out the characteristic of slip surface at Wonoreja Village. Data acquisition were conducted using geoelectrical method with Wenner-Schlumberger array. The data were obtained at 3 lines with 210 length. The data were processed using RES2DINV ver 3.54 software. We found that the slip surface for first line is around 12 meter depth with clay material that forms an inclined plan pattern which can be potential to cause translational landslide. On the second line we found two slip surfaces around 20 to 25 meter depth, the first one forms a concave pattern which can be potential to cause a rotational landslide, another slip surface on second line forms an inclined plane which can be potential to cause a translational landslide. On the third line we found an inclined plan slip surface at 15 meter depth which can be potential to cause translational landslide.

1. Introduction

Karanganyar Regency located in Central Java Province with coverage area between 110°40' - 110°70' E and 7°28' - 7°46' S. Based on RPJP Kabupaten Karanganyar 2005-2025 [1], more than 60% of the land is in the landslide-prone zone. This is due to the condition of the land that has a fairly steep slope starting from a slope of 20 degrees even worse. There are 327 landslide-prone points in Karanganayar Regency. In the last 5 years there have been 300 cases of landslides in Karanganyar Regency.

There are 14 sub-districts in Karanganyar Regency with medium vulnerability of landslides spesifically Jenawi, Ngargoyoso, Tawangmangu, Jatiyoso, Jumapolo, Jumantono, Jatipuro, Karangpandan, Matesih, Karanganyar, Kerjo, Kebakkramat, Gondangrejo, and Mojogedang. Reported from the official website of BPBD Karanganyar, there were landslides in Jatiyoso, Jumapolo, Kerjo, and Tawangmangu in 2016 which caused a total of nine building houses to be damaged. The latest, reported by Solopos.com, landslides struck in two villages in Jatiyoso District, Karanganyar Regency on April 14, 2021, which is Berusk and Wonorejo with seven houses affected. It was raining heavily came from 16.00 until the next day when the landslide struck in. With that series of cases, it is necessary to conduct a scientific study to provide information and improve understanding of landslide disasters.

This research apply the geoelectrical method to find and characterize the slip surface that has been suspected to be the cause of the landslide based on the resistivity value from each layer. It also to determine the structure and subsurface layers of the research site. This information can be used to determine the boundaries of soil instability which can be used as a reference in regional development in the Karanganyar area and its surroundings. Therefore, to determine the structure and layering of the soil this research was carried out using geoelectrical method of resistivity with Wenner-Schlumberger array.

1.1 Landslide and Slipsurface
Landslide is a displacement of slope-forming components such as rock, sediment, and debris that move down and out of the slope [2]. There are two factors that affect landslides, namely driving factors, this is the factors that affect material conditions, and the other is triggering factors, this is the factors that cause soil displacement [3]. Landslides are common disaster in high rainfall areas. With a lot of water infiltration, the potential for the water to penetrate to impermeable layer (slip subrface) is even greater. This causes the soil layer above it to move.

There are 6 types of landslides according to PVMBG (2015)[4] :

1. Translational landslide: It is a type of landslide that often occurs in Indonesia, where this type of landslide occurs due to the movement of soil and rock masses contained in a flat slip surface.
2. Rotational lanslide: is a type of landslide that occurs because there is a movement of soil and rock masses on a concave slip surface.
3. Blok landslide: This is a type of landslide that occurs due to rock shifting on a flat slip surface. This type of landslip is often referred to as a stone block translational landslip because it is only rock that undergoes displacement
4. Rock collapse: is a type of terrible landslide where this type of landslide occurs due to the movement of rock that moves downward by means of free fall. In general, this type of landslide occurs on steep hills near the coast.
5. Soil creep: A type of slow-moving landslide. This type of landslide is almost undetectable. Over time, this type of termite landslide can cause telephone poles, trees, or houses to tilt downwards.
6. Debris: Usually occurs in valleys and volcanoes. This can occurs when the soil mass moves and then pushed by water. The speed of it flow depends on the level of the slope, volume and water pressure.

1.2 Rock Resistivity
The electrical properties of rocks are related to the response of rocks to electric currents. Rocks are formed by various processes and composed by various minerals. This makes rocks have various physical characteristics, so these factors make the resistivity values of rocks so vary. The electric current propagates under the earth's surface is spread through the constituent rocks as a conducting medium. Rock resistivity means the ability of a rock to conduct electricity. Rock electric current is divided into 3 [5]:

1. Electronic Conduction
   utilizes free electrons in rock constituent minerals as a transport conductor of electric current.
2. Electrolytic Conduction
   Utilizes electrolyte ions in water that settle in the pores of the rock as a conductor. So that its conductivity and resistivity are also determined by the porosity and water content stored in it.
3. Dielectrical Conduction
   Dielectric conduction utilizes the influence of an external electric field to separate electrons from the nucleus so the additional free electrons are obtained.

Table 1. Rock resistivity values

Rock type	Resistivity (Ωm)
Andesite	$1{,}7x10^2$ - $45x10^4$
Bassalt	200 – 100.000
Quartz	500 – 800.000
Tuff	$2x10^3$ - 10^5
Sandstone	200 - 8000
Sand	1 - 1000
Clay	1 - 100
Wet clay	10 - 35
Sandy clay	30 - 215

Source : [6]

1.3 Geoelectrical Method

Geoelectric method is geophysical method which studies the properties of electric currents that occur under the earth's surface. By assuming the earth as an isotropic homogeneous medium, so when there is a current flowing under surface the current flow will be the same and continuous. The resistivity measured in this case is a true resistivity. But in fact the earth is not a homogeneous medium, so the measured resistivity is the average resistivity value or referred to be apparent resistivity [6].

This method uses two current electrodes (C1 and C2) and two potential electrodes (P1 and P2). An electric current is injected into the earth and will produce a measurable potential as a result of the injection of a given current. The result of the measurement is a potential value that it can be used to determine the apparent resistivity value of the subsurface results [7]. In the processing, the apparent resistivity value is inverted with data processing software to get the true resistivity value. With the help of the geological study of the research area, this value is then used to interprate the subsurface conditions [8].

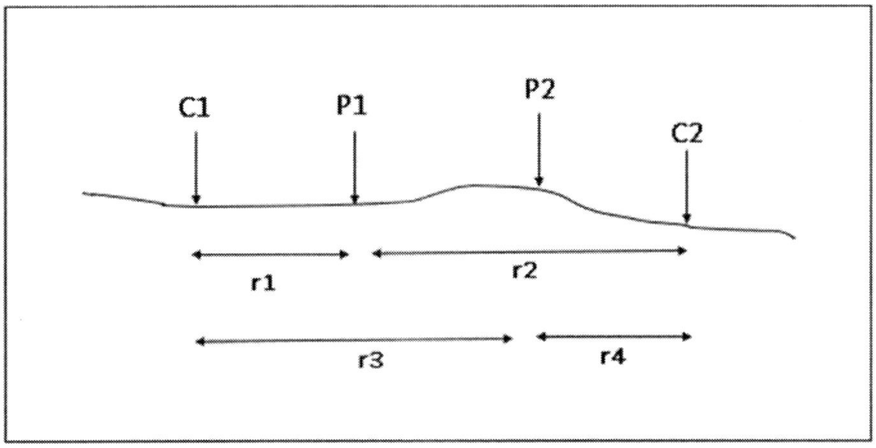

Figure 1. Geoelectric Configuration concept

1.4 Wenner-Schlumberger Array

The resistivity mapping method aims to study the horizontal variation of the resistivity of the subsurface layer [9]. Electric current is injected into the earth by using two electrode currents (C1 and C2), and the measurement of potential difference using two potential electrodes (P1 and P2). The Wenner-Schlumberger configuration is often used to measure resistivity, in a system with electrodes arranged at a constant distance as shown in Figure 2.2 [10].

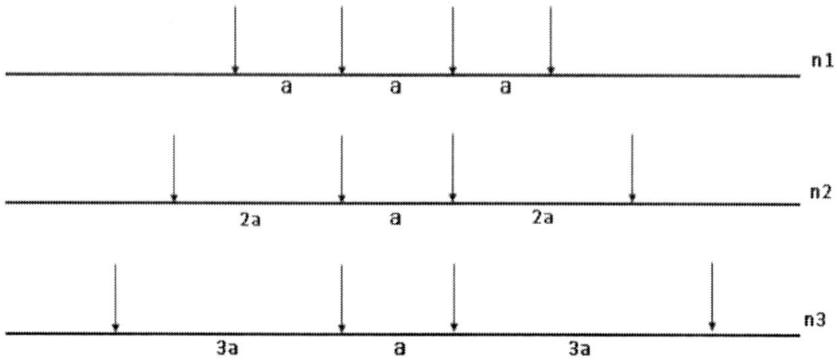

Figure 2. Wenner–Schlumberger electrode arrangement

This configuration is a combination of Wenner alpha array and the Schlumberger array. Wenner alpha-Schlumberger configuration can be seen in Figure 2.3

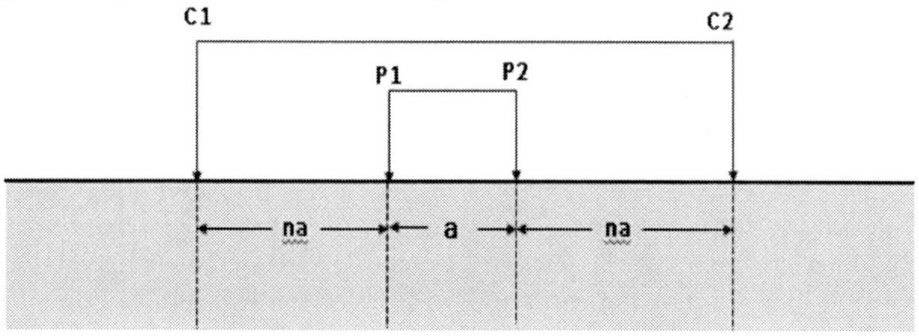

Figure 3. Wenner–Schlumberger configuration

The value of the geometry factor of the Wenner alfa-Schlumberger configuration is [11]:

$$k = \pi(n^2a + na) \quad (1)$$

The apparent resistivity can be calculated based on the following equation.

$$\rho a = k\frac{\Delta V}{I} \quad (2)$$

With the geometry factor (k), the apparent resistivity value (ρa) will be grouped according to the layer depth (n). Then the data that has been arranged is processed using the Res2dinv software. In order to obtain an image of the subsurface conditions based on the difference in resistivity values [12].

2. Method

The study was conducted on 8 – 11 June 2021 in Wonorejo Village, Jatiyoso District, Karanganyar Regency. Determination of research points was carried out by reviewing landslide-prone areas that endanger public facilities, residents' houses, and agricultural land. The study was conducted at three lines, the set up lines shown on figure 4.

Figure 4. Research Lines Set Up

Measurements were made using the Geoelectric resistivity method to estimate the slip surface that has the potential to conduct a movement. The configuration used in this research is Wenner-Schlumberger array with 210 meters track length on each line and 15 meters of electrode spacing unit. The main equipment used in the measurements there are OYO McOHM-EL resistivity meter, electrodes, cables, meters, hammers, GPS, calculators and other supporting equipment. Data processing uses RES2DINV Software to perform the inversion and display the 2D sections of the subsurface resistivity distribution. The 2D model of each research line can be interpreted by reviewing several references to identify the slip surface.

3. Result and Discussions

The research was conducted in Wonorejo Village, Jatiyoso District, Karanganyar Regency. The study was conducted to determine the potential for ground movement by identifying the presence and structure of the slip surface at the research site. The slip surface is a layer below the soil surface which has the potential to cause the layer above it to make a movement.

Wonorejo Village is a hilly area with a slope of more than 20 degrees. The stratigraphic setting of Wonorejo village is included in Lava Jobolarangan formation which is an andesite lava flow from Mount Jobolarangan towards the southwest which contains andesine, quartz, feldspar, and a bit of hornblend [13].

The measurement was conducted on three tracks with a stretch of 210 meters each. The primary data obtained during the acquisition is then inverted with RES2DINV Software to obtain a 2D sections of subsurface resistivity model and its contours.

3.1. Line 1

The first line is located at 70 42' 2" S - 1110 7' 59.8" E (first electrode) to 70 42' 2.2" S - 1110 7' 53.2" E (last electrode) with an average elevation of 1217 meters above the surface sea. The first line is located on a hill ridge in a West-East direction. Based on the results of the RES2DINV inversion, the distribution of the subsurface resistivity values in the first track ranges from 33.9 Ohm meters to more than 1500 Ohm meters.

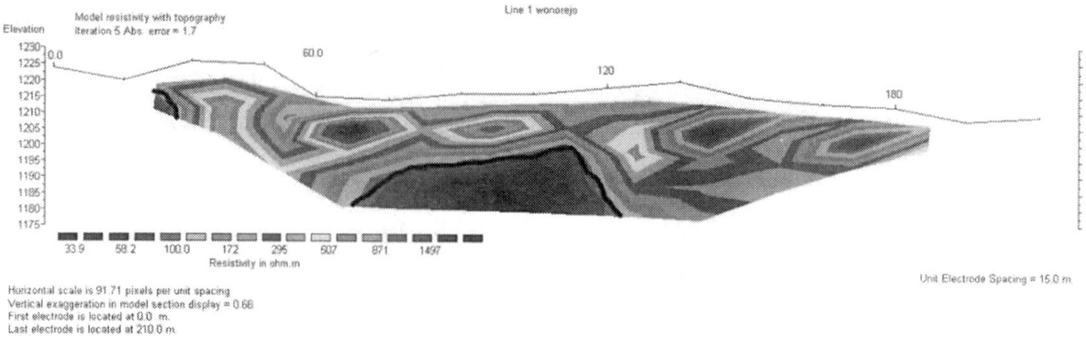

Figure 5. 2D Section of 1st line

The layers on the first line are classified into three, they are low resistivity layers between 33.9 - 100 Ohm meters with a dominant blue on color scheme, medium resistivity ranges from 136 - 400 Ohm meters with a dominant color of green, and high resistivity, which is above 500 Ohm meters predominantly yellow to red.

3.2. Line 2

The second line is located at 70 41' 44.5" S - 1110 7' 48.5" E (first electrode) to 70 41' 50.6" S - 1110 7' 46.7" E (last electrode) with an average elevation of 1003 meters above sea level. The second line intersects a steep ridge in a northeast-southwest direction. Based on the results of the RES2DINV inversion, the distribution of the subsurface resistivity values in the first track ranges from 2.86 Ohm meters to more than 1900 Ohm meters.

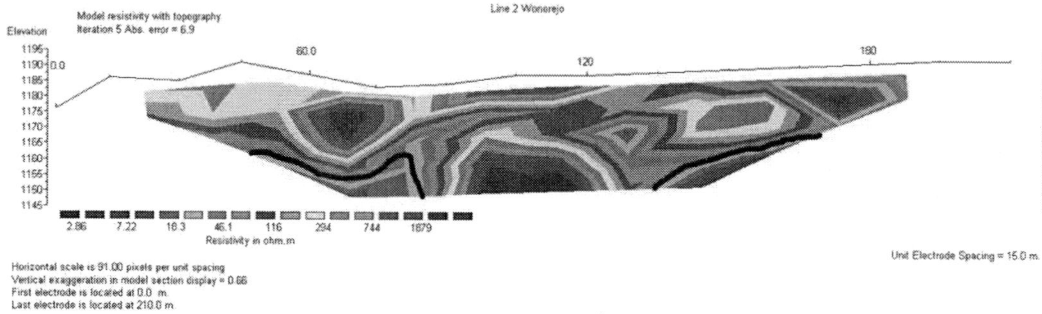

Figure 6. 2D Section of 2nd line

The layers on the second line are classified into three, they are low resistivity layers, ranges between 2.86 - 33.2 Ohm meters with a dominant color of blue, then medium resistivity that is ranges from 46.1 - 294 Ohm meters with dominant colors of green to yellow, and high resistivity, which values above 520 Ohm meters with the dominant color of red.

3.3. Line 3

The third line is located at 70 41' 54.5" S - 1110 7' 47.1" E (first electrode) to 70 41' 48.7" S - 1110 8' 10.4" E (last electrode) with an average elevation of 1186 meters above sea level. The third line is on the ridge with a northwest-southeast direction. Based on the results of the RES2DINV inversion, the distribution of the subsurface resistivity values in the first track ranges from 22.1 Ohm meters to more than 544 Ohm meters.

IOP Conf. Series: Earth and Environmental Science **986** (2022) 012030 doi:10.1088/1755-1315/986/1/012030

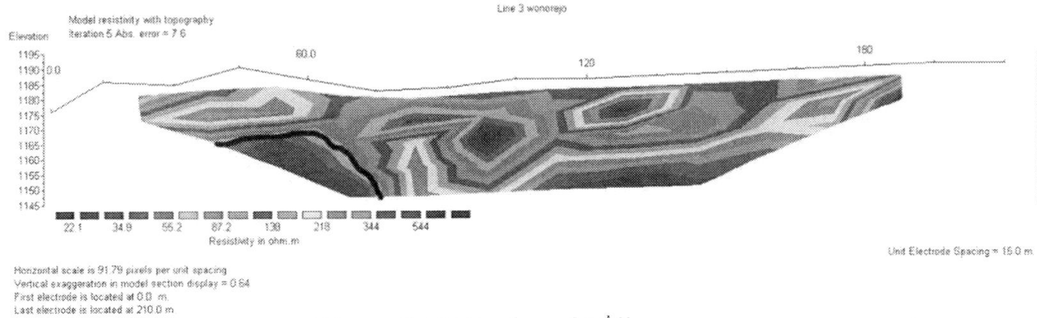

Figure 7. 2D Section of 3rd line

The layers on the third line are classified into three, the are low resistivity layers that ranges between 22.1 - 71.2 Ohm meters with a dominant color of blue, medium resistivity which ranges from 87.2 - 178 Ohm meters with a dominant image color of green, and high resistivity that varies from 344 to above 544 Ohm meters with a dominant color of yellow to red.

3.4. Slip Surface Identification

Table 2. Layers type of each line

Location	Low resistivity	Medium resistivity	High resistivity
Line 1	Clay	Sandy clay	Igneous rock
Line 2	Wet clay	Sandy clay	Igneous rock
Line 3	Clay	Sandy clay	Igneous rock

The slip surface under the three lines is clearly clay layer with the highest water saturation among the layers above it. In the first line the slip surface is in the form of quite thick clay at a depth of 12 meters which is located below the 5th electrode to the 9th electrode (Figure 5). The slip surface forms an inclined plane pattern that has the potential to cause translational landslides [14].

On the second line there are two slip surfaces in the form of a layer of wet clay (figure 6) where one of them has to be a concave pattern along the 4th to 7th electrode with a depth of 25 meters. The slip surface in the form of a concave has the potential to cause rotational landslides [15]. Another slip surface forms an inclined plane pattern at the 10th to 12th electrodes with a depth of about 20 meters. The slip surface has the potential to cause translational landslides [14].

The slip surface under the third line is a layer of clay located under the 3rd electrode to the 6th electrode at a depth of about 15 meters (Figure 7). The slip surface forms an inclined plane so that it has the potential to cause translational landslides.

Clay layers with low resistivity have high water saturation but they are not good enough to pass the waters. Clay Layers with these characteristics can be classified as an aquiclud layers based on their role in groundwater distribution [16]. The sandy clay layer above it can be potential to move because the clay underneath with a high enough water saturation causes the surface to become slippery so it is not good enough to withstand the volume of the load on the layer above it. This situation is reinforced by a fairly steep slope.

If the frequency of rain highly increase, this layer of clay has the potential to become a slip surface. This is because the water that passed by the layer above it stops and is accommodated there, so that it becomes very slippery for the layer above it. Planting vegetation with good water absorption may be a solution to help absorb the water so it doesn't quickly pass to the clay layer.

4. Conclusion

The slip surface formed on the first line forms an inclined plane pattern at a depth of 12 meters which has the potential to cause translational landslides. On the second line there are two slip surfaces, the first forming a concave pattern at a depth of 25 meters. This has the potential to cause rotational landslides. The second one is at a depth of 20 meters which forms an inclined plane, this has the potential to cause translational landslides. In the third line, the slip plane forms an inclined plane pattern at a depth of 15 meters which has the potential to cause translational landslides.

With high rainfall intensity, the water will be accommodated in the slip surface (clay layer) which makes it slippery and has the potential to move. As a first prevention step, it is possible to plant some vegetation with good water absorption and do some early warning system installment as a part of mitigation.

References

[1] Bappeda Kabupaten Karanganyar 2015 *Rencana Pembangunan Jangka Panjang (RPJP) Kabupaten Karanganyar 2005-2025* (Kabupaten Karanganyar)

[2] Ottowitz D, Jochum B, Pfeiler S, Gruber S, Supper R and Kim J-H 2017 Pechgraben Landslide: Evaluation of Geophysical/Geotechnical Methods in Terms of Remediation Support (Springer, Cham)

[3] Supper R, Ottowitz D, Jochum B, Kim J H, Römer A, Baron I, Pfeiler S, Lovisolo M, Gruber S and Vecchiotti F 2014 Geoelectrical monitoring: An innovative method to supplement landslide surveillance and early warning *Near Surf. Geophys.* **12** 133–50

[4] Pusat Vulkanologi dan Mitigasi Bencana Geologi 2015 *Perkiraan wilayah potensi terjadi gerakan tanah / tanah longsor dan banjir bandang di seluruh Indonesia* (Bandung)

[5] Telford W M, Geldart L P and Sheriff R E 1990 *Applied geophysics* (Cambridge: Cambridge University Press)

[6] Milsom J 2003 *Field geophysics* vol 11 (England: John Wiley & Sons Ltd)

[7] Sugito, Irayani Z and Jati I P 2015 Investigasi Bidang Gelincir Tanah Longsor Menggunakan Metode Geolistrik *J. Berk. Fis.* **13** 49–54

[8] Hapsari S B 2013 *Interpretasi Data Geolistrik untuk Memetakan Potensi Air Tanah dalam Menunjang Pengembangan Data Hidrogeologi di Kabupaten Jombang, Jawa Timur* (Universitas Sebelas Maret)

[9] Sudha, Tezkan B, Israil M, Singhal D C and Rai J 2010 Geoelectrical mapping of aquifer contamination: A case study from Roorkee, India *Near Surf. Geophys.* **8** 33–42

[10] Fitrikayanti Hasibuan, Susilawati and Rahmatsyah 2014 Studi Intrusi Air Laut Dengan Menggunakan Metode Resistivitas Konfigurasi Dipole- 1–6

[11] Priambodo I C, H P, N R and Juanda 2011 Aplikasi Metode Geolistrik Konfigurasi Wenner-Schlumberger pada Survey Gerakan Tanah di Bajawa, NTT *Bul. Vulkanol. dan Bencana Geol.*

[12] REYNOLDS J M 1997 *An Introduction to Applied and Environmental Geophysics* (England: John Wiley & Sons Ltd)

[13] Sampurna and Samodra 1997 *Peta Geologi Lembar Ponorogo* (Bandung)

[14] Ling C, Xu Q, Zhang Q, Ran J and Lv H 2016 Application of electrical resistivity tomography for investigating the internal structure of a translational landslide and characterizing its groundwater circulation (Kualiangzi landslide, Southwest China) *J. Appl. Geophys.* **131** 154–62

[15] Pratiwi E S, Sartohadi J and Wahyudi 2019 Geoelectrical Prediction for Sliding Plane Layers of Rotational Landslide at the Volcanic Transitional Landscapes in Indonesia *IOP Conf. Ser. Earth Environ. Sci.* **286** 0–9

[16] Darsono D 2016 Identifikasi Akuifer Dangkal dan Akuifer Dalam dengan Metode Geolistrik (Kasus: Di Kecamatan Masaran) *Indones. J. Appl. Phys.* **6** 40

Soil erodibility mapping for soil susceptibility in the upstream of Kelara Subwatershed in Jeneponto Regency

A Ahmad[1], M Farida[2], N Juita[1] and N Amin[1]

[1] Departement of Soil Science, Faculty of Agriculture, Hasanuddin University, Jalan Perintis Kemerdekaan km. 10, 90245, Makassar, Indonesia
[2] Departement of Geology, Faculty of Engineering, Hasanuddin University, Makassar, Indonesia

*Email: asmita.ahmad@agri.unhas.ac.id

Abstract. Landslides and flash floods in Rumbia Village, Rumbia District, Jeneponto Regency on June 11-12, 2020, have caused material and non-material losses to the local population. The incident occurred very quickly with an area with a disaster impact on seven sub-districts and 18 villages. This study aims to map soil erodibility to assess soil susceptibility to landslides in the Upper Kelara Sub-watershed. Calculate soil erodibility using the Wischmeier and Smith method, texture with hydrometer method, c-organic with Walkley and Black, mapping of soil erodibility with the kriging approach, and expert judgment for soil susceptibility category. The results showed that c-organic value content (1.19 to 2.47%) has low in landslides areas, with soil permeability ranging from 0.23 to 1.16 cm/hour and soil texture dominated silty clay. Soil erodibility in the landslides area reaches a value of 0.4 with the high category of soil susceptibility. Soil erodibility is in line with soil susceptibility value, the high erodibility value, the high soil susceptibility category. Soil erodibility mapping showed a distribution of erodibility index increase in the bottom part of the Kelara Subwatershed. The Mitigation actions through government assistance and socialization of disaster-aware communities need to be carried out immediately so that incidents can be minimized and prevented in the future.

1. Introduction

Based on BNPB data for 2020, it is known that the risk index value for disaster events in 2015-2020 is still relatively high at 159.49 [1]. The number of landslides and flash floods in South Sulawesi is increasing every year. Although the incident data in Jeneponto Regency is classified as a moderate risk class, the events cause losses and casualties [2]. The data of landslides and flash floods in Rumbia Village, Rumbia District, Jeneponto Regency on June 11-12, 2020, have caused material and non-material losses to the local population. The incident occurred very quickly with an area with a disaster impact on seven sub-districts and 18 villages.

Disaster events are triggered by various things, such as changes in land use from forest to plantation [3], extreme rainfall anomalies [4,5], tectonic earthquakes, and volcanic earthquakes [6]. Various regional susceptibility maps related to landslides and flash floods have been made by ESDM [7] to take early mitigation actions against disaster events. However, the potential and magnitude of the event cannot be estimated because the geobiophysical conditions of each region are different [8] and very dynamic [9], so the extent of the impact cannot be evaluated at this time. So it is important to make a map of the ability of the land to not experience landslides which can trigger flash floods. The purpose of this study is to be able to determine the level of erodibility to predict the upstream flash flood hazard of the Kelara Sub-watershed, Jeneponto Regency.

2. Material and methods

Soil sampling was carried out on several slope classes in the Kelara watershed (Figure 1) in Jeneponto Regency at a 0-40cm depth at 24 soil sampling points. Soil analysis was carried out following the procedures of the Soil Research Institute [10] include; analysis of soil texture (sand, fine sand, silt and clay) with the hydrometer method, analysis of c-organic Walkley and Black soil, permeability with permeameter, and direct observation for soil structure. Map created with Arc-GIS 10.3 software.
The calculation of the value of the soil erodibility index follows the procedure of Wischmeier and Smith [11], with the equation:

$$K = \frac{1{,}292[2{,}1M^{1.14}(10^{-4})(12-a)+(3{,}25(b-2)+2{,}5(c-3)]}{100} \tag{1}$$

where:
K = index of soil erodibility
M = $percent\ of\ fine\ sand + percent\ of\ silt\ \times (100 - percent\ of\ clay)$
a = percent of organic matter
b = the code of soil structure
c = the code of permeability class

Soil erodibility class use of Arsyad classification [12]. The category is presented in table 1.

Table 1. Classification of soil erodibility value

Class	K Value	Category
1	0,00-0,10	Very low
2	0,11-0,21	Low
3	0,22-0,32	Moderate
4	0,33-0,44	Slightly high
5	0,44-0,55	high
6	0,56-0,64	Very high

Source : [12]

Soil erodibility mapping in the Kelara sub-basin was carried out by kriging interpolation with semivariogram value [13], with the equation:

$$\gamma * (h) = \frac{1}{2Nh}\sum_{i=1}^{Nh}[z\ (xi) - Z\ (xi + H)]^2 \tag{2}$$

where:
Nh = The number of pairs
H = lag distance

Validation with root mean square error (RMSE), with the equation:

$$RMSE = [\frac{\sum_{i=1}^{n}(z(xi)-\check{z}(xi))^2}{n}]^{1/2} \tag{3}$$

where:
z(xi) = the height value at reference data
ž = predited height of value

The susceptibility assessment is carried out using the method of expert judgment through a questionnaire to provide categories related to the level of soil susceptibility, taking into account factors; erodibility value and slope. Rating categories are divided into very high, high, medium, and low.

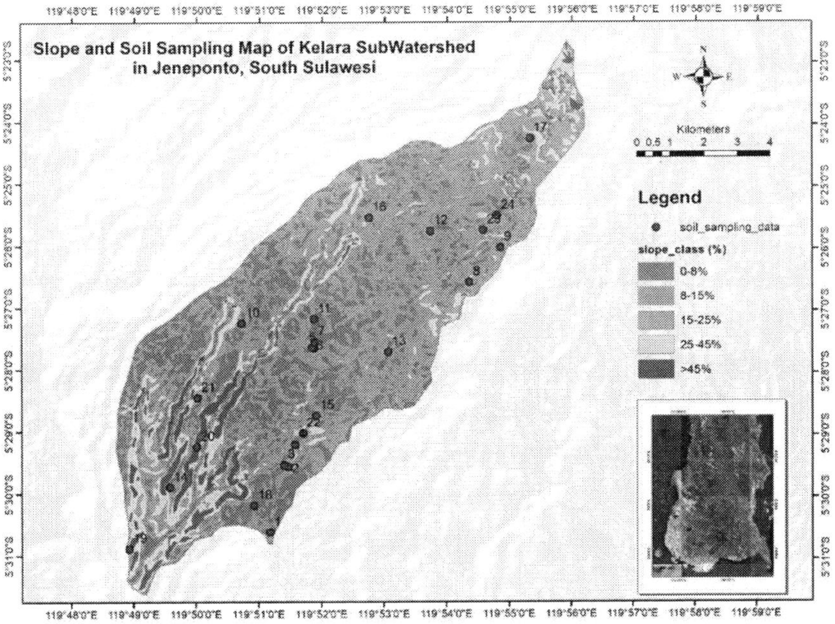

Figure 1. Slope and soil sampling map of the Kelara Subwatershed in Rumbia District, Jeneponto Regency

3. Results

The results showed that c-organic value content (1.19 to 2.47%) has low in landslides areas (table 2). The soil permeability ranged from 0.23 to 1.16 cm/hour in the very slow to slow category of permeability [14]. The soil texture dominated silty clay, with a silt fraction average is 49%; this value will reduce soil cohesion [15]. Slope classes vary widely and spread locally from the bottom to the upstream part of the SubWatershed. The dominant slope classes are 8-15% and 15-25% (figure 1).

Soil erodibility index in the landslides area reaches a value of 0.4 at slope 15-45% at sampling point 4 and sampling point 22, while on gentler slopes with higher erodibility values, no landslide triggers flash floods (figure2). The model of soil erodibility index in the upstream of Kelara Subwatershed in Jeneponto Regency showed the distribution of the erodibility index increases in the northeastern and eastern part of the subwatershed locally and increases at the bottom of the Subwatershed so that the accumulation of rain in the upstream area can cause landslides and flash flood at the bottom of the Subwatershed in line with the increase in the value of soil erodibility and the percent of the slope.

Validation data with RMSE show a value of 0.106, close to 0, which indicates the data used is accurate and precise to predict the spread of the erodibility index value in the upstream area of the Kelara sub-watershed in Rumbia District, Jeneponto Regency. The normal distribution of data value with plot distribution close to the model (straight line) can be seen in Figure 3.

Based on soil erodibility value and slope data, soil susceptibility increases in line with the increase in soil erodibility value (table 2). Landslides occurred at sampling points 4 and 22 on riverbanks that triggered flash floods, with a high susceptibility status (table 2).

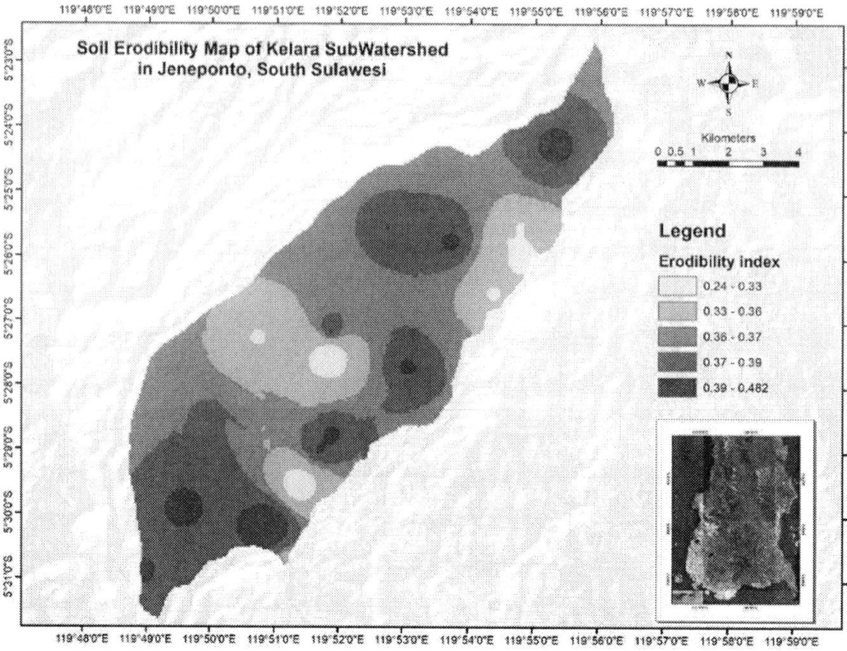

Figure 2. Soil erodibility map of the Kelara Subwatershed in Rumbia District, Jeneponto Regency

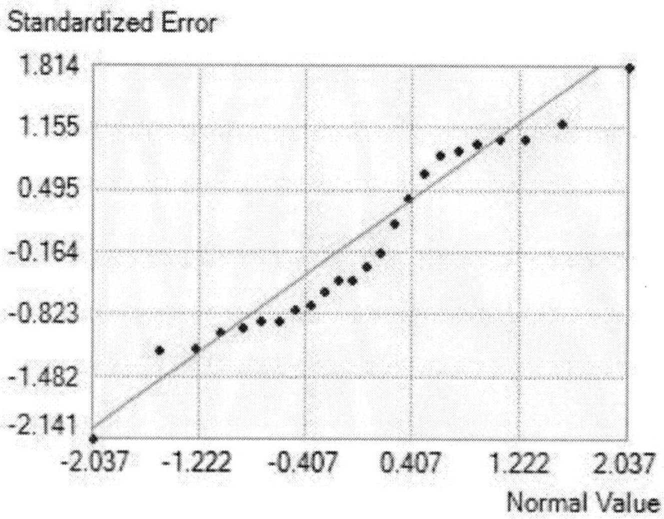

Figure 3. Normal distribution of plot data value closed to the model

Table 2. Soil characteristic in Kelara Subwatershed in Rumbia District of Jeneponto Regency

Soil sampling	Texture				Texture Class	Soil organic matter	Value of Soil Texture	Soil Structure	Code of soil structure	Soil permeability	Class of soil permeability	Code of soil permeability	Soil Erodibility	Soil Susceptibility category
	sand	fine sand	silt	clay		%	M			cm/hour			K	
1	10	5	43	42	silty clay	3.07	2524	granule medium-coarse	3	0.99	slow	5	0.29	Moderate
2	5	8	34	54	clay	4.26	1575	blocky	4	0.42	very slow	6	0.27	Moderate
3	6	4	12	78	clay	3.45	268	blocky	4	0.71	slow	5	0.16	Low
4	6	6	35	53	clay	2.05	1651	blocky	5	0.27	very slow	6	0.40	High
5	7	0	55	38	silty clay loam	4.22	3426	granule fine-medium	2	0.99	slow	5	0.29	Moderate
6	12	0	30	65	clay	3.41	1050	blocky	4	1.39	slow	5	0.21	Low
7	4	8	40	48	silty clay	3.62	2088	granule fine-medium	2	1.16	slow	5	0.20	Low
8	7	8	45	40	silty clay	3.26	2708	granule fine-medium	2	0.76	slow	5	0.26	Moderate
9	7	0	47	46	silty clay	3.43	2518	granule fine-medium	2	1.07	slow	5	0.24	Low
10	5	0	43	52	silty clay	2.48	2064	granule fine-medium	2	0.31	very slow	6	0.25	Moderate
11	4	3	67	26	silt loam	2.97	4961	granule fine-medium	2	0.56	slow	5	0.46	High
12	2	3	61	34	silty clay loam	2.65	4029	granule medium-coarse	3	0.41	very slow	6	0.47	High
13	6	5	62	26	silt loam	2.24	4593	granule fine-medium	2	0.5	slow	5	0.46	High
14	9	10	60	21	silt loam	2.09	4750	granule fine-medium	2	0.51	slow	5	0.48	High
15	3	0	60	37	silty clay loam	2.34	3780	granule medium-coarse	3	0.36	very slow	6	0.45	High
16	4	0	57	39	silty clay loam	2.09	3477	granule medium-coarse	3	0.36	very slow	6	0.43	High
17	8	11	63	18	silt loam	2.74	5177	granule fine-medium	2	0.94	slow	5	0.50	Very high
18	7	5	71	17	silt loam	2.31	5898	granule fine-medium	2	0.61	slow	5	0.59	Very high
19	8	13	59	20	silt loam	3.45	4733	granule coarse	3	1.16	slow	5	0.47	High
20	5	7	57	31	silty clay loam	3.14	3940	granule medium-coarse	3	0.51	slow	5	0.41	High
21	12	9	47	32	clay loam	3.34	3205	blocky	4	0.48	very slow	6	0.41	High
22	7	13	43	37	silty clay loam	2.09	2722	blocky	5	0.28	very slow	6	0.44	High
23	7	0	47	46	silty clay	2.10	2518	granule fine-medium	2	0.31	very slow	6	0.30	Moderate
24	3	7	43	47	silty clay	2.09	2286	granule fine-medium	2	0.23	very slow	6	0.28	Moderate

4. Discussion

The high slope classes at the bottom of the upstream Kelara Subwatershed trigger landslides that enter the river and induce flash floods. This is supported by the results of interviews with local communities, where before the flash flood, it was preceded by hillside landslides from riverbanks that entered the striped river flow and obstructed river flow. Continuous rain events had resulted in the overflow of river flows carrying material as flash floods and caused fatalities on June 12, 2021.

The decrease in soil carbon content in sloped areas of >15% occurred in all study locations. This triggers an increase in the value of soil erodibility. In this study, soil erodibility values that reach an index value of 0.4 have triggered landslides and flash floods. The decrease in soil carbon content and the high fraction of clay and soil dust are the causes of increased soil erodibility and increased soil saturation, as well as lowering the cohesion value of the soil so that the soil is easily dispersed [16,17].

The increase in the value of soil erodibility in areas with small slopes (<15%) in hilly regions (highlands) is caused by an increase in land-use intensity. This situation results in a decrease in soil carbon content, which binds the soil fraction to compact the soil quickly [18], inhibits infiltration, increases runoff, and triggers flooding in lower areas. The decrease in soil carbon content can trigger flooding in flat areas, and this is following the results of research from Saint-Laurent [19] in southern Quebec, where floods often happen in lower areas with less organic carbon content.

The increase in the erodibility value of the soil is in line with the increase in erodibility; the higher the probability, the more sensitive the soil to landslide events, especially on slopes above >15% with intensive land use. Intensive land use for agricultural land, cropland, and plantation by the community has increased the incidence of landslides in Western Ghat, India [20]. The occurrence of landslides mainly occurred at the erodibility value reaching an index value of 0.4 with a high susceptibility category on land use for agriculture. The increase in the value of soil vulnerability is mostly due to human land use activities for agriculture and settlement [21].

Soil erodibility has been able to predict soil susceptibility so that soil erodibility value mapping can be a solution in disaster mitigation. The government can immediately implement the mitigation process, especially for areas with high erodibility values, by applying vegetative and mechanical conservation techniques in the form of civil buildings to prevent the collapse of riverbank walls, especially in the lower area of the Kelara watershed. Mitigation actions can also be carried out through socialization to the community to be aware of disasters so that incidents can be minimized and prevented in the future.

5. Conclusions

Soil erodibility in the landslides area reaches a value of 0.4. This value impacts the increased value of soil erodibility at the bottom of the upstream of Kelara Sub-Watershed map to increase soil susceptibility, and it's connected to the slope of $\geq$15%. The Mitigation actions through government assistance and socialization of disaster-aware communities need to be carried out immediately so that incidents can be minimized and prevented in the future.

Acknowledgment

Thanks to RISTEK-BRIN and LP2M Hasanuddin University for the financial assistance provided through research grants.

References

[1] BNPB 2020 *Indeks Resiko Bencana Indonesia* (Badan Nasional Penanggulangan Bencana Indonesia)

[2] BNPB 2020 Info bencana *Badan Nas. Penanggulangan Bencana Indones. (National Agency Disaster Manag.* 1–7

[3] Reichenbach P, Busca C, Mondini A C and Rossi M 2014 The Influence of Land Use Change on Landslide Susceptibility Zonation : The Briga Catchment Test Site (Messina , Italy) *Environ. Manage.* **54** 1372–84

[4] Chen C W, Saito H and Oguchi T 2015 Rainfall intensity–duration conditions for mass movements in Taiwan *Prog. Earth Planet. Sci.* **2**

[5] Ahmad A, Lopulisa C, Imran A and Baja S 2019 Rainfall erosivity in climate changes and the connection to landslide events Rainfall erosivity in climate changes and the connection to landslide events *Earth Environ. Sci.* **280** 1–8

[6] Lacroix P, Berthier E and Maquerhua E T 2015 Earthquake-driven acceleration of slow-moving landslides in the Colca valley, Peru, detected from Pl??iades images *Remote Sens. Environ.* **165** 148–58

[7] ESDM 2009 *Peta Zona Kerentanan Gerakan Tanah Provinsi Sulawesi Selatan (Map of Soil Susceptibility of South Sulawesi Province)*

[8] Skrzypczak I, Kokoszka W, Zientek D, Tang Y and Kogut J 2021 Landslide Hazard Assessment Map as an Element Supporting Spatial Planning : The Flysch Carpathians Region Study *Remote Sens.* **13** 1–20

[9] Turconi L, Luino F, Gussoni M, Faccini F, Giardino M and Casazza M 2019 Intrinsic Environmental Vulnerability as Shallow Landslide Susceptibility in Environmental Impact Assessment *Sustainability* **11** 1–22

[10] BPT 2005 *Analisis kimia tanah, tanaman, air dan pupuk*

[11] Wischmeier W H and Smith D D 1978 *Predicting rainfall erosion losses* (United States Department of Agriculture)

[12] Arsyad S 2011 Konservasi Tanah Dan Air 1–19

[13] Setiyoko A, Basaruddin T and Arymurthy A 2020 Minimax Approach for Semivariogram Fitting in Ordinary Kriging *IEEE Access* **8** 82054–65

[14] Arsyad S 2010 *Konservasi Tanah & Air (Soil and Water Conservation)* (IPB Press)

[15] Ahmad A, Lopulisa C, Imran A M and Baja S 2018 Soil physicochemical properties to evaluate soil degradation under different land use types in a high rainfall tropical region : A case study from South Sulawesi , Indonesia *Earth and Environmental Science* (IOP Publishing) pp 1–7

[16] Shahangian S 2011 Variable Cohesion Model for Soil Shear Strength Evaluation *Geotech. Conf.* 1–9

[17] Abidin M H Z, Saad R, Wijeyesekera D C, Ahmad F, Baharuddin M F T, Tajudin S A A and Madun A 2017 The Influences of Basic Physical Properties of Clayey Silt and Silty Sand on Its Laboratory Electrical Resistivity Value in Loose and Dense Conditions *Sains Malaysiana* **46** 1959–69

[18] Błońska E, Lasota J, Piaszczyk W, Wiecheć M and Klamerus-Iwan A 2018 The effect of landslide on soil organic carbon stock and biochemical properties of soil *J Soils Sediments* **18** 2727–37

[19] Saint-laurent D, Paradis R, Drouin A and Gervais-beaulac V 2016 Impacts of Floods on Organic Carbon Concentrations in Alluvial Soils along Hydrological Gradients Using a Digital Elevation Model (DEM) *Water* **8** 2–17

[20] Sujatha E R and Rajamanickam G V 2012 Analysis of Landslide Susceptibility to Different Land Use Patterns in a Part of Kodaikkanal , Tamilnadu **2** 53–5

[21] Chen L, Guo Z, Yin K, Pikha Shrestha D and Jin S 2019 The influence of land use and land cover change on landslide susceptibility: A case study in Zhushan Town, Xuan'en County (Hubei, China) *Nat. Hazards Earth Syst. Sci.* **19** 2207–28

IOP Conf. Series: Earth and Environmental Science **986** (2022) 012032 doi:10.1088/1755-1315/986/1/012032

Land degradation due to tourism gentrification in Tawangmangu, Karanganyar, Indonesia

S Sholeh[1*] and Y Yusuf [1*]

[1] Geography Education Study Program, Faculty of Teacher Training and Education, Sebelas Maret University, Surakarta, Indonesia

*Email: sholehsofyan@gmail.com, yasinyusup@staff.uns.ac.id

Abstract. It is important to study the case of tourism gentrification which has an impact on human resources, ecosystem services and the role of actors who play planning and socio-economic development of the region. The mountainous landscape at the foot of Mount Lawu offers aesthetic views that other regions don't offer. This has resulted in increased demands and pressures on the emergence of new cities, where the area that should be conservation land and recharge area becomes built up land. From these pressures and demands, an in-depth analysis of the right tools or policies in sustainable landscape management is needed. This study focuses on the problem of anthropogenic pressure caused by land use change taking into account socio-economic and political factors. The methodology used consists of making maps, calculating and analyzing indicators, linking geospatial and socio-economic data to explore landscape changes and identify the drivers of rapid development. The findings show that the urbanization of tourism at the foot of Mount Lawu is increasing, while the drivers and impacts are different. The development carried out by local communities who do not have permits is very rapid, while unsustainable regional planning and development policies will threaten the aesthetics of the existing natural landscape. In addition, the rapid development in the recharge area has resulted in an increase in surface runoff, which will threaten the iconic tourism, namely gerojogan sewu. To stop land degradation, it is necessary to have a policy that is synergized between socio-economic needs and the sustainability of ecosystem services

1. Introduction

Gentrification is an urbanization of an area, where there is a transfer of property ownership from working people who have low incomes to people who have middle-upper income [1]. The phenomenon of the transfer of ownership of the property is also accompanied by a revitalization of the quality of settlements. Revitalization will have an impact on increasing tax rates, and property prices. This will result in indigenous people with low incomes shifting or moving to areas that have low property prices [2–4]. In addition, natives who know property prices are increasing will mass sell their properties, so they feel they have made high profits after selling their properties [5].

Tawangmangu Village, Kalisoro Village and Blumbang Village are villages in Tawangmangu District that are experiencing the phenomenon of gentrification. The findings related to the gentrification phenomenon are based on the identification of gentrification markers. According to Prayoga (2013: 2) that the characteristics of gentrification include changes/shifts in population, social changes, segregation and revitalization of areas within a society. The factor that causes gentrification in 3 sub-districts of Tawangmangu to take place quickly is tourism or what is called Tourism Gentrification. According to

Gotham (2005) Tourism gentrification refers to the transformation of the middle class environment into a space characterized by the proliferation of various entertainment companies and tourism places.
In any area, the development of a tourism causes a change in a space/place and brings a significant transformation to the community. Many investors (from outside Tawangmangu District) invest in Tawangmangu District, which have high prospects in the investment world, both in the form of tourist attractions and lodging services. The cool atmosphere with views of the mountains and pine trees that spoil the eye is an attraction for people from outside Tawangmangu District to visit.

Tourism gentrification is able to produce all changes in the landscape of a region. The characteristics of villages with local cultures in Tawangmangu, Kalisoro and Blumbang villages will fade, mainly because the displacement of the indigenous population by immigrant communities and increased access have made the villages closer to a central location, namely the city of Surakarta [6]. Improved road accessibility provides more opportunities for migrant residents to have a secondary residence in 3 sub-districts of Tawangmangu sub-district.

The gentrification of tourism also poses a threat to the livelihoods of the indigenous people who mostly work as farmers due to the impact of tourism development. According to Hines (2012) that there is a big shift from those that characterize changes in the production landscape (social and physical) to consumption. The community will change from livelihood in agriculture to shift to tourism or trade services. Even the natives will feel profit if they sell their property and move to another area.

2. Methods

Exploratory qualitative research is intended to describe and illustrate how people who are affected by changes in the physical and socio-economic environment are able to adapt to the phenomenon of gentrification, as well as how those in positions of power, namely the government and gentrifiers perceive these changes [7].

The research approach used is a spatial approach. The spatial approach has aspects of geographical space including location, natural conditions and socio-economic conditions of the community. The spatial approach in its implementation must also have the principles of distribution, interrelation and interaction of these aspects of geographic space. Tawangmangu Village, Kalisoro Village and Blumbang Village are an area located at the foot of Mount Lawu and has cool natural conditions, fertile soil, and beautiful scenery. This makes Tawangmangu Village, Kalisoro Village and Blumbang Village one of the tourist destinations.

3. Result And Discussions

3.1. Process and Form of Gentrification in Tawangmangu Village, Kalisoro Village and Blumbang Village

The natural beauty of the mountains and the cool air is a factor in the rapid gentrification in Tawangmangu Village, Kalisoro Village and Blumbang Village, so it is called Tourism Gentrification [8]. The 3 urban villages are based on 4 gentrification characteristics, namely displacement/shift of indigenous people, area revitalization, social change, and community segregation [9]. There are 3 forms of gentrification [1,10,11] (Figure 1):

a. State-led Gentrification

   This gentrification process has been going on since the era of Mangkunegara's control between 1917 and 14 July 1946 [12]. Where the Mangkunegara Government uses Dutch heritage equipment in the form of hydroelectric power plants to supply electricity in 3 villages in Tawangmangu District, namely Tawangmangu Village, Kalisoro Village and Blumbang Village. For people who want to enjoy electricity, they must submit a girik (proof of paying land tax). In addition to land tenure using girik, the government in the Mangkunegara era also carried out a land swap policy with indigenous people, especially in the Kalisoro Village community and part of the Tawangmangu Village. The land swap policy was an effort to develop a hydroelectric power plant that was built during the Mangkunegara era. From this policy, many people in Kalisoro Village and Tawangmangu Village built houses in Nglurah Hamlet, Tawangmangu Village.

In 1946 there was the handover of parcels of land to the government, wherein the policy of leasing plots of land emerged. 1950 was the year when many immigrants from outside Tawangmangu with medium to upper income looked for a secondary house/villa/business in the lodging sector, so they rented government land managed by Perusda Citra Mandiri Central Java (PDCMJT) and built inns and villas - villa on government land. The phenomenon of land rental is only able to be done by immigrant communities because only people who have middle and upper income are able to rent it (see Figure 1). The development policy process that involves government intervention is referred to as State-gentrification or gentrification initiated by the government [1].

After the issuance of Law Number 5 of 1960, land must be certified. The certificate must have valid evidence, such as girik or other proof of land ownership. Due to the constraints of legal evidence, land that should be owned by indigenous peoples cannot become proprietary land, while land leased by immigrant communities turns into land with building rights (HGB).

Then in 1997 a new policy was issued for people who want to change land, the government must pay compensation with 3 factors, namely the price based on the NJOP, the price based on the BPN and the price based on the market, then divided based on the number of these factors. This has become one of the agrarian conflicts between the indigenous people and the Citra Mandiri Regional Company in Central Java (PDCMJT) through the Tawangmangu Tourism Company (PPT) as the manager of the parcels of land. Indigenous people assume that they have occupied the land since their parents or ancestors, which until now is still occupied as a private house. Legally, the indigenous people lost because there was no legal evidence to prove that the land was their land. So that mediation was carried out between the indigenous people and the government with the final result that the community had to pay compensation for half of the actual price offered by the government, but the price was still burdensome for the indigenous people.

b. Gentrifier-led Gentrification

1970 was the year where many inns began to be built. The early pioneers in the development of lodging in the Tawangmangu sub-district were immigrant communities who tried to do business in the luxury inns [1]. In acquiring property in Tawangmangu District, there are several ways, namely, immigrant communities directly come to the location or through a broker. If you go through a direct seller, you will benefit from a cheaper price, but not necessarily the newcomers who want to buy the property will get the location, price, and view you want. It is different if the buyer through a broker will get a choice of location, the price that the buyer wants but at a higher price. This is what causes indigenous people to become impromptu brokers.

c. Self-Gentrification

In addition to buying cheap properties in other areas, the factor that causes people to sell their properties is as capital for people who want to renovate their homes into lodging/home stays for out-of-town tourists who want to stay in Tawangmangu District [1,8]. This is referred to as Self-Gentrification or Gentrification led by the natives. The self-gentrification process has a desire to improve the fate of the indigenous population, but also pays attention to the community they belong to. When immigrants came to invest in lodging with a modern style, this caused the indigenous people to start copying these ideas and create inns/homestays.

IOP Conf. Series: Earth and Environmental Science **986** (2022) 012032 doi:10.1088/1755-1315/986/1/012032

Figure 1. Villa Building on Government Land (Persil Land) with Building Use Rights Status.

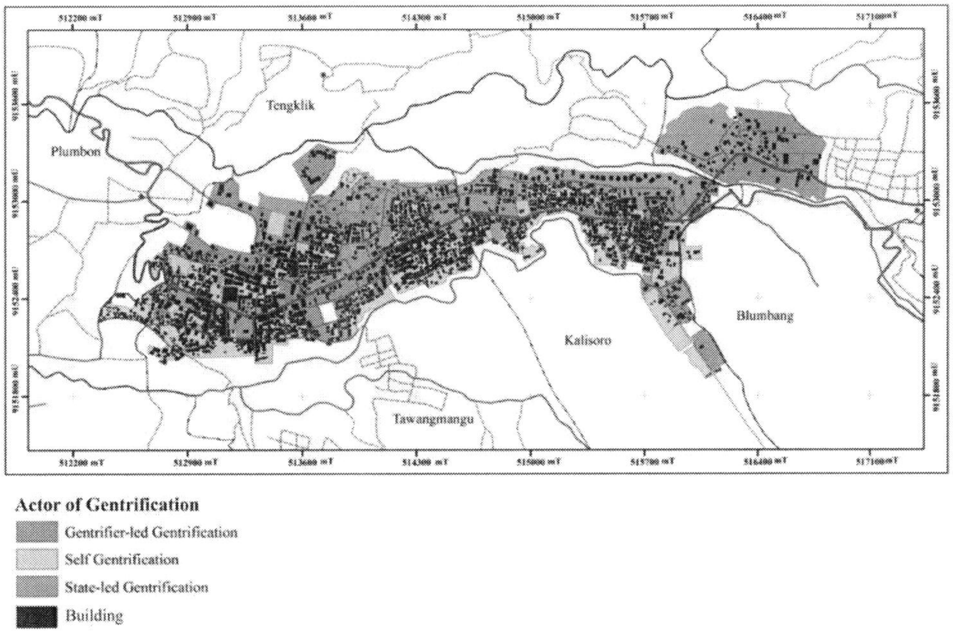

Figure 2. Map of Gentrification Actors in Tawangmangu Village, Kalisoro Village and Blumbang Village in 2020.

3.2. The Impact of Gentrification on the Socio-Economic Community and Land Degradation in Tawangmangu Village, Kalisoro Village and Blumbang Village

a. Physical Environment Transformation in Tawangmangu Village, Kalisoro Village and Blumbang Village

This is what happened in Tawangamngu Village, Kalisoro Village and Blumbang Village where when indigenous people have sold property to gentrifier actors, they as gentrifier actors will transform the physical environment of their area by doing area refreshes in accordance with the form of the modern area they brought from their home. previous environment [1,2,4,10]. Environmental transformation does provide competition between the development of commercial areas between immigrant communities (gentrifiers) and indigenous communities, one of which is the development of modern shops by immigrant communities. With a strong capital economy, the immigrant group is able to provide and provide complete facilities to provide services to consumers when shopping. The habit pattern of urban tourism groups is to

IOP Conf. Series: Earth and Environmental Science **986** (2022) 012032 doi:10.1088/1755-1315/986/1/012032

look for modern shops that are the same or similar to what is in their environment, so they feel that it is easier and more comfortable to buy in modern shops (Figure 3).

Revitalization of the area is also carried out by the government, which used to be a shopping area for indigenous people that was built on government/illegal land then the government revitalized the area so that it was redeveloped into cafes that could be rented by the community. However, because the rental price is set too high, only immigrants who have high capital are able to rent the building. So the policy is not very concerned with the local community.

Figure 3. Gentrifier-led Gentrification: on the left (Google Earth, 8 August 2015) before revitalization and on the right after revitalization (primary data, 8 September 2020)

b. Increase in property prices in Tawangmangu Village, Kalisoro Village and Blumbang Village
The Karanganyar Regency Government also has an impact on price increases by strengthening the effect through various land zoning, development financing, and fiscal practices [7]. The practice of zoning and development financing really affects a property that is highly valued, where the practice of zoning is related to access to a road while development financing leads to future government development planning. Easy accessibility to housing locations and the construction of facilities that have been carried out by policy makers are the causes of the increase in property prices. This was proven when the policy of developing Halloween Village tourism in Kalisoro Village resulted in an increase in land prices. From this policy, the development and widening of the road that was originally managed by the kelurahan is now a road that is managed by the Karanganyar Regency.

Fiscal practices are also a factor in increasing property prices in Tawangmangu Village, Kalisoro Village and Blumbang Village, where the process of buying and selling land is very high. This massive sale and purchase provides benefits for the Karanganyar Regency Government in making a fiscal policy, namely an increase in the NJOP starting in 2019 (Karanganyarkab, December 21, 2018). Although the NJOP increase, this does not affect PBB-P2 that must be paid by land owners, where the policy of the increase only affects Land and Building Rights Acquisition Fees (BPHTB) so as to increase the Regional Original Income (PAD) of Karanganyar Regency when the transaction occurs. When the NJOP increases, the receivables that must be paid by the land owner will be subsidized by the Karanganyar Regency Government with the difference between before the increase and after the increase.

Speculators are also influential in rising property prices where they neglect land/property until they realize that changing a policy can generate extraordinary profits (Goezt, et al 2019). This phenomenon will be detrimental to certain groups who need land. When the demand for a property grows larger.

Figure 4. Productive land that is neglected/abandoned by the owner

c. Degradation of the Quality of Settlement of Indigenous People in Tawangmangu Village, Kalisoro Village and Blumbang Village

In this study, the impact of gentrification was found, namely the degradation of the quality of the settlements of indigenous people. By using the unit of analysis of the physical appearance of the environment, the researchers delineated the density of settlements so as to produce a map of the density of settlements in 3 sub-districts of Tawangmangu Sub-district, namely Tawangmangu Village, Kalisoro Village and Blumbang Village. To determine the density of settlements, delineation of residential buildings is then carried out. The results of the delineation show that the density in Tawangmangu Village has a percentage of 84%, Kalisoro Village has a density percentage of 84% and Blumbang Village 88% (Figure 5).

The factor causing the density of settlements is the indigenous people who do not have strong economic power in providing land for their sustainability. They only rely on the land inherited from their parents/descendants which they have lived in until now. From the results of the inheritance, they then distribute it to the next generation. This is what causes the degradation of the residential environment due to inheritance land which is arguably not too wide, making the house buildings narrow and dense.

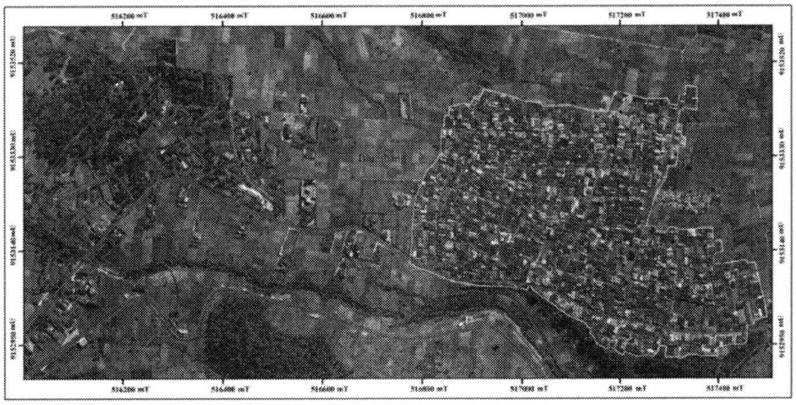

Figure 5. Residential Density in Blumbang Village in 2020

d. Social – Economic Community in Tawangmangu Village, Kalisoro Village and Blumbang Village

Gentrification really affects the socio-economic community in the Tawangmangu sub-district. Indigenous people who used to work in agriculture gradually began to change in the commercial field, especially in the field of lodging/homestay services and trade [2,7]. They think that the lodging and trading services business is very profitable compared to agriculture, which is sometimes high fertilizer prices and unprofitable harvests or even crop failures. From these problems, it encourages people to switch to lodging services as a support for the progress of tourism in Tawangmangu District.

In addition to having a positive impact, gentrification also has a negative impact on the social environment of the community. Due to the proliferation of businesses in the lodging sector and increasingly fierce competition between innkeepers, many inn owners rent out to young people who are not married but want to have a relationship with a partner like husband and wife. This is as a result of the socio-economic dynamics that have developed in the field of tourism construction. in this case, they charge between IDR 50,000 – IDR 70,000 per room at a price that can be negotiated according to the agreement before renting. This phenomenon will pollute the continuity of tourism in Tawangmangu District regarding the values and norms that exist in the tourist area.

In addition, the role of gentrifier actors in revitalizing their land into modern-style shophouses encourages the advancement of indigenous people to develop their trading businesses. This has an impact on the reduction of slum buildings built by indigenous people to sell on the roadside. Although only some indigenous people who have high capital are able to rent these shophouses. At the beginning of opening, the shophouse was rented out at a price of 800,000 per/month, but as more tourists came to eat and drink, it was increased to 1,000,000 per/month

Another impact of gentrification is the increasing number of farm workers because they lose their agricultural land. People who sell their land are forced to sell their land because of urgent needs such as a sick family member, the need to pay for their children to continue their education to a higher level. There are several interesting phenomena in the process of buying and selling agricultural land, namely an agreement to manage the land for planting crops by the seller of the land to the buyer. The seller (which here is the indigenous people) makes an agreement with the buyer regarding the land they buy, where the seller may work on the land before the buyer builds on the agricultural land. There are some land owners who rent out the opponents to be planted by indigenous people for free, but there are also those who if they want to work on the land have to pay rent.

e. Land conversion in Tawangmangu Village, Kalisoro Village and Blumbang Village

The increase in land demand that is not matched by the availability of land causes land use change or land use changes, while the land itself is permanent (Rindarjono, MG, 2017). Land conversion in Tawangmangu District is the result of the gentrification phenomenon, where agricultural land is the victim of the ferocity of gentrification [4]. In Tawangmangu Village, Kalisoro Village and Blumbang Village, environmental degradation has occurred, where agricultural land has turned into built-up land, even though the attraction of Tawangmangu tourism is the view of vegetable plantations and not towering villas, restaurants or hotels. The conversion of land to built-up land is a factor that affects the increase in runoff (Figure 6).

This will have an impact on the availability of river water in the rainy season, it will have abundant river discharge, while in the dry season it will decrease and even no water flows / dies. The sustainability of the Gerojokan Sewu Waterfall will be a negative impact that will be felt in the future. Where the tour is very dependent on the availability of water that is always flowing. The interesting thing that happened here is that 65% of the indigenous people do not have a building permit and own a business. So that this phenomenon is very detrimental to the government as local revenue (PAD) and will have an impact on future regional planning.

Figure 6. Runoff flow due to loss of water catchment areas from increased built environment.

4. Conclusions

Gentrification of tourism that has an impact on human resources, ecosystem services and the role of actors who play planning and socio-economic development of the region. The mountainous landscape at the foot of Mount Lawu offers aesthetic views that other regions don't offer. This has resulted in increased demands and pressure on the emergence of new cities, where the area should be conservation land (Land Conservation).

References

[1] Chan J H, Iankova K, Zhang Y, McDonald T and Qi X 2016 The role of self-gentrification in sustainable tourism: Indigenous entrepreneurship at Honghe Hani Rice Terraces World Heritage Site, China *J. Sustain. Tour.* **24** 1262–79

[2] Davidson M 2009 Displacement, space and dwelling: Placing gentrification debate *Ethics, Place Environ.* **12** 219–34

[3] Goes T, Heights B and Angeles L 2017 There Goes the Barrio : Measuring Gentefication in Boyle Heights ,

[4] Jackson Candidate R 2006 Bruny on the Brink: Governance, Gentrification and Tourism on an Australian Island *Isl. Stud. J.* **1** 201–22

[5] Slater T, Curran W and Lees L 2004 Guest Editorial *Environ. Plan. A Econ. Sp.* **36** 1141–50

[6] Ocejo R E 2011 The early gentrifier: Weaving a nostalgia narrative on the lower east side *City Community* **10** 285–310

[7] Goetz E G, Lewis B, Damiano A, Calhoun M, Murray K, Mehta N and Wood M 2019 THE DIVERSITY OF GENTRIFICATION : Multiple Forms of Gentrification in Minneapolis the important contributions to this work made by *Cent. Urban Reg. Aff. Univ. Minnesota - Twin Cities*

[8] Gotham K F 2013 Tourism gentrification: The case of New Orleans' vieux carre (french quarter) *The Gentrification Debates* **42** 145–76

[9] Prayoga I N T 2011 Permukiman Pinggiran Kota Semarang *Skripsi Urusan Perenc. Wil. Dan Kota Fak. Tek. Univ. Diponegoro Semarang*

[10] Bouman M, Smith C, Ward K, Coles T, Stainer L, Wheeler J O, Bourne L S, Smith N and Champion A G *Urban Geography*

[11] Um J and Yoon S 2021 Evaluating the relationship between perceived value regarding tourism gentrification experience, attitude, and responsible tourism intention *J. Tour. Cult. Chang.* **19** 345–61

[12] Septiningsih I 2014 Ismawati Septiningsih , SH , MH Dosen Fakultas Hukum - Universitas Surakarta *RECHSTAAT Ilmu Huk. Fak. Huk. UNSA* **8** 1–13

Analysis changes of carrying capacity base ecological footprint in surrounding Yogyakarta international airport

Chatarina Muryani[1,2,*] and Fitria Dewi Kartika[1,2]

[1]Disaster Research Center, Universitas Sebelas Maret Surakarta
[2]Geography Education, Universitas Sebelas Maret Surakarta

Email: chatarinamuryani@staff.uns.ac.id

Abstract. The Yogyakarta International Airport (YIA) construction has caused changes in land use both at the project site and in the surrounding area. This study aims to determine the environmental carrying capacity assessment of the area around YIA, and there are Temon District, Kulon Progo Regency, and in Purwodadi District, Purworejo Regency. Data analysis is spatial-temporal using the GIS approach. The land use data in 2009 was taken from the LANDSAT-7 image data, while the land use data for 2017 and 2019 was taken from the LANDSAT-8 image. Analysis of land-use change by overlaying three land use maps. Changes in environmental carrying capacity are calculated based on the ecological footprint and biocapacity. The results of land-use changes analysis indicate that the most decreased land is agricultural, there are 86.28% in Temon District, Kulon Progo Regency, and 85.82% in Purwodadi District, Purworejo Regency. Meanwhile, land use that significantly increased in the area was settlement land (76.87% in Purwodadi District and 86.01% in Temon District) and YIA development covering 587.28 hectares). Based on the analysis of ecological footprint and biocapacity, there has been a decrease in environmental carrying capacity in Purwodadi District, Purworejo Regency from 0.908389 in 2009 to 0.847857 in 2019 and also a reduction of ecological carrying capacity in Temon District, Kulon Progo Regency from 0, 750097 in 2009 to 0.642614 in 2019).

1. Introduction

Population growth impacts the natural environment, primarily affecting land cover and land use changes, mainly related to agricultural expansion and urbanization [1]. The growth rate in Indonesia in 2010-2019, with a span of 9 years, reached 1.31% [2]. The population growth will affect the dynamics of land use in Indonesia because land needs are increasing. The main factor of land use is humans, who influence the environment [3]. Land use is based on land cover features for life, be it natural or artificial features. The need factor gives rise to the form of land use [4]. Land use patterns are influenced by socio-economic and natural factors, and their use by humans in space and time [5]. It can be seen historically that changes in land use made by humans are forest land clearing for settlement development, including coastal areas [3]. Changes in land use caused by climatic, physical, and socio-economic factors directly affect socioeconomic status [6]. Currently, the demand for land exceeds the level sustainably provided by the earth, causing more long-term severe sustainability problems and the coastal regions around the world [6,7]. Two important trends need to be noted. First, the total area of land that is being reduced due to human use (e.g., settlement, agriculture, forestry, etc.) has grown significantly. Secondly, increased production of goods and services has increased land use and tenure [8].

The Indonesian government's policy regarding the development of the South Coast of Java is to balance the progress of the development of the North Coast of Java as one factor that attracts people to move to the coastal area. Efforts made by the government in the development of the South Coastal area of Java are the construction of the Yogyakarta International Airport (YIA) in Kulon Progo, the development of the Southern Cross Line or Daendels Road, and the development of coastal tourism areas. The three coastal area developments impact coastal land use in the Kulon Progo Regency and Purworejo Regency. In addition, the Government of Purworejo Regency plans to build a border city on an area of 2,000 hectares with the concept of an industrial city that is integrated with Aero City in Kulonprogo Regency [9]. Thus, human activities will gradually affect the site (e.g., housing development, agriculture, etc.) [10,11]. Dahuri et al. (2013)[12] mention that in some coastal areas that have been utilized, indications appear that the carrying capacity or sustainability capacity (sustainable potential) of coastal and marine ecosystems has been exceeded.

This study aims to calculate environmental carrying capacity changes caused by the development of interrelated areas in Kulon Progo Regency's coastal regions and Purworejo Regency's, especially in Temon District, Kulonprogo Regency and Purwodadi District, Purworejo Regency. The coastal areas of Purworejo Regency and Kulonprogo Regency are dominated by agricultural land use. The development of YIA potentially changes the economic pattern from agriculture and fisheries to industry and trade. In addition, the development of the tourism sector and urbanization significantly affects the expansion of agricultural land, reducing the capacity of the population and the carrying capacity of land that exceeds the maximum population limit [13]. Urban expansion in agricultural land significantly affects agricultural production and food supply [14]

Calculation of changes in environmental carrying capacity in the study area using the method of ecological footprint and biocapacity. The right method to measure sustainability. The Ecological Footprint was discovered by Mathis Wackernagel in 1992 as a new theory and method for quantitatively measuring the state of sustainable development [4]. According to Swiader (2018) [15], The environmental carrying capacity (ECC) assessment method uses ecological indicators such as biocapacity (BC) dan Ecological Footprint (EF), ECC is a parameter to determine policy planning and sustainable development [15]. The use of EF and BC will be necessary for the successful implementation of Ecological Carrying Capacity [16–18].

2. Methods

This study uses a quantitative descriptive method with a spatial-temporal approach. Multi-temporal and multispectral data are used to measure the amount, location and type of land cover [3]. Land use data is known through time series analysis of land use data from LANDSAT-7 image for 2009 and LANDSAT-8 for 2017 and 2019. The year selection is based on phenomena that support land-use changes. Processing land use data using Geographic Information Systems (GIS) with digital delineation using the Supervised classification method. ECC is known from calculating EF and BC, the calculation structure is shown in Figure 1:

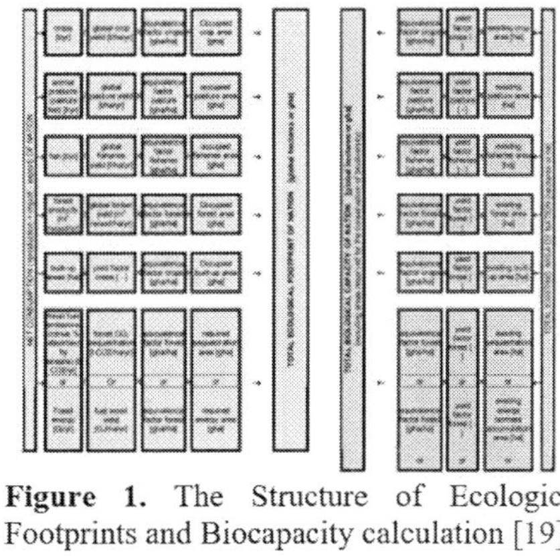

Figure 1. The Structure of Ecological Footprints and Biocapacity calculation [19]

Equivalence factors and yield factors to calculate the ecological footprint and required biocapacity are shown in table 1.

Table 1. Equivalence factors and yield factor

Land Use	Equivalence Factor (Gha/Ha)	Yield Factor
Cropland	2,51	0,98
Forest	1,26	0,82
Grazing Land	0,46	1,81
Marine & Inland Water	0,37	2,96
Buil-up Land	2,51	0,98

Source: Ewing et al., 2010 [20]

How to calculate the carrying capacity of the environment with the following formula:

$$ECC = BC/EF$$

Notes:
ECC = Ecological Carrying Capacity
BC = Biocapacity
EF = Ecological Footprint

Based on this formula, then if:

1. ECC > 1 means that there is a surplus condition, where the ecosystem can support the people living in it (ecological debt)
2. ECC < 1 This means that there is an overshoot condition where the ecosystem cannot support the population's life (ecological deficit).

3. Result and Discussion

3.1 Land-use changes

Yogyakarta International Airport (YIA) stands on a land area of about 600 hectares and has a terminal area of 210,000 square meters with a capacity of 20 million passengers per year. Its use was inaugurated on August 28, 2020. The existence of this new airport in Kulonprogo Regency is an urban catalyst that will have a positive and increase the economic impact of urban areas to improve people's quality of life [21]. The construction of this airport causes changes in land use at the project site also causes changes inland in the surrounding area.

The results of land-use changes analysis in Temon District, Kulon Progo Regency, the location where YIA was built, and Purwodadi District, Purworejo Regency, which is the closest sub-district on the west side of YIA using Landsat 7 imagery for 2009 and Landsat 8 imagery for 2017 and 2019 are as follows (see Table 2).

Tabel 2. Land use changes at Temon District Kulon Progo Regency and Purwodadi District Purworejo Regency Year 2009, 2017 dan 2019

No	District	Regency	Land use	Area (Ha) 2009	Area (Ha) 2017	Area (Ha) 2019
			Farm	1052,12	733,07	694,24
			Empty land	77,96	0	8,91
			Settlemet	797,9	1356,54	1398,71
1	Purwodadi	Purworejo	Ricefield	3250,77	2943,02	2937,8
			Bush	23,63	45,3	39,6
			Fishpound	4,65	174,37	169,49
			Shrubs	365,56	320,27	323,84
			Airport	0	0	587,28
			Farm	582,92	670,07	379,83
			Empty land	215,67	13,41	9,38
			Settlement	274,6	841,65	1044,46
2	Temon	Kulonprogo	Ricefield	2055,34	1321	1245,24
			Bush	25,07	42,82	17,87
			Fishpound	26,84	354,54	246,69
			Shrubs	442,24	391,04	112,26
		Total		**9281,54**	**9281,54**	**9281,54**

Source: processing results of LANDSAT Image 7 (2009) and LANDSAT Image 8 (2017 and 2019)

it was found that:

- In Temon District, Kulon Progo Regency: the land uses that have decreased in the area are rice fields (-810.10 ha), dry fields (-329.98 ha), empty land (-206.29 ha), farms (-203.09 ha), and bush (7.2 ha). Meanwhile, the land use that is increasing is settlement (+ 769.86 ha, YIA (+587.28 ha), and fishpond (219.85 ha).
- In Purwodadi Subdistrict, Purworejo Regency: the land uses that have decreased in the area are farm (- 357.88 ha), ricefields (312.97 ha), empty land (-69.05) and, a moor (-41.72 ha). Meanwhile, the land uses that have increased in the area are settlement (+ 600.81), fishponds (+164.84) , and shrubs (+ 15.97 ha).

The decline in land use in the area is mostly (86.28% in Temon District and 85.82% in Purwodadi District from reduced land area) is agricultural land (rice fields, agriculture, and dryland agriculture) and most of it has changed. Being a residential area (76.87% in Purwodadi District) and 86.01% in Temon District turned into an airport and settlement from the increasing land use. And if you look at the distribution, the agricultural land turned into settlements is around YIA and along the southern causeway (Jandals Street). This means a significant reduction in productive land and an increase in humans that the environment must support because the increase in settlements is identical to the rise in population.

Figure 2. Example of Land-use Change from Agricultural to Built-up Land

IOP Conf. Series: Earth and Environmental Science **986** (2022) 012033 doi:10.1088/1755-1315/986/1/012033

Figure 3. Land-use changes along Deanndels Road

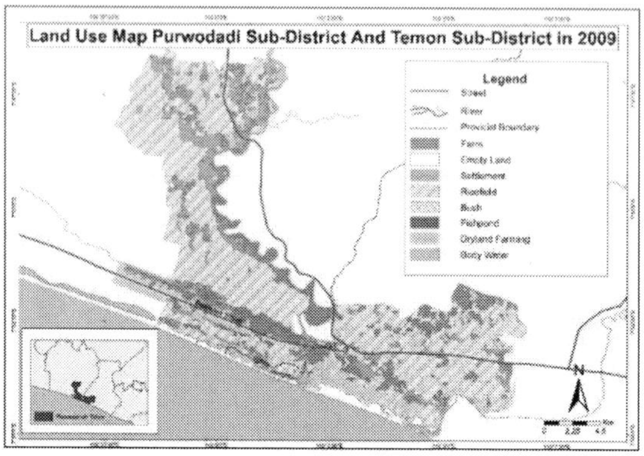

Figure 4. Landuse map Purwodadi and Temon Sub-district year 2009

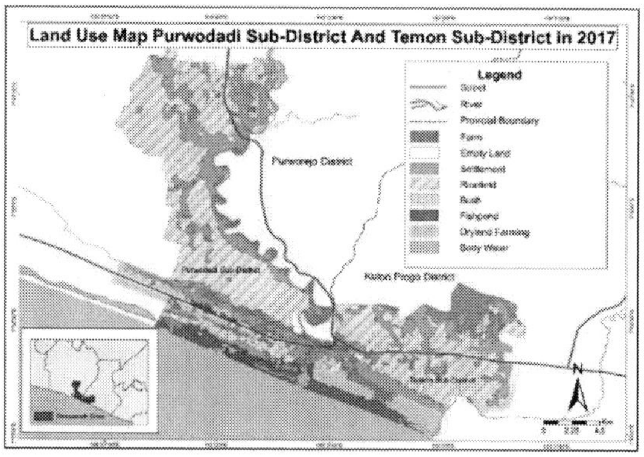

Figure 5. Landuse map Purwodadi and Temon Sub-district year 2017

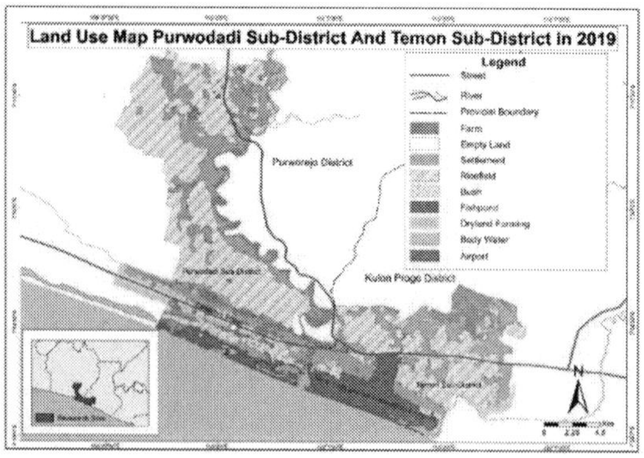

Figure 6. Landuse map Purwodadi and Temon Sub-district year 2019

Thus, the reduction in agricultural land and the expansion of residential land, which means an increasing number of residents, can reduce the carrying capacity of land in the area. LCC is the largest human population that an area can support under certain environmental conditions and levels of economic development [22]. The impact of extraordinary land-use changes, LCC research has experienced a drastic change from the carrying capacity of the land population to the integrated carrying capacity of the land [23].

3.2 Ecological Carrying Capacity (ECC) Changes

ECC assessment uses two factors, namely the Ecological Footprint and Biocapacity. ECC is the heaviest load environment that can sustain sustainably [24]. Furthermore, Vitousek (1986) states that the appropriate carrying capacity is the land needed to absorb waste and provide resources [25].

The results of the environmental carrying capacity calculation are

Table 3. The Ecological Carrying Capacity of Temon District in 2009. 2017 and 2019

	Land use	Population	Area (Ha)	Ecological Footprint (Gha/person)	Biocapasity (Gha/person)	Ecological Carrying Capa	Description
ECC 2009	Agricultural	24471	3080,50	0,441009	0,309649	0,702138	Overshoot
	Built Land	24471	490,27	0,027602	0,049282	1,785418	Surplus
	Fishpond	24471	26,84	0,003924	0,002698	0,687395	Overshoot
	Bush	24471	25,07		0,001395	0	
	Water	24471	71,75		0,001964	0	
	Total			**0,486586**	**0,364987**	**0,750097**	**Overshoot**
ECC 2017	Agricultural	26960	2382,11	0,330967	0,217341	0,656684	Overshoot
	Built Land	26960	855,06	0,061708	0,078015	1,264266	Surplus
	Fishpond	26960	354,54	0,033008	0,032348	0,98	Overshoot
	Bush	26960	42,82		0,002163	0	
	Water	26960	59,90		0,001488	0	
	Total			**0,425683**	**0,331354**	**0,778406**	**Overshoot**
ECC 2019	Agricultural	29446	1737,34	0,31821	0,14513	0,456082	Overshoot
	Built Land	29446	1641,12	0,108644	0,137092	1,261853	Surplus
	Fishpond	29446	246,69	0,047498	0,020607	0,433853	Overshoot
	Bush	29446	17,87		0,000826	0	
	Water	29446	51,42		0,00117	0	
	Total			**0,474352**	**0,304825**	**0,642614**	**Overshoot**

Ecological carrying capacity in Temon District as a whole, since 2009, the ECC has overshot. In 2017 and 2019, it remains in the overshoot class even the value has decreased (from 0.750097 in 2009 to 0.642614 in 2019). This decrease was mainly due to the conversion of agricultural land into settlements that were quite large in the study area. This is also indicated by a decrease in the agricultural sector's ecological carrying capacity, from 0.44-gram hectare/person to 0.32 gha/hectare. Specifically, the ecological footprint of the fishery sector has increased even though it is small (0.003924 in 2009 to 0.047498 in 2019) due to the increase in the pond area.

Table 4. The Ecological Carrying Capacity of Purwodadi District in 2009. 2017 and 2019

	Land use	Population	Area (Ha)	Ecological Footprint (Gha/person)	Biocapasity (Gha/person)	Ecological Carrying Capa	Description
ECC 2009	Agricultural	36435	4668,44	0,383992	0,315176	0,820789	Overshoot
	Built Land	36435	797,90	0,021971	0,053868	2,451745	Surplus
	Fishpond	36435	4,65	0,005118	0,000314	0,061336	Overshoot
	Bush	36435	101,60		0,003797		
	Water	36435	14,52		0,000267		
	Total			**0,411081**	**0,373421**	**0,908389**	**Overshoot**
ECC 2017	Agricultural	37357	3996,37	0,373507	0,263144	0,704522	Overshoot
	Built Land	37357	1398,71	0,047804	0,092099	1,926591	Surplus
	Fishpond	37357	174,37	0,003487	0,011482	3,292614	Surplus
	Bush	37357	48,51		0,001768		
	Water	37357	14,52		0,00026		
	Total			**0,424798**	**0,368753**	**0,868066**	**Overshoot**
ECC 2019	Agricultural	37485	3955,89	0,378191	0,259589	0,686397	Overshoot
	Built Land	37485	1407,61	0,048428	0,092369	1,907335	Surplus
	Fishpond	37485	169,49	0,003616	0,011122	3,075864	Surplus
	Bush	37485	39,60		0,001438		
	Water	37485	14,52		0,000259		
	Total			**0,430235**	**0,364778**	**0,847857**	**Overshoot**

Source: Calculation results

The calculation of the change in the carrying capacity of land in Purwodadi Subdistrict, Purworejo Regency from 2009 to 2019 in an overshoot condition and the ecological carrying capacity is increasing even though it is decreasing even though it was small from 0.908389 in 2009 to 0.847857 in 2019. This is also mainly due to the conversion of agricultural land into residential land and built-up land (YIA). Even though it belongs to the overshoot class, the LCC number is relatively high and close to 1 (one). However, the existence of YIA will have the potential to become a catalyst for regional development. Therefore it is necessary to control land-use change. In Temon Sub-district, there is also an increase in fishery land with an increase in the pond area, and based on field observations in this area, there is also mangrove conservation. LCC in Temon District is higher

If a comparison is made between LCC in Temon District, Kulon Progo Regency, and in Purwodadi District, Purworejo Regency, it can be seen that LCC in Purwodadi District is higher than LCC in Temon District. Based on field observations, it is true that environmental conditions in Purwodadi District are dominated by the use of agricultural land and economic activities of the population, especially in the agricultural and fishery sectors. From the results of the LCC calculation, it can also be concluded that changes in land use from agriculture to settlements and built-up land have reduced the LCC value significantly.

4. Conclusion

1. The analysis of land-use changes shows that the land that has been dramatically reduced in the area is agricultural (86.28% in Temon District, Kulon Progo Regency, and 85.82% in Purwodadi District, Purworejo Regency). Meanwhile, land use that significantly increased in

the area was residential land (76.87% in Purwodadi District and 86.01% in Temon District) and YIA development, covering an area of 587.28 hectares).

2. Based on the analysis of environmental carrying capacity based on ecological footprint and biocapacity, there has been a decrease in ecological carrying capacity in Purwodadi Subdistrict, Purworejo Regency, from 0.908389 in 2009 to 0.847857 in 2019 and also a reduction of environmental carrying capacity in Temon District, Kulon Progo Regency from 0.750097 in 2009 to 0.642614 in 2019).

Acknowledgments
This article is a research output article titled "Improving Branding of Research Groups and Cooperation of the Center for Disaster Studies at Sebelas Maret University LPPM" with the Institutional Strengthening Research Scheme (PPI-UNS). Thank you to the Purworejo Regency Regional Disaster Management Agency, which has helped a lot in providing data, and thanks also to the Purworejo coastal thesis team: Fitria, Tanasiva, Abdul, and Melyda, who have helped carry out the survey

References

[1] Coulter L L, Stow D A, Tsai Y, Ibanez N, Shih H, Kerr A, Benza M, Weeks J R and Mensah F 2016 Remote Sensing of Environment Classi fi cation and assessment of land cover and land use change in southern Ghana using dense stacks of Landsat 7 ETM + imagery *Remote Sens. Environ.* **184** 396–409
[2] BPS 2020 Kabupaten Purworejo Dalam Angka Tahun 2020 *Badan Pus. Stat.* 1–226
[3] Ac-chukwuocha N B 2015 Analysis of Structural Changes in Land Use and Land Cover of Otamiri Watershed in Owerri , South East Nigeria **5** 51–60
[4] Ritohardoyo S 2013 *Penggunaan dan Tata Guna Lahan* (Yogyakarta: Penerbit Ombak)
[5] Rawat J S and Kumar M 2015 Monitoring land use/cover change using remote sensing and GIS techniques: A case study of Hawalbagh block, district Almora, Uttarakhand, India *Egypt. J. Remote Sens. Sp. Sci.* **18** 77–84
[6] Kaliraj S, Chandrasekar N, Ramachandran K K, Srinivas Y and Saravanan S 2017 Coastal landuse and land cover change and transformations of Kanyakumari coast, India using remote sensing and GIS *Egypt. J. Remote Sens. Sp. Sci.* **20** 169–85
[7] Xiuwan C 2002 Using remote sensing and GIS to analyse land cover change and its impacts on regional sustainable development **1161**
[8] Fazal S, Banu N and Azharuddin 2017 Bundelkhand Region : A Geographical Analysis Of Land Use Dynamics *Punjab Geogr.* **13** 77–90
[9] Ridarineni N and Assidiq Y 2017 Purworejo Berencana Bangun Kota Perbatasan dengan DIY
[10] Kolios S and Stylios C D 2013 Identi fi cation of land cover / land use changes in the greater area of the Preveza peninsula in Greece using Landsat satellite data *Appl. Geogr.* **40** 150–60
[11] Serra P, Pons X and Saurı D 2008 Land-cover and land-use change in a Mediterranean landscape : A spatial analysis of driving forces integrating biophysical and human factors **28** 189–209
[12] R D, J R, Ginting S . and Sitepu M . 2013 *Pengelolaan Sumberdaya Wilayah Pesisir dan Lautan Secara Terpadu* (Balai Pustaka)
[13] Sriartha I P and Giyarsih S R 2015 Spatial zonation model of local irrigation system sustainability (A case of subak system in Bali) *Indones. J. Geogr.* **47** 142–50
[14] Jiang L, Deng X and Seto K C 2013 The impact of urban expansion on agricultural land use intensity in China *Land use policy* **35** 33–9
[15] Świąder M, Lin D, Szewrański S, Kazak J K, Iha K, van Hoof J, Belčáková I and Altiok S 2020 The application of ecological footprint and biocapacity for environmental carrying capacity assessment: A new approach for European cities *Environ. Sci. Policy* **105** 56–74
[16] Liu X, Fu J, Jiang D, Luo J, Sun C, Liu H, Wen R and Wang X 2018 Improvement of ecological footprint model in national nature reserve based on net primary production (NPP) *Sustain.* **11**

IOP Conf. Series: Earth and Environmental Science **986** (2022) 012033 doi:10.1088/1755-1315/986/1/012033

[17] Volodya E, Yeo M J and Kim Y P 2018 Trends of Ecological Footprints and policy direction for sustainable development in Mongolia: A case study *Sustain.* **10**
[18] Cheng C, Liu Y, Chen Y, Liu Y, Zhang Y, Shen S, Yang R, Xu Z and Hong Y 2019 Diagnosing cropland's allowable range and spatial allocation in China's typical mountainous plateau area: An evaluation framework based on ecological carrying capacity *Sci. Total Environ.* **685** 1255–68
[19] Wackernagel M, Monfreda C, Moran D, Wermer P, Goldfinger S, Deumling D and Murray M 2005 National Footprint and Biocapacity Accounts 2005 : The underlying calculation method *Land use policy* **21** 231–46
[20] Ewing B, Moore D, Goldfinger S H, Oursler A, Reed A and Wackernagel M 2010 Ecological Footprint Atlas 2010 *Glob. Footpr. Netw.* 1–111
[21] Prasetyo H E and Trijeti 2020 Analisis Dampak Pembangunan Bandara Udara Yogyakarta International Airport (YIA) terhadap Kapasitas Jalan Nasional di Kabupaten Kulonprogo *Semin. Nas. Penelit. LPPM UMJ*
[22] Qian Y, Tang L, Qiu Q, Xu T and Liao J 2015 A comparative analysis on assessment of land carrying capacity with ecological footprint analysis and index system method *PLoS One* **10** 1–17
[23] Ding-ding L, Su-qiong W and Zhi-qiang C 2007 Primary Study of Carrying Capacity of Urban Land Resources–Fuzhou as a Case *J. Shenyang Norm. Univ.* **25** 252–6
[24] Catton W . 1986 Carrying Capacity and the Limits to Free- dom," Paper Prepared for Social Ecology Session 1 *XI World Congress of Sociology* (New Delhi)
[25] Vitousek P M, Ehrlich P R, Ehrlich A H and Matson P A 1986 Human Appropriation of the Products of Photosynthesis *Bioscience* **36** 368–73

IOP Conf. Series: Earth and Environmental Science **986** (2022) 012034 doi:10.1088/1755-1315/986/1/012034

Identification of mangrove forest changes used Geographic Information System (GIS) in South Banawa District, Donggala Regency

David Tri Budi[1], Ida Arianingsih[2], Iwan Alim Saputra[1], and Rendra Zainal Maliki[1]

[1]Geography Education Department, Faculty of Teacher Training and Education, Universitas
Tadulako, Palu, Indonesia

[2]Forestry Department, Faculty of Forestry, Universitas Tadulako, Palu, Indonesia

Email: zainalrendra@untad.ac.id

Abstract. This research was motivated by changes in mangrove forest areas in Central Sulawesi Province, Donggala Regency, especially South Banawa District. The objective of the research is to analyze the change of mangrove forests in the district of South Banawa years 2002-2018 and to determine the dominant factors influencing the change of mangrove forests. This study uses a quantitative approach combining geographic information systems and remote sensing, using overlay and descriptive analysis. Results from research this showing that the mangrove forest in the district South Banawa years 2002-2018 decline area amounting to 47.59 hectares, which was originally large mangrove forest 261.32 hectares to 213.73 hectares on in 2018. The Changing of mangrove forests to fishponds is a dominant factor in the change of mangrove forests when compared to other changes.

1. Introduction

Indonesia is an archipelagic country with around 17,508 islands and has a coastline of 81,000 kilometers. It makes Indonesia has very potential coastal resources such as agricultural resources, fisheries, coral reefs, mangroves, and tourism. Based on the Decree of the Minister of Maritime Affairs and Fisheries Number KEP.10/MEN/2002, coastal areas are violations of interacting land and sea ecosystems, where the direction of the provincial coastline and one-third of the sea area is for the district/city and the land direction is the administrative boundary of the city district.

Mangrove forest is one of the ecosystems in coastal areas and has an important role like mangroves have high productivity to support the surrounding environment and can prevent and reduce the impact of disasters such as strong coastal wind, tsunamis, torrential storms, and other natural disasters [1–3]. Furthermore, it provides a variety of ecological services for human beings, pollutant absorption, and economical [4–6].

Mangrove forest is the main ecosystem that supports life in coastal areas. It has various very complex ecological functions, including as a reservoir and natural waste treatment (bioremediation) or natural biofilter which is very effective in tackling pollution. Excessive use of mangroves has resulted in the reduction of the area of mangrove forests. The current pattern of utilization of mangrove forests is mostly based on the interests of certain individuals or groups, so that many mangrove forests have shifted their functions into settlements, ponds, and even vacant land due to massive logging. In general, the community in using mangroves ignores environmental aspects and is destructive. The impact of excessive use of mangroves results in coastal abrasion, flooding, sedimentation, and reduced diversity of marine natural resources [7].

One of the areas in Indonesia that has experienced changes is the area of mangrove forests in Central Sulawesi Province. According to the Environment Agency of Central Sulawesi Province in 1988 it reached 46 thousand hectares but decreased to 29 thousand hectares in 2010, and even then most of the area is in a damaged condition. This proves that the mangrove forest area in Central Sulawesi has also experienced a shift in function, one of which is in South Banawa District, which is currently changing the function of the mangrove forest in the area to become a fish pond. Land conversion occurs in many areas of the mangrove ecosystem into cultivation, development, or other areas. The main problem with mangrove habitat comes from various economic pressures that cause changes in function between settlement activities, ponds, or various community activities on the coast [8].

Detection using GIS and remote sensing is an alternative to determine shoreline changes. So this can also be used to detect and map changes and monitor mangrove conditions [9–12]. The first step to prevent mangrove forests from changing their function is monitoring mangrove forests, but an effective method for obtaining and analyzing information is needed, one of which is remote sensing. Remote sensing imagery can be used to easily utilize land use around the coastline [13]. Utilization of satellite image data in terms of natural advantages has been carried out, one of which is by changing the mangrove area of the coastline [14]. Several previous studies have been carried out related to changes in land use and land cover. For example, studies on land cover change in the territory of Indonesia [14–16]. Identification of land cover changes is important to minimize the occurrence of land cover changes so that land can be avoided [17]. The purpose of this study was to analyze changes in mangrove forests in South Banawa District in 2002-2018 and to determine the dominant factors influencing changes in mangrove forests in South Banawa District.

2. Methods

2.1 Study Area

The research location is in South Banawa District, Donggala Regency, Central Sulawesi Province, which is located on the west coast of Sulawesi Island. The research sample is villages located on the South Banawa Coast which has 7 villages, namely Tosale, Helpano, Lembasada, Tanahmea, Salusumpu, Lalombi, and Surumana. The reason for choosing this area as the research site due to there is the largest mangrove forest in Donggala Regency other than Balaesang District.

2.2 Analysis Data and Processing Techniques

The type of research used is a quantitative research using descriptive analysis. The method used is a combination of geographic information system (GIS) and remote sensing. The ordered data sources are primary namely changes in mangrove forests and secondary data namely Landsat 7, Landsat imagery in 2002, Landsat 7 imagery in 2010, and Landsat 8 imagery in 2018, RBI Sheet Watatu Map, RBI Sheet Bambarimi, and RTRW Donggala Regency. Primary data is carried out through image interpretation, while secondary data collection is data that does not come directly from any source. Data analysis was carried out by two methods, namely the overlay and descriptive analysis. The overlay method is used to determine changes in the mangrove forest at the research site and to explain further the two image interpretation methods and the overlay method, both of which are closely related in this study. The descriptive analysis explains the results of changes in mangrove forests in 2002, 2010, and 2018 in the South Banawa District area and the factors that influence changes in mangrove forests.

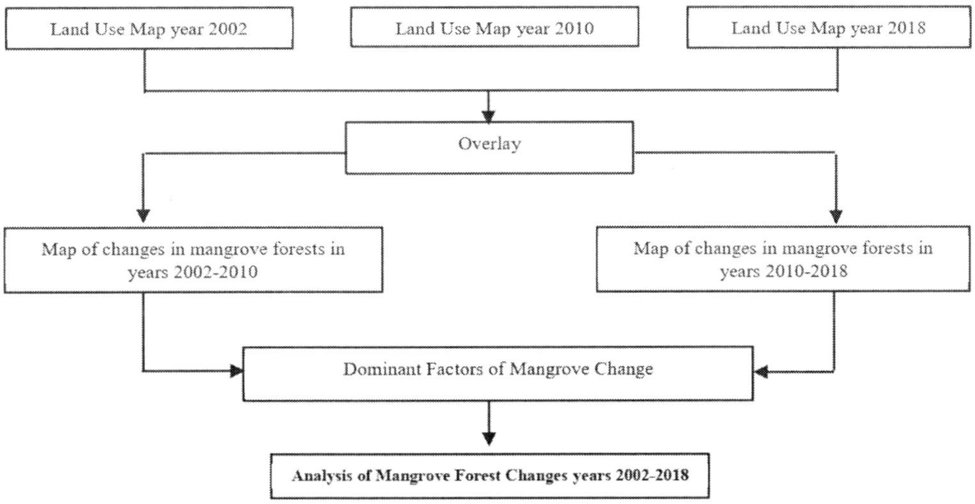

Figure 1. Step Research

3. Result and Discussion

3.1 Changes in Mangrove Forest 2002-2010

Based on the interpretation of Landsat 7 imagery in 2002 and 2010, the land uses in South Banawa District are lowland forest, mangrove forest, settlements, plantations, rice fields, and ponds. This refers to the 2014 SNI regarding land cover. Changes in the area of each land use and its changes over for 8 years in South Banawa District can be seen in Table 1 as follows.

Table 1. Changes in Mangrove Forest Years 2002-2010

No	Land Use	2002		2010		Change
		Large (Ha)	%	Large (Ha)	%	Large (Ha)
1	Lowland Forest	2733.38	46.87	2675.43	45.90	-57.95
2	**Mangrove Forest**	**261.32**	**4.48**	**234.64**	**4.03**	**-26.67**
3	Settlement	151.09	2.59	161.09	2.73	+10.00
4	Farm	1450.07	24.87	1508.70	25.88	+58.63
5	Field	367.51	6.34	369.61	6.30	+2.10
6	Fishpond	866.05	14.85	884.15	15.17	+18.11
	Total	5831.52	100	5831.52	100	

Mangrove forest in South Banawa District is one of the land uses that have decreased by 26.67 hectares in 8 years. This shows that mangrove forest degradation has occurred in a multitemporal manner. The decrease in the area of mangrove forests from 2002-2010 was followed by an increase in the use of residential land, plantations, and ponds. The area of mangrove forest has decreased by 26.67 hectares, from 261.32 hectares in 2002 to 234.64 hectares in 2010. The area of mangrove forest has changed by 234.64 hectares, which turned into 18.11 hectares of ponds and 8.39 hectares of plantations. Map of changes in mangrove forests in 2002-2010 can be seen in Figure 2 below:

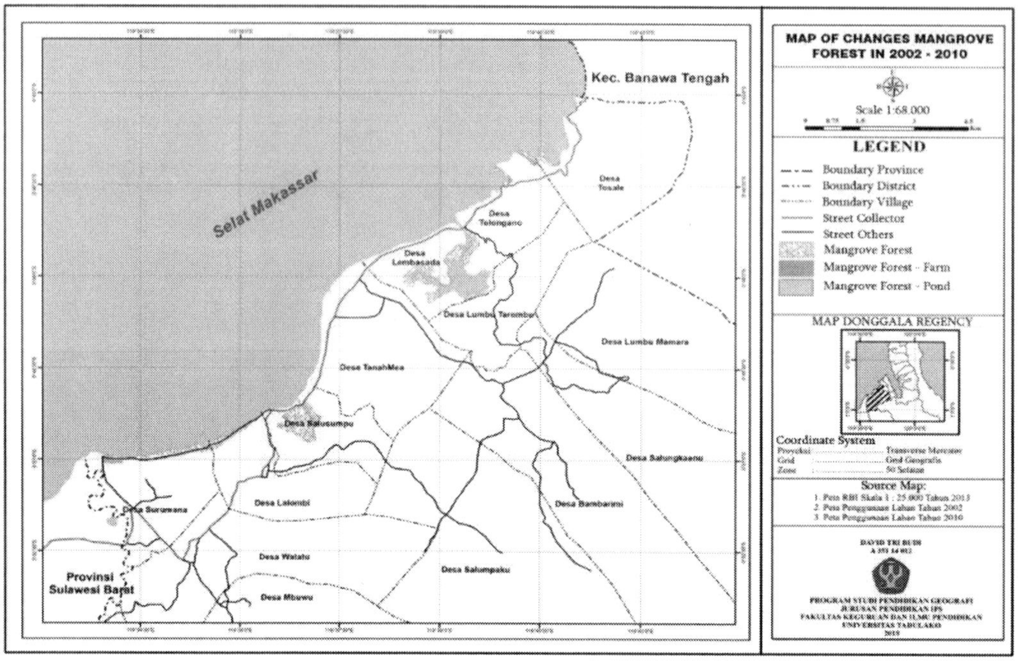

Figure 2. Map of changes in mangrove forests in 2002-2010

Most of the changes in mangrove forests into ponds due to mangrove forests are directly adjacent to ponds so that every year people will cut down to open new land for ponds. Changes in mangrove forests into plantations also occur due to the majority of people who live or own land around mangrove forests work as farmers. So, most farmers who own land will open mangrove forests to be used as agricultural land. Based on the interpretation of Landsat 7 in 2010 and Landsat 8 in 2018, the land uses in South Banawa District are lowland forest, mangrove forest, settlements, open land, plantations, rice fields, and ponds, this refers to the 2014 SNI regarding land cover. Changes in the area of each land use and its changes over for 8 years in South Banawa District can be seen in Table 2 below:

Table 2. Changes in Mangrove Forest Years 2010-2018

No	Land Use	2010		2018		Change
		Luas (Ha)	**%**	**Luas (Ha)**	**%**	**Luas (Ha)**
1	Lowland Forest	2675.43	45.90	2633.05	44.57	-42.38
2	**Mangrove Forest**	**234.64**	**4.03**	**213.73**	**3.17**	**-20.91**
3	Settlement	161.09	2.73	162.36	2.78	+1.27
4	Farm	1508.70	25.88	1531.53	26.32	+22.83
5	Field	367.51	6.30	401.59	6.86	+34.08
6	Fishpond	884.15	15.17	891.36	15.77	+7.21
7	Open Field	0.00	0.00	7.90	0.23	+7.90
	Total	**5831.52**	**100**	**5831.52**	**100**	

3.2 Changes in Mangrove Forest 2010-2018

Types of land use in South Banawa District from 2010-2018 was increased and decreased. Land uses that experienced an increase in the area were settlements by 1.27 hectares, plantations 22.83 hectares, ponds 7.21 hectares, rice fields 34.08 hectares, and open land 7.90 hectares, while those experiencing a reduction in the area were lowland forests by 42.38 hectares and mangrove forests by 20.91 hectares.

In 2010 the forest area of 2675.43 hectares was reduced in 2018 to 2633.05, of mangrove forests in 2010 was 234.64 reduced in 2018 to 213.73 hectares. The types of land use that increased, namely settlements in 2010 increased 161.79 hectares in 2018 to 180.36, plantations in 2010 the area of 1508.70 increased to 1531.53, ponds which were originally in 2010 the area of 884.15 hectares increased to 891.36 hectares, rice fields which were originally in 2010 the area of 367.51 increased to 401.59 and open land which in 2010 did not exist but in 2018 had an area of 7.90 hectares.

Mangrove forests in 2010-2018 decreased in area of 21.91 hectares. The reduced area of mangrove forests is caused by turning into ponds, plantations, and vacant land. This is evidenced by the increase in the area of 7.21 hectares of ponds, 22.83 plantations, and open land which initially did not exist but in 2018 the area reached 7.90. The total area of mangrove forests has changed, namely 20.91 hectares. Changes in mangrove forest to open land 5.99 hectares, changes to plantations 2.31 hectares, and changes to ponds by 10.81 hectares while those that do not experience changes are 213.76 hectares. The map of mangrove forest change in 2010-2018 can be seen in Figure 3.

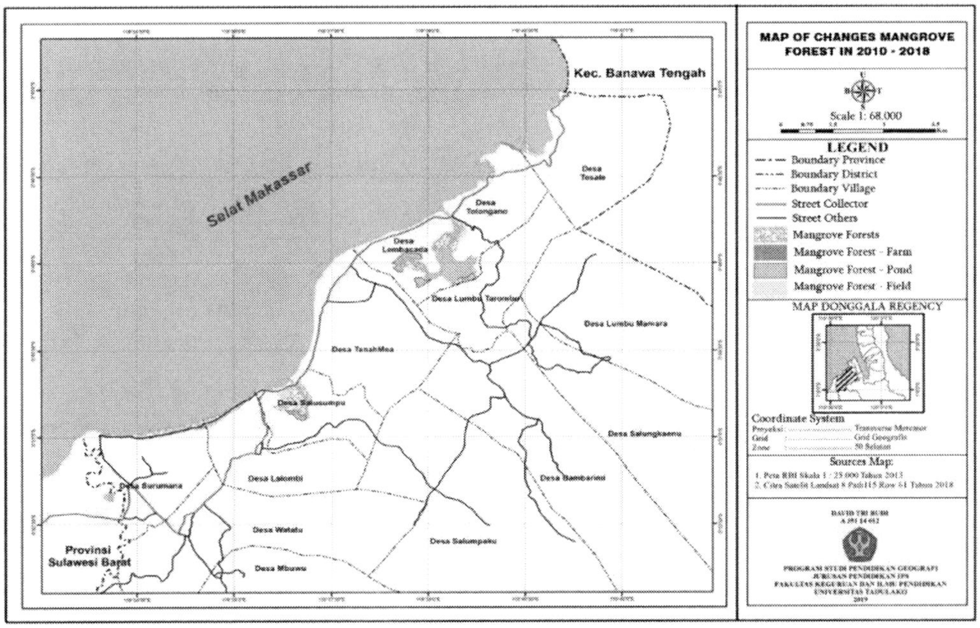

Figure 3. Map of changes in mangrove forests in 2010-2018

The change of mangrove forest into open land, in 2010 there was no open land but in 2018 the area of open land was 7.95 hectares and 5.99 hectares came from mangrove forests. The change in mangrove forests into open land is caused by many people. They use mangrove forests as building materials or firewood, logging for building materials such as house poles is mostly done by people around mangrove forests. Based on information communities in several areas, especially in Tosale Village, mangrove forests are damaged due to pollution. Changes in mangrove forests into vacant land are also caused by development by the government. For example, the construction of environmental roads in mangrove forest areas so that mangrove forests are cut down.

3.3 Dominant Factors of Mangrove Change

The decreasing area of mangrove forests from 2002-2018 is influenced by various factors, but the main factor is human activity. The dominant direction of mangrove forest change can be seen in Table 3 below.

Table 3. Dominant Factors of Mangrove Change

No	Types of Changes in Mangrove Forest	2002-2010 (Ha)	2010-2018 (Ha)
1	Mangrove Forest-Pond	18.11	10.87
2	Mangrove Forest-Farm	8.39	2.31
3	Mangrove Forest-Open land	-	5.94
4	Mangrove Forest-Low Land Forest	-	-
5	Mangrove Forest-Settlement	-	-
6	Mangrove Forest-Field	-	-
	Total	26.67	20.91

Table 3 data shows forest changes for 2 periods, namely 2002-2010 and 2010-2018, mangrove forests into ponds, mangrove forests into plantations, and mangrove forests into open land. The change of mangrove forests into ponds has become a dominant change when compared to other changes. It is accordance with interviews with local residents that the cause of the reduction in the area of mangrove forests is the change of mangrove forests into ponds. In 2002-2010 mangrove forests became ponds of 18.11 hectares and in 2010-2018 it was 10.87 hectares. Changes in mangrove forest to plantations in 2002-2010 amounted to 8.39 hectares and decreased in 2010-2018 to 2.31 hectares. Changes in mangrove forests into open land in 2002-2010 were not found, but in 2010-2018 the area changed was 5.94 hectares.

Mangrove forests in South Banawa District from 2002-2018 have always decreased. It is caused by community activities in the mangrove forest area. The increasing due to higher population growth marked by the rate of population growth which is increasing every year, based on South Banawa District 2018 population growth experienced an increase of 1.03% from the previous year. Population growth which increases every year is also accompanied by an increase in residential areas of 11.27 hectares from 2010-2018.

Figure 4. Condition Mangrove Forest. (a) Mangrove Forest bordering the Community Pond (b) Mangrove forest that became vacant land

Communities in South Banawa Sub-district who live around mangrove forests have knowledge not only limited to utilization but also the causes of the reduced area of mangrove forests. This is due to the conversion of mangrove forests into ponds or agricultural land. Currently, people are starting to realize the impact of converting mangrove forests into ponds or agricultural land. Many communities in Tosale and Surumana villages in South Banawa sub-district have started replanting mangrove forests in their areas. Many people are aware of the benefits of mangrove forests, one of which is preventing the beach in their village from being damaged from abrasion or protecting it in the event of a tsunami natural disaster. The results of research by [18] that there has been a change in the function of pond land because land use has changed its function into housing, public facilities, offices, and commercial buildings built along with coastal areas.

4. Conclusions

Mangrove forest in South Banawa Sub-district in 2002-2018 experienced a decrease in the area of 47.59 hectares, from 261.32 hectares of mangrove forest to 203.73 hectares in 2018. The reduced mangrove forest area turned into 28.98 hectares of ponds, turned into plantations of 10.7 hectares, and changes to 5.94 hectares of vacant land. The conversion of mangrove forests into ponds is the dominant factor in changing mangrove forests in South Banawa District.

Acknowledgments

Ramlin, S.E as Sub-district Head of South Banawa District, Donggala Regency who has allowed researchers to conduct research so that this research can be resolved.

References

[1] Souhoka J 2009 Kondisi Dan Keanekaragaman Jenis Karang Batu Di Pulau Nusalaut, Maluku Tengah *J. Perikan. Univ. Gadjah Mada* **11** 54–65
[2] Omar H, Afizzul Misman M and Musa S 2019 GIS and Remote Sensing for Mangroves Mapping and Monitoring *Geogr. Inf. Syst. Sci.* 1–15
[3] Kamil E A, Takaijudin H and Hashim A M 2020 Distribution of Mangroves in Kedah, Malaysia: A Remote Sensing and Ground-Truth Based Assessment *IOP Conf. Ser. Earth Environ. Sci.* **549** 0–9
[4] Liu K, Li X, Shi X and Wang S 2008 Monitoring mangrove forest changes using remote sensing and GIS data With decision-tree learning *Wetlands* **28** 336–46
[5] Hamzah M L, Amir A A, Maulud K N A, Sharma S, Mohd F A, Selamat S N, Karim O, Ariffin E H and Begum R A 2020 Assessment of the mangrove forest changes along the pahang coast using remote sensing and gis technology *J. Sustain. Sci. Manag.* **15** 43–58
[6] Hashim N I, Zaharin N F, Hashim N H and Abdul Halim M 2019 Temporal Changes of Mangrove Distribution in Mukim Kuala Selangor Using GIS Approach *IOP Conf. Ser. Earth Environ. Sci.* **385** 0–8
[7] Carong S R, Atjo A A, Arbit N I S, Saru A and Omar S B A 2019 Analysis of mangrove forest ecosystem damage in polewali mandar regency through environmental science approach *IOP Conf. Ser. Earth Environ. Sci.* **280** 0–7
[8] Bengen D G 2001 Ekosistem dan Sumberdaya Pesisir dan Laut serta Pengelolaan Secara Terpadu dan Berkelanjutan *Pros. Pelatih. Pengelolaan Wil. Pesisir Terpadu. 29 Oktober-3 Novemb. 2001* 28–55
[9] Azhar R, Rusdi M, Irham M and Fuadi A 2021 Spatial distribution of mangrove using a geographic information system in Aceh Besar *IOP Conf. Ser. Earth Environ. Sci.* **674** 0–8
[10] Khairul Abdullah Halim M, Hidayah Halid N, Ahmad A, Mohd Suhaimi H and Hidayat Jamal M 2019 Monitoring mangrove forest cover declination at kilim karst geoforest park, Langkawi from 2005 to 2017 using geospatial technology *IOP Conf. Ser. Earth Environ. Sci.* **220** 0–8
[11] Pham T D and Yoshino K 2016 Characterization of mangrove species using ALOS-2 PALSAR in Hai Phong city, Vietnam *IOP Conf. Ser. Earth Environ. Sci.* **37** 0–8
[12] Rudiastuti A W, Yuwono D M and Hartini S 2018 Mangrove Mapping Using SPOT 6 at East Lombok Indonesia *IOP Conf. Ser. Earth Environ. Sci.* **165** 0–14

[13] Parman S 2010 Deteksi Perubahan Garis Pantai Melalui Citra Penginderaan Jauh Di Pantai Utara Semarang Demak *J. Geogr.* **7** 30–8
[14] Putra D M, Akbar M R A and Habiburrahman F 2021 Deteksi Perubahan Luasan Mangrove Dan Garis Pantai Menggunakan Data Penginderaan Jauh Di Area Tambak Ujung Pangkah Kabupaten Gresik Provinsi Jawa Timur *Semin. Nas. Geomatika* 727
[15] Wijaya N 2015 Deteksi Perubahan Penggunaan Lahan Dengan Citra Landsat Dan Sistem Informasi Geografis: Studi Kasus Di Wilayah Metropolitan Bandung, Indonesia *Geoplanning J. Geomatics Plan.* **2** 82–92
[16] Dwi Agung Pramono 2020 Sistem Informasi Geografis untuk Studi Perubahan Tutupan Lahan Vegetasi dan Non-Vegetasi di Desa Benhes *Bul. Loupe* **16** 54–9
[17] Yulius Y, Tanto T A, Ramdhan M, Putra A and Salim H L 2015 Landcover Changes in Coastal Bungus Region of Kabung Bay, West Sumatra From 200-2013 Based on Geographic Information System Application *J. Ilmu dan Teknol. Kelaut. Trop.* **6**
[18] Rahmi M M, Sugianto S and Faisal 2017 Analisis Perubahan Lahan Tambak Di Kawasan Pesisir Kota Banda Aceh *Pros. Semin. Nas. Pascasarj. Unsyiah* 98–103

IOP Conf. Series: Earth and Environmental Science **986** (2022) 012035 doi:10.1088/1755-1315/986/1/012035

Analysis of relationship between economic sectors of Dki Jakarta towards land use changes and erosion in Ciliiwung Hulu watershed, Bogor Regency

Y Arifien[1*], J Sinurat[1] and T Susdiyanti[1]

[1] Universitas Nusa Bangsa

*Email: yunus@unb.ac.id

Abstract. Land resources are increasingly important as the population continues to grow, which will put pressure on demand, so land use is increasingly competitive. The purpose of this study is to determine the interrelationships between economic sectors between DKI Jakarta and Bodetabek and the impact of DKI Jakarta's economic development on the economy and changes in land use in Bodetabek. Research using IRIO approach and analysis of Landsat imagery. The sector in DKI Jakarta that has an effect on Bodetabek is the trade sector. Food industry, building, restaurant and hotel, communication, building and service business,. On the other hand, the sectors in Bodetabek that affect DKI Jakarta are the chemical, wood, and metal goods industries. According to the IRIO table, an area of 2,561.86 ha of built land is needed in DKI Jakarta and 251,167.57 ha in DKI Jakarta and 251,167.57 ha while the results of the interpretation of TM7 images in 2010 and 2019 mean that the built area of Bodetabek has changed from 183,681.86 ha to 249,867.91 ha. or has increased to 66,186 ,05 ha. There is a difference in calculations with the IRIO table and image interpretation of 1,299.66 ha. The land cover of the Ciliwung Hulu watershed was obtained through analysis of Landsat imagery data and 2019 was dry land agriculture of 4435 ha in 2019. A sizeable reduction also occurred in paddy fields and forest cover of 2% each. The reduction in forest area to dry land agriculture causes disruption of hydrological functions and higher erosion

1. Introduction

1.1. Background

DKI Jakarta as the nation's capital and the main gateway to Indonesia has carried out a series of development activities with the characteristics of being the center of government, trade and service center. The ongoing development has caused the relationship and interrelationship between various economic sectors in DKI Jakarta to move in a higher and more complex direction. Changes that occur in one particular sector have a reciprocal effect on various other sectors. In fact, changes that occur in the economic sector in DKI Jakarta have affected the surrounding areas (hinterland), including Bogor, Depok, Tangerang and Bekasi (Bodetabek).

In accordance with the Instruction of the President of the Republic of Indonesia No. 13 of 1976, Bogor, Tangerang and Bekasi worked as buffer zones for DKI Jakarta, and indirectly stated that these areas should be able to accommodate an abundance of activities that were not accommodated by DKI Jakarta, including: population, industry and trade overflows[1]. In addition to accommodating settlements, Bogor also functions as an air bag to meet the need for clean water for residents who live in the DKI Jakarta and Bodetabek areas, while Tangerang and Bekasi accommodate residential and industrial activities. To support the implementation of this function, it is necessary to regulate the land used with the proportion of available land to allocate the physical development of the facilities and infrastructure needed and use it for agriculture[2]

Space is something dynamic, for example, which contains roads on the one hand will interfere with the pattern of human life, but on the other hand it is humans who intervene on roads because humans need accessibility for their mobility needs. The existence of globalization, where humans can be anywhere and anytime. Space changes are natural. But that does not mean that all changes in space are beneficial for humans. Spatial transformation is a process of spatial change from being characterized by rural areas to urban areas or it can also be a process of urban change or the development of suburban areas to become modern[3].

The development emphasis in the urban sector has been shown to increase change in sectors and locations with high productivity levels. Big cities are growing rapidly as regional growth centers that often experience changes in their functions to provide services to regional villages [4]. DKI Jakarta as the nation's capital and the main gateway to Indonesia has carried out development activities with its characteristics as the center of government, trade and service center. The ongoing development has led to the relationship and interrelationships between various sectors in DKI Jakarta moving towards a higher and more complex Bodetabek. Changes that occur in one particular sector have a reciprocal effect on other sectors. In fact, changes in the economic sector in DKI Jakarta have affected changes in land use in the surrounding areas (hinterland), including Bogor, Depok, Tangerang and Bekasi (Bodetabek).

Land resources are becoming increasingly important because a growing population will put pressure on demand. Meanwhile, in terms of the available land supply, it can be said to be constant, but its value increases from time to time, even with the development of the economy and the increase in population and the expansion of economic activities outside of agriculture, the use of land is increasingly competitive. The land supply market system also often does not work well because it experiences institutional constraints such as many regulations that cause land availability to become increasingly scarce. The results of a study conducted by the Research Institute of IPB in Jakarta and in one of the buffer zones of Jakarta, Bekasi Regency, show that there is a link between the migration process, economic growth and land conversion.

Land resources are becoming increasingly important because a growing population will put pressure on demand. Meanwhile, in terms of the available land supply, it can be said to be constant, but its value increases from time to time, even with the development of the economy and the increase in population and the expansion of economic activities outside of agriculture, the use of land is increasingly competitive [5]. The land supply market system also often does not work well because it experiences institutional constraints such as many regulations that cause land availability to become increasingly scarce. The results of a study conducted by the Research Institute of IPB[6] in Jakarta and in one of the buffer zones of Jakarta, Bekasi Regency, show that there is a link between the migration process, economic growth and land conversion.

The Ciliwung Watershed (DAS) located in two provinces, namely West Java Province and DKI Jakarta, is one of 13 watersheds in a very critical condition due to changes in land use. The upstream Ciliwung watershed covers an area of 14,860 ha located in Bogor Regency and is a mountainous area with an elevation between 300 m to 2,040 m above sea level. This area is generally characterized by fast flowing rivers, especially during the rainy season, and high slope variations above 45% [7]. And the description above, it can be formulated to what extent the economic development that occurs in DKI Jakarta affects the economy and changes in land use in Bogor Regency, especially the Ciliwung Hulu watershed.

IOP Conf. Series: Earth and Environmental Science **986** (2022) 012035 doi:10.1088/1755-1315/986/1/012035

1.2. Research purposes
In general, this study aims to determine the economic structure of DKI Jakarta and Bodetabek as well as the impact of economic development of DKI Jakarta on the economy and changes in land use in Bogor Regency, especially the Ciliwung Hulu watershed.

2. Research methods

2.1. Writing Analysis Design
In general, this analysis is descriptive-analysis with secondary data base for analysis. The secondary data in question is Regency/City data in the Jabodetabek area in time series using the IRIO table[8]. The areas studied are DKI Jakarta and Bogor Regency, especially the upstream Ciliwung watershed.

2.2. Interregional Input Output Analysis (IRIO) And Landsat Image
The Jabodetabek IRIO table is used as an analytical tool to see the structure of economic linkages between sectors in an economy and the multiplier effect of a sector on the sector or the economy of DKI on land use change in Bogor Regency.

To prove the type of land use change through IRIO with the location of the land use change in the upstream Ciliwung watershed, an analysis was carried out through Landsat imagery and then calculated the impact of erosion that occurred

3. Results and Discussion

3.1. Economic Condition
In 2020, the total final demand for goods and services in Jakarta is Rp. 2,772.38 trillion. This amount is used to meet the needs of households, non-profit institutions, the government, investment and traded to other provinces and abroad. DKI Jakarta's economy contributes 17.56 percent to the national economy. The development of DKI Jakarta's GRDP based on current prices from 2016 to 2020. The average GRDP during that period was Rp. 2,541.24 trillion. The average increase in the value of GRDP is Rp. 156.66 trillion per year. When viewed by component, the nominal household consumption expenditure (PKRT) figure has the largest number among other components, meaning that household consumption expenditure has the largest contribution to the Jakarta economy.

Economic growth can also describe an increase in goods and services as measured by an increase in real output. Jakarta's economic growth during the 2016-2020 period. The average economic growth of DKI Jakarta in that period was 4.33 percent. If seen every year, the economic growth rate of DKI Jakarta is always above five percent. However, the Covid-19 pandemic caused DKI Jakarta's economic contraction by minus 2.36 percent

Based on the contribution of inputs and outputs in DKI Jakarta and Bodetabek, it can be shown that the output of DKI Jakarta used for activities (inputs) in Bodetabek is 1.22 percent of the total national output. Meanwhile, the Bodetabek output used for activities (input) in DKI Jakarta is 1.06 percent of the total national output. For the utilization of output in its own region, in DKI Jakarta it is 56.66 percent and Bodetabek is 81.53 percent of the total national output.

For DKI Jakarta, the contribution of household consumption is 56.34 percent, government consumption is 5.25 percent, investment is 20.85 percent, and foreign exports are 17.78 percent of the final national demand. Meanwhile in Bodetabek, household consumption contributed 52.27 percent, government consumption 1.94 percent, investment 18.25 percent, and foreign exports 20.65 percent of the final national demand.

Based on the total national final demand, the proportion of household consumption in DKI Jakarta is 14.51 percent of the total national household consumption. The absorption of investment in DKI Jakarta is 20.44 percent of the absorption of national investment. This is very striking when compared to the area of DKI Jakarta which is only 0.03% of the total area of Indonesia. Meanwhile, the proportion of household consumption in Bodetabek is 5.96 percent of the total national household consumption.

The proportion of government consumption in Bodetabek is 1.96 percent of the total national government consumption. The absorption of investment in Bodetabek is 7.92 percent of the absorption of national investment. Likewise with Bodetabek, the area of Bodetabek is only 0.32% of the total area of Indonesia.

3.2. Sectoral Economic Linkages and Land Use Change

The inequality of development in Indonesia is a regional development problem and needs more attention. On a national scale, the development process that has been carried out so far has in fact caused quite complex development problems, and tends to lead to a very large development gap between regions [1]. In this study, the DKI Jakarta area is the center of development. Community activities are dominated by the secondary and tertiary sectors, while the Bodetabek area which is a buffer zone for DKI Jakarta is dominated by the primary and secondary sectors. Investment and resources are absorbed and concentrated in urban areas and growth centers, while the hinterland areas experience excessive resource depletion.

This gap in the end creates problems which in a macro context are very detrimental to the development process to be achieved as a nation. The imbalance of development between regions on the one hand occurs in the form of poor distribution and allocation of resource use which creates inefficiency and suboptimal economic systems. Inequality of development results in the structure of relations or linkages between regions that form an interaction that weakens each other, the hinterland becomes weak due to excessive draining of resources (backwash) which results in net flow and accumulation of added value in development centers. massively and excessively. But on the other hand, the occurrence of capital accumulation or added value in the growth center area. It can be seen that the DKI Jakarta area is dominated by the banking sector and other financial institutions, but this sector does not become an output multiplier for Indonesia as a whole, which indicates that mediation for the national economic node is not functioning as it should, in fact there is only capital accumulation in Indonesia, DKI Jakarta[9].

In simple terms, from the results of the interregional input and output analysis, the contribution of inputs and outputs in DKI Jakarta, Bodetabek with weak linkages. Utilization of output for its own region is higher than for other regions, in DKI Jakarta at 56.66%, Bodetabek at 81.53%, and the rest of Indonesia at 96.57% of the total national output.

The output of DKI Jakarta used for activities (input) in Bodetabek is 1.22%, while the output of Bodetabek used for activities (input) in DKI Jakarta is 1.06%. in the industrial sector, trade, banks and other financial institutions, as well as building businesses and corporate services.

The output from DKI Jakarta which is used as input by Bodetabek, the sectors that have the highest value are the industrial, trade, restaurant and hotel sectors, building businesses and corporate services, as well as social and community services.

The sector with the highest income multiplier value in DKI Jakarta is the mining and quarrying sector. The sectors that have the strongest inter-sector linkages in DKI Jakarta are the building business sector and corporate services. For Bodetabek, the strongest inter-sector linkages are the industrial and trade sectors. Wikarya's research shows[10] that the asymmetric pattern of economic linkages between islands in Indonesia results in a counter-productive situation with equalization efforts, namely the increase in final demand for output produced in Java and Bali only has a slight impact on increasing output, income, and income. labor on other islands. On the other hand, the increase in final demand for output produced outside Java and Bali has a significant impact on increasing output, income, and employment opportunities in Java and Bali.

Inequality in development results in a structure of relations or linkages between regions that weaken each other. In the hinterland, there is an excessive drain of resources (backwash) which results in a net flow and accumulation of capital or added value in development centers in excess. The accumulation of capital or added value occurs in the growth center areas which in turn leads to the process of poverty and underdevelopment in the back/rural areas. Finally, this situation encourages the migration of people out of the villages to the cities, so that cities and growth centers eventually become weakened, due to the emergence of various extraordinary "urbanization diseases".

Economic development in DKI Jakarta as indicated by the increase in Gross Regional Domestic Product (GRDP) which has caused changes in the sectoral economic structure and affects final demand, has affected economic development and land use in Bodetabek. The increase in the Bodetabek economy can be seen from the increase in output and income and its relation to changes in the use of agricultural land to non-agricultural land.

The phenomenon of urbanization that weakens the development of cities can be seen in the development of big cities in Indonesia which are experiencing "over-urbanization". A person's motivation to migrate is due to economic motives. This motive arises because of the gap between regions. Therefore, population migration reflects an economic balance between rural and urban areas. This status will trigger a migration pattern that tends to go to cities or to villages depending on the strength of the pull and push factors (puff-push factors). The factors that encourage and attract someone to migrate are; (1) demographic factors (gender, education), (2) income expectations that are greater than what is actually obtained in the village, (3) narrow and low land resources reflected in productivity, (4) limited agricultural job opportunities for labor work with higher education. In addition, other pull factors are: (1) the existence of employment opportunities in the industrial and service sectors in the region

In addition, the development of the urban area of DKI Jakarta is very necessary to pay attention to the problem of social capital (social capital). These developments are characterized by the occurrence of various forms of inefficiency and problems, such as the expansion of slum areas, high levels of pollution, traffic jams, the spread of crime, and so on. In addition, according to Ibrahim (2012) the process of development and development of DKI Jakarta which results in new activity centers spread across various areas of DKI Jakarta, results in spatial inequality. This is why DKI Jakarta cannot be treated the same as a homogeneous settlement pattern, both at the macro and micro levels.

From the IRIO table, 2,561.86 ha of built land is needed in DKI Jakarta and 251,167.57 ha in Bodetabek. Based on the results of the interpretation of TM7 images in 2010 and 2019, the area of Bodetabek built up area has changed from 183,681.86 ha to 249,867.91 ha. or has increased to 66,186.05 ha. The difference in calculations with the IRIO table and image interpretation of 1,299.66 ha is due to increasing output not always increasing the land area to maximum production or increasing this area by building multi-storey buildings.

Data on land cover in the Upper Ciliwung watershed was obtained through analysis of Landsat image data in 2010 and 2019 through a guided classification method with the area of each type of land cover showing that dry land agriculture was the land that experienced a lot of reduction, which was 9% of the 5623 ha area in 2015. 2010 to 4435 ha in 2019. A sizable reduction also occurred in paddy fields and forest cover of 2% each. Along with the reduction in the area of land cover, there was an increase in other land areas, namely on built up land by 11% and productive tea gardens by 2%.

The results of the overlapping analysis resulted in 8 patterns of land cover change in the Ciliwung Hulu watershed area. The resulting change pattern is divided into two characteristics, namely one-way and two-way. The pattern of change that is one-way means that the land change does not change back to its original use. On the other hand, those that have two-way properties will return to their original use. From the results of the analysis, a one-way pattern of change occurs in: (1) forest – dry land agriculture, (2) dry land agriculture – productive tea gardens, (3) dry land agriculture – built up land, and (4) rice fields – land. awakened. While the two-way ones are: (1) productive tea gardens - fallow tea gardens, (2) fallow tea gardens - productive tea gardens, (3) rice fields - dry land agriculture, and (4) dry land agriculture - rice fields.

Changes in land cover patterns that occur in the Upper Ciliwung watershed have a tendency to increase from year to year towards uses with smaller infiltration characteristics and result in reduced conservation functions of the upstream Ciliwung area. The reduction in forest area to dry land agriculture causes disruption of hydrological functions and higher erosion

4. Conclusion

Based on the results of an analysis of the economy of DKI Jakarta and Bodetabek using the Jabodetabek IRIO Model, the following results are obtained:

4.1. The sector in DKI Jakarta that has an effect on Bodetabek is the trade sector. Food industry, building, restaurant and hotel, communication, building and service business. On the other hand, the sectors in Bodetabek that affect DKI Jakarta are the chemical, wood, and metal goods industries.

4.2. According to the IRIO table, 2,561.86 ha of built land are needed in DKI Jakarta and 251,167.57 ha in DKI Jakarta and 251,167.57 ha while the results of the interpretation of TM7 images in 2010 and 2019, the area of Bodetabek built land changes from 183,681.86 ha to 249,867.91 ha. or has increased the area of 66,186.05 ha. There is a difference in calculations with the IRIO table and image interpretation of 1,299.66 ha

4.3. Land cover data for the Upper Ciliwung watershed obtained through analysis of Landsat image data in 2010 and 2019 shows that dry land agriculture is the land that has experienced a lot of reduction, which is 9% from an area of 5623 ha in 2010 to 4435 ha in 2019. Sufficient reduction large areas also occurred in paddy fields and forest cover, each of which was 2%. Along with the reduction in the area of land cover, there was an increase in other land areas, namely on built up land by 11% and productive tea gardens by 2%. The reduction in forest area to dry land agriculture causes the hydrological function to be disturbed and the erosion that occurs is higher

References

[1] Arifien Y and Rustiadi E Change Pattern of Land Use of Jabodetabek Area Through Irio and Dimamic System Approach **8713**

[2] Sejati L B, Arifien Y and Maad F 2020 Economic valuation of rice agricultural land in Bogor regency *Journal of Physics: Conference Series* vol 1517

[3] Ginsburg, Kopped M and Gee The Extended Metropolis: Settlement Transition Is Asia - Google Books

[4] Rustiadi E, Saefulhakim. S and Panuju D 2017 Perencanaan dan Pengembangan Wilayah - Google Books

[5] Arifien Y, Badri I and Fitriani A 2020 Optimization of farmer pattern as an effort to improve farmers' income and environmental friendliness in Bogor, Indonesia *IOP Conf. Ser. Mater. Sci. Eng.* **830**

[6] E. Rustiadi et all Perencanaan dan Pengembangan Wilayah 87–102

[7] Soewandita H and Sudiana N 2018 Studi Dinamika Kualitas Air Das Ciliwung *J. Air Indones.* **6** 24–33

[8] Miller R and P B 1997 *Input Output Analysis*

[9] Nazara S 1997 *Analisis input-output* (Lembaga Penerbit, Fakultas Ekonomi, Universitas Indonesia)

[10] Wikarya 2003 Keterkaitan Ekonomi Sektoral dan Spasial di Indonesia

IOP Conf. Series: Earth and Environmental Science **986** (2022) 012036 doi:10.1088/1755-1315/986/1/012036

Assessment of soil degradation potency for biomass production and the strategy for its management in Giriwoyo-Indonesia

Mujiyo [1,*], T Hardian [2], H Widijanto [1] and A Herawati [1]

[1] Department of Soil Scince, Agriculture Faculty, Sebelas Maret University, Surakarta, Indonesia

[2] Department of Agrotechnology, Agriculture Faculty, Sebelas Maret University, Surakarta, Indonesia

*E-mail: mujiyo@staff.uns.ac.id

Abstract. Biomass production supports humans need for food so needs to be supported by optimal soil conditions. The study related to the potential for soil degradation in land characteristics is the first step for sustainable agricultural control. Potential soil degradation is compiled from the number of scores for each land map unit (soil type, slope, rainfall, land use) using technical guidelines for the preparation of soil degradation, and the Pearson correlation test to examine the causal relationship of soil degradation from each land characteristic. The results show that the land with the highest potential is with moderate level potential of 4,671.83 hectares (46.44% area) and the other part of 3,953.13 hectares (39.30% area) has low potential level. The main factors for potential soil degradation are the slopes, rainfall, and land use. The steeper the slopes and the high rainfall carried out on intensive agricultural land, the greater the degradation to the soil. Suggested land management strategies include soil conservation with minimal tillage, mulch application, terrace construction, and planting parallel to the soil contour. In addition, erosion research needs to be carried out to determine in detail the magnitude of the erosion prediction in this area.

1. Introduction

Land is defined as a territorial unit including soil, climate, hydrological activity, and the variety of vegetation in it. Biomass production is inseparable from land use as a growing medium. Biomass production supports the community's need for food, so it needs to be supported by optimal soil conditions. Land that is not managed optimally will experience a decrease in quality and/or soil degradation. Agricultural land becomes the identity of an area which is the result of the complexity of natural interactions that occur and human actions in management [1]. Soil is the key to biomass production and an important role in ecosystems and agricultural systems [2]. The study related to the potential for soil degradation by identifying land characteristics is the first step for sustainable agricultural control.

Land characteristics are environmental factors that have a major influence on the quality and resilience of a land. Environmental factors that play an important role are the morphology of the area including climatic conditions, parent form material, and land relief in the area. Environmental factors in an area will indirectly change the condition of the land. Changes in soil conditions towards decreasing are the result of natural events including soil and water erosion [3], reduced population of soil biodiversity [4], and topographic mismatches of land for agricultural land such as the length of the slopes in the soil, the area of cultivated land and the steepness of the slopes in the landscape [5].

The other substantial factors that linked to degradation is for what purpose the land is used, how land management is carried out by farmer or stakeholders, and farming systems. Tillage for the purpose of crop production will significantly impact on soil characteristics such as groundwater conservation, soil temperature, infiltration, and evapotranspiration [6]. Even in areas with dry and semi-arid land, excessive use of agricultural land is included in resource exploitation which often has negative impacts on agroecosystems and widespread soil degradation [7]. Often the land experiences nutrient exploitation, due to the lack of discipline of farmers in post-harvest soil management so that soil nutrients are transported by plant products and lost slowy [8], and water resources/irrigation for agricultural land are of less concern for quality [9].

Wonogiri Regency is one of the many areas in the province of Central Java which is a hilly area, and there are at least 20% of the limestone hills area. The hilly area has a varies land topography so that it has limitations and a large risk of being used for agriculture, and requires more input in its management. Even so, Wonogiri Regency has very promising agricultural development potential to support the economy of the community and local government. Almost every District in Wonogiri Regency consists of agricultural land. One of the districts that has agricultural land with an area of more than 50% of its area is Giriwoyo District. BPS Wonogiri data in Giriwoyo District produces a lot of food crops including rice (paddy field and upland paddy), soybeans, corn, cassava, peanuts and tubers. These food crops can not only be consumed personally, but can also be a source of income for the local community to fulfill economic needs.

Based on the description above, the agricultural is a prospect sector to be developed and managed wisely in Giriwoyo. Besides being able to prosper the local community, it can also increase the percentage of the regional economy. Aims of the mapping of potential soil degradation needs to be done as a support for protecting and planning agricultural land management. Sustainable agricultural systems need to be based on the selection of land management decisions [10]. The purposes of information related to potential soil degradation can later be used as guidelines for local governments in controlling agricultural activities so in land management they prioritize aspects of the sustainability and can create long-term sustainable agriculture in the area.

2. Material and Method

2.1. Research area description

This research was conducted in Giriwoyo District, Wonogiri Regency, province of Central Java. Giriwoyo has 10,060.13 hectares area consist of 16 village, Tirtosuworo, Guwotirto, Ngancar, Bulurejo, Gedongrejo, Pidekso, Tukulrejo, Bumiharjo, Selomarto, Sejati, Sendangagung, Platarejo, Sirnoboyo, Tawangharjo, Giriwoyo, and Girikikis. Giriwoyo located at the coordinates of 7° 58' 52.82" - 8° 5' 27.26" South Latitude, and 110° 52' 0.08" - 111° 1' 52.19" East Longitude. The research area is mostly in the form of limestone mountains, especially in the southern part which is a row of a thousand mountains and the springs of Bengawan Solo. The boundaries of the Giriwoyo District include Baturetno District in the north, Batuwarno District in the east, Giritontro District in the south, and Eromoko District in the west. Giriwoyo District has a varies topography with slopes ranging from 0%-15%, and 15%-40%. The research area is classified as dry land agriculture in 8,624.96 hectares, with an area of land for biomass production including 3,922.91 hectares of plantation and moor, and 4,702.05 hectares of irrigated and non-irrigated paddy fields.

2.2. Tools and Materials

Working map using a land map unit (Figure 1) which was created by overlaying several thematic maps of regional characteristics including soil type maps, slope maps, rainfall maps, and land use maps, compiled using the GIS ArcView 3.3 software. The tools used in this study include GPS, clinometer, computer, ArcGIS 3.3 software, and SPSS version 16.0.

IOP Conf. Series: Earth and Environmental Science **986** (2022) 012036 doi:10.1088/1755-1315/986/1/012036

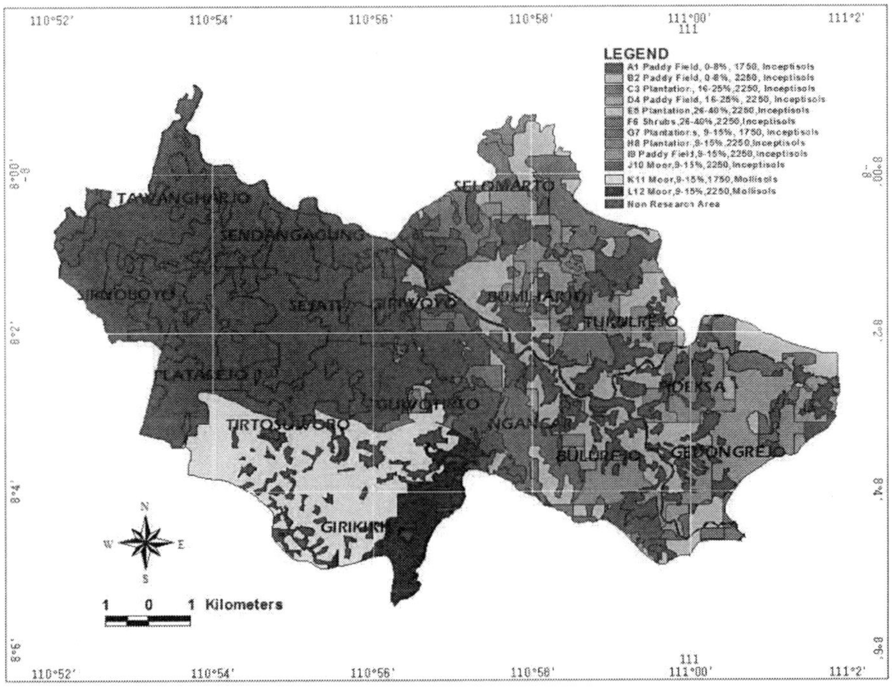

Figure 1. Land Mapping Unit for Biomass Production in Giriwoyo District

2.3. Methods

Research is using an exploratory survey method. Data collection is by observing conditions by maps information, field verification, and statistical tests of observational data. The work map is in the form of a land map unit (LMU). The land map unit of the research area is divided into 12 land units (Table 6). The methods for study about evaluation of soil degradation is using individual assessment of the countries. Indonesia has the basic procedure of soil degradation assessment on Regulation of The Government of Indonesia Number 150 on Soil Degradation Control for Biomass Production and [11] Regulation of the Minister of Environment No. 7 of 2006 on Procedures Measurement Criteria of Soil Degradation for Biomass Production.

Potential soil degradation is compiled from the number of scores on each land map unit. The assessment of each thematic map is different, from the rating value and score, each can be seen in the assessment of the soil type map (Table 1), slope map (Table 2), rainfall map (Table 3), and land use map (Table 4). After that, matching to the weight scores (the result of multiplying the rating and scores) of each thematic maps in each LMU. Then, classified to the accumulation score of the LMU (Table 5). A map of the soil degradation potential (SDP) was made according to the category. The maps gives information the details of level soil degradation by the characteristics itself not only about soil characteristics but also the morphology, the climates in the region, and land use system.

Table 1. Assessment for soil degradation potential (SDP) based on soil type

Soil Type	Soil Degradation Potency	Score	Rating	Score Weight (score x rating)
Vertisols	Very Low	2	1	2
Oxisols	Low	2	2	4
Alfisols, Mollisols, Ultisols	Moderate	2	3	6
Inceptisols, Entisols, Histosols	High	2	4	8
Spodosol, Andisol	Very High	2	5	10

Source: [12]

Table 2. Assessment for soil degradation potential (SDP) based on slope

Slope (%)	Soil Degradation Potency	Score	Rating	Score Weight (score x rating)
1-8	Very Low	3	1	3
9-15	Low	3	2	6
16-25	Moderate	3	3	9
26-40	High	3	4	12
>40	Very High	3	5	15

Source: [12]

Table 3. Assessment for soil degradation potential (SDP) based on rainfall

Rainfall (mm)	Soil Degradation Potency	Score	Rating	Score Weight (score x rating)
<1000	Very Low	3	1	3
1000-2000	Low	3	2	6
2000-3000	Moderate	3	3	9
3000-4000	High	3	4	12
>4000	Very High	3	5	15

Source: [12]

Table 4. Assessment for soil degradation potential (SDP) based on land use

Land Use	Soil Degradation Potency	Score	Rating	Score Weight (score x rating)
Natural forest, Paddy field, Pure fertile reed	Very Low	2	1	2
Mix plantation, Shrubs, Savanna	Low	2	2	4
Production forest, dry land	Moderate	2	3	6
Moor	High	2	4	8
Open field	Very High	2	5	10

Source: [12]

IOP Conf. Series: Earth and Environmental Science **986** (2022) 012036 doi:10.1088/1755-1315/986/1/012036

Table 5. The level of SDP based on accumulation score

Symbol	Soil Degradation Potency	Accumulation Score
PR.I	Very Low	< 15
PR.II	Low	15 – 24
PR.III	Moderate	25 – 34
PR.IV	High	35 – 44
PR.V	Very High	45 – 50

Source: [12]

Table 6. Land mapping unit and SDP total scores in Giriwoyo

Land Mapping Unit (LMU)	Soil Type	Slope(%)	Rainfall (mm/year)	Land Use	Accumulation Degradation Score	Area (ha)
A1	Inceptisols	0-8 %	1750	Paddy field	19	2599.16
B2	Inceptisols	0-8 %	2250	Paddy Field	22	807.44
C3	Inceptisols	16-25 %	2250	Plantation	30	486.04
D4	Inceptisols	16-25 %	2250	Paddy field	28	644.75
E5	Inceptisols	26-40 %	2250	Plantation	33	434.70
F6	Inceptisols	26-40 %	2250	Shrub	33	358.39
G7	Inceptisols	9-15 %	1750	Plantation	24	546.53
H8	Inceptisols	9-15 %	2250	Plantation	27	310.99
I9	Inceptisols	9-15 %	2250	Paddy field	25	650.70
J10	Inceptisols	9-15 %	2250	Moor	31	295.83
K11	Mollisols	9-15 %	1750	Moor	26	1144.62
L12	Mollisols	9-15 %	2250	Moor	29	345.81

2.4. Data analysis

The observational data in the form of accumulated scores of potential soil degradation were then tested statistically to examine the relationship between the characteristics of land-forming and the potential for degradation, using the Pearson's correlation test, processed with SPSS 16.0 software.

3. Result and Disscuccion

3.1. Soil degradation potency

Land-forming characteristics in the research area of Giriwoyo District consist of 2 types of soil, namely Mollisols and Inceptisols, topography ranging from 2-40%, 4 types of land use for agriculture, namely paddy fields with an area of 4,702.05 hectares with very low degradation potential (Table 5), plantation 1,778.26 hectares and 358.38 hectares of shrubs with low potential for degradation, and 1,786.26 hectares of moor with high potential for degradation, as well as climatic conditions with rainfall of 1,750 mm/year classified as low (Table 4) in the west, and 2,250 mm/year classified as moderate in the eleventh east of Giriwoyo District. Based on Figure 1, Mollisols with moderate potential (presented in Table 2) are located in the west of Giriwoyo District, namely Tirtosuworo and Girikikis villages, while Inceptisols soils with high degradation potential dominate the Giriwoyo area. Wonogiri Regency is known for its hilly areas so that its topography is very diverse, but what is used as agricultural land is divided into a slope of 0-8% with a very light potential (presented in Table 3), a slope of 9-15% with a mild potential, a slope of 16-25% with a moderate potential, and slopes of 26-40% high potential, and some of them >40% as un-used land.

Table 7. Area of Soil Degradation Potency in Giriwoyo District

Soil Degradation Potency	Symbol	Area (hectares)	Area (%)	LMU
Low	SDP I	0.00	0	0
Slightly	SDP II	3,953.13	39.30	1, 2, 5
Moderate	SDP III	4,671.83	46.44	3, 4, 6, 7, 8, 9, 10, 11, 12
High	SDP IV	0.00	0	0
Very High	SDP V	0.00	0	0

The results of mapping the potential for soil degradation in Giriwoyo District (Figure 2) the land with the highest potential, located in the villages of Selomarto, Bumiharjo, Pideksa, Gedongrejo, Ngancar, Bulurejo, Platarejo, and Guwotirto with moderate degradation potential (SDP III) covering an area of 4,671.83 hectares (46.44%) (show in Table 7). Some other areas covering an area of 3,953.13 hectares or about 39.30% potential at slightly level (SDP II) are distributed in the villages of Tawangharjo, Sirnoboyo, Sendangagung, the northern part of Tukulrejo, the southern part of Giriwoyo, and the western part of Selomarto. Some of the land that is not agricultural land is not included in the observation area, and is land used for settlement, which is in 1,435.17 hectares area.

The lowest soil degradation for biomass production occurred in LMU A1 covering an area of 2599.16 hectares with the current type of land use in the form of paddy fields, located on flat land (0-8%), low annual rainfall of 1,750 mm/year, and Inceptisols soil type. Meanwhile, the greatest potential for soil degradation occurs in LMU E5 and LMU F6 (793.09 ha) with the current land use were being plantation and shrubs, located on land with a steep slope (26-40%) and 2,250 mm/year of rainfall. Plantation land is a land with a variety of annual vegetation, such as mahogany, teak, and sengon, while shrubs land is a land with a variety of undergrowth, such as shrubs and reeds.

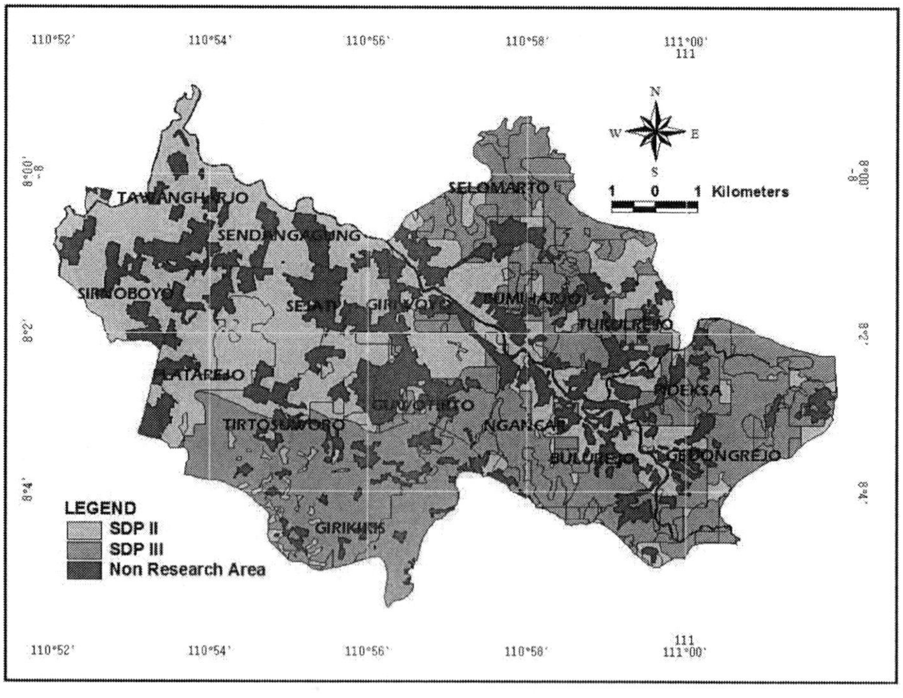

Figure 2. Map of SDP in Giriwoyo District

Table 8. Land characteristics factor of SDP in Giriwoyo District

Land Characteristics	Correlation with SDP
Slope	0.857**
Rainfall	0.595**
Land Use	0.415*
Soil Type	-0.027

Remark: A number followed by (**) correlated in alpha 1%, and (*) correlated in alpha 5%.

The correlation test results show (Table 8) that slope, rainfall, and land use have a very significant linked with the potential for soil degradation. Potentially degradation land can be caused by artificial factors, one of which is land management activities carried out by stakeholders and farmers. The current condition of land use in the research area has a very significant cause-and-effect linked (Sig. = 0.012, R = 0.415*) with the potential for soil degradation. Agriculture that is intensively used to produce food crops and horticulture is paddy fields and dry fields (moor). Paddy fields in the study area consist of irrigated paddy fields and non-irrigated paddy fields, both of which are used for paddy production one to two times a year. The dry land is used to grow corn, soybeans, beans, and tubers. In addition, paddy fields and moor are carried out on land with gentle to steep slopes. Dry land has the highest potential degraded score among other land uses and is used continuously throughout the year in the dry season and rainy season.

The use of land by the surrounding community aims to fulfill economic needs, because the livelihoods of the majority of the population of this area are farmers. The need for food and the economy makes the land continuously given management for reaching production results with better quality [13]. One of them is the use of inorganic fertilizers. The use of inorganic NPK fertilizers is widely used in paddy fields and dry fields (moor) when cultivating with the aim of high production, because fertilizers play a role in increasing plant dry weight per 100 seeds [14], dry crop yield [15], and accelerating the growth of cereal crops [16]. However, the continuous use of inorganic fertilizers will certainly have an impact on the soil and the environment [17].

The impact of land management that is not in accordance with land conditions and continuous use to achieve economic value is considered a land exploitation activity. Chemical fertilizers with high content such as sodium, potassium, and acid-forming in a sustainable manner actually have a negative impact on soil conditions including degrade to soil structure, liming the soil, decreasing soil efficiency for plants, and triggering soil degradation [13]. In addition, it can lead to a decrease in soil quality and an imbalance of soil organic carbon [18], reduced land productivity as a vegetation medium [19], significantly reduce the quality to quantity of groundwater in the area [20], and have an impact on meeting the socio-economic needs of the surrounding community.

In addition to land use factors, the results also show the characteristics of land-forming research areas that have the potential to degraded soil, including slope conditions (Sig. = 0.000, R = 0.857**), and rainfall (Sig. = 0.000, R= 0.595**). The results of the study show that both them are very significant positively correlated, where the higher the percentage of slope and rainfall, the greater the potential for degraded to the soil. Giriwoyo District has a hilly topography and an average slope of up to 30°, where in this structural condition the land is naturally formed by thin solum and has hard rock not far from the top layer [21], and has the potential to degraded the soil. Agricultural land with a fairly steep slope (26-40%) dominates in the eastern part of Giriwoyo District where the rainfall is 2,250 mm/year, including in Ngancar village, west of Bulurejo village, north of Tukulrejo and Selomarto villages, and east of Pideksa and Gedongrejo villages.

[22] stated that erosion is a natural event that is naturally triggered by the slope angle of a land and regional rainfall. On a slope of 16-40% and conditions of high annual rainfall distribution this triggers erosion in which the topsoil is transported from the upper slope to the lower area (catchment area) by water [23]. The transportation of the topsoil on the slopes is strongly influenced by rainfall, because in the early period of the rainy season, splashing water that falls to the ground causes the most soil loss

[24], and resulting in a decrease in soil organic carbon content [25]. Land use for intensive agricultural activities on fairly steep slopes in the study area should be minimized, and converted to biomass production land that does not require much processing to reduce the risk of soil erosion. The rate of water flow increases in the erosion process so that the organic matter content of the soil and soil nutrients is transported and will later inhibit the vegetative productivity of plants [26]. Research of [27] proves that soil erosion on erosive agricultural land will result in loss of crop productivity, and plants that experience the highest productivity loss are secondary crops such as paddy, wheat, and corn with a total of 3 million tons. So that not only will there be thinning of the surface/topsoil, but it will also not to be optimal for biomass production.

3.2. Strategy for land management

Climatic conditions, land use differences, as well as topography or land forms, if not used in appropriate proportions will result in a decrease in land productivity, land degradation, and an increase in critical land. Efforts to prevent the risk of degradation due to topography, rainfall, and land use factors need to be implemented. This prevention can be done by land management that considers the scoring of soil degradation in each land map unit, and reduces the intensity on lands with high potential for degradation. [28] stated that in order to achieve the socio-economic development of the community, it is necessary to protecting the land use according to the characteristics of the land and region as an effort to preserve the environment.

Management of agricultural land in the topography of the research area to prevent serious soil degradation requires special soil conservation. Conservation measures that can be taken are minimal tillage or no tillage system. The no-tillage system is a land management system where at the time of planting preparation, the soil is not given excessive treatment such as hoeing or plowing by a tractor, only processed in the planting hole with the aim of reducing the breakdown of soil aggregates. Land preparation with no tillage techniques or minimum tillage can increase soil density, reduce the risk of disturbance to the soil ecosystem, and increase the load of organic matter on the soil surface and around planting holes.

In addition, it is possible to make terraces and use mulch in plant management [29]. The results of the study [30] that the application of mulch on hilly land in dryland significantly reduced the amount of loss of soil nutrients such as nitrogen and organic matter, where this mulch plays an important role in resisting the erosion process when it rains. Making terraces and planting following the contour lines of the slopes is to prevent soil erosion with terraces made in a certain area. Soil conservation by making terraces according to contours is very suitable to be applied to hilly topographic conditions (for steep slopes vary up to 40% or more) to help manage short and long term economic crops, such as woody plants [31]. Prevention of soil erosion can also be done by increasing the vegetation cover around the plots [32]. It aims to reduce the number of direct water collisions with the soil.

4. Conclusion

4.1. Conclusion

Potential degradation to land for biomass production in Giriwoyo District is in the moderate level degraded potential (SDP III) covering an area of 4,671.83 hectares (46.44%), and some other areas covering an area of 3,953.13 hectares or about 39.30% potential at a low level (SDP II). Factors that linked to soil degradation in the research area are land topography (slope), climate (rainfall), and land use. The steeper slope and the higher rainfall, also the intensive land management will increase the potential soil degradation. Thus, in this area it is necessary to apply land management according to the characteristics of the land and region as an effort to preserve the environment. The recommended land management strategies for the Giriwoyo area include land with gentle to slightly steep slopes that need to be implemented without tillage, the use of mulch during cultivation/planting, terrace construction and planting parallel to the contour, and increasing the amount of the annual vegetation around the cultivated land.

4.2. Suggestion
Authors suggest the need for research related to the estimation of erosion hazard to determine in detail the amount of erosion on agricultural land in the Giriwoyo District.

Acknowledgment
Authors thanked to research team, and Novi Rahmawati for the participation.

References
[1] Tarolli P and Straffelini E 2020 Agriculture in hilly and mountainous landscapes: Threats, Monitoring and Sustainable Management *Geogr. Sustain.* **1** 70–6
[2] Lehmann J, Bossio D A, Kögel-Knabner I and Rillig M C 2020 The concept and future prospects of soil health *Nat. Rev. Earth Environ.* **1** 544–53
[3] Muchena F N, Onduru D D, Gachini G N and Jager A De 2005 Turning the tides of soil degradation in Africa : capturing the reality and exploring opportunities *Land use policy* **22** 23–31
[4] Rickson R J, Deeks L K, Graves A, Harris J A H, Kibblewhite M G and Sakrabani R 2015 Input constraints to food production: the impact of soil degradation *Food Secur.* **7** 351–64
[5] Sharma A, Tiwari K N and Bhadoria P B S 2011 Effect of land use land cover change on soil erosion potential in an agricultural watershed *Environ. Monit. Assess.* **173** 789–801
[6] Busari M A, Kukal S S, Kaur A, Bhatt R and Dulazi A A 2015 Conservation tillage impacts on soil, crop and the environment *Int. Soil Water Conserv. Res.* **3** 119–29
[7] Rahmanipour F, Marzaioli R, Bahrami H A, Fereidouni Z and Bandarabadi S R 2014 Assessment of soil quality indices in agricultural lands of Qazvin Province, Iran *Ecol. Indic.* **40** 19–26
[8] Zingore S, Mutegi J, Agesa B, Tamene L and Kihara J 2015 Soil Degradation in sub-Saharan Africa and Crop Production Options for Soil Rehabilitation *Better Crop.* **99** 24–6
[9] Widodo T, Budiastuti M T S and Komariah 2019 Water Quality and Pollution Index in the Grenjeng River, Boyolali Regency, Indonesia *Caraka Tani J. Sustain. Agric.* **34** 150–61
[10] Bhattacharyya R, Ghosh B N, Mishra P K, Mandal B, Rao C S, Sarkar D, Das K, Anil K S, Lalitha M, Hati K M and Franzluebbers A J 2015 Soil Degradation in India: Challenges and Potential Solutions *Sustainability* **7** 3528–70
[11] Minister of Environment 2006 Regulation of Environment Minister's Number 7 in 2006 ordinance of measuring the raw criteria of soil degradation in biomass production *Peraturan Menteri Lingkungan Hidup Nomor 07 Tahun 2006 Tentang Tata Cara Pengukuran Kriteria Baku Kerusakan Tanah Untuk Produksi Biomassa* vol 3 pp 2063–81
[12] Minister of Environment 2009 GUIDELINES FOR MAPPING AND DETERMINING LEVEL OF SOIL DEGRADATION ON BIOMASS PRODUCTION *Pedoman Teknis tentang Tata Cara Pengukuran Kriteria Baku Kerusakan Tanah Untuk Produksi Biomassa* pp 1–22
[13] Savci S 2012 Investigation of effect of chemical fertilizers on environment *APCBEE Procedia* **1** 287–92
[14] Purba R S, Irwan S N R and Putra E T S 2020 The Effect of Spent Bleaching Earth Filler-Based NPK Fertilization on Proline, Growth and Yield of Maize *Caraka Tani J. Sustain. Agric.* **35** 44–53
[15] Wisnubroto M P, Putra□ E T S and Kurniasih B 2021 Effects of Spent and Deoiled Bleaching Earth Filler-Based NPK Fertilization on the Soil Nutrient Status and Growth of Soybean (Glycine max L . Merrill) *CarakaTani J. Sustain. Agric.* **36** 213–26
[16] Wijanarko A, Taufiq A and Harnowo D 2016 Effect of liming, manure, and NPK fertilizer application on growth and yield performance of soybean in swamp land *Degrad. Andmining Landsmanagement* **3** 527–33
[17] Supriyadi, Vera I L P and Purwanto 2021 Soil Quality at Rice Fields with Organic, Semi Organic and Inorganic Management in Wonogiri Regency, Indonesia *Caraka Tani J. Sustain. Agric.* **36** 259–69

[18] Zeraatpisheh M, Bakhshandeh E, Hosseini M and Alavi S M 2020 Assessing the effects of deforestation and intensive agriculture on the soil quality through digital soil mapping *Geoderma* **363** 114139

[19] Baskan O, Dengiz O and Demirag İ T 2017 The land productivity dynamics trend as a tool for land degradation assessment in a dryland ecosystem *Env. Monit Asses* 189–212

[20] Aher S and Deshmukh K 2019 Identifying the impact of intensive agriculture practices on groundwater quality using GIS and multi-tracer techniques around Sangamner city, Maharashtra, India *Geocarto Int.* **36** 1136–60

[21] Ashari A 2013 Erosive level studies of several soil types in Baturagung mountainous Putat and Nglanggeran Village, Patut District, Gunungkidul Regency *Informasi* **39** 15–31

[22] Guerra A J T, Fullen M A, Jorge M Do C O, Bezerra J F R and Shokr M S 2017 Slope Processes, Mass Movement and Soil Erosion: A Review *Pedosphere* **27** 27–41

[23] Sholagberu A T, Mustafa M R U, Yusof K W and Ahmad M H 2016 Evaluation of rainfall-runoff erosivity factor for Cameron highlands, Pahang, Malaysia *Ecol. Eng.* **17** 1–8

[24] Zhang X, Li Z, Tang Z, Zeng G, Huang J, Guo W, Chen X and Hirsh A 2013 Effects of water erosion on the redistribution of soil organic carbon in the hilly red soil region of southern China *Geomorphology* **197** 137–44

[25] Olson K R, Al-kaisi M, Lal R and Cihacek L 2016 Impact of soil erosion on soil organic carbon stocks *Featur. J. Soil Water Conserv.* **71** 61–7

[26] Jain P K, Husain T, Expressway Y, Noida G, Pradesh U, Expressway Y, Noida G, Pradesh U and Id- E 2021 Impact of soil erosion on agriculture *Int. J. Mod. Agric.* **10** 2053–60

[27] Panagos P, Standardii G, Montanarella L, Bosello F, Borrelli P and Lugato E 2018 Cost of agricultural productivity loss due to soil erosion in the European Union : From direct cost evaluation approaches to the use of macroeconomic models *L. Degrad. Dev.* **29** 471–84

[28] Lestrelin G, Bourgoin J, Bouahom B and Castella J 2011 Measuring participation : Case studies on village land use planning in northern Lao PDR *Appl. Geogr.* **31** 950–8

[29] Liu X-E, Li X G, Hai L, Wang Y P and Fu T-T 2014 Film-Mulched Ridge–Furrow Management Increases Maize Productivity and Sustains Soil Organic Carbon in a Dryland Cropping System *Soil Water Manag. Conserv.* **78** 1434–41

[30] Chalise D, Kumar L, Sharma R and Kristiansen P 2020 Assessing the Impacts of Tillage and Mulch on Soil Erosion and Corn Yield *Agronomy* **10** 1–13

[31] Fatimah S and Dibyosaputro S 2015 Evaluasi Praktek Konservasi Tanah Cara Teras Di Das Secang Kecamatan Kokap Kabupaten Kulonprogo Daerah Istimewa Yogyakarta *J. Bumi Indones.* 1–10

[32] Han J, Ge W, Hei Z, Cong C, Ma C, Xie M, Liu B, Feng W, Wang F and Jiao J 2020 Agricultural land use and management weaken the soil erosion induced by extreme rainstorms *Agric. Ecosyst. Environ.* **301** 107047

IOP Conf. Series: Earth and Environmental Science **986** (2022) 012037 doi:10.1088/1755-1315/986/1/012037

Determination of land conservation policy using the analytical hierarchy process method in Weninggalih Village, Jonggol District, Bogor Regency, West Java

C Setiawan[1*], O S Hardi[1], F R A'Rachman[1], O Ariyanti[1], R M Fattah[1], M Baihaqy[1] and Z Abidin[1]

[1] Departement of Geography Education, Faculty of Social Sciences, Universitas Negeri Jakarta, Jakarta

*Email: cahyadi-setiawan@unj.ac.id

Abstract. Weninggalih Village is one of the villages that experiences repeated droughts every year and the conversion of land functions into built areas, this has an impact on the decline in village rice production. While most of the land cover in Weninggalih Village is rice fields. This study aims to determine the land conservation policy in Weninggalih Village, Jonggol District, Bogor Regency, and West Java. The survey method was used to collect data from the field. The sample was determined by proportional sampling as many as thirty farmer households. The Analytical Hierarchy Process method was used to determine the appropriate land conservation policy for farmer households in Weninggalih Village. Based on the results of the study, it is known that those that influence land conservation policies are education, farming experience, conservation attitudes, expansion education, and cost sharing. The land in Weninggalih Village requires land conservation considering that the physical condition of the village land tends to be dry, based on observations. Land conservation policy efforts can be used as material for consideration in planning appropriate land use.

1. Introduction

Land is an important need for people to produce food, provide a place for shelter, as raw material and other activities from various economic, social, cultural sectors. Incorrect land use management causes new problems in the land [24]. In the mid-1990s, the suburbs of Jakarta were mostly controlled by the private sector and new urban and industrial areas would be built [9]. Even in the Jonggol area, there are plans to build a new city of 30,000 ha, which will later be referred to as Bukit Jonggol Asri. However, the plan was hampered by violations of laws and regulations of land use and Indonesia is in an economic crisis [8].

This causes various problems in the form of water shortages for agriculture and industry, decreased soil quality, decreased food supplies, soil erosion, decreased function of water catchment areas, loss of agricultural land to land conservation [6]. Soil erosion can be controlled using innovative agricultural technologies that involve the use of organic farming, bio fertilizers, and biological pest control [18]. Critical land causes land productivity to be low and gradually degraded [23]. Efforts to over-come land problems are land conservation which of course uses water resources efficiently [3].

Weninggalih is one of the many villages that have often hit by a water crisis over the last five years. This is due to a significant climate change that makes the Cipamingkis River and Cihoe River experience a reduction in quantity and are unable to irrigate Weninggalih Village. Based on observations throughout 2020, peak of the dry season start from July to September causing the region to have a water crisis [17]. In addition to water crisis, there are other phenomena in the form of erosion on agricultural land. This phenomenon begins with cracks in rice field in the dry season, but during the rainy season the rainwater fills the cracked soil. This causes the soil to move according to the pattern of water flow.

Land degradation seen from the physical, chemical, and biological aspects is caused by natural and artificial influences [5]. Land improvement is the main goal of land conservation, in the form of preventing soil damage, repairing critical land, and maintaining soil productivity to achieve production. Soil management needs to focus on restoring the physical and biological qualities of the soil [25]. Efforts to preserve dry land in the form of planting hedges, maintaining soil moisture by harvesting rainwater, and increasing soil fertility with green manure [11]. In addition to physical land improvement, community participation in land conservation efforts is very important in terms of knowledge and skills in carrying out land development [10]. Another study states that the development factors for land conservation are through education, farming experience, conservation attitudes, extension education, and cost sharing [1].

In 2008, Adjaye studied using developer factors in an effort to conserve land using the Analytical Hierarchy Process (AHP) method [1]. This method is used to determine the priority weights of the five developer factor parameters in determining land conservation policies in Weninggalih Village. The results of the study aim to determine the appropriate policies to be applied in Weninggalih Village in overcoming land degradation, by utilizing the AHP analysis method.

2. Methods

Weninggalih is one of the villages located in Jonggol District, Bogor Regency or astronomically at -6.464312 South Latitude 107.102556 East Longitude. The distance from Weninggalih village to Jakarta is 57.6 km or to spend approximately 1 hour 23 minutes. The area of Weninggalih Village reaches 444,506 ha with almost 80% of the area being rice fields. Weninggalih is bordered by several villages, namely Sirnajati in the north, Sirnagalih and Karang Indah in the south, Babakan Raden and Sukagalih in the east, and Jonggol and Sukamanah in the west, as shown in Figure 1 the majority of the land cover in Weninggalih is agriculture, making most of the people farmer's livelihoods.

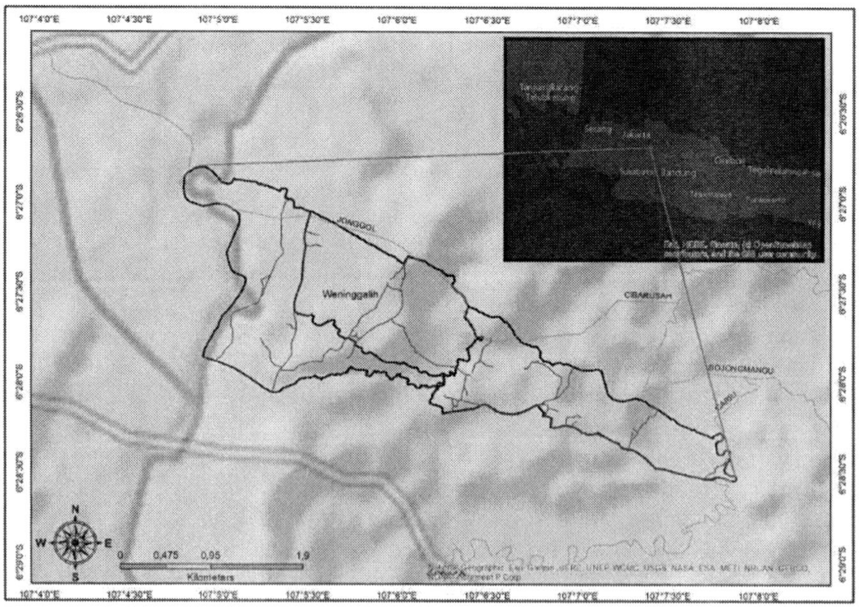

Figure 1 Administrative Map of the Research Area

This research starts from the condition of the area experiencing drought and the conversion of land into a factory since 2019. This problem certainly causes land degradation, so a land conservation development program is needed. There are two development factors, namely personal factors and institutional factors. Personal factors come from education, farming experiences, conservation attitudes. Meanwhile, for institutional factors, there are extension education and cost sharing [7].

In the initial stage, we collected raw data related to community perceptions of soil erosion, decisions to take conservation actions, and soil conservation efforts based on two development factors. Data collection was carried out using a questionnaire that provided several questions about land conservation policies to respondents. The number of samples taken were 30 community respondents from three hamlets in Weninggalih Village. In addition, we also use drone mapping and geospatial information agency to create village administration maps.

The second stage, the available data focused on processing raw data, using the AHP analysis method for questionnaire data while overlaying for land cover data and village boundaries. The overlay method was used to produce village administrative maps, using spatial analysis. While on AHP analysis method to determine the weight of each factor resulting from developers of land conservation in Figure 2.

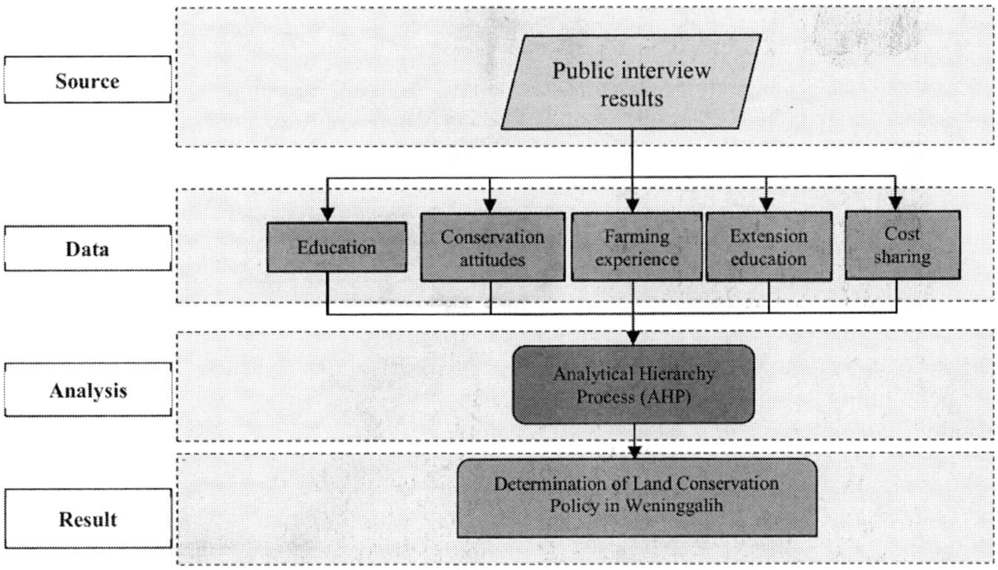

Figure 2 Research Design

In this study, five factors were used to develop rural agricultural land conservation is education, farming experience, conservation attitudes, extension services, and cost sharing. Data were obtained from the results of questionnaires from community representatives from area one to area three in Weninggalih Village. The AHP method has principles for complex decision making [22] especially to determine decisions that are suitable for the public [15,4]. The AHP method can be used as a balance to produce weights for each criterion, and can be used as a tool for decision making in conservation issues [12].

Table 1. AHP comparison standard table

Comparison Syntax	. Value
Extreme importance	9
Very strong importance	7
Essential or strong importance	5
Moderate importance of one over another	3
Equal importance	1
The middle value between two adjacent values	2,4,6,8

Source: Saaty (1987)

The weighting will later determine the dominant influence of the development of land conservation with a comparison index of the highest value (9 = very important) and the lowest value (1 = equally important) contained in Table 1 [20].

Consistency ratio stages are used to validate the research model used, because this model comes from an assessment, validation needs to be done. If the consistency index (CI) = 0, then the matrix value is consistent or perfect [21]. While the value of the consistency of the suggested ratio is not greater than 0.1 [25,19].

The initial step is to determine the consistency ratio value in each comparison, namely by calculating the weighted sum vector, consistency vector, and consistency index values. The weighted sum vector stage is multiplying the matrix comparison with the eigenvalues. The results of the two matrices are determined by the average value to produce a consistency vector. The results of the consistency vector will be used to calculate the consistency index value, using the equation: [2].

$$\mathrm{CI} = \frac{(p-n)}{(n-1)} \quad (1)$$

n = number of alternative

Consistency ratio requires a random index (RI) value based on the Oarkridge Table, with the equation:

$$\mathrm{CR} = \frac{CI}{RI} \quad (2)$$

RI value = 1.12 (based on Oarkridge Table)

Table 2. Pairwise Comparison Matrix

	MATRIX					Priority	Parameter	Value	Result
	A	B	C	D	E				
A	0.18	0.18	0.18	0.17	0.17	0.18	Max. Eigen Value	5.00	Consistency Ratio is 0% consistency is acceptable
B	0.15	0.16	0.15	0.16	0.16	0.16			
C	0.30	0.32	0.31	0.30	0.32	0.31	CI	0.00	
D	0.16	0.14	0.15	0.15	0.14	0.15	RI	1.12	
E	0.22	0.20	0.20	0.23	0.21	0.21	CR = CI/RI	0.00%	
S.O.R	1.00	1.00	1.00	1.00	1.00	1.00			

Provisions CR < 0.10% with consistent ratings criteria. Based on the data shown that calculation of the CR value on the criteria for land conservation development factors is namely < 0.10% (Table 2). These results indicate that the consistency ratio value is acceptable.

3. Result and Discussion

In the process of determining land conservation development policies in Weninggalih Village, there are five criteria needed, namely education, farming experience, conservation attitudes, education extension and cost sharing. In a similar study, using several more complex variables such as age and sex composition, family size and educational status, social position and farming experience, farm size and source of farmland, distance of the land from residential area, slope gradient of the farm plots, fertility status of the plots, and land tenure issues [16].

Table 3. Result of Eigen Value and Weight Value

Criteria	Iteration-1 Eigenvalue	Iteration-2 Eigenvalue	Criteria Weight
Education	0.18	0.18	17.63%
Farming experience	0.16	0.16	15.56%
Conservation attitudes	0.31	0.31	30.96%
Extension education	0.15	0.15	14.70%
Cost sharing	0.21	0.21	21.15%

The Table 3 can be seen that the survey of respondent's perceptions show conservation attitudes with a weight of 30.96%. Another supporting factor is cost sharing with a weight of 21.15%, followed by education 17.63%, farming experience by 15.56%, and the last is the education factor of extension workers at 14.70%. Whereas in previous studies, factors that significantly affect land conservation are extension and cost sharing. Providing counseling to the community through counseling has a positive impact on public perceptions regarding the problem of soil erosion. Meanwhile, the economic variable of farm income greatly affects cost sharing, but the status of land ownership has no significant effect [1]. In 2018 the findings of Siles, the factors that have a significant influence are development, the education level of the head of the household, and farmers' perceptions of the erosion problem [23]. Another study uses four significant factors that influence land conservation benefits, environmental benefits from soil, and labor efficiency [13]

In 1982, the findings of Ervin & Ervin's research shown that the most influential factor is the extension education variable from personal factors and the distance variable to the research location from economic factors [7]. The factor that distinguishes Weninggalih Village from other locations is that there are no people who have received counseling or socialization about land conservation from the government, while 18% of Fijian farmers have participated in an extension program on land conservation [1]. This is also influenced by the low level of education of Weninggalih farmers, most of whom are elementary school graduates.

The second factor that affects the development of land conservation in Weninggalih is cost sharing, where the community is constrained by the costs incurred to carry out land improvement efforts. The absence of a special extension program regarding the improvement of agricultural land makes the level of experience of farmers in Weninggalih in dealing with agricultural land problems tends to be low. In 2016, the study conducted by Mena shows that there are 55% of respondents who are involved and have experience in agricultural activities around land improvement.

Table 4. Calculation of Sub-Criteria Weights Using the AHP Method

			Education			
	A	**B**	**C**	**S.O.C**	**2nd Eigen Vector**	**Weight**
A	0.22	0.22	0.22	0.67	0.22	22.29%
B	0.36	0.36	0.36	1.09	0.36	36.32%
C	0.41	0.41	0.41	1.24	0.41	41.40%
	Amount			**3.00**	**1.00**	**100.00%**
			Farming Experience			
	A	**B**	**C**	**S.O.C**	**2nd Eigen Vector**	**Weight**
A	0.35	0.35	0.35	1.04	0.34	34.55%
B	0.35	0.35	0.35	1.04	0.34	34.55%
C	0.31	0.31	0.31	0.93	0.31	30.90%
	Amount			**3.01**	**1.00**	**100.00%**
			Conservation Attitudes			
	A	**B**	**C**	**S.O.C**	**2nd Eigen Vector**	**Weight**
A	0.28	0.28	0.28	0.84	0.28	28.09%
B	0.29	0.29	0.29	0.86	0.29	28.76%
C	0.43	0.43	0.43	1.29	0.43	43.14%
	Amount			**2.99**	**1.00**	**100,00%**
			Extension Education			
	A	**B**	**C**	**S.O.C**	**2nd Eigen Vector**	**Weight**
A	0.26	0.26	0.26	0.77	0.26	25.67%
B	0.38	0.38	0.38	1.15	0.38	38.33%
C	0.36	0.36	0.36	1.08	0.36	36.00%
	Amount			**3.00**	**1.00**	**100.00%**
			Cost Sharing			
	A	**B**	**C**	**S.O.C**	**2nd Eigen Vector**	**Weight**
A	0.25	0.25	0.25	0.74	0.24	24.58%
B	0.34	0.34	0.34	1.03	0.34	34.22%
C	0.41	0.41	0.41	1.24	0.41	41.20%
	Amount			**3.01**	**1.00**	**100.00%**

[A] Perception of soil erosion problem
[B] Decision to use soil conservation measures
[C] Soil conservation effort

Based on the results of alternative calculations, each sub-criteria produces education with the dominant factor being conservation efforts with a percentage weight of 41.40%, while the decision factor in land conservation actions has a weight of 36.32%. On the conservation attitude factor, the dominant weight is conservation efforts with a weight of 43.14%, the percentage value is superior to the education factor. In addition to the cost sharing factor, the dominant weight value is found in soil conservation efforts with a weight of 41.20% (see Table 4). Seeing these results, Weninggalih tends to be dominant in land conservation efforts. In contrast to the findings of Adjaye (2008), a significant alternative factor is the perception of the problem of soil erosion [1]. So Adjaye provides an alternative by providing counseling on appropriate erosion land conservation practices.

The findings of Mena research with the chi-square test in 2016 show that the factor that has a significant influence is education status, while the factor that does not have a significant effect is farming experience [16]. This is similar to the case in Weninggalih, where farming experience (15.56%) had no significant effect on land conservation.

Based on the results of the weight of filling out the community questionnaire, it is necessary to have a conservation attitude in an effort to prioritize soil conservation. The lack of community attitudes in terms of land conservation care is also constrained by costs with a weight gain of 21.15%. In addition, education is needed about soil conservation efforts to provide additional insight in understanding land conservation practitioners. This problem was also expressed by community leaders, that the community still lacks insight into conservation attitudes in land conservation efforts.

4. Conclusion

The application of AHP proves that in determining the development of land conservation, it is necessary to look at the attitude of the community in land conservation and the issue of costs, based on the results of data from the answers to the questionnaire from the people of Weninggalih Village. Whereas in Weninggalih the education level of village farmers was only completed at the elementary school level. So most people lack will knowledge about land conservation. Another factor that hinders farmers in making efforts to improve land is constrained costs. Most of the farmers complained about the costs incurred in land improvement efforts. Moreover, the Weninggalih people have not participated in activities related to land improvement, so they do not know for sure in carrying out land conservation in their agriculture. With regard to the appropriate policy on Weninggalih in terms of land conservation, it is necessary to consider several of the above variables, so that it can be implemented effectively. In this case study, farmers need to be given counseling or farming experience about land conservation, provide training or socialization about land conservation, agricultural land management efficiency, and empowering modern farmers.

Acknowledgments

We would like to thank Mr Abdul Rahim and all respondents in Weninggalih Village who have provided opportunities for researchers and have provided encouragement and enthusiasm for us to carry out this research.

References

[1] Adjaye, J. A. (2008). Factors Affection the Adoption of Soil Conservation Measures: Case Stucy of Fijian Cane Farmers. *Journal of Agricultural and Resources Economics Vol. 33 No. 1*, 99-117.

[2] Alonso, J. A., & Lamata, T. (2006). Consistency in the Analytical Hierarchy Process: A new Approach. *International Journal of Uncertainty*, 445-459.

[3] Bashir, S., Javed, A., Bibi, I., & Ahmad, N. (2018). Soil and Water Conservation. *Agriculture*, 263-284.

[4] Bonito, M. d., Clarkson, J., Wienhold, M. L., & Pecchia, L. (2013). Integrating Analytical Hierarchy Process (AHP) and Geographical Information System (GIS) for Priotising and Planning Conservation Choices in Wales. *Proceedings of the International Symposium on the Analytical Hierarchy Process 2013* (pp. 1-10). Nottingham: -.

[5] Bossio, D., Noble, A., Molden, D., & Nangia, V. (2008). Land Degradation and Water Productivity in Agricultural Landscapes. *Conserving Land, Protecting Water*, 20-31.

[6] de Vries, F. P., Acquay, H., Molden, D., Valentin, C., Cofie, O., & Scherr, S. (2008). Learning from Bright Spots to Enchane Food Security and to Combat Degradation of Water and Land Resources. *Conserving Land, Protecting Water*, 1-19.

[7] Ervin, C. A., & Ervin, D. A. (1982). Factors Affecting the Use of Soil Conservation Practices: Hypotheses, Evidence, and Policy Implications. *Land Econ 58*, 277-292.

[8] Firman, T. (1997). Land conservation and urban development in the northern region of West Java, Indonesia. *Urban Studies*, 1027-1046.

[9] Firman, T. (2004). New Town development in Jakarta Metropolitan Region: a prepective of sppatial segregation. *Habitat International*, 349-368.

[10] IUCN Bangladesh Country Office. (2015). *Bangladesh National Conservation Strategy (2016-2031.* Bangladesh: Ministry of Environment and Forest Government of the People's Republic of Bangladesh.

[11] IUCN Cancade Development Project Information. (2016, January _). *Conservation of Soil Resources of the Kapiriggama Village Tank Cascade System.* Retrieved from Water Programme: https://www.iucn.org/sites/dev/files/content/documents/conservation_of_soil_re-sources_brief_6_jan29_2016.pdf

[12] Kamps, C., & Zargani, R. J. (2021). Decision Making in Dynamic Environments: An Application of Machine Learning too the Analytical Hierarchy Process. *International Journal of the Analytical Hierarchy Process Vol. 13 Issue 1*, 21-50.

[13] La, C., Halbrendt, C. C., Halbrendt, J., Naik, D., & Ray, C. (2012). Farmers Preference of Conservation Agricultural Practices in Kendujhar, Odisha using Analytical Hierarchy Process. *Appendix 2*, 1-13.

[14] Leaf, M. (1991). Land regulation and housing development in Jakarta, Indonesia. *Ph.D Dissertation*, Universty of California, Berkeley.

[15] Liberatore, M. J., & Nydick, R. L. (2008). The Analytical Hierarchy Process in Medical and Health Care Decision Making: A literature review. *European Journal of Operational Research vol. 189*, 194-207.

[16] Mena, M. M. (2016). Determinants of Choice Decision for Adoption of Conservation Intervention Practice: The Case of Mt. Damota Sub-Watershed, Wolaita Zone, Ethiopia. *Global Journal of Human-Social Science: B Vol. 16 Issue 6*, 32-41.

[17] Meteorological, Climatological, and Geophysical Agency. (2020). *Seasonal Climate and Vegetation Monitoring Outlook. Special Focus: Potential Impacts of La Nina.* Jakarta: Meteorological, Climatological, and Geophysical Agency.

[18] National Institute of Open Schooling. (2021, Agustus 13). *Environmental Science*. Retrieved from Secondary Courses: https://www.nios.ac.in/online-course-material/sr-secondary-courses/envi-ornmental-science-(333).aspx

[19] Qurnain, S. S., Sattanathan, M., & Sankaranarayanan, B. (2020). Analysis of social inequality factors in implementation of building energy conservation policies using Fuzzy Analytical Hierarchy Process Methodology. *International Journal of Sustainable Energy Planning and Management*, 153-170.

[20] Saaty, R. W. (1987). The Analytical Hierarchy Process-What it is and How it is Used. *Mathi Modelling Vol. 9*, 161-176.

[21] Saaty, T. L. (1977). A scaling method for priorities in hierarchial structures. *Journal of Mathematical Psychology*, 234-281.

[22] Saaty, T. L., & Zoffer, H. J. (2011). Negotiating the Israeli-Palestinan Controversy from a New Case Study in the Lake Byes_ehircatchment area. *Waste Management*, 20-37-2046.

[23] Sarminah, S., Karyati, Karmini, Jhonatan, S., & Erikson, T. (2018). Rehabilitation and soil conservation of degraded land using sengon (Falcataria moluccana) and peanut (Arachis hypogaea) agroforestry system. *Biodiversitas Vol. 19, No. 1*, 222-228.

[24] Ulger, N. E., Aydemir, S., Iban, C. M., & Akbulut, H. (2018). Land Use Problems and Land Management: A Land Inventory Study in Istanbul. *Land Governance in an Interconnected World*, 1-26.

IOP Conf. Series: Earth and Environmental Science **986** (2022) 012037 doi:10.1088/1755-1315/986/1/012037

[25] Zheng, G., Zhu, N., Tian, Z., Chen, Y., & Sun, B. (2012). Application of a trapezoidal fuzzy AHP method for work safety evaluation and early warning rating of hot and humid environmments. *Safty Science*, 228-239.

Study of bathymetry and sedimentation in Tondano Lake

A T Moningkey[1*], M M F Rampengan[2], A A Tumengkol[1] and J C Kumaat[3,1]

[1] PROGEO Indonesia, Tataaran 1, Lingkungan 3, Kecamatan Tondano Selatan Kabupaten Minahasa 95618, Indonesia

[2] Department of Biology, Manado State University, Jl. Kampus UNIMA, Kelurahan Tonsaru, Kecamatan Tondano Selatan, Kabupaten Minahasa 95618, Indonesia

[3] Department of Geography, Manado State University, Jl. Kampus UNIMA, Kelurahan Tonsaru, Kecamatan Tondano Selatan, Kabupaten Minahasa 95618, Indonesia

*Email: atmoningkey@gmail.com

Abstract. Tondano Lake is the largest lake on the northern peninsula of Sulawesi, which is naturally formed at an altitude of 680 m above sea level. Tondano Lake extends from north to south, while the width of the lake leads from west to east, the width of the lake varies from 3 to 5 kilometers, and the length of the lake in the longest part is 12.8 kilometers. The sedimentation or silting of Tondano Lake is one of the problems that can disrupt the ecological function and socio-economic life of the people who live and live around the lake area. The purpose of the study was to determine the rate of silting of the lake from 2015 to 2020. The study was carried out from June to October 2020, at Tondano Lake, Minahasa Regency, North Sulawesi. The analytical method used is the descriptive research method that describes the variables and phenomena that occur. Bathymetry data for 2015 was obtained from the bathymetry map 2015 by the Sulawesi River Region I. While the bathymetry data for 2020 was obtained from the results of a research survey from December 2019 to June 2020. The results of the analysis of bathymetry maps in 2015 and 2020 showed that there was a changing depth, in 2015 the maximum depth was 31.81 m and the average depth was 12.88 m. In 2020, the maximum depth is 30.22 m and the average depth is 12.66 m. Based on the results of spatial analysis, it is known that there is a siltation rate of 4.4 cm per year in Tondano Lake.

1. Introduction

Tondano Lake is the largest lake on the northern peninsula of the island of Sulawesi which is naturally formed at an altitude of 680 m above sea level. According to Sudrajat and Bintoro [1] there are two versions of the formation of Tondano Lake, the first formed as a result of an ancient volcanic eruption (Lake Creater), and the second that occurred due to the damming of the drainage system as a result of the Minahasa geanticline, namely the emergence of two volcanoes Soputan and Mahawu. Furthermore, Sorey et al [2] explained that Tondano Lake extends from the north to the south, while the width of the lake leads from the west to the east, the width of the lake varies from 3 to 5 kilometers, while the length of the lake in the longest part is 12.8 kilometers.

The silting of a lake is characterized by changes in the depth of the lake periodically. According to Koperberg in Hikmatullah et al [3], the depth of Tondano Lake from the results of a survey by the Dutch in 1928 was 28 meters in the southern part of Remboken-Telap, a depth of 15 m in the north, 3.5 km south of the mouth of the lake. Furthermore, Zen and Muziel [4] stated that Tondano Lake has a maximum length of 12 km and an average width of 4 km. The depth between Remboken and Telap is not more than 28 m. Then according to Murdiyanto and Sulastriningsih [5], in 2010 the State Ministry for the Environment from the WIP Bathymetry Survey on May 2010 the depth of the deepest lake ranged from 22 - 34 m.

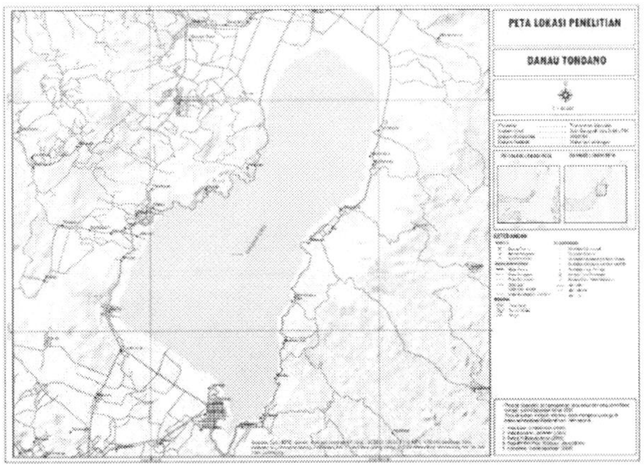

Figure 1. Map of Tondano Lake

The silting of Tondano Lake is one of the problems that can disrupt the ecological function and socio-economic life of the people who live and reside around the lake area. Therefore, this study aims to determine the rate of silting of Tondano Lake from 2015 to 2020 through bathymetry and sedimentation studies.

2. Methods

This research uses a positivistic/quantitative approach, which is a rational linear process. The analytical method used is a descriptive research method that describes the variables and phenomena that occurred, Aprilliyana [6]. The data used are primary and secondary, obtained through field observations and surveys, and collected through lake depth measurements (bathymetric surveys).

The bathymetry survey or sounding implementation uses the Garmin Echosounding GPS Map 585 Plus, while the projection system for coordinates using UTM (Universal Transverse Mercator) and the WGS 84 ellipsoid, Tendean et al [7]. The device emits sound waves vertically to the bottom of the water and is reflected back to the echosounder through planned paths on a digital pre-plot map, which can be read on a computer. Measurement of the depth of the lake using an echosounder was carried out from a motorboat, with a maximum ship speed of 5 (five) knots (2.5 m/second) and the condition of the ship was stable K [8]. Downloading field data from GPS with the Mapsource program and changing the format to be compatible with the Arcgis 10.7.1 program, then making lake depth contours based on the principle of interpolation.

Bathymetry data for 2015 was obtained from the bathymetry map 2015 by the Sulawesi River Region I (Balai Wilayah Sungai Sulawesi I), while the bathymetry data for 2020 was obtained from the results of a research survey from December 2019 to January 2020. Both bathymetry data were processed using Arcgis 10.7 software. For the purposes of analyzing the morphometric aspects required in this study. The morphometric aspects observed in this study are the surface area of the lake water (ha), the average depth (m), the maximum depth (m), and the total volume of the lake water. The surface area is the lake's water surface area, which is measured based on the polygon area on the bathymetry map. The average depth is expressed in meters, i.e. volume (V) divided by surface area. the total volume of lake water is calculated with Arcgis 10.7 software using 3d analyst tools. The four morphometric aspects are mapped with the same scale. According to Murdiyanto and Sulastriningsih [5], the analysis of changes in lake morphometry is carried out based on bathymetry maps with overlay techniques. The change information is then used as a basis for describing the rate of siltation of the lake.

3. Results and Discussion

Surveys and mapping of bathymetry Tondano Lake are carried out as basic data in analyzing changes in lake water depth and determining the rate of lake silting. To obtain bathymetric information, a sounding method is used, using an echosounder that is integrated with GPS.

The 2015 bathymetry data was obtained from the bathymetry map 2015 by the Sulawesi River Region I (Balai Wilayah Sungai Sulawesi I). The analysis results from the 2015 Tondano Lake Bathymetry map showed a maximum depth of 33.88 m located at coordinates 1°13'23.04"N - 1°13'23.04"N, depth an average of 12.88 m.

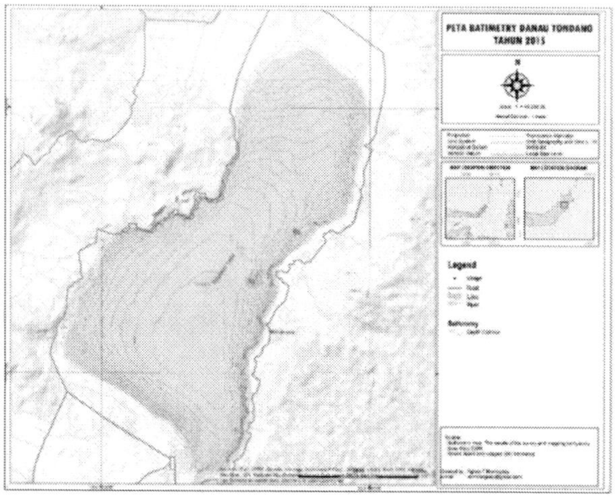

Figure 2. Tondano Lake Bathymetry Map 2015

The source of the bathymetry data for Tondano Lake in 2020 is the result of a survey and mapping conducted by the research from December 2019 to June 2020. The analysis results from the bathymetry map of Tondano Lake 2020 show a maximum depth of 31.81m located at coordinates 1°13'23.04"N - 1 °13'23.04"N. The average depth is 12.66 meters.

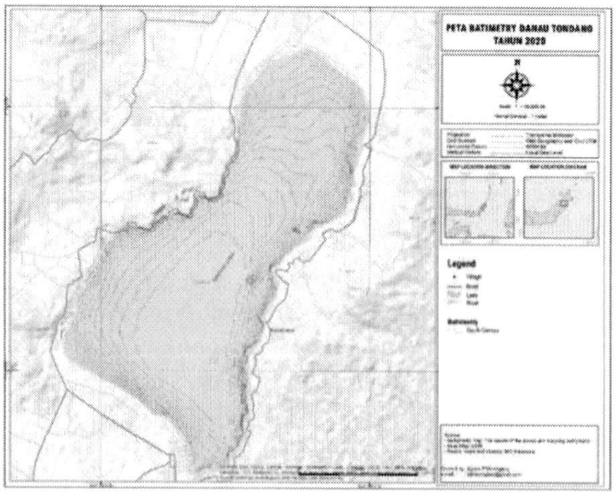

Figure 3. Tondano Lake Bathymetry Map 2020

The results of the calculation of changes in the depth of Tondano Lake are presented in the bar chart image below (Figure 4).

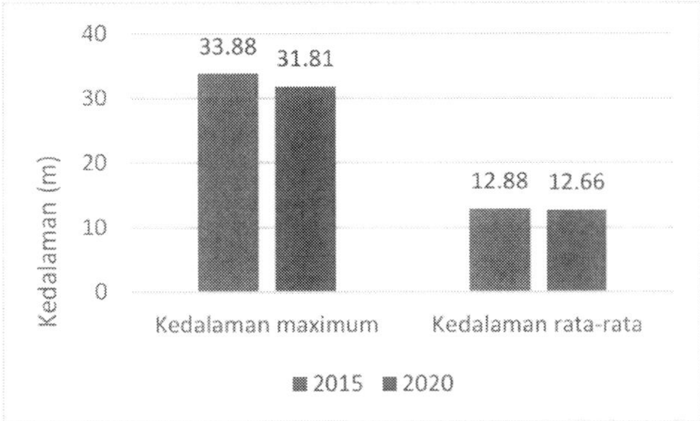

Figure 4. The Maximum and Average Depth Changes of Tondano Lake

Figure 4 shows a change in depth, in 2015 the maximum depth was 31.81 meters and the average depth was 12.88 meters, while in 2020, the maximum depth was 30.22 meters and the average depth was 12.66 meters. The results of the bathymetry map study in 2010 by the Director-General of Natural Resources-BWS, in Wantasen et all [9], the maximum depth is 32.70 m and the average depth is 15.11 m.

4. Conclusions

The spatial analysis of bathymetry maps in 2015 and 2020, it is known that the rate of silting of Tondano Lake is 4.4 cm every year. The results of previous studies explained that the rate of siltation was 13.41 cm per year (2010) in the analysis of the bathymetry map of Tondano Lake 2004-2010 (Directorate General of Natural Resources-BWS in Wantasen et all) [9]. Murdiyanto and Sulastriningsih in 2018 [5], found that there was no significant change in the depth of the lake bottom in the last 5 years, as predicted by previous researchers who stated that the process of silting the lake was 0.43 m per year.

5. References

[1] Sudrajat A and Bintoro A 2016 Pengukuran konsentrasi ortofosfat di danau tondano *Bul. Tek. Litkayasa* **14** 127–33
[2] Sorey T D, Bisri M and Dian S 2016 Kajian Penanganan Sedimentasi Danau Tondano *J. Tek. Pengair.* **7** 259–67
[3] Hikmatullah, H S, U K and I A L 2000 Evaluasi Erosi dan Siltasi Danau Tondano Berdasarkan Sifat Hidrologi dan Tanah dari Daerah Aliran Sungainya *J. Tanah dan Iklim* **18** 13–25
[4] Zen M T and A M 1974 Tondano Pyroclastics 7–19
[5] Murdiyanto and Sulastriningsih S 2018 Detection of Sediment Material Sources Based on the Morphometry Changes of Lake Tondano , Sulawesi Utara *Pros. Semin. Nas. Lingkung. Lahan Basah* **3** 342–6
[6] Aprilliyana D 2015 Pengaruh Perubahan Penggunaan Lahan Sub DAS Rawapening terhadap Erosi dan Sedimentasi Danau Rawapening *J. Pembang. Wil. Kota* **11** 103
[7] Tendean M, Moningkey A T and Kumaat J C 2020 Pemanfaatan Data Hidro Oseanografi dan Batimetri Untuk Penataan Pantai Tatapaan, Minahasa Selatan *J. Episentrum* **1** 1
[8] Kumaat J C, Haluan J, Wiryawan B, Hariwisudo S and Monintja D 2012 Model Perencanaan Kawasan Perikanan Tangkap Berbasis Konservasi Laut di Kabupaten Kepulauan Sitaro *Sci. Rep.*
[9] Kementerian Lingkungan Hidup Republik Indonesia 2014 *Gerakan Penyelamatan Danau (GERMADAN) Tondano*

Evaluation of land suitability for Coffee (*Coffea sp.*) in Nawangan District, Pacitan Regency, East Java

D E Rahmatika[1]*, Supriyadi[1], A Herawati[1], H Widijanto[1], M Nufus [2] and G Masyithoh[2]

[1] Department of Soil Science, Faculty of Agriculture, Universitas Sebelas Maret, Surakarta, Indonesia

[2] Department of Forest Management, Faculty of Agriculture, Universitas Sebelas Maret, Surakarta, Indonesia

*Email: Devvesya@gmail.com

Abstract. Nawangan District has been hit by clove crop failures for years. Due to this incident, alternative plant replacements are needed following the land conditions in the area. Coffee (*Coffea sp.*) is one of the relevant choices. Land evaluation needs to be done to determine the land suitability class. This study aims to determine the land suitability class of coffee plants in Nawangan District. The type of coffee used in this research is *Coffea arabica* and *Coffea canephora*. This study uses the matching method with the minimum law on each land unit. Based on data obtained from field surveys and laboratory analysis, the actual land suitability for the two types of coffee is S3 (marginal), with various limiting factors. *Coffea arabica* has a limiting factor: temperature, water availability, rooting condition, oxygen availability, nutrient retention, and nutrient availability. The limiting factors of *Coffea canephora* are rooting condition, oxygen availability, nutrient retention, and nutrient availability. The improvement to decrease the limiting factors are irrigation, liming, composting, fertilizing, and adding organic matter. The potential land suitability class for *Coffea arabica* after improvement efforts was S3, while for *Coffea canephora* was S2 and S3. So based on the smallest limiting factor, the appropriate type of coffee in Nawangan District is *Coffea canephora*.

1. Introduction

Climate change is an unavoidable result of global warming and has a real impact on all aspects of life, especially the agricultural sector. From time to time, it has been predicted that the development of the agricultural sector will face major problems, including: 1) declining productivity, 2) degradation of land and water resources, 3) diversity and climate change [1] . One of the impacts of climate change occurs in Nawangan District, Pacitan Regency. In the area, there has been a decline in production and crop failure of clove plants since 2017, which at that time became a leading commodity. This incident certainly caused a lot of losses, for that it's necessary to develop alternative plants that are in accordance with the current climate and land characteristics in the area.

Coffee is a plant with great potential in the economic aspect, which is Indonesia's leading export commodity. Based on these, coffee is one of the plants that is recommended to be an alternative crop option for areas in Nawangan District, Pacitan Regency. Indonesia has several types of coffee that are *Coffea canephora*, *Coffea arabica*, and *Coffea liberica* [2]. This is because each type of coffee has different characteristics and growing conditions. *Coffea arabica* grows at 1000 to 1500 meters above sea level; *Coffea canephora* can grow at an altitude of 40 to 900 meters above sea level, while *Coffea canephora* is generally grown on peatlands [3]. In general, coffee can be grown in areas with rainfall of 1500 to 3500 mm in a year and a maximum dry month of three months. Some of the things that distinguish the requirements for growing *Coffea arabica* and *Coffea canephora* are temperature and nutrient retention. The appropriate temperature for *Coffea arabica* is 16-20^{0}C, while for *Coffea canephora* is 20-24^{0}C. Moreover, soil acidity suitable for *Coffea arabica* plants is 5.6 to 6.6, while for *Coffea canephora* is from 5.3 to 6.0 [4]. Soil characteristics including soil texture, CEC, soil acidity, soil organic carbon, base saturation, soil nitrogen total, soil available potassium, and soil available phosphorus influence the quality of the coffee, so this is important to pay attention to in cultivation [5–7].

Pacitan Regency is an agricultural area with various plant commodities [8]. This research was conducted in Nawangan District, Pacitan Regency, East Java. In this case, coffee becomes an alternative crop that allows it to be empowered in the region. Based on the condition of the research area, the types of coffee that are possible to be planted are *Coffea arabica* and *Coffea canephora*. Knowing land suitability in land use is the first step to designing a land management system so that the potential and limitations of the land can be defined [9,10].

The development of alternative crops needs the necessary data on land characteristics that can be determined by evaluating land suitability. Appropriate and careful planning is needed to make decisions based on plant suitability so that optimum production is obtained. Land suitability class assessment is carried out based on the criteria for the physical and chemical properties of the land. The purpose of this study was to determine the land suitability class for *Coffea arabica* and *Coffea canephora* in Nawangan District.

2. Materials and methods

2.1. Study site

This research was conducted in Nawangan District, Pacitan Regency. Nawangan District is located at 7^{0}59'26.50" S dan 111^{0}10'37.89" E. There are 9 villages in Nawangan District, that are Nawangan, Tokawi, Penggung, Gondang, Mujing, Sempu, Pakis Baru, Jetis Lor, dan Ngromo Villages [11]. Nawangan District is a highland and mountainous area located at an altitude of 500-1000 meters above sea level. Determination of the sampling location based on the Land Unit Map. The land unit map is determined using the overlay method from thematic maps, including RBI maps, slope maps, geological maps, land use maps, and soil types in Nawangan District. From the results of overlaying these maps, fifteen land units were obtained. The area used in the research is plantation and moorland, with slope of about <8 and 8-15%. The type of soil in the research site is litosol and complex of reddish latosol and litosol. At the same time, the soil analysis was carried out in the physics and chemistry laboratory of Soil Science Study Program, Faculty of Agriculture, Sebelas Maret University.

2.2. Materials

The tools used are a digital camera, roll meter, Munsell Soil Colour Chart, compass, GPS, clinometer, paper pH indicator, hoe, flacon, pipette, soil boring, ring sample, drik, and paper for field observations. The tools used to compose the Land Unit Map are a computer that has ArcMap GIS 10.4 installed on it, RBI maps, rainfall maps, land use maps, slope maps, geological maps, and soil type maps. The chemicals used are H_2O; NaF; HCl 1,2 N; H_2O_2 10%; KCNS 10%; HCl 10%; $K_4Fe(CN)_6$ 0,5%; for soil analysis in field. Furthermore, H_2O; alcohol; ammonium acetate; concentrated NaOH; NaCl; HCl 0.1 N; Zn; H_3BO_3 storage solution; $K_2Cr_2O_7$; Concentrated H_2SO_4; Olsen's solution; $CuSO_4.K_2SO_4$; ammonium molybdate; $SnCl_2$; Bray and Kurt 1 extraction and standard solution for soil analysis in the laboratory.

2.3. Soil Analysis

The research was conducted with an exploratory descriptive through a survey approach. Methods survey was conducted by direct observation in the field and taking soil samples. The technique of determining the sample point in this study was carried out by purposive random sampling, taking samples randomly based on certain adjusted criteria [12]. Before conducting laboratory soil analysis, soil wind dried and sieved in accordance with the requirements in advance. Soil pH analysis using an electrometric method, pH determination in this method is based on weak acid or weak base [13]. Determination of the percentage of C-organic using the Walkey and Black method by oxidizing organic carbon in soil samples in an acidic state [14,15]. Analysis of total soil N using the Kjeldahl method this method is based on the color reaction between $NH4^{+}$ and a weak base mixture of Na salicylate and chlorine [16]. Available P analysis uses Olsen and Bray 1 methods. The Olsen method is used for slightly acidic to alkaline soils. The Bray and Kurt 1 method is used for neutral to acidic soils. [17]. Available K, cation exchange capacity, and base saturation analysis using the extraction of ammonium acetate 1 N pH 7,0 method by knowing the results of the extraction of ions in the soil [18]. Soil texture analysis using the pipette method uses a micropipette based on deposition at a certain time [19].

2.4. Data Analysis
Data analysis used matching system by matching land characteristics with land use requirement for *Coffea arabica* and *Coffea canephora*. After knowing the limiting factors on each land unit, then the improvement efforts that can be applied are determined. Actual land suitability uses land characteristic data based on environmental conditions and soil analysis, while potential land suitability uses land characteristic data that has been applied improvement efforts. The analysis showed land suitability classes in high (S1), moderate (S2) and marginal (S3) suitability, as well as unsuitable (N) [20]. Based on an assessment of actual and potential land suitability, land suitability maps created by using overlay method with GIS technology [21].

3. Result and discussion

3.1. Land characteristic
Land characteristics were observed in the study include temperature, water availability, oxygen availability, rooting condition, nutrient retention, erosion, flood, and land preparation. The table below (Table 1.) shows the results of observations and soil analysis on each land unit.

The research location is located at an elevation of 645-1,034 m above sea level with an air temperature of 20.10 ^{0}C to 22.43 ^{0}C. Water availability is determined by rainfall, long dry months, and humidity. Nawangan District has an average rainfall of 1,767mm/year, the dry month for 4.6 months, and humidity of 81.86%. Climate elements such as temperature, rainfall and humidity can affect agricultural yields qualitatively and quantitatively [22,23].

The drainage conditions at the research site are medium well-drained and somewhat poor drainage because the soil is dominated by micropores, so that the movement of water and air in the soil is slow, but that's more able to hold water. Soil drainage class indicates the extent and occurrence of when the soil is saturated with water [24].

The soil in all land units has a fine and slightly fine texture and has a rough material of <15%. Soil texture has an effect on a nutrient circulation, biomass activity, and differences in the ability to store nutrients. This is caused by differences in the nature of the particle surface area of different soil [25]. The good percentage of rough material for plants is <15%, too much coarse material causes a decrease in the ability of the soil to hold water [26]. Land units 2, 3, and 6 have a 50-75 cm soil depth, while other land units have a 75-100 cm soil depth. Effective soil depth is related to plant roots, and coffee plants need a soil depth of more than 50 cm because it has a deep root.

Nutrient retention was observed are cation exchange capacity (CEC), base saturation, pH, and C-organic. The soil has a CEC value of 8.77 cmol/kg to 22.12 cmol/kg. This means that the research location has a low to moderate CEC value. CEC describes the ability of the soil to bind soil cations on the colloidal surface of the soil. The bound cations include aluminum (Al), hydrogen (H), ammonium (NH4), calcium (Ca), potassium (K), sodium (Na) and magnesium (Mg). CEC is influenced by the amount of clay, organic matter, and mineral composition [27].

IOP Conf. Series: Earth and Environmental Science 986 (2022) 012039 doi:10.1088/1755-1315/986/1/012039

Table 1. Land characteristics in each land unit

Land Characteristics	Land Unit														
	1	2	3	4	5	6	7	8	9	10	11	12	13	14	15
Temperature coefficient (tc)															
Temperature (°C)	23,38	22,40	22,40	22,11	22,11	22,33	22,08	22,08	22,10	22,10	22,26	22,15	22,43	20,13	22,12
Elevation (masl)	653	650	650	698	662	703	703	1,034	1,034	1,034	674	692	645	1,029	1,030
Water availability (wa)															
Rainfall (mm)	1767.2	1767.2	1767.2	1767.2	1767.2	1767.2	1767.2	1767.2	1767.2	1767.2	1767.2	1767.2	1767.2	1767.2	1767.2
Dry month (month)	4,6	4,6	4,6	4,6	4,6	4,6	4,6	4,6	4,6	4,6	4,6	4,6	4,6	4,6	4,6
Humidity (%)	81,86	81,86	81,86	81,86	81,86	81,86	81,86	81,86	81,86	81,86	81,86	81,86	81,86	81,86	81,86
Oxygen availability (oa)															
Drainase	MWD	MWD	MWD	SPD	SPD	MWD	MWD	MWD	MWD	MWD	MWD	MWD	MWD	MWD	MWD
Rooting condition (rc)															
Texture	Clay	Clay	Clay	Clay	Clay	Clay	Clay	Clay	Silty clay loam	Silty clay loam	Silty clay loam	Silty clay loam	Silty clay loam	Clay	Clay
Rough material	<15	<15	<15	<15	<15	<15	<15	<15	<15	<15	<15	<15	<15	<15	<15
Soil depth (m)	75-100	50-75	50-75	75-100	50-75	75-100	75-100	75-100	75-100	75-100	75-100	75-100	50-75	75-100	75-100
Nutrion retention (nr)															
CEC (cmol/kg)	14.73	12.81	13.73	8.77	10.57	16.14	15.39	14.29	14.94	12.64	22.12	20.29	14.93	12.77	13.25
BS (%)	33.32	46.30	41.67	63.50	52.24	52.05	38.19	50.16	35.72	39.44	52.69	49.89	54.03	37.49	35.49
pH	5.20	5.36	5.54	6.61	6.67	6.51	5.51	5.90	5.8	5.8	6.3	6.7	6.6	6.2	6.1
C-Organic (%)	0.61	1.6	0.18	0.29	0.29	1.86	0.24	0.6	1.12	1.01	0.22	0.52	1.14	0.87	1.91
Nutrion availability (na)															
N total (%)	0.36	0.27	0.22	0.19	0.18	0.41	0.21	0.23	0.19	0.18	0.20	0.25	0.24	0.28	0.43
P availability (ppm)	0.33	0.32	0.94	0.73	0.73	0.74	0.73	0.77	0.77	0.70	0.75	0.75	0.79	0.74	0.76
K availability (cmol/kg)	0.13	0.09	0.08	0.11	0.10	0.14	0.08	0.08	0.10	0.09	0.07	0.10	0.08	0.20	0.08
Erosion hazard (eh)															
Slope (%)	8-15	8-15	8-15	8-15	8-15	<8	<8	<8	8-15	8-15	<8	8-15	8-15	8-15	8-15
Erosion hazard	Light-modium	Light-modium	Light-modium	Light-modium	Light-modium	Light-modium	Light-modium	Light-modium	Light-modium	Light-modium	Light-modium	Light-modium	Light-modium	Light-modium	Light-modium
Flood hazard (fh)															
Height (cm)	-	-	-	-	-	-	-	-	-	-	-	-	-	-	-
Long (day)	-	-	-	-	-	-	-	-	-	-	-	-	-	-	-
Land preparation (lp)															
Surface rocks (%)	<5	<5	<5	<5	<5	<5	<5	<5	<5	<5	<5	<5	<5	<5	<5
Rock outcrop (%)	<5	<5	<5	<5	<5	<5	<5	<5	<5	<5	<5	<5	<5	<5	<5

The soil has a low to high base saturation with a value of 33.32% to 63.50%. Base saturation is related to soil pH. Soil with a low pH generally has a low base saturation and otherwise. This is because, at low base saturation, the base cations will be reduced and be replaced by H^+ ions which cause the soil pH to decrease [28,29]. Based on Table 1, the pH of the soil is 5.20 to 6.7. This means Soil at the research site is acidic, slightly acidic, and neutral. The cause of soil acidity is H^+ and Al^{3+} ions concentration in the soil solution, H^+ and OH^- ions concentration, soil minerals, rainwater, and source rock [30]. Soil pH influences soil organic matter, controlling microbial activity in the soil and the availability of soil carbon for microbes [31].

Organic C content in soil is 0.6% to 1.91%, which can be categorized as very low to low. The low organic c content indicates the low production of organic matter at the research site. Organic C content may vary due to differences in the type and amount of vegetation on the land [32]. The high and low organic C content is also influenced by the decomposition process of organic matter [33].

The availability of nutrients were observed are N total, available P, and available K. N total in the soil is 0.18 to 0.43% which can be categorized as low to moderate. N total is influenced by the type and amount of organic matter. Organic matter will undergo further decomposition and mineralization, thereby affecting the levels of C and N in the soil [34]. The organic matter decomposed by soil microorganism to be converted into organic nitrogen [35]. The low content of N can occur because of the process of NO_3 leaching, denitrification NO_3 into N_2, $NH4^+$ volatilization into NH3-, fixed by clay minerals or consumed by soil microorganisms [36].

The available P at the research site is 0.32 ppm to 0.94 ppm, which is categorized as very low. The low content of available P is due to the source material that forms the soil, which is poor in P elements, and low P content in an organic matter [32]. The low P-availability is also caused by phosphate ions bound by Al and Fe. [37].

The available K is 0.07 cmol/kg to 0.20 cmol/kg, which is categorized as low to very low. Available K means the availability of potassium that can be exchanged and absorbed by plants. The high and low availability of K is influenced by soil mineral content. Weathering soil minerals will release the potassium that contained and release it for plant growth [27].

Erosion hazard is determined based on the slope and the potential for erosion. Based on Table 1, the research location has a slope of 3-8% and 8-15%. Erosion can be affected by slope, rainfall, soil type and land use[22,38]. Based on field observations, all land units have a mild to medium erosion hazard potential. Entisols and inceptisols with annual crops land use have a high risk of erosion [39].

The flood hazard is determined by the height of the inundation and the duration of the flood. The research location is a highland area, so that floods rarely occur. Despite the high rainfall, some water can be absorbed and retained by the soil because the soil texture is dominated by micropores, while some others flow into the runoff.

Land preparation is influenced by surface rocks and rock outcrops. Surface rock is the volume of rock found on the ground surface. Meanwhile, a rock outcrop is the volume of rock that appears on the ground surface. Overall the volume of surface rocks and rock outcrops in all land units is only about <5%.

3.2. Land suitability analysis

Land suitability for *Coffea arabica* and *Coffea canephora* land use in this study was determined by using a traditional system, that is by matching the quality of the land and with the needs of plants with qualitative diagnostic factors defined in the very suitable, quite suitable, marginal and not suitable categories [40]. Based on soil sample analysis and field observations, it is known that all land units have a suitability class of S3 for *Coffea arabica* and *Coffea canephora* with different limiting factors in each land unit. The table below (Table 2) shows the actual and potential land suitability level for land use of *Coffea canephora* and *Coffea arabica* in each land unit.

Table 2. Land suitability of *Coffea canephora* and *Coffea Arabica*

Land Unit	Caffea Arabica		Coffea canephora		Area (ha)
	Actual	Potential	Actual	Potential	
1	S3-tc, wa, nr, na	S3-tc, wa	S3-nr, na	S2-wa, rc	33.51
2	S3-tc, wa, rc, nr, na	S3-tc, wa, rc	S3-nr, na	S3-rc	1,880.17
3	S3-tc, wa, rc, nr, na	S3-tc, wa, rc	S3-rc, nr, na	S3-rc	289.23
4	S3-tc, wa, oa, nr, na	S3-tc, wa	S3-oa, nr, na	S2-wa, rc	814.30
5	S3-tc, wa, oa, nr, na	S3-tc, wa	S3-oa, nr, na	S2-wa, rc	646.89
6	S3-tc, wa, rc, na	S3-tc, wa, rc	S3-rc, na	S3-rc	143.01
7	S3-tc, wa, nr, na	S3-tc, wa	S3-nr, na	S2-wa, rc	26.27
8	S3-tc, wa, nr, na	S3-tc, wa	S3-nr, na	S2-wa, rc	94.14
9	S3-tc, wa, na	S3-wa	S3-na	S2-wa, rc	367.60
10	S3-tc, wa, na	S3-wa	S3-na	S2-wa, rc	238.80
11	S3-tc, wa, nr, na	S3-wa	S3-nr, na	S2-wa, rc	175.20
12	S3-tc, wa, nr, na	S3-tc, wa	S3-nr, na	S2-wa, rc	156.01
13	S3-tc, wa, rc, na	S3-tc, wa	S3-rc, nr, na	S3-rc	77.58
14	S3-tc, wa, na	S3-wa	S3-na	S2-wa, rc	32.80
15	S3-tc, wa, na	S3-wa	S3-na	S2-wa, rc	23.78

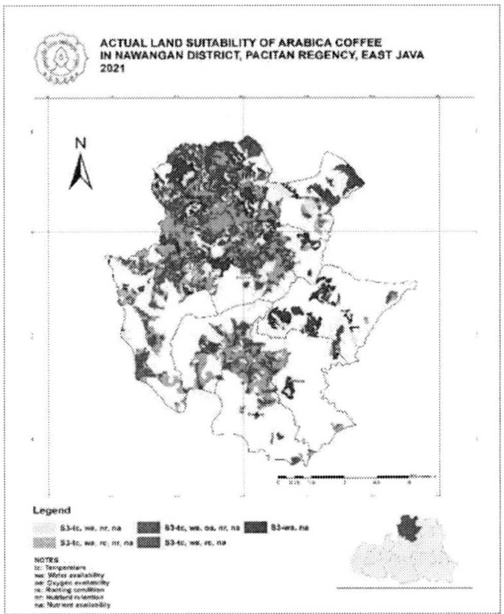

Figure 1. Actual land suitability map for *Coffea arabica*

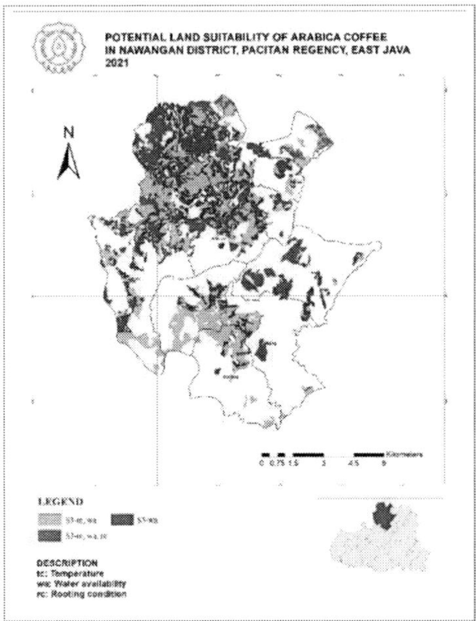

Figure 2. Potential land suitability map for *Coffea arabica*

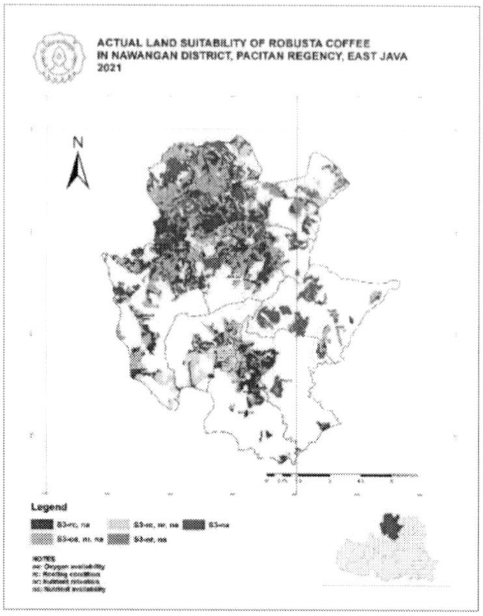

Figure 3. Actual land suitability map for *Coffea canephora*

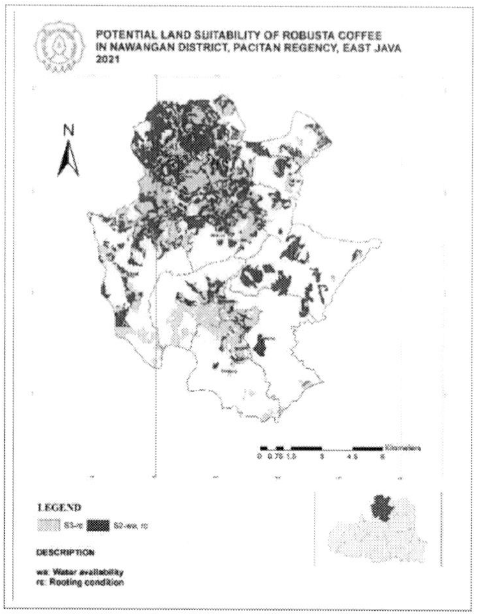

Figure 4. Potential land suitability map for *Coffea canephora*

The actual land suitability of *Coffea arabica* is marginally suitable (S3), that are S3-tc,wa,nr,na with an area of 485.13 ha, S3-tc,wa,rc,nr,na with an area of 2,169.4 ha, S3-tc,wa,oa,nr,na with an area of 1,461.19 ha, S3-tc,wa,rc,na with an area of 220.59 ha and S3-wa na with an area of 662.98 ha. The actual land suitability of *Coffea canephora* is also marginally suitable (S3), S3-rc,na with an area of 2,023.18 ha, S3-oa,nr,na with an area of 1,461.19 ha, S3-rc,nr,na with an area of 366.81 ha, S3-nr,na with an area of 329.12 ha and S3-na with an area of 662.98 ha. After being given recommendations for improvement efforts, the result of potential land suitability for *Coffea arabica* is marginally suitable (S3), that are S3-tc wa with an area of 1,848.7 ha, S3-tc,wa,rc with an area of 2,312.41 ha dan S3-wa with an area of 838.18 ha. Meanwhile, the potential land suitability class for *Coffea canephora* are moderate suitable (S2) and marginally suitable (S3), that are S2-wa,rc with an area of 2,609.3 ha and S3-rc with an area of 2,398.99 ha.

3.3. Improvement efforts for decrease the limiting factors

Improvement efforts are applied to improve the quality of the land so that it is in accordance with the requirements for plant growth. Limiting factors are divided into permanent and non-permanent limiting factors. The limiting factor that can be improved is the non-permanent limiting factor; there are dry months, oxygen availability, nutrient retention, and nutrient availability. Irrigation canals or irrigation that we made can overcome the limiting factor for the length of the dry period. Irrigation can accommodate rainwater in the wet months so that the availability of water in the dry period is sufficient. Irrigation can also increase oxygen availability by improving the drainage system. Efforts to increase soil CEC can be made by fertilizing compost and adding zeolite. Soil pH, which tends to be acidic, can be improved by liming. The low content of C-organic can be increased by adding organic matter (compost). Available nutrient content includes N total, available P, and available K. Improvement efforts that can be applied are fertilizing with single or compound fertilizers, according to the needs of the land. To increase N nutrients can be done by fertilizing manure.

4. Conclusion

Based on the results, it can be concluded that the actual land suitability class for *Coffea arabica* is S3 with limiting factors temperature, water availability, rooting condition, oxygen availability, nutrient retention, and nutrient availability. While actual land suitability class *Coffea canephora* is S3 with limiting factors root media, nutrient retention, oxygen availability, and nutrient availability. The potential land suitability for Arabica coffee is S3, while for robusta coffee is S2 and S3. Improvement efforts were applied to decrease the limiting factors (dry months, oxygen availability, nutrient retention, and nutrient availability). The improvements that can be applied are making irrigation, liming, composting, fertilizing, and adding organic matter. The improvement efforts are adjusted to the needs of each land. According to environmental conditions (elevation, temperature, and humidity) and potential suitability, Nawangan District is more suitable for *Coffea canephora* than actual land suitability class. The results of the evaluation and recommendations for improvement can be used as a reference for the sustainable development of coffee plantations in Nawangan District.

Acknowledgment

The author would like to thank the community and government of Nawangan District, and all parties from Sebelas Maret University who have helped make this research successful, so that this publication can be carried out properly.

References

[1] Surmaini E, Runtunuwu E and Las I 2010 Upaya sektor pertanian dalam menghadapi perubahan iklim *J. Litbang Pertanian,* **30** 1–7
[2] Martauli E D 2018 Analisis Produksi Kopi Di Indonesia **01** 112–20
[3] Hulupi R and Martini E 2013 *Pedoman Budidaya dan Pemeliharaan Tanaman Kopi di Kebun Campur* (Bogor, Indonesia)
[4] Djaenudin D, H. M, H. S and Hidayat A 2011 *Petunjuk Teknis Evaluasi Lahan untuk Komoditas Pertanian*
[5] Mintesnot A, Dechassa N and Mohammed A 2015 Association of Arabica Coffee Quality Attributes with Selected Soil Chemical Properties *East African J. Sci.* **9** 73–84
[6] Yao R, Yang J and Zhang T 2013 Short-term effect of cultivation and crop rotation systems on soil quality indicators in a coastal newly reclaimed farming area *J. Soils Sediments* **13** 1335–50
[7] Castro-tanzi S, Dietsch T, Urena N, Vindas L and Chandler M 2012 Analysis of management and site factors to improve the sustainability of smallholder coffee production in Tarrazú, Costa Rica *"Agriculture, Ecosyst. Environ.* **155** 172–81
[8] Nita I, Putra A N and Fibrianingtyas A 2020 Analysis of drought hazards in agricultural land in Pacitan Regency , Indonesia *SAINS TANAH – J. Soil Sci. Agroclimatol.* **17** 7–15
[9] Nurfadila J S, Baja S, Neswati R and Rukmana D 2020 Evaluation of land suitability for coffee plants based on fuzzy logic in Enrekang district *IOP Conf. Ser. Earth Environ. Sci.* **486**
[10] Endar Hidayat, Asmak Afriliana, Gusmini H H 2020 LAND SUITABILITY EVALUATION OF COFFEE IN TOKUNOSHIMA ISLAND, JAPAN *J. Appl. Agric. Sci. Technol.* **4** 146–54
[11] Badan Pusat Statistik Kabupaten Pacitan 2019 *Kecamatan Nawangan Dalam Angka 2019* ed BPS Kabupaten Pacitan (Pacitan: BPS Kabupaten Pacitan)
[12] Cholil M, Anna A N and Setyaningsih N 2016 Analisis Kesadahan Air Tanah Di Kecamatan Toroh Kabupaten Grobogan Propinsi Jawa Tengah *3rd Univ. Res. Colloq.* 88–98
[13] Thomas G W 1996 Soil pH and Soil Acidity *Methods of Soil Analysis Part 3 - Chemical Methods* ed M E S D.L. Sparks, A.L. Page, P.A. Helmke, R.H. Loeppert, P. N. Soltanpour, M. A. Tabatabai, C. T. Johnston (Lexington, Kentucky: SSA Book) pp 475–90
[14] Pramod Jha, A. K. Biswas, Brij Lal Lakaria, R. Saha M S & A S R 2014 Predicting Total Organic Carbon Content of Soils from Walkley and Black Analysis Predicting Total Organic Carbon Content of Soils from Walkley and Black Analysis *Communications in Soil Science and Plant*

Analysis vol 45, ed Taylor & Francis (Bhopal) pp 713–25

[15] Darmawan D 2019 Evaluasi Status Bahan Organik Dan Sifat Fisik Tanah (Bulk Density, Tekstur, Suhu Tanah) Pada Lahan Tanaman Kopi (Coffea Sp.) di Beberapa Kecamatan Kabupaten Dairi *J. Chem. Inf. Model.* **53** 1689–99

[16] Baethgen W E and Alley M M 1989 A manual colorimetric procedure for measuring ammonium nitrogen in soil and plant Kjeldahl digests *Communications in Soil Science and Plant Analysis* vol 20 (Blacksburg: Taylor & Francis) pp 961–9

[17] Irving G C J and McLaughlin & M J 1990 Analysis A rapid and simple field test for phosphorus in Olsen and Bray No . 1 extracts of soil *Communications in Soil Science and Plant* vol 21 (Canberra: Taylor & Francis) pp 2245–55

[18] Raij B Van, Quaggio J A and Silva N M 1986 Extraction of phosphorus , potassium , calcium , and magnesium from soils by an ion - exchange resin procedure *Communications in Soil Science and Plant Analysis* vol 17 (Campinas) pp 547–66

[19] Miller W P and Miller D M 1987 A micro - pipette method for soil mechanical analysis *Communications in Soil Science and Plant Analysis* vol 18 (Georgia: Taylor & Francis) pp 1–15

[20] Ritung S, Suryani K N, Mulyani A and Suryani E 2011 *Petunjuk Teknis Evaluasi Lahan Untuk Komoditas Pertanian* (Bogor: Balai Besar Penelitian dan Pengembangan Sumberdaya Lahan Pertanian, Badan Penelitian dan Pengembangan Pertanian)

[21] Ennaji W, Barakat A, Baghdadi M El and Oumenskou H 2018 GIS-based multi-criteria land suitability analysis for sustainable agriculture in the northeast area of Tadla plain (Morocco) *J. Earth Syst. Sci.*

[22] Pusponegoro I H, Herawati A and Widijanto H 2019 Planning of banana plant development based on the land conservation aspect in Jenawi District *J. Degrad. Min. Lands Manag.* **5** 1319–26

[23] Herawati A, Rahayu, Herdiansyah G, Supriyadi and Wijayant R 2021 The impact of climate change on land suitability and cocoa productivity in Tulakan District , Pacitan Regency *IOP Conf. Ser. Earth Environ. Sci.* **824** 1–9

[24] Asgari N, Ayoubi S, Demattê J A M, Jafari A, Safanelli J L and Silveira A F D Da 2020 Digital mapping of soil drainage using remote sensing, DEM and soil color in a semiarid region of central Iran *Geoderma Reg.* **1** e00302

[25] Hamdi H, Hechmi S, Naceur M, Rahma I, Benzarti S, Mokni-tlili S, Hassen A and Jedidi N 2019 Repetitive land application of urban sewage sludge : Effect of amendment rates and soil texture on fertility and degradation parameters *Catena* **172** 11–20

[26] Mahfut T, Buchari H and Cahyono P 2015 Kandungan bahan kasar dan sifat fisik tanah ultisol di lahan perkebunan nanas terbanggi besar lampung tengah *J. Agrotek Trop.* **3** 155–9

[27] Riwandi, Prasetyo, Hasanudin and Cahyadinata I 2017 *Kesuburan Tanah Dan Pemupukan* ed A Munawar (Bengkulu: Yayasan Sobat Alam Rafflesia)

[28] Soewandita H 2008 Studi kesuburan tanah dan analisis kesesuian lahan untuk komoditas tanaman perkebunan di Kabupaten Bengkalis *J. Sains dan Teknol. Indones.* **10** 128–33

[29] Wilson W, Supriadi S and Guchi H 2015 Evaluasi Sifat Kimia Tanah Pada Lahan Kopi Di Kabupaten Mandailing Natal *J. Agroekoteknologi Univ. Sumatera Utara* **3** 104299

[30] Prabowo R and Subantoro R 2017 Analisis Tanah Sebagai Indikator Tingkat Kesuburan Lahan Budidaya Pertanian Di Kota Semarang *J. Ilm. Cendekia Eksakta* **2** 59–64

[31] Stark S, Männistö M K and Eskelinen A 2014 Nutrient availability and pH jointly constrain microbial extracellular enzyme activities in nutrient-poor tundra soils **383** 373–85

[32] Rahmi A and Biantary M P 2014 Karakteristik Sifat Kimia Tanah Dan Status Kesuburan Tanah Lahan Pekarangan Dan Lahan Usaha Tani Beberapa Kampung Di Kabupaten Kutai Barat *Ziraa'ah* **39** 30–6

[33] Wang X, Butterly C R, Baldock J A and Tang C 2017 Science of the Total Environment Long-term stabilization of crop residues and soil organic carbon affected by residue quality and initial soil pH *Sci. Total Environ.* **8** 502–9

[34] Farrasati R, Pradiko I, Rahutomo S, Sutarta E S, Santoso H and Hidayat F 2020 C-organik Tanah di Perkebunan Kelapa Sawit Sumatera Utara: Status dan Hubungan dengan Beberapa Sifat Kimia Tanah *J. Tanah dan Iklim* **43** 157

[35] Mujiyo, Setyawan Y Y, Herawati A and Widijanto H 2021 The effect of land use on soil quality in Giriwoyo Sub-district , Wonogiri Regency *J. Degrad. Min. LANDS Manag.* **8** 2559–68

[36] Rusdiana O 2012 Pendugaan Korelasi Antara Karakteristik Tanah Terhadap Cadangan Karbon (Carbon Stock) Pada Hutan Sekunder *J. Silvikultur Trop.* **3** 14–21

[37] Lemanowicz J 2018 Dynamics of phosphorus content and the activity of phosphatase in forest soil in the sustained nitrogen compounds emissions zone *Environ. Sci. Pollut. Res.* **25** 33773–82

[38] Herawati A, Mujiyo, Widijanto H, Isnaini and Romdhati F 2018 Land conservation on banana cultivation based on erosion risks level at Land conservation on banana cultivation based on erosion risks level at Jenawi District-Karanganyar , Indonesia *Int. Conf. Clim. Chang.* **200** 1–8

[39] Herawati A, Suntoro, Widijanto H, Pusponegoro, Sutopo N R and Mujiyo 2018 Soil degradation level under particular annual rainfall at Jenawi District – Karanganyar , Indonesia *IOP Conf. Ser. Earth Environ. Sci.* **129** 1–7

[40] Akpoti K, Kabo-bah A T, Zwart S J, Rice A and Ivoire C 2019 Agricultural land suitability analysis : State-of-the-art and outlooks for integration of climate change analysis *Agric. Syst. J.* **173** 172–208

Is psychological adaptability needed for high school students in disaster-prone areas of Mount Merapi Eruption?

S H N Hafida[1*], N K M Isa[2], Jumadi[3] and N Khotimah[4]

[1] Department of Geography Education, Faculty of Teacher Training and Education, Universitas Muhamamdiyah Surakarta, Indonesia
[2] Department of Geography, Faculty of Human Science, Universiti Pendidikan Sultan Idris, Malaysia
[3] Faculty of Geography, Universitas Muhamamdiyah Surakarta, Indonesia
[4] Department of Geography Education, Faculty of Social Sciences, Universitas Negeri Yogyakarta, Indonesia

*Email: shnh421@ums.ac.id

Abstract. One of the disaster events that has a clear cycle is the eruption of Mount Merapi. Despite having a clear cycle, students are still one of the vulnerable groups during the eruption of Mount Merapi. The vulnerability of students will ultimately affect the psychological condition of students. Many researchers have found that the psychological condition of students who experience a disaster will return to normal in a long period of time, around 4-5 years. Therefore, this study aims to determine whether psychological adaptation skills are needed for high school students in areas prone of the Mount Merapi. The case study design was used in this study to determine students' perceptions of their psychological adaptability. The number of respondents used was 364 students that spread across 15 high schools and located in disaster-prone areas 1 to 3 of Mount Merapi. Questionnaires were used as a data collection technique, and data analysis was carried out using descriptive analysis. The analysis result shows that about 56.20% of respondents feel the need to have disaster adaptation capabilities. The existence of disaster adaptation is expected to help respondents who do not know good stress management when dealing with the eruption of Mount Merapi. Psychological disaster adaptation will affect the readiness of students t deal with disasters. The existence of psychological adaptation will encourage high school students to become a generation that is resilient to disaster because with psychological adaptation, students will more easily adjust their psychological conditions when a disaster occurs.

1. Introduction

Students are one of the vulnerable groups in natural disasters. This vulnerability is because students' knowledge and understanding of natural disasters are low [1]. According to [2], students in Indonesia still have an inadequate disaster preparedness and ultimately encourages them to become victims of natural disasters. This is certainly not in accordance with the location of Indonesia which is in a disaster-prone area. According to the calculation of the disaster index by Maplecroft in 2010, Indonesia is one of the countries most vulnerable to natural disasters after Bangladesh [3]. Despite being in an area with the highest level of disaster vulnerability, the implementation of disasters for school students in Indonesia is still limited.

One of the natural disasters that often occurs in Indonesia is volcanic eruptions. Indonesia is located at the confluence of three active tectonic plates, thus encouraging many active volcanoes. Several catastrophic volcanic eruptions in Indonesia have even affected the lives of the world's people, such as the eruptions of Mount Tambora (1815) and Mount Krakatau (1883). Volcanic eruptions have a clear cycle compared to other types of natural disasters; therefore, students should be able to have better resilience to eruption disasters. One active volcano with a clear cycle (4 to 6 years) is Mount Merapi [4,5]. Even though it is located in an area prone to eruption disasters, the slopes of Mount Merapi are still actively used by the community to carry out their daily activities. Given the frequent occurrence of disasters at Mount Merapi, the community must have a better understanding of disasters because disasters can impact all age groups, including school students [6].

One sector of activity that has experienced the impact of the eruption of Mount Merapi is the education sector. There are 15 Senior High Schools (SMA) located on the slopes of Mount Merapi [7] and are included in the disaster-prone areas of Mount Merapi I to III. Disaster-prone areas are a grouping of areas based on their vulnerability to the eruption of Mount Merapi. Disaster-prone area I has a slight eruption impact, and disaster-prone area II has a moderate hazard risk due to the eruption of Mount Merapi. In contrast, disaster-prone area III is an area with the highest hazard risk. Each of these areas has different characteristics, so there needs to be an understanding to adapt to its characteristics.

Many high schools on the slopes of Mount Merapi ultimately encourage various efforts to improve disaster preparedness for school students on the slopes of Mount Merapi. The efforts that have been made so far have not been able to encourage disaster impact reduction optimally. This is due to the many changes that occur, both in terms of policies, environmental conditions, or community conditions; therefore, students must adapt to any changes that exist. This adjustment is referred to as the disaster adaptation process. This is in accordance with [8] statement that adaptation is vital because of the many changes that occur, such as government and non-government policies, socio-economic conditions, and cultural values in the context of disasters.

Disaster adaptation is different from disaster risk reduction or disaster resilience; however, the three aspects have a very close relationship. Disaster risk reduction efforts are part of the adaptation process [9], while resilience is often unstable, constantly changing, and requires adaptation to external and internal resources to minimize the impact of disasters. Therefore, adaptation is a process of adjustment of each individual [10]. The same thing was stated by [11], which says that adaptation is a process, action, and ability of an individual or a system to adapt to changes in the environment through social learning. Disaster adaptive capacity will not be realized suddenly; it takes a long time to achieve good adaptive capacity. This process is a complex process of activities, actions, decisions, and attitudes that reflect existing social norms and processes [12,13].

The adaptability of each individual will vary depending on the location of the region [14,15]. A good understanding of adaptive capacity will have a long-term effect on each individual to help realize better mitigation and prevention efforts against volcanic activity [16]. With adaptive practices, students will have anticipatory actions carried out to reduce the impact of disasters and increase their ability to respond, recover, and return to a better condition [8]. [17] also stated that good adaptation would benefit the mobilization of knowledge and a more comprehensive institutional network for each individual.

There are two forms of disaster adaptation, namely behavioral and psychological adaptation. Behavioral adaptation is the first response that humans will make in the face of change [18,19] so that behavioral adaptation becomes an exclusive adaptation for someone who has a high level of knowledge so that they can act according to their understanding [20]. In contrast, psychological adaptation will be related to processing negative attitudes during disaster events [20]. Psychological adaptation can help individuals describe and explain the causes of a disaster event and inform efforts to minimize the impact of disasters [21].

There are many forms of behavioral adaptation that communities can carry out in eruption-prone areas [16], while psychological adaptation is still very rarely considered. This is because it is easier to adapt behavior compared to psychological adaptation, whereas on the other hand, psychological adaptation is more important for people living in disaster-prone areas [20]. High school students often experience psychological and emotional trauma [22] due to eruption disasters; therefore, psychological adaptation aims to improve students' psychological abilities because, during disaster situations, students' psychological conditions will be vulnerable to disturbances [23–25]. The trauma experienced by students during a disaster will encourage various negative impacts on students. A psychological-based approach to disaster needs to be considered for disaster development, especially the eruption of Mount Merapi. Although students have gained a lot of understanding about eruption disasters, it is often difficult for students to make decisions because they do not have suitable adjustments to the surrounding conditions. This study aims to analyze the level of psychological adaptation needs based on high school students' perceptions in disaster-prone areas of Mount Merapi.

2. Method

This research was conducted using a quantitative approach and survey design. According to [26], survey research is conducted to generalize the data results from a sample to the population so that the conclusions that will be obtained will be more valid. This study uses a survey design to determine students' perceptions of their psychological adaptability before, during, and after the eruption of Mount Merapi occurred. This study was conducted on high school students in 15 selected schools (Table 1). The selection of schools was based on the location of schools in disaster-prone areas I to III of Mount Merapi. Three hundred sixty-four high school students are respondents in this study.

Table 1. Research Location

No	School Name	Disaster Prone Areas	No	School Name	Disaster Prone Areas
1	SMA Muhammadiyah 1 Muntilan	I	9	SMK Dirgantara Putra Bangsa	I
2	SMK N 1 Salam	I	10	SMA Islam Al Azhar 9 Yogyakarta	I
3	SMA Maarif Srumbung	I	11	SMA N 1 Dukun	II
4	SMK Muhammadiyah 1 Salam	I	12	SMK Muhammadiyah Dukun	II
5	SMK Muda Patria Sleman	I	13	SMA Sunan Kalijogo	III
6	SMA N 1 Kalasan	I	14	SMK Muhammadiyah Cangkringan	III
7	SMK Muhammadiyah Mlati	I	15	SMK Kanisius 1 Pakem	III
8	SMK Kesehatan Binatama	I			

Data collection techniques in this study were conducted using a questionnaire. According to [27], questionnaires can be used to collect various information simultaneously. Therefore, the researcher used a questionnaire as the main instrument in this study. Questionnaires are used to determine respondents' responses to their psychological adaptations. In contrast, the analysis technique uses descriptive statistics. [26] states that descriptive statistics is an analytical technique that does not depend on the relationship of the data in the analysis. The researcher feels that this analytical technique is suitable for use in this study. Descriptive statistical analysis technique was used to determine the respondents' answers to the problem of psychological adaptation for students in disaster-prone areas of Mount Merapi.

3. Result and Discussion

Psychological adaptation ability for high school students is essential to understand because of the many psychological and emotional impacts experienced by students after experiencing a disaster. Several questions related to psychological adaptability can be seen in Table 2. These questions focus on the mental and social conditions of a student before, during, and after the eruption of Mount Merapi occurred.

Table 2. Respondents' Responses Regarding Psychological Adaptation

No	Questions	Percentage (%)				
		Strongly Disagree	Disagree	Doubtful	Agree	Strongly Agree
1	I feel confident in my ability to deal with the disaster situation of the eruption of Mount Merapi	2.47	7.97	32.97	32.69	23.90
2	I do not feel afraid when facing the eruption of Mount Merapi	5.77	10.71	32.69	28.57	22.25
3	I can maintain a good/ positive feeling state during the disaster situation of the eruption of Mount Merapi	2.20	7.14	29.12	34.07	27.47
4	I stay calm when I am in a disaster situation of the eruption of Mount Merapi	5.49	11.54	27.75	28.57	26.65
5	I know what I can do to calm myself down during the disaster situation of Mount Merapi eruption	1.65	8.24	28.30	30.49	31.32
6	If I am in a disaster situation of the eruption of Mount Merapi, I know how to respond to that situation	1.37	5.77	29.95	35.16	27.75
7	If people around me experience stress due to the eruption of Mount Merapi, I can calm them down	4.40	15.66	35.44	23.63	20.88
8	I realized when I was experiencing anxiety and stress due to the eruption of Mount Merapi	4.12	13.19	32.69	26.10	23.90
9	I have mentally prepared for the worst situation that I will encounter during the eruption of Mount Merapi	1.65	6.04	29.95	32.42	29.95
Total		29.12 (3.24)	86.26 (9.58)	278.85 (30.98)	271.70 (30.19)	234.07 (26.01)

Based on Table 2, it can be seen that many respondents answered the question with a doubtful answer (30.98%). This certainly shows that high school students still do not understand their psychological condition well. The highest percentage of uncertain answers (35.44%) is at question 7, where students do not know how to calm the people around them when a disaster occurs. The percentage of other doubtful answers is indicated in point 1 (confidence in their ability when dealing with disaster situations), point 2 (fear when facing disasters), and point 8 (anxiety and stress due to catastrophe). The high ignorance of students in calming the people around them during a disaster situation can be influenced by their ignorance of their mental abilities. When students do not understand their mental state well, they will find it difficult to help those around them. This is in accordance with [28] statement that reflective adaptation will be necessary because of individual fears and anxieties that need to be resolved in the communication process.

Individuals who have a good psychological adaptation capacity will pay attention to relationships in a particular community [20] so that it will be easier for them to calm the people around them when the disaster situation of the eruption of Mount Merapi occurs. In accordance with the opinion of [29] then, with social support, subjective support, and support utilization, people affected by disasters will find it easier to recover psychologically. In addition, adaptability will be realized if there is consistent role allocation, sustainable role relationships, and improvised behavior from each individual [30]. This can be the basis for the development of psychological adaptability. A clear division of roles between each individual (including school students) will help them carry out a better role relationship. Thus, to realize the ability of students to calm the people around them, it can be done with an understanding of the appropriate division of roles first.

Related to point 1, namely confidence in their abilities when facing disaster situations, high school students in disaster-prone areas of Mount Merapi still do not have confidence in their abilities. This is contrary to [31] that people who live in disaster-prone areas will have high confidence in their abilities; this is due to the experience they have while living in the area. To increase confidence in their abilities, students need to be accustomed to working together in a group so that they can increase knowledge [32] and familiarize students with solving problems around them [33].

Point 2 in table 2 shows that students do not understand well the fear they have when facing disaster situations. This will encourage the psychological adaptation of students will not to be realized. [34] stated that one of the negative emotions that school students will feel is fear, so that students who understand their fear well will find it easier to adapt. The same thing was conveyed through the [33] that one of the normal reactions seen in students is fear if disaster will happen again. Based on this, high school students in disaster-prone areas of Mount Merapi have not understood their mental conditions well, especially fear.

Related to point 8, when students do not know the anxiety and stress they experience, the ability to adapt has not been realized. This is in accordance with [17] statement that adaptation must always pay attention to various causes of stress, both economic, social, cultural, and technical factors. The existence of an understanding of the factors that cause pressure will undoubtedly help students to more easily control their emotions; however, if students do not know the factors that cause stress, they will find it difficult to control their emotions.

High school students in disaster-prone areas of Mount Merapi have an excellent perception on point 3 (the ability to maintain a good mood during a disaster situation), point 4 (calm in dealing with disaster situations), point 5 (how to calm down), point 6 (understanding in responding to a disaster situation), and point 9 (mental preparation for the worst condition). The excellent ability of these students is influenced by the habits that have been done for years. The frequent occurrence of Mount Merapi eruptions certainly encourages various efforts that the government and the community have made to minimize the impact of disasters. These habits can encourage the realization of psychological adaptation capacity. This is in accordance with the statement of [9,16] that adaptation will be achieved if there are repeated learning efforts about the surrounding disasters.

Different things were conveyed by [10] that psychological adaptability must be adjusted to the learning capacity and depth of the material so that students are easier to adjust. In addition, adaptive capacity is not always related to an individual's level of education and years of schooling [35]. What is needed is good knowledge. [20] stated that knowledge has a close relationship with psychological adaptation to process negative emotions and behavioral changes related to the causes of disasters. In addition to the knowledge factor, adaptability can be obtained through access to various sources, networks, and positive perspectives related to justice in society [17]. [36] also added that relationships with other organizations would facilitate the transfer of knowledge to develop the flexibility to deal with disasters. Adaptive capacity in disasters needs to integrate a set of outcomes and process-oriented characteristics to more adequately measure social resilience [35].

To find out the value and percentage of psychological adaptation needs for high school students in disaster-prone areas of Mount Merapi, see Table 3.

Table 3. Percentage of Psychological Adaptation Needs for Students

Category	Value	Percentage (%)
No need	46.67	12.82
Doubtful	112.78	30.98
Need	204.56	56.20

Table 3 shows that high school students in disaster-prone areas of Mount Merapi have a high percentage of the need for psychological adaptation, reaching 56.20%. The existence of this increased need is in accordance with [37], which states that it will be easier to realize adaptation efforts from the needs of an individual. This high need also shows that adjustment to psychological conditions does need to be considered, especially for students. Students are often victims who need protection in disasters [38]. One of the impacts of disasters that students will feel is a sense of trauma. [39] stated that students who experienced psychological trauma during a catastrophe were more likely to experience high blood pressure for up to 4 years after the disaster occurred. Seeing this, students need to pay attention to their psychological conditions to survive into adulthood. The existence of disaster adaptability will be beneficial for the community before, during, and after a disaster occurs [8], mainly psychological adaptation.

Psychological adaptation can help individuals process negative emotions into more positive ones [20] to change society towards sustainability [8,40]. This level of adaptation at the individual level will be closely related to the outcome of the mixing of self-interest and prosocial motives (care) [20]. Psychological adaptation will help students to better understand the feelings and conditions of others when a disaster situation occurs. Adaptation focuses on reducing the adverse effects of a disaster and maximizing potential opportunities in response to existing disturbances [11]. Therefore, psychological adaptation is expected to minimize psychological disruptions felt by students during the eruption of Mount Merapi. Psychological adaptation will not happen suddenly; it needs a long process of continuous response to create resilience in disasters [11]. This is in accordance with [41] statement that adaptation at the individual level is one of the effective efforts in increasing disaster resilience.

4. Conclusion

The need for disaster adaptability will increase in line with the external changes that occur. Efforts to minimize the impact of disasters cannot be carried out statically; however, there need to be adjustments to existing modifications through adaptation. Disaster adaptability is an essential part of the disaster management process, so that this disaster adaptation will encourage people to more easily achieve disaster resilience. Disaster adaptation consists of two types, namely behavioral and psychological adaptation. However, psychological adaptation capacity becomes essential to discuss because the impact of disasters felt by students is related to fear and trauma. Both of these effects will not be eliminated suddenly without the ability to adapt psychologically. The existence of psychological adaptability will help students understand the mental condition and feelings of themselves and others so that trauma can be more easily minimized.

Based on the results of data analysis, it is known that 56.20% of students feel the need for psychological adaptation skills for students in areas prone to the eruption of Mount Merapi. With the psychological adaptation ability, it is hoped that students will have more sympathy and empathy for the people around them. Psychological adaptation will not be realized in a short time; it needs a gradual process until a student can adapt psychologically to the eruption of Mount Merapi. Therefore, a fundamental role in developing psychological adaptation for students needs to be carried out by the government, schools, and society in general.

References

[1] Hermon D, Ganefri, Erianjoni, Dewata I, Iskarni P and Syam A 2019 A policy model of adaptation mitigation and social risks the volcano eruption disaster of sinabung in Karo regency - Indonesia *Int. J. GEOMATE* **17** 190–6

[2] Amri A 2017 *Pendidikan Tangguh Bencana* (Sekretariat Nasional Satuan Pendidikan Aman Bencana Kementerian Pendidikan dan Kebudayaan)

[3] Indonesian National Agency for Disaster Management 2013 *IRBI (Indeks Risiko Bencana Indonesia)*

[4] Surono, Jousset P, Pallister J, Boichu M, Buongiorno M F, Budisantoso A, Costa F, Andreastuti S, Prata F, Schneider D, Clarisse L, Humaida H, Sumarti S, Bignami C, Griswold J, Carn S, Oppenheimer C and Lavigne F 2012 The 2010 explosive eruption of Java's Merapi volcano-A "100-year" event *J. Volcanol. Geotherm. Res.* **241–242** 121–35

[5] Darmawan H, Walter T R, Troll V R and Budi-Santoso A 2018 Structural weakening of the Merapi dome identified by drone photogrammetry after the 2010 eruption *Nat. Hazards Earth Syst. Sci.* **18** 3267–81

[6] Almazan J U, Cruz J P, Alamri M S, Albougami A S B, Alotaibi J S M and Santos A M 2019

Coping Strategies of Older Adults Survivors Following a Disaster: Disaster-Related Resilience to Climate Change Adaptation *Ageing Int.* **44** 141–53
[7] BAPPENAS and BNPB 2011 *Ringkasan Eksekutif Rencana Aksi Rehabilitasi Dan Rekonstruksi Pascabencana Erupsi Merapi Di Wilayah Provinsi Di Yogyakarta Dan Provinsi Jawa Tengah, Tahun 2011-2013*
[8] Christine Wamsler Ebba Brink 2015 The Role of Individual Adaptive Practices for Sustainable Adaptation *Int. J. Disaster Resil. Built Environ.* **6** 102–16
[9] Davis M and Vulturius G 2014 *Disasters, climate change and development: Reducing risk by tackling the drivers of vulnerability*
[10] Sharpe J 2016 Understanding and unlocking transformative learning as a method for enabling behaviour change for adaptation and resilience to disaster threats *Int. J. Disaster Risk Reduct.* 1–22
[11] Lei Y, Wang J, Yue Y, Zhou H and Yin W 2014 Rethinking the relationships of vulnerability, resilience, and adaptation from a disaster risk perspective *Nat. Hazards* **70** 609–27
[12] Nelson D R, Adger W N and Brown K 2007 Adaptation to Environmental Change: Contributions of a Resilience Framework *Annu. Rev. Environ. Resour.* **32** 395–419
[13] Sridarran P, Keraminiyage K and Amaratunga D 2018 Enablers and Barriers of Adapting Post-disaster Resettlements *Procedia Eng.* **212** 125–32
[14] Pilli-Sihvola K, Harjanne A and Haavisto R 2018 Adaptation by The Least Vulnerable: Managing Climate and Disaster Risks in Finland *Int. J. Disaster Risk Reduct.* **31** 1266–75
[15] Adhikari S, Baral H and Nitschke C 2018 Adaptation to Climate Change in Panchase Mountain Ecological Regions of Nepal *Environments* **5** 42
[16] Sword-Daniels V, Wardman J, Stewart C, Wilson T M, Johnston D and Rossetto T 2011 *Infrastructure impacts, management and adaptations to eruptions at Volcán Tungurahua, Ecuador, 1999-2010*
[17] Bennett N J, Dearden P, Murray G and Kadfak A 2014 The Capacity to Adapt?: Communities in a Changing Climate, Environment, and Economy on the Northern Andaman Coast of Thailand *Ecol. Soc.* **19**
[18] Jain S and Cohen A K 2013 Behavioral Adaptation Among Youth Exposed to Community Violence: A Longitudinal Multidisciplinary Study of Family, Peer and Neighborhood-Level Protective Factors *Prev. Sci.* **14** 606–17
[19] Wong B B M and Candolin U 2015 Behavioral Responses to Changing Environments *Behav. Ecol.* **26** 665–73
[20] Oakes L E, Ardoin N M and Lambin E F 2016 "I know, therefore I adapt?" Complexities of individual adaptation to climate-induced forest dieback in Alaska *Ecol. Soc.* **21**
[21] Swim J K, Stern P C, Doherty T J, Clayton S, Reser J P, Weber E U, Gifford R and Howard G S 2011 Psychology's Contributions to Understanding and Addressing Global Climate Change *Am. Psychol.* **66** 241–50
[22] Bhebhe S, Runhare T and Monobe R J 2019 Strategic Approaches for Developing A Culture of Safety Management in Schools: Indications from Literature Studies *Jàmbá J. Disaster Risk Stud.* **11** 1–7
[23] Shaufa A 2018 *Enhancing individual knowledge of disaster resilience among children and public based on best practices of japan and other countries*
[24] Proulx K and Aboud F 2019 Disaster Risk Reduction in Early Childhood Education: Effects on Preschool Quality and Child Outcomes *Int. J. Educ. Dev.* **66** 1–7
[25] Torani S, Majd P M importance of education on disasters and emergencies: A review article, Maroufi S S, Dowlati M and Sheikhi R A 2019 The Importance of Education on Disasters and Emergencies: A Review Article *J. Educ. Health Promot.* **8** 85
[26] Creswell. J W 2014 *Research Design : Qualitative, Quantitative, and Mixed Methods Approaches* (California: SAGE Publications, Inc.)
[27] Macdonald S and Headlam N 2008 *Research methods handbook: Introductory guide to research*

methods for social research (Manchester: Centre for Local Economic Strategies.)
[28] Fearnley C J, Bird D K, Haynes K, McGuire W J and Jolly G 2018 *Observing the Volcano World Volcano Crisis Communication* ed K Nemeth (Barcelona, Spain: International Association of Volcanology and Chemistry of the Earth's Interior – IAVCEI)
[29] Dai W, Chen L, Tan H, Wang J, Lai Z, Kaminga A C, Li Y and Liu A 2016 Association Between Social Support and Recovery from Post-traumatic Stress Disorder After Flood: A 13-14 Year Follow-up Study in Hunan, China *BMC Public Health* **16** 194–203
[30] Lindell M K 2013 Disaster Studies *Curr. Sociol.* **61** 797–825
[31] Becker J S, Paton D, Johnston D M, Ronan K R and McClure J 2017 The Role of Prior Experience in Informing and Motivating Earthquake Preparedness *Int. J. Disaster Risk Reduct.* **22** 179–93
[32] Musiyam M, Suharjo S, Prayitno H, Hafida S and Widiyatmoko W 2019 The Effectiveness of the Use of Learning Strategies in Delivering the Materials of Disaster Guide Book in Senior High Schools and Vocational High Schools *Proceedings of the 4th Progressive and Fun Education International Conference*
[33] ASEAN Safe School Initiative (ASSI) 2016 *School Disaster Risk Management Guidelines for Southeast Asia*
[34] O'Toole V M and Friesen M D 2016 Teachers as First Responders in Tragedy: The Role of Emotion in Teacher Adjustment Eighteen Months Post-earthquake *Teach. Teach. Educ.* **59** 57–67
[35] Saja A M A, Teo M, Goonetilleke A and Ziyath A M 2018 An Inclusive and Adaptive Framework for Measuring Social Resilience to Disasters *Int. J. Disaster Risk Reduct.* **28** 862–73
[36] Orsato R J, Ferraz de Campos J G and Barakat S R 2018 Social Learning for Anticipatory Adaptation to Climate Change: Evidence From a Community of Practice *Organ. Environ.*
[37] Hafida S H N, Setiawan B and Fajariyah R A A 2018 The Effectiveness of Community Radio Infrastructure to Support Disaster Preparedness (Case Study of Community Radio in Merapi Volcano, Yogyakarta, Indonesia) *MATEC Web Conf.* **229** 6–11
[38] Haynes K and Tanner T M 2015 Empowering young people and strengthening resilience: youth-centred participatory video as a tool for climate change adaptation and disaster risk reduction *Child. Geogr.* **13** 357–71
[39] Watanabe M, Hikichi H, Fujiwara T, Honda Y, Yagi J, Homma H, Mashiko H, Nagao K, Okuyama M and Kawachi I 2019 Disaster-related trauma and blood pressure among young children: a follow-up study after Great East Japan earthquake *Hypertens. Res.*
[40] Eriksen S and Brown K 2011 Sustainable adaptation to climate change *Clim. Dev.* **3** 3–6
[41] Muller V M, Burke R V, Berg B M, Lin A C and Upperman J S 2014 A Mixed-methods Pilot Study of Disaster Preparedness and Resiliency Among Faith-based Organizations *Prehosp. Disaster Med.* **29** 127–33

Capacity building: lessons learned Kampung Sewu Surakarta

Rita N[1,2*]

[1] Faculty of Teacher Training and Education, Universitas Sebelas Maret, Surakarta 57126, Indonesia
[2] Disaster Research Center, Research institutions and community service, Universitas Sebelas Maret

*Email: ritanoviani@staff.uns.ac.id,

Abstract. Persistent flood disasters in Surakarta City almost every year pose diverse challenges to the region's sustainable development. However, existing institutional policies on flood control had succeeded to adequately complement known structural measures with the relevant non-structural standards of flood risk management. In this paper, we present an assessing institutional capacities Sibat for flood disaster risk reduction. This review article provides a deep understanding of flood vulnerability capacity building as an often neglected non-structural measure in flood-prone regions of developing countries. An extensive systematic review was conducted on flood vulnerability capacity building related to peer-reviewed journal articles and documents domiciled in the Web of Science, Scopus, and field survey. The study reveals the underlying potentials and procedure of flood vulnerability capacity building as a robust non-structural strategy that is reliable and should precede structural measures in managing flood risks and flood hazards in the Bengawan solo Riverbank.

1. Introduction

The frequency and severity of hazards have increased over the past decade, including floods, droughts, and hurricanes. Practically, capacity building is not a new paradigm that stands alone but is an important component to reduce the impact of disasters.

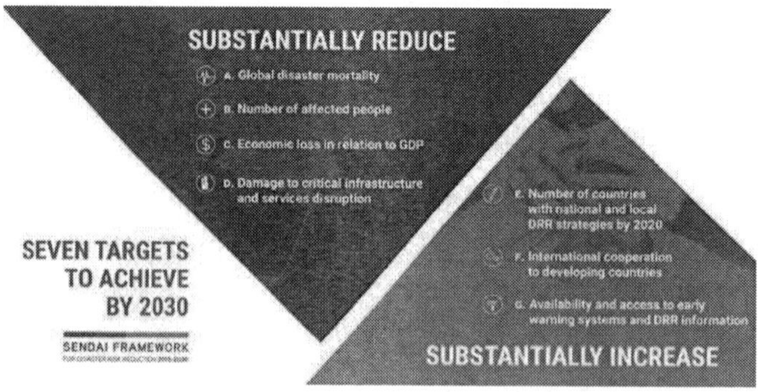

Figure 1. Sendai Framework 2015-2030[1]

The Sendai Framework for disaster risk reduction 2015-2030 is the successor to the Hyogo Framework for Action 2005-2015. The framework is an important agreement of the post-2015 development agenda. It also provides concrete steps to protect development from disaster risk in member countries.

The Sendai Framework for Disaster Risk Reduction 2015-2030 outlines seven clear targets (Figure 1) and four priorities for action to prevent new and reduce existing disaster risks: (i) Understanding disaster risk; (ii) Strengthening disaster risk governance to manage disaster risk; (iii) Investing in disaster reduction for resilience and; (iv) Enhancing disaster preparedness for effective response, and to "Build Back Better" in recovery, rehabilitation, and reconstruction [1].

It aims to substantially reduce disaster risks and losses in the lives, health livelihoods, livelihoods, and physical, economic, cultural, social and environmental assets of people, communities, businesses and nations over the next 15 years. The Sendai Framework for Disaster Risk Reduction 2015–20130 and the Sustainable Development Goals (SDGs) recommend enhancing holistic disaster risk reduction (DRR) policies and plans to increase disaster resilience globally.

Disaster risk reduction (DRR) – *"The concept and practice of reducing disaster risks through systematic efforts to analyze and manage the causal factors of disasters, including through reduced exposure to hazards, lessened vulnerability of people and property, wise management of land and the environment, and improved preparedness for adverse events*[2]. " Consequently, building DRR capacity can be defined as "efforts to strengthen the competencies and skills of a target organization, group or community so that the target could drive DRR efforts, or in a broader sense development, in a sustainable way in the future" [3].

Capacity is to prepare for and manage natural disasters to take the necessary actions and prepare for the most appropriate response. It is very important to establish policies regarding public awareness with disaster management. This leads to education and training, information programs, and research on disaster risk reduction.

In disaster risk reduction, capacity building can be achieved by strengthening infrastructure and enhancing organizational abilities [4]. Various elements of capacity building[2,5]

a. Education on disaster prevention and response
   It includes educating the vulnerable communities and the population of the state for the possible hazards and their impacts in their area. It also covers the preventive measures and response strategies in case of a disaster strike to the surface.
b. Training to vulnerable communities
   They held basic disaster management training for the community and local volunteers. This is intended so that the community and volunteers know how to help and survive when a disaster occurs.
c. Collaboration with relief agencies
   This includes collaboration between various government agencies and other non-governmental organizations involved in relief work through the states for all types of disasters.
d. Mock drill
   It is a method of practising how a building would be evacuated in a fire or other emergency. Usually, the building's existing fire alarm system is activated, and the building is evacuated as if the crisis had occurred
e. Household preparation
   This includes preparation by securing valuables in waterproof bags, specifying minimal clothing, and being prepared with an evacuation plan
f. Understanding warning/de-warning messages
   Dissemination of EWS messages to vulnerable communities achieved through training and involvement of all stakeholders.
g. First aid preparedness
   Along with all necessary things, it's important to have a well-stocked first-aid kit to deal with minor accidents and injuries

Capacity building factors include the relationship between disaster and development, participation of all stakeholders, post-disaster psychosocial support, training approaches, training manuals, sustainable development, and local wisdom [2]

While Principal For DRM Capacity Development, IFRC is Focus on holistic risk reduction, Plan and design programs carefully for long lasting impact, Strengthen functional capacity for DRM, Grow ownership and Pay attention to context. The capacity building process must be carried out by all levels in all sectors [6].

2. Methods

Surakarta City is one of the regencies in Central Java Province, The city of Surakarta is located in the intermountain basin to the west of Mount Merapi and Merbabu, to the east of Mount Lawu, to the east, to the south of Mount Sewu, and the north of Mount Kendeng. The Bengawan Solo river crosses the solo city, the longest river in Java across provinces. Besides Bengawan Solo, several rivers include the Premulung River, Anyar river, Pepe river, etc. (**Figure 2**).

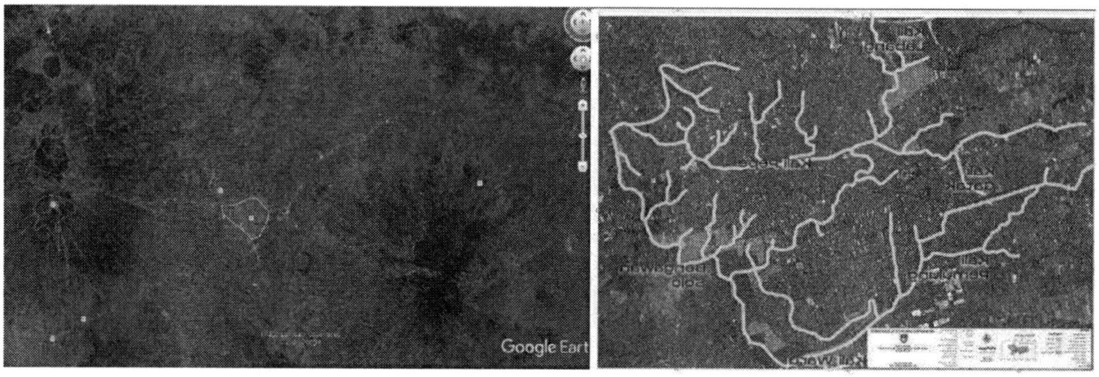

Figure 2. Location and number of rivers in the city of Surakarta

3. Results and Discussion

Surakarta City is one of the cities in Java Island which is prone to flooding. According to its history, in 1966, the city of Surakarta experienced a great flood and inundated the city square of Surakarta. As a result, solo has an increasing level of vulnerability to flooding. Increased vulnerability cannot be separated from unplanned urbanization. This can be seen from the increasingly dominant urban land use on the one hand; on the other hand, the narrowing of open space (open space); and the decreasing ability of drainage channels to drain surface runoff due to lack of maintenance, large sedimentation including accumulation of garbage, and inhabited riverbanks.

BBWS Bengawan Solo (2011) explain that Flooding in Solo, apart from being caused by the Bengawan Solo river, other rivers, and high rainfall, caused by (1) changes in land use, (2) irregularities in the granting of IMB, (3) narrowing and silting of the river due to the use of riverbanks as illegal settlements, (4) the habit of some people throwing garbage into the river, (5) the Bengawan Solo River embankment is broken/damaged, and (6) broken floodgates[7]. According to data from the Regional Disaster Management Agency (BPBD) of Solo City, 25 sub-districts out of 51 villages are included in flood-prone areas. Almost every year, when the rainy season arrives, the areas drained by the river experience flooding, both on a medium and large scale. As a result, very large losses must be borne by all components, the community, the private sector, and local governments.

People living in the southern-eastern part of Surakarta, Semanggi, Sangkrah dan Sewu are generally aware that they live in flood-prone areas, especially during the rainy season. They stay in the location because they have lived in the location for generations, inherited from their parents, close to work, and the family's economic condition forces them only to rent or occupy a house in the area. Hence, the impact of flooding in several aspects: (1) Population Aspect, in the form of fatalities/deaths, drifting, drowning, injuries, missing victims, displacement, outbreaks of epidemics, and isolated residents. (2) Government Aspects, in the form of damage or loss of documents, archives, office equipment, and supplies and disruption of the running of the government. (3) Economic Aspect, in the form of loss of livelihood, non-functioning of traditional markets, damage, loss of property, livestock, and disruption of the community's economy. (4) Facilities/Infrastructure aspects, in the form of damage to people's houses, bridges, roads, office buildings, social facilities, public facilities, electricity installations, drinking water, and communication networks. And (5) Environmental Aspects, in the form of damage to ecosystems, tourism objects, rice fields/agricultural land, clean water sources, and damage to dams/irrigation networks.

One of the problems in handling floods in Surakarta City is the high vulnerability to flooding hazards, both individual, group, and community vulnerabilities. Individually, the elderly have an adequate understanding of flooding, but their weak health can affect their ability to protect themselves. A child can better take care of his physical needs, but he is vulnerable due to failure to understand when a flood warning is issued. Women are a vulnerable community because they have physical and temporal resources that limit their abilities during a flood disaster.

In the aspect of group vulnerability, poor minority groups are vulnerable because they do not have adequate resources to buy safe shelter from flooding, such as the poor living on the banks of the Bengawan Solo River. In terms of community vulnerability, it is known that the location of residence around the Bengawan Solo River is very vulnerable to the danger of flooding. In addition, the large number of people living on the banks of the river also causes narrowing and silting of the river.

As a response to this vulnerable situation, a community-based disaster preparedness village (SIBAT) was formed. SIBAT is a PMI program that aims to overcome disasters. The role of SIBAT in disaster management reduction.

Before the disaster

1. SIBAT does as a River Worker regularly every day cleaning the river from garbage that hampered the flow of the river at several bridge points that residents often used to dispose of garbage.
2. SIBAT also provides environmental and disaster education to residents about the importance of caring for and protecting the environment to minimize disaster risk. This activity is done continuously, both directly and through social media like Facebook
3. Making bio pure holes as a diminishing catchment area in urban areas, saving rainwater as an alternative water source,
4. Planting Akar Wangi as a water barrier on riverbanks, managing waste, and planting vertimina as an additional food source for residents.

During Disaster

1. SIBAT has the responsibility to monitor river water discharge directly at the viewing post,
2. providing information via radio and social media and utilize the early warning system technology developed by the BBWS Bengawan Solo.
3. Its establishment of public kitchens, first aid teams, and evacuation teams.

After Disaster

SIBAT did emergency response, recovery, and trauma healing for residents and ensured that the affected residents could do.

Figure 3. SIBAT activities and disaster risk reduction in Kampung Sewu

4. Conclusion

The Bengawan Solo Riverbank is an environmentally sensitive area, thus, highly vulnerable to various ecological and environmental hazards. Flood vulnerability mapping and capacity building are non-structural measures that complement structural measures in disaster risk reduction and are recommended for the Bengawan Solo riverbank, especially Kampung Sewu. Community participation as one of the stakeholders in Kampung Sewu is very good. Emergency disaster management activities, before the disaster, during the disaster, and after the flood disaster, community participation is balanced with other stakeholders. The community's funding sources as a spontaneous humanitarian step have developed in the emergency response stage (intervention). One strategy for non-structural disaster risk reduction is recognizing the importance of learning for building and maintaining social and ecological resilience [8][5].

References

[1] United Nations Office for Disaster Risk Reduction 2015 *Sendai Framework for Disaster Risk Reduction 2015 - 2030*

[2] Anil K, Gupta, Sreeja S. Nair, Shiraz A. Wajih, Shashikant chopde G G and G A 2019 *Training Module Mainstreaming Climate Change Adaptation and Disaster Risk Reduction*

[3] Walker J and Cooper M 2011 Genealogies of resilience: From systems ecology to the political economy of crisis adaptation *Secur. Dialogue* **42** 143–60

[4] UN/ISDR 2004 *Living with risk: a global review of disaster reduction initiatives* vol 1
[5] Wong K K and Zhao X 2001 Living with floods: Victims' perceptions in Beijiang, Guangdong, China *Area* **33** 190–201
[6] Magel H and Kötter T 2006 *The contribution of the surveying profession to disaster risk management* vol 66
[7] Toto Cahyono, M. Pramono Hadi D M 2015 Pemodelan Spasial Untuk Pembuatan Peta Rawan Banjir Dan Peta Tingkat Risiko Banjir Bengawan Solo Di Kota Surakarta *Maj. Geogr. Indones.* **29** 32
[8] Adger W N, Hughes T P, Folke C, Carpenter S R and Rockström J 2005 Social-ecological resilience to coastal disasters *Science (80-.).* **309**

IOP Conf. Series: Earth and Environmental Science **986** (2022) 012042 doi:10.1088/1755-1315/986/1/012042

Community social capital in facing landslide hazards in Banjarnegara

Ferani Mulianingsih[1*], Arif Purnomo[1], Sofyan Soleh Nur Hidayatulloh[1] and Irma Mulia Setyarini[2]
[1] Social Studies Program, Universitas Negeri Semarang, Indonesia
[2] English Education, Pascasarjana Universitas Negeri Semarang, Indonesia

[*]Email: ferانigeographer@mail.unnes.ac.id

Abstract. This research was conducted in Karangkobar District, Banjarnegara. Karangkobar District had 2 (two) moderate landslide threat levels covering an area of 2,419.96 hectares or 75%, while the high landslide threat level is 789.29 hectares or 25% spread over 11 villages. There had not been a disaster preparedness village, there had been no socialization or training on disaster reduction, the community still used kentongan and mosque speakers as media to convey information. The focus of this research was the social capital of the community, with data collection methods, namely documentation and observation. The data analysis used was descriptive analysis. The results of the study showed that there were 3 villages that had a moderate level of community social capital, with the characteristics of having regulations and institutions, early warning and disaster risk studies, reduction of basic risk factors, and development of preparedness on all lines, but disaster education was not yet available. There had not been a disaster preparedness village in all villages, only in Sampang Village, socialization and training on disaster reduction had been carried out, the community used kentongan and starts using mobile phones as a medium to convey information. Meanwhile, the village with a high level of landslide threat that had a high level of community social capital was Pagerpelah Village, which had the characteristics of having all parameters of community capacity in the form of rules and institutions, early warning and disaster risk studies, disaster education, reduction of basic risk factors, and preparedness development at all levels.

1. Introduction

Banjarnegara Regency according to the geomorphology of Central Java is part of the central zone of Central Java which is part of the North Serayu Mountains mandala whose topography is relatively undulating with semi-steep to steep slopes, where the mountains have slopes between 15-40% which are at risk of landslides [1]. The statigraphy of Banjarnegara Regency consists of the oldest rocks, namely molion (metamorphic) rocks consisting of crystalline schist, slate, phyllite, quartzite, and limestone [2].

Banjarnegara Regency has high to very high rainfall intensity, which is between 3000->3500 mm/year [3]. This high rainfall causes the soil to become saturated with water, and as a result it will weaken the bonds of soil particles. Water molecules infiltrate the soil particles and become a catalyst for the sliding process between the particles and cause landslides [4].

Banjarnegara Regency is traversed by many fault lines, especially around areas where landslides occur, this is a controlling factor that can cause landslides, so that it has a hilly land texture that has steep and upright slopes. These fault lines cause the bond between the layers of rock supporting the soil to split and become brittle [5].

From the various types of disasters that have occurred in Central Java Province, landslides are the most frequent types of natural disasters compared to other types of natural disasters , which are 485 times or about 29.6% of the total natural disasters [6]. Banjarnegara Regency is an area that has a frequency of occurrence of areas that are prone to center movement which can be found in the northern Banjarnegara area, this is indicated by morphology or topography, most of which have slope angles >30^0 [7]. The types of constituent rocks consist of claystone, breccia, sandstone from propagation formations which are also very influential on the occurrence of soil movements in this area [8].

Karangkobar District is one of the areas in Banjarnegara Regency that is prone to landslides. Throughout 2018 there have been 23 landslides or 15% of landslides occurred in Karangkobar District out of a total of 152 landslides that occurred in Banjarnegara Regency. Where the number of landslides that occurred was the highest compared to other areas in Banjarnegara Regency. Landslides that occurred in 2018 caused losses in the form of 6 people being injured, 4 families in danger, 2 houses being heavily damaged, 1 house being lightly damaged, and 8 houses being moderately damaged. Landslides also damaged 10 infrastructure in the form of roads and bridges [9].

The landslide disaster that occurred caused the community to suffer losses both in physical and non-physical aspects. Losses from the physical aspect are in the form of damage to residential buildings, closed main road access, damage to rice fields and gardens owned by residents, and closed rivers due to landslides. Meanwhile, the loss of landslides when viewed from a non-physical aspect is in the form of losses from the psychological aspect of the community. The high level of losses experienced by the community due to natural disasters is due to the lack of information obtained by the community about the possibilities of disasters that occur in the vicinity, so that public awareness of disasters is very minimal.

The National Disaster Management Agency has a strategy to create a disaster-resilient society in the form of developing disaster mitigation as an effort to reduce disaster risk. The disaster mitigation process directs the community to participate in disaster management management to reduce the risks and impacts of disasters. Indirectly, the understanding of disaster mitigation as a means to overcome or reduce the community's vulnerability to a disaster [10]. Disaster management through mitigation can be analyzed using the concept of social capital that enables communities to respond to disasters and create community attitudes that are disaster-resistant (disaster resilience). Therefore, the role of community social capital in dealing with the danger of landslides in Banjarnegara must be able to maintain and increase existing social capital, in order to achieve mutual success. This article aims to explain that the role of social capital in landslide disaster mitigation in Banjarnegara is very important and needs to be maintained as a form of realization in the landslide disaster recovery process for the community.

2. Method

The method of data collection in this study was using documentation and observation. This research was a quantitative descriptive study, namely research that only analyzed to the level of description, facts were analyzed and presented systematically so that it could be more easily understood and concluded. In this study, descriptive analysis was used to describe the social capital of the community in dealing with the danger of landslides in Banjarnegara.

3. Result and Discussion

Karangkobar sub-district is astronomically located between $7^0$14'50"LS-$7^0$19'15"LS and is located between $109^0$40'05"BT-$109^0$44'45"BT. Karangkobar District is one of the sub-districts in Banjarnegara Regency which is located in the mountainous area of North Serayu. The topography of the area is mostly hilly with various slopes, ranging from plain to very steep. Karangkobar District is located in the northern part of Banjarnegara Regency which is a hill with an altitude ranging from 710-1,025 meters above sea level. The administrative boundaries are bordered on the north by Kalibening and Wanayasa

Districts, in the east by Wanayasa District, in the south by Banjarmangu District, and in the west by Banjarmangu District.

Karangkobar sub-district consists of 13 villages, 46 hamlets, 50 RW, and 186 RT. The total area of Karangkobar District is 3,209.252 ha or 3.65% of the total area of Banjarnegara Regency. The village that has the largest area in Karangkobar District is Slatri Village with an area of ± 468,264 ha with a percentage of 14.59% of the total area of Karangkobar District. Meanwhile, the village that has the smallest or narrowest area in Karangkobar District is Gumelar Village, which is 128,679 ha with a percentage of 4.01% of the total area of Karangkobar District.

Karangkobar District based on the results of documentation and observations, there are several locations of landslides that have occurred in Karangkobar District. From several locations of landslides, there are 5 points or 5 locations of landslides that threaten residents in Karangkobar District. The area in Central Java Province that is most prone to landslides is Banjarnegara. While in Banjarnegara the most prone to landslides is Karangkobar Regency. So that Karangkobar Regency is included in the red zone.

The location of the first landslide occurred in Diwek Hamlet, Karangkobar Village. Residents whose houses were damaged have been relocated to safer places around Diwek Hamlet, Karangkobar Village. Currently the location of the avalanche has been turned into agricultural fields and cattle pens. This landslide location point in Diwek Hamlet, Karangkobar Village is one of the areas that has a high level of landslide threat. The condition of the former landslide in Diwek Hamlet, Karangkobar Village can be seen in Figure 1.

Figure 1. Location of Landslide in Diwek Hamlet, Karangkobar Village

The second point of the landslide occurred in Jemblung Hamlet, Sampang Village, Karangkobar District. Currently, the location where landslides have occurred has become a conservation area. However, there are also many residents who still use the location of the former landslide as agricultural fields. Based on the analysis of the landslide threat level in Jemblung Hamlet, Sampang Village, this is an area that has a high level of landslide threat. The condition of the landslide location in Jemblung Hamlet, Sampang Village can be seen in Figure 2 as follows.

Figure 2. Landslide Location in Jemblung Hamlet, Sampang Village

The landslide at point 3 occurred in Gintung Hamlet, Binangun Village. This landslide location point in Gintung Hamlet, Binangun Village is one of the areas that has a high landslide threat. The location of the landslide in Gintung Hamlet, Binangun Village, can be seen in Figure 3 as follows.

Figure 3. Landslide Location in Gintung Hamlet, Binangun Village

The landslide at the 4th point occurred in Jlegong Village, Karangkobar District. The landslide location point in Jlegong Village is an area that has a high level of landslide threat. The landslide at point 5 occurred in Ambal Hamlet, Ambal Village, Karangkobar District, which is a high landslide threat area.

In the view of economics, capital is anything that can be profitable or produce. Capital itself can be divided into: (1) capital in the form of materials such as money, buildings or goods; (2) cultural capital in the form of quality education, local cultural wisdom; and (3) social capital in the form of togetherness, social obligations instituted in the form of common life, role, authority, responsibility, reward systems and other linkages that m enghasilkan collective action [11].

Communities in areas with low levels of social capital still use traditional tools in the form of kentongan as a medium of communication to convey information to the public. Kentongan is used by the community as a medium to provide information about natural disasters, one of which is landslides. In addition to kentongan, the community also uses mosque speakers as a medium to convey information or to gather people.

Regional disaster risk studies are already available in Slatri Village and Sampang Village in the form of availability of landslide threat maps and landslide risk assessment documents. Meanwhile, Ambal Village is not yet available. Slatri Village and Sampang Village already have an early warning system for landslides and socialization as well as early warning simulation training, while in Ambal Village it is not yet available.

Pagerpelah village has high social capital, because the parameters of disaster management rules and institutions are already available. Such as national or local legal and policy frameworks for disaster risk reduction, resources allocated specifically for disaster risk reduction activities, community participation and decentralization are established. Relevant information about disasters that can be accessed by all interests of disaster education parameters is available in Pagerpelah Village, namely there has been an archive of disaster event data for the last 5 years. In the field of school curricula, relevant educational and training materials covering concepts and practices of disaster reduction and recovery already exist in Pagerpelah Village, in the form of increased skills in dealing with disaster emergencies in schools, namely at the elementary level. In the area of availability of research methods for multi-disaster risk studies and cost benefit analysis, Pagerpelah Village is not yet available. Meanwhile, strategies to build awareness of the entire community in implementing disaster-resistant cultural practices that are able to reach the community are available in Pagerpelah Village, in the form of media publishing public awareness building to carry out disaster risk reduction practices and initiatives to build disaster-resilient villages.

Disaster risk reduction is one of the objectives of policies and plans related to the environment, natural resource management, land use, and adaptation to climate change. In these villages, they have started planning the formation of disaster-resilient villages (DESTANA), which are villages that have the independent ability to adapt and deal with potential disaster threats, and to recover quickly from the adverse effects of disasters.

4. Conclusion

The community of Karangkobar District, Banjarnegara Regency is in an area that has a high potential for landslides. The potential for landslides makes the community must have skills in disaster mitigation, in order to minimize the occurrence of disasters. Theoretically, the study of social capital in disaster mitigation based on local wisdom in Karangkobar District, Banjarnegara Regency is a form of implementing social capital studies. The community is able to build a social network by starting the formation of a disaster-resilient village Desa Tangguh Bencana (DESTANA). Destana is a village that has the independent ability to adapt and face the threat of disasters, as well as recover quickly from the adverse effects of disasters. Thus, the village has the ability to recognize threats in its area and is able to organize community resources to reduce vulnerability and at the same time increase capacity to reduce disaster risk.

Reference

[1] Keswara and Panji R 2015 *Geomorfologi Jawa Tengah*
[2] Anon 2018 *Pusat Vulkanologi dan Mitigasi Bencana Geologi*
[3] Badan Meteorologi Klimatologi dan Geofisika 2015 *Data Curah Hujan Tahunan Kabupaten Banjarnegara*
[4] Suranto J P 2008 Kajian Pemanfaatan Lahan Pada Daerah Rawan Bencana Tanah Longsor di Gununglurah, Cilongok, Banyumas *Pemanfaat. Lahan Pada Drh. Rawan Bencana Tanah Longsor* 1–165
[5] Ayudya V R and Rofi A 2015 Strategi Adaptasi Masyarakat dalam Menghadapi Bencana Tanah Longsor di Desa Sampang Kecamatan Karangkobar Kabupaten Banjarnegara *J. bumi Indones.* **6**
[6] Badan Nasional Penanggulangan Bencana 2014 *Rekapitulasi Bencana Nasional*
[7] Hutomo I A and Maryono M 2016 Model Prediksi Kawasan Rawan Bencana Tanah Longsor di Kecamatan Karangkobar *J. Pembang. Wil. Kota* **12** 303
[8] Warnadi W 2014 Inventarisasi Daerah Rawan Longsor Kabupaten Banjarnegara Jawa Tengah *Spat. J. Wahana Komun. dan Inf. Geogr.* **12** 35–45
[9] Badan Penanggulangan Bencana Daerah Kabupaten Banjarnegara 2019 *Rekapitulasi Kejadian Bencana Tahun 2018*
[10] Samuda S 2016 bari fola sebagai modal sosial dan instrumentasi masyarakat tangguh bencana *J. Penelit. Hum.* **21** 109–18
[11] Mawarni A Pentingnya Modal Sosial dalam Pembangunan Pasca Bencana

Liveable city from the perpective of disaster management

P Iskandar[1*] and W Prasetyo[2]

[1]Jl. Prof. Sudarto No.13, Tembalang, Kota Semarang
[1]Program Magister Arsitektur Departemen Arsitektur, Universitas Diponegoro
[2]Program Studi Teknologi Konstruksi Bangunan Air, Politeknik Pekerjaan Umum

*Email: pu3isk4nd4r04@gmail.com

Abstract. Research purpose is provide an overall image of what a *liveable city* from the perspective of disaster management. Disaster is seen as an event that can disrupt and threaten people's lives, while a liveable city is an idea to create a city that is economically qualified. environment dan equity of live. The initial stage begins with the theoretical aspect and the final stage with the practical aspect, is attempt to provide answers to the following questions: can liveable cities be applied to disaster-risk areas and what kind of concept will be applied to disaster-vulnerable areas. The elements of liveable cities which are generally different in each country because they are adapted to the geographical conditions and demographics of each region. The method of this study uses literature studies and case studies. Disaster management from the beginning must be taken into consideration in planning a city, so as to minimize losses due to disasters and the people who live in it can live side by side with nature. The results of this study explain that the application of the concept of a *liveable city* seen from disaster management is not only the urgency of fulfilling the requirements of a *liveable city* that is ready to face disasters but also views that liveable urban planning as a city ecosystem that has appropriate values and culture and is comfortable to live in so that it creates sustainable city.

1. Introduction

A disaster is a series of events that threaten and disrupt lives and livelihoods caused by natural factors and factors of non-natural and human factors that resulted in the onset of human casualties, environmental damage, loss of property, and the psychological impact [1]. The magnitude of the disaster risk can be expressed in the most significant losses incurred (the treasure, the soul, injury) to the importance of certain events. In 3 (Three) Years recorded catastrophic natural disasters in Indonesia was among the flood as much as 19.675 disaster, landslides as much as 10.246 disasters, earthquakes as much as 10.115 disasters, drought as much as 8.587 disaster and as much as 24,89% are some of the natural disasters that hit Indonesia, such as tsunamis, forest fires, volcanic eruptions, angina tornadoes, tidal waves of the sea and floods [2]. The risk of a disaster in an area is dependent on several factors of Nature/geography/geology (the possibility of the occurrence of the phenomenon of danger), the Vulnerability of society to the phenomenon (the condition and number of buildings), the Vulnerability of the physical area (the condition and number of buildings) as well as the readiness of the local community for emergency response and rebuilding, In general, the risk of disaster. Disaster management is required in order to provide protection from disasters and reduce the risk of loss of life and injury with the evacuation plan is good, environmental planning and design standards. Disaster management basically sought to prevent the public from disasters both by reducing the possibility of the emergence of a hazard (Danger) and resolves vulnerabilities. Disaster management can be defined as a series of activities consisting of various aspects of disaster management before, during, and after catastrophic events. Some disasters in Indonesia is not impossible to overcome, and minimize the im pact of such events is required appropriate policies in handling the disaster, the resistance of the public in the face of disaster, and the process of development as a community effort in improving the quality of life to realize the sustainable City. Improvement in quality of life which requires physical and habitat of social to make it happen is the concept of a liveable City. [3] Disaster is seen as an event that can disrupt and threaten people's lives. At the same time, a *Liveable City* is an idea to create an economically qualified city—environment and equity of life. *Livability* reflects the well-being of society and consists of many characteristics that make it a place where people want to live now and in the future [4]. According to IAP (Association of Planning Experts), that *Liveable City* reflects a comfortable environment and city atmosphere as a place to live and work by looking at various aspects both physically (urban facilities, infrastructure, spatial planning, etc.), relationships, economic activities, etc.). From the beginning, disaster management must be considered in planning a city to minimize losses due to disasters, and the people who live in it can live side-by-side with nature.

2. Liveable City Consep and Disaster Management

The review of literature is needed to facilitate an understanding of the research focus. In realizing a *Liveable City* or *Liveable City* must have basic principles, these principles must be owned by cities that want to make their city a liveable and comfortable city for the city community. The basic principle for a *Liveable City* is the availability of various basic needs of the community, public and social facilities, public spaces and places, safe, supporting economic, social, and cultural functions, and environmental sanitation and the physical environment [5]. *Liveable City* can be said to be based on 4 (four) pillars, namely: (1) improving the life opportunity system for the welfare of the community, (2) providing employment opportunities, (3) a safe and clean environment for health, and to maintain economic growth, and (4) good governance [6]. *Livability* is the highest value of new urbanism, focusing on growth conflict management by integrating livability and economic matters through urban design [7]. The following is a livability prism model (see Figure 1):

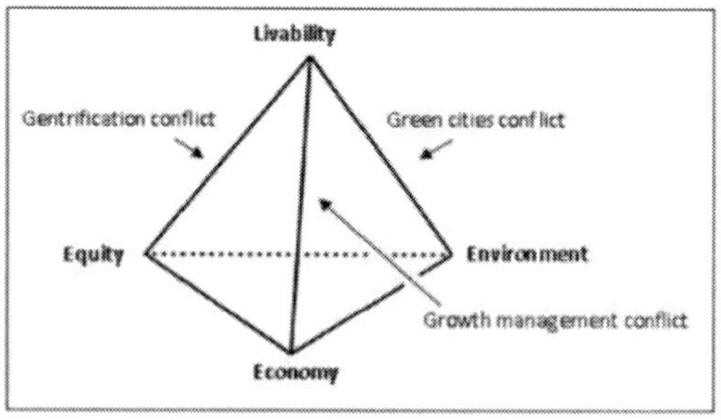

Figure 1. Model Prisma Livablility
Source: Godschalk, 2004

In the prism of livability model, the achievement of livability is highly dependent on other aspects, namely the economy, equality, and the environment, where the three elements must be balanced. These three aspects require comprehensive management and integration, both planning and development. According to [8], *Liveable City* factors include security, housing, education, employment and income, health and public services, transportation, public open space, social integration and democracy, culture, local food and goods, natural environment. Talking about the natural environment is also related to geological, hydrometeorological, and social disasters, so disaster management is needed in tackling disasters. Disaster management from the beginning must be taken into consideration in planning a city

IOP Conf. Series: Earth and Environmental Science **986** (2022) 012043 doi:10.1088/1755-1315/986/1/012043

so as to minimize losses due to disasters, and the people who live in it can live side by side with nature. The objectives of disaster management are:

a. Protecting the community from disaster hazards.
b. Straighten the existing statutory regulations.
c. Ensure the implementation of disaster management in a planned, integrated, coordinated, and comprehensive manner.
d. Respect local culture.
e. Develop public and private participation and partnerships.
f. Encouraging the spirit of cooperation, solidarity, and generosity.
g. Creating peace in the life of society, nation, and state.

With this point, it can be concluded that the application of the concept of a liveable city seen from disaster management is not only the urgency of fulfilling the requirements of a liveable city that is ready to face disasters but also views that liveable urban planning as a city ecosystem that has appropriate values and culture and is comfortable to live in so that it creates a sustainable city.

3. Methods

This study uses literature studies and case studies. From the beginning, disaster management must be considered in planning a city to minimize losses due to disasters, and the people who live in it can live side-by-side with nature. This research method is very suitable with the research objective, namely provide an overall image of a liveable city from the perspective of disaster management. Disaster is seen as an event that can disrupt and threaten people's lives. At the same time, a liveable city is an idea to create an economically qualified city: environment and equity of life. Most of the data and information used in this study came from secondary sources such as literature reviews, journals, statistical data, institutional annual reports, and others.

4. Result and Discussion

4.1. Liveable City From The Perspective of Disaster Management

The rate of growth has a significant contribution in influencing the design of a city. Water crisis, lousy waste management, air pollution that increasingly pollutes cities, natural disasters are urban problems that begin with uncontrolled population growth. The liveable city concept requires integrated and comprehensive planning and design from all sectors. They consider that natural disasters cannot be predicted when they arise, and the effects of environmental crises add to the thinking of planners and designers to think universally in realizing a liveable city. In principle, Liveable City and disaster management have the same goal, namely the need for the environment, both housing, and a safe social environment that can assist economic, social, and cultural functions, environmental sanitation, and the beauty of the physical environment towards a Liveable City and sustainable city. Disaster management is divided into disaster, which is severe damage due to extraordinary natural phenomena and caused by human activities that cause casualties, material losses, and environmental damage whose impacts exceed the ability of the local community to cope and require external assistance. Disaster consists of 2 (two) components, namely, Hazard and Vulnerability, and risk is the possible adverse impact caused by a hazard and vulnerability [9]. Several approaches are needed, but most of those who analyze them agree that a city is less dependent on the laws of the country it belongs to (developed or developing) to be liveable. However, this feature refers to the city's attractiveness due to its disaster management, social activities, economic resilience, physical aspects, environment, safety, space, etc. Specific general factors: savings, quality of life, happiness, fun, etc. Over plans are needed as part of the critical path to nocturnal sources, shallow oil, more sustainable, low carbon cities. In addition, many cities will be directly threatened by the effects of climate change. Some of the world's largest cities under stress are particularly at risk. A combination of sea-level rise, coastal erosion, saltwater pollution, and increased storms worsens the environment and improves people's living conditions in coastal areas.

4.1.1 Singapore

Singapore is one of the Asian countries where liveability is progressing rapidly. The application of the principles of elements related to liveability is more focused and is the main dimension. Liveability is one of the main elements of urban development; the other is good governance, the process of economic and environmental development. Quality of life Reflects behavior towards the environment, but discipline is needed in carrying it out—understanding liveability in urban management. Singapore is continuously and consistently tackling challenges and turning them into the opposite condition. The rules made include building energy efficiency and information technology in various sectors (waste, transportation, security, and others). (see Figure 2)

Figure 2. Singapore Smart City Progame
Source: BBC.com

The area of Singapore is about 728.6 km2, with a population of 5.68 million/person and a density of 7,810 km2. Singaporeans' ethnic and cultural patterns are Chinese, Malay, Indian, Pakistani, Bangladeshi, and Sri Lankan. Singapore's population is dominated by Chinese as much as 77%, Malays by 15%, Indians by 8%, and other ethnic groups by 1%. Singapore has an area with high humidity and rainfall every year, a tropical climate, and the influence of the monsoon season. The flood disaster also hit the State of Singapore in 1954; it is undeniable that the geographical condition of Singapore with its coastal mangrove swamps [10]. Floods and high tides are a combination of inevitable disasters, especially when there is high rainfall. Singapore goes through two monsoons, with the peak being from November to January. Extreme weather cannot be fully legalized because it all returns to sustainable environmental management. People living in low-lying, swampy, and flood-prone areas have a higher risk of being affected by disasters. Singapore designed the city of disaster resilience as a concept that starts from policymakers to the application of policies that have been made. After the big flood hit in 1954, Singapore began to rehabilitate the Bedok river (near the coast of Singapore), rehabilitate drainage, involve and support the community in dealing with flood disasters. (see Figure 3-5). The real challenge in the real world is disaster management, while technology is problem-solving. Albert Einstein said that Knowledge is not an attribute of intelligence but imagination. Technology can deal with urban problems, especially disasters. Singapore has developed an IoT system capable of monitoring disaster-prone areas. By placing sensors at various points in every corner of Singapore to minimize disaster victims.

Figure 3. Bedok Singapore 1994
Source: https://www.roots.gov.sg/

Figure 4. Bedok Resevoir Singapore 1980
Source: https://www.roots.gov.sg/

Figure 5. Bedok Resevoir Singapore 2020
Source: https://www.roots.gov.sg/

Singapore is one of the countries in Asia that has the title of a Liveable City. Singapore uses the Liveable City concept as shown in the Liveable City framework diagram. (see Figure 6). The Liveable City concept applied by Singapore shows the following outputs:

a. Competitive economy
b. Classy environment
c. High quality of life (level of welfare and psychology of society)

Efficiency and effectiveness of design and planning is a long-term strategy in increasing productivity coupled with flexibility and application of innovative systems. Singapore manages its city dynamically, starting from the government, community, stakeholders and making local wisdom a hallmark of sustainable development.

IOP Conf. Series: Earth and Environmental Science **986** (2022) 012043 doi:10.1088/1755-1315/986/1/012043

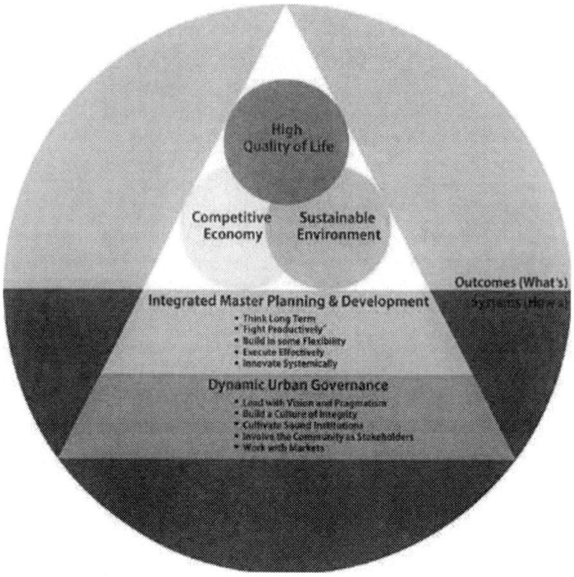

Figure 6. Diagram Framework Liveable City
Source: Center for Liveable City Singapore, 2012

Center Liveable City (CLC) in 2012, Singapore divided it into the following:

a. Based on the level of homeownership
b. building safety quality
c. park user satisfaction percentage
d. A number of people working in urban areas
e. A number of programs aimed at reforestation
f. Busy hours
g. Percentage of population who are satisfied with life, work, and the environment
h. Comparison of the proportion of green open space
i. Percentage of public transportation users
j. Public transportation user satisfaction
k. Clean water affordability
l. Sanitation access
m. State-owned land use rate
n. Air pollution level
o. Household water consumption rate per capita
p. Drinking water quality
q. Amount of water and air pollution
r. Recycling rate
s. Access to clean water sources
t. Flood-prone area
u. Energy consumption rate

Singapore designed its city starting from the ideals of Liveable City to the implementation of Smart City with a resilience disaster management approach. Dynamic governance and an integrated planning system are the keys to a Liveable City. In the Liveable City framework diagram, there is a sustainable environment section as the basis for forming a city's Liveable City. This means that a sustainable environment is very important and becomes a guide in planning and designing a city towards a Liveable

City. It is necessary to ensure that the sustainable background is carefully calculated so that the survival of the people of Singapore is guaranteed. Control of disasters such as pollution has been very tightly guarded so as not to impact the future. Singapore is also zoning its territory. Regional zoning needs to be done to minimize and facilitate problem-solving. A clean and beautiful city is an ideal that must be realized as support for Liveable City. The reciprocal of that is investor confidence in urban development in the economic sector.

Community support is very important, so the Singapore government educates its people with the sense of place, community involvement in landscape planning, information related to the quality of life by considering the local wisdom heritage board. To maintain the availability of natural resources and Singapore's economic development to streamline water and waste management, the process reduces unemployment and impacts water availability in the long term. Clean water is currently a rare item and will run out at any time if it is not managed properly. The close relationship between water management and waste management with community involvement in maintaining a sustainable environment impacts the visuals of the city being clean, beautiful, increasing attractiveness (tourists), maintaining public health, and improving the quality of life of the community.

4.1.2 Australia

The fire disaster that hit Australia in the 2019s was the worst in history. Climate change that causes heat waves and oscillations in rainfall dramatically affects the environment. Burnt forests produce substances that are harmful if inhaled by residents [11]. Mitigation of fire disasters is needed in a comprehensive and integrated manner to adapt an area to areas with extreme climatic conditions [12]. Poor air quality has an impact on public health conditions as much as 80%. Wildfires are a feature of Australia's natural environment, but climate change pushes for longer fire conditions. February is known as "Black Saturday" because of the worsening fires. Acres of burned forest created a "Black Summer" of severe air pollution and affected countries around Australia. The forest fire disaster claimed the lives of humans and animals, vegetation and ecosystems had a heavy impact. Starting from the Aboriginal tribes since centuries ago, they have understood and implemented their nature by managing the gifts of nature with local cultural values. Australia designed a disaster mitigation strategy to strengthen the settlement infrastructure sector and land use management. Being in a bad climate, Australia tries to develop a disaster management strategy, namely:

a. Integrated and sustainable urban planning minimizes the risk of forest fires
b. Using local wisdom as a fire management approach
c. Community resilience is a key element in dealing with disasters (ease of property insurance process)
d. Improving the quality of materials in buildings in fire-prone areas
e. Restoration and protection of animals, vegetation, and ecosystems
f. Improving the quality and quantity of firefighters and involving the fire emergency response community (see Figure 7).
g. Disaster mitigation sensor technology system update
h. Traditional forest management is the right solution in disaster management

For generations, people in Australia have been taught how to handle fire disasters, so disaster mitigation is part of their culture. Realizing that the area is prone to fire disasters, the Australian population indirectly changes the city's development towards low density. It is essential to educate the public about the dangers of fire and its handling, turning weaknesses into strengths. This shows that disasters should not be avoided but rather adapted to regional conditions.

Figure 7. Australian fire conditions
Source: Cape association Picture, 2019

5. Conclusion
The purpose of this study is to provide an overall image of what is a liveable city from the perspective of disaster management. Disaster is seen as an event that can disrupt and threaten people's lives. At the same time, a liveable city is an idea to create an economically qualified city: environment and equity of life. Starting from the theoretical aspects presented in the first part of this study and ending with the practical ones, an attempt was made to answer the following question: can liveable cities be applied to disaster-risk areas, and what kind of concept will be used? To disaster-vulnerable areas. Albeit, there is no generally accepted concept of Liveability which can be applied equally from one country to another. Still, the most important thing is the elements of liveable cities that are generally different in each country because they are adapted to each region's geographical conditions and demographics. The application of the concept of a liveable city seen from disaster management is not only the urgency of fulfilling the requirements of a liveable city that is ready to face disasters but also views that liveable urban planning as a city ecosystem that has appropriate values and culture and is comfortable to live in so that it creates a sustainable city.

Acknowledgements
Authors wishing to acknowledge assistance or encouragement from colleagues who have helped and supported the preparation of this paper as well as resource persons and relevant agencies who have participated who provided time and input for this research.

References
[1] Badan Nasional Penanggulangan Bencana 2007 *UNDANG-UNDANG REPUBLIK INDONESIA NOMOR 24 TAHUN 2007 TENTANG PENANGGULANGAN BENCANA* (Jakarta)
[2] Badan Pusat Statistik 2018 *Upaya Antisipasi/Mitigasi Bencana Alam 2014 dan 2018*
[3] Peter E 2002 *Liveable Cities? The Politics of Urban Livelihood and Sustainability* (Berkeley: University of California Press)
[4] Victorian Competition and Efficiency Commission 2008 *A State of Liveability: An Inquiry into Enhancing Victoria's Liveability*
[5] Hahlweg D 1997 *"The City as a Family" In Lennard, S. H., S von Ungern Sternberg, H. L. Lennard, eds. Making Cities Liveable. International Making Cities Liveable Conferences* (California, USA: Gondolier Press)
[6] Douglass M 2002 From global intercity competition to cooperation for livable cities and economic resilience in Pacific Asia *Environ. Urban.* **14** 53–68
[7] Goldschalk 2014 *Urban Land Use Planning (5th Ed)*
[8] Lowe M, Whitzman C, Badland H, Davern M, Hes D, Aye L, Butterworth I and Giles-Corti B 2013 *Liveable, healthy, sustainable: What are the key indicators for Melbourne neighbourhoods?*
[9] Purnama S G 2017 Modul Manajemen Bencana *Fak. Kedokt. Univ. Udayana* 1–89
[10] Williamson F 2018 The politics of disaster: The Great Singapore Flood of 1954 *Environ. Plan. E*

Nat. Sp. **1** 251484861877687
[11] Hunter A 2020 *The Limits of Livability*
[12] Borchers Arriagada N, Horsley J A, Palmer A J, Morgan G G, Tham R and Johnston F H 2019 Association between fire smoke fine particulate matter and asthma-related outcomes: Systematic review and meta-analysis *Environ. Res.* **179** 108777

Public risk perception and public acceptance of the existing flood and drought mitigation measure in Bandung city

Linggar Y Asmara_[1], Saut Sagala[1], Danang Azhari[2*] and Elisabeth Rianawati [2]

[1] School of Architecture, Planning and Policy Development, Institut Teknologi Bandung, Bandung 40132, Indonesia
[2]Resilience Development Initiative, Bandung 40135, Indonesia

[*]Email: dana.azhari@rdi.or.id

Abstract. Flood and drought are hydrometeorological hazard that annually occurred in Bandung City. This problem occurs due to static natural conditions such as geographical, topographical conditions, as well as dynamic natural conditions such as climatological matters that are exacerbated by climate change and human activities. Flood and drought risk management in urban areas generally emphasizes physical development by ignoring various social dimensions. Therefore, this research aims to understand people's knowledge and attitudes towards disasters, represented by the public risk perception towards flooding and drought, as well as public acceptance of the existing programs that have been provided by the municipality. To identify public risk perception and their acceptance of existing programs use statistical descriptive methods. While the data collection use questionnaire with 99 samples. This research shows the affected communities can assess the risks, emotion, and expectation for the future risk, and the implemented program by municipality is mostly accepted by the community. Public risk perception and public acceptance of a program are prominent factors that determine the success or failure of a program by ensuring the compatibility of the program and the community. Thus, this research is critical to give the public perspective on implementing a community-based disaster mitigation program.

1. Introduction

In the recent years, Indonesia has been dealing with Hydrometeorological hazard that linked to climate change. Hydrometeorological hazards are the most dominant natural dangers occurring worldwide the impact of extreme climate and weather events like droughts, floods, tornadoes, landslides, or hurricanes [1]. Floods and droughts are common hydrometeorological dangers in Indonesia and have a complex and serious impact on various sectors of human life. In Indonesia, disaster has been increased rapidly from 2000-2010, and hydro-meteorological disasters account for more than 70% of the disasters in Indonesia [2]. Moreover, hydrometeorological hazards can also cause from static natural conditions such as geographic, topographical conditions, and dynamic natural conditions such as climate-related problems that are worsened by climate change and human activity [3,4].

West java is one of the province that has the highest vulnerability on flood and drought in Indonesia [5]. Bandung city as the capital of West Java is also prone to floods and droughts resulting from its static and dynamic natural conditions. From the side of the static natural condition, Bandung was situated in a Bandung basin surrounded by mountains and hills that served the upper part of Citarum watershed and had an important role in water catchment. Whereas, from a side of dynamic nature, it also deals with climate change that affects precipitation and the effect human activity on the environment. In Bandung, land conversion due to urbanization pressure significantly reduced infiltration that caused water shortages or clean water crises during the dry season and increasing frequency and intensity flood in Bandung City. Moreover, the increase in the proportion of built up land in Bandung City due to urbanization pressures has caused 95% of rainwater to be unable to infiltrate which has an impact on land subsidence and drainage failure, as well as increased pollution and sedimentation in waterways [6,7]. This characteristic makes the city exposed to various climate change issues and hydro-meteorological hazards.

Floods and droughts can be controlled through structural and non-structural efforts. In Indonesia structural efforts are still likely to be prioritize [6]. Furthermore, Indonesia government also adapted the top-down approach in disaster management [8]. In accordance with Indonesia, the structural efforts and top-down policy also dominate flood and drought management in Bandung, which is reflected in various efforts taken by the Bandung city government in the last 5 years. Since 2015, Bandung City has made various structural efforts flood and drought mitigation measures in the form of the construction of 6 retention ponds, normalization of rivers watershed, the construction of water basements (toll air), *Drumpori* (a larger bio pores) programs, and others. Unfortunately, the previous efforts that have been implemented still cannot overcome the problem of floods and droughts in Bandung and continue year after year with quit a big loss.

The various infrastructure solutions and top-down plans related to flood, and drought management can be effective when the use is adapted to the characteristics of the implementation areas. The complexity of interactions in social, ecological, and physical aspects of floods and droughts issue can produce the challenges in the control the issue, especially regarding its understanding, modelling, and management [4,9]. Furthermore, the management of flood and drought risks in urban areas including Bandung City generally prioritize on physical development by ignoring various social dimensions. The public's perception of the risk they face is an important part of determining social vulnerability and guaranteeing compatibility between programs and community. Researchers believe that risk perception has a direct impact on risk prevention and countermeasures in the public [10,11]. Moreover, the knowledge regarding how community understands and responds must be considered because individual beliefs about hazards are a crucial think about individual risk-making decisions.

In addition to risk perception, in implementing a top-down perspective it is also commonly need public acceptance. Public acceptance explains whether the public in general can accept the efforts that have been made in disaster management. It is also one of the determining factors in the success or failure of a project, program, and policy [12,13]. It is crucial to understand whether the community already know about the flood and drought measures and the expectation for the measures, so the existing measures can be improved if it is not in accordance with people's expectation or the target. Therefore, this research aims to understand people's knowledge and attitudes towards disasters, represented by the public risk perception towards flooding and drought, as well as public acceptance of the existing programs that have been provided by the municipality.

2. Literature Review

2.1. Public Perception

Risk perception is defined as complex and subjective process for an individual in assessing uncertainty (risk), which is reflected by their attitudes, intuition, expectations, experiences, as well as information related to the context [10,11,14]. Briefly, risk perceptions are often referred to as judgments made after they evaluate and characterize hazardous technologies and activities [15]. Risk perception leads to people's actions, decisions, values, emotions related events, and researchers admit that the direct explanation for flood prevention and response is the perception of flood risk. [16]. The main factor behind individual risk-taking decisions is the individual's belief about the danger with reference to the purchase of insurance and self-protection [17].

Risk perceptions are often used to assess individuals or the community's vulnerability to an environmental hazard [18,19]. Perception of each individual aims to assess the vulnerability to disaster by approaching the risk of disaster. Since risk perception can describe how people understand the physical environment and interactions therein [20]. Risk perceptions can be identified through the community assessments of the risk, their emotions related to the risk, and their future views regarding the risk. The process of forming risk perceptions is influenced by various factors. It consists of the individual's background, characteristics or contexts of the conditions, and the characteristics of something to be perceived [21]. Vulnerability can be identified through information on socio-economic as well as exposure and sensitivity that is related to the environment and self-condition of the people [10]. Information about previous disaster is also needed to determine the characteristics of the disaster according to community's view, and how the impact due to the disaster physically, materially, and psychologically. The understanding of public perception among the populations allows agencies to get wider information and views to refine and determine the risk communication, priority, modes of delivery related to the measure of risk disaster [20,22]. It also ensures the suitability and effectiveness of a measure by considering the specific needs of the targeted community.

2.2. Public Acceptance

Public acceptance can be conceptualized as a passive evaluation by the public regarding policy or a measure that has been implemented [23]. Public acceptance portrays the public action towards a new policy/technology which has been implemented. It is prevalently used in a top-down perspective, in which policies or measures are given by authorities to be accepted by the public. Despite the technological/policy promise, the implementation is limited by public acceptance. Public acceptance is a key point in successful technology/policy implementation [23–26]. Moreover, previous research discovered that even public acceptance can be an obstacle in the implementation of a policy, no matter how it has been chosen [13,27]. Basically, efforts to overcome the disaster must be done by all parties, including the community. Government runs bureaucracy and provides public services, and community's a target group of government's policy and services [28]. As a common hydrometeorological hazard that occurs in all regions in the worlds, floods and drought are naturally impossible to prevent. Mitigation measures will help to be more prepared to deal with disaster, build more stable ecosystems, recover from disasters by increasing resilience, and reducing the impact of disasters. [29].

Since public acceptance is a matter in the implementation of a program, it's necessary to find out the influencing factors or shape public acceptance. Public acceptance could represented by the public support of a program, as it has been assessed by Firestone et al. [30] and Petrova [31], despite Zverinova et al. [23] stated that public support has a further meaning. Public acceptance is influenced by various factors and characteristics in society, such as personal factors, socio-psychological factors, as well as other contextual factors [23,32]. Furthermore, knowledge, expectation, and public trust related to a program also need to identify to point out psychologically the public understanding dan awareness of a program. While public involvement and perceived responsibility related to a program can be identified to know how sense of belonging and responsibility has something to do with public decision to accept a program. Drought and flood impact mitigation is one of the main water management issues that should be approached on the basis of a proactive strategy which needs to be planned in advance in terms of long-term and short-term strategies for the response to hydrometeorological hazard.

3. Methodology

The data collection method used in this study was carried out through secondary and primary methods. Secondary data collection is done through data collection from media popular, journal, paper, as well as the authorities related to floods and drought, such as Diskar PB and DPU of Bandung City. The primary data collection was conducted by online self-completion questionnaire surveys for 99 samples who have been affected by floods and droughts in Bandung City before. Determination of the number of samples has done with random sampling method from the entire population in Bandung, which results a total of 99 samples. This research uses the qualitative methods to understand phenomena or occurrences affecting individuals, which in this research are the public risk perception on floods and drought and public acceptance on flood and drought mitigation measures in Bandung City. The quantitative research approach is used to measure quantitative data or objective statistics through scientific calculations [33]. Descriptive quantitative chemical analysis wont to descriptively interpret statistical data and aims to elucidate and summarize various conditions, situations or various variables that arise and become the thing of the research supported what happened. Previously, desk study is also used to determine the variables of public risk perception and public acceptance.

4. Result and Discussion

4.1. Flood and Drought Mitigation in Bandung City

To reduce the runoff volume, Bandung Municipality has carried out various programs to solve flood and drought disasters. Either specific measures aimed at a disaster or aimed at both disasters such as retention ponds and *drumpori*. Retention ponds are pond designed to attenuate surface runoff and improve water quality by promoting pollutant removal through sedimentation. In Bandung city, it was designed to release water through infiltration and groundwater recharge as well. *Drumpori* is an adapted infiltration trench into a smaller scale, using drums filled with certain materials as a medium to infiltrate water. It was an improvement of the bio pores program that was launched previously in 2013 [34]. In specifically coping with the floods, Bandung city has also launched several measures as providing water basements, maintaining the retaining wall (*TPT),* as well as dredging of sedimentation to normalize the waterbodies' function and capacity. Meanwhile to cope with drought problem, the Government of West Java in collaboration with Ministry of Public Works and Human Settlements established a drinking water supply system in January 2019. These measures are aimed to support achieving the water demand by providing access to clean water to 140,000 inhabitants with details of 250L/s for 80,000 inhabitants within the City of Bandung.

4.2. Characteristics of Respondents and Disasters

Based on the survey, respondents were distributed in various regions of Bandung City, with the majority being in Gedebage sub-district. Although it has a low population that is only about 0.015% of the total population, these sub-districts experienced severe flooding and drought in Bandung City. Forty-two percent of respondents' homes were semi-detached, 32% single homes, and the rest of the respondents lived in detached homes. Meanwhile, the pavement in which most respondents lived is dominated by asphalt or concrete (59%).

Most respondents experienced flooding for less than three times a year (62%), where the rest is more. The majority are flooded at an altitude of fewer than 0.5 meters (69%), followed by 0.5 - 1 meter (22%), with duration varying from 30 minutes - 1 hour (23%), to less than 30 minutes (37%), more than 3 hours (15%), and the rest were between 2 and 3 hours. This has the same flood characteristics as documented in DISKAR PB and DPU of Bandung City reports. In terms of drought, 48.5% of the respondent's experience drought for once a year. Whereas, only 11.1% of the respondent's experience drought for 5 times or more. The duration of drought that is felt by the respondents is mostly less than 1 month (64.6%), 18.2% for 1-2 months, and 6% for more than 4 months. Based on the surveys done, 60.6% of the respondents have easy access to water, yet 11.1% of the respondents have difficulty in getting water. For more than 35% of the respondents have their water streams all day long, 12.1% for more than 12 hours a day, 13.1% for 7-12 hours a day, 18.2% for less than 3 hours a day and 4% for no water at all.

4.3. Public Risk Perception on Flood and Drought in Bandung City

Knowledge can be defined as awareness as well as an information leading to understanding [35]. In terms of floods, most respondents viewed heavy rains and the drainage blockage due to trash as the major cause of floods. For the drought, dryness in the main water source and lack of groundwater are believed by most of the respondents to be the main causes of drought. The result indicates that public knowledge on flood risk source is likely focus on common knowledge, the same case also applies to the respondents' knowledge on drought. It reflected on fact that the absence of infiltration areas due to the increased impermeable surfaces is less understood as one of the most significant causes of urban flooding as well as drought, including in Bandung city. In accordance with Venkataramanan et al. whereas from the 85 studies around the world, communities could accurately identify that the cause of the flooding is incomplete drainage, extreme weather, and the increased impervious surfaces in neighbourhoods. For the drought, based on the document by Mediterranean EU Water Initiative [36], drought is not only cause by natural factors (precipitation deficiency, shifts in atmospheric circulation), but also it is caused by anthropogenic factors including population growth, land use and water demand. The community's understanding on floods and drought may lead to reduce the vulnerability of the community by encouraging changes in the community's behaviour [37].

Public awareness related to the possibility of risk in their environment is an essential element that determined how they act or adapt to the risk [38]. Regarding public awareness, respondents assessed that the current possibility of floods in their neighbourhoods is very low to quite high (72%). In terms of drought, respondents estimated that the current possibility for occurrence of drought is quite high to very high (77%). Nevertheless, respondents show a more pessimistic answer that both floods (48%) and drought (77%) might happen in their future neighbourhoods. Inline, most of the respondents viewed that the severity of floods (49%) as well as drought (47%) in the future will get worsen. Public awareness of future possibility of floods and drought is shown a positive signal. Since if public don't think that they might ever be suffering from floods or drought in their area, either present or within the future, they could not willing to require any action to decrease the risk [14]. Therefore, public awareness does contribute in achieving disaster risk reduction as it will increase the resilience of the community [39].

Experience with hazards has a significant effect on how people consider risks [40,41]. There is a condition referred to as "familiarity with risk source" in risk perception assessment related to the publics' experience. This condition is related to the publics' ability to tolerate and cope with risks resulting from the frequency of disasters [42]. Based on respondents' experience, most respondents felt that they rarely experienced floods (51%), which is less than three times a year. They also confessed that they feel no effect of floods as well. Further, 72% of respondents assessed that the floods led to minor losses which the losses average varies from 0 (55%) to IDR 500,000 (24%). Likewise, based on respondents' experience in flood, most respondents (48.5%) only experience drought for once a year. Only as many as 11.1% of the respondent's experience drought for 5 times or more a year. When drought happens, the effects of drought is perceived by the respondents in various levels. 62.6% of respondents feel the effects of drought, while only 6% feel no effects of drought. From these answers, it is seen that despite the lower frequency of drought compared with floods, the effects of drought bring more indicative consequences to the life of respondents. It indicated that their ability to tolerate and cope with the most frequent disaster that is flooding is increased. In the other side, public experience in varying severity time and again will influence public awareness, thus they will tend to be better informed as well as prepared [14].

Emotional reactions of individuals exposed to risk are the definition of worry [38]. Worry in the risk perception context is essential element because it serves as a normative value of awareness. Since, individual might understand the risks, but they may not be afraid of underestimating the consequences [38,40]. For the occurrence of both floods and drought, respondents' concern more on their health/safety, assets, as well as their economic condition tends to be extremely high with means of 46% on floods and 67% on drought. From these answers, respondents' concerns that tend to be high can be a potential emotion that leads them to act regarding risk but remain controlled through risk communication to avoid the excess of public fear.

When disaster appears, public's vulnerability takes a big part in determining the survival ability of the societies. The vulnerability is affected by various factors, including the receiving of disaster warnings, ability to evacuate in response to disaster warnings and ability in accessing post-disaster aid [43]. The majority of respondents felt capable of survive during the floods (92%), while only 41% of them capable of surviving during the drought. In dealing with both disasters, less than 8% of respondents have a specifically reserved fund to cope with floods or drought risk, even though 96% of them agreed that they do need prevention efforts to deal with that disaster risk. But most of them can provide the fund from their savings or other expense if it is necessary. However, this phenomenon reflected "optimism bias" which is a condition where the individual has the power and ability to address a hazard, incomprehensibly they have small basic to do so [22,44], which is in this context is reflected in how they willing to reserve their funds specifically to a disaster risk.

4.4. Public Acceptance on Flood and Drought Mitigation Measures in Bandung City

According to Zvěřinová et al. [23] conceptualized as a inactive person assessment, whereas bolster highlights a encourage meaning with respect to the behavioral response. In spite of these quick thinks about, acknowledgment and back are still utilized traded to work out whether people can acknowledge the endeavors that are made. Assessing the social acceptability of drought and flood management measures will contribute to the progress of drought and flood mitigation measures by selectors, thus it will lead to better community resilience. In terms of floods, several measures that have been launched tends to be accepted by the respondents (90%). Likewise, most of the respondents (93%) supports the provided programs related to drought. The public acceptance is reflected that most of the respondents supported to develop the same program in another area of Bandung city to deal with floods and drought.

According to Watanabe et al [45], public awareness and education related to the programs provided are the key elements for the disaster management. Awareness of the programs will contribute to the reduction of vulnerability thus cases of flood and drought can be minimized [46]. From the survey, Rancabolang Retention Pond is the closest program to where most respondents live (34%), compared to other measures related to both the flood risk as well as drought risk. However, most respondents stated that they were previously unaware of the program both for floods (54%) and for drought (59%). It is related to the absence of information regarding floods program 48% as well as drought program (54%), although some respondents received information from local governments (27%) and other sources such as social media, NGOs, and other popular media. In contrast, most of them felt that the closest program had decreased the impact of flooding that occurred in their neighborhoods (83%). Accordingly, most respondents are satisfied with the program (56%). According to Kholis et al. [47], satisfaction is an individual feeling resulting from a process that involves individual expectations and the performance of a program. Thus, the floods and drought measures in Bandung city performances have been in line with the publics' expectations reflected by the public's satisfaction with the programs.

According to Ross et al. [25], trust is a critical factor in acceptance regarding risk. Trust can be defined as a psychological state where individuals have the intention of accepting something based on their positive expectations. Public trust in the given program by the municipality in Bandung city tends to be high (90%). Besides, respondents also tend to trust the institution responsible for floods management (87%). Similarly, with floods, majority of respondents trust the program provided in dealing with drought (91%) as well as trust to the institution handling the program (87%). Although public trust is a prominent element, previous studies stated that many individuals only rely on their trust in the authority to make decisions because of their limited resources [48,49]. Moreover, excessive public trust can lead the individual intention to take any action regarding risk independently decreased (Cannon, 2020).

Respondents asked to choose a sufficient measure scale to reduce the flood and drought risk. As each community is unique and has their own characteristics, the measure scale of mitigation efforts by the respondents is perceived at a different level effected by one's capability in mitigating the hazards [50]. The result pointed out that large scale and small-scale measure are the most measures selected by the respondents. Whereas in the drought program, the top two scales of drought program that respondents believe will have the best fit are large scale (45%) and communal scale (28%). Respondents also assessed that the responsibility regarding flood risk is upon themselves as well as the government. In terms of drought, aside from government's responsibility in overcoming the drought as their most selected answer, the respondents pointed out the urgency of PDAM to participate as it becomes their second most-selected answer. Most of them assessed that it's a necessity to involve society in all stages of development from the planning process, the implementation, and the management of a program (53% for floods and 64% for drought). Individually, most of the respondents felt that they have a responsibility to carry out flood (97%) as well as drought mitigation efforts (95%). It is a positive indication since a greater sense of responsibility will leads to their intention to take any actions regarding risk. Although public involvement in flood and drought management programs requires more investment of time, efforts, as well as money, it will ensure the sustainability of measures adopted and build community resilience [4,51,52]. Public involvement, especially in the first stage of planning, will lead to the best decisions and plans, with the shared of information between the experts and the local that can also improve implementation that ensures the priorities of affected parties are addressed [4]. According to Heldt et al. [27] the most efficient way to engage the general public and find a generally recognized arrangement is to join open collaboration early in the design process.

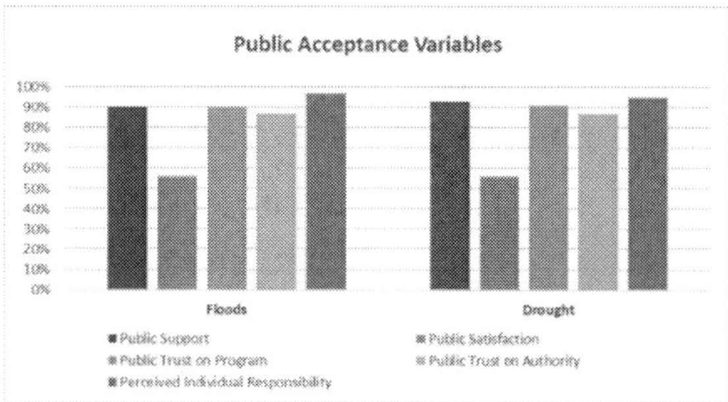

Figure 1. Rate of respondents who chose positive answers to floods and drought risk perception

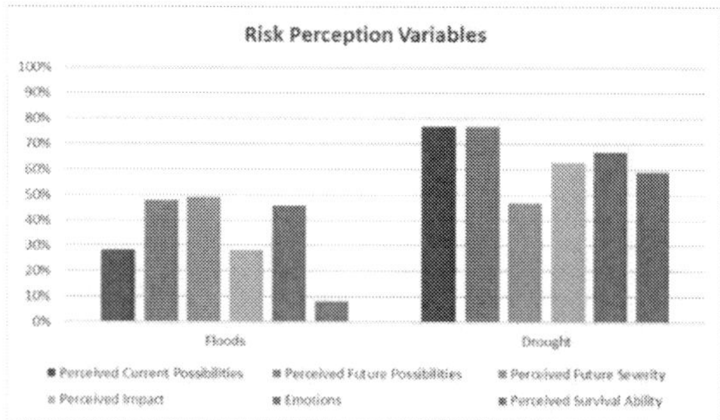

Figure 2. Rate of respondents who chose positive answers to floods and drought

The occurrences of floods and drought in Bandung City has shaped the publics' risk perception and has driven the municipality to provide several measures to cope with the risk. According to the survey, affected communities could assess the risks [see figure 1], although, its shown that the publics' flood risk perception is lower than the publics' drought risk perception. Furthermore, the affected communities supported the implemented program by the municipality that reflected the publics' acceptance of the programs. It also has been driven by the publics' trust both in the program and the authority as well as the perceived responsibility of the individual to take part in mitigation measures, even though the community satisfaction is still not shown a high number [see figure 2]. Nonetheless, all the programs provided could not be able to prevent both floods and drought yet.

5. Conclusion

The failure of structural measures which is expensive and publicly funded to cope with the risks has driven a paradigm shift in various countries. The core principle of these paradigm shifts is that the responsibility must be shared through engaging a wide range of stakeholders. Regarding the survey, the affected community in Bandung City has the potential to be involved in floods and drought coping efforts. It has reflected by their knowledge and awareness of the responsibility to take action as well as the view that it is necessary to involve the community in every stage of program development. Even though there is still a low-risk perception for floods and high public acceptance of government-provided programs that contradicts protective motivation theory, which would reduce motivation to take action. But these potentials can be developed to encourage shared responsibilities in coping with the risks. By involving the public could be the right way to improve risk information, raise awareness regarding risks, and increase community resiliency, in addition to spreading the cost of risk reduction measure itself. For this reason, it leads to the conclusion that a community-based disaster mitigation program should be enhanced in Bandung City.

Acknowledgments

This research was funded by Kurita Water and Environment Foundation through Kurita Overseas Research Grant 2019, as a part of research output entitled Stakeholder Acceptance Behaviour of Infiltration Well and Rainwater Harvesting System as Runoff Water Management in Bandung City. We would also thank Deffany Rosa and Silvia R. Dewi for the writing or editing support for this manuscript.

References

[1] Wu H, Huang M, Tang Q, Kirschbaum D B and Ward P 2016 Hydrometeorological Hazards: Monitoring, Forecasting, Risk Assessment, and Socioeconomic Responses *Adv. Meteoro*
[2] UNISDR 2017 *Country Assessment Report for Indonesia: Strengthening of Hydrometeorological Services in Southeast Asia*
[3] BNPB 2016 RBI Risiko Bencana Indonesia
[4] Challies E, Newig J, Thaler T, Kochskämper E and Levin-Keitel M 2016 Participatory and collaborative governance for sustainable flood risk management: An emerging research agenda *Environ. Sci. Policy* **55** 275–80
[5] Bencana B N P 2010 Buku Risiko Bencana Indonesia *Jakarta BNPB*
[6] Sagala S, Yamin D and Wimbardana R 2014 Adaptasi Non Struktural Penduduk Penghuni Permukiman Padat terhadap Bencana Banjir: Studi Kasus Kecamatan Baleendah, Kabupaten Bandung *Work. Pap. Ser.* **5**
[7] Hutasoit L 2009 Kondisi Permukaan Air Tanah dengan dan tanpa peresapan buatan di daerah Bandung: Hasil Simulasi Numerik *Indones. J. Geosci.* **4** 177–88
[8] Efendi D, Agustiyara and Putra H A 2019 Natural Disasters Management and the Challenge of Governability in Indonesia *Indian J. Public Adm.* **65** 627–45
[9] Jha A K, Bloch R and Lamond J 2011 *A Guide to Integrated Urban Flood Risk Management for The 21st Century*
[10] Adelekan I O and Asiyanbi A P 2016 Flood risk perception in flood-affected communities in Lagos, Nigeria *Nat. Hazards* **80** 445–69
[11] Miceli R, Sotgiu I and Settanni M 2008 Disaster preparedness and perception of flood risk: A study in an alpine valley in Italy *J. Environ. Psychol.* **28** 164–73
[12] Takagi K, Otaki M, Otaki Y and Chaminda T 2019 Availability and public acceptability of residential rainwater use in Sri Lanka *J. Clean. Prod.* **234** 467–76
[13] Gu Q, Chen Y, Pody R, Cheng R, Zheng X and Zhang Z 2015 Public perception and acceptability toward reclaimed water in Tianjin *Resour. Conserv. Recycl.* **104** 291–9
[14] Messner Frank and Meyer V 2006 Flood Damage , Vulnerability and Risk Perception – Challenges for Flood Damage *Environ. Res.* 149–67
[15] Slovic P and Peters E 2006 Risk Perception and Affect **15** 322–5
[16] Wang Z, Wang H, Huang J, Kang J and Han D Analysis of the Public Flood Risk Perception in a Flood-Prone City: The Case of Jingdezhen City in China **10** 1577
[17] Botzen W J W, Aerts J C J H and van den Bergh J C J M Dependence of flood risk perceptions on socioeconomic and objective risk factors **45** 15

[18] Birkholz S, Muro M, Jeffrey P and Smith H M 2014 Rethinking the relationship between flood risk perception and flood management *Sci. Total Environ.* **478** 12–20
[19] Armaş I and Avram E 2009 Perception of flood risk in Danube Delta, Romania *Nat. Hazards* **50** 269–87
[20] Adomah Bempah S and Olav Øyhus A 2017 The role of social perception in disaster risk reduction: Beliefs, perception, and attitudes regarding flood disasters in communities along the Volta River, Ghana *Int. J. Disaster Risk Reduct.* **23** 104–8
[21] Harliani F 2014 Persepsi Masyarakat Kampung Cieunteung, Kabupaten Bandung tentang Rencana Relokasi Akibat Bencana Banjir (Cieunteung Village Community Perception on *Journals.Itb.Ac.Id* **25** 37–57
[22] Bodoque J M, Díez-Herrero A, Amerigo M, García J A and Olcina J 2019 Enhancing flash flood risk perception and awareness of mitigation actions through risk communication: A pre-post survey design *J. Hydrol.* **568** 769–79
[23] Zvěřinová I, Ščasný M and Kyselá E 2014 What influences public acceptance of the current policies to reduce GHG emissions *Prague Charles Univ. Environ. Center. Retrieved April* **9** 2015
[24] Dolnicar S and Hurlimann A Water alternatives-who and what influences public acceptance? n/a--n/a
[25] Ross V L, Fielding K S and Louis W R 2014 Social trust, risk perceptions and public acceptance of recycled water: Testing a social-psychological model *J. Environ. Manage.* **137** 61–8
[26] Stenekes N, Colebatch H K, Waite T D and Ashbolt N J Risk and Governance in Water Recycling: Public Acceptance Revisited **31** 107–34
[27] Heldt S, Budryte P, Ingensiep H W, Teichgräber B, Schneider U and Denecke M 2016 Social pitfalls for river restoration: How public participation uncovers problems with public acceptance *Environ. Earth Sci.* **75**
[28] Susanti G, Halwatiah S, Rukmana N S, Yani A A, Hidayat A R and Ahmad S Public Service Performance and Public Trust in Government -- An Indonesian Case Study *Proceedings of the International Conference on Administrative Science ({ICAS} 2017)* (Atlantis Press)
[29] Solh M and Van Ginkel M 2014 Drought preparedness and drought mitigation in the developing world's drylands *Weather Clim. Extrem.* **3** 62–6
[30] Firestone J, Kempton W, Lilley M B and Samoteskul K 2012 Public acceptance of offshore wind power across regions and through time *J. Environ. Plan. Manag.* **55** 1369–86
[31] Petrova M A 2013 NIMBYism revisited: Public acceptance of wind energy in the United States *Wiley Interdiscip. Rev. Clim. Chang.* **4** 575–601
[32] Devine-Wright P 2007 Reconsidering public attitudes and public acceptance of renewable energy technologies : a critical review *Architecture* **Working Pa** 1–15
[33] Creswell J W 2014 *Research Design: Qualitative, Quantitative and Mixed Methods Approaches* (California, United States of America: SAGE publications)
[34] Rianawati E and Sagala S 2014 *Communal based flood mitigation measures in Bandung city* (Working Paper Series of Resilience Development Initiative, 10)
[35] Venkataramanan V, Lopez D, Mccuskey D J, Kiefus D, Mcdonald R I, Miller W M, Aaron I and Young S L 2020 Knowledge, Attitudes, Intentions, and Behavior Related to Green Infrastructure for Flood Management: A Systematic Literature Review *Sci. Total Environ.* 137606
[36] MED Joint Process WFD/EUWI 2006 *Water Scarcity Management in the Context of WFD*
[37] UNESCO Office Jakarta and Regional Bureau for Science in Asia and the Pacific 2004 *Flood Mitigation: a Community-based Project; maximizing knowledge to minimize impacts*
[38] Bradford R A, O'Sullivan J J, Van Der Craats I M, Krywkow J, Rotko P, Aaltonen J, Bonaiuto M, De Dominicis S, Waylen K and Schelfaut K 2012 Risk perception - Issues for flood management in Europe *Nat. Hazards Earth Syst. Sci.* **12** 2299–309

[39] Joyce C, Olivia K and Jordaan A 2013 Public awareness campaigns, a disaster risk reduction strategy for fire and flood hazards in the Western Cape, South Africa
[40] Bubeck P, Botzen W J W and Aerts J C J H 2012 A Review of Risk Perceptions and Other Factors that Influence Flood Mitigation Behavior PERCEPTIONS AND PRIVATE
[41] WMO 2016 Public Perception of Flood Risk and Social Impact Assessment
[42] Figueiredo E, Valente S, Coelho C and Pinho L 2009 Coping with risk: Analysis on the importance of integrating social perceptions on flood risk into management mechanisms - The case of the municipality of Águeda, Portugal *J. Risk Res.* **12** 581–602
[43] Substance Abuse and Mental Health Services Administration 2017 *Disaster Technical Assistance Center Supplemental Research Bulletin. Greater Impact: How Disasters Affect People of Low Socioeconomic Status*
[44] Harclerode M A, Lal P, Vedwan N, Wolde B and Miller M E 2016 Evaluation of the role of risk perception in stakeholder engagement to prevent lead exposure in an urban setting *J. Environ. Manage.* **184** 132–42
[45] Watanabe T, Cullmann J, Pathak C S, Turunen M, Emami K, Ghinassi G and Siddiqi Y 2018 Management of Climatic Extremes with Focus on Floods and Droughts in Agriculture *Irrig. Drain.* **67** 29–42
[46] United Nations International Strategy for Disaster Reduction 2005 *Hyogo Framework for Action 2005-2015: Building the Resilience of Nations and Communities to Disasters*
[47] Kholis N, Ratnawati A and Farida Y N 2018 Customer Satisfaction On The Performance Of Social Security Administrator (BPJS) Health in Central Java-Indonesia *Int. J. Organ. Innov.* **10** 150
[48] Ross V L, Fielding K S and Louis W R Social trust, risk perceptions and public acceptance of recycled water: Testing a social-psychological model **137** 61–8
[49] Cannon C, Gotham K F, Lauve-Moon K and Powers B 2020 From the general to the specific: the influence of confidence and trust on flood risk perception *J. Risk Res.* **0** 1–19
[50] Centers for Disease Control and Prevention 2020 Community Mitigation Framework
[51] The Intergovernmental Authority on Development 2020 *Community participation in drought management*
[52] Wang Y, Sun M and Song B 2017 Public perceptions of and willingness to pay for sponge city initiatives in China *Resour. Conserv. Recycl.* **122** 11–20

The role of the *Paguyuban Masyarakat Peduli Hutan* (PMPH) community as a form of disaster management in the highlands of Kudus Regency

Arif Purnomo[1,*], Ferani Mulianingsih[1], Amirudin Fahry[1] and Irma Mulia Setyarini[2]

[1] Social Studies Program, Universitas Negeri Semarang, Indonesia

[2] English Education, Pascasarjana Universitas Negeri Semarang, Indonesia

*Email: arifpurnomo32@mail.unnes.ac.id

Abstract. This article discussed the role of the *Paguyuban Masyarakat Peduli Hutan* (PMPH) community as a form of disaster management in the highlands of Kudus Regency. The purpose of this article was to analyze nature conservation activities as a form of disaster management which carried out by the *Paguyuban Masyarakat Peduli Hutan* (PMPH) community in Rahtawu Village, Gebog District, Kudus Regency. The focus of the study was to reveal the role of the PMPH community by using the phenomenological paradigm as a form of disaster management. The results of this paper found that Gebog District is one of the districts that was included in the landslide-prone area. Among the villages that are prone to landslides are Rahtawu Village, Colo Village, Menawan Village, Kedungsari Village, and Jurang Village. Rahtawu village was located in the highlands with an altitude of ± 1,627 m above sea level. This village physically had an undulating topography and was located in an area where the rainfall was quite high, namely 2,500-3,000 mm/year, with an average daily temperature of 290C. Rahtawu village in terms of morphology was an area located in the zone of moderate to high ground movement vulnerability. This village was also a protected area for the area below, because it was located in the upstream area of the Gelis sub-watershed and made Rahtawu Village a water catchment area that fills the Kudus groundwater basin. Rahtawu Village was designated as a protected forest area, community forest area, land movement prone area, cultural heritage area, local protected area around springs, and groundwater absorption area. , *Paguyuban Masyarakat Peduli Hutan* (PMPH) Community in Kudus Regency has carried out several activities (living Qur'an) which are actually Qur'anic values regarding human duty to protect and preserve the universe in order to bring benefits to the people.

Keywords: *Highlands, Paguyuban Masyarakat Peduli Hutan (PMPH) Community, Disaster Management*

1. Introduction

Various natural disasters have recently become the main news in various Indonesian media, both prints, electronic and social. During 2018-2021 more than five major natural disasters hit Indonesia. A number of earthquakes, tsunamis, volcanic eruptions, and the phenomenon of liquefaction claimed many victims. Indonesia, which is located in the Pacific ring of fire, has a high potential for natural disasters.

Administratively, the Muria Mountains area is located in three districts and 20 sub-districts. The Muria area in Kudus Regency includes three sub-districts, namely: Dawe District, Gebog District, and

Jekulo District. The Muria area in Jepara Regency includes nine sub-districts, namely: Bangsri District, Bale Alit District, Donorojo District, Keling District, Mayong District, Numsari District, Pakishaji District, Pecangaan District and Tahunan District. Pati Regency includes eight sub-districts, namely: Cluwak District, Dukuhseti District, Gembong District, Gunungwungkal District, Margorejo District, Margoyoso District, Tayu District, and Tlogowungu District.

Kudus Regency is a district in Central Java that has a variety of topographical structures, namely in the north is the highlands of Mount Muria with an altitude of 1,600 meters above sea level and sharp slopes that are prone to landslides. The southern part of Kudus Regency is in the form of lowlands with an altitude of 5 meters above sea level which is a basin area, which is prone to disasters, either in the form of floods, droughts, and hurricanes. Earth movement events that occur every year in Rahtawu Village, Gebog District, Kudus Regency make this village an attractive location for research.

Gebog District is a sub-district in Kudus district which is located at an average height of 155 meters above sea level. Gebog District is one of the sub-districts that is included in the area prone to landslides. Landslides are disasters that can occur due to ground movement in unstable soil conditions. Among the villages in it which are prone to landslides are Rahtawu Village, Menawan Village, Kedungsari Village, and Jurang Village [1].

Rahtawu village in terms of morphology is an area located in the zone of moderate to high ground movement vulnerability. This village is also a protected area for the area below it, because it is located in the upstream area of the Gelis sub-watershed (DAS) and makes Rahtawu a water catchment area that fills the Kudus groundwater basin [2].

Rahtawu village is designated as a protected forest area, community forest area, land movement prone area, cultural heritage area, local protected area around springs, and groundwater infiltration area [2]. The community needs to understand the efforts to deal with landslide disasters that can threaten safety. The purpose of this article is to analyze nature conservation activities as a shape of disaster management done by *Paguyuban Masyarakat Peduli Hutan* (PMPH) Community in Rahtawu Village, Gebog District, Kudus Regency.

Research on the PMPH community of the Ratawu Village community in the study of the living Qur'an has not been carried out. The phenomenon of the living Qur'an is a social phenomenon, so the research method model used is a social research model. Information about the activities of the PMPH community was collected using in-depth interview techniques. In addition, the technique of observing the actions of the PMPH community is also used.

2. Method

Qualitative research methods with a phenomenological approach are used in this study. The reason for using this method is that the focus of the research is to reveal the role of the PMPH community. On the other hand, the phenomenological approach aimed to describe the life experiences meaning experienced by several individuals, regarding certain concepts or phenomena, by exploring the structure of human consciousness.

The elements of the research method described in this paper were location, approach and perspective, data collection techniques, data analysis units, data collection strategies and data presentation. Information about the activities of the PMPH community was collected using in-depth interview techniques. The primary data sources for the study included the PMPH (*Paguyuban Masyarakat Peduli Hutan*) community and some village communities. Secondary data were obtained from various sources, such as books, interpretations that have relevance to the title, articles, magazines, journals, internet, and other media that support the research.

3. Result and Discussion

Rahtawu village is astronomically located at 6^{0}36'20"South Latitude to 6^{0}40'55"South Latitude and 110^{0}50'40"East Longitude to 110^{0}54'20"East Longitude. Rahtawu Village is located in an area designated for the protected area below it. This village is the most upstream village of the Gelis sub-watershed. The typology of Rahtawu Village is mostly highlands with an altitude of 500-1,200 meters above sea level and undulating land surface. The type of soil found in Rahtawu Village is regosol soil which is sensitive to erosion with soil thickness varying between 0.5 m-10m.

One of the villages in Kudus Regency that frequently occurs landslide disasters is Rahtawu Village, which is located in the administrative area of Gebog District, as shown in Figure 1. In this village, in every rainy season there are frequent landslides, both on river cliffs, road cliffs. or the cliffs that surround the village.

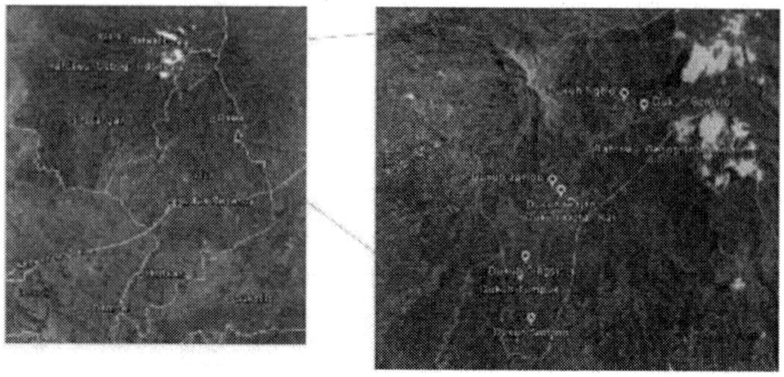

Figure 1. Rahtawu Village, Gebog District, Kudus Regency

This village is the most upstream village of the Gelis Sub-watershed, which has a function as a strategic area for environmental functions and carrying capacity that needs to be limited in development to maintain the sustainability of the area, determine a protected function and should not be converted as a cultivation area that has a designation as a protected forest area. , community forest areas, land movement prone areas, cultural heritage areas, local protected areas around springs, and groundwater infiltration areas [3].

The village of Rahtawu in the long dry season experiences drought and in the rainy season it experiences ground movement, therefore efforts are needed to keep water from being absorbed into the soil to fill the aquifer, but the water content in the soil must also be kept stable.

Regarding the Muria forest, the village community has a fairly high level of awareness to preserve the forest. This is evidenced by the existence of local groups who independently strive to maintain and preserve Muria's protected forest, namely the PMPH community or group (Paguyuban Masyarakat Protecting Forest), and the Kuncen Rejo Farmer's group which was inaugurated by the Tamzil Regent in 2018.

The initial background for the emergence of the PMPH community was from the hereditary instincts of Sunan Muria about 490 years ago. Sunan Muria has spread Islam in the middle of the wilderness. The people of Colo as the children and grandchildren of Sunan Muria feel they should have an active role in preserving the forest, protecting and caring for the forest in the Mount Muria area.

The Living Al-Qur'an anthropologically basically views this phenomenon as a socio-cultural phenomenon, namely as a symptom in the form of individual behavior patterns that emerge from the basis of their understanding of the Qur'an. Viewing this, the phenomenon that later becomes the object of study is no longer the Qur'an as a book, but the human treatment of the Qur'an and how patterns of behavior that are considered based on an understanding of the Qur'an are realized. The research object here is how the various meanings of the Qur'an above are present, practiced and take place in human daily life [4].

Placing the meaning of the Qur'an and its embodiment in human daily life as the central point of study, the paradigms needed are no longer the same as those used to study the Qur'an as a book. There are several anthropological paradigms that are considered suitable for studying the living Al-Qur'an. Of the many paradigms, the paradigms that are considered usable and can provide satisfactory results if applied properly are the acculturation paradigm, the structural paradigm, the functional paradigm, the phenomenological paradigm, and the hermeneutic paradigm [5].

In the level of thinking of the community and members of the Village PMPH community, in explaining the definition of preserving the environment or forest, it is not much different. Because in

IOP Conf. Series: Earth and Environmental Science **986** (2022) 012045 doi:10.1088/1755-1315/986/1/012045

the PMPH community itself the guidelines "Forest is a mandate, protecting and caring for it is worship", that is a slogan that is firmly held and believed by its members. From the slogan, PMPH members independently without any tendencies, Lillahi ta'ala want to maintain and care for the protected forest in the Muria Mountains. The meaning of "conserving nature" is often identified with the slogan of PMPH.

To maintain and save the environment, many efforts are taken by humans. But the most important thing here is starting from the improvement of human behavior towards the environment. And commonly referred to as ethics in the environment. Ethics is rooted in a person's awareness and morals. Environmental ethics is basically what actions are considered good for the environment and what is not good for the environment [6]. Environmental ethics can be realized in our daily lives.

Various efforts have been made by the PMPH group in conserving the Muria Mountain Forest. First, invite the community to be aware of the forest environment. The PMPH community is trying to make people aware of the environment around them. Sometimes by giving material in Friday sermons when Friday prayers take place in mosques.

Second, recommend not throwing garbage in the forest. If someone invites other people to talk about nature or environmental conservation, then from small things you have to apply it yourself, for example by requiring you to dispose of garbage in its place, then you must be able to reduce the production of plastic waste.

Third, protecting water sources. Forests have an important role in the availability of water. Forests that are not deforested will remain sustainable. Large trees with their roots function as absorbers of water that falls to the ground. So that the source of water will also increase. This water can be used as a reserve during the dry season. And forests as the lungs of the world are not just words and icons. Because if there are no trees, it will be fatal to the springs which can lead to drought.

Fourth, conducting forest patrols. Every year several forest patrols are held in the districts of Kudus, Jepara, and Pati. PMPH community members who attended were also self-supporting, before going on a forest patrol trip, a briefing was held.

Fifth, doing reforestation (replanting). Reforestation (replanting) is an alternative way to conserve forests. Reforestation of damaged forests can also prevent deforestation. With reforestation, the forest will be maintained and can still be used by humans.

Sixth, protecting forest areas from illegal logging and forest encroachment activities. Illegal logging is an act that violates the law. Encroaching on the forest by destroying it is not justified in religion, especially in the government because there is a law that regulates illegal logging. Here, the government must be involved in forest conservation.

Seventh, maintaining biodiversity in the Muria forest. Let wildlife live freely in nature. Don't hunt it wild. Because in the Muria forest there are still many rare animals in it. In the protected forest of Muria, once someone researched and installed a trep camera. In the trep camera, it is evident that there are still many rare animals that live in the Muria forest, such as tigers and tigers, wild boars, deer and deer, monkeys and langurs, snakes, and various kinds of birds.

Eighth, developing productive plants. Coffee plants are treated in such a way by being processed by PMPH. In fact, now coffee has recovered like it used to be with better quality with the presence of the PMPH community.

Ninth, developing ecotourism areas. PMPH has promoted ecotourism, although it has not been opened on a large scale. The existence of ecotourism packages, providing education about coffee plants. Every time there is an activity, the package offered is about education on preserving the environment of the Muria Forest. This is in accordance with PMPH's vision for a campaign on nature conservation

The PMPH community hopes that with the promotion of forest conservation, people can take advantage of the forest without having to destroy it by logging or forest encroachment. Restoring public awareness becomes very important at a time when humans have lost their morals and morals.

IOP Conf. Series: Earth and Environmental Science **986** (2022) 012045 doi:10.1088/1755-1315/986/1/012045

4. Conclusion

Building the trust of the people of Rahtawu Village is a challenge that must be answered by the parties, both the people of Rahtawu Village themselves and related parties related to the preservation of the Muria forest, namely Perhutani, Kudus Regency Government, Educational Institutions, and existing non-governmental organizations. It must be understood together that social capital will not run out if it is used, but will only increase. PMPH has done activities/behaviors (living Qur'an) that are exactly Qur'anic values regarding human duty to protect and preserve the universe so that it can bring benefits and maslahah to others. Nature conservation also has a status as a manifestation of piety, faith and worship to Allah SWT. Caring for the forest environment means showing a caring attitude, not only to nature, but also a form of concern for the sustainability of future generations.

Reference

[1] Badan Penanggulangan Bencana Daerah 2017 *Kudus*
[2] Riani M, Prabandiyani S and Izzati M 2013 Pemetaan Kondisi Tanah dan Vegetasi Sebagai Upaya Mengurangi Terjadinya Bencana Gerakan Tanah di Desa Rahtawu Kecamatan Gebog Kabupaten Kudus *Pros. Semin. Nas. Pengelolaan Sumberd. Alam dan Lingkung. 2013* 283–8
[3] Badan Perencanaan Pembangunan Kabupaten Kudus 2012 *Rancangan Review Rencana Tata Ruang Wilayah Kabupaten Kudus 2013-2023*
[4] Abdul Mustaqim 2007 *Metode Penelitian Living Qur'an Model Penelitian Kualitatif, in Metodologi Penelitian Living Qur'an dan Hadis* (Yogyakarta: Teras)
[5] Ahimsa-Putra H S 2012 the Living Al-Qur'an: Beberapa Perspektif Antropologi *Walisongo J. Penelit. Sos. Keagamaan* **20** 235
[6] Khafidhoh M 2013 Teologi Bencana Dalam Perspektif M. Quraish Shihab *ESENSIA J. Ilmu-Ilmu Ushuluddin* **14** 37

IOP Conf. Series: Earth and Environmental Science **986** (2022) 012046 doi:10.1088/1755-1315/986/1/012046

Analysis of run-off potential for flash flood management in Rahakolowu Watershed

R D Mazalena[1*], Ahmad [1, 2] and Y Yusup [1, 2]

[1] Master of Geography Education, Sebelas Maret University, Surakarta
[2] Lecturers of Geography Education, Faculty of Teacher Training and Education, Sebelas Maret University, Surakarta

*Email: mazalenarahma08@gmail.com

Abstract. Rahakolowu Watershed is an important and strategic issue area for the South Buton Regency. The characteristic of this region is in the form of hills and having wavy until the very steep slope which is very potential of having high *run-off*. The purpose of this study was: 1) to determine the potential for surface *run-off* that occurs in the Rahakolowu Watershed; 2) to map the potential for flash floods that occur in the Rahakolowu Watershed. The approach used in this study was a modified HRU (Hydrological Response Unit) with *a run-off* occurrence land unit as its unit analysis. Based on the result, the highest *run-off* water discharge produced in each sub-basin was an average of more than 0.50 m3/sec, both on forest, moor, residentials, mix garden and shrubs. The high slope and the area of each land unit in a flow occurrence is the factor causing the high *run-off* in the Rahakolowu Watershed. Flash floods occur along the main river channel and become inundation floods when they reach downstream. The accumulation of run-off in the southern sub-basin has more potential as a flood contributor in the Rahakolowu Watershed.

1. Introduction

Run-off is part of the rainfall that flows over the land surface to rivers, lakes, or oceans [1,2]. *Run-off* occurs when the amount of rainfall exceeds the amount of infiltration. After fulfilling the infiltration rate and basin, the water will flow freely on the ground surface [3]. The increase in run-off volume will result in flooding problems in the downstream watershed [2]. Surface run-off is affected by many factors simultaneously. In general, the affecting factors of run-off grouped into two are the meteorological factors and the characteristics of the watershed, which include rain intensity, duration of rain [4], distribution of rainfall, width, and shape of the watershed, and topography [5].

The watershed functions as a system of run-off processes that can be described as a conceptual model. Therefore, a watershed-based erosion study is crucial to examine the characteristics of the run-off relationship between spatial units [6]. Rahakolowu Watershed is an important and strategic issue area for the South Buton Regency because the South Buton Regency Government will build an office center in that part of the Watershed. According to the location, the head office of the South Buton Regency government is at the foot of the Rahakolowu Watershed.

Based on the landform, the office center is under construction on a fluvial landform. It means that area of the building is very vulnerable to flooding and sedimentation problems, both short-term and long-term. The characteristic of the region is in the form of hills and having wavy until the very steep slope, which is very potential for a high *run-off*. The run-off water acts as a force that causes erosion and landslides, then transported and deposited to a lower place as sedimentation. At the same time, the accumulation of run-off water can cause flooding.

Rainfall data from the Betoambari observation station shows that in the last 11 years, the highest monthly rainfall intensity occurred in December, reaching 628 mm/month (classified as very high-intensity rainfall). If the monthly rainfall is assumed to be 50% (approximately 314 mm) falling within 24 hours, the resulting run-off can cause flash floods in the upstream area. Based on the assumption, a very high run-off occurred in the Rahakolowu Watershed. It proved by the cross-sectional width of the river after arriving at the office construction reaching 60 meters, whereas, when there is no intensive rain, the width is only about 7 meters.

Therefore, *run-off* analysis in the Rahakolowu Watershed is a top priority as the basis for formulating policies to control potential flash floods and inundation floods in the short, medium, and long term in the South Buton Regency office area. The aim of this research was: (1) to determine the potential for surface *run-off* that occurs in the Rahakolowu Watershed; (2) to map the potential for flash floods that occur in the Rahakolowu Watershed.

2. Methods

As the research location, the Rahakolowu Watershed has an area of approximately 2,128, 62 ha, with quite diverse location characteristics. Therefore, to make it easier to organize data, it is done by modifying the Hydrologic Response Unit (HRU) approach with a unit analysis unit of land flow occurrences as the object of research. It means that all data to be collected and analyzed, patterned in object units, are called land units of flow occurrences.

Hydrologic Response Unit (HRU) is a land unit with sub-watershed characteristics that affect erosion [7] and surface run-off. Hydrologic Response Unit (HRU) formed from a watershed or sub-watershed into smaller parts connected by a river network. The smallest part of the watershed is the smallest unit, where all hydrological processes were simulating [8]. The concept of HRU is according to an integrated systems approach which is a requirement to analyze and dynamically model the hydrology of the heterogeneous drainage structure of the watershed and its interaction with soil, geology, and land cover [9]. HRU is the smallest land unit obtained by overlaying soil, land use, and slope data [10]. Meanwhile, HRU analysis requires land-use data and soil maps that overlaid with watershed boundaries [11].

The flow occurrence land unit map was made from 3 (three) maps: the land unit map, HRU map, and contour map. The map illustrates the distribution of land units in flow occurrence. It means that each land unit polygon can only consist of one flow occurrence but can also consist of two or more flow occurrences. Therefore, the border of observation and unit analysis for each land unit is the border flow occurrences (Figure. 1).

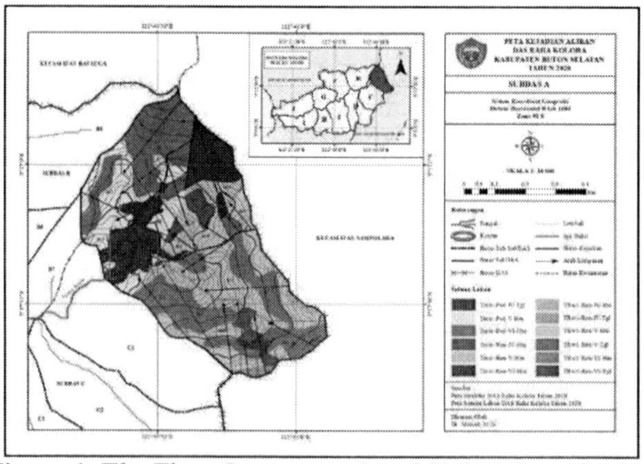

Figure 1. The Flow Occurrence Land Units Map of Sub-watershed A

Based on the aims which have been described, the steps of analysis were:

2.1. Run-off Potential Analysis

2.1.1 Land Unit Characteristics
The means annual rainfall per month and annual rainy days per month are calculated based on the amount of precipitation and rainy days in the same month for 10 years divided by 10. The maximum 24-hour rain taken was the highest 24-hour rain for 10 years.

Soil characteristics obtained through laboratory analysis results from samples of soil types in the research location. The slope characteristic was determined by using equation (1):

$$(\%) = (tg\alpha / 45) \times 100 \quad (1)$$

Types of land use and land management were obtained from field observations. Categorizing of land management was determined according to the kind of conservation effort by the community on each land use. The unit area of the flow occurrence was analyzed digitally using the ArcGIS program..

2.1.2 Run-off Potential. Run-off
Potential was calculated by using equation (2):

$$Q = 0,0028CIA \quad (2)$$

Where Q is *a run-off*, C is a *run-off* coefficient, I is rainfall intensity, and A is the area of flow occurrence land units. The run-off coefficient value (C) uses the C value from the U.S. Forest Service (1980 in Asdak, 2014).

Rainfall intensity was calculated by using equation (3):

$$I = \frac{R_{24}}{24}\left(\frac{24}{tc}\right)^{2/3} \quad (3)$$

Where I is the average rain intensity for t hours (mm/hour), R24 is the daily rainfall or 24 hours of rain (mm), Tc is the concentration-time or flow arrival time (hours).

Tc was calculated using the Kirpich equation (4):

$$Tc = 0,0195L^{0.77}S^{-0.385} \quad (4)$$

Where L is the maximum length of water travel in one land unit of run-off occurrence and S is the height difference between the highest and lowest points of one of flow occurrence land units divided by the length of run-off.

2.2. Flood Potential Analysis
Analysis of flood potential calculated by using (i.e., (2)) multiplied by the area of each land unit included in one flow occurrence. The results of the overall run-off discharge at each flow occurrence compared with the river's capacity. When the accumulation of run-off discharge exceeds the river's flow capacity, whether caused by the flow occurrences in the sorrounding or caused by the addition in the intersection point of the tributary branches, at that moment, it is included in the flood category. Flood categorization is analyzed based on the location of the flood.

Flash floods occur when the accumulation of run-off discharge in the upstream river exceeds the capacity. At the same time, it has called inundation flood when it arrives downstream.

3. Result and Discussion

3.1 Run-Off Potential

Rahakolowu Watershed is a unitary land area with the Rahakolowu River, and the tributaries always give the input that then enters the Rahakolowu River and out to one outlet. The boundaries are ridges or mountain peaks from upstream to downstream with an area of 2,182.62 ha. Rahakolowu Watershed was divide into 10 sub-watershed, the distribution of which consists of 3 sub-watershed on the upper, 5 sub-watersheds on the middle, and 2 sub-watersheds on the lower. Furthermore, the distribution of sub-watersheds in the Rahakolowu Watershed can be seen in Table 1. and Figure. 2.

Table 1. Distribution Of Rahakolowu Sub-Watersheds

No	Sub Watershed	Coordinat	Sub Sub Watershed	Type of Flow Event		Hectare
				Singular	Dual	
1.	A	5°36'44,16"–5°37'45,74" LS dan 122°40'13,44"–122°41'6,56" BT	3	2	20	165,85
2.	B	5°36'43,02" – 5°37'39,59" LS dan 122°39'22,57" – 122°40'34,20" BT	5	11	41	219,56
3.	C	5°37'25,04" – 5°38'22,10" LS dan 122°39'51,85" – 122°40'47,65" BT	3	30	44	201,22
4.	D	5°37'38.82" - 5°39'7.75" LS dan 122°39'25.64" - 122°40'24.50" BT	4	15	38	294,15
5.	E	5°37'55.55" - 5°39'14.87" LS dan 122°38'41.98" - 122°39'35.68" BT	8	13	37	251,95
6.	F	5°36'53.72" - 5°38'3.59" LS dan 122°38'27.57" - 122°39'34.79" BT	7	11	43	248,04
7.	G	5°37'29.90" - 5°38'22.50" LS dan 122°38'14.29" - 122°39'6.19" BT	4	28	37	172,31
8.	H	5°38'18.47" - 5°39'17.18" LS dan 122°38'14.98" - 122°38'58.62" BT	4	10	38	153,54
9.	I	5°38'15.50" - 5°39'8.22" LS dan 122°37'0.89" - 122°38'26.01" BT	4	14	44	216,59
10.	J	5°37'42.71" - 5°38'40.03" LS dan 122°36'45.26" - 122°38'17.57" BT	5	16	42	259,61
	Total		**47**	**150**	**384**	**2182, 82**

Source : Data Analysis, 2020

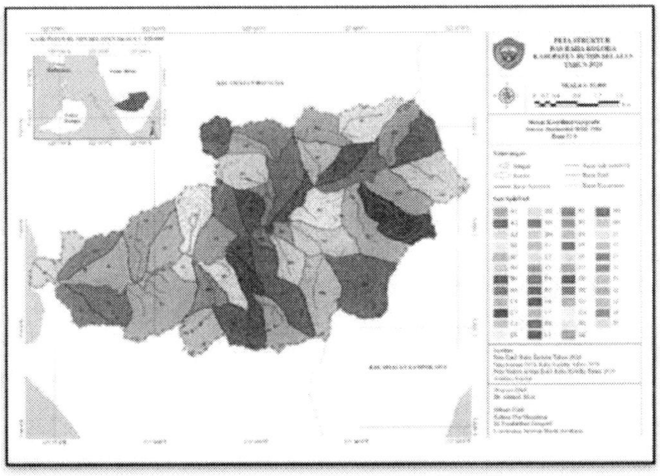

Figure 2. Distribution Of Rahakolowu Sub-Watersheds

The results show that the highest run-off discharge produced in each sub-watershed is an average of more than 0.50 m3/sec, both on forest land use, dry land, residential, and shrubs. The highest run-off discharge generated by land units with residential, which is located on a slope of 8-15%, reaching 24.52 m3/s (Table 2).

Table 2. The Highest Peak Run-Off According to Land Use

No.	Landuse	Highest Run-off		Sub Watershed
		Q (m3/sec)	Slope (%)	
1.	Forest	1,88	>140	F
2.	Mix Garden	0,24	2 - 7	J
3.	Moor	4,22	7 - 15	I
4.	Residential	24,52	7 - 15	J
5.	Shrubs	4,24	7 - 15	H

Source : Data Analysis, 2020

Based on Table 2, each land use has a different flow rate based on the slope level. The steep slope can cause high peak discharge, volume of the *run-off* [12], *run-off* erosion [13]. The potential for the *run-off* in the Rahakolowu Watershed tends to be caused by high slope. The slope in the Rahakolowu watershed is dominantly steep (30%-70%) to very steep (70%-140%). That slope level is spread evenly from the upstream to the downstream of the Rahakolowu Watershed. However, along an area adjacent to the main river from the middle to the downstream, the slope level is gentle (7%-15%) to moderately steep (15%-30%). It is necessary to know that the outermost watershed edge of the northern tends to be steeper than the southern.

There are 5 types of land use in the Rahakolowu Watershed: forest, shrubs, gardens, residential, and moorland. Each land use has a different flow rate based on the slope level. Land use is an indicator when determining the value of the *run-off* coefficient (C). The method of determining the *run-off* coefficient value for each land use is by considering 3 (three) indicators in each sample, including slope, soil texture, and land cover. Then the median value is taken from the range of C values determined by the U.S. Forest Service (1980) [1]. In the Rahakolowu Watershed, the dominant land use is forest. Forest only has a *run-off* coefficient value of 0.15, but the highest peak *run-off* discharge produced is 1.88 m3/s at a slope of >140% (Table 2). The falling power of *run-off* will be higher when *run-off* occurs on very steep slopes. It has a tremendous risk of accumulation of water flowing in the Kolowu tributaries which, then causes the river channel to be unable to accommodate the flow rate, resulting in overflow. Another risk that arises from high *run-off* can result in massive power as a scraper and soil carrier along the area traversed. However, the slope factor has a lack effect on residential. The *run-off* discharge produced by the residential is 24.52 m3/s at a slope of 7-15% (Table 2). Residential land use has the highest coefficient value, which is 0,40. Under certain conditions, the *run-off* coefficient (C) can affect the amount of the generated *run-off*. The higher the value of the *run-off* coefficient (C) at that time, the greater the *run-off* discharge produced.

Management *run-off* is carried out based on the river capacity. The principal determination to arrange management *run-off* in this study is similar to the erosion management that has been conducted in the Keduang Watershed Wonogiri, Central Java [6]. The difference is that this research is not concentrated on the number of erosive indexes but only on the flow sources. The flow capacity that enters the river comes from the accumulation of *run-off* contributions by several flow events in sub-sub-watershed. To calculate the controlled flow, we should calculate the difference between the total flow contribution and the river's capacity. Furthermore, the number of flow occurrences is required to be a controller of the *run-off* discharge. That's because each flow occurrence has the same capacity to manage the excess *run-off* generated in the sub-sub-watershed (Table 3).

Table 3. Comparison Of Peak Run-off With Tributary Capacity in Rahakolowu Watershed

No.	Sub Watershed	Sub Sub Watershed	Total of Flow Occurence	Amount of Peak *Run-off* (m^3/ sec)	Tributary Capacity (m^3/ sec)
1.	A	A1	6	53,33	50
		A2	8	26,98	
		A3	8	49,51	
2,	B	B1	16	60,94	50
		B2	9	34,53	
		B3	9	40,62	
		B4	8	39,43	
		B5	10	45,15	
3.	C	C1	22	27,20	50
		C2	28	33,01	
		C3	24	42,10	
4.	D	D1	9	41,48	50
		D2	12	53,08	
		D3	19	59,60	
		D4	13	85,24	
5.	E	E1	1	6,44	70
		E2	6	28,82	
		E3	1	0,58	
		E4	19	76,83	
		E5	1	9,98	
		E6	7	21,92	
		E7	9	58,80	
		E8	6	25,51	
6.	F	F1	5	21,02	70
		F2	9	39,63	
		F3	10	36,93	
		F4	6	25,32	
		F5	5	35,84	
		F6	13	51,91	
		F7	6	7,47	
7.	G	G1	13	21,75	70
		G2	12	31,27	

		G3	18	35,51	
		G4	22	35,92	
8.	H	H1	12	31,94	70
		H2	9	36,46	
		H3	12	40,52	
		H4	15	59,16	
9.	I	I1	2	19,27	100
		I2	37	101,64	
		I3	10	26,29	
		I4	9	7,23	
10.	J	J1	11	29,71	100
		J2	22	30,69	
		J3	12	43,82	
		J4	10	7,05	
		J5	3	34,60	

Source : Data Analysis, 2020

For example, in Sub-Watershed A, Sub-sub-watershed A1 produces the highest *run-off* by 53.33 m3/s, while the tributary capacity in Sub-watershed A is only 50 m3/s. It means we should control the flow by about 3,33 m3/s. There are 6 (six) contributing flow occurrences in section A1. So that must be managed 16-17% of the total controlled *run-off* by each occurrence. The high *run-off* in Sub-Sub-Watershed A1 is due to the contributed by dual flow occurrences. Dual flow occurrences are *run-off* that passes through two or more land units within an occurrence border. Each land unit within an occurrence border has a different area. That area affects the accumulation of the flow. If the *run-off* passes through two large land units, then more *run-off* will be generated. When *run-off* occurs in a dual flow occurrence, the *run-off* discharge from the land unit in the lower gets additional *run-off* from the land unit above. So when the flow enters the river channel is formerly the accumulation of *run-off* discharge by the sorrounding flow occurrences. It means that the amount of *run-off* discharge is also affected by the area of the land unit.

Land management needs to be directed well at all land uses in the Rahakolowu Watershed. A high *run-off* potential on land units throughout the Rahakolowu Sub-watershed can result in massive power as soil scrapers and transporters if the land is not well-managed. Therefore, land management also needs to be directed well at all land uses in the Rahakolowu watershed. Regarding the flash flood potential in the Rahakowu Watershed will be explained in the next point.

3.2 Mapping the Flash Flood Potential

The construction of a civic center in the Rahakolowu Watershed area can be a magnet that can attract local people to bring their activities closer. It is possible because during development will be followed by adequate facilities and infrastructure. Therefore, the surrounding area may experience expansion, mainly residential, both planned and sporadically in the long term. The office center is a vital facility for the South Buton Regency government, so it has to be secured from the hazard of potential flash floods.

Flash floods are floods caused by peak flows that occur in a flash time after extreme rain [14]. Another opinion explains that flash floods are universal natural disasters that are dangerous and cause economic losses [15,16]. Every year around the world, the lives of millions of people are threatened by floods [17]. Flash floods are characterized by rapid and intense runoff formation, making them difficult to predict [18].

Rainfall and run-off data are of the essential hydrological elements in the flood mapping of watershed systems [19]. Based on the number of sub-watersheds in the Rahakolowu watershed area, sub-watershed E has the most sub-sub-watersheds, namely 8 sub-basins. Sub-watersheds A and C are sub-watersheds that have the least number of sub-watersheds, which are 3 sub-watersheds each. Based on the number of flow events, it can be explained that the Subwatershed that has the most flow events is Subwatershed

C, which is 74 flow events consisting of 30 single flow events and 44 double flow events. Sub-watershed A has the least number of flow events, namely 22 cash flow events consisting of 2 single flow events and 20 double flow events. This means that sub-watersheds A and C receive flow contributions from 22 and 74 flow events that are patterned in the three sub-watersheds (Table 3).

Prediction of potential flooding in the Rahakowu watershed is calculated based on the potential for accumulation of runoff water originating from each flow event that enters through the sub-sub-watershed which then enters the sub-watershed and then enters the main river flow. The results of the total runoff water discharge in each sub-watershed are then compared with the capacity of the tributaries. This is done so that it can be seen which sub-sub-watersheds have the potential to be a source of flooding.

The measurement of the carrying capacity of the creek was not carried out thoroughly because the dominant terrain was steep, so the measurements were only carried out in easily accessible locations. According to the measurement results, the width of the Rahakowu tributary reaches 10 m while its height reaches 5 - 10 m. The results in Table 3 show that several sub-watersheds produce a total peak run-off that exceeds the capacity of their tributaries, so they have the potential to contribute to flooding when the flow enters the main river.

Sub-watershed A is the uppermost sub-watershed in Rahakowu watershed. It should be noted that there are 9 sub-watersheds under Sub-watershed A, including Sub-watersheds B, C, D, E, F, G, H, I, and J. These 9 sub-watersheds function as reservoirs for peak flow. runoff from its sub-watersheds. The water flowing from the uppermost sub-watershed will accumulate with the flow coming from the sub-watershed below it, so that the water discharge will increase even though the capacity of the river in the downstream area is getting bigger. The amount of water discharge that collects in the main river is certainly influenced by the accumulation of run-off generated from each sub-watershed. Based on the analysis of the run-off potential in the previous discussion, it is stated that the run-off potential in the Rahakowu watershed is high. The high run-off if it is concentrated in one place can cause the river channel to be unable to accommodate the peak discharge, thus encouraging flash floods in the main river channel.

Flash floods occur when the accumulation of flow discharge in the upstream area exceeds the capacity of the main river and will become inundation when it reaches the downstream area. The next thing to note is the condition of the main river that flows along the watershed from upstream to downstream. The position of each sub-watershed greatly determines the amount of runoff accumulated in the main river. The table below presents the results of the analysis after calculating the peak runoff accumulation in each sub-watershed (Table 4).

Table 4. Comparison of Accumulation Peak Run-off with Main River Capacity

No	Sub Watershed	Hectare	Sub Sub Watershed	Type of Flow Occurrence		Amount of Peak *Run-off* (m^3/ sec)	Accumulate of Amount Peak *Run-off* (m^3/ sec)	Main River Capacity (m^3/ sec)
				Singular	Dual			
1.	A	165,85	3	2	20	129,82	129,82	50
2.	B	219,56	5	11	41	220,67	350,49	150
3.	C	201,22	3	30	44	102,31	452,80	150
4.	D	294,15	4	15	38	239,40	692.20	170
5.	E	251,95	8	13	37	228,88	921.08	200
6.	F	248,04	7	11	43	218,12	1.139,20	200
7.	G	172,31	4	28	37	124,45	1.263,65	250
8.	H	153,54	4	10	38	168.08	1.431,73	360
9.	I	216,59	4	14	44	154,43	1.586,16	360
10.	J	259,61	5	16	42	145.87	1.732,03	360
	Total	**2182, 82**	**47**	**150**	**384**			

Source : Data Analysis, 2020

The accumulated peak runoff that flows from Subwatershed A has exceeded its capacity when it reaches the main river. Sub-watershed A1 resulted in accumulated run-off of 53.33 m3/s, while the tributary channels in sub-sub-watershed A1 were only able to accommodate 50 m3/s. This means that river water begins to overflow in the A1 sub-watershed. The overflow is sourced from runoff water produced in the A1 sub-watershed area. The capacity of the river channel in the A2 and A3 sub-watersheds is still greater than the accumulated peak discharge generated (river water is still flowing normally). However, things started to turn around when the streams met at a fork near the main river. The total peak discharge flow generated by the Sub-DAS reaches 2x the capacity of the main river. In other words, banjir bandang has started to occur when it passes through the main river in Subwatershed A. More specifically, the explanation is mapped in full in Figure 3.

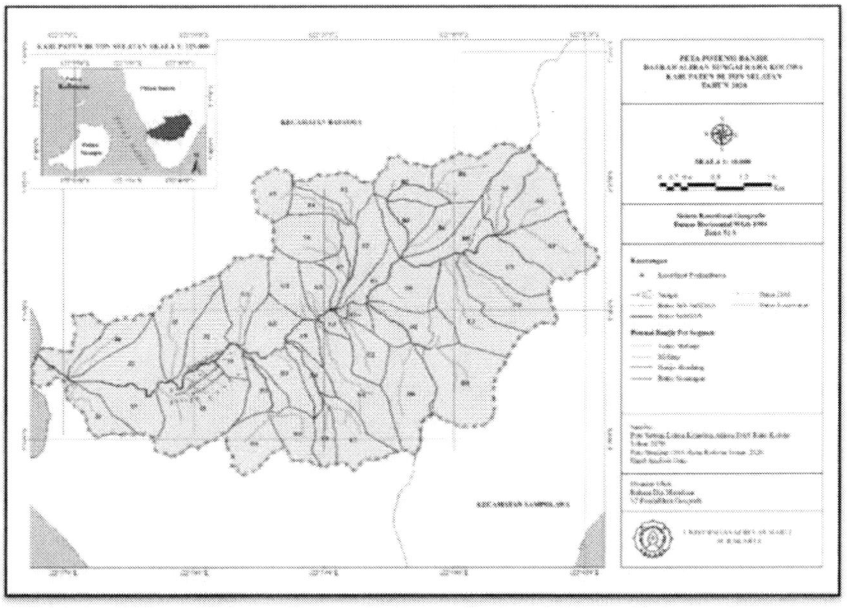

Figure 3. Map of Flood Potential According Section of the Rahakolowu Watershed

The color of the river flow line in Figure 3 has different meanings from each other. The green color means that the flow rate is still flowing normally (not exceeding capacity). The orange color means that there has been an overflow caused by the accumulation of flow events in a sub-watershed. While the red color means that a flash flood has occurred. Runoff events in the southern part of the Rahakowu watershed are more likely to be a contributor to flooding. This is because the number of tributaries that have begun to overflow is greater than in the northern part of the watershed. The overflow of the tributaries causes flash floods before reaching the main river. If this happens, then when it reaches the main river flow, the resulting damage will be even greater, so that it is able to erode the walls of the river cliffs in its path.

The thing that needs to be considered is the flow of the creek in the office area (sub-DAS I). Grooves or valleys are outlets of water flow from the highest areas of the rain catchment area, such as watersheds and/or sub-watersheds. Based on the results of field observations, the depth of the channel in subwatershed I is only 10 meters, while the distance from the main river to the office location is approximately 50 meters. In sub-watershed I there has been an overflow caused by local flow events, so it needs to be controlled so that flooding does not occur. The proximity of the location to the main river flow (stream) that is too close can also threaten the existence of office construction. River flow is a very dynamic area depending on soil conditions and river discharge. Based on Table 4, the accumulation of peak discharge in sub-watershed I has exceeded the capacity of the main river. At that time, there was a widening of the river's flow because the water that flowed extended beyond the main river channel. Another thing that might happen is when the river channel is blocked, the river water can find a way outside the main channel (shift into the surrounding area).

Figure 4. a) and b) : The Condition of the main river in Sub-watershed I; c) Sedimentation in Sub-watershed I; d) Cliff Erosion in Sub-watershed I

4. Conclusion

Based on the research as explained, the following conclusions are the highest run-off water discharge produced in each sub-watershed is an average of more than 0.50 m3/s, both on forest land use, dry land, settlements, mixed gardens, and shrubs. The determination of management run-off is based on the river's capacity by involving the number of contributory run-off occurrences in each Sub-watershed. Flash floods occur along the main river channel and become inundation floods when they reach downstream. It happens because the accumulation of peak discharge increases when it flows downwards, thus exceeding the capacity of the main river. The quantity of run-off occurrences in the southern sub-basin has more potential as a flood contributor in the Rahakolowu Watershed.

So far, management of watershed environmental problems in various regions has not been effective because it prioritizes the intervention control model or forces nature to accept the applied model, so it is inefficient for long-term control. This modified Hydrologic Response Unit (HRU) concept can determine the characteristics of the flow when it rains on each flow unit which affects the amount of power generated. The approach of this modification reflects the character of the watershed where the rain falls that regulates the natural flow of water. So, in formulating flash flood management, we can involve natural characteristics more. The benefit of this method can manage run-off as the primary cause of flooding in all flow segments, so it is pretty effective for the implementation of long-term management.

Impact prevention is necessary, but don't ignore the source of the problem. Flood management must be carried out thoroughly and simultaneously from upstream to downstream. By controlling the runoff discharge from the upstream, the flash floods potential in the Rahakolowu Watershed can be decreased. So, it can secure the middle and downstream of the watershed, especially the location of office center construction.

Acknowledgments
The authors would like to thank the Regional Government of South Buton Regency for facilitating this research and several members of the Masiri Village community who were willing to help during the research.

References

[1] Asdak C 2009 *Hidrologi dan Pengelolaan Daerah Aliran Sungai* (Yogyakarta: Gadjah Mada University Press)
[2] Verrina G P, Anugrah D D and sarino 2013 Analisa Runoff Pada Sub Das Lematang Hulu *J. Tek. Sipil dan Lingkung.* **1** 23–31
[3] Sari S 2010 Studi Limpasan Permukaan Spasial Akibat Perubahan Penggunaan Lahan (Menggunakan Model Kineros) *J. Pengair. UB* **2** 13
[4] Guo B, Xie T and Subrahmanyam M V 2021 Rainfall runoff in small watershed after wenchuan earthquake in western china based on remote sensing technology – a case study of jianping gully watershed *Polish J. Environ. Stud.* **30** 2543–52
[5] Budinetro H S, Fatchan A K and Sahid M N 2012 Pengendalian Aliran Permukaan Akibat Perubahan Tata Guna Lahan Dengan Konsep Low Impact Development *Semin. Nas. Tek. Sipil UMS 2012* 100–11
[6] Ahmad 2016 *Kajian spasial ekologi erosifitas hujan dan limpasan untuk modifikasi prediksi erosi di DAS Keduang Wonogiri* (Gadjah Mada University)
[7] Alexander M C and Sapei A 2018 Analisis Potensi Air Sungai Pada Embung 190 Di PG. Bungamayang PTPN VII, Lampung *J. Tek. Sipil dan Lingkung.* **3** 77–84
[8] Ferijal T 2012 KRUENG JREU MENGGUNAKAN MODEL SWAT Predicting Runoff and Erosion Rate from Krueng Jreu Subwatershed Using SWAT Model *Agrista* **16(1)** 29–38
[9] Emiyati ., Kusratmoko E and Sobirin . 2017 Spatial Pattern of Hydrologic Response Unit (Hru) Effect on Flow Discharge of Ci Rasea Watershed Using Landsat Tm in 1997 To 2009 *Int. J. Remote Sens. Earth Sci.* **13** 39
[10] Christanto N, Setiawan M A, Nurkholis A, Istiqomah S, Sartohadi J and Hadi M P 2018 Analisis Laju Sedimen DAS Serayu Hulu dengan Menggunakan Model SWAT *Maj. Geogr. Indones.* **32** 50
[11] Suhaila M R N, Zuhairi A, Azlyn A N S and Zaini M M 2021 Soil data definition for hydrologic response unit analysis in swat model of Langkawi Island, Malaysia *Eurasian J. Soil Sci.* **10** 68–76
[12] Abuzied S M and Mansour B M H 2019 Geospatial hazard modeling for the delineation of flash flood-prone zones in Wadi Dahab basin, Egypt *J. Hydroinformatics* **21** 180–206
[13] Wang N, Yao Z, Liu W, Lv X and Ma M 2019 Spatial variabilities of runoff erosion and different underlying surfaces in the Xihe River basin *Water (Switzerland)* **11** 1–15
[14] Toukourou M, Johannet A, Dreyfus G and Ayral P A 2011 Rainfall-runoff modeling of flash floods in the absence of rainfall forecasts: The case of "cévenol flash floods" *Appl. Intell.* **35** 178–89
[15] Group M R 2005 Annual review : natural catastrophes 2005 *Knowl. Ser. Top. Geo* 56
[16] Gökçekuş H, Kassem Y, Alijl N and Tawalbeh M 2020 Flash flood risk mitigation plan: Zarqa Ma'in Basin, along the dead sea in Jordan *Int. J. Sci. Technol. Res.* **9** 4089–95
[17] Khaleghi S, Mahmoodi M and Karimzadeh S 2015 Integrated application of HEC-RAS and GIS and RS for flood risk assessment in Lighvan Chai River *Int. J. Eng. Sci. Invent.* **4** 38–45
[18] Water I and Management R 2014 RS & GIS Based Spatial Mapping of Flash Floods in Karamana and Vamanapuram River Basin , Thiruvananthapuram District , Kerala 1236–43
[19] El Maghraby M, Masoud M and Niyazi B 2014 Assessment of surface runoff in arid, data scarce regions; an approach applied in Wadi Al Hamd, Al Madinah Al Munawarah, Saudi Arabia *Life Sci. J.* **11** 271–89

IOP Conf. Series: Earth and Environmental Science **986** (2022) 012047 doi:10.1088/1755-1315/986/1/012047

Empirical study on energy consumption, carbon footprint, and economic growth in Indonesia

Shella Elly Sritrisniawati[1], Ayya Agmulia A.I[2], Evi Gravitiani[3]
Faculty of Economics and Business, Universitas Sebelas Maret, Surakarta, Indonesia

*Email: shellasritrisniawati@gmail.com

Abstract. Energy is a basic material that can be used for economic development. Increasing growth requires continuous energy exploration, so it requires new energy sources and technological innovations in energy. Energy consumption makes a big contribution in influencing climate change in the presence of carbon. In Indonesia, with a fairly dense population, it certainly increases the most energy consumption. Based on this, the purpose of this study is to analyze the relationship between variables of energy consumption, carbon footprint as an indicator of climate change and economic growth. The energy consumption used in this study is the consumption of electrical energy, oil, gas and energy innovation as alternative energy that can be realized. The method used is VARX analysis. The use of this method is to analyze the relationship between variables and analyze the resulting impact on each variable. The results show that there is a correlation between energy consumption and economic growth. In addition, there is a significant impact of shocks on the energy consumption variable. The higher the level of energy consumption, the higher the carbon footprint. energy consumption cannot prevent climate change, while energy consumption provides benefits in minimizing carbon. so that energy consumption in Indonesia needs to be controlled.

1. Introduction

The consumption sector is a major driving factor in a country's economic growth. As in Indonesia, economic growth is formed from high and high levels of consumption. The Ministry of Finance of the Republic of Indonesia projects economic growth that increased 5 percent in the first quarter of 2019 when compared to last year. But in 2020 it coincides with a pandemic of economic growth to depreciate. Despite the economic slowdown, consumption behavior will continue to be needed. Global energy consumption continues to grow mainly in developing countries such as India and China, besides that the country is also the most populous country like Indonesia and includes the top 10 energy consumers in the world [1]. Energy consumption continues to increase due to economic activities and activities that are still running despite restrictions on activities due to the COVID-19 pandemic.

Energy is a basic material that can be used for economic development. Growth requires continuous energy exploration, so it needs new energy sources and technological innovations in energy. Indonesia's economy before the pandemic, total energy consumption increased greatly in 2018 and 2019. Although it can encourage economic development, Indonesia is also a dilemma to increase economic growth or reduce the negative impact of energy consumption during pandemics such as today. Huang and Le Huang's research (2019)[2] showed between individual energy consumption and economic growth in China with the ARDL method from 2004 to 2017. The study resulted in individual energy consumption having a positive impact on economic growth. The analytical tools used are granger cointegration and causality. The result is a two-way causality between energy consumption and economic growth. So it can be interpreted that economic growth can affect energy consumption, and vice versa.

In contrast to research conducted by Susanto and Laksana (2013)[3] on energy consumption and economic growth in six ASEAN countries in 1998-2010. The results showed a one-way relationship between economic growth and energy consumption. Economic growth leads to energy consumption, but energy consumption does not result in changes in economic growth. In line with research conducted by Shaari, *et al* (2013)[1] which examined the relationship between energy consumption and economic growth in Malaysia. Using Johansen's cointegration analysis tool. The results showed that petroleum and coal consumption had no impact on economic growth. The relationship of turbidity occurs only one way from gas consumption to economic growth. So if the government's policy to reduce gas use will have a negative impact on Malaysia's economic growth. Natural resources are also used as energy sources, then will be utilized by consumers [4]. In addition, another opinion that energy consumption has more influence on environmental degradation. Increased energy consumption that results in carbon dioxide emissions that can directly damage the environment. The carbon footprint produced by energy consumption has a negative impact on the environment and health. There are many environmental problems that occur because the energy consumed increases [1,5,6] also explained about environmental problems that can hamper the economic development process if the government does not intervene in reducing energy and the impact of carbon emissions. So there needs to be efforts to reduce the carbon footprint produced by applying environmentally friendly energy consumption.

In response to this problem in 1980, the *United Nations* Conference on New and Renewable Energy was defined as a new technology and new materials, modernized development and satisfaction from traditional renewable energy sources. Energy is essential to achieving interrelated economic, social, and environmental goals in sustainable development. So that in developing countries such as Indonesia face challenges and problems in increasing economic growth and reducing the negative impact of energy consumption on the environment [1,7]. Energy consumption has a role in improving the economy, therefore in testing the relationship between energy consumption and economic growth and in addressing policy issues regarding energy consumption.

With an interest in issues regarding global energy, it triggers the development of research that links the energy sector with the economic sector in a particular country or region. In this study raises the problem of how the relationship between energy consumption and economic growth in Indonesia and the second how the impact of the *shock* of each variable and which variables have a greater influence. The novelty of this research uses energy consumption variables from various sides, including the consumption of electricity, oil, gas and energy innovation as alternative energy and includes CO2 as an indicator of the carbon footprint left by energy use. In terms of research methods, this research uses the VARX model with the latest research data in 2020.

2. Method

2.1 Types and Sources of Data

This research is a type of quantitative research. Based on this type of research, the study emphasizes more testing and measurement of research variables in the form of numbers, as well as conducting data analysis using statistical tools. Research data used from 1985 to 2020. In analysis is also based on theories and literature studies. This study uses secondary data obtained from: World Bank, BPS Indonesia, Bank Indonesia, INDO DAPOER, economic reports and other literature studies.

2.2 Operational Definition

Table 1. Operational definition

	Variable Name	Proxy	Information
Dependent Variables	Economic Growth (EG)	GDP Growth	Percent (%)
Independent Variables	Electric Energy Consumption (ELEC)	Electricity from oil, gas and coal sources	Percent of total consumption energy (%)
	Oil Energy Consumption (OIL)	Energy use of oil	Kg per capita
	Gas Energy Consumption (GAS)	Fossil Fuel Energy Consumption	Percent of total consumption energy (%)
	Renewable Energy (RE) Consumption	Alternative and nuclear energy	Percent of total energy use (%)
	Carbon Footprint (CF)	CO2 Emissions	Total CO2 Emission from consumption energy (kt)

Source: Multiple sources, processed

2.3 Research Model

The study adopted model specifications from Chaudhry *et al* (2012)[8] and Shaari et *al* (2013)[1] relating to energy consumption to economic growth. The equations in this study consider functional and econometric models. Empirically the selection of research models that describe the variables taken include eg dependent variables*(Economic Growth)*and independent variables namely ELEC (Electricity*Consumption)*,OIL (Energy*Use of Oil)*,GAS (Fossil*Fuel Energy Consumption)*,RE (Renewable Energy) and CF (Carbon*Footprint)*. Based on previous research models, this study can be derived as follows:

$$EG = f(ELEC, OIL, GAS, RE, CF) \quad (3.1)$$
$$EG = \beta_0 + \beta_1 ELEC + \beta_2\, LOIL + \beta_3 GAS + \beta_4 RE + \beta_5 LCF + U_t \quad (3.2)$$

Information:

EG = Gdp Growth,
LOIL = Log energy *use of oil*,
GAS = *Fossil fuel energy consumption*,
ELEC = *electricity consumption*,
RE = Renewable Energy,
LCF = Log carbon *footprint*.
U_t = *error term* (error factor)

The equation describes the relationship between consumption, oil, gas, renewable energy and carbon footprint that affects economic growth.

2.4 Vector Auto Regression Exsogen (VARX)

Vector *Auto Regression Exsogen* (VARX) method is a simultaneous relationship that is also determined by exogenous variables that have an influence on other exogenous variables, but are difficult to influence by endogenous variables. VARX is a structural model of VAR with exogenous variables [9]. There are three types of exogenous variables, namely weak, strong and very strong. In this case it is assumed, that in the model there are endogenous variables (Y) and exogenous (X). Explanations of these three types include:

1. Weak exogenous variables occur if variable Y (endogenous) cannot explain X (exogenous), this state is required for estimation.
2. Strong exogenous variables occur if the lag value Y does not explain X (no *feedback*relationship), this state is required for*forecasting*.
3. Very strong exogenous variables occur when variable Y does not change at all when variable X changes.

In this model it is assumed that in the model there are endogenous () and $Y_t(X_t)$exogenous variables, the stationary value of the variable can be at the level or level 1st *difference.* These conditions are adjusted to the data used assuming the basis of no cointegration between variables. Common forms of VAR models include the following:

$$X_t = \beta_0 + \beta_n X_{t-n} + E_t \quad (3.4)$$
$$EG_t = \beta_0 + \beta_1 ELEC_{t-n} + \beta_2 LOIL_{t-n} + \beta_3 GAS_{t-n} + \beta_4 RE_{t-n} + \beta_5 LCF_{t-n} + E_t \quad (3.5)$$

Which $X_t$is the vector element of the EG model = Electrical Energy Consumption (ELEC), Oil Consumption (OIL), Gas Consumption (GAS), Renewable *Energy* (RE) and Carbon Footprint (CF). $\beta_0$is the notation for *the vector* constant n x 1. Is the coefficient of n is the length of lag. It is the $\beta_n X_t E_t$*vector* of each *variable's* error. From the above variables are transformed into VARX analysis. Based on the VAR equation above it is derived into the basic model of VARX as adopted from Bierens (2004):

$$X_t = \alpha_0 + A_1 Y_{t-1} + \cdots + A_P Y_{t-p} + B_1 X_{t-1} + \cdots + B_1 X_{t-P} + E_t \quad (3.6)$$

Where $Y_t€Ꝁ^{k}$ $€ꝀX_t^{k}$ is an exogenous variable, $\alpha_0€Ꝁ^k$ is a slop *intercept,* is the metric coefficient k x A_pk, is the metric coefficient k x B_pm, and $\alpha_0€Ꝁ^k$ is the error *term.* So that derived varx equation model, among others:

$$E\ [e_t|EG_t = \alpha + \beta_1 ELEC_{t-n} + \beta_2 LOIL_{t-n} + \beta_3 GAS_{t-n} + \beta_4 RE_{t-n}\{LCF_{t-n}\}_{n=1,2,3,4,5}] = 0\ (€Ꝁ^k) \quad (3.7)$$

It was later assumed that the VARX model for LCF_{t-n} is as follows:

$$LCF_{t-n} = c_t + C_1 X_{t-1} + \cdots + C_r X_{r-1} + v_t$$
$$E\ [v_t|EG_{t+1} = \alpha + \beta_1 ELEC_{t+1} + \beta_2 LOIL_{t+1} + \beta_3 GAS_{t+1} + \beta_4 RE_{t+n} + \beta_5 LCF_{t+n} + \beta_6 ELEC_{t-n} + \beta_7 LOIL_{t-n} + \beta_8 GAS_{t-n} + \beta_9 RE_{t-n}\{LCF_{t-n}\}_{n=1,2,3,4,5}] = 0\ (€Ꝁ^m) \quad (3.8)$$

The equation implies that EG, ELEC, LOIL, GAS, and RE do not explain LCF, thus indicating exogenity. This exogenity arises because the LCF variable is a resulting impact can also be called the carbon footprint in energy consumption that arises due to energy use by humans. But in equation (8), $\beta_5 \neq 0$ it is possible if EG, ELEC, LOIL, GAS, and RE have indirect effects on LCF through the interdependence effect on and. The VARX model, which reflects the relationship between endogenous and exogenous variables that interdependence indirectlyby echoing$e_t v_t$equations (3.5) and (3.8)is obtained:

$$EG_t = \beta_0 + \beta_1 ELEC_t + \beta_2 LOIL_t + \beta_3 GAS_t + \beta_4 RE_t + \beta_5 LCF_t * EG_t + E_t \quad (3.9)$$

IOP Conf. Series: Earth and Environmental Science **986** (2022) 012047 doi:10.1088/1755-1315/986/1/012047

2.5 Impluse Response Function (IRF) Test

The Impluse Response Function is used to look at the movement of the impact or effect of a*shock*that occurs in one of thevariables and its effect on other variables in the present and future periods. IRF can find the impact of a shock on endogenous variables. In IRF a good time span is to know the effect or impact of shocks at several periods.

3. Results And Discussion

3.1 VARX Analysis Results

The second method in this study is to conduct an analysis of varx model estimates. The VARX model shows the relationship between exogenous variables and endogenous variables according to the assumptions that have been formed. The exogenous variables in this study are the impact or carbon footprint resulting from energy consumption (LCF) as a result of which carbon components are harmful to the environment and human health. The carbon footprint used in this study is carbon dioxide (CO2 Emissions) emissions. Based on the results of VARX estimates are indicated by the results of VARX *estimates* and Impulse Response Function (IRF). The results of the varx model estimates are as follows:

$$EG_t = \beta_0 + \beta_1 ELEC_t + \beta_2 LOIL_t + \beta_3 GAS_t + \beta_4 RE_t + \beta_5 LCF_t * EG_t + E_t$$

$$EG_t = 3.58198 + 2.01438\ ELEC_t + 0.171343\ LOIL_t + 1.39185\ GAS_t + 0.446551\ RE_t + LCF_t * EG_t + E_t.$$

Overall the equation of VARIABLES ELEC, LOIL, GAS and RE at time has a probability of 0.000 significant at the level of $\alpha < 0.05$ (5%) with a positive influence on EG or economic growth in Indonesia. But the results are different when it has included time lag. The time lag used in this study is in lag 2. The results of these estimates include the following:

Table 2. VARX Estimated Results with lag 2 Energy Consumption, Carbon Footprint and Economic Growth in Indonesia

Dependent Variables	Exsogenous Variables	Independent Variables	Coefficient value	Prob.
EG		EG_{t-1}	0.2791051	0.119
		EG_{t-2}	-0.1737441	0.261
		$ELEC_{t-1}$	-0.6879292	0.039**
		$ELEC_{t-2}$	-0.3056224	0.250
		$LOIL_{t-1}$	1.44938	0.740
		$LOIL_{t-2}$	4.508329	0.353
		GAS_{t-1}	-0.4407251	0.381
		GAS_{t-2}	0.4851397	0.419
		RE_{t-1}	0.2381165	0.859
		RE_{t-2}	-0.7417548	0.581
	LCF	-	4.530145	0.188
	Consanta	-	-10.11467	0.783

* significant α=1%, ** significant α=5%, *** significant α=10%.

Based on the results of the estimated VARX model at lag 2 ya ng indicated bytheprobability value with the value of α as the category of significance level. The table above obtained the result that the EG dependent variable has a significant effect only on the dependent variable ELEC t-1 with Prob. α = 0. 039 where less than $\alpha<5\%$. While in other dependent variables in lag 2 does not have a significant influence indicated by a Prob value of more than $\alpha > 5\%$ (0.05). Likewise, exogenous lcf variables have no significant effect on EG (economic growth) with prob values of 0.188 and 0.783

more than 0.05. But when viewed from the coefficient value in LCF t-1 has a positive value it can be interpreted that LCF (carbon footprint C02) has a positive direction. So that if LCF (CO2 carbon footprint) increases, increasing it can increase EG (economic growth). This is due to the large number of economic activities that use energy consumption.

These results are in line with Huang and Le Huang's (2019)[2] research that provides individual energy consumption results have a positive impact on economic growth. Supported also by the results of research conducted by Qiang Hou (2009)[10] in his research in China that economic growth can affect energy consumption, and vice versa. The results of this study also showed that energy consumption that has a big influence is the consumption of electrical energy, then oil and gas energy. This is because during the covid-19 pandemic, electricity consumption has increased because many activities are carried out at home by utilizing the internet network and electricity energy. With the restriction of community activities outside activities that require oil energy becomes reduced.

3.2 Impulse Response Function (IRF) Test Results on endogenous variables

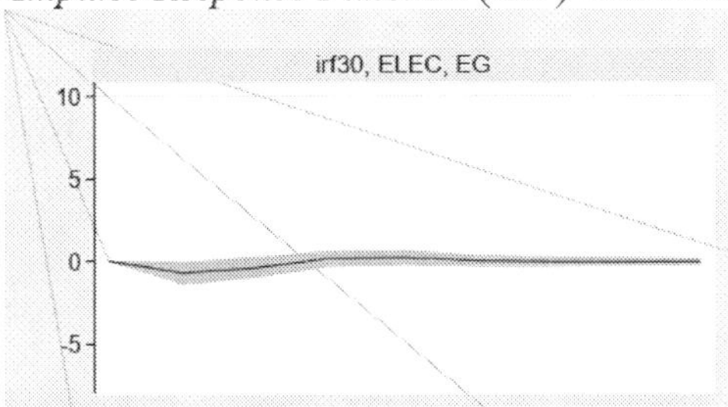

Figure 1. IRF between ELEC and EG

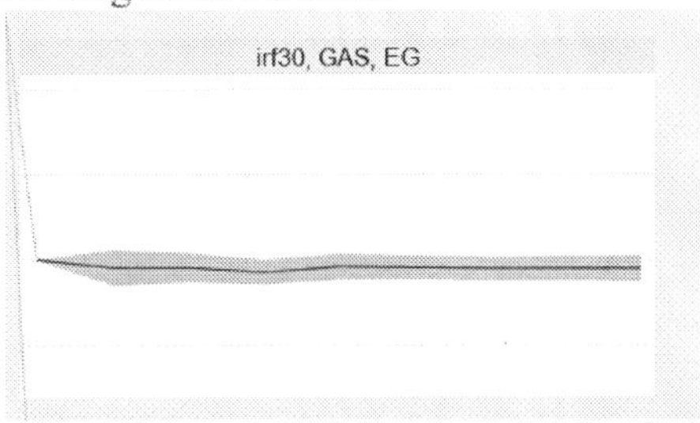

Figure 2. IRF between GAS and EG

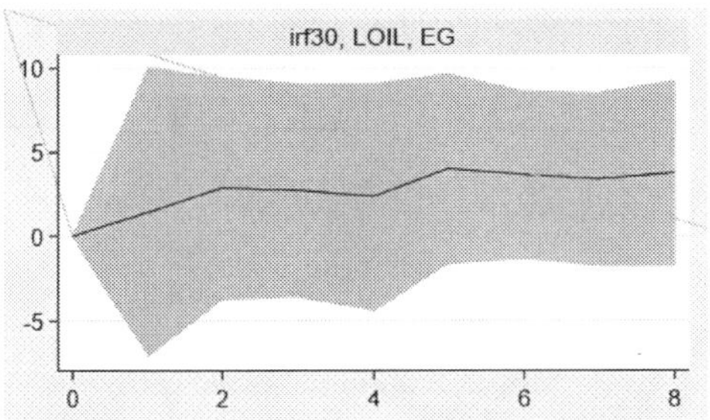

Figure 3. IRF between LOIL and EG

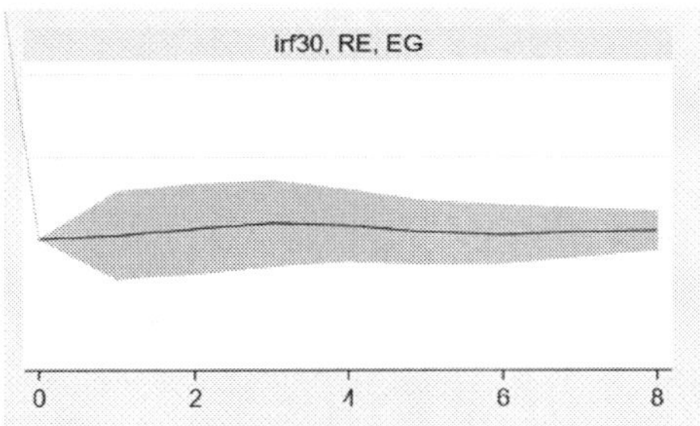

Figure 4. IRF between RE and EG

IRF testing of Energy Consumption, Carbon Footprint and Economic Growth in Indonesia shown in Figures 1 to 4 shows that there is a refreshment between variables. The response that occurs in the economic growth variable (EG) caused by shocks in independent variables at 8 periods. The shock condition in Figure 1 with ELEC impulse to EG occurs a stable shock with a surprise that occurs only at the beginning of the period. There are fluctuations in *shock* at the beginning of period to period 4. After that it regains balance (stable). The *Fossil Fuel Energy Consumption* (GAS) variable seen in Figure 2 explains that the occurrence of shocks is almost the same as the previous variable and the EG variable responds constantly. Only the increase at the beginning of the period, in the next period is lined balance.

Impulse variable shock Log *Energy Use of Oil* (LOIL) to response by EG is shown in Figure 3. Shocks from the LOIL variable elicit a very variable response at the beginning of the period to the 8th period. The shock response is very strong and leads to a positive trend. This can be interpreted that if there is a shock in oil consumption will have an impact on fluctuations in economic growth in Indonesia. From Figure 4 explained that the shocks of variable *impulse* RE (Renewable Energy) to EG tend to be stable and do not give a significant response in the event of a shock. Direction remains constant and precise on the balance line. This condition allowsthat the existence of alternative energy used will not be easily influenced by theuncertain economy.

The results of the IRF in this study are in line with Costantini *et al.* (2009)where shocks due to oil and gas use will negatively impact economic growth in Malaysia. In addition, another opinion Fong *et al* (2007) that non-renewable energy consumption produces carbon dioxide emissions that can directly damage the environment (environmental degradation). The carbon footprint generated by energy consumption has a negative impact on the environment and health and many environmental problems that will occur. Energy consumption such as oil, gas, and electricity which is non-renewable energy will cause dependence so that it will be more easily affected by shocks / shocks that come from outside. In contrast to the use of renewable energy that is environmentally friendly and can self-sufficiency energy independently.

4. Conclusion

VARX test results showed that the overall equations of variables ELEC*(electricity consumption)*, LOIL (Log energy *use* of *oil)*, GAS (Fossil *fuel* energy *consumption)*and RE (Renewable Energy) at times are not significant to EG (economic growth) with a positive influence on EG (Economic Growth) or growth. economy in Indonesia. But the results are different when it has included the presence of time lag in lag 2. Where the EG dependent variable has a significant effect only on the ELEC dependent variable at time t-1, while in other dependent variables it has no significant effect. Likewise with exogenous LCF variables which have no significant effect on EG (economic growth). The IRF results in the endogenous independent variable impulse to the dependent variable shock response that has the most shock effect, namely on the LOIL variable. While in the ELEC variable, GAS and RE are volatile but not variable and tend to be stable.

The results of this study are expected to encourage alternative energy innovations that can be used for more efficient management and use efforts that have greater potential. So Indonesia can be self-sufficient in energy. Utilization of alternative energy is still very low in Indonesia. So that there is a need for renewable energy innovations, so that with this energy can create environmentally friendly energy.

References

[1] Shaari M S, Hussain N E and Ismail M S 2013 Relationship between Energy Consumption and Economic Growth : Empirical Evidence for Malaysia *Bus. Syst. Rev.* **2**

[2] Huang Z and Huang L 2020 Individual new energy consumption and economic growth in China *North Am. J. Econ. Financ.* **54** 101010

[3] Susanto J and Laksana D H 2013 Uji Kausalitas Antara Konsumsi Energi Dan Pertumbuhan Ekonomi Di Asean *Bul. Ekon.* **11** 1–6

[4] Costantini V and Martini C 2010 The causality between energy consumption and economic growth: A multi-sectoral analysis using non-stationary cointegrated panel data *Energy Econ.* **32** 591–603

[5] Apergis N and Payne J E 2009 CO2 emissions, energy usage, and output in Central America *Energy Policy* **37** 3282–6

[6] Taherdoost H 2018 Decision Making Using the Analytic Hierarchy Process (AHP); A Step by Step Approach *Int. Journel Econ. Manag. Syst.* **2** 244–6

[7] Cassim R, Jackson W, Gavera L and Robertson P 2004 International Trade in Services and Sustainable Development : The Case of Tourism in South Africa Contributing authors : About the Trade Knowledge Network

[8] Chaudhry I S, Safdar N and Farooq F 2012 Electricity consumption and economic growth empirical evidence from Pakistan *Pakistan J. Soc. Sci.* **32** 371–82

[9] Ocampo S 2012 An Introductory Review of a Structural VAR-X Estimation and Applications 479–508

[10] Hou Q 2009 The Relationship between Energy Consumption Growths and Economic Growth in China *Int. J. Econ. Financ.* **1** 232–7

Land cover changes impact on lst between 1990-2005-2020 in Bandar Lampung City, Indonesia

Pramudiyasari T[1,2*], Wibowo A[1] and Adi Suko P[2*]

[1] Department of Geography, Faculty of Mathematics and Natural Sciences, University of Indonesia, 16424 Depok, Indonesia

[2] Indonesian Agency for Meteorological, Climatological and Geophysics

*Email: tia.pramudiyasari@ui.ac.id, adi.w@sci.ui.ac.id

Abstract. Urban areas with low vegetation trigger an increase in LST, while areas with dominant vegetation trigger a decrease in LST. Green space can produce shadows covering the earth's surface to prevent heat radiation directly from the sun, while land expansion for residential, office, industrial, urban purposes, causing an increase in LST >30°C. The study is to analyse the distribution and relationship between NDBI, NDVI and LST in Bandar Lampung City using Landsat 5, Landsat 7 and Landsat 8 data from 1990 to 2020. There has been an increasing distribution of LST, which tends to the outskirts of the central city to the northwest, north and east from the centre of Bandar Lampung City. Although it is not significant to the maximum temperature, thus saw from the wide distribution of the LST region >30°C. Landcover changes based on NDVI and NDBI in 1990 tended to be low to moderate, while 2005 was moderate to strong and 2020 was strong. This relation indicates that the land cover change from vegetated land to built-up land affects the expansion of the LST area in 2020 compared to 1990 and 2005. The result concluded that the average LST trend from 1990 to 2020, increasing 4,5°C.

1. Introduction

Changes in surface coverage and landscape patterns caused by rapid urbanisation result in land Surface Temperature. It is clearly defined as the higher temperatures of the central urban location compared to the nearby areas of similar elevation (Dorigon & Amorim, 2019). The United States Environmental Protection Agency (EPA) (2008) describing two types of heat islands: surface heat islands (SUHI) can be found at any time of day or night, although they are most prevalent during the day when the sun is shining and atmospheric heat islands when an urban area is warmer than nearby rural surroundings. Surface UHI (SUHI) based on LST and atmospheric UHI was based on air temperature (AST) (Priyankara et al., 2019).

A study by Pal and Ziaul (2017) showed that Land Surface Temperature (LST) significantly had a positive correlation with the Normalized Difference Built-up Index (NDBI) and a negative correlation with the Normalised Difference Vegetation Index (NDVI). Other studies have also shown that urban areas with low vegetation trigger an increase in LST, while areas with dense vegetation trigger a decrease in LST (Sun & Chen, 2017; Wibowo & Salleh, 2018). Green space area can produce shadows that cover the earth's surface from the direct radiation, while the expansion of the built-up area for residential, office, industrial, infrastructure and road in the downtown area can cause an increase in LST >30°C (Zhou et al., 2011; Li et al., 2012; Tursilowati et al., 2012). The increase in LST is an indication of the UHI phenomenon as has occurred in several major cities in Indonesia, such as Bandung (Ningrum & Narulita, 2018; Sasky et al., 2017), Surabaya (Sobirin & Fatimah, 2015), Tangerang (Wibowo et al., 2016), and Jakarta (Estoque et al., 2016).

Bandar Lampung City is the capital city of Lampung Province. Besides being the centre of government, social, political, educational and cultural activities, this city is also the centre of economic activity in Lampung Province. Bandar Lampung City is located in a strategic area because it is a transit area for economic activities between the islands of Sumatra and Java, making it profitable for the growth and development of Bandar Lampung City as a centre for trade, industry and tourism. The Spatial Planning Regulation (RTRW) of Bandar Lampung City 2011-2030 explained that the existing potential and development trends regarding a strategic location, natural potential, population, and potential supporting areas, Bandar Lampung City could be developed as a growth centre for Southern Sumatra as well as the centre of national production commodity (sippa.ciptakarya.pu.go.id). This condition impacts the population in Bandar Lampung City, with a population growth rate in 2018-2019 reaching 1.71% (BPS Bandar Lampung City, 2020). Population growth and land-use change can cause severe UHI in Bandar Lampung City.

Therefore, UHI mitigation has become one of the essential topics in urban ecology, urban landscapes, urban geography, and urban meteorology. The development of urban areas in Bandar Lampung City can expand the UHI and negatively impact the urban climate. To research was conducted within a spatial pattern of NDVI, NDBI, LST, and UHI values in Bandar Lampung City from 1990-2020 and the Relationship between NDBI and NDVI values on LST and UHI in Bandar Lampung City.

2. Methods

The research used Landsat 5, Landsat 7 and Landsat 8 data and used NDBI, NDVI and LST to answer the research. The research conduct compares the spatial pattern of Land Surface Temperature in 1990, 2005 and 2020, so that changes and directions of surface temperature, vegetation density, and building density will be seen on Urban Heat Island in the Bandar Lampung area. In addition, it is also compared with the observation air temperature for 30 years and is associated with Urban Heat conditions in Bandar Lampung. Geographic Information System (GIS) is used through the tools of NDVI (Normalized Difference Vegetation Index), NDBI (Normalized Difference Building Index) and LST (Land Surface Temperature). NDVI was used to determine the vegetation density classification based on remote sensing data. NDBI is used to determine the density class of buildings based on remote sensing data. Based on the density difference, the type of building density could be interpreted. LST analysis is used to determine the classification of Land Surface Temperature based on remote sensing data to the land surface temperature can be interpreted base on the temperature difference. To answer the spatial relationship between the dependent and independent variables using the Linear Regression Model (Figure.1).

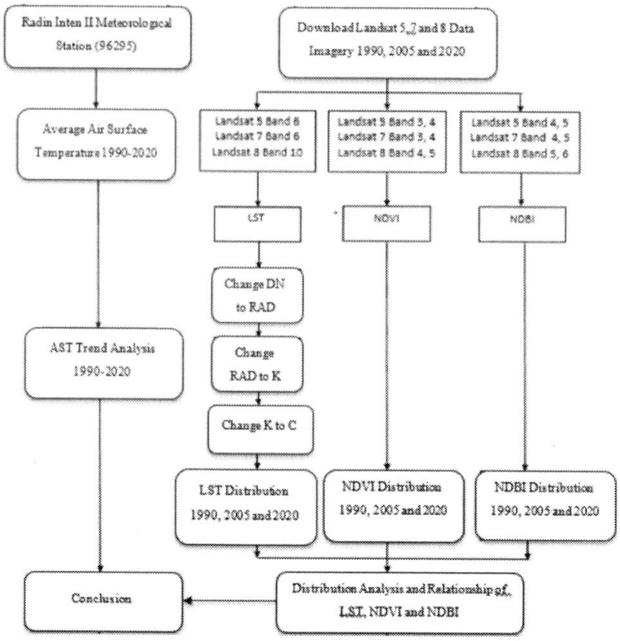

Figure 1. Diagram of Research Methodology

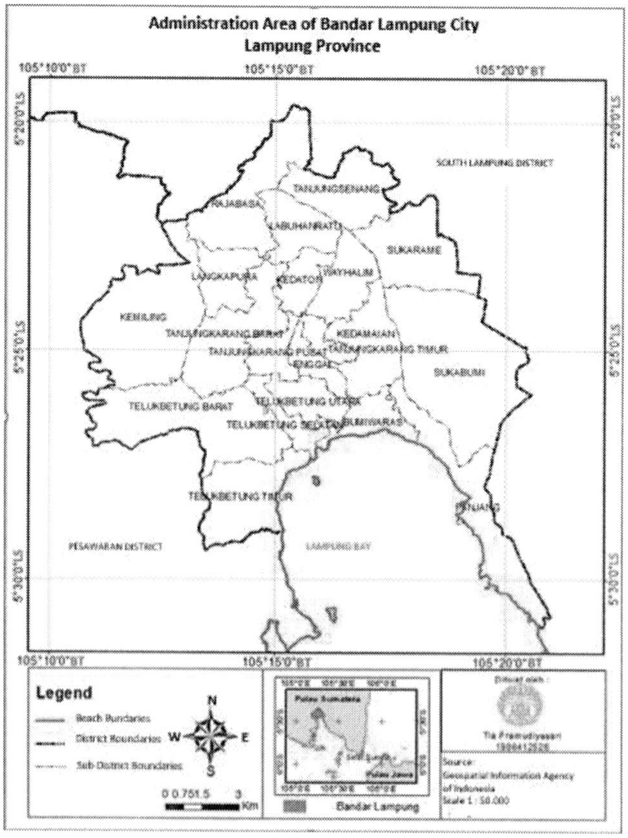

Figure 2. Administration Area of Bandar Lampung City

The variables in this study divided into two: Land Surface Temperature (LST), and the independent variables are vegetation density and building density. The data processed are Landsat satellite images of 1990, 2005, and 2020. The samples were taken in months with shallow cloud cover, mostly taken when Bandar Lampung was in the dry season, with sunny weather and no clouds or wind. Data will be analysed using qualitative and quantitative methods. Qualitative analysis was carried out with a spatial approach. The spatial approach is used to analyse the spatial pattern of Land Surface Temperature and its changes and analyse the Relationship between LST, vegetation density, and building density. In addition to qualitative analysis, quantitative analysis is also carried out in the form of statistical calculations. Quantitative analysis was carried out by analysing the trend of changes in Air Surface Temperature in 1990-2020. Statistical analysis using Pearson Product Moment correlation test to find out the Relationship between LST, vegetation density and building density to Urban Heat Island.

3. Result and discussion

3.1. LST, NDVI and NDBI Distribution Analysis

Based on Figure 3 saw that between 1990 to 2020, vegetation with a very low-vegetation category to non-vegetation is expanding while the area with a moderate to a high-vegetation category from 1990 to 2005 and 2020 decreases in size.

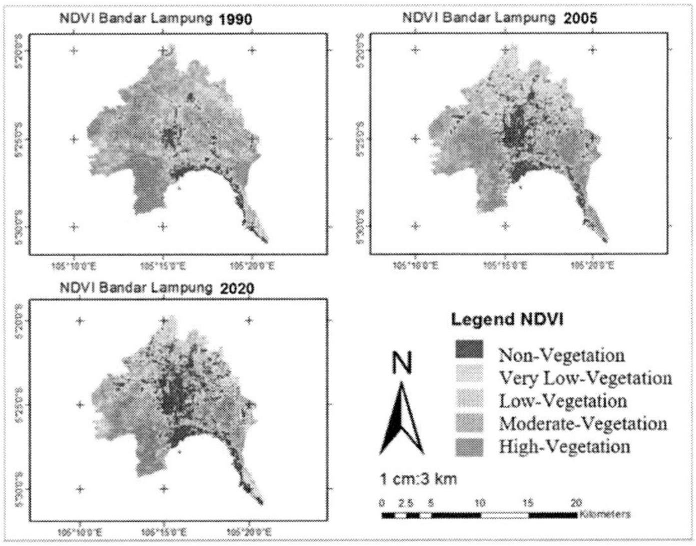

Figure 3. Bandar Lampung NDVI Distribution Map 1990, 2005 and 2020

From table 1, vegetation in 1990, the city of Bandar Lampung had the broadest category of medium-high vegetation density compared to 2005 and 2020. This percentage of vegetation that 53.63% of the total area of Bandar Lampung in 1990 consists of vegetation with medium to high density. Meanwhile, if we look at 2005, the area from medium to high density has decreased to 42.41% of the total area of Bandar Lampung City. Based on data 2020, the area with medium to high vegetation density is decreasing, which has 35.03% of the total area of Bandar Lampung City. Therefore, 2020 has the lowest high-vegetation area category with 2518.37 ha and the largest non-vegetation area with 3318.60 ha compared to 1990 and 2005.

Table 1. Vegetation Density Area of Bandar Lampung in 1990, 2005 and 2020

NDVI Category	1990		2005		2020	
	Area (Ha)	Percent (%)	Area (Ha)	Percent (%)	Area (Ha)	Percent (%)
Non-vegetation area	1246.25	6.94	2592.53	14.44	3318.60	18.49
Very low-vegetation area	2561.16	14.27	3575.05	19.92	4263.30	23.75
Low-vegetation area	4514.74	25.15	4168.98	23.23	4079.47	22.73
Medium-vegetation area	5401.37	30.09	4056.75	22.60	3769.54	21.00
High-vegetation area	4226.06	23.54	3554.81	19.81	2518.37	14.03

Source: data analysis

Table 2. Changes in Vegetation Density Area in Bandar Lampung

NDVI Category	Area Change					
	1990 - 2005		2005 - 2020		1990 - 2020	
	Δ (Ha)	%	Δ (Ha)	%	Δ (Ha)	%
Non-vegetation area	1346.28	7.50	726.07	4.04	2072.35	11.55
Very low-vegetation area	1013.89	5.65	688.25	3.83	1702.14	9.48
Low-vegetation area	-345.76	-1.92	-89.51	-0.50	-435.28	-2.42
Medium-vegetation area	-1344.62	-7.49	-287.22	-1.60	-1631.84	-9.09
High-vegetation area	-671.25	-3.74	-1036.44	-5.78	-1707.69	-9.51

Source: data analysis

Table 2 saw the largest vegetation density reduction in the medium-vegetation category from 1990 – 2005, which decreased by 1344.62 ha and an increase in a non-vegetation category, which increased by 1346.28 Ha. Meanwhile, the largest vegetation density reduction in the high-vegetation category was in 2005 - 2020, which reduced to an area of 1036.44 Ha. In the last 30 years (1990 to 2020), the high-vegetation category decreased by 1707.69 ha with 9.51%, and the medium-vegetation category decreased by 1631.84 ha or 9.09%. An increase followed this condition in the area of the non-vegetation category, which increased by 2072.35 Ha or 11.55%.

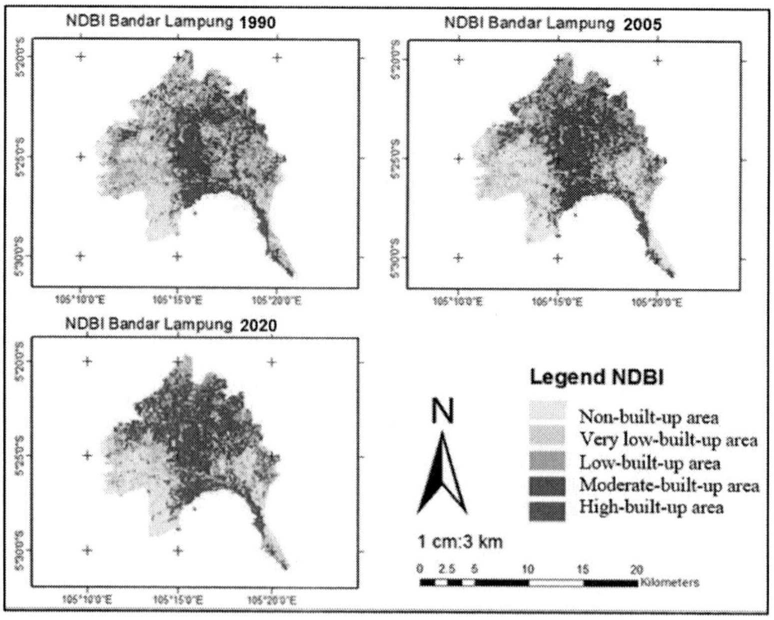

Figure 4. Bandar Lampung NDBI Distribution Map 1990, 2005 and 2020

Figure 4 explained that developed land density in the medium to the high category has expanded from 1990-2020. On the other hand, an area with shallow vegetation changed into a non-built category between 1990 and 2005. In the year 2020, it has decreased in size. Furthermore, Figure 3 seen that the built-up land in 1990 was more likely to be in the central and southern parts of Bandar Lampung City, while in 2005, it expanded to the northwest – east of Bandar Lampung City. This condition continued to expand until 2020.

Table 3. The density of Built-Up Area in Bandar Lampung 1990, 2005 and 2020

NDBI Category	Area					
	1990		2005		2020	
	Ha	%	Ha	%	Ha	%
Non-built area	2986.84	16.64	3352.54	18.68	3470.85	19.34
Low-built-up area	5361.77	29.87	4353.97	24.26	3966.12	22.10
Moderate-built-up area	4559.58	25.40	4365.00	24.33	3506.10	19.53
High-built-up area	3286.14	18.31	3622.36	20.19	4008.94	22.34
Very high-built-up area	1754.55	9.78	2249.62	12.54	2996.85	16.70

Source: data analysis

Table 4. Changes in Built-Up Area Density in Bandar Lampung 1990, 2005 and 2020

NDBI Category	Area Change (Ha)					
	1990 – 2005		2005 - 2020		1990 - 2020	
	Δ	%	Δ	%	Δ	%
Non-built area	365.69	2.04	118.31	0.65	484.01	2.70
Low-built-up area	-1007.79	-5.61	-387.85	-2.17	-1395.64	-7.78
Moderate-built-up area	-194.58	-1.08	-858.90	-4.79	-1053.47	-5.87
High-built-up area	336.21	1.88	386.58	2.15	722.80	4.03
Very high-built-up area	495.07	2.76	747.24	4.16	1242.30	6.92

Source: data analysis

Table 3, reported in 1990, Bandar Lampung City had the lowest built-up area density in a non-built category with a total area of 8348.61 Ha. This largest area was compared between 2005 and 2020, wherein 2005 it was reduced to 7706.51 ha and in 2020 changed to 7436.97 ha. The total area of built-up area density in change high to a very high category in 1990 had the lowest area 5040.69 ha with 28.09% of Bandar Lampung City. Thus condition then increased between 2005 and 2020, becomes 7005.79 ha with 39.04% of Bandar Lampung City. This condition will lead the density of built-up land in Bandar Lampung City to expand in 2020.

Table 4 shows the change in the built-up area density from 1990 to 2020 in Bandar Lampung City. Judging from the changes from 1990 to 2005, the density of built-up area in the non-built category experienced an increase in 365.69 ha. However, the area was significantly reduced to 1007.79 Ha, followed by an increase in the area of high to very high-built-up area category, which was a total of 831.28 Ha. Changes in the density of built-up area from 2005 to 2020 experienced an expansion in the non-built area category, which was 118.31 Ha, however, the density of low-built-up category decreased by 387.85 Ha, followed by the expansion of built-up area in the high to very high category with a total of 1133.82 Ha. From the past 30-year (1990 to 2020), the change in the density of built-up area in the low category decreased by 1395.64 Ha, followed by the expansion of built-up area in the high and very high categories, with a total of 1965.10 Ha or 10.95%, which means that during the past 30 years period, Bandar Lampung City experienced an expansion of built-up area.

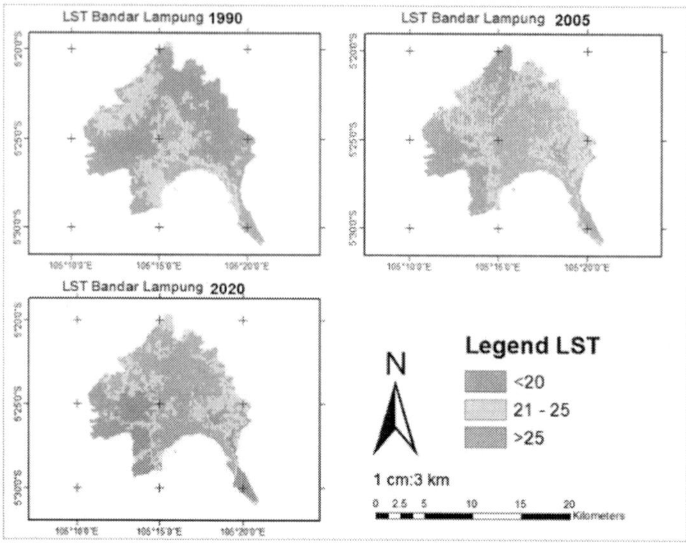

Figure 5. Bandar Lampung LST Distribution Map 1990, 2005 and 2020

Based on figure 5, in 1990, LST with temperatures >25^{0}C tended to be in the central and southern parts of Bandar Lampung City, while in 2005 areas with LST >25^{0}C expanded to the northwest and east of Bandar Lampung City, in 2020 the areas with LST >25^{0}C is increasingly widespread in the northwest, north and east of Bandar Lampung City. Meanwhile, areas with LST <25^{0}C was decrease from 1990 to 2020. Based on Figure 5, we agree that during the past 30-year period, LST in Bandar Lampung city experienced an expansion.

Table 5. Changes in the Density of Built-Up Area in Bandar Lampung

LST	Area					
	1990		2005		2020	
	Ha	%	Ha	%	Ha	%
<20	11723.92	65.29	6281.18	34.98	4893.35	27.25
21-25	5017.26	27.94	5930.78	33.03	4935.20	27.48
>25	1214.53	6.76	5742.28	31.98	8127.56	45.26

The widespread of LST in Bandar Lampung City is strengthened by the processed data of LST according to table 5. It can be seen that in 1990, there were 65.29% or an area of 11723.92 Ha from Bandar Lampung City which had LST <20^{0}C. Meanwhile, LST >25^{0}C covering an area of 1214.53 Ha or 6.76% which become warmer in 2005 and continued to increase until 2020. In 2020, LST >25^{0}C significantly increase to 45.26% of the total area of Bandar Lampung city.

3.2. Relationship of LST, NDVI and NDBI

In graph 1 (a), it can be seen that LST and NDVI have a coefficient of determination R2 = 0.0854, which means that the independent variable NDVI is able to explain the variance of the dependent variable LST of 8.54%, and has a correlation of R = -0.292 which means it has a negative correlation. where the greater NDVI value, the lower LST value but with a low relationship. While graph 1 (b) has a coefficient of determination of R2 = 0.1959, which means that NDBI is able to explain the variance of the LST of 19.59%, and has a correlation of R = 0.4426 which means it has a positive correlation where the greater NDBI value, the greater LST value with a moderate relationship.

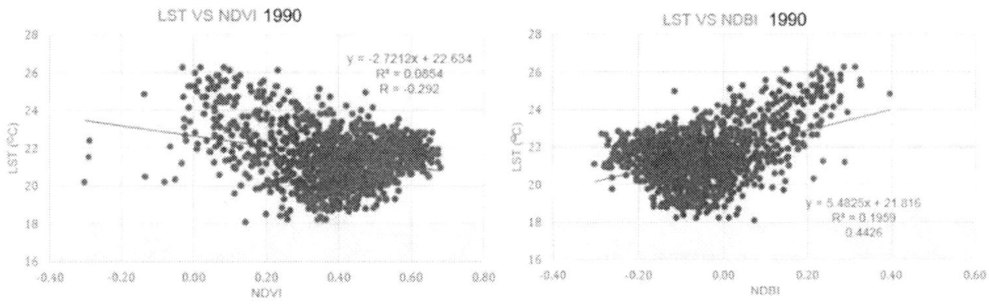

(a) (b)

Graph 1. (a) LST vs NDVI 1990, (b) LST vs NDBI 1990

In graph 2 (a), it can be seen that LST and NDVI have a coefficient of determination of R2 = 0.2668, which means that the independent variable NDVI is able to explain the variance of the dependent variable LST of 26.68%, and has a correlation of R = -0.517 which means it has a negative correlation. where the greater NDVI value, the lower LST value with a moderate relationship. While graph 2 (b) has a coefficient of determination of R2 = 0.3886, which means that NDBI is able to explain the variance of LST by 38.86%, and has a correlation of R = 0.623 which means it has a positive correlation where the greater NDBI value, the greater LST value with a strong relationship.

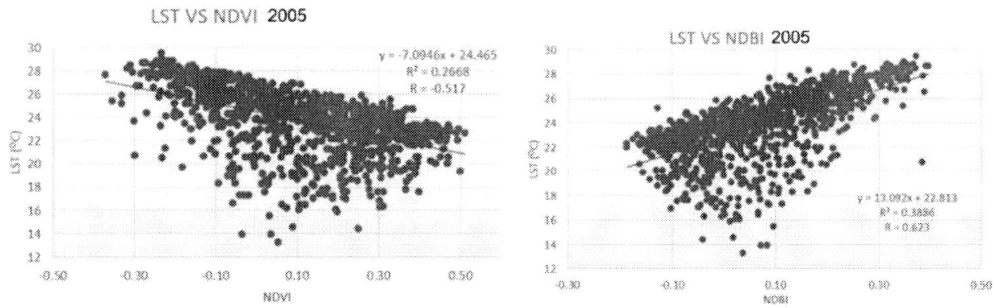

(a) (b)

Graph 2. (a) LST vs NDVI 2005, (b) LST vs NDBI 2005

In graph 3 (a), it can be seen that LST and NDVI have a coefficient of determination of R2 = 0.3681, which means that the independent variable NDVI is able to explain the variance of the dependent variable LST of 36.81%, and has a correlation of R = -0.607 which means it has a negative correlation. where the greater NDVI value, the lower LST value with a strong relationship. While graph 3 (b) has a coefficient of determination of R2 = 0.5112, which means that NDBI is able to explain the variance of LST by 51.12%, and has a correlation of R = 0.715 which means it has a positive correlation where the greater NDBI value, the greater LST value with a strong relationship.

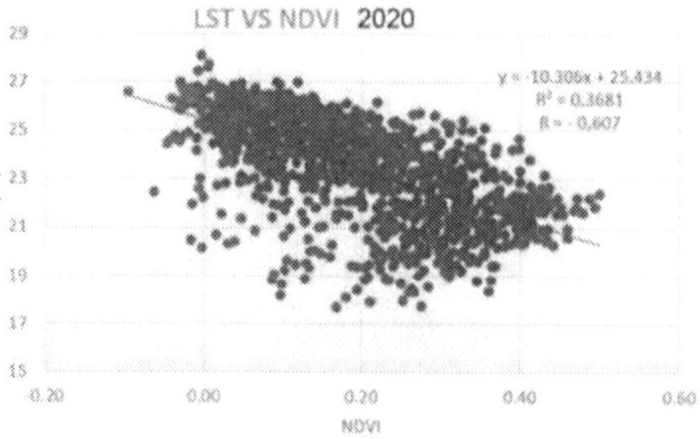

(a)

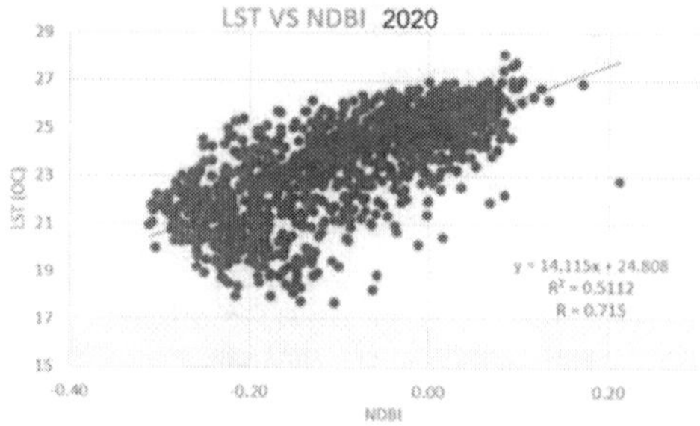

(b)

Graph 3. (a) LST vs NDVI 2020, (b) LST vs NDBI 2020

From Graph 4 (a), the average air temperature throughout 1990 obtained from the Radin Inten II Lampung Meteorological Station experienced a negative trend with the equation y = -0.0017x + 26,625, indicating that there was a decrease in average of air surface temperature 0.0017^{0}C/day. Graph 4 (b) shows a positive trend with the equation y = 0.0018x + 26,286, which means that there is an average increase in air surface temperature 0.0018^{0}C/day. Graph 4 (c) shows a negative trend with the equation y = -0.0011x + 27,177, which means that there is a decrease in average of air surface temperature 0.0011^{0}C/day.

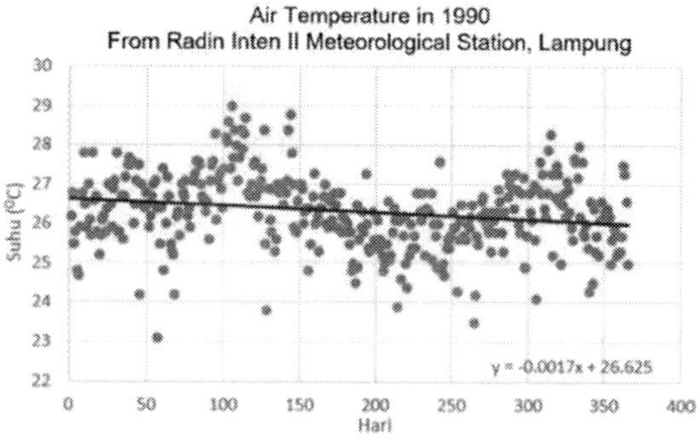

(a)

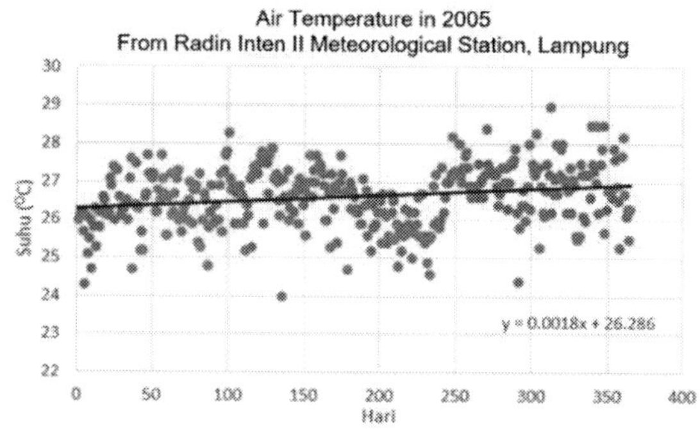

(b)

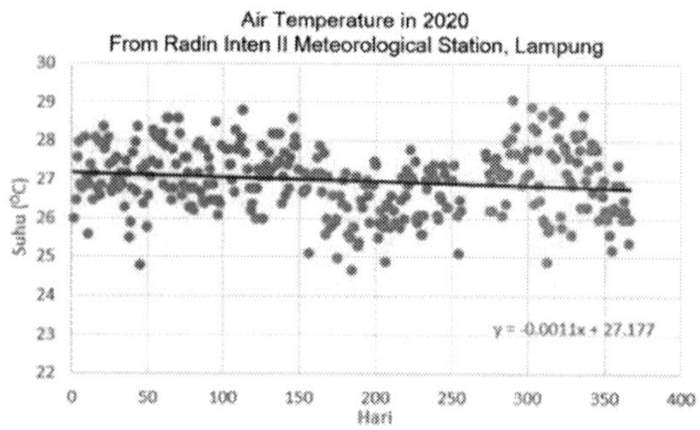

(c)

Graph 4. (a) Air Temperature in 1990, (b) Air Temperature in 2005, (c) Air Temperature in 2020

4. Conclusion

The conclusion this research on land cover changes to LST 1990-2005-2020 are:

1. There has been an increasing distribution of LST which tends to the periphery of the center of Bandar Lampung City, namely to the northwest, north and east areas of the center of Bandar Lampung City. Although not significant to the maximum temperature, it is clearly visible from the wide distribution of the LST region > 25^{0}C.
2. Land changes in the form of a decrease in vegetation density (NDVI) and an increase in built-up area (NDBI) will cause an expansion area of LST >25°C in Bandar Lampung City in 2020.
3. The Relationship between LST, NDVI and NDBI in 1990 tends to be low - moderate, while in 2005 the relationship tends to be moderate to strong and looks significant in 2020 where the relationship tends to be strong. This means that the effect of building and land density on the widespread increase in ground surface temperature in Bandar Lampung City in 2020 is greater than 1990 and 2005. And periodically the relationship tends to increase from 1990 to 2020.

Acknowledgements

The authors are deeply grateful to Indonesian Agency for Meteorological, Climatological and Geophysics and Department of Geography, Faculty of Mathematics and Natural Sciences, University of Indonesia for support of the research. Research Grant supported the publication from Faculty Mathematics and Natural Sciences of Universitas Indonesia year 2021 based on contract number: NKB-002/UN2.F3/HKP.05.00/2021.

References

[1] Akbari, H., & Kolokotsa, D. (2016). Three Decades Of *Urban Heat Islands* And Mitigation Technologies Research Three Decades Of *Urban Heat Islands* And Mitigation Technologies Research. Energy & Buildings. https://doi.org/10.1016/j.enbuild.2016.09.067

[2] Alfraihat, R., Mulugeta, G., Gala, T. S. (2016). Ecological Evaluation of Urban Heat Island in Chicago City, USA. Journal of Atmospheric Pollution, 2016, Vol. 4, No. 1, 23-29. Department of Geography, Chicago State University, Chicago, IL., USA.

[3] Amorim, M.C.C.T., Dubreuil, V. (2017). Intensity of Urban Heat Island in Tropical and Temperate Climates. MDPI Journal of Climate 2017, 5, 91; doi:10.3390/cli5040091

[4] Andana, E.K. (2015). Pengembangan Data Citra Satelit Landsat 8 Untuk Pemetaan Area Tanaman Hortikultura Dengan Berbagai Metode Algoritma Indeks Vegetasi (Studi Kasus: Kabupaten Malang dan Sekitarnya). Prosiding Seminar Nasional Manajemen Teknologi XXII. Program Studi MMT-ITS: Surabaya. ISBN: 978-602-70604-1-8.

[5] Andrade, M.J., Alcoforado, H. (2008). Global warming and the urban heat Island. In: En Marzluff, J. (Ed.), Urban Ecology. Springer, USA, pp. 249–262.

[6] Ariani, R. D., & Harini, R. (2012). Tekanan Penduduk terhadap Lahan Pertanian di Kawasan Pertanian Kasus Kecamatan Minggir dan Moyudan. Yogyakarta: Universitas Gajah Mada.

[7] Bai, Y., Wang, X.Y., Jiang, H.M., Liu, S.D. (2013). Progress of urban heat island effect. J. Meteorol. Environ, 2, 101–106.

[8] Bhatti, S.S., & Tripathi, N.K. (2014). Built-up area extraction using Landsat 8 OLI imagery, GIScience & Remote Sensing, 51:4, 445-467, DOI: 10.1080/15481603.2014.939539

[9] Bonafoni, S., Baldinelli, G., Verducci, P. (2017) Sustainable strategies for smart cities: Analysis of the town development effect on surface urban heat island through remote sensing methodologies. Sustain. Cities Soc. 29, 211–218.

[10] BPS Kota Bandar Lampung, Kota Bandar Lampung Dalam Angka 2020. ISSN/ISBN 0215-4102

[11] Bryson, R.A., Ross, J.E. (1972). The Climate of the City. Urbanisation and Environment. Duxbury Press: Belmont, CA, USA.

[12] Cardoso, R.S., & Amorim, M.C.C.T. (2018). Urban Heat Island Analysis Using the 'Local Climate Zone' Scheme in Presidente Prudente, Brazil. Investigaciones Geográficas, (69), 107-118. from https://doi.org/10.14198/INGEO2018.69.07

[13] Chandler, T.J. (1965). The Climate of London. Hutchison of London: London, UK, 1965; p. 287.
[14] Chapman, S., Watson, J.E.M., Salazar, A. et al. (2017). The impact of urbanisation and climate change on urban temperatures: a systematic review. Landscape Ecol 32, 1921–1935. https://doi.org/10.1007/s10980-017-0561-4
[15] Cole-Dai, J., Ferris, D., Lanciki, A., Savarino, J., Baroni, M., and Thiemens, M. H. (2009), Cold decade (AD 1810–1819) caused by Tambora (1815) and another (1809) stratospheric volcanic eruption, Geophys. Res. Lett., 36, L22703, doi:10.1029/2009GL040882.
[16] Danoedoro, P. (2012). Pengantar Penginderaan Jauh Digital Edisi 1. ISBN : 9789792931129
[17] Delarizka, A., Sasmito, B., & Hani'ah, H. (2016). Analisis Fenomena Pulau Bahang (Urban Heat Island) Di Kota Semarang Berdasarkan Hubungan Antara Perubahan Tutupan Lahan Dengan Suhu Permukaan Menggunakan Citra Multi Temporal Landsat. Jurnal Geodesi Undip, 5(4), 165–177.
[18] Dorigon, L.P., Amorim, M.C.C.T. (2019). Spatial Modelling of An Urban Brazilian Heat Island in A Tropical Continent Climate. Urban Climate, Volume 28 (2019) 100461. ISSN 2212-0955, https://doi.org/10.1016/j.uclim.2019.100461
[19] Du, H., Ai, J., Cai, Y., dkk. (2019). Combined Effects of the Surface Urban Heat Island with Landscape Composition and Configuration Based on Remote Sensing: A Case Study of Shanghai, China. MDPI Sustainability 2019, 11, 2890; doi:10.3390/su11102890
[20] Fardani, I., Adisurya, I.A., dan Saraswati. (2018). Penggunaan Citra Satelit Landsat Untuk Analisis Urban Heat Island. Seminar Nasional Geomatika 2018: Penggunaan dan Pengembangan Produk Informasi Geospasial Mendukung Daya Saing Nasional. Perencanaan Wilayah dan Kota, Universitas Islam Bandung.
[21] García, M.C.M., (1999). Climatologia Urbana, in: Departament de geografia Física i Anàlisi, Geográfica Regional. Col-lecció Textos Docents, Universitad de Barcelona, pp. 71.
[22] Guha, S., Govil, H., Dey, A & Gill, N. (2018). Analytical study of land surface temperature with NDVI and NDBI using Landsat 8 OLI and TIRS data in Florence and Naples city, Italy, European Journal of Remote Sensing, 51:1, 667-678, DOI:10.1080/22797254.2018.1474494. https://doi.org/10.1080/22797254.2018.1474494
[23] Hoan, N.T et al. (2018). Assessing the Effects of Land-Use Types in Surface Urban Heat Islands for Developing Comfortable Living in Hanoi City. Remote Sensing. 2018, 10, 1965; doi:10.3390/rs10121965
[24] Ilhamsyah, Y. (2012). A Mesoscale Meteorological Model of Modified Land Cover to the Effect of Urban Heat Island in Jakarta, Indonesia. Aceh International Journal of Science and Technology, 1 (2): 60-66 ISSN:2088-9860
[25] Kirchner, I., Stenchikov, G. L., Graf, H.-F., Robock, A., and Antuña, J. C. (1999), Climate model simulation of winter warming and summer cooling following the 1991 Mount Pinatubo volcanic eruption, J. Geophys. Res., 104(D16), 19039– 19055, doi:10.1029/1999JD900213.
[26] Lillesand dan Kiefer. (1998). Penginderaan Jauh dan Interpretasi Citra. Dulbahri (Penerjemah). Gajah Mada University Press. Yogyakarta.
[27] Landsberg, H.E. (1981). The Urban Climate, 1st ed.; Academic Press: New York, NY, USA.
[28] Lowry, W.P. (1977). Empirical estimation of urban effects on climate: A problem analysis. J. Appl. Meteorol, 16, 129–135.
[29] Malik, M.S., Shukla, J.P., & Mishra, S. (2018). Relationship of LST, NDBI and NDVI using Landsat-8 data in Kandaihimmat Watershed, Hoshangabad, India. Indian Journal of Geo Marine Sciences Vol. 48 (01), January 2019, pp. 25-31.
[30] Mohajerani, A., Bakaric, J., & Jeffrey-Bailey, T. (2017). The Urban Heat Island effect, its causes, and mitigation, with reference to the thermal properties of asphalt concrete. Journal of Environmental Management, 197, 522– 538. https://doi.org/10.1016/j.jenvman.2017.03.095
[31] Nadira, C., Saraswati, R., dan Wibowo, A. (2019). Pengaruh Perubahan Tutupan Lahan Terhadap Fenomena Urban Heat Island di Kecamatan Cikarang Utara, Kabupaten Bekasi Tahun 2007 – 2018 Menggunakan Citra Landsat 5 dan 8. Depok : Departemen Geografi, FMIPA UI.

[32] Ningrum, W., Narulita, I. (2018). Deteksi Perubahan Suhu Permukaan Menggunakan Data Satelit Landsat Multi-Waktu. Jurnal Teknologi Lingkungan Vol. 19, No 2.
[33] Nurhuda, A., Huda, D. N., dan Adhisurya, S. (2019). Penginderaan Jauh Untuk Analisis Spasial Temporal Suhu Permukaan Daratan di Kota Manado Tahun 2015 dan 2018. Seminar Nasional Penginderaan Jauh ke-6 Tahun 2019. Depok : Universitas Indonesia.
[34] Oke, T, R., (1978). Boundary Layer Climates. Metheuen & Co, London.
[35] Oke, T.R. (1987). Boundary Layer Climates, 2nd ed. Methuen and Co.: New York, NY, USA, 1987; pp. 262–303.
[36] Pal, S., Ziaul, Sk. (2017). Detection of land use and land cover change and land surface temperature in English Bazar urban centre. The Egyptian Journal of Remote Sensing and Space Science, Volume 20, Issue 1, Pages 125-145, ISSN 1110-9823, https://doi.org/10.1016/j.ejrs.2016.11.003.
[37] Priyankara, P., Ranagalage, M., Dissanayake, D. M. S. L. B., Morimoto, T., & Murayama, Y. (2019). Spatial process of surface urban heat island in rapidly growing Seoul Metropolitan area for sustainable urban planning using landsat data (1996–2017). Climate, 7(9), 110.
[38] Sheng, L., Tang, X., You, H., Gu, Q., Hu, H. (2017). Comparison of the urban heat island intensity quantified by using air temperature and Landsat land surface temperature in Hangzhou, China. Ecological Indicators, Vol.72, Pages 738-746, ISSN 1470-160X, https://doi.org/10.1016/j.ecolind.2016.09.009.
[39] Sobirin & Fatimah, R. N. (2015). Urban Heat Island Kota Surabaya. Depok : Departemen Geografi, FMIPA – Universitas Indonesia.
[40] Sun, R.; Chen, L. (2017). Effects of green space dynamics on urban heat islands: Mitigation and diversification. Ecosyst. Serv., 23, 38–46.
[41] Tursilowati, L., Sumantyo, J. T. S., Kuze, H., dan Adiningsih, E. S. (2012), relationship between urban heat island phenomenon and land use/land cover changes in Jakarta, Indonesia. Journal of Emerging Trends in Engineering and Applied Sciences, 3(4), 645-653.
[42] Wang, S, Ma, Q, Ding, H., & Liang, H. (2016). Detection Of Urban Expansion and Land Surface Temperature Change Using Multi-Temporal Landsat Images. Resources, Conservation and Recycling. https://doi.org/10.1016/j.resconrec.2016.05.0
[43] Wibowo, A., Kuswantoro, Ardiansyah, Rustanto, A., & Shidiq, I.P.A. (2016). Spatial temporal analysis of urban heat hazard in Tangerang City. IOP Conference Series: Earth and Environmental Science, 47(1), [012039]. https://doi.org/10.1088/1755-1315/47/1/012039
[44] Wibowo, A., & Salleh, K.O. (2018). Landscape features and potensial heat hazard threat: a spatial-temporal analysis of two urban universities. Nat Hazards. 92:1267–1286 https://doi.org/10.1007/s11069-018-3363-3
[45] Wibowo, A., dkk. (2020). IOP Conf. Ser.: Earth Environ. Sci. 500 012009. doi:10.1088/1755-1315/500/1/012009
[46] Winkler, J. A., Skaggs, R. H., & Baker, D. G. (1981). Effect of temperature adjustments on the Minneapolis-St. Paul urban heat island. Journal of Applied Meteorology, 20(11), 1295-1300.
[47] Yang, C., He, X., Yan, F., dkk. (2017). Mapping the Influence of Land Use/Land Cover Changes on the Urban Heat Island Effect—A Case Study of Changchun, China.
[48] Zahro, H., Sobirin, S., & Wibowo, A. (2018). Variasi spasiotemporal urban heat island di kawasan perkotaan Yogyakarta tahun 2015-2017. Jurnal Geografi Lingkungan Tropik, 2(1).
[49] Zha, Y., Gao, J., & Ni, S. (2003). Use of normalised difference built-up index in automatically mapping urban areas from TM imagery. International Journal of Remote Sensing, 24(3), 583–594.

Analysis of social vulnerability index in Watershed Bengawan Solo downstream

Suryanto[1*], A H Juwita[1] and D R Wicaksono[1]

[1] Faculty of Economics and Business, Universitas Sebelas Maret

*Email: suryanto_feb@staff.uns.ac.id

Abstract. Purpose of this research to know 1) the level of social vulnerability, 2) socio economic description, and 3) mapping of comparison of social vulnerability in Watershed of Bengawan Solo covering Bojonegoro, Tuban, Lamongan and Gresik. The method used in this research by using Social Vulnerability Index (SoVI) method as social vulnerability analysis tool and Geography Information System (GIS) analysis tool to map the area and its vulnerability level. The results showed the highest level of social vulnerability with the highest score of SoVI score obtained by Tuban Regency with SoVI score of 7, then Bojonegoro with SoVI score of 6, Lamongan with SoVI score of 4, and Gresik with SoVI score of 1. Bojonegoro And Kabupaten Tuban have the same highest score on the components of social vulnerability level, then Lamongan regency, and the lowest level of social vulnerability is in Gresik regency. For the grade of social vulnerability, Kabupaten Tuban and Bojonegoro are categorized into high social vulnerability classes, Lamongan District with moderate social vulnerability, and Gresik Regency with the lowest social vulnerability.

1. Introduction

Climate change is an environmental issue that has recently become a problem discussed in various countries in the world. Global climate change occurs due to the greenhouse effect caused by increasing greenhouse gas emissions and decreasing green forest area and increasing industrial and vehicle exhaust gas pollution [1][2], and [3]. Disasters are disturbances or disturbances that occur in the normal pattern of life. The impact on humans is quite large, such as loss of life, loss of residence, injuries, and loss of property. The impact of social and economic structures such as damage to infrastructure in the form of road systems, clean water, electricity, communications and other important services [4][5].

In Indonesia, there are three types of climates in Indonesia, including seasonal climates, hot climates, and marine climates. High rainfall can result in crop failure for farmers who plant rice in the fields because heavy rains can create strong winds or cause flooding of rice fields due to overflowing irrigation channels from the river. Prolonged dry season can result in dry rice fields.

Table 1. Natural Disaster Events in East Java 2018 Number of Natural Disasters in East Java Province by Regency/City, 2018

Regencies	Floods	Drought
Bojonegoro	8	1
Tuban	3	1
Lamongan	3	-
Gresik	5	-
East Java	87	20

Source: Central Bureau of Statistics Republic of Indonesia (BPS)

In Table 1, the data from the East Java National Disaster Management Agency (BNPB) explains that natural disasters that often occur in the East Java area in the 2018 period are floods. Bojonegoro Regency became a city that experienced relatively more flood events in 2018 with a frequency of 8 times, Gresik Regency 5 times, while Tuban and Lamongan Regency experienced 3 times of flood events. [6] stated that the watershed (DAS) that often occurs in disasters in Indonesia is the Bengawan Solo watershed. The Bengawan Solo River is the longest river on the island of Java with a watershed area of about 16.100 km2 [7]. The Bengawan Solo Downstream sub-watershed which includes the regencies of Bojonegoro, Tuban, Lamongan, and Gresik often has frequent flooding due to the overflow of the Bengawan Solo River every year the rainy season arrives due to siltation. This is due to river sedimentation that occurs due to soil surface erosion, tree felling in water catchment areas, and damage to watersheds which are tidal lands. On the other hand, climate change in the dry season also resulted in several disasters that occurred in the Bengawan Solo Hilir Sub-watershed area such as soil drought and water shortages. It is the local community who can get the impact if climate change occurs in the area. In [8] in their journal, in February 2015 in Gresik Regency a 30-100 cm high flood submerged 1,245 houses in Benjeng Subdistrict, 655 houses in Cerme Subdistrict, and 581 houses in Menganti Subdistrict. the souls of three people died. The overall material loss is estimated at IDR 18 billion, with 3,000 ha of agricultural land being submerged and 378 ha of them declared crop failure. In addition to material losses, floods inundated schools which resulted in disrupting accessibility so that people had difficulty getting clean water and food.

Community welfare is the output of the process of managing family resources and overcoming problems faced by the community. Welfare can change in the event of a natural disaster, of course, the community must face social vulnerability in the event of a natural disaster due to the impact of climate change. Social vulnerability is an indicator of vulnerability that is measured based on the social conditions of the community. Social vulnerability is the level of community's ability to deal with disasters that occur in an area. Natural conditions that cannot be controlled, with this condition the community becomes threatened. Thus, it is necessary to measure the level of social vulnerability of the community to face climate change.

In principle, vulnerability analysis is used as: (1) a diagnostic tool to understand the problems and factors that cause vulnerability, (2) a planning tool as a basis for determining the priority of activities and the sequence of planned activities, (3) a risk measurement tool to assess risk comprehensively. specific, and (4) tools for empowerment and mobilization of vulnerable community groups [9][10]. Vulnerability analysis is part of risk analysis aimed at disaster management in managing disaster risk.

According to [10], vulnerability analysis develops and is used in various sectors. Currently, there are 20 to 25 definitions of vulnerability will have an impact on various indicators and instruments of vulnerability. Similarly, the components of physical and environmental vulnerability. Social experts agree on several main factors that influence social vulnerability including lack of access to resources (information, knowledge, and technology), limited access to political power and representation, social capital, connections and social networks, customs and cultural values [11][12]. The research will complement the research results that base the analysis on the level of physical vulnerability of an area.

2. Research methods

2.1. Research sites

This research is located in the Bengawan Solo Watershed (DAS) downstream. This area has a fairly developed agricultural sector. There are four districts in the downstream Bengawan Solo watershed, namely Tuban Regency, Bojonegoro Regency, Gresik Regency, and Lamongan Regency.

2.2. Data types and sources

This study uses secondary data obtained from agencies or institutions related to this research. Secondary data is used to analyze the level of social vulnerability of the community in the Bengawan Solo watershed. Secondary data in this study were obtained from the Central bureau of Statistics (BPS), the Population and Social Family Planning Agency (BKKBN), the One Stop Integrated Investment and Licensing Service (DPMPTSP), the Financial Revenue and Asset Management Office (DPPKA), the Social Service (DinSos), and the General Election Commission (KPU) from Tuban Regency, Bojonegoro Regency, Gresik Regency, and Lamongan Regency.

2.3. Methods and data analysis

2.3.1. Socio-economic description

The socio-economic description carried out using descriptive statistical methods. This analysis conducted to determine the socio-economic description of the community in the Bengawan Solo Hilir watershed. The data used is the result of calculating the z-score carried out in the step of obtaining the SoVI score. Each main component explained in a comparison of each region so that it will be easier to see the socio-economic description.

2.3.2. Social Vulnerability Index (SoVI)

The SoVI method or the social vulnerability index has developed by Cutter in 2003. The study used 11 independent variables used to calculate the Social Vulnerability Index (SoVI). According to Cutter (2003), among others: a) Socio-economic status, b) Gender, c) Race and ethnicity, d) Age, e) Industrial development, f) Unemployment, g) Rural/urban, h) Property, i) Population with special needs, j) Tenants, k) infrastructure, l) Family structure, m) Education, n) Population growth, o) Health facilities, p) Social dependence, q) employment. The seventeen variables used consisted of several components which were developed based on the results of a literature review of each of the main variables called sub-components. Each sub-component in the SoVI uses an additive model (Cutter et al, 2003:68), in this approach, each factor contributes to the overall vulnerability in the district. Measurement of the social vulnerability that occurred in 1990 in the United States developed by these components.

2.3.3. Analysis of Geographic Information Systems (GIS)

This study uses Geographic Information System (GIS) analysis which will be used to see which areas are the most vulnerable in 4 districts, namely Bojonegoro Regency, Tuban Regency, Gresik Regency, and Lamongan Regency. The calculation of the score generated from the calculation of the Social Vulnerability Index (SoVI) is the data used in calculating the GIS.

3. Results and Discussion

3.1. Socio-economic description

3.1.1. Bojonegoro Regency

Socio-economic descriptions are discussed regarding administration, population, and the number of flood events in each research area. The discussion started from the regencies of Bojonegoro, Tuban, Lamongan, and Gresik.

Administratively, Bojonegoro currently divided into 28 k ecamatan with 419 villages and 11 urban villages. The total area of Bojonegoro Regency is 230,706 Ha. Of the area, 40.15 percent is state forest, while 32.58 percent is used for rice fields.

According to BPS data, the population development of Bojonegoro Regency until 2020 has increased every year. The population in 2020 is 1,344,038 people. When compared with the total population in 2019 of 1,331,077 people, it experienced an increase of 12,961 people within 1 year. So the population growth of Bojonegoro Regency is 0.96 percent. This increase in population is thought to be caused by births, migration and economic growth.

Statistically, flood events in Bojonegoro Regency can be seen in Table 2 below:

Table 2. Number of Flood Events in Bojonegoro Regency 2014, 2015 and 2016

Frequency, Impact	2014	2015	2016
Floods	7	4	8
Crop Failure	744 Ha	30 Ha	91 Ha
The number of people	1765 people	250 people	0

Source: National Disaster Management Agency (BPNB)

Table 2 describes the number of flood events in 2014 as many as 7 times, in 2015 as many as 4 times, in 2016 as many as 8 times. Land damage caused by floods in 2014 was 744 hectares, in 2015 it was 30 hectares, and in 2016 it was 91 hectares. The victims suffered from floods in 2014 amounted to 1765 people, and in 2015 there were 250 people.

3.1.2. Tuban Regency

Administratively, Tuban Regency is divided into 20 sub-districts consisting of 311 villages and 17 sub-districts. The total area of Tuban Regency is 183,994,562 Ha, the sea area is 22,068 km2, and the coastal area is 65 km stretching from the east of Palang sub-district to the west of Bancar sub-district. The area of agricultural land in Tuban Regency is 183,994,562 Ha consisting of 54,860,530 Ha of rice fields and 129,134,031 Ha of dry land.

The total population in Tuban Regency in 2015 was recorded at 1,304,080 people, consisting of 653,413 male and 650,667 female. The data shows that the number of men in Tuban district is more than the female population with a *sex ratio of* 100.42 percent. Statistically, flood events in Tuban Regency can be seen in Table 3 below:

Table 3. Number of Flood Events in Tuban Regency 2014, 2015 dan 2016

Frequency, Impact	2014	2015	2016
Floods	5 kali	4 kali	1 kali
Crop Failure	75 ha	0 ha	200 ha
The number of people	0	150 people	0 people

IOP Conf. Series: Earth and Environmental Science **986** (2022) 012049 doi:10.1088/1755-1315/986/1/012049

Source: National Disaster Management Agency (BPNB)

Table 2 explains the number of flood events in Tuban Regency during 2014, 2015, and 2016. Flood events in Tuban Regency have decreased since the 2014 flood, 5 times in 2015 and 4 times in 2016 and once in 2016. Land damage caused by floods in 2014 damaged about 75 ha of land, and in 2016 damaged 200 ha of land. The population who suffered from flooding was recorded only in 2015 around 150 people.

3.1.3. Lamongan Regency

Administratively, Lamongan Regency is divided into 27 sub-districts, covering 462 villages and 12 sub-districts. Lamongan Regency has an area of approximately 1,812.80 Km2 equivalent to 181,280 Ha or + 3.78% of the total area of East Java Province with a coastline of 47 Km. The agricultural area in Lamongan Regency as a whole covers an area of 91,458.91 Ha, with details of wetland agriculture (paddy fields) covering an area of 79,320 Ha and dry land agriculture of 12,138.91 Ha.

The population of Lamongan Regency in 2015 was 1,342,266 people with 672,636 male gender and 669,630 female sex. The population density in Lamongan district is 740 people/km2. The average age of the entire population in Lamongan Regency is 81,455 people. Statistically, flood events in Tuban Regency can be seen in Table 4 below:

Table 4. Number of Flood Events in Lamongan Regency 2014, 2015 dan 2016

Frequency, Impact	2014	2015	2016
Floods	4	1	2
Crop Failure	0 ha	0 ha	0 ha
The number of people	0	0	0

Source: National Disaster Management Agency (BPNB)

Table 4 describes the number of flood events in Lamongan Regency from 2014 as much as 4 times, in 2015 as much as 1 time, in 2016 as much as 2 times. There is no data related to land damage and victims suffering from flooding.

3.1.4. Gresik Regency

Administratively, Gresik Regency consists of 18 sub-districts, with 330 villages and 26 urban villages. The area of Gresik Regency is around 1,191.25 km2 or equivalent to an area of 125,193.64 Ha. Part of Gresik Regency is a coastal area with a beach length of 140 km on the mainland of Java Island extending from Kebomas, Gresik, Manyar, Bungah, Sidayu, Ujungpangkah, and Panceng sub-districts and 71 km in Sangkapura and Tambak sub-districts located on Bawean Island.

The total population in Gresik Regency in 2015 was 1,303,773 people consisting of 655,363 male residents and 648,410 for female *sex* with a *sex ratio of* 101 which means that where the number of women is 100, the male population is 101. soul. The average age of the population in Gresik Regency is the population aged between 35-39 years.

Table 5. Number of Flood Events in Gresik Regency 2014, 2015 dan 2016

Frequency, Impact	2014	2015	2016
Floods	2	3	5
Crop Failure	745 ha	2383 ha	60 ha
The number of people	0	6477	0

Source: National Disaster Management Agency (BPNB)

Table 5. Describes the number of flood events in Gresik Regency during 2014 that occurred 2 times, in 2015 it occurred 3 times, in 2016 it occurred 5 times. This flood incident caused damage to an area of 745 hectares in 2014, an area of 2383 hectares in 2015, an area of 60 hectares in 2016. The number of victims suffering was only recorded in 2015 as many as 6477 people.

3.2. Social Vulnerability Index (SoVI)

The main components and sub-components of each region contained in table 6 explain that the highest SoVI score is Tuban Regency with a score of 7, then Bojonegoro Regency with a score of 6, Lamongan Regency with a score of 4, and the lowest SoVI score is Gresik Regency with a score of 1.

The resulting SoVI score is then included in the vulnerability class that has been created. Bojonegoro Regency with a score of 6 is included in the class with a high level of vulnerability, Tuban Regency with a score of 7 is included in the class of high vulnerability level, Lamongan Regency with a score of 4 is in the category of moderate level of vulnerability, and Gresik Regency with a score of 1 is classified as an area. with a low level of social vulnerability. Areas with a higher level of social vulnerability indicate more scores of components that increase vulnerability and fewer scores of components that decrease vulnerability levels, so that areas with low levels of vulnerability mean fewer scores of components that increase vulnerability and more scores of components that reduce vulnerability.

Table 6. Main Component Score, Sub-component and SoVI Score

No	Vulnerability Indicators	Sub Vulnerability Indicators	(+/-)	Bojonegoro		Tuban		Lamongan		Gresik	
				Skor SK	Skor KU	Skor SK	Skor KU	Skor SK	Skor KU	Skor SK	Skor SU
1	Socio-economic status	Income per capita	-	-3	3	-2	4	-2	2	-3	2
		Government dependence on DAU*)	+	3		3		2		3	
		Number of pre-prosperous families	+	3		3		2		2	
2	Age	Average age	+	3	8	2	7	2	8	2	7
		Percent of elderly	+	3		2		3		2	
		Percentage of children under 5 years	+	2		3		3		3	
3	Property	Number of companies	-	-2	-4	-3	-6	-2	-5	-2	-4
		agriculture GRDP	-	-2		-3		-3		-2	
4	Health	Number of doctors	-	-3	-6	-2	-5	-2	-4	-3	-6
		Number of health facilities	-	-3		-3		-2		-3	
5	Population	Percent change in population	+	2	5	2	4	3	5	3	5
		Population density	+	3		2		2		2	
6	Unemployment	Unemployment number	-	-2	-2	-2	-2	-3	-3	-3	-3
7	Rural/Urban	Land used by farmers	-	-2	1	-2	0	-3	0	-2	0
		Percentage of farmer population in rural areas	+	3		2		3		2	
8	Profession	Percentage of population in the labor force	-	-3	-2	-2	2	-3	1	-2	-2

No	Vulnerability Indicators	Sub Vulnerability Indicators	(+/-)	Bojonegoro		Tuban		Lamongan		Gresik	
				Skor SK	Skor KU	Skor SK	Skor KU	Skor SK	Skor KU	Skor SK	Skor SU
		Percentage of women in the labor force	+	2		3		3		2	
		Number of people working in agriculture, fisheries, mining and forestry	+	2		3		3		2	
		Percentage of people working in transportation, communication and other public facilities	-	-3		-2		-2		-3	
		Percentage of people working in the community service sector	-	-2		-3		-3		-3	
		Number of TKI	+	2		3		3		2	
9	Family Structure	Percentage of women as heads of household*****)	+	3	3	3	3	2	2	2	2
		SoVI score			6		7		4		1

Source: Calculation of SoVI

3.3. SIG Vulnerability Map.

Spatial analysis is needed when the results of non-spatial analysis are considered insufficient to answer some questions. As well as to find out the location and distribution of a disaster problem. Spatial analysis in the field of disaster itself is the result of a synergy between geography and disaster science with various attributes. Then with the process of geomatics and geo information produced a result of spatial analysis as well as the regionalization of a problem both formally and functionally. Technological advances have an impact in performing spatial analysis, namely the discovery of a tool to analyze spatial data known as a Geographic Information System.

Based on the premise that there is a relationship between the magnitude of the impact or loss due to disasters and public health phenomena, the conditions that cause differences in the level of vulnerability to floods in East Java caused by the number of events, the magnitude of the impact, and population density. Therefore, to estimate the potential for disaster in the future so that flood prevention can be done more effectively and efficiently, it is necessary to do the analysis of exploratory against the region - the area is prone to flooding index distribution of each district / city in East Java. Efforts in disaster management begin at the time before the disaster (pre-disaster) in the form of prevention, mitigation (softening/taming impacts), and preparedness activities. Activities carried out during a disaster in the form of temporary emergency response activities after a disaster occur in the form of recovery/rehabilitation and reconstruction activities.

IOP Conf. Series: Earth and Environmental Science **986** (2022) 012049 doi:10.1088/1755-1315/986/1/012049

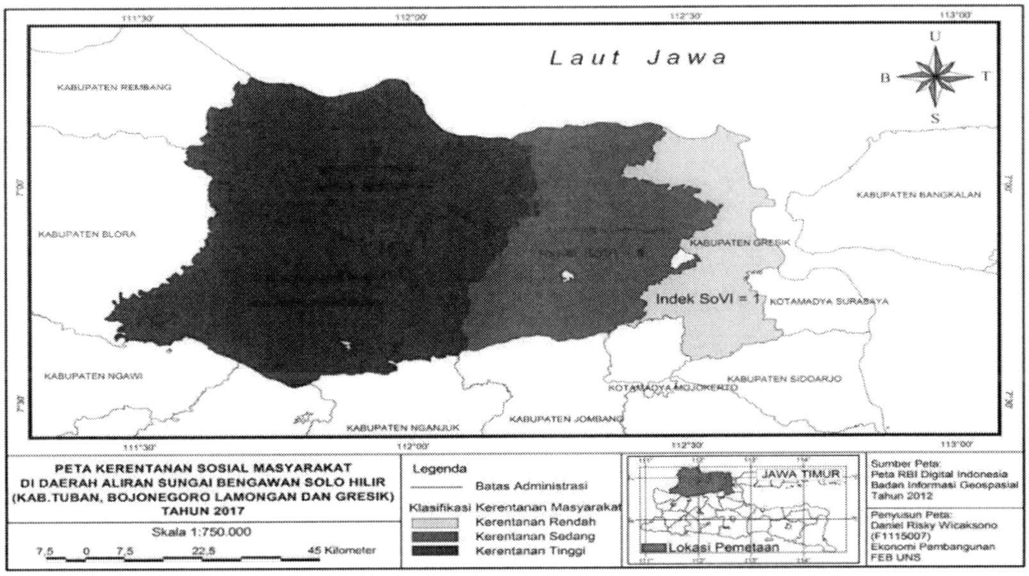

Source: Calculation of SoVI

Figure 1. Mapping of Social Vulnerability

4. Conclusion and Implication

4.1. Conclusion
Based on the data obtained to determine the level of social vulnerability, look at the description of socio-economic life, and mapping the comparison of social vulnerability in the Bengawan Solo River Basin (DAS) Lower part, this study can be concluded that:

The calculation using the Social Vulnerability Index (SoVI) method explains that the highest level of social vulnerability is Tuban Regency with a SoVI score of 7, then Bojonegoro Regency with a SoVI score of 6, Lamongan Regency with a SoVI score of 4, and the lowest level of social vulnerability is Gresik Regency with a score of 4 Based on the socio-economic component used, Bojonegoro Regency and Tuban Regency have the highest score on socio-economic components, then Lamongan Regency and Gresik Regency have the same score with a low vulnerable class.The results of the mapping using the Geographic Information System (GIS), the resulting map illustrates that Tuban Regency and Bojonegoro Regency are included in the highest vulnerability class, Lamongan Regency has a medium vulnerability class, and Gresik Regency has a low vulnerability class.

4.2. Implications
Based on the results of the research that has been obtained, the suggestions that can be given relating to this research are:

People in Bojonegoro, Tuban, Lamongan, and Gresik districts should be more concerned with the condition of the rivers in their respective areas not to throw garbage on the riverbanks so that they are free from flooding. Local governments pay more attention to the sub-components that affect social vulnerability so that the level of social vulnerability of the area can be lower. For example, the sub-component of the agricultural sector is the sector that is most vulnerable to vulnerability to flooding. The government can provide various trainings and briefings for farmers regarding disaster response, so that when a flood disaster occurs, farmers are more prepared and responsive, so the potential losses experienced by farmers can be minimized.

For further researchers, more areas should be used as research objects, especially in the districts that are passed by the Bengawan Solo River Basin so that all areas that reach the area that the Bengawan Solo River flows can be compared for the level of social vulnerability.

References

[1] Opiyo F, Wasonga O, Nyangito M, Schilling J and Munang R 2015 Drought Adaptation and Coping Strategies Among the Turkana Pastoralists of Northern Kenya *Int. J. Disaster Risk Sci.*

[2] Guan D, Hubacek K, Weber C L, Peters G P and Reiner D M 2008 The drivers of Chinese CO2 emissions from 1980 to 2030 *Glob. Environ. Chang.*

[3] Runtunuwu E and Syahbuddin H 2007 Perubahan Pola Curah Hujan dan Dampaknya Terhadap Periode Masa Tanam *J. Tanah dan Iklim* **26** 1–13

[4] Alhassan S and Hadwen W L 2017 Challenges and opportunities for mainstreaming climate change adaptation into WaSH development planning in Ghana *Int. J. Environ. Res. Public Health* **14**

[5] Khalid A and Rudianto I 2017 Kajian Tingkat Kerentanan Bencana Kekeringan Pertanian Di Kabupaten Demak *J. Tek. Sipil dan Perenc.* **19** 9–16

[6] Auliyani D and Wahyuningrum N 2020 POLA HUJAN DI BAGIAN HULU DAERAH ALIRAN SUNGAI BENGAWAN SOLO DALAM PERENCANAAN PEMANFAATAN SUMBER DAYA AIR *JPPDAS* **4** 53–62

[7] Wijaya R C 2021 Modeling Bengawan Solo River to Predict the Area Inundation of Modeling Bengawan Solo River to Predict the Area Inundation of Flood *J. Eng. Appl. Sci.* **11** 14416–30

[8] Fristyananda M A and Idajati. H 2017 TINGKAT BAHAYA BENCANA BANJIR DI KALI LAMONG KABUPATEN GRESIK *J. Tek. ITS* **6**

[9] Weichselgartner J and Kelman I 2015 Geographies of resilience: Challenges and opportunities of a descriptive concept *Prog. Hum. Geogr.*

[10] Masud M M, Azam M N, Mohiuddin M, Banna H, Akhtar R, Alam A S A F and Begum H 2017 Adaptation barriers and strategies towards climate change: Challenges in the agricultural sector *J. Clean. Prod.* **156**

[11] Shirley W L, Cutter S L and Boruff B J 2003 Social Vulnerability to Environmental Hazards *Soc. Sci. Q.* **84** 242–261

[12] Cutter S L 2003 The vulnerability of science and the science of vulnerability *Ann. Assoc. Am. Geogr.*

[13] Jariyah N A and Pramono I B 2018 Kerentanan sosial ekonomi dan biofisik daerah aliran sungai solo (socio-economic and biophysical vulnerability of solo watershed) *J Penelitian Pengelolaan Daerah Aliran Sungai (Journal of Watershed Management Research)*, **2** 2 89-110

[14] Handayani T D and Mataburu I B 2021 The Level of Social Vulnerability for Flood Disaster in the City of Tangerang, Banten Province *J SPATIAL Wahana Komunikasi dan Informasi Geografi*, **21** 1, 29-36.

IOP Conf. Series: Earth and Environmental Science **986** (2022) 012050 doi:10.1088/1755-1315/986/1/012050

Assessing the impact of spatio-temporal land cover changes on land surface temperature using satellite data in Beas Valley, Himachal Pradesh, India

R Lata[1] and S Ghosh[1*]

[1] G.B. Pant National Institute of Himalayan Environment, Himachal Regional Centre, Mohal, Kullu- 175126, Himachal Pradesh, India

*Email: visittosayanta@gmail.com

Abstract. Here, the impact of spatio-temporal Land Use Land Cover Changes (LULC) on the variations of Land Surface Temperature (LST) has been estimated using Landsat 7 ETM+, Landsat 8 TIRS/OLI and Sentinel-2 data for Beas basin of Kullu district, Indian Himalayan Region (IHR). Also, the relationship of various remote sensing indices such as NDVI, NDSI, NDBI and NDMI with LST were established using coefficient of correlation to monitor the role of natural ecosystem alteration on LST. LST was calculated using Single Window algorithm (SW) for eight major land cover categories extracted from Landsat-7 ETM+ and Landsat-8 (Path-147 and Row-38) Thermal Infrared Sensor (TIRS) data for October, 2000 and October, 2020 respectively. LULC change detection revealed that there was a huge increase in agricultural land including orchard expansion of 123 % during the year 2020 than that of 2000. Also there's a sharp increase of 40.63 % in settlement areas which includes the tourism activities. These are the significant factors for the changes in LST. Further, it shows that negative correlation with the strong correlation coefficients of $R^2 = 0.7072$ and $R^2 = 0.5642$ between NDSI and LST in Pre-winter season of 2000 and that of 2020 respectively. Whereas, the correlation between NDVI and LST showed positive correlation with the coefficient of $R^2 = 0.2577$ in 2000, which increased to $R^2 = 0.5959$ in 2020. This positive relationship highlighted the fact that an increase in LST provides favourable conditions for the vegetation growth in the valley during the pre-winter season 2000 and 2020 respectively. The main outcome of the paper, we believe, will be helpful in analysing the dynamics of land cover changes and sustainable environmental planning in the Beas Valley, which is crucial for livelihood sustainability of the people residing in the already fragile IHR.

1. Introduction

Land Surface Temperature (LST), is the main subject in geophysical parameters that have functions in monitoring climate change, hydrological cycle, soil moisture, land use dynamics, natural resource management, weather modeling, etc. LST variations in anthropogenic and naturally changing land cover complexity [1–4]. On the other hand, land cover and land use changes mainly influence the increase in the earth's surface temperature [5]. Temporal change detection of LULC reveals the variation in composition of vegetation, waterbodies and built-up areas as well as anthropogenic activities [6]. This variation has an important role in the temporal changes of LST and climate change [7,8]. Geographic Information System (GIS) and Remote Sensing data are systems used to investigate LST derivation and LULC changes [9–11]. Application of Thermal Remote Sensing technique by capturing ESG land cover on a local scale because ESG comes from solar radiation [12]. There are many studies on the impact of land use change on LST around the world. Imran et al. (2021)[13] monitored the Impact of Land Cover Changes on LST for Dhaka City of Bangladesh, whereas Dhar et al. (2019) [14] observed the impact of LULC variations on LST for Rajarhat Block of North 24-Parganas District in West Bengal. Kumar et al. (2018)[15] carried out the same studies for Spiti valley of Himachal Pradesh. On the other hand, Guha et al. monitored LST-NDVI relationship for Raipur City of India [16]. The association of LST, NDBI and NDVI using Landsat-8 data has been carried out in the Kandaihimmat Hoshangabad watershed, India by Malik et al. (2018) [17]. This study attempts to assess the impact of Spatial-temporal Land Cover Change on Land Surface Temperature (LST) using Landsat 7 ETM+, Landsat 8 TIRS/OLI, and Sentinel-2 data for the Beas Kullu Valley district, Indian Himalayan Region (IHR). The spatial variation in LST over different LULC classes in the catchment area of river Beas in Kullu District over the period of time where anthropogenic activities including rapid expansion agricultural land, tourism activities, Hydro-electric Power projects etc. alter natural ecosystems in relation to land and water use have been investigated. Moreover, the relationship of vegetation (NDVI), snow cover (NDSI), settlements (NDBI) and vegetation moisture content (NDMI) with LST were also established using coefficient of correlation for the study area. Outcome of the paper, we believe, will be helpful in monitoring the trend of land cover changes and sustainable environmental planning in the Beas Basin, which is very important for the sustainability of local people's livelihoods.

2. Methods

2.1 Study Area

The river Beas originated from the southern face of Rohtang Pass situated at an altitude of 4,361 meters above mean sea-level in Kullu district and flows for some 470 kilometers to the Sutlej River in Punjab, India. The drainage basin of the river covers an area of approximately 20,303 km^2[18]. In Kullu district, the drainage basin of river Beas covers an area of 3274 km^2 which consist of the entire Kullu except the extreme south. The topographic elevation of the water basin varies from 959 meters to over 6500 meters (Fig. 1a). The basin is extended between latitudes 31° 43' 30" N to 32° 25' 0" N and longitudes between 76° 58' 26.4" E to 77° 51' 44.399" E. Location map and Contours map of the study area with altitude gradient has been developed in ArcGIS 10.8 (Fig. 1a & 1b). The valley has climate variability with the elevation from Hot & sub-tropical in the lower elevation (959 – 1400 m) to warm temperate (1400-1900m), cool temperate (1900-2500m) and cold alpine to glacial (2500-6000m) in the northern & eastern mountain ranges. The Beas valley is rich in plant diversity and dense forest because of high precipitation in the form of rainfall in July- September and heavy snowfall during winter months of January- March [18]. Spatial distribution maps of annual Rainfall distribution, Relative Humidity (RH), average Air Temperature for the year 2020 have been developed to analyze the hydro-meteorological status using the Inverse Distance Weighting (IDW) interpolation method in ArcGIS 10.8 for Beas Basin (Fig. 1c, 1d, 1e, 1f, 1g). Rainfall, Average Relative Humidity, Average Air Temperature data have been collected from KVK Bajaura, Meteorological Centre, Shimla, District level survey document Kullu and NASA POWER Data Access Viewer.

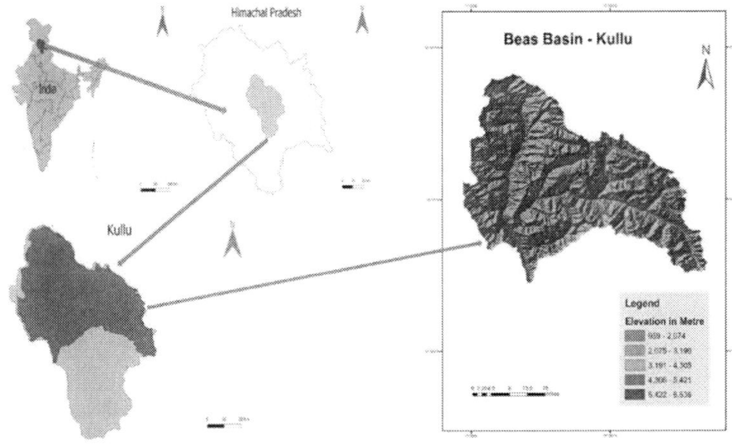

Figure 1a. Location of the study are with altitude gradient

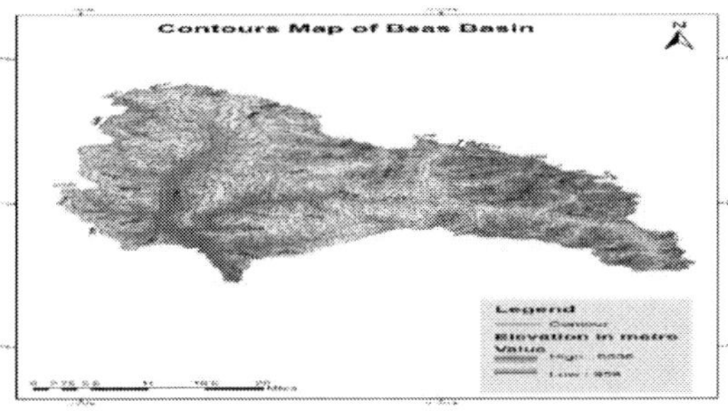

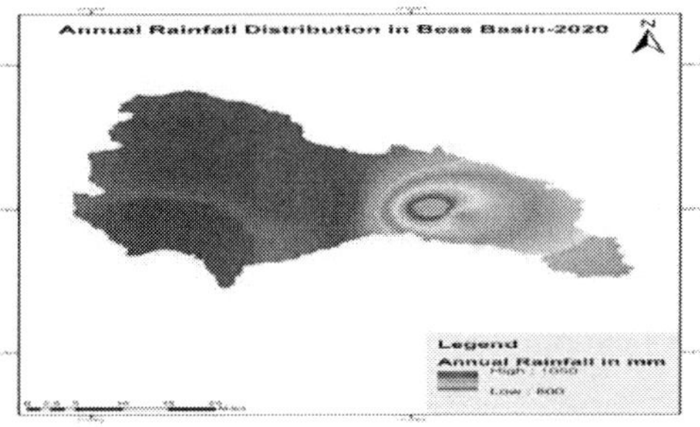

Figure 1b. Contours map of Beas Basin; **Figure 1c.** Annual Rainfall distribution pattern of Beas Basin (2020)

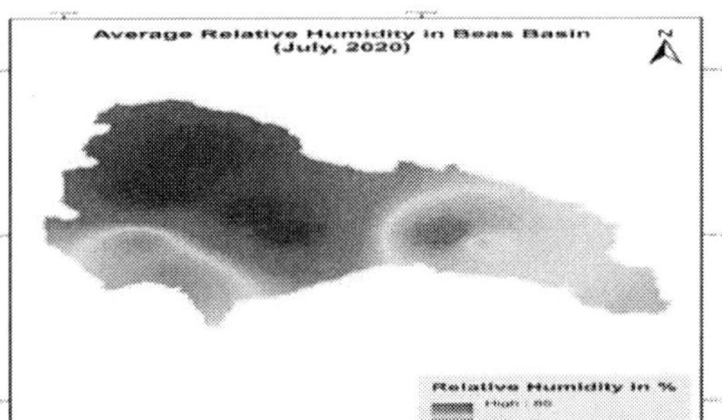

Figure 1d. Average RH distribution map of July, 2020; **Figure 1e.** Average RH distribution map of Desember 2020

IOP Conf. Series: Earth and Environmental Science **986** (2022) 012050 doi:10.1088/1755-1315/986/1/012050

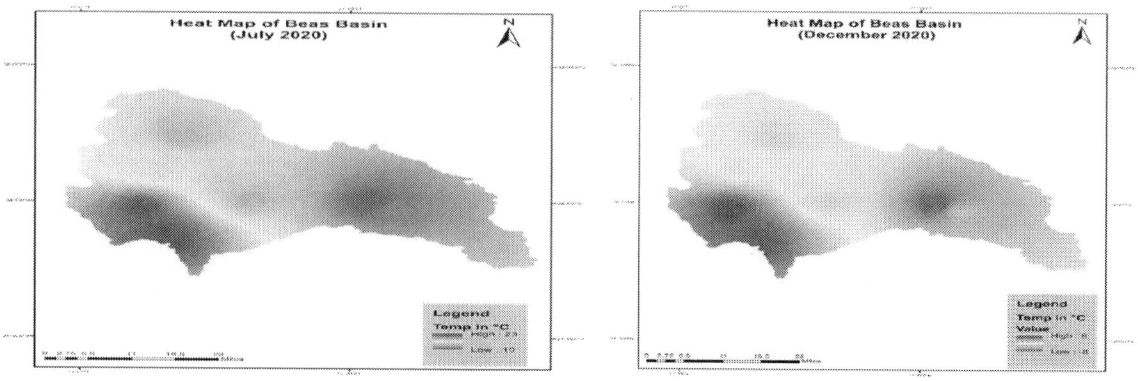

Figure 1f. Average Heat distribution map of July, 2020; **Figure 1g.** Average Heat distribution map of December, 2020

2.2. Methodology

To analyze the spatio-temporal changes in Land Surface Temperature (LST), LST products have been derived for Beas basin (Path-147 and Row-38) using single window/ single channel algorithm from Landsat-7 ETM+ and Landsat 8 TIRS/OLI satellite images of 30 m spatial resolution for 15th October, 2000 and 30th October, 2020 respectively (Fig. 2a & Fig. 2b).

LULC map of the Beas basin of Kullu district for October, 2000 (Fig. 3a) and October, 2020 (Fig. 3b) have been generated using Landsat 7 ETM+ at 30 m resolution, ESA Sentinel-2 imagery at 10 m resolution satellite data. Eight major LULC classes have been pointed out and used here, namely (i) Glacier/Snow covered area (ii) Open/Rocky Land (iii) Water Bodies (iv) Forest Green area (v) Vegetation/Scrub/Grass land (vi) Barren land (vii) Built-up/Settlements (viii) Agriculture/Plantation Area.

The Normalized Difference Vegetation Index (NDVI) is a vegetation index that is widely used to observe vegetation globally. Relatively healthy vegetation reflects in the near-infrared (NIR) region of the electromagnetic spectrum. While vegetation has a low reflectance in the red light of the electromagnetic spectrum. This index is calculated by subtracting the Red band from the NIR band and dividing by the number of Red bands and NIR bands (Equation 1). The NDVI value varies from 1 to +1, where a negative value indicates a body of water and a higher positive value indicates dense green leafy vegetation. Medium NDVI values (0.2 to 0.5) represent old shrubs and meadows or plants while lower NDVI values represent barren rock, sand, or snow. Estimated NDVI using Landsat 7 (ETM+) data for October 15, 2000, and Landsat 8 (OLI/TIRS) data for October 30, 2020, using Eq. 1 has been shown in figures 4a & 4b.

$$\text{NDVI} = (\text{NIR band} - \text{Red band}) / (\text{NIR band} + \text{Red band}) \qquad \text{...Eq. (1)}$$

For Landsat 7, NIR band is band 4 and Red band is band 3 whereas in case of Landsat 8 data, NIR band is represented by band 5 and that of Red band is band 4.

Here, the NDVI values for 15th October, 2000 and 30th October, 2020 have been calculated using raster calculator in ArcGIS 10.8.

The Normalized Difference Snow Index (NDSI) describes the snow cover area and is used as a parameter in hydrological forecasting. This index is calculated by subtracting the SWIR band from the Green band and dividing by the number of Green bands and SWIR bands (Equation 2). Snow mapping is important because snow is not only highly reflective in the visible part especially in the green band of the electromagnetic spectrum, but is also highly absorbent in the short-wave infrared (SWIR) spectrum. Since snow absorbs most of the incident radiation in SWIR while clouds do not, this allows NDSI to distinguish snow from clouds. An NDSI value > 0.4 usually indicates the presence of snow in the area.

Here, the Beas Basin NDSI values for October 15, 2000 (Fig. 5a) and October 30, 2020 (Fig. 5b) have been calculated using the raster calculator in ArcGIS 10.8.

$$NDSI = (Green\text{-}SWIR) / (Green + SWIR) \quad \ldots Eq. (2)$$

For Landsat 7, Green band is Band 2 and SWIR band is represented by Band 5 whereas in case of Landsat 8 data Band 3 and Band 6 represent the Green band and SWIR band of the electromagnetic spectrum respectively.

NDBI is the most common indexes for analysis the built-up areas. The build-up areas and bare soil reflects strongly in SWIR than that of NIR. NDBI can be calculated by using this following formula (Eq. 3):

$$NDBI = (SWIR - NIR) / (SWIR + NIR) \quad \ldots Eq. (3)$$

For Landsat 7, SWIR band is Band 5 and NIR band is represented by Band 4 whereas in case of Landsat 8 data Band 6 and Band 5 represent the SWIR band and NIR band of the electromagnetic spectrum respectively.

Also, the Normalize Difference Build-up Index value lies between -1 to +1. Negative value of NDBI indicate water bodies as water body doesn't reflect in the Infrared spectrum region. Whereas higher value represents built-up areas. Lower NDBI values indicate vegetation. Normalized Difference Built-up Index (NDBI) maps of Beas basin of Kullu district using Landsat 7 (ETM+) data for 15th October, 2000 and Landsat 8 (OLI/TIRS) data for 30th October, 2020 using Eq. 3 have been shown in figure 6a & 6b.

Normalized Difference Moisture Index (NDMI) is used to determine the water content of vegetation. This formula uses shortwave infrared (SWIR) and near infrared (NIR) to capture moisture variations in vegetated areas. NDMI is applicable in applications that monitor drought and changes in vegetation moisture conditions. NDMI is useful for fire hazard assessment.

NDMI can be calculated by subtracting SWIR band from NIR band and then dividing by the sum of NIR band and SWIR band (Eq. 4).

$$NDMI = (NIR - SWIR) / (NIR + SWIR) \quad \ldots Eq. (4)$$

In case of Landsat 4 to 7, Band 4 and Band 5 represent the NIR and SWIR band respectively whereas in case of Landsat 8, NIR and SWIR bands are represented by Band 5 and Band 6 respectively.

Normalized Difference Moisture Index (NDMI) maps of Beas basin of Kullu district using Landsat 7 (ETM+) data for 15th October, 2000 and Landsat 8 (OLI/TIRS) data for 30th October, 2020 using Eq. 4 have been shown in figure 7a & 7b.

3. Results and Discussion

3.1. Assessment of Spatio-Temporal Changes of Land Surface Temperature in Beas Basin (Over the Period of 20 Years)

Spatial-temporal variations of Land Surface Temperature (LST) were investigated using Landsat 7 ETM+, Landsat 8 TIRS/OLI data for the Beas valley in Kullu district, Indian Himalayan Region (IHR). LST products were deployed in the Beas basin (Path-147 and Row-38) using a single-window/single-channel algorithm from Landsat 8 TIRS/OLI and Landsat-7 ETM+ satellite images for October 30, 2020, and October 15, 2000, respectively. Respectively. The Beas Basin ESG map, dated 15 October 2000, shows the highest value of around 34.3°C, which is shown at the bottom of the Beas Basin which is a tourist and residential area, and an area where hydroelectric projects and other anthropogenic activities are located. Meanwhile, the lowest ESG value was observed at -24.7°C in the Beas basin glacier cover area (Fig. 2a). The ESG Beas Basin map of October 30, 2020, showing an elevation of approximately 36.23°C, which is also observed in and around the lower Beas Basin which consists of barren land and tourist and residential areas where development activities occur. Although the lowest ESG value observed was -27.26°C in the glacier cover area of the Beas basin (Fig. 2b). Spatial-temporal land cover changes over a 20-year period were also analyzed in the follow-up study stage and the relationship of vegetation (NDVI), snow cover (NDSI), settlements (NDBI), and vegetation moisture conditions (NDMI) with ESG was also determined using correlation coefficients. The results show that there has been a rapid increase in agricultural land including expansion of gardens, residential areas, and other anthropogenic activities over the years, which is a significant factor in changes in soil surface temperature.

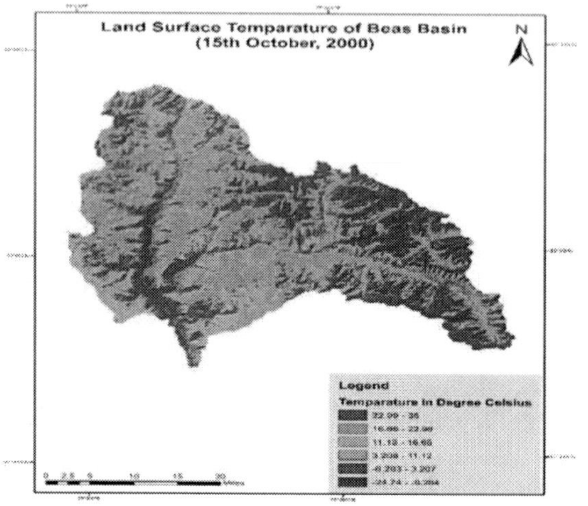

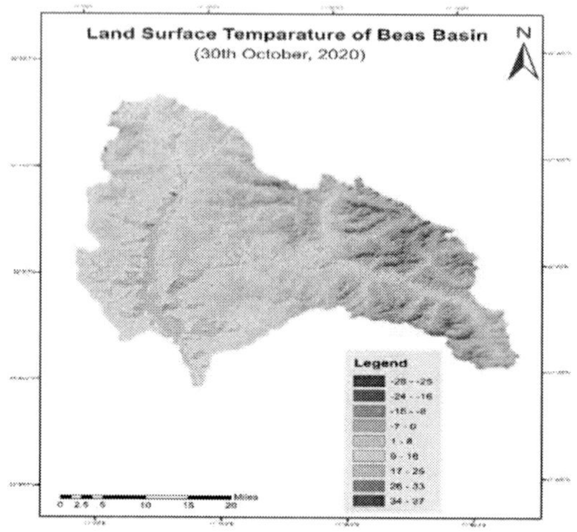

Figure. 2a: LST map of Beas Basin (15/10/00); **Figure. 2b:** LST map of Beas Basin (30/10/20)

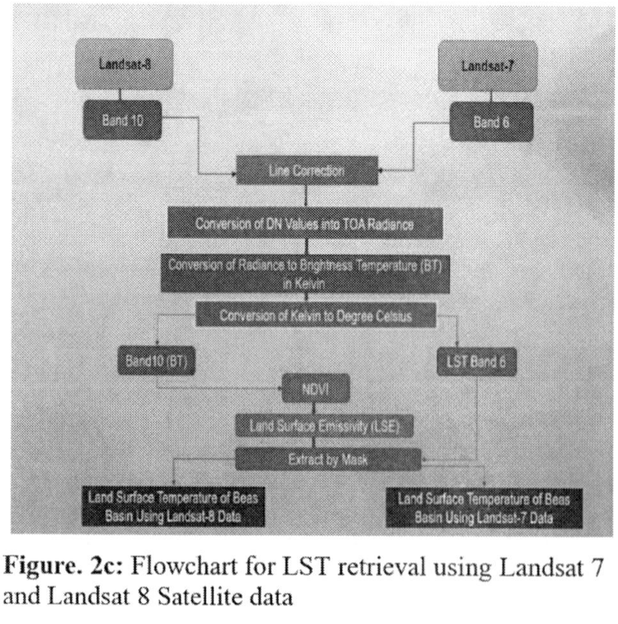

Figure. 2c: Flowchart for LST retrieval using Landsat 7 and Landsat 8 Satellite data

3.2. Temporal Change Detection for Eight Major Land Cover Types In Beas Basin (Over The Period Of 20 Years)

Land Use Land Cover (LULC) map of the Beas for October, 2000 (Fig. 3a) and October, 2020 (Fig. 3b) respectively has been generated using Landsat 7 ETM+ at 30 m resolution, ESA Sentinel-2 imagery at 10 m resolution satellite data. Eight LULC categories have been identified and used in this study, namely (i) Glacier/Snow covered area (ii) Open/Rocky Land (iii) Water Bodies (iv) Forest Green area (v) Vegetation/Scrub/Grass land (vi) Barren land (vii) Built-up/Settlements (viii) Agriculture/Plantation Area. The results revealed that there was a huge increase in Agricultural land including orchard expansion of 123 % during the year 2020 than that of 2000. Also there's a sharp increase of 40.63 % in settlement areas which includes the tourism activities such as hotels, restaurants etc. in the year 2020 than that of 2000. Moreover, it is also observed from LULC change detection that there's a sharp decline in the shrub vegetation and grass land about 45 % during the year 2020 than that of 2000.

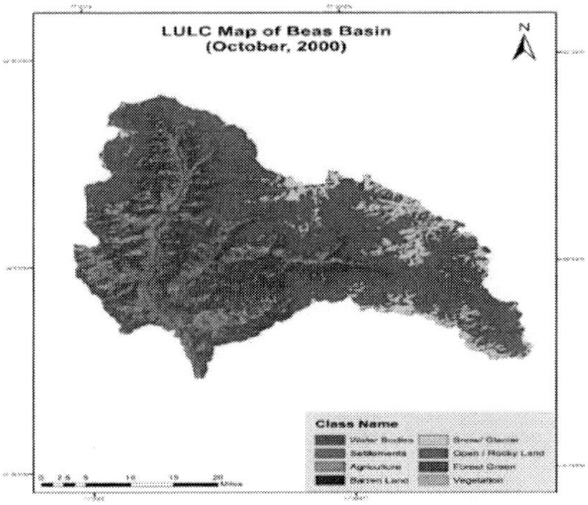

Figure. 3a: Status of LULC of Beas Basin (15/10/00);

Figure. 3b: Status of LULC of Beas Basin (30/10/20);

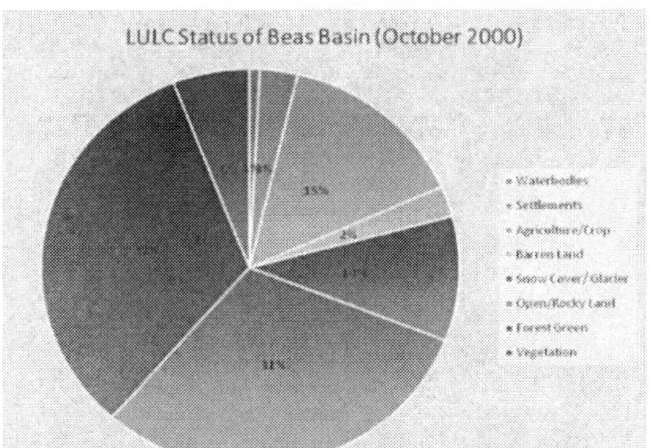

Figure. 3c: Area in % under different LULC categories (15/10/00);

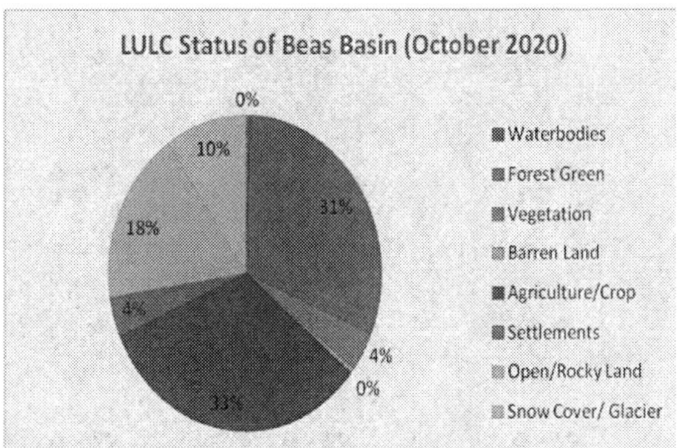

Figure. 3d: Area in % under different LULC categories (30/10/20);

Table 1: Change in LULC categories (Over the period of 20 years)

LULC Categories	Area in Sq. Km. (Oct, 2000)	Area in Sq. Km. (Oct, 2020)
Glacier/Snow covered area	331.6	312
Open/Rocky Land	1002.4	584.2
Water Bodies	26.3	15.6
Forest Green area	1063.2	1006.7
Vegetation/Scrub/Grass land	194.8	134.4
Barren land	81	16.9
Built-up/ Settlements	96	135
Agriculture/Plantation Area	480	1070

3.3. Monitoring the Relationship of Vegetation (NDVI) With LST (Over the Period Of 20 Years)

The relationship of vegetation (NDVI) with LST was monitored using coefficient of correlation in ArcGIS 10.8 for 15th October, 2000 (Fig. 4c) and 30th October, 2020 (Fig. 4d) respectively. The result indicates a positive correlation between NDVI and LST with the value of correlation coefficient $R^2 = 0.2577$ in 2000, which increased to $R^2 = 0.5959$ in 2020. This positive relationship suggests that an increase in LST values provides favourable conditions for the growth of vegetation in the Beas basin during the pre-winter season 2000 and 2020 respectively.

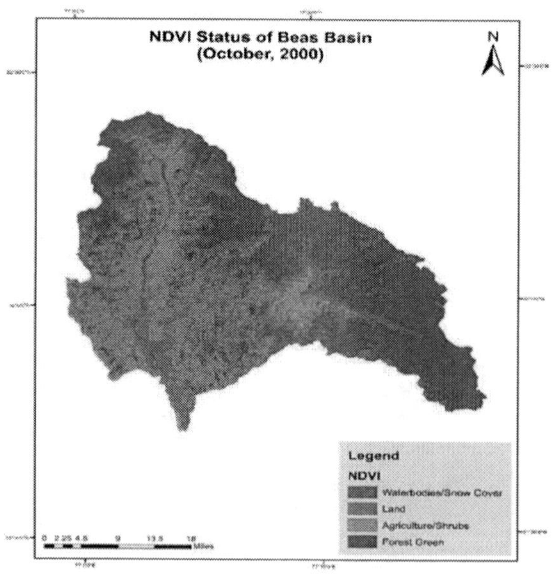

Figure. 4a: NDVI map of Beas Basin (15/10/00)

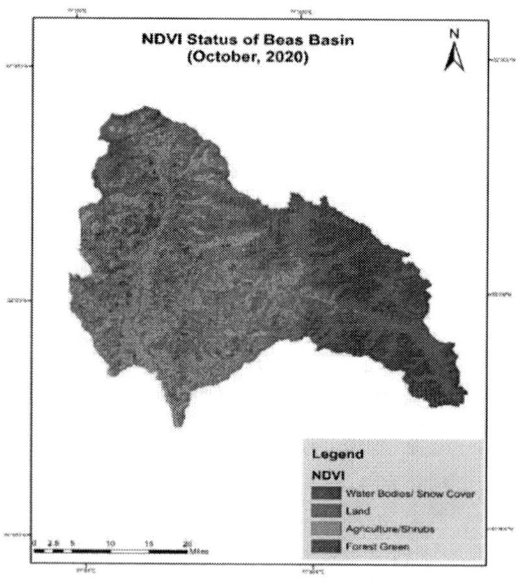

Figure. 4b: NDVI map of Beas Basin (30/10/20)

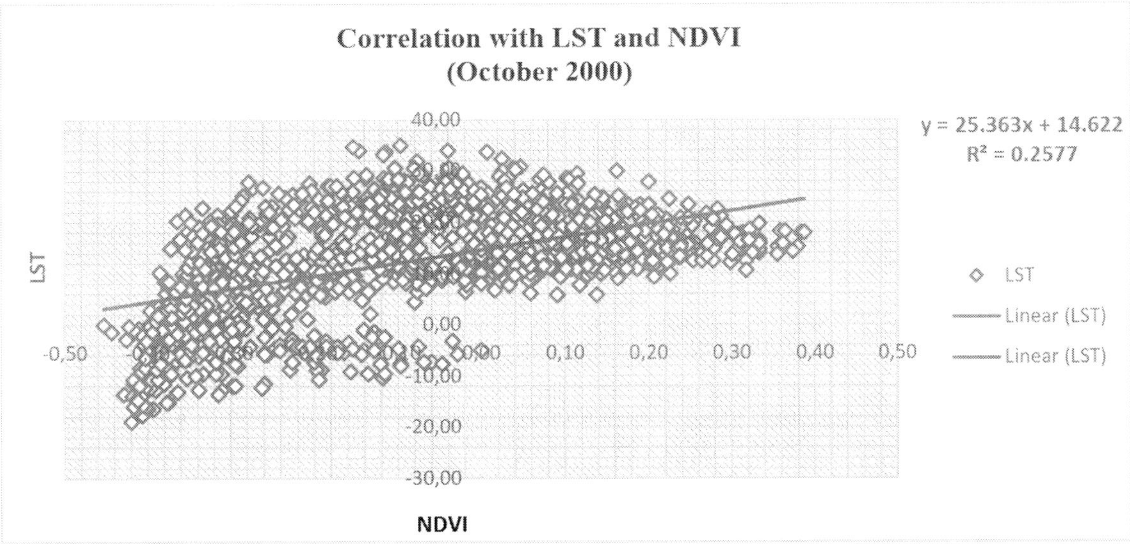

Figure. 4c: Relationship of vegetation (NDVI) with LST for 15th October, 2000

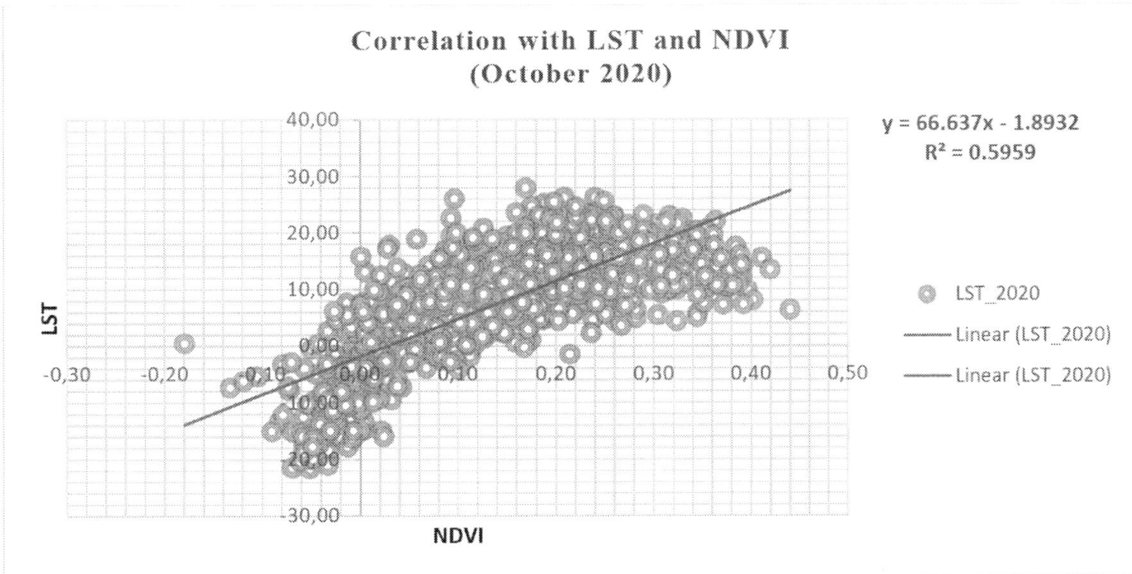

Figure. 4d: Relationship of vegetation (NDVI) with LST for 30th October, 2020

3.4. Monitoring the Relationship of Snow Cover (NDSI) With LST (Over the Period Of 20 Years)
The relationship of snow cover (NDSI) with LST was observed by using coefficient of correlation in ArcGIS 10.8 for 15th October, 2000 (Fig. 6c) and 30th October, 2020 (Fig. 6d) respectively. It indicates that negative correlation with the strong correlation coefficients of $R^2 = 0.7072$ and $R^2 = 0.5642$ between NDSI and LST exist in Pre-winter season of 2000 (15th October, 2000) and that of 2020 (30th October, 2020) respectively.

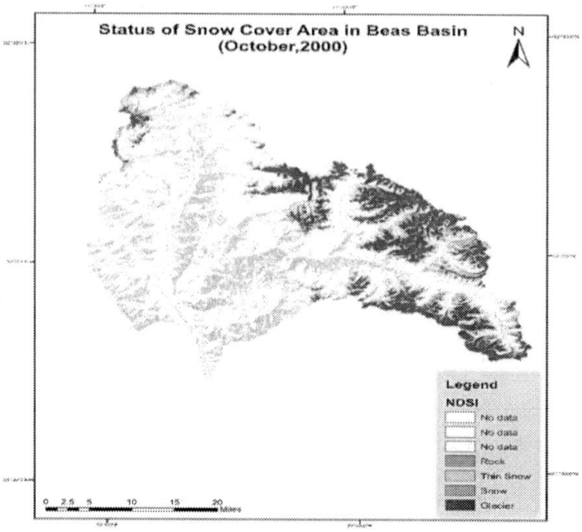

Figure. 5a: Snow Cover status of Beas Basin (15/10/00)

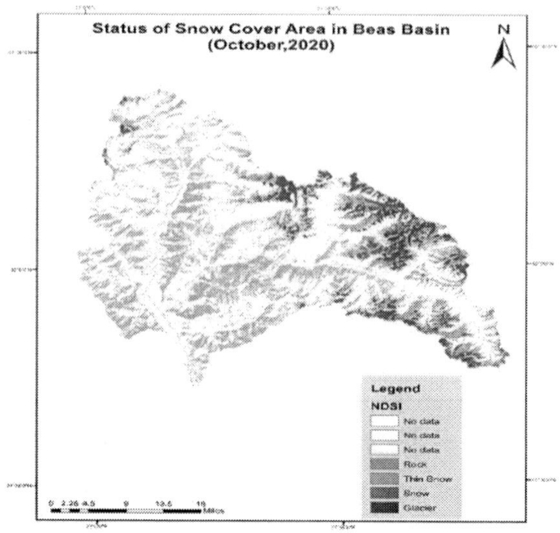

Figure. 5b: Snow Cover status map of Beas Basin (30/10/20)

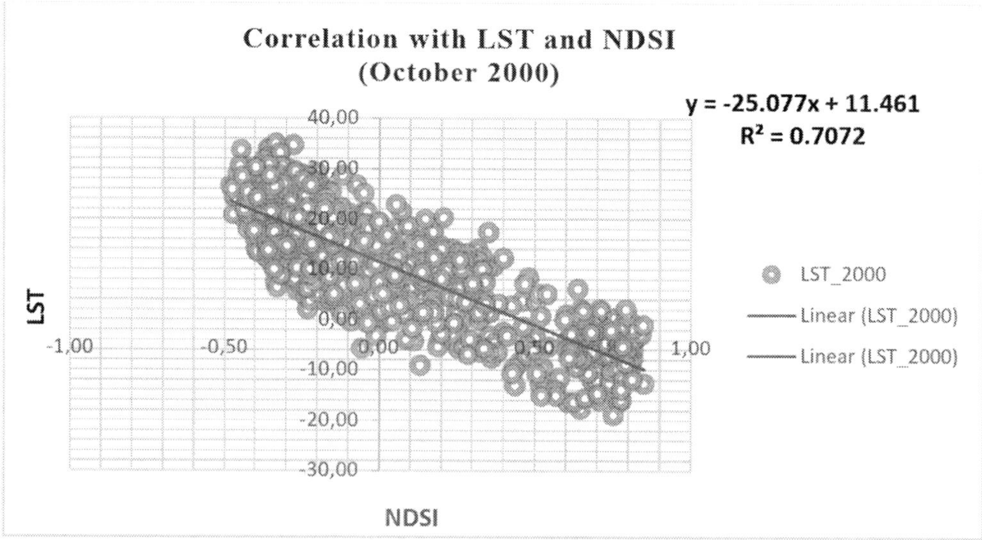

Figure. 5c: Relationship of snow cover (NDSI) with LST (15/10/00)

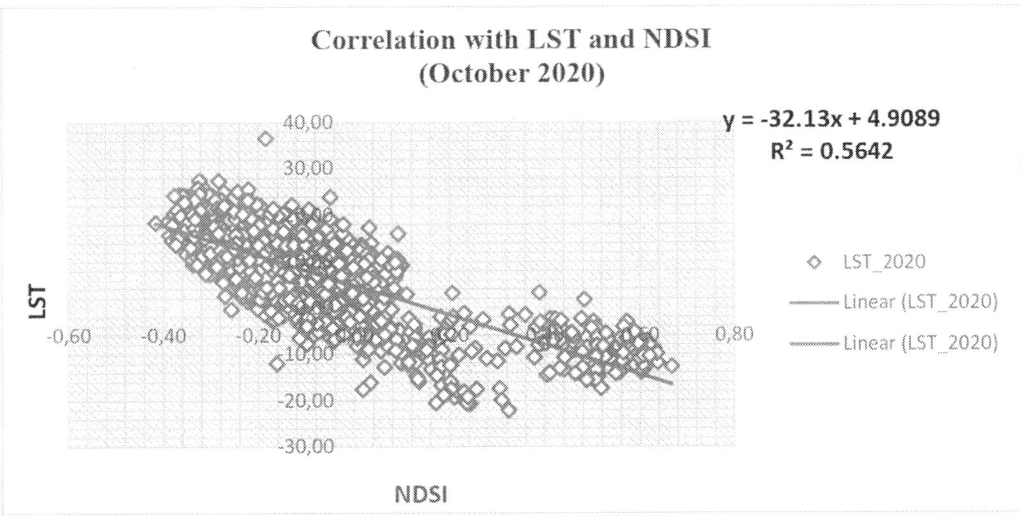

Figure. 5d: Relationship of snow cover (NDSI) with LST (30/10/20)

3.5. Monitoring the Relationship of Settlements (NDBI) With LST (Over the Period Of 20 Years)
The relationship of settlements (NDBI) with LST shows that positive correlation with the moderate correlation coefficients of R^2 = 0.4382 and R^2 =0.2081 between NDBI and LST exist in Pre-winter season of 2000 (15th October, 2000) and that of 2020 (30th October, 2020) respectively, indicating that, development activities, increasing built-up areas and LST have positive correlation.

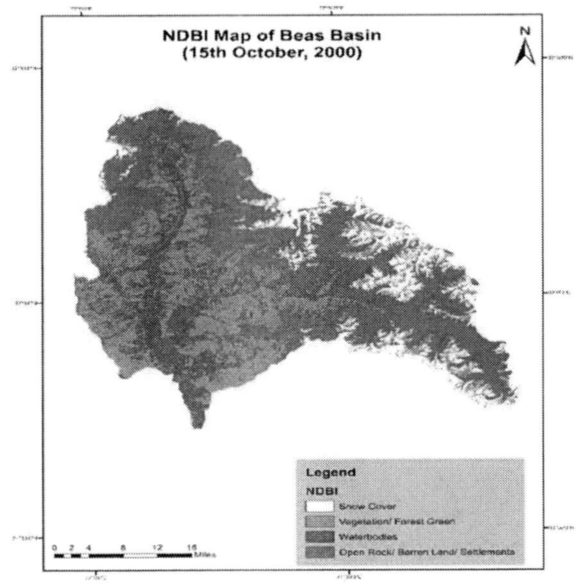

Figure. 6a: NDBI map of Beas basin (15/10/00)

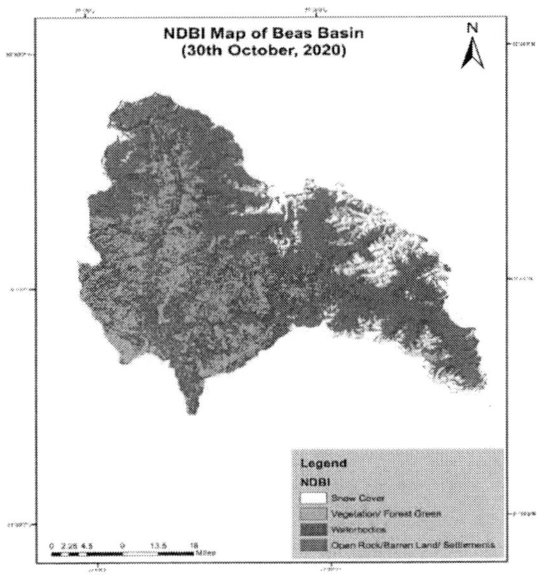

Figure. 6b: NDBI map of Beas basin (30/10/20)

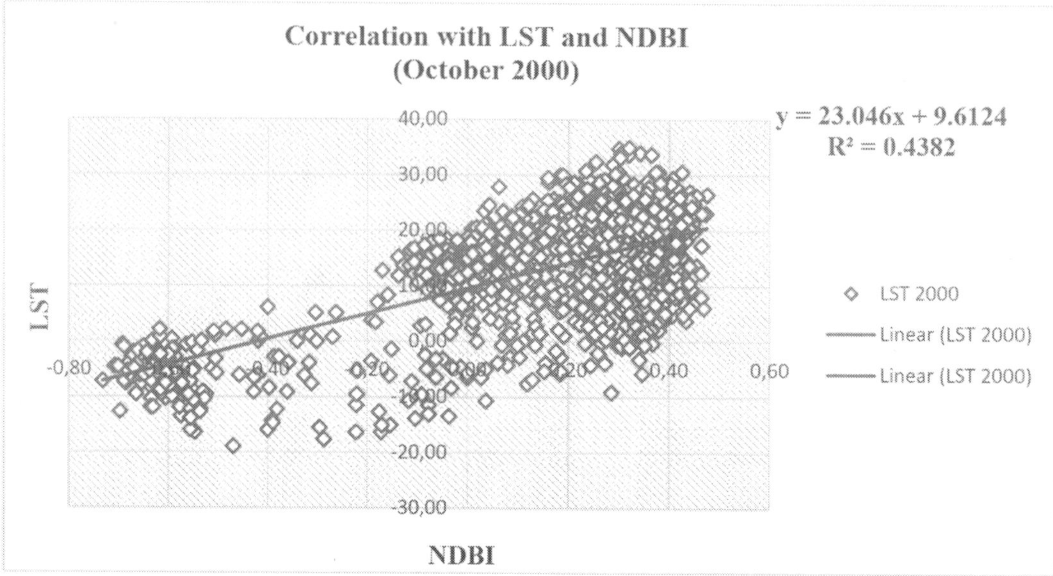

Figure. 6c: Relationship of settlements/ built-up areas (NDBI) with LST (15/10/00)

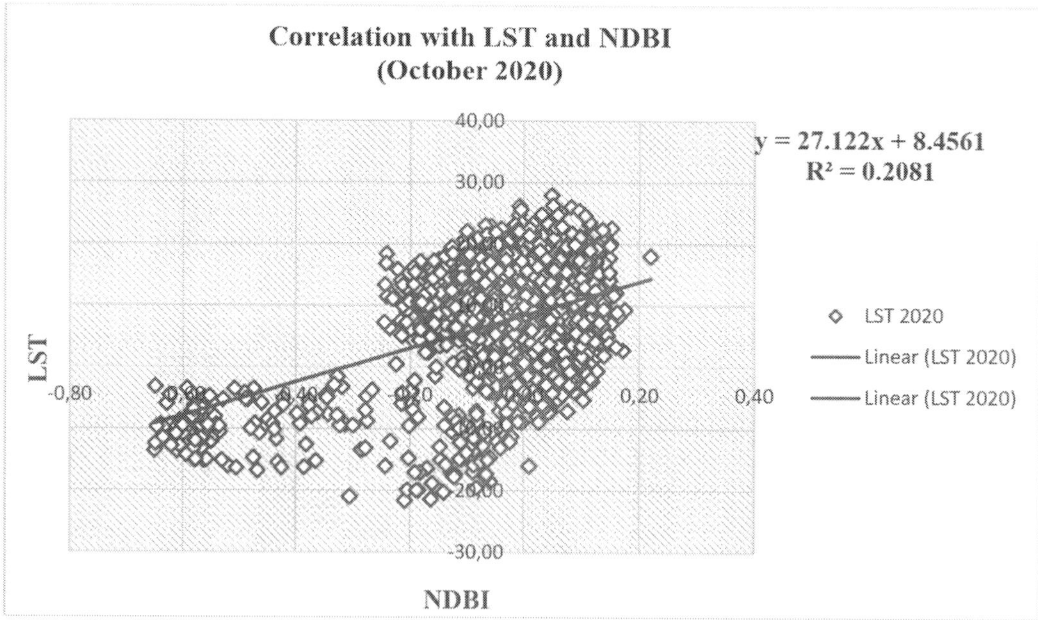

Figure. 6d: Relationship of settlements/ built-up areas (NDBI) with LST (30/10/2020)

3.6. Monitoring the Relationship of Normalized Difference Moisture Index (NDMI) With LST (Over The Period of 20 Years)

The relationship of Vegetation Moisture Content (NDMI) with LST indicate negative correlation exist between moisture content in the vegetation and LST with moderate correlation coefficients of R^2 = 0.4382 and R^2 = 0.2266 during the Pre-winter season of 2000 (15th October, 2000) and that of 2020 (30th October, 2020) respectively.

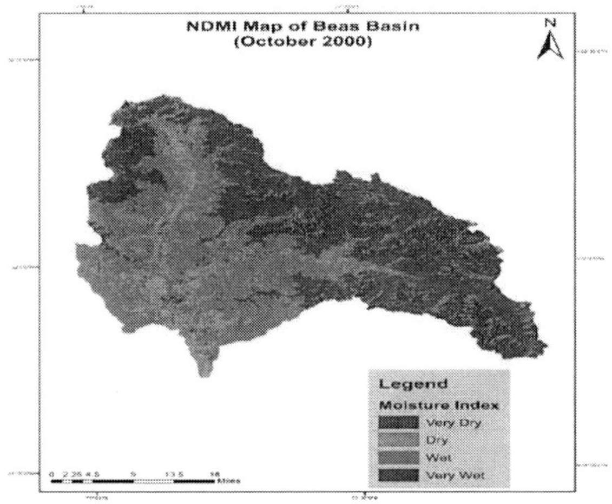

Figure. 7a: NDMI maps of Beas basin (15/10/00)

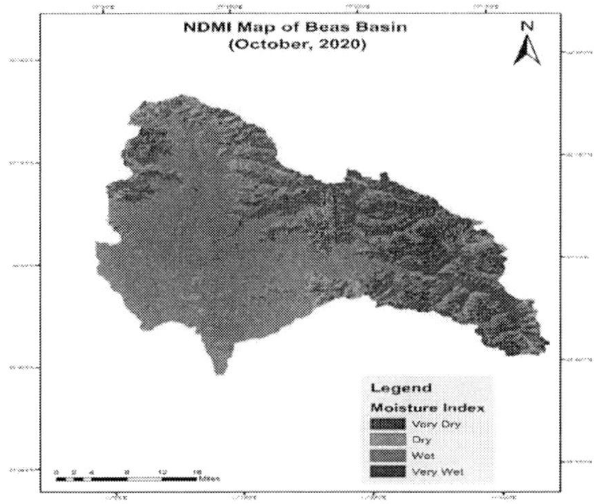

Figure. 7b: NDMI maps of Beas basin (30/10/20)

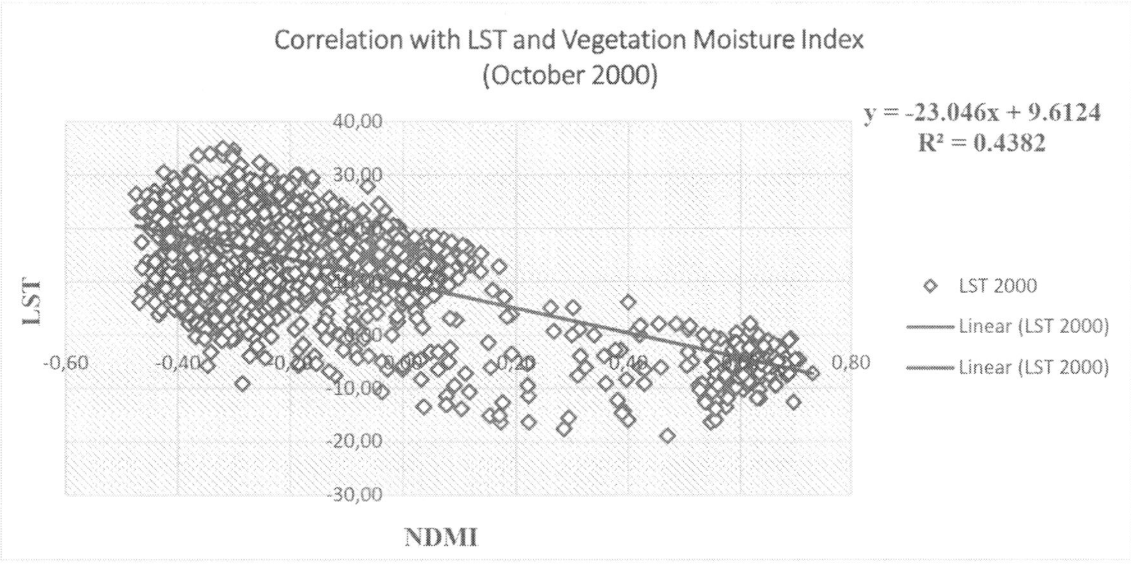

Figure. 7c: Relationship of vegetation moisture index (NDMI) with LST for 15th October, 2000

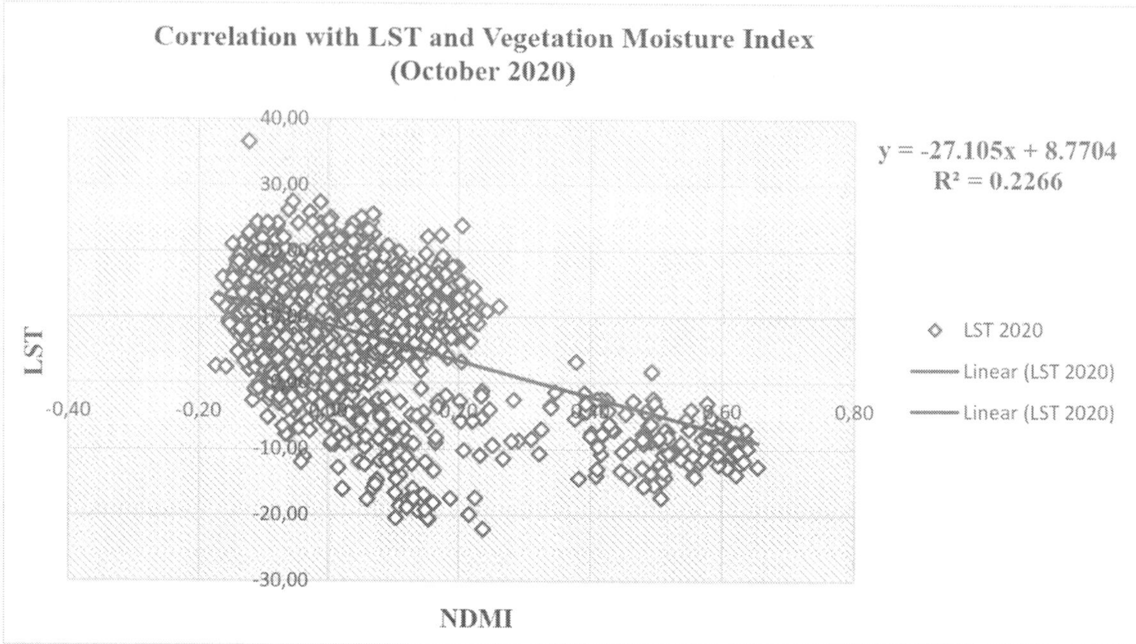

Figure. 7d: Relationship of vegetation moisture index (NDMI) with LST for 30th October, 2020

4. Conclusion

The study was conducted in the Beas watershed in Kullu District, Himachal Pradesh of the Indian Himalayan Region (IHR), an attempt was made to examine ESG changes in different LULC classes in the area over a 20 years period in which anthropogenic impacts played a role. An important role in changing natural ecosystems. An increase in mean ESG was observed at 2° Celsius in the study area over a 20 years period. Increased development, population settlements, increased agricultural activities, and decreased forest area are the factors causing the increase in ESG. The relationship between vegetation (NDVI) and ESG shows a positive correlation between NDVI and LST with a correlation coefficient value of R2 = 0.2577 in 2000, which increases to R2 = 0.5959 in 2020. This positive relationship indicates that an increase in ESG provides favorable conditions. profitable. Favorable for vegetation growth in the valley during the pre-winter season of 2000 and 2020. The relationship between snow cover (NDSI) and ESG shows that the correlation is negative with a strong correlation coefficient R2 = 0.7072 and R2 = 0.5642 between NDSI and LST in Pre-winter 2000 and 2020. The relationship between the Normalized Difference Built-up Index (NDBI) and ESG shows a positive correlation between settlements and ESG, implying that an increase in residential area contributes to an increase in ESG. The relationship between Vegetation Moisture Content (NDMI) and ESG shows a fairly negative correlation between vegetation moisture content and ESG.

Acknowledgement

Author Thanks to Er. Kireet Kumar, Managing Director, G.B. Pant National Institute of Himalayan Environment, Kosi-Katarmal, Almora, Uttarakhand for providing the facilities. Thanks to Er. R. K. Singh, Head of Himachal Regional Center Center, Himalayan G. B. Pant National Environmental Institute, for his support and providing necessary permits and facilities at the regional center during the study period. The authors also thank the funding agencies - National Mission for Himalayan Studies (NMHS), MoEF & CC, Govt. India to provide the necessary funds to carry out the work.

References

[1] Karnieli A, Agam N, Pinker R T, Anderson M, Imhoff M L, Gutman G G, Panov N and Goldberg A 2010 Use of NDVI and land surface temperature for drought assessment: Merits and limitations *J. Clim.* **23** 618–33

[2] Nikam B R, Ibragimov F, Chouksey A, Garg V and Aggarwal S P 2016 Retrieval of land surface temperature from Landsat 8 TIRS for the command area of Mula irrigation project *Environ. Earth Sci.* **75**

[3] Ramanathan V, Crutzen P J, Kiehl J T and Rosenfeld D 2001 Atmosphere: Aerosols, climate, and the hydrological cycle *Science (80-.).* **294** 2119–24

[4] Corresponding Z W and Wang P 2010 International Journal of Remote Using MODIS Land Surface Temperature and Normalized Difference Vegetation Index products for monitoring drought in the southern Great Plains , USA *Int. J. Remote Sens.* **25** 37–41

[5] Purwanto, Utomo D H and Kurniawan B R 2016 Spatio Temporal Analysis Trend of Land Use and Land Cover Change Against Temperature Based on Remote Sensing Data in Malang City *Procedia - Soc. Behav. Sci.* **227** 232–8

[6] Pal S and Ziaul S 2017 Detection of land use and land cover change and land surface temperature in English Bazar urban centre *Egypt. J. Remote Sens. Sp. Sci.* **20** 125–45

[7] Igun E and Williams M 2018 Impact of urban land cover change on land surface temperature *Glob. J. Environ. Sci. Manag.* **4** 47–58

[8] Nzoiwu C P, Agulue E I, Mbah S and Igboanugo C P 2017 Impact of Land Use/Land Cover Change on Surface Temperature Condition of Awka Town, Nigeria *J. Geogr. Inf. Syst.* **09** 763–76

[9] Jahan K, Pradhanang S M and Bhuiyan M A E 2021 Surface runoff responses to suburban growth: An integration of remote sensing, gis, and curve number *Land* **10** 1–18

[10] Bhuiyan M A E, Witharana C and Liljedahl A K 2020 Use of Very High Spatial Resolution Commercial Satellite Imagery and Deep Learning to Automatically Map Ice-Wedge Polygons across Tundra Vegetation Types *J. Imaging* **6** 137

[11] Hegazy I R and Kaloop M R 2015 Monitoring urban growth and land use change detection with GIS and remote sensing techniques in Daqahlia governorate Egypt *Int. J. Sustain. Built Environ.* **4** 117–24

[12] John J, Bindu G, Srimuruganandam B, Wadhwa A and Rajan P 2020 Land use/land cover and land surface temperature analysis in Wayanad district, India, using satellite imagery *Ann. GIS* **26** 343–60

[13] Imran H M, Hossain A, Islam A K M S, Rahman A, Bhuiyan M A E, Paul S and Alam A 2021 Impact of Land Cover Changes on Land Surface Temperature and Human Thermal Comfort in Dhaka City of Bangladesh *Earth Syst. Environ.* **5** 667–93

[14] Dhar R B, Chakraborty S, Chattopadhyay R and Sikdar P K 2019 Impact of Land-Use/Land-Cover Change on Land Surface Temperature Using Satellite Data: A Case Study of Rajarhat Block, North 24-Parganas District, West Bengal *J. Indian Soc. Remote Sens.* **47** 331–48

[15] Kumar P, Husain A, Singh R B and Kumar M 2018 Impact of land cover change on land surface temperature: A case study of Spiti Valley *J. Mt. Sci.* **15** 1658–70

[16] Guha S, Govil H and Diwan P 2020 Monitoring LST-NDVI Relationship Using Premonsoon Landsat Datasets *Adv. Meteorol.* **2020**

[17] Malik M S and Shukla J P 2018 Retrieving of Land Surface Temperature Using Thermal Remote Sensing and GIS Techniques in Kandaihimmat Watershed, Hoshangabad, Madhya Pradesh *J. Geol. Soc. India* **92** 298–304

[18] Anon 2012 *Survey Document District Kullu* (Kullu)

Exploring conservation education and learning activities in Bajulmati Sea Turtle Conservation area

L Y Irawan[1*], M M R Devy[1], Ditian[1], W E Prasetyo[1], M F M Kamal[1], A Sahrina[1], Sumarmi[1] and M Lelitawati[2]

[1] Department of Geography, Faculty of Social Science, Universitas Negeri Malang, Jl. Semarang 5, Malang, Indonesia

[2] Deparment of Biology, Faculty of Mathematics and Science, Universitas Negeri Malang, Jl. Semarang 5, Malang, Indonesia

*Email: Listyo.fis@um.ac.id

Abstract. Bajulmati Beach is one of the most well-known tourist destinations in the coastal area of the southern Malang Regency. It has several unique tourism destinations to visit. A community-based sea turtle conservation facility is one of them, which is managed by Bajulmati Sea Turtle Conservation (BSTC). This paper aims to describe the environmental conservation education and learning activities of BSTC. This is qualitative research conducted from primary data. Data collection was done using three methods, namely 1) in-depth interview with key informants, 2) focus group discussion (FGD) with LMDH, POKMASWAS, and BSTC, and 3) field survey. The result shows that BSTC has been consistently practicing some main environmental conservation actions in Bajulmati Beach. They are natural school (*Sekolah Alam*) twice a week, sea turtle's hatchling nest protection, clutch relocations, post-hatchling care, baby sea turtle releases, and environmental conservation education for the locals and tourists. In general, education and learning activities are increasing the environmental knowledge of the targeted audience. Furthermore, it is slowly shaping a more positive environmental attitude towards the locals.

1. Introduction

The southern coasts of East Java are ideal regions for sea turtle habitat and its conservation. One of the great examples is located in the coastal area of Malang Regency. It is supported by the geomorphological features of the area, such as sandy beach with gentle slope and rounded particles [1–3]. Bajulmati Beach is the frontier of sea turtle conservation action in the area.

The administrative location of Bajulmati coastal area is in Gajahrejo Village, Malang Regency, East Java. Bajulmati Beach has been successfully conserving the sea turtles and its habitat for the last seven years. Accordingly, the sea turtle conservation effort potentially acts like the seed of a greater conservation tree—there is a high chance for the action to be developed into one of eco-tourism activities. Eco-tourism is a tourism approach that incorporates nature safeguarding, the collaboration among the community and its nature conservation, and include vacationers with the neighborhood local area to acquire the potential benefit [4,5]. Through eco-tourism activities, the advantages acquired are the expanding economies of the seaside local area, acquiring information on sightseers, and the possibility to be more capable in protecting sea turtles.

Conservation can be translated into the preservation or maintaining the balance of the natural carrying capacity, quality, function, and ability of an environment. The act of conservation has two main objectives. Firstly, it is to support and improve the welfare and quality of human life better. Second, to utilize the natural resources and its ecosystem without disrupting the dynamic equilibrium of the ecology.

Bajulmati Beach is constantly developing as tourism destinations, yet still manage to uphold its nature conservation spirit [6,7]. The uniqueness therefore offers a great tourism prospect to the area. If the conservation area is classified as unique and rare, this will attract the attention of wider communities and potential tourists.

Bajulmati Sea Turtle Conservation (BSTC)—a local community group that focus on sea turtle conservation—create a Nature School (*Sekolah Alam*) to support eco-tourism activities through education in the area. Because the school is located in the area of tourism, BSTC also targeting to increase the environmental knowledge and awareness of the tourists regarding Bajulmati Beach ecological condition in general [1,2]. Accordingly, the Nature School aims to foster environmental character education in the community through the programs offered.

Nature School is a place of environmental education in BSTC facilities that are intended for the general public for all ages and backgrounds. Nature School offers a holistic understanding of nature and nature conservation to its students. The school uses natural environment of Bajulmati Beach as the learning media itself. Therefore, the students can directly experience the related topics that Nature School has provided before. In addition, learners can indirectly participate in conservation actions of eco-tourism by increasing the economy of the coastal community. Eco-tourism also able to increase environmental knowledge of the tourist and provide further potential better attitude change towards the environment, and sea turtle conservation in particular.

2. Methods

This is qualitative-descriptive research. Descriptive research methods have two main characteristics [8,9]: (1) anchoring on current or actual issues at the research period and (2) describing the facts of the issues after interpretation. The discussion of this research is presented in the form of a description of the words (description). The subjects of this research are sea turtle conservation actors in Bajulmati.

There is only one type of data in this study, namely primary data. Primary data were obtained by using in-depth semi-structured interviews, open-ended questions, and neutral questioning based on a qualitative approach. Semi-structured interviews were conducted by asking pre-prepared questions. However, there are questions that were developed from the answers of the informants to be able to dig deeper information.

The stages of the research data collection process can be detailed as follows. Firstly, literature Study on the implementation of conservation and development of Nature School in the Bajulmati Beach environment. Second, in-depth interviews and discussions with informant. Third, observations of conservation activities carried out by informants as validation of interview data. The combination of various data collection techniques is referred to as data collection triangulation.

The data obtained from the literature study will be confirmed and compared with the data obtained from the informants. The data collection technique used in this study is the most likely technique to obtain as much data as possible. The amount of data in question is used for the analysis stage.

The data analysis process starts from the beginning of the research (literature study). The data is organized so that data that can support research activities and data acquisition in the field can be grouped. After that, it was continued with the process of synthesizing data from interviews in the field and literature study data. This analytical technique makes it easier for researchers to identify the causes and implications of environmental conservation activities carried out by Bajulmati Sea Turtle Conservation (BSTC). Then, the findings are presented descriptively to explain the development of natural schools in BSTC facilities.

3. Results and Discussion

3.1. Initiation and beginning of the establishment of a nature school by Bajulmati Sea Turtle Conservation (BSTC)

Attention to assure the insurance of sea turtles are increasing, one of which occurred in Malang. The coastal area of Southern Malang is one of the natural habitats for sea turtles. One of them is located in Bajulmati Beach. The local community therefore marked a small patch of area for sea turtle conservation. Eggs hunting triggered the local residents to establish a conservation area as they became knowledgeable and aware of the crucial role of sea turtle in the ecology. If one of the ecological chains is broken, it will affect the entire way of life that develops in the biological system. Therefore, in order to adjust and reduce poaching, the surrounding area is starting to build shelters.

Bajulmati Sea Turtle Conservation (BSTC) is a local-based community group that concerns in the sea turtle population and ecosystem conservation. Based on the results of interviews, this conservation effort started from the presence of physical and anthropogenic threats to the sea turtles and the environment of Bajulmati Beach itself. The main physical or natural threat is high waves which is common in the southern Java. It is because the winds and currents from the Indian Ocean move with high speed and strength. As implications, the coastal area of southern Java faces high erosion (abrasion) rate that threat the sea turtle nesting habitats. However, the threat to sea turtle population is exacerbated by egg poaching actions. Therefore, BSTC undertakes conservation efforts with protection, preservation, and sustainable utilization strategies.

BSTC already has plans for sea turtle conservation efforts for the short and long term. One of the long-term conservation efforts is environmental education through the Nature School (*Sekolah Alam*). The establishment of the Nature School aims to introduce the actual conditions in the field, everything related to the environment at Bajulmati Beach. This includes lessons related to the phenomenon of high waves. In another sense, the education in the Nature School is not only limited to sea turtle conservation education.

Through the Nature School, BSTC also wants to increase knowledge and raise awareness of the surrounding community to be able to understand the importance of preserving the marine and coastal environment. It is hoped that this awareness can continue to involve the participation of the surrounding community to be able to carry out conservation efforts. In addition, the Nature School also targets tourists and visits from educational institutions to learn about the environment directly in nature. Nature School is expected to provide environmental education more intensively and effectively to the community and tourists. Accordingly, environmental knowledge and awareness also increases. Furthermore, positive behavior change towards the environment can occur.

The meaning of the name of *Sekolah Alam* is education or knowledge obtained directly from nature. In simple terms, BSTC defines a natural school as a means of learning from and in nature. This principle was used by Mr. Sutari—the head of BSTC—as the basis for building the Nature School. He obtained his knowledge from the observation of the natural phenomenon in the Bajulmati Beach environment. In another sense, knowledge related to the environment and disaster is not obtained by Mr. Sutari from formal education.

Based on the results of the interview, BSTC hopes that there will be further theoretical research related to the knowledge that he get directly from nature. The goal is that the knowledge that will be given at the Natural School has received scientific validation. In addition, further research activities are also expected to be a separate lesson for managers. In the long term, the BSTC natural school can be used as one of the leading tourist attractions on Bajulmati Beach. Local tourism actors plan to cooperate in the management of eco-edutourism at Bajulmati Beach.

3.2. The learning model for environmental conservation composed by Bajulmati Sea Turtle Conservation (BSTC)

BSTC uses outdoor learning as the learning model in Nature School. Learnings are mostly done in open spaces, by utilizing the natural environment at Bajulmati Beach. Visitors who will learn about sea turtle conservation can explore the environment of Bajulmati Beach and the entire ecosystem in it. This is intended to widen the knowledge and understanding of visitors about the natural potential of Bajulmati Beach. Furthermore, it is intended to hone their ability to find the meaning of environmental conservation in a broader term [10–12].

There are several reasons that make the Bajulmati coastal environment very important in conservation studies, including:

1. As a place of learning
   Natural environment is an object to be explained to visitors or in this case the individual or group will learn about conservation. The learning activities held every Monday and Thursday at the facility of BSTC.
2. As a learning resource
   The environment is a real learning resource that is used by managers and visitors to know, understand, and implement environmental conservation measures, especially in the effort to conserve sea turtles. BSTC also use social media as a publication tool for the activities as well as sea turtle conservation information. The contents include: sea turtle species, life circle of sea turtle, sea turtle nesting and hatchling process, sea turtle's egg relocation and preparation for releasing the individuals back to the sea, as well as the ecological condition in the Bajulmati Beach area.

The learning implementation carried out by BSTC has represented environmental-based learning that contains comprehensive elements in learning including: empirical, caring, aesthetic, and social such as experts' descriptions [13,14].

1) Empirical, which provide the widest possible opportunity for learners to interact with the environment directly. Here students can observe, understand, analyze, and interpret all phenomena and resources found in the environment.
2) Caring, is to provide information and specific stimulus. The aim is to raise the awareness of environmental sustainability importance. In this case, the students were directed in order to understand that all the elements that exist in the environment interact and influence each other.
3) Aesthetic, which provide insight to students about the existence of natural resources and beautiful tourist attraction to be enjoyed. Learners are given the understanding that the resources that are able to meet the needs of non-physical human needs, such as landscapes, order a comfortable and beautiful environment, give a sense of tranquility. This also instills a sense of responsibility learners to natural resources and the environment.
4) Social, providing knowledge about the social, cultural, and economic life of the community in the Bajulmati coastal environment.

3.3. Learning facilities at nature school

The Nature School activities are carried out at the sea turtle conservation facility owned by BSTC. These facilities are in the form of sea turtle conservation facilities consisting of 1) sea turtle nesting tanks from the relocation, 2) temporary ponds for hatched eggs, 3) ponds where adult sea turtles live, 4) sea turtle graves, and 5) infographic boards containing information related to sea turtles and their conservation efforts. More specifically, BSTC also provides library containing books related to the environment. The Nature School library opens donations for tourists, the general public, or agencies who wish to donate related books. If the learning material requires a more detailed explanation visually, Mr. Sutari as the tutor uses the blackboard as a learning medium.

3.4. Challenges in developing the nature school

BSTC has two obstacles, namely funding and publication. Because BSTC is an independent community group, funding related to conservation activities that are carried out continuously is a major obstacle. In addition, publications related to these conservation activities have not been published optimally. As a result, many of the people and prospective tourists of Bajulmati Beach are not aware of the existence of

BSTC. In fact, BSTC can be used as an object of potential tourist attraction if it is developed with careful planning and management.

Funding problems can be handled if the publication of sea turtle conservation efforts can be handled properly. Widespread publications, especially online publications, will increase the exposure of BSTC to various levels of society. It is possible that in the future there will be CSR (corporate social responsibility), NGO (non-governmental organizations), government agencies, educational institutions, communities, and the general public who want to help.

In addition to obstacles, BSTC also has challenges in managing sea turtle conservation efforts at Bajulmati Beach. The challenge is the lack of synergy between BSTC and tourism managers related to tourist management and tourist attraction objects. There are still differences in the vision of tourism management in Bajulmati Beach between BSTC and tourism actors. This is because the community does not yet have an awareness of the urgency of turtle conservation efforts on the Bajulmati beach.

4. Conclusion

The implementation of sea turtle conservation by Bajulmati Sea Turtle Conservation (BSTC) has provided meaningful lessons to the community about the importance of continuously preserving nature. Environmental-based learning in the form of eco-edutourism provided by Bajulmati beach has a role as a place of learning and as a learning resource. Furthermore, the implementation of learning from the nature has a positive regarding the empirical, caring, aesthetic, and social experiences. On the other hand, there is a need for synergy between various policy makers and the community to continue to manage the sustainability of conservation, especially sea turtles in the Bajulmati coastal environment.

Acknowledgement

This research would have been impossible without the support of the Research and Community Services Institute (LP2M), UM under the PNBP UM 2021 grant.

References

[1] Irawan L Y, Sumarmi, Bachri S, Rosbella Devy M M, Faizal R and Prasetyo W E 2021 Coastal Community Resilience Planning Toward Disaster: A Case Study on Coastal Area in Malang Regency, East Java, Indonesia *IOP Conf. Ser. Earth Environ. Sci.* **747**

[2] Sunarto, Purnama I L S, Mardiatno D, Malawani M N, Prasetyaningrum A, Safitri A E and Sasongko M H D 2020 Integration of participatory and aerial mapping for sea turtle conservation zoning in Malang coastal area *IOP Conf. Ser. Earth Environ. Sci.* **451**

[3] Sunarto, Malawani M N, Sasongko M H D, Elvira I, Nastiti I P, Arlita P and Arinda E S 2019 A Geomorphological Evaluation of Sea Turtles Nesting in the Southern Sea of West Java *IOP Conf. Ser. Earth Environ. Sci.* **256**

[4] Sumarmi, Bachri S, Irawan L Y, Sholeha A W and Aliman M 2021 Ecotourism Development Strategies of Pulau Merah Beach, Banyuwangi, Indonesia *IOP Conf. Ser. Earth Environ. Sci.* **747**

[5] Rudianto 2018 Restorasi Ekosistem Mangrove Desa Pesisir Berbasis Co-Management Restoration of Mangrove Ecosystem in Coastal Village Based on Co-Management *J. Perikan. Univ. Gadjah Mada* **20** 1–12

[6] Rudianto R and Bintoro G 2018 Future Turtle Management: Opportunities for Habitat Restoration Governance in East Java, Indonesia *Int. J. Environ. Agric. Biotechnol.* **3** 1721–31

[7] Rudianto R, Susilo E and Yamindago A 2015 Coastal Ecosystem Restoration Based on "Good Governance": Case Study Gresik and Malang Regency, East Java Province *Res. J. Life Sci.* **2** 161–72

[8] Bhat A 2018 Descriptive Research: Definition, Characteristics, Methods, Examples and Advantages. QuestionPro

[9] Wong A 2020 Descriptive Research Design | Definition, Methods and Examples

[10] Siswoyo A A, Setyawan A, Citrawati T, Bendriyanti R P and Dewi C 2020 Management of

outdoor learning models for environmental education courses *Univers. J. Educ. Res.* **8** 5036–43

[11] Aliman M, Budijanto, Sumarmi and Astina I K 2019 Improving environmental awareness of high school students' in Malang city through earthcomm learning in the geography class *Int. J. Instr.* **12** 79–94

[12] Wayan Sukarjita I, Ardi M, Rachman A, Supu A and Dirawan G D 2015 The integration of environmental education in science materials by using MOTORIC learning model *Int. Educ. Stud.* **8** 152–9

[13] Corte E D 2019 Learning design: Creating powerful learning environments for self-regulation skills *Vopr. Obraz.* **4** 30–46

[14] Krott M and Giessen L 2014 Learning from practices - implications of the "practice based approach" for forest and environmental policy research *For. Policy Econ.* **49** 12–6

IOP Conf. Series: Earth and Environmental Science **986** (2022) 012052 doi:10.1088/1755-1315/986/1/012052

Urban Flood Resilience through Spatial Plan in Bandung City, Indonesia

D Afriyanie [1,*], M M Julian [2,3], and H A Nugraha [4]
[1] Lokahita, Research Center on Ecology and Geospatial
[2] Hydrography Research Group, Faculty of Earth Sciences and Technology, Institut Teknologi Bandung (ITB), Bandung, Indonesia
[3] Geographic Information Science, Friedrich-Schiller-Universität Jena (FSU-Jena), Jena, Germany
[4] Study Program Urban and Regional Planning, School of Architecture, Planning, and Policy Development, Bandung Institute of Technology, Bandung City, Indonesia

*Email: dian.afriyanie@gmail.com

Abstract. According to New Urban Agenda, the spatial plan has been promoted as an instrument to strengthen urban resilience. In the context of urban flood resilience, flood management measures could be incorporated into the spatial plan to address inundation and flood risks. However, the existence of a spatial plan does not necessarily reduce inundation and flood risks due to the ineffectiveness of spatial plan implementation and/or insufficient flood management measures. This study aims to examine flood management measures within Bandung's spatial plan in achieving urban flood resilience in the future. We combine flood management measures framework and characteristics of urban system resilience as an analytical framework. The effectiveness of flood management measures is highly associated with the topographic characteristics and the combination of measures selected in suitable locations. We use thematic analysis to identify flood management measures in Bandung's Spatial Plan and compare the result with the spatial distribution of future inundation and flood risks data from the previous study. We reveal that flood management measures in Bandung's spatial plan have not been selected and located based on ecological constraints and future inundation information. This study demonstrates the importance of flood risks information to formulate an effective spatial plan for urban flood resilience.

1. Introduction

Floods are the most frequent disaster for more than 30 years in Asia, and their trend increases significantly [1]. The number of populations at risks and losses affected by floods highest in Asia compared to any other continents [1,2]. Indonesia is the third country in Asia that experienced the most frequent flood events between 1973-2002 [1]. Based on Indonesian Disaster Data and Information (DIBI) from National Agency on Disaster Management (BNPB), floods are the most frequent disasters in Indonesia between 2010 and 2019. DIBI recorded that in 2019 there were 343 flood events in Indonesia and resulted in 1,045 people being injured and 94,493 houses flooded.

Many researchers have agreed that land-use change and climate change are among drivers or factors contributing to flood hazards and risks [1,3–7]. At a global scale, the impacts of climate change on the hydrological processes are stronger than land-use change. However, at the regional scale, climate change worsens the impact of land-use change on the hydrological process [8]. The changes in precipitation were more dominant in the annual cycle (wet season become wetter, dry season become drier) than in the long-term annual trend. Land-use change enhances the seasonal changes due to climate changes on hydrological responses [9].

Land-use zoning and regulation are one of the flood management measures to address flood hazards and risks, especially to manage land-use change as a contributing driver. In Indonesia, land-use zoning and regulation are embedded within a detailed spatial plan ("*Rencana Detail Tata Ruang* – RDTR"), which is formulated based on a general spatial plan ("*Rencana Tata Ruang Wilayah* – RTRW"). According to [10], spatial planning refers to "*the processes used by the agencies (in both public and private sectors) in deliberate attempts to influence the spatial distributions of humans and non-humans and of various land-using activities*". Today, spatial plan has been promoted as one of the instrument to strengthening the resilience of cities as stated in New Urban Agenda (2017) as follows: "*We commit ourselves to strengthening the resilience of cities ..... through the development of quality infrastructure and spatial planning, by adopting and implementing ... ecosystem-based approaches in line with the Sendai Framework for Disaster Risk Reduction 2015-2030 and by mainstreaming holistic and data-informed disaster risk reduction and management at all levels to reduce vulnerabilities and risk....*"

There are opportunities to integrate various flood management measures into each spatial plan's contents, such us: spatial structures and patterns, spatial programs, and spatial utilization control and/or zoning regulation. However, the existence of spatial plan does not necessarily reduce inundation and flood risks in the future. due to the ineffectiveness of spatial plan implementation and/or insufficient flood management measures [11]. Researchers argue that spatial planning process has been overlooked environmental concerns as a primary concern [12–14]; and replaced by "techno-managerial, expert management and administration" [15]. As a result, hydrological responses and ecological substances are often overlooked in the spatial planning processes and lead to ineffective quality of spatial plan to accommodate flood management measures in reducing future inundation and flood risks [11].

In Indonesia, integrating environmental concerns into spatial planning has been established since the Spatial Use of Management Act 1992 through the hierarchical concept of spatial planning, from a national level to provincial level and local level [16]. Today, a strategic environmental assessment (SEA) has been promoted to integrate environmental concerns into the spatial planning process at all plan levels. During the spatial planning process, SEA requirements ensure that space allocation for development activities could still maintain environmental carrying capacity. SEA recommendation for spatial plan integrates into all contents, including spatial structures, spatial pattern, strategic areas, and spatial utilization control. However, the implementation of SEA in Indonesia faces some challenges such as : "limited capacity, insufficient data, lack of funding, the perception of SEA as a burden" [17–19]. Despite the challenges, SEA assessments could provide an opportunity to explore tools, models and methods in integrating environmental concerns into spatial plan [20].

This paper aims to examine the effectiveness of flood management measures within Bandung's spatial plan in achieving urban flood resilience in the future, based on ecological constraints, future inundation and flood risks conditions within resilience system characteristics framework. We reveal that Bandung's spatial plan has accommodated flood management measures. However, the measures have not selected and located properly according to ecological constraints and future inundation in achieving urban flood resilience in the future.

2. Methods

The location of the study is in Bandung City, the capital of West Java Province. Bandung has a long history of floods since the late 19th century, followed by the second and third largest flood events in 1917 and 1931 [21]. Bandung is a naturally flood-prone area due to its location on the basin area formed by a prehistoric lake over 55,000 years ago, which blocked the course of Citarum River and created a natural dam [22,23]. However, flood events appeared to have become more severe and resulted in high risk, especially on people and assets (e.g., public infrastructures, buildings, and schools). Based on data from the Indonesian National Board for Disaster Management (BNPB), floods risks in Bandung City between 1998 and 2018 have 1,608 people affected and 1,480 buildings damaged. Recent study predicts that in year 2035, 173,420 people will be affected by flood and damage value could reach USD 272,383 in Bandung City [11].

This paper combines flood management measures framework [24] and characteristics of urban system resilience [25] to analyze the effectiveness of flood management measures within Bandung's spatial plan in achieving urban flood resilience in the future (figure 1). The effectiveness of flood management measures is highly associated with the topographic characteristics, the combination of measures selected and their location to achieve urban flood resilience [24]. In the context of socio-ecological systems, disaster management, and urban sustainability; resilience is understood to require the following characteristics: flexibility, adaptive capacity, transformation, learning and change [26–32]. Therefore, two principles are applied during the analysis:

1. Understanding the role of various measures in reducing risk parameters helps facilitate choosing an appropriate option under site/location constraints, the examples of suitable measures for different topographical characteristics are as follows [24]:
   - Suitable measures at catchment, consists of the following: catchment management, meteorological and hydrometric telemetric monitoring as part of an early warning system, watershed protection through soil conservation, forest conservation.
   - Suitable measures at the river, consist of shielding banks from erosion, river conveyance, and river sediment excavation.
   - Suitable measures at floodplain include dykes (in densely populated and high economic areas; but dyke is not recommended in wetlands and marshes areas and regions with clay soil type), floodproofing or relocation (small town or few houses), water storage, and artificial groundwater recharging basins
2. Understanding the characteristics of urban system resilience helps choosing the combination of measures to achieve urban flood resilience. The following characteristics are contributing to the resilience of the systems [25]:
   - Flexibility and diversity: "ability to perform essential task and/or deliver services under a wide range of conditions; key elements are spatially distributed and have functional diversity".
   - Redundancy and modularity: "Spare capacity for contingency situations to accommodate increasing or extreme surge pressures or demand; multiple pathways and various options for service delivery".
   - Safe failure: "Ability to absorb sudden shocks in ways that avoid catastrophic failure. Failures in one structure or linkage being unlikely to result in cascading impacts across other systems".

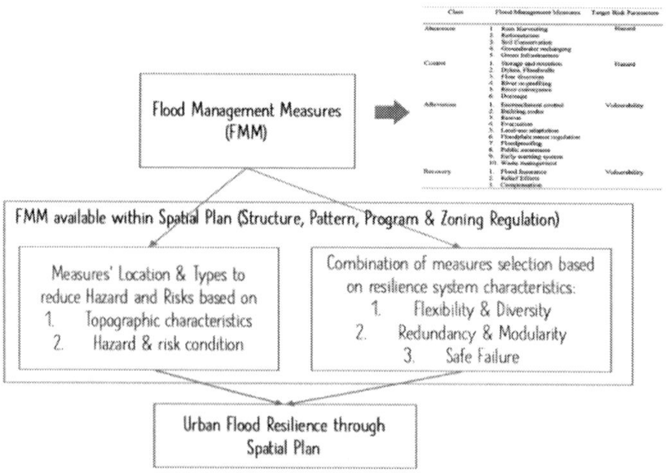

Figure 1. Analytical Framework

We uses qualitative method through thematic analysis. Thematic analysis is a method to identify, analyze and report the patterns (themes) through interpreting meanings and concepts in a data set with a specific theme [33–35]. Thematic analysis was used to: a). Identify the availability of flood management measures in Bandung's spatial plan; b). Propose flood management measures based on future inundation and flood risks data in Bandung City calculated by [11]. The comparison between existing and proposed measures; and the analyzes of resilience system characteristics of the flood management measures helps uncover the effectiveness of flood management measures within the spatial plan to reduce future inundation and flood risks in Bandung City (Figure 2). Data and information on inundation and flood risks in Bandung City from [11] are calculated in two periods: baseline (2015) and future (2035), as shown in table 1.

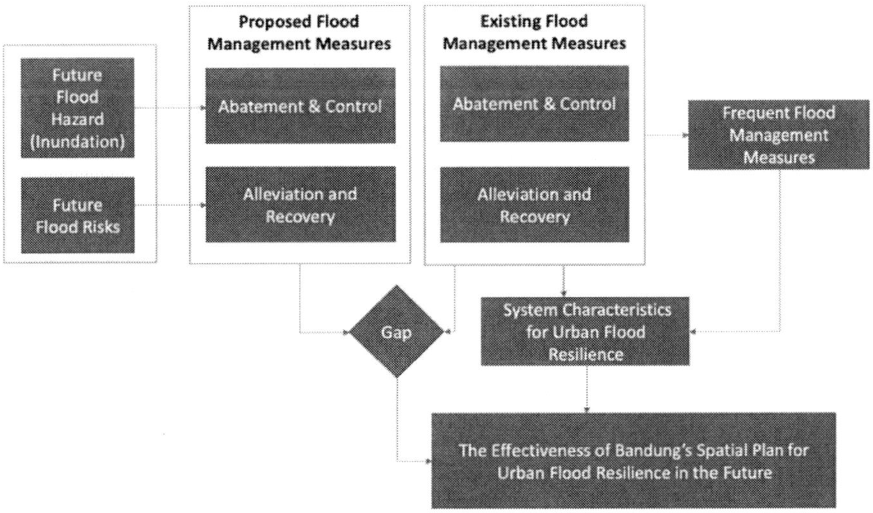

Figure 2. Analytical Procedures

Table 1. Flood Inundation and Flood Risks in Bandung City

City sub-area	Flood Inundation Areas (Ha)		Flood Risks			
			Population at Risk (People)		Damage Value (USD)	
	Baseline	Future	Baseline	Future	Baseline	Future
Bojonagara	0,00	0,81	0	0	0	12.507
Cibeunying	0,00	7,33	0	329	0	64.268
Tegallega	0,00	0,00	0	0	0	0
Karees	41,80	217,10	3.424	61.188	10.393	55.674
Arcamanik	33,01	423,09	6.393	80.477	25.954	48.632
Ujungberung	0,00	80,91	0	6.118	0	22.468
Kordon	138,92	158,01	5.665	21.144	20.844	45.701
Gedebage	66,77	150,49	4.723	4.164	12.903	23.133

Source: [11].

IOP Conf. Series: Earth and Environmental Science **986** (2022) 012052 doi:10.1088/1755-1315/986/1/012052

3. Results and Discussions

Most Frequent Flood Measures within Bandung's Spatial Plan

Bandung City's government has considered flood management measures to be included in its spatial plan. The measures covers from abatement, control and alleviation; , but there is no evidence of recovery measures included in the plan. Most measures are available in spatial program of the plan (Figure 3). A spatial program is a guideline for space utilization, specifically indicating various activities to be implemented in each location according to its land-use, which the spatial structure and pattern have determined. The measures in the spatial structure are limited to control measures through drainage development and maintenance; and alleviation measures through evacuation route and waste management network development. Available measures in the spatial pattern are abatement and alleviation through the allocation of protection areas (such as conservation areas and green infrastructure) and land-use adaptation to increase water infiltration capacity. Both measures are also available in zoning regulation through encroachment control, green infrastructures, and the basic building coefficients (KDB – Koefisien Dasar Bangunan). The application of KDB is aim to provide space for water infiltration in each land parcel; the lower KDB is the bigger space for water infiltration.

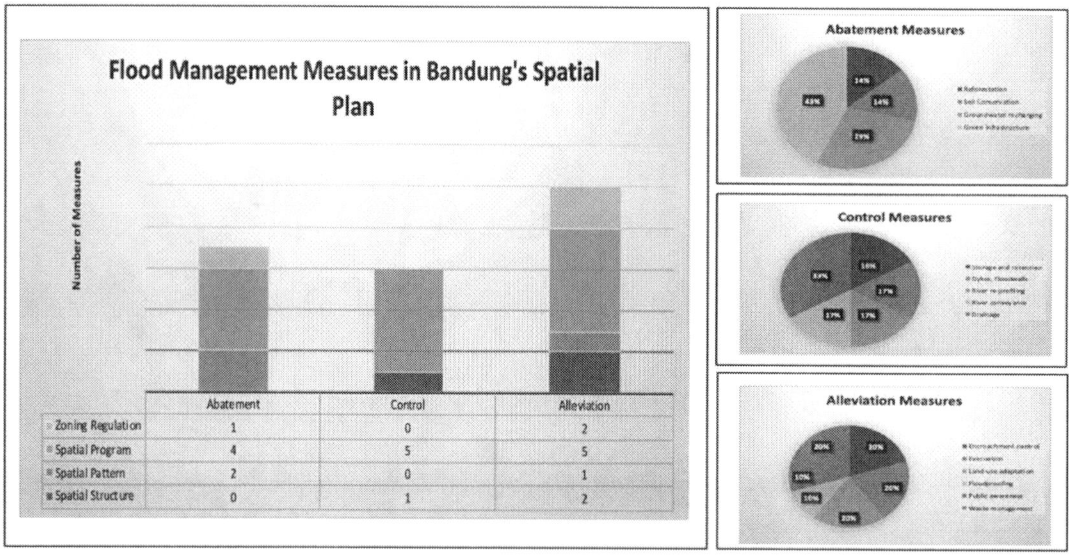

Figure 3. The Availability of Flood Management Measures in Bandung's Spatial Plan

Alleviation measures are the most available measures (6 types) in Bandung's spatial plan; followed by control measures (5 types) and abatement measures (4 types). Alleviation measures are consist of encroachment control, evacuation, land-use adaptation, floodproofing, public awareness and waste management. Control measures are consist of storage and retention, dykes and floodwalls, river re-profilling, river conveyance, and drainage. Abatement measures are consist of reforestation, soil conservation, groundwater recharging, and green infrastructure. However, recovery measures seem missing from the plan. The possible recovery measures such us relief and compensation could be part of a spatial program, while flood insurance could be part of zoning regulation in flood prone zones as incentives-disincentives instruments. However, insurance premium costs need to be paid by the party who occupied the land in water catchment areas. This instrument and its mechanism need to be explored in order to ensure the equality of social justice principles.

IOP Conf. Series: Earth and Environmental Science 986 (2022) 012052 doi:10.1088/1755-1315/986/1/012052

The Effectiveness of Bandung's Spatial Plan for Urban Flood Resilience in the Future.
The analysis shows that Bandung's spatial plan has accomodated various flood management measures in its contents. However, we found that flood management measures in the spatial plan are similar in each city sub-areas, despite the differences of topographic characteristics, future inundation, flood risks in each locations (figure 4 and 5). This condition implies that the selection and location of the measures have not considered the ecological constraints, future inundation and flood risks information. For instances, Tegallega with zero inundation in the future has similar measures with Arcamanik where the future inundation is the highest. In Bojonagara and Cibeunying where upstream watershed areas are located and have fewer inundated areas, the proportion of urban green spaces as abatement measure is still below minimum standard provision (20% of total urban areas). Moreover, land-use in flood prone sub-districts in Bojonagara and Cibeunying covered by high-density housing and commercial areas; and they have fewer green spaces (for water infiltration) in each land parcel indicated by high KDB (table 2). Control measures provision such us water storage and retention, dykes, and floodwalls are more preferable than abatement measures with sufficient capacity in Bojonagara and Cibeunying.

Inundated areas in Arcamanik and Karess located in high-density housing and commercial areas. Both areas have high KDB which have fewer space for water to infiltrate. This condition indicates that the provision of green space in flood-prone areas is not a priority measure compare to control measures for the Bandung city government. Some researchers [20,36,37] argue that economic activities will usually dominate the use of urban space compared to conservation activities, which is partly regulated through land-use planning.

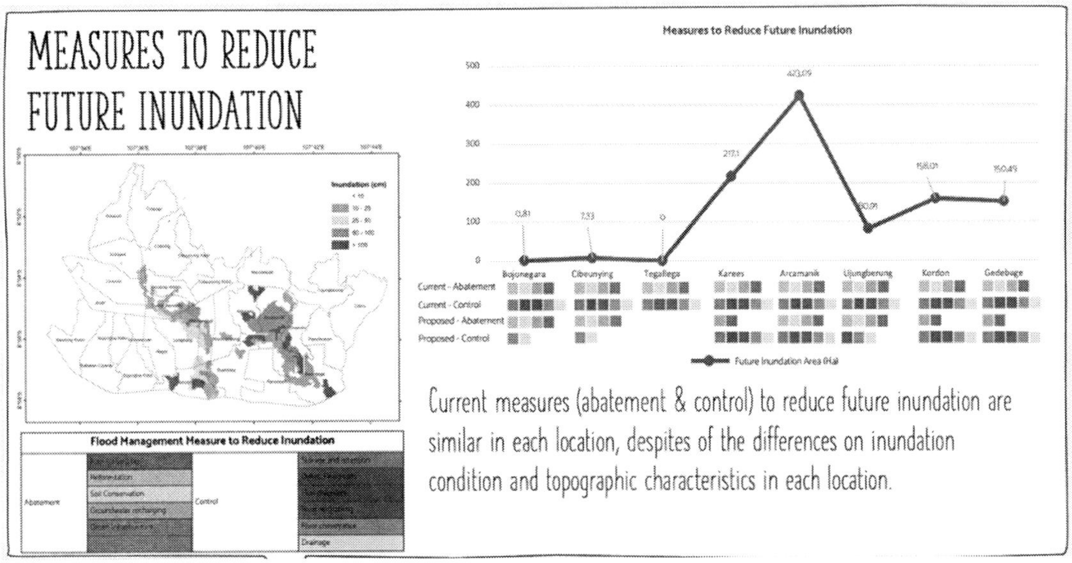

Figure 4. Current and Proposed Measures to Reduce Future Inundation

Table 2. Abatement Measures in Future Inundated Sub-districts Based on Bandung's Spatial Plan

City Sub-area	Green Space Areas (%)[a]	Future Inundated Sub-districts[b]	Zoning Regulation in Future Inundated Sub-districts[a]	
			Land use Type	KDB
Bojonagara	8,28%	Sukajadi	High density housing	60%-80%
			Road	N/A
Cibeunying	4,90%	Coblong	Public facility	50%-60%
			Road	N/A
			Medium density housing	50%-60%
		Bandung Wetan	High density housing	40%-80%
			Service and trading areas	50%-70%
			Road	N/A
		Sumur Bandung	Offices	40%-50%
			Road	N/A
			Service and trading areas	50%-70%
Tegallega	3,01%	N/A	N/A	N/A
Karees	6,24%	Regol	Medium density housing	50%-60%
			Buffer areas for high voltage electricicy network	2%
			High density mixed-use	
		Lengkong	Defence and security areas	50%-60%
			Service and trading areas	50%-70%
			Road	N/A
		Batununggal	High density housing	40%-80%
			Service and trading areas	50%-70%
			Road	N/A
Arcamanik	13,67%	Arcamanik	Housing	40%-80%
			Service and trading areas	50%-70%
			Road	N/A
		Mandalajati	Road	N/A
		Antapani	Housing	40%-80%
			Road	N/A
			Service and trading areas	50%-70%
Ujungberung	15,43%	Cinambo	Housing	40%-80%
			Industry and warehouse area	40%-60%
			Mixed-use	N/A
Kordon	2,78%	Bandung Kidul	Housing	40%-80%
			Road	N/A
			Service and trading areas	50%-70%
		Buahbatu	Housing	40%-80%
			Service and trading areas	50%-70%
			Road	N/A
Gedebage	7,64%	Gedebage	Housing	40%-80%
			Mixed-use	N/A
			Urban green open space	2%

[a] Source: RDTR dan PZ Kota Bandung 2015-2035
[b] Source: [11]

Bandung's spatial plan recognized flood-prone areas as floodplain zones which stipulated in spatial pattern. Sub-districts with high flood risk in the future are mostly considered as floodplain zones in Bandung's spatial plan (table 3). Measures in floodplain zones are vary from abatement, control and alleviation measures. Alleviation measures that are available in Bandung's spatial plan are: monitoring and information dissemination on flood hazards and risks (public awareness), illegal building demolition around buffer areas (encroachment control), evacuation, and settlement relocation (floodproofing). However, table 3 indicates that those measures are similar in each location. Floodproofing and/or settlement relocation are the suitable measures for small town where few houses located, but they applied in all locations includes Karees, Arcamanik, and Kordon, where densely populated and commercial activities are located. Moreover, flood insurance is not available in any locations, even though it could be applied through zoning regulation in floodplain zones as incentives-disincentives instruments.

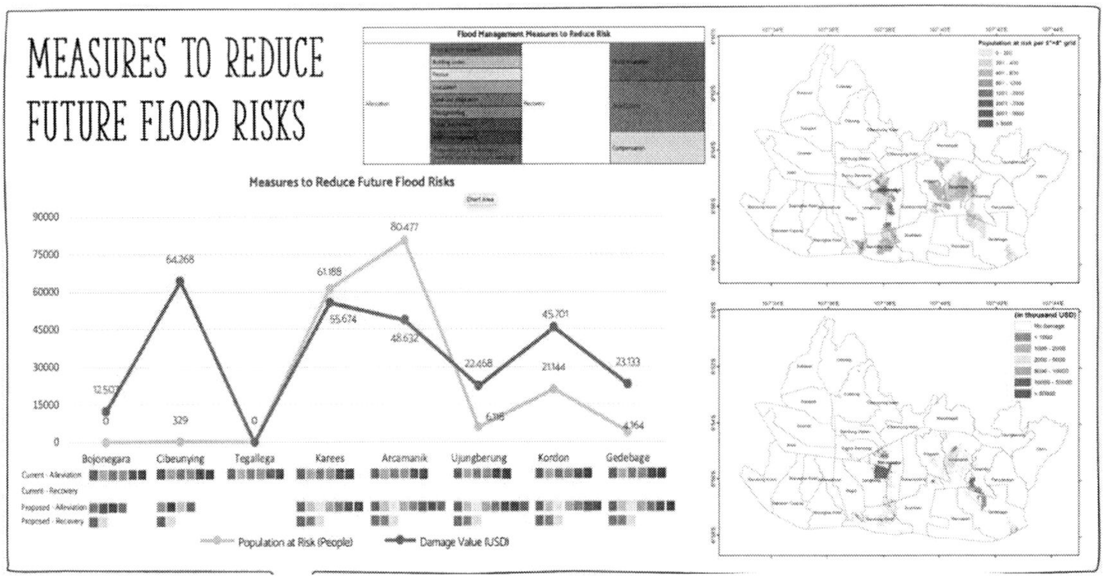

Figure 5. Current and Proposed Measures to Reduce Future Flood Risks

Table 3. Flood Management Measures in Floodplain Zones based on Bandung's Spatial Plan

City Sub-area	Sub-district with High Damage and Population at Risk[a]	Location of Floodplain Zones based on Bandung Spatial's Plan[b]	Flood Management Measures in Flood-Prone Zones[b]
Bojonegara	Sukajadi	• Cicendo • Andir • Sukajadi • Sukasari	1. Public awareness 2. Encroachment control 3. Evacuation 4. Floodproofing 5. Drainage development 6. River conveyance 7. Reforestation 8. Groundwater re-charging 9. Dykes and floodwalls
Cibeunying	• Bandung Wetan • Sumur Bandung	• Cidadap • Coblong • Bandung Wetan • Sumur Bandung • Cibeunying Kidul • Cibeunying Kaler	
Tegallega	N/A	• Bojongloa Kidul (0,6 Ha) • Bojongloa Kaler (0.8 Ha)	
Karees	• Regol • Lengkong • Batununggal	• Regol • Lengkong • Batununggal	
Arcamanik	• Arcamanik • Antapani	• Mandalajati • Antapani • Arcamanik	
Ujungberung	Cinambo	• Panyileukan (0,8 Ha in Cisalatri & 0,5 Ha in Cipamulihan). • Cinambo (0,2 Ha).	
Kordon	• Bandung Kidul • Buahbatu	• Bandung Kidul • Buahbatu	
Gedebage	Gedebage	Rancasari	

[a] Source: [11]
[b] Source: RDTR dan PZ Kota Bandung 2015-2035

In the context of resilience system characteristics, the availability of various measures in each city sub-areas indicates that flexibility and diversity of the resilience system have been adopted by Bandung's spatial plan to achieve urban flood resilience in the future. However, in terms of redundancy of the system, abatement measures have fewer redundancy than control and alleviation measures. The proportion of urban green spaces in Bandung City is below 20%, as mandatory by the regulation. Moreover, the building coefficient (KDB) value in flood-prones areas is high, meaning less water infiltration space. On the other hand, redundancy in control measures has been designed to accommodate unexpected demand or extreme weather climate events. This indicates that various types of control measures are available in each city sub-areas. Redundancy is also designed for alleviation, for instances, a waste management program is available in each city sub-areas. However, both redundancies in control and alleviation measures need to be followed by a detailed program plan such as: drainage system masterplan and waste management system masterplan in order to ensure sufficient space capacity to accommodate unexpected demand in the future.

Bandung's city government's choice to prioritise control measures instead of nature-based solution through alleviation measures could lead to cascading failures when control measures are not followed by a detailed program plan. Therefore, prioritizing control measures in urban areas should be decided based on examining services needed; in this case, information on future inundation and flood risks need to be considered in the spatial planning process. Unfortunately, the spatial planning practices in Bandung City have not considered that information [11,20].

4. Conclusions

An increasing future inundation and flood risks in Bandung city have been shown by several studies [11,20,38]. Yet, few studies have analysed the effectiveness of flood management measures within Bandung's spatial plan in achieving urban flood resilience in the future based on resilience system characteristics. This paper advances flood management measures framework and resilience system characteristics to fill the gap. The flood management measures framework and resilience system characteristics could provide information on specific measures needed in each location to achieve urban flood resilience based on the inundation and flood risks information.

We compared flood management measures within Bandung's spatial plan with proposed measures based on future inundation and flood risks, and evaluate the gap based on the resilience system characteristics to examine the effectiveness of Bandung's spatial plan in achieving urban flood resilience in the future. Our findings suggest several points: 1). The selection of flood management measures in Bandung's spatial plan has not been considered a topographic condition, ecological constraints, inundation and flood risks characteristics in each location; indicated by similar measures in all locations; 2). Control measures are the most favorable choices to reduce future inundation compared to abatement measures; and alleviation measures are preferable choice instead of recovery measures; 3). The selection of flood management measures in Bandung's spatial plan has been considered diversity and redundancy of system characteristics to achieve urban flood resilience in future, especially for control and alleviation measures; but not for abatement measures; 4). The effectiveness of Bandung's spatial plan in achieving urban flood resilience in the future depend on detailed program plan on abatement, control and alleviation measures which consider ecological contraints, future inundation and flood risks as well as the implementation of those plans in order to prevent cascading failures of those measures. This research confirmed findings from previous study that spatial planning process has been overlooked environmental concerns as a primary concern [12–14]; and SEA implementation in Indonesia to integrate environmental concerns into spatial planning process still faces some challenges [17–19].

This research uses secondary data and information from previous studies and Bandung's spatial plan documents and maps. Whereas primary data from the interview and/or focus group discussion with relevant stakeholders are needed to confirm and corroborates the findings. Therefore, further research should include primary data from the interview and/or focus group discussions. Furthermore, this research shows that the application of flood management framework and resilience system characteristics could inform the effectiveness of the spatial plan in accommodating flood management measures to achieve urban flood resilience.

Acknowledgments

The study was funded by a research program 2021 (FITB.PN-6-04-2021) from "*Lembaga Penelitian dan Pengabdian kepada Masyarakat (LPPM) Institut Teknologi Bandung (ITB)*". The authors also wish to acknowledge Bandung City Government in providing spatial plan document and its maps and Dian Afriyanie in providing the inundation and flood risks information and their maps for this research.

References

[1] Dutta D and Herath S 2004 Trend of Floods in Asia and Flood Risk Management with Integrated River Basin Approach
[2] Ward P J, De Moel H and Aerts J C J H 2011 How are flood risk estimates affected by the choice of return-periods? *Nat. Hazards Earth Syst. Sci.* **11** 3181–95
[3] Marconi M, Gatto B, Magni M and Marincioni F 2016 A rapid method for flood susceptibility

mapping in two districts of Phatthalung Province (Thailand): present and projected conditions for 2050 *Nat. Hazards* **81** 329–46
[4] Budiyono Y, Aerts J, Brinkman J J, Marfai M A and Ward P 2015 Flood risk assessment for delta mega-cities: a case study of Jakarta *Nat. Hazards* **75** 389–413
[5] Caljouw M, Nas P J M and Pratiwo 2009 Flooding in Jakarta: Towards a blue city with improved water management *Bijdr. tot taal-, land-en volkenkunde/Journal Humanit. Soc. Sci. Southeast Asia* **161** 454–84
[6] Steinberg F 2007 Jakarta: Environmental problems and sustainability *Habitat Int.* **31**
[7] Poerbandono, Julian M M and Ward P J 2014 Assessment of the effects of climate and land cover changes on river discharge and sediment yield, and an adaptive spatial planning in the Jakarta region *Nat. Hazards* **73** 507–30
[8] Giambelluca T W 2015 Land Use and Water Resources under a Changing Climate *Encyclopedia of Hydrological Sciences Edition* (New Jersey: John Wiley & Sons Ltd Hoboken)
[9] Julian M M, Brenning A, Kralisch S and Fink M 2019 Hydrological responses in the upper citarum basin based on the spatial plan of west java province 2029 and climate change *Int. J. Technol.* **10** 866–75
[10] Hillier J 2010 Introduction *The Ashgate Research Companion to Planning Theory: Conceptual Challenges for Spatial Planning* ed J Hillier and P Healey (Ashgate Publishing Limited) pp 1–34
[11] Afriyanie D 2020 *Re-framing urban green spaces planning towards flood resilience through socio-ecological resilience* (Institut Technology Bandung)
[12] Wilkinson C 2011 Social-ecological resilience: Insights and issues for planning theory *Plan. Theory* **11** 148–69
[13] Davoudi S and Strange I 2009 Introduction *Conception of Space and Place in Strategic Spatial Planning* ed S Davoudi and I Strange (Taylor & Francis) pp 7–42
[14] Eraydin A, Taşan-kok T, Stead D, Taşan T, Lu P, Dias L F, Costa J P T a, Schmitt P, Harbo L G, Diş A T and Henriksson A 2013 *Resilience Thinking in Urban Planning* vol 106
[15] Swyngedouw E 2010 Trouble with Nature:'Ecology as the New Opium for the Masses' *The Ashgate Research Companion to Planning Theory: Conceptual Challenges for Spatial Planning* ed J Hillier and P Healey (Ashgate Publishing Limited and Ashgate Publishing Company) pp 299–318
[16] Suroso D S A 2000 *Spatial Planning and Environmental Assessment in Indonesia* (LAP LAMBERT Academic Publishing AG & Co. KG)
[17] Dusik J 2010 *SEA as a dialogue and planning support tool: Lessons from pilot projects in Indonesia I*
[18] Susilowardhani A 2014 The Potential of Strategic Environmental Assessment to Address the Challenges of Climate Change to Reduce the Risks of Disasters: A Case Study from Semarang, Indonesia *Procedia - Soc. Behav. Sci.* **135** 3–9
[19] Victor D and Agamuthu P 2014 Policy trends of strategic environmental assessment in Asia *Environ. Sci. Policy* **41**
[20] Afriyanie D, Julian M M, Riqqi A, Akbar R, Suroso D S A and Kustiwan I 2020 Re-framing urban green spaces planning for flood protection through socio-ecological resilience in Bandung City, Indonesia *Cities* **101**
[21] Kunto H 1986 *Semerbak Bunga di Bandung Raya* (Granesia)
[22] M.A D 1994 *The late quaternary evolution of the Bandung basin, west Java, Indonesia* (VU Amsterdam)
[23] Tarigan A K M, Sagala S, Samsura D A A, Fiisabiilillah D F, Simarmata H A and Nababan M 2016 Bandung City, Indonesia *Cities* **50** 100–10
[24] Tariq M A U R, Farooq R and van de Giesen N 2020 A critical review of flood risk management and the selection of suitable measures *Appl. Sci.* **10**
[25] Tyler S and Moench M 2012 A framework for urban climate resilience *Clim. Dev.* **4** 311–26

[26] Adger W N, Hughes T P, Folke C, Carpenter S R and Rockström J 2005 Social-ecological resilience to coastal disasters *Science (80-.).* **309**

[27] Antrobus D 2011 Smart green cities: From modernization to resilience? *Urban Res. Pract.* **4**

[28] Berkes F, Johan C and Folke C (Eds.). 2003 Navigating Social-Ecological Systems: Building Resilience for Complexity and Change. Cambridge University Press. *J. Fam. Issues* **39**

[29] Fields B 2009 From green dots to greenways: Planning in the age of climate change in post-Katrina New Orleans *J. Urban Des.* **14**

[30] Prasad N and Ranghieri F 2008 *Climate resilient cities; A Primer on Reducing Vulnerabilities to Disasters* vol June

[31] Miller F, Osbahr H, Boyd E, Thomalla F, Bharwani S, Ziervogel G, Walker B, Birkmann J, Van der Leeuw S, Rockström J, Hinkel J, Downing T, Folke C and Nelson D 2010 Resilience and vulnerability: Complementary or conflicting concepts? *Ecol. Soc.* **15**

[32] Twigg J 2007 Characteristics of a disaster-resilient community: a guidance note *Nat. Hazards Rev.* **1** 1–40

[33] Braun V and Clarke V 2006 Using thematic analysis in psychology *Qual. Res. Psychol.* **3**

[34] Javadi M and Zarea K 2016 Understanding Thematic Analysis and its Pitfall *J. Client Care* **1**

[35] Neuendorf K A 2019 Content analysis and thematic analysis. In P. Brough *Research methods for applied psychologists: Design, analysis and reporting*

[36] Ernstson H 2008 The social production of ecosystem services: lessons from urban resilience research *Ecol. Res.*

[37] Ernstson H 2013 The social production of ecosystem services: A framework for studying environmental justice and ecological complexity in urbanized landscapes *Landsc. Urban Plan.* **109** 7–17

[38] Tollenaar D, R D and Wageenar D 2016 *Flood Risk Assessment as Input to Informed Decision Making for Future Cities: The case of Bandung, Indonesia*

IOP Conf. Series: Earth and Environmental Science **986** (2022) 012053 doi:10.1088/1755-1315/986/1/012053

Adaptation strategy community from exposure floods lake in the face of floods in Semayang Village, East Kalimantan, Indonesia

Yaskinul Anwar *, **Iya Setyasih, and Dika Oktoberdinata**
Department of Geography Education, Faculty of Teacher Training and Education, Mulawarman University

*Email: yaskinul.anwar@fkip.unmul.ac.id

Abstract. Semayang Village is one of the villages located on the lake of Semayang which is inundated every flood season. Semayang lake is a flood exposure lake, where the increase in standing water can reach >10 meters during a flood with a flood duration of >2 months. The purpose of this study was to determine the adaptation strategy of the semayang villagers in dealing with floods. Data collection techniques used the Rapid Rural Appraisal (RRA) method and interviews. Interviews were collected from 30 respondents in Semayang Village that were affected by the flood. The results showed that the overall adaptation carried out by the community in Semayang Village using houses on stilts with almost the same model, and some residents make terraced houses. In addition, each house has a backup board to raise the floor of the house when the flood crosses the floor of the house. there is also a floating house system attached to a tree or building around the floating house. In order to protect the house buildings from waves, a wave barrier was also built at the village entrance gate from the lake. This strategy makes the villagers of Semayang never evacuate and move from Lake Semayang even though they are always affected by flooding every year.

1. Introduction

Disaster is a natural and non-natural phenomenon that is very disturbing to the community because it can disturb the community's comfort, peace, and security [1]. It was recorded that throughout 2020 4,650 disaster events occurred in Indonesia, which was dominated by hydrometeorological disasters. Where this is the highest number of disaster events in the last 10 years [2].

Hydrometeorological disasters (meteorological natural disasters) are natural disasters related to climates such as floods, landslides, cyclones, tidal waves, and droughts [3]. Throughout 2020, floods occurred 1143 times and in 2021 there were 447 flood events in Indonesia caused by various natural phenomena such as high rainfall, rising sea levels, land degradation, and overflowing lake water [4].

East Kalimantan Province is also inseparable from flooding, the Mahakam river which flows for 980 kilometers from Mahakam Ulu Regency to the Makassar Strait in Kutai Kartanegara Regency. Along with the flow, of course, some areas are affected by the river's overflow, causing the area to flood. From 2019 to December 2020, there was 214 times the number of flood events in East Kalimantan, where floods were caused by high rainfall, damaged mine embankments to overflowing lake water [5].

The phenomenon of flooding that occurred in East Kalimantan will certainly have an impact on people's lives, especially in flood-prone areas. The adaptations made by the community include making houses on stilts, raising the foundations or floors of their houses, and making embankments so that their houses are not affected by flooding [6].

The Semayang Village area which is located on the shores of Semayang Lake is one of the areas in Kutai Kartanegara Regency which is an area exposed to flooding. The flood that occurred in this area as a result of the overflow of Semayang Lake. Semayang Lake is a collection of lakes from Mahakam Cascade Lake. Mahakam Cascade Lake (MCL) is a lake exposed to flooding, which makes the area of the lake and its borders largely determined by the water level, and the catchment area is very wide [7] [8].

Generally, a lake has an important role as a reservoir for overflowing river water during the flood season. However, flooding will still occur when the lake water has an excess volume so that the water capacity of the lake increases and causes flooding around the lake area [9]. Since 2000, in Semayang Village almost every rainy season there have been extreme floods that are very long like in 2006 went on for 63 days. Typically, significant floods happen throughout an extensive stretch of time (over two months) [7].

The extreme flood conditions that often occur in Semayang Village make people's lives in the village certainly disrupted. However, there were no residents who moved or evacuated due to flooding. According to the people living around Lake Semayang, flooding is not a disaster, but a fortune because fish catches will increase, so many expect floods but do not expect extreme floods [10]. Based on these conditions, the aim of this research is to define the adaptation strategy of the semayang villagers in dealing with floods.

2. Method

2.1. Location

Semayang Village is one of village from Kenohan Sub-district, Kutai Kartanegara District, East Kalimantan Province, Indonesia (**Figure 1**). Semayang Village located in Semayang Lake, one of Cascade Mahakam Exposure Lake.

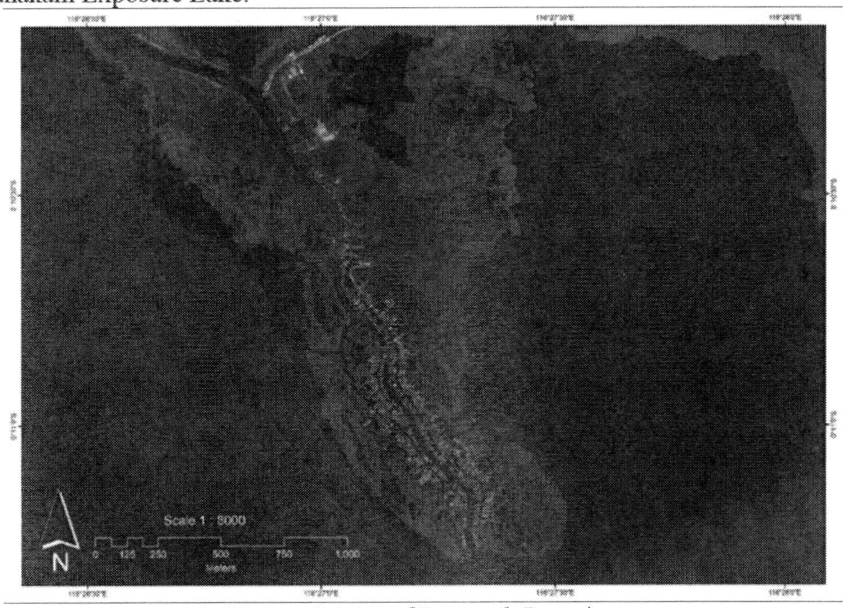

Figure 1. Map of Research Location.

Topography of semayang village are lowland that part of flood area Semayang Lake. Ethnic of Semayang Village are dominated by kutai. People live in Semayang Village are 1.423 people. The livelihoods of the people of Semayang village are fishermen, from capture fisheries and aquaculture. The flood disaster in Semayang Village is an annual phenomenon, where not only floods are faced by the community, but also hurricanes and high waves in the lake which also causes damage to buildings in Semayang Village.

2.2. Collection and Analysis Data

To explore the community adaptation strategy in this study, qualitative research was conducted that focused on the observation of the Semayang village community, both in terms of overall behavior and actions. Withdrawal of informants in the village of Semayang using purposive sampling on the villagers affected by the flood. The number of samples in this study was 30. This study also used observation and interview methods as follows:

a. Field observation

Observation of the adaptation of the Semayang village community was carried out using the Rapid Rural Appraisal (RRA) method. In this method, the introduction of the Semayang village area is carried out quickly to get an overview of the research area. Observations in this study were carried out on:

- House Model in Semayang Village,
- Flood disaster risk reduction activities by community in Semayang Village.

b. Interview

This interview was conducted to explore adaptation strategies and how to survive the Semayang village community in the face of flooding. informants were randomly selected from flood-affected communities. Information collected includes:

- Impact of floods,
- Knowledge of flood, and
- Forms of adaptation for life in Semayang Village.

Data analysis was done by the triangulation method. The process was carried out since field data were obtained with observations, interviews, and documents from BPBD Kutai Kartanegara, Semayang Village government, with the aim to simplify the data and easier to understand.

3. Result and Discussion

The residents of Semayang Village are residents who have been side by side with the flood disaster from Semayang Lake. Floods are considered a disaster only during extreme floods because people can still function normally during regular floods (Figure 2). Residents are looking forward to flood events (but not extreme floods), because after receding fish in the lake will be abundant and can increase the income of the villagers of Semayang. The flood model in Mahakam Cascade Lake rises slowly so that residents can prepare themselves when the flood comes. The occurrence of extreme flooding in Semayang Lake does not occur every year, based on the record of the highest flood water level for the last 20 years, which occurred in 2017 which caused many houses to be flooded. Even though the flood was extreme and made people's houses flooded (houses on stilts), no residents were evacuated and remained in their respective homes. The community's strategy can be seen from the shape of the house and the efforts made to reduce disaster risk and still be able to survive in Lake Semayang. The forms of community adaptation strategies in Semayang Village are:

Figure 2. Educational Activities in Semayang Village During Ordinary Floods.

3.1. House model in Semayang Village

Disasters often cause damage and cause losses if disaster risk reduction strategies are not implemented [11] [12]. To reduce and survive in disaster-prone areas, humans often have to adapt to survive in these environments [13]. One of the adaptation models for the community in Semayang Village is the adaptation of the house model and the road model. The house models to withstand flooding in Semayang Village are 2 house models: a stilt house and a raft house. The shape of this house is a typical house shape in Indonesia which is near the waters [14] [15] [16]. The stilt house model is a house built in an area outside the lake's grooves. The shape of this house can be in the form of a one-story house or more than one floor (Figure 2). The main material of this stilt house is wood. The wood used mainly for house foundations is ironwood. The design of this stilt house is a house model that can be modified for the height of the house (house foundation), where this house can be raised again by adding wood to the foundation of the house. The form of this stilt house is quite tough against waves during a flood [14] [17]. This is because the shape of the building's foundation is hollow so that it can break waves and transmit wave power.

This adaptation strategy for stilt houses requires a large amount of money but has stronger resistance in the face of high waves during floods. The position of the building foundation attached to the ground and the hollow foundation of the building make this building more resistant to wave impact. Based on the results of interviews, 50% of respondents said that they had raised the foundation of their house following the increase in water during ordinary floods. This is done so that the house does not get flooded during a flood. In addition, if this stilt house still sinks during the flood, the people of Semayang Village adapt by raising the floor by adding wood from the bottom of the house floor. This wood is obtained from the forest or the nearest sawmill or market from Semayang Village.

Figure 3. Stilt House in Semayang Village

In addition to the model of the house on stilts, the community in this area adapts to making raft houses. This raft house is a house that floats in the area of the lake's grooves which is always flooded by water throughout the year (Figure 3). This house has a cheaper manufacturing cost than a stilt house and is easy to move around. This raft/floating house model is a house with a system that follows the water level so that water never enters the house. Based on the results of interviews with residents in Semayang Village, this house model has a drawback, namely that it cannot withstand the blows of high waves during floods. But this house model is easy to move, so the raft house can be moved to a safer place from the waves when the waves are high. In addition, raft house owners are not too worried during floods because high waves only occur during extreme floods that occur once every five years, and many houses on stilts have helped break the waves before reaching their homes.

Figure 4. Raft House / Floating House in Semayang Village.

3.2. Strategies for reducing flood risk
The residents of Semayang Village are residents who have been side by side with the flood disaster since they were born in Semayang Village. Floods are no longer included in disasters if the floods are not extreme. The way residents reduce the risk when water starts to enter the house floor is to raise the floor of the temporary house with wooden planks (especially for houses on stilts). This floor elevation uses wood from the forest around Lake Semayang and also from sawmills and markets in Kota Bangun District. Based on interviews 100% of respondents do not have wood reserves to deal with floods, because extreme floods are rare, during the last 20 years there have only been three extreme floods, namely 2005, 2007, and 2017 [18]. The wood is usually gets from the forest around Semayang and also in the market.

In addition to the modification of temporary houses, the residents and the village government have also raised the road in Semayang Village to be able to carry out normal activities during ordinary floods (Figure 5). This road elevation was carried out in 2016 although during extreme floods the road would still sink. The added road height is around 1.5 meters from the previous road height.

To reduce wave power during flooding, a breakwater has been built at the entrance to Semayang Village (Figure 6). This breakwater not only functions as a breakwater but also functions as a weed net that enters residential areas during flooding. Based on interviews with residents of Semayang village, the waves will get higher following the water level and wind speed, so that before there was a breakwater, many houses and fish cage were damaged by the waves from the lake. Adaptation efforts made by the community in dealing with disasters are an effort to survive humans in the face of disasters [19] [20]. Likewise, the adaptation efforts carried out by the Semayang village community who survived the flood from the Semayang Lake were good by building stilt houses, raft houses/floating houses, and building breakwaters. In addition, the community also maintains the forest around their settlements to help protect against waves during floods. This is done because of the high cost of constructing a breakwater, so people choose an alternative to natural protection.

Figure 5. Raise the road in Semayang Village

Figure 6. Breakwater in Semayang Village

4. Conclussion

In Conclusion, this form of adaptation can be seen from the shape of the house in the form of a house on stilts and a raft house. In addition, residents also have a strategy to reduce flood risk by elevating the floors of temporary houses, elevating roads, building breakwaters, and maintaining the forest around settlements. In addition to the advantages, disadvantages of the adaptation strategy must be explained, and further research plans.

References

[1] Hermon D 2012 *Hydrometeorological Disaster Mitigation* (Padang: UNP Press)
[2] Arifin D 2020 A total of 2,925 natural disasters occurred in 2020 in the country, hydrometeorological disasters dominate *BNPB*
[3] Qodriyatun S N 2013 Hydrometeorological Disasters and Climate Change Adaptation Efforts *Info*

Singk. Kesejaht. Sos. **5** 9–12
[4] BNPB 2021 Information Data Indonesia *BNPB*
[5] Bappeda Kalimantan Timur 2021 East Kalimantan Natural Disaster Data *Bappeda Kalimantan Timur*
[6] Anwar Y, Setyasih I and Ningrum M V R 2021 Multi-ethnic communities adaptation to flooding in the north samarinda sub-district, samarinda city, east kalimantan province, indonesia *IOP Conf. Ser. Earth Environ. Sci.* **683**
[7] Directorate of Inland Water Damage Control 2019 *Mahakam Cascade Lake Management Plan* (Samarinda: Kementrian Lingkungan Hidup dan Kehutanan)
[8] Kalimantan B K I and B E 2013 *Collaborative Management of Lake Regions in the Mahakam Watershed* (Samarinda)
[9] Liu Y, Yang S Q, Jiang C, Sivakumar M, Enever K, Long Y, Deng B, Khalil U and Yin L 2019 Flood mitigation using an innovative flood control scheme in a large lake: Dongting Lake, China *Appl. Sci.* **9** 1–16
[10] Sucipto 2020 Fishermen Restless Waiting for Lake Flood to Return *Kompas.id* 1
[11] UNISDR 2020 What is Disaster Risk Reduction? *UNISDR* 1
[12] European Commission 2013 *Disaster Risk Reduction: Increasing resilience by reducing disaster risk in humanitarian action* (Brussels: European Commission)
[13] Bronfman ID N C, Cisternas P C, Repetto P B and Castañeda J V 2019 Natural disaster preparedness in a multi-hazard environment: Characterizing the sociodemographic profile of those better (worse) prepared
[14] Isfa Sastrawati 2009 the Characteristics of the Self-Support Stilt-Houses Towards the Disaster Potentiality At the Cambaya Coastal Area, Makassar *Dimens. (Jurnal Tek. Arsitektur)* **37** pp.33-40
[15] Naing N, S H R and Soemarno I 2009 Sustainability of the floating house at Tempe Lake , Sulawesi Island *43rd Annual Conference of the Architectural Science Association* (Tasmania: University of Tasmania) pp 1–7
[16] Kusliansjah K and Suriansyah Y 2013 RAFTA2011 , the Innovation Of The Manufactured Floating House Model : A New Concept Of Waterfront Settlements For Flood Risk Reduction In Indonesia *Int. J. Eng. Sci.* **2** 18–29
[17] UNISDR 2008 *Indigenous Knowledge for Disaster Risk Reduction: Good Practices and Lessons Learned from Experiences in the Asia-Pacifi c Region* ed and J B Rajib Shaw, Noralene Uy (Bangkok: UNISDR)
[18] BWS IV Kalimantan 2019 *Mahakam Cascade Lake Boundary Delimitation Plan* (Samarinda)
[19] Preston J, Chadderton C, Kitagawa K and Edmonds C 2015 Community response in disasters: an ecological learning framework *Int. J. Lifelong Educ.* **34** 727–53
[20] United Nations Office for Disaster Risk Reduction 2020 *Integrating Disaster Risk Reduction and Climate Change Adaptation in the UN Sustainable Development Cooperation Framework* (Geneva: United Nations Office for Disaster Risk Reduction)

IOP Conf. Series: Earth and Environmental Science **986** (2022) 012054 doi:10.1088/1755-1315/986/1/012054

The impact of climate change to livelihood vulnerability for smallholders farmers in Wonogiri, Indonesia

Suryanto[1*], E S Rahayu[2], O P Astirin[3] and F Susilowati[4]

[1] Faculty of Economics and Business, Universitas Sebelas Maret, Surakarta
[2] Faculty of Mathematics and Natural Sciences Universitas Sebelas Maret, Surakarta
[3] Faculty of Agriculture, Universitas Sebelas Maret, Surakarta
[4] Faculty of Business, Universitas PGRI Yogyakarta

*Email: Suryanto_feb@staff.uns.ac.id

Abstract. Climate change has affected human life, especially farmers in rural areas. Their livelihoods are very dependent on the climate so that their socio-economic life becomes vulnerable. If there is a risk of crop failure, the community becomes vulnerable to poverty and hunger. This study determines the livelihood vulnerability of smallholders in Wonogiri District, Indonesia. Purposive sampling was used to select respondents in the sub-districts of Girimarto, Eromoko, and Tirtomoyo. The number of respondents was 120 households which were selected for farmers who only own less than 2 hectares of land. Descriptive statistics and modified social vulnerability index (SoVI) were used in data analysis. We measured the level of vulnerability for economic and social. The results show that climate change to social vulnerability index with SoVI scores of 0.68 for Tirtomoyo, 0.61 for Girimarto, and 0.55 for Eromoko, respectively. The differences in components between the three sample areas include regional gross domestic product (GDRP), poverty rate, population density, population growth rate, sex ratio, children, and elderly. The results of the study conclude that climate change has the potential to cause poverty levels in the three sub-districts to increase. In terms of resilience, they are quite resilient because they do not depend on government assistance. However, when climate change causes continuous crop failure, the level of resilience of farmers will also decline.

1. Introduction

The effect of climate change causes the dry season to be longer than usual or the rainy season to be longer than usual. Changes in rain scheme can be initiated from the start of the season earlier than usual. The dry season will last longer which will cause drought disasters, reduce productivity and planting area. If the intensity of rainfall is high and exceeds the environmental capacity, it can cause flooding or landslides [1,2].

For farmers, climate change causes the level of vulnerability to livelihood to be higher. The dependence of their livelihood on the agricultural sector means that if the risk of failure in the agricultural sector will have an impact on the economic livelihood of farmers. Currently, farmers are difficult to determine the right time to do farming, for example when is the right time to start planting or determine the kind of plant they will cultivate. According to [3] and [4] mention that farmers, especially sharecroppers are not able to adapt to climate change.

The risk of climate change not only affects national food security, but also decreases the welfare of farmers. A report from the Ministry of Agriculture stated that an estimated 3 million Indonesians lived below the poverty between October and December 2016, 1.2 million among them depended on rain for food production and livelihoods [5].

Research on the level of drought vulnerability in Wonogiri as carried out by [6,7] showed that when the dry season hits, some areas in Wonogiri Regency would be dry certainly. The drought would have an impact on the economic condition of farmers. However, research that further investigates the level of community vulnerability in several sub-districts has not been carried out. This study aims to measure the level of livelihood vulnerability due to climate change. By this research, it is expected that appropriate adaptation or mitigation can be formulated for small farmers.

Measurement of vulnerability has evolved from the measurement of Livelihood Vulnerability Index (LVI), Climate vulnerability index (CVI), Livelihood Index (LI) to Social Vulnerability Index (SoVI). From various measurements but they agreed that there is a specificity or uniqueness related to the location. Therefore, it is important to use measurements that include local elements. This is necessary to gain a better understanding of the basic features that underlying the vulnerability along with appropriate targeting of adaptation policies for relevant agencies at local, national and international sites [8–10].

The preparation of indicators of vulnerability level of or adaptive capacity will be more appropriate if it is based on local values [11] and [8]. According to [9], the more effective approach is the commonly used variables in the research area. Among villages can have different vulnerability factors and these factors also significantly influence the level of vulnerability, such as effective adaptive options for rural communities with unique characteristics.

2. Research Method

The benefits of measuring SoVI as stated by Bolow [10] to measure the level of vulnerability, they used an activity-based adaptation index (AAI). Based on the AAI they explained the correlation between farmers' adaptation and some social economic variables. They concluded that with the AAI model the difference in vulnerability level between locations could be seen more clearly. In addition, the cooperation of countries that are members of the International Panel for Climate Change (IPCC) has exchanged idea on the importance of including local elements in measuring vulnerability. Gbetibouo [12] who conducted research in South Africa concluded that it was not always area that was exposed to climate change always have a high level of vulnerability. It can be concluded that vulnerability measurement requires a specific and unique approach depending on each region.

SoVI measurements have been carried out for areas that have a vulnerability level of flooding [13], landslides [14,15], or even drought [9,16]. Based on this research, it is known that the types of hazard faced by the community can be different. SoVI can be applied by identifying physical, social, economic, and cultural characteristics to make appropriate indicators.This study relies on secondary data to be analyzed using the SoVI approach. Data were collected from three sub-districts that were used as samples for this study, there were Eromoko District, Tirtomoyo District, and Girimarto District. Data were obtained from various sources including the Central Bureau of Statistics and related technical offices (OPD). This research was conducted with a quantitative approach.

The location choice is based on Wonogiri Regency with the consideration that most of community in Wonogiri depends on the agricultural sector. The second consideration is that Wonogiri area has vulnerability level of drought and landslides in several sub-districts.

The variables used include Gross Regional Domestic Product at Current Prices (PDRB ADHB) Sub-district, population growth rate, sex ratio, population density data, number of population at children vulnerable age, number of population at elderly vulnerable age, number of poor residents. The operational definition of each data is as follows:

2.1. Gross Regional Domestic Product (GRDP) at the District level

Gross Regional Domestic Product (GRDP) is one of the important indicators in the regional economic balance that can be measured and assessed from its economic activity. The calculation of GRDP at the sub-district level is important because with this calculation, it will be able to know the economic structure of each sub-district, the distribution of added value of each sector by sub-district and GRDP per capita at the sub-district level within a certain period of time. The GRDP used is GRDP at Current Prices to show the ability of economic resources, shifts, and the economic structure of a region.

2.2. The Poor
The poor are people who have an average monthly per capita expenditure below the poverty (BPS, 2020). Poverty not only has an impact on the economic aspect, but poverty indicators will also have a serious impact on the political aspect related to the right decision-making process in accordance with the conditions of the community. So the poverty level also becomes indicator for local governments and stakeholders to measure the level of community ability in dealing with various types of disaster threats in the future, including natural disaster such as landslide that can occur at unexpected times.

2.3. Population Density
The population distribution in an area is related to the population density. The higher population density in an area, the more densely populated the area is, but otherwise if the population density is smaller, then the population distribution will be less. Population density needs to be considered, this is related to the carrying capacity of the environment.

2.4. The rate of population growth
The rate of population growth can be used to determine the rate of increase in population per year in a certain period of time. The higher the population growth rate affects to the low per capita growth which will then have big impact on the community's burden to meet primary, secondary and tertiary needs such as food, clothing, housing, education, health, and other needs. This will certainly affect the level of vulnerability of an area to disaster risks that occur in the future, especially areas that are very vulnerable to disasters.

2.5. Sex Ratio
Sex ratio is an indicator used for the gender-oriented development planning in an area. There is an assumption that women are more vulnerable when exposed to the threat of disaster compared to men, both in terms of physical limitation during disaster evacuation and decision making. The higher number of female residents in an area, the more vulnerable the area is to the threat of disaster, so the government needs to pay attention to the sex ratio to help in making policy on handling natural disasters that can occur at any time.

2.6. Vulnerable group
Vulnerable group is a group of people who are classified as most vulnerable to the threat of landslides. Who include in the vulnerable group are the children under 15 years and the elderly over 65 years. These two groups are included in the vulnerable group because of their inability to meet their own needs, as well as to protect or save themselves in the event of a disaster threat.

The weight value of each variable is determined by the researcher based on considerations such as the ability to describe the socio-economic condition of the community. The weight of the number of poor people gets a higher portion because it can better describe the level of economic vulnerability than GRDP. The level of population density is considered to better describe the level of demand for goods and services in an area. The following table presents the weighting of each variable.

Table 1. Weight of Compiler Indicator of Economic and Social Vulnerability

Variabel	Indicator	Weight
Ekonomic Aspect	Gross Regional Domestic Product (GRDP) at the District	40 %
	The poor	60 %
Sosial Aspect	Population density	30 %
	Population growth rate	25 %
	Sex ratio	15 %
	Children	15 %
	Elderly	15 %

Source: Professional judgement

Classification of the vulnerability level of economic and social is known based on the scoring result that have been assigned to each indicator. The placement of each indicator refers to low, medium, and high vulnerability classification that has been determined, which can be seen in the following table:

Table 2. Indikacor Score of economic and social vulnerability

Indicator	**Unit**	**Score**		
		Low (1)	**Medium (2)**	**High (3)**
Standardization Value		**< 0,33**	**0,33 – 0,66**	**> 0,67**
Ekonomic Aspect				
Gross Regional Domestic Product (GRDP) at the District	Percent	< 3	3 – 6	> 6
The Poor	People	< 2.600	2.600 – 4.910	> 4.910
Social Aspect				
Population Density	People/Km	< 420	420 – 2.234	> 2.234
Population Growth Rate	Percent	< 0,5	0,5 - 1	> 1
Sex Ratio	Percent	> 99,86	99,86 – 104,06	< 104,06
Children	People	<6.500	6.500 – 18.295,2	> 18.295,2
Elderly	People	< 2.200	2.200 – 5.711,2	> 5.711,2

Source: [17] Modification

The steps for calculating the Social Vulnerability Index (SoVI) method are started by collecting data for each of the variables used in the study. The second step is to find the z score or min-max value. The purpose of normalization is that each data that has different unit can be compared so that it can be used to draw the conclusion.

When using Max-min, normalization is done as follows:

$$Z = \frac{X - Xmin}{Xmax - Xmin} \quad (1)$$

X= value of each data
Xmin= X minimum value
Xmax= X maximum value

The max-min method is not substantially different from the Z-score method. When using the Z-Score, the calculation will make negative and positive number in an infinite range, which sometimes makes it difficult to draw conclusions. By using the max-min method, the Z-Score value will be changed to between 0 and 1, so that the max-min method is actually intended to convert the Z-Score value to make it easier to draw conclusions [9,11].

The result of the processing above is in the form of score or index, the closer to 1 the more to have socio-economic vulnerability. To provide a clearer explanation, calculations will be carried out in sub-districts in Wonogiri Regency.

The third step is to compare between sub-districts that min-max values have been made. From the results of the min-max value, it will be known that the sub-district that has a value of 1 as the most vulnerable sub-district on certain variable and value of 0 as the sub-district that is relatively not vulnerable.

The fourth step is to score components that have represented all components and then do the scoring. Scoring is done by entering the average min-max value into the vulnerability classification table released by the Head of the The National Agency for Disaster Countermeasure (Table 2).

3. Results and Discussion

The following is the result of the calculation of the value of social vulnerability by using the social vulnerability index (SoVI) approach. Specific comparison is addressed to the three selected sub-districts, these are in Eromoko District, Tirtomoyo District, and Girimarto District. The analysis started from mapping the drought vulnerability and landslide susceptibility level, indicators of economic vulnerability, indicators of social vulnerability, and level of socioeconomic vulnerability.

3.1. Mapping of Disaster Vulnerabilty Level

Wonogiri Regency had a level of disaster vulnerability these were drought and landslides. These two types of disasters were triggered by climate change. If there was an increase in weather shift, it was predicted that several sub-districts which were categorized as drought-prone areas would experience a higher probability. The second type of disaster was landslide, this type was opposite to drought. If climate change caused less rainfall, it would cause drought, so if the rainfall increased it, increased the vulnerability to landslide. Figure 1 shows the level of drought vulnerability in Wonogiri Regency.

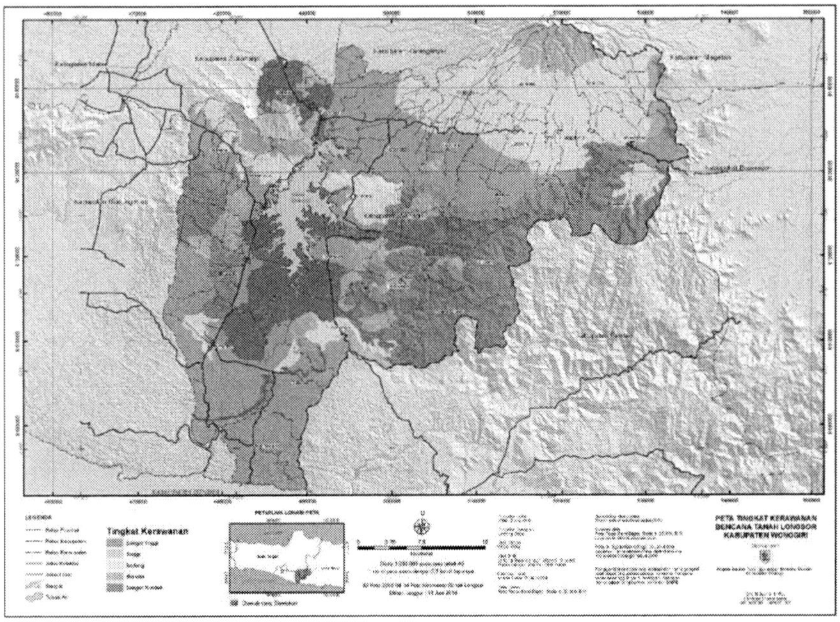

Source: Regional Disaster Management of Wonogiri Region

Figure 1. Landslide Vulnerability Level in Wonogiri Regency

Based on Figure 1, it is known that the Districts of Bulukerto, Purwantoro, Jatisrono, Slogohimo, Tirtomoyo, are areas that have a high level of vulnerability. Meanwhile, Eromoko and Girimarto sub-districts have a relatively low level of vulnerability. Geographically, areas that have a high level of vulnerability are on the east side and a small part is on the west side of Wonogiri Regency.

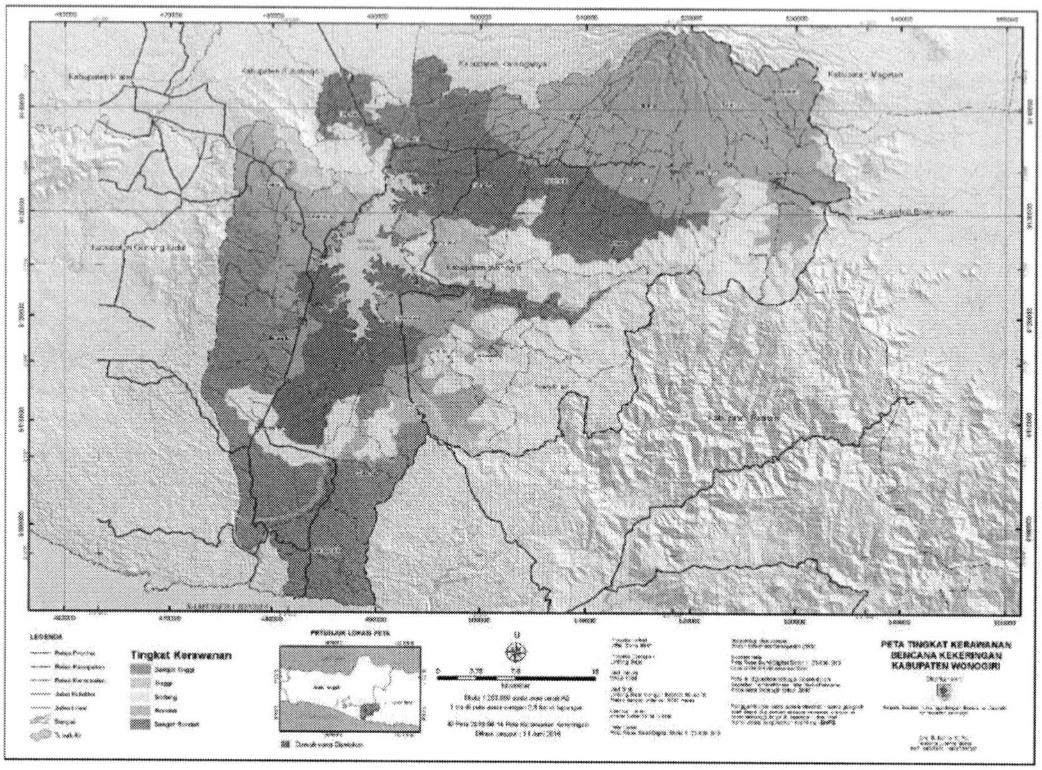

Source: Regional Disaster Management of Wonogiri Region

Figure 2. Drought Vulnerability Level in Wonogiri Regency

Based on Figure 2, it is known that the areas which have high level of vulnerability to drought. Pracimantoro, Paranggupito, and Eromoko sub-districts are identified as areas that have high level of drought vulnerability. The areas of Girimarto and Tirtomoyo sub-districts have relatively lower level of vulnerability.

Among the three sub-districts these are Tirtomoyo, Eromoko, and Girimarto, only Girimarto has relatively lower level of vulnerability. Meanwhile, Tirtomoyo represents landslide vunerable area, and Eromoko represents drought area.

3.2. *Economic Vulnerability Indicator*

In Table 3. inform about indicators of economic vulnerability. In this study, the variable of economic vulnerability is represented by Gross Regional Domestic Product (GRDP) and poverty rate. The weight or proportion of each variable is 40 percent for GRDP and 60 percent for the poverty rate.

Table 3. Economic Vulnerability Indicator

Indicator	Unit	Score			Value
		Eromoko Sub-district	Tirtomoyo Sub-district	Girimarto Sub-district	
Gross Regional Domestic Product (GRDP)	Percent	Medium (2)	Medium (2)	Low (1)	40 %
Poverty	People	Medium (2)	High (3)	Medium (2)	60 %

Source: data processing

1. Economic Vulnerability Level in Eromoko Sub-district
   Based on the calculation of the economic vulnerability index using the SoVI equation, it was known that the economic vulnerability index in Eromoko District was 0.67. The number was the result of multiplying the GRDP indicator scores and the poor population with a predetermined weight. If it was adjusted to the existing classification, the level of economic vulnerability in Eromoko Sub-district was included in the category of high economic vulnerability.
2. Economic Vulnerability Level in Tirtomoyo Sub-district
   Based on the calculation of the economic vulnerability index using the SoVI equation, it was known that the economic vulnerability index in Tirtomoyo District was 0.87. The number was the result of multiplying the GRDP indicator scores and the poor population with a predetermined weight. If it was adjusted to the existing classification, the level of economic vulnerability in Tirtomoyo Sub-district was included in the category of high economic vulnerability.
3. Economic Vulnerability Level in Girimarto Sub-district
   Based on the calculation of the economic vulnerability index using the SoVI equation, it was known that the economic vulnerability index in Girimarto District was 0.54. The number was the result of multiplying the GRDP indicator scores and the poor population with a predetermined weight. If it was adjusted to the existing classification, the level of economic vulnerability in Girimarto Sub-district was included in the category of medium economic vulnerability.

3.3. Social Vulnerabilty Indicator

Table 4. presents indicators of social vulnerability in three sub-districts. In this study, the proportion of social vulnerability variables, population density with 30 percent, population growth rate with 25 percent, sex ratio with 15 percent, children with 15 percent, and the elderly with 15 percent.

Table 4. Social Vulnerabilty Indicator

Indicator	Unit	Score			Weight
		Eromoko Sub-district	Tirtomoyo Sub-district	Girimarto Sub-district	
Population Density	People/Km	Low (1)	Low (1)	Medium (2)	30 %
Population Growth Rate	Percent	Low (1)	Low (1)	Low (1)	25 %
Sex Ratio	Percent	Medium (2)	High (3)	High (3)	15 %
Children	People	Low (1)	Low (1)	Medium (2)	15 %
Elderly	People	Medium (2)	Medium (2)	High (3)	15 %

Source: data processing

1. Social Vulnerability Level in Eromoko Sub-district
   Based on the calculation of the social vulnerability index using the SoVI equation, it was known that the social vulnerability index in Eromoko Sub-district was 0.44. This number was the result of multiplying the indicator scores for population density, population growth rate, sex ratio, population of children, and the elderly population. If it was adjusted to the existing classification, then the level of social vulnerability in Eromoko Sub-district was in the category of medium social vulnerability.
2. Social Vulnerability Level in Tirtomoyo Sub-district
   Based on the calculation of the social vulnerability index using the SoVI equation, it was known that the social vulnerability index in Tirtomoyo Sub-district was 0.49. This number was the result of multiplying the indicator scores for population density, population growth rate, sex ratio, population of children, and the elderly population. If it was adjusted to the existing classification, then the level of social vulnerability in Tirtomoyo Sub-district was in the category of medium social vulnerability.
3. Social Vulnerability Level in Girimarto Sub-district
   Based on the calculation of the social vulnerability index using the SoVI equation, it was known that the social vulnerability index in Girimarto Sub-district was 0.69. This number was the result of multiplying the indicator scores for population density, population growth rate, sex ratio, population of children, and the elderly population. If it was adjusted to the existing classification, then the level of social vulnerability in Girimarto Sub-district was in the category of high social vulnerability.

3.4. Social Vulnerability Index

Based on calculations using the SoVI method, it was seen that the overall economic and social vulnerability index in Eromoko Sub-district was 0.55 and it was included in the medium category. For Tirtomoyo Sub-district, the overall social and economic vulnerability index was 0.68 and it was included in the high category. Meanwhile, for Girimarto Sub-district, it was known that the overall economic and social vulnerability index was 0.61 and it was included in the medium category. It could be concluded that of the three sub-districts, Tirtomoyo sub-district was the sub-district that has the highest level of economic and social vulnerability. The result of the SoVI calculation in Wonogiri also strengthened the result of the study [12]. In theory, the Girimarto area had a low level of vulnerability, but in fact, in terms of the level of socio-economic vulnerability, it was higher than the Tirtomoyo sub-district. Tirtomoyo sub-district consists of 12 sub-districts which are geographically bordered on the north by Jatiroto and Sidoharjo sub-districts, in the south with Batuwarno sub-district, in the west with nguntoronadi sub-district and in the east with East Java Province. The area of Tirtomoyo sub-district is 9,301.08 hectares consisting of rice fields and fields of 5,099.22 hectares, buildings and yards of 2,402.55 hectares, state forest of of 1,572.39 hectares, grasslands of 56.20 hectares and other land uses of 170.74 hectares.

The area of Eromoko sub-district is 12,035.87 hectares consisting of rice fields and fields of 7,590.04 hectares, buildings and yards of 1,784.85 hectares, state forest of 1,302.92 hectares and other land uses of 1,358.06 hectares. If it is compared to Tirtomoyo sub-district, the area is 9,301.08 hectares consisting of rice fields and fields of 5,099.22 hectares, buildings and yards of 2,402.55 hectares, state forest of 1,572.39 hectares, grasslands of 56.20 hectares and other land uses of 170.74 hectares.

Girimarto sub-district has a higher population density than the other two sub-districts. If in Girimarto District the population density reaches 618 people per square km, then in Tirtomoyo District and Eromoko District the density level is 537 people per square km and 345 people per square km.

IOP Conf. Series: Earth and Environmental Science **986** (2022) 012054 doi:10.1088/1755-1315/986/1/012054

4. Conclusion

From the results of the SoVI calculation, it can be concluded that areas that have a high level of vulnerability to the danger of drought or landslides are directly proportional to the level of social and economic vulnerability. Areas that have a high vulnerability to drought have a high probability of experiencing crop failure, but if it is observed from the socio-economic vulnerability there is higher than other region. On the other hand, areas that have a low level of vulnerability but if there is a drought, it will cause the community to be more vulnerable to suffer losses.

Suggestions for further research can be developed for mapping the disaster risk in Klaten Regency. By a disaster risk map, it will be known which sub-districts have vulnerability level and socio-economic vulnerability level in a map.

References

[1] Runtunuwu E And Syahbuddin H 2007 Perubahan Pola Curah Hujan Dan Dampaknya Terhadap Periode Masa Tanam Persoalan Global Yang Melibatkan Banyak Negara Dan Latitude Perubahan Pola Hujan Yang Mengakibatkan Pergeser- Waktu Singkat Dengan Kecenderungan Intensitas Beberapa Organisme Pengganggu *J. Tanah Dan Iklim* **26** 1–12

[2] Putri F A Dan S 2012 Strategi Adaptasi Dampak Perubahan Iklim (Climate Change) Terhadap Sektor Pertanian Tembakau (Studi Kasus : Kecamatan Bulu , Temanggung) *J. Ekon. Dan Stud. Pembang.* **13** 33–42

[3] Salampessy Y L . 2018 Menakar Kapasitas Adaptasi Perubahan Iklim Petani Padi Sawah (Kasus Kabupaten Pasuruan Jawa Timur) *J. Ilmu Lingkung.* **16** 25

[4] Kifli F W, Mulyo J H And Sugiyarto 2015 Analisis Kerentanan Perubahan Iklim Terhadap Ketahanan Pangan Dan Kesejahteraan Rumahtangga Tani Di Propinsi Riau *The 2nd University Research Coloquium 2015* Pp 489–95

[5] Suryanto S 2017 Pemetaan Dan Valuasi Ekonomi Kerugian Banjir Di Karesidenan Surakarta *Piramida* **XIII** 1–8

[6] Karlina K 2017 Meteorological Drought Assessment In Wonogiri District *J. Civ. Eng. Forum* **2** 159

[7] Yudhanti F Y 2017 Drought Risk Assessment For Resource Management Towards Resilient Development In Eromoko District , Wonogiri Regency , *Prosiding Seminar Nasional Geografi UMS 2017 Pengelolaan Sumberdaya Wilayah Berkelanjutan* Pp 233–40

[8] Shirley W L, Cutter S L And Boruff B J 2003 Social Vulnerability To Environmental Hazards *Soc. Sci. Q.* **84** 242–261

[9] Razafindrabe B H N, Parvin G A, Surjan A, Takeuchi Y And Shaw R 2009 Climate Disaster Resilience: Focus On Coastal Urban Cities In Asia *Asian J. Environ. Disaster Manag. - Focus. Pro-Active Risk Reduct. Asia* **01** 101

[10] Below T B, Mutabazi K D, Kirschke D, Franke C, Sieber S, Siebert R And Tscherning K 2012 Can Farmers' Adaptation To Climate Change Be Explained By Socio-Economic Household-Level Variables? *Glob. Environ. Chang.* **22** 223–35

[11] Lawal O And Arokoyu S B 2015 Modelling Social Vulnerability In Sub-Saharan West Africa Using A Geographical Information System *Jamba J. Disaster Risk Stud.* **7** 1–11

[12] Gbetibouo, G.A., Ringler, C. And Hassan R 2010 Mapping South African Farming Sector Vulnerability To Climate Change And Variability. *Nat. Resour. Forum* **34** 175–187

[13] Hapsoro A W And Buchori I 2015 Kajian Kerentanan Sosial Dan Ekonomi Terhadap Bencana Banjir *J. Tek. PWK* **4** 542–53

[14] Putri T D And Muhammad F 2019 Analisis Kerentanan Sosial Masyarakat Dan Adaptasi Perubahan Iklim Kampung Gemblakan Atas , Kota Yogyakarta Analysis Of Social Vulnerability And Climate Change Adaptation In Gemblakan Atas Village , Yogyakarta City *Proceeding Biol. Educ. Conf.* **16** 256–64

[15] Gariano S L And Guzzetti F 2016 Landslides In A Changing Climate *Earth-Science Rev.* **162** 227–52

IOP Publishing
doi:10.1088/1755-1315/986/1/012054

[16] Ullah R, Shivakoti G P And Ali G 2015 Factors Effecting Farmers' Risk Attitude And Risk Perceptions: THE Case Of Khyber Pakhtunkhwa, Pakistan *Int. J. Disaster Risk Reduct.* **13** 151–7
[17] BNPB 2012 *PERKA BNPB No. 2 Tahun 2012 Tentang Pedoman Umum Pengkajian Risiko Bencana*

IOP Conf. Series: Earth and Environmental Science **986** (2022) 012055 doi:10.1088/1755-1315/986/1/012055

Sustainable Urban Drainage System (SUDS) as Nature Based Solutions Approach for Flood Risk Management in High-Density Urban Settlement

Saut Sagala [1], Arini Murwindarti [1], Belia Ega Avila [2,*], Arief Rosyidie [1] and Danang Azhari [2]

[1]School of Architecture, Planning and Policy Development, Bandung Institute of Technology (ITB), 40132, Bandung, Indonesia
[2]Resilience Development Initiative (RDI), 40135, Bandung, Indonesia

*Email: belia.ega@rdi.or.id

Abstract. Nature-based Solution (NBS) is an umbrella for ecosystem-based approaches to prevent or mitigate the impacts of hydro-meteorological hazards. The increasing hydro-meteorological disaster occurrence such as floods in Bandung City because of both climate change phenomenon and rapid growth of Bandung City has caused a big impact on people live in slums and high-density settlement. Realizing this risk, Bandung City has implemented strategies to reduce flood risk, including the Sustainable Urban Drainage System (SUDS) as stated on the Bandung City Spatial Plan (RTRW) 2011-2031. However, this measure had not effectively reduced flood risk in Bandung City. Therefore, this research focuses to analyse the implementation of SUDS and how it can capture the concept of NBS, with the study case Cibadak Administrative Village, one of the poor and high-density urban settlements in Bandung City. The result was developed using a mixed method consisting of a descriptive quantitative approach, spatial analysis for satellite imagery, and content analysis technique to analyse each challenge of SUDS implementation. The result indicates the lack of capacity in scale, budget, natural characteristic, and social demographic for SUDS implementation in poor and high-density urban settlement in Bandung, which aligns with the challenge of NBS implementation in the global south.

1. Introduction

Over half of the Indonesian population now is living in urban area. The number will be increasing to 67.1% in 2045 based on the BPS projection [1]. Rapid urban population growth brings a number of challenges, such as lack of basic infrastructures, traffic jams[2], environmental problems, such as water and/or air pollution [3,4]. This is also happening with the loss of urban natural spaces in pursuit of space for built up areas [5]. In addition, the population growth will increase a lot of problems and lead to disasters in urban areas [6]. Besides that, disasters in Indonesia have been increased rapidly from 2000-2010, and hydro-meteorological disasters induced by climate change such as flood account for more than 70% of the disasters in Indonesia [7]. Inadequately managed and planned rapid growth in urban area will create threats that undermine existing development accomplishments. Besides that, uncontrollable rapid growth area can cause poverty, resulting to have no access to clean drinking water, no sewage system and can lead to water scarcity, especially in the dry season [8]. Moreover, poor urban management can also increase vulnerability and risk of flooding [9].

Bandung City as the fourth dense city in Indonesia with a population of 2.5 million [10], meanwhile, population in Bandung Metropolitan Area is 8,790,308 [11] and it has faced similar problems with the urban population, including the provision of settlement and basic needs. Based on BNPB data (2021), almost 90% of Bandung City area is highly prone to flooding, mostly in the southern part of the city. In 2020, floods in Bandung City have inundated 25 locations, most of which are high-density settlements. Floods in Bandung City often occur because of climate change and land-use conversion, thus leading to water catchment areas shrinkage, affecting the increase of infiltration level in the area, and creating a high flow of run-off water [12,13]. Furthermore, the inadequacy of drainage channels to drain excessive run-off water causes floods. This significantly affects the community's risk of hydro-meteorological hazards in poor and high-density settlements.

Cibadak Administrative Village is located in the centre of Bandung City with an area of 0.473 km^2 and 13,370 population [10]. Cibadak is also crossed by the Citepus River, which is the stream of the Citarum River. Water from the river always overflowed during the rainy season and caused flooding in that area that made several houses inundated. Besides that, there have been many lands use changes into built-up areas in Cibadak, causing a reduction in infiltration areas. Moreover, there have been several programmes from the Bandung City Government to mitigate the flood which is mainly focused on structural conventional measures such as building water embankment. However, the effectiveness of this measure is still questionable since some hamlets are still often flooded. Realizing this challenge, flood risk reduction innovations are needed that take into account the hydrological system and physical characteristics of the settlement and prioritize the ecosystem. Meanwhile, Bandung City has implemented strategies to reduce flood risk, including the Sustainable Urban Drainage System (SUDS) based on the Bandung RTRW 2011-2031. However, this measure had not effectively reduced flood risk in Bandung City.

NBS is an umbrella for ecosystem-based approaches to prevent or mitigate the impacts of hydro-meteorological hazards. NBS brings more sustainability and addressing societal challenges as well as providing human well-being and biodiversity benefits. SUDS as a promising NBS type have various benefits ranging from flood risk management to water management [14]. Based on the prior research, the research about SUDS still focused on discussing the types of SUDS that are suitable to be applied in urban areas [15], identifying SUDS performance and the factors that influence it [16], as well as the importance of SUDS in stormwater management [17]. Therefore, this research aims to fill the research gap by analysing the challenge implementing SUDS in high-density settlement and slum area and how it can capture the concept of NBS, with the study case of Cibadak, one of the poor and high-density urban settlements in Bandung City.

2. Literature Review

2.1. Hydro-meteorological Hazard and High-Density Settlement

Globally, hydrometeorological disasters reckon for more than 75% of damage, including loss of life, economic loss, infrastructure damage, and regular life disruption [18]. The UNISDR [19] defines a *hydrometeorological hazard* as a phenomena that can damage infrastructure, environmental aspects, economic losses, and social disturbances. Of all hydrometeorological disasters, floods are one of the most frequent and expensive hydrometeorological disasters [20]. From the estimation, the average population per year affected by floods is expected to rise from 1 million in 1990 to 25 million in 2050 due to global warming climate change [21]. Floods strike all countries and prompt further destructiveness and goods and property harm than some different kind of hazard. Floods can disturb water purifying and residue disposal systems and produce toxic water waste sited to overflow, primarily in high-density settlements.

Settlements with high density are easy to identify through satellite imagery because of the proximity of residential buildings [22]. High-density urban settlements refer to areas located in urban areas. Housing is the primary land use, with a building area of less than 90 per cent. According to European Commission, high-density settlements are designated as all cells including the local community density of at slightest 1,500 residents/km^2 or with a locally built area division of about 0.50 and clustered in 4 connectivity objects of at least 50,000 people or all cells following from median filtering of 3x3 or with gap-filling <15 km^2 (median filtering and gap-filling are referred to a group of cells that provide the prior logic).

In most countries, high-density settlements are amongst the most defenceless to flood. The natural characteristics and high-density settlements risk scale change because of the dynamic land-use patterns, unplanned growth, and climate change impacts. In addition, the floods impact increment on high-density settlements is because of the area of residences in the floodplain area, the construction of dams that join the rivers across which transportation routes overlap, the decline in current drainage into ineffective hydro-technical products, and more.

With widespread urbanisation and high-density settlements, the potential for damage to human life and infrastructure is enormous. Vulnerability of flood includes parts at the chance, such as the inhabitants of a flood-prone area, a built environment, or an ecosystem revealed to flood hazard [23]. Meantime, vulnerability remains commonly recognised by many researchers to consist of three components: degree of exposure, susceptivity and resilience or response capacity of a population in a remarkable area [24].

2.2. Sustainable Urban Drainage System (SUDS) in High Density Settlement

Urbanisation phenomenon has led to the impermeabilization of the soil, causing a greater volume of surface runoff. When it converged to lower regions, the greater volume of runoff becomes one of the main causes of flooding [25]. The efficiency of traditional drainage in managing the increasing amount of surface runoff have significantly decreased because they have designed according to a determined return period and how many users that use them. Therefore, the SUDS was introduced as it includes different technologies and techniques in a more sustainable manner [26]. SUDS is an approach to decelerate and decrease surface water runoff which aims to reduce downstream flood risk and possibilities of pollution caused by the runoff. It includes reaping, penetrating, decelerating, reserving, transporting and handling runoff on site [27].

SUDS imitates natural processes to protect water cycles from negative impacts of development [26]. The way SUDS work makes it appear as one of the most promising form of NBS implementation. SUDS tackles water quality problems and, at the same time, enhances green spaces within developments, supporting the ideal condition for habitats and places for wildlife to live and flourish [27]. For urban areas that are highly populated, these co-benefits of SUDS implementation are very much needed because urban development has replaced well-drained soil with impermeable or imprevious surfaces; hence, disturbing the water cycle in urban ecosystems [28,29]. Moreover, High-density urban areas are faced with a lot of problems regarding the quantity and quality of urban water. Hence, the SUDS implementation in high-density urban areas is crucial so that surface runoff can be well-managed.

Successful integration of SUDS into urban design will require an early consideration of surface water as part of the design and planning process, and also a collaborative and interdisciplinary team that [27]. However, implementing SUDS in urban areas can face several challenges [30]. Placing a combined SUDS components in sequences and spread around the development are can help to create an effective system (good water quality and hydraulics) that does not require too much of a cost [27]. In general. there are four factors that contribute to the implementation of SUDS, namely scale, natural characteristics, social demographic, budget, and planning (see Table 1 for more details).

Table 1. Factors, Variables, and Indicators that Affect the Implementation of SUDS

Factor	*Variable*	*Indicator*	*Parameter*	*Source*
Scale	Type/use of open space	Public space	The existence of parks, squares, roads dividers, sidewalks, or parking lots.	(Jiménez Ariza et al., 2019)
		Private space	The existence of open space of residential, commercial, and industrial areas, or public facilities.	
	Spatial aspect	Land value	High/low land value due to competing development objectives.	(Kuller et al., 2017)
		Land zoning	SUDS implementations require a specific drainage plan	
Natural characteristics	Soil	Type	Soil with good infiltration capacity	(Kuller et al., 2017)
	Slope	Slope rate	Steep slopes are prohibitive to most SUDS types	
	Hydrology	Topography	Natural drainage channel availability	
		Proximity to water body	The closer to the water body, the easier to implement SUDS	
	Climate	Rainfall averages	Rainfall averages determine the volume and condition of water managed by SUDS	
		Temperature	Local temperature determines plant selection, and it is important to several types of SUDS	
	Urban Fabric	Roof surface areas	Flat rooftop will facilitate green roof implementation	
		Open space	Some SUDS technologies favour or require open space.	
		Permeable pavement	Permeable pavements facilitate infiltration	
	Ecosystem Type	Available vegetation	Vegetation improves infiltration	
	Flood hazard	Depth of inundation	Inundation leads to floods and material loss	(Minister of Public Works and Public Housing, 2014)
		Duration of inundation	The longer an area being drowned, the higher the losses	
		Inundation area	Bigger inundation area means bigger or higher losses	
		Inundation frequency	More frequent floods mean bigger losses	
	Flood impacts	Tangible impact	Tangible impact on public or private properties influences the type and location of SUDS	(Hammond et al., 2015)
		Intangible impact	Intangible impact on health increases the community awareness	
Social Demographic	Demographic	Income	Micro approach of the SUDS is more suitable for community or area with low average income	(Kuller et al., 2017)
		Density	High population density usually has higher levels of imperviousness	
	Public awareness and values	Risk awareness and value	Public awareness and values determine the degree of community acceptance towards SUDS	
	Social cohesion	Volunteering	The existence of volunteering activities indicates community initiative.	
		Community organization membership	Participation in community organization enhance coordination and mutual assistance	
Budget and Planning	Political stability	Political responsiveness	SUDS requires strong, ongoing, and responsive commitment from government	(Kuller et al., 2017)
	Development opportunity	Organizational capacity	Opportunities for development of SUDS may arise from high organisational capacity and coordination	
	Economic cost	Initial cost	Initial costs are estimated to determine the SUDS type that fits the budget	(Gogate et al., 2017)
		Operation and Maintenance (O&M) Cost	O&M costs per year are estimated to maintain the quality of SUDS	

3. Methods

Cibadak administrative village was chosen as the location for this research because it is one of the densely populated settlements in Bandung City that is very prone to flood disaster. This research aims to identify the challenges in the implementation of SUDS in high-density settlement and how it captures the concept of Nature-based Solution.

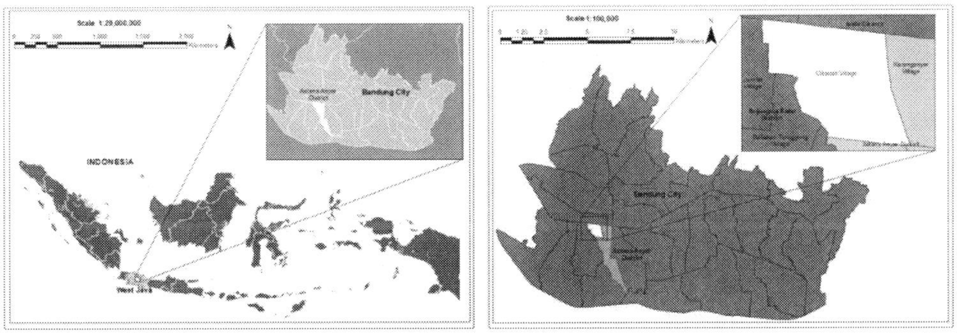

Figure 1. Map of Study Location

It uses mixed-method approach consisting of quantitative, qualitative, and spatial analysis. The quantitative approach is used to analyse the statistical data, including slope rate, temperature, rainfall averages, proximity to receiving water body, and flood hazards (depth of inundation, duration of inundation, inundation area, inundation frequency). Besides that, the spatial analysis is also used to describe the characteristics of the land use zone, urban fabric, availability of vegetation, and density. Meanwhile, content analysis as qualitative approach is carried out to explore the historical causes, conditions and impacts of flooding in Cibadak administrative village as well as the challenge in implementing SUDS in that area. Furthermore, Secondary data collection is conducted through literature review related to the previous SUDS research, online news about flood in Cibadak administrative village, regulation, programs, and policies about flood risk management in Bandung city or Cibadak administrative village. On the other hand, primary data collection is utilised through in-depth interview with local stakeholder and observation from worldview image and field documentation. Stakeholders for in-depth interview are selected based on purposive sampling to obtain the first key informants who have better understanding about flood condition in that area and other key informants are explored through snowball sampling from the recommendation of the first key informants.

4. Results and Discussions

As stated on the Bandung City Spatial Plan (RTRW) 2011-2031, Bandung City has implemented SUDS as one of the strategies to reduce flood risk although further details are needed regarding the locations and the types of SUDS implemented in the city. This study explores some challenges in implementing SUDS in high density settlement. The results will be elaborated in four sections based on the factors that contribute to the implementation of SUDS.

4.1. Scale of Implementation

The first factor is scale of implementation, which refers to the site selection process of where the SUDS infrastructure will be built. The scale of implementation determined by type/use of the open space so SUDS can be built in either public open spaces, private open spaces, or both. There are several steps or procedures that should be taken to determine or design the most feasible and suitable SUDS that can be implemented [31]. In Cibadak administrative village, most of the land use area is private space, which indicates from the previous observation and the spatial analysis. 64,56% (31,06

Km^2) of the area is private and build up area. This makes various implementations of large-scale SUDS more difficult to implement, due to limited public land. On the other hand, micro or household scale SUDS would be an appropriate option to implement in Cibadak Administratie Village.

"The difficulty to implement the SUDS is the availability of space. It is because the location is very dense, densely populated. Even for the previous programme to find a location to create local garden was very difficult. Then, we had another option to look for plants that can absorb water and we looked for the yard of the community's house which is wide enough to grow the plant."
M. Dio Wana Paratama, Chief of *Karang Taruna* of Cibadak Administratie Village

Figure 2. Type of Build-up Area of Cibadak

The second variable of spatial aspect refers to the land value and land zoning or regulation in the area where the SUDS will be built. In Cibadak, the land value has been increasing due to the rapid development in this area. From the interview with the local government, they stated that currently the land value in Cibadak is about IDR 13.000.000 – 25.000.000 per m^2. Hence, transforming the existing land use to SUDS would be a big challenge. This situation aligns with Woods Ballard et al., (2015) research that mentioned about sites within urban areas are often confined and restricted, and land is often more valuable, which means introducing SUDS will be challenging because it is faced with competing development purposes. Furthermore, land zoning refers to the classification of urban areas based on the different land uses. Based on Bandung Spatial Plan, Cibadak classified as high-density settlement, trade and services centre, and some part of river watershed. Current land use mostly has the same classification as the spatial plan. However, there is the lack of river watershed area and proper drainage system in Cibadak since there is no specific legislation or drainage plan in that area [32].

Figure 3. Conventional Drainage System in Cibadak administrative village (Left) and the Citepus River (Right)

4.2. Natural Characteristics

Natural characteristics can be observed through the site characterises and the flood conditions. Based on the Bandung RTRW 2011-2031, generally the soil type in Bandung City consists of alluvial soil and latosol soil, while in Cibadak administrative villages, most of the soil is alluvial. This type of soil is formed from dry silt deposits through water flow and gravity force, as a result, this soil is prone to erosion. Meanwhile, the slope in that area is around 0 – 15%, meaning that the slope is flat or sloping and it caused the area to be more prone to flooding. Besides that, for hydrology conditions, there is the Citepus River as a natural drainage channel that crosses Cibadak Administrative Village. This river often overflowed when it rained heavily in the north, carrying black-colored water because it was polluted with garbage from the upstream. The conditions around the river are also the factors in the occurrence of flooding because there are many settlements established in river border areas so that there is no open space to absorb stormwater runoff. Whereas, it has been regulated in Government Regulation No. 38 of 2011 concerning Rivers that the borderline of the embankment river in the urban area is at least 3 meters from the outer edge of the embankment foot along the river channel.

Furthermore, regarding climatic conditions, the average temperature in Cibadak administrative village is between 22°C – 25°C, while for the rainfall, the heaviest rain occurred in December at around 313.5 mm, related to severe flooding in December 2020. Meanwhile, through observing the urban fabric, it is noticeable that most of the building roofs are not flat, meaning that the implementation of green roofs as a type of SUDS is not possible. Besides that, the availability of open space is also few, only 1,28 km2 as the total area. In addition, the road pavement in Cibadak generally uses asphalt and very little vegetation so that the absorption of rainwater runoff becomes slow.

Based on the interview, floods always occur during the rainy season, especially consignment floods originating from the north. The tributaries (7 - 8 tributaries) in the Citepus river cause water to overflow faster during heavy rain.

"As a matter of fact, the flood that happened in Cibadak was consignment flood from the north. Meawnhile, the community here actually has already been aware to throw the rubbish into the dustbin. However, there are approximatelly 7 tributaries that enter the Citepus River. Thus, it caused flood and carried black-colored water."

Ceppy Setiawan, Chief of Hamlet 2, Cibadak Administrative Village

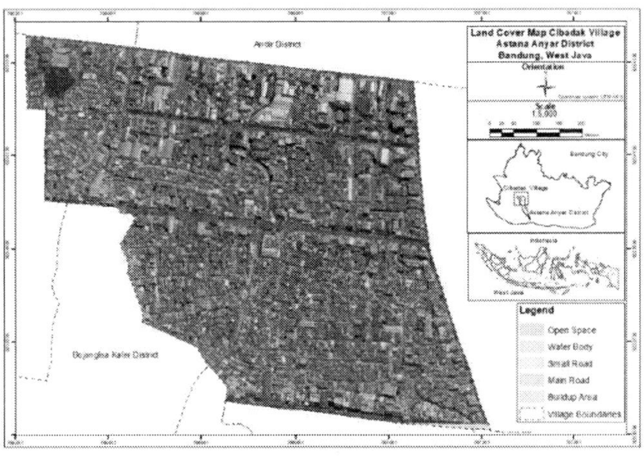

Figure 4. Map of Land Cover in Cibadak Administrative Village

The area is also inundated easier because of the lower soil structure conditions in Hamlet 5, 6, 7 and 8. The local stakeholder also mentioned that the average inundation depth was 1 meter, while the highest inundation which reached 2 meters occurred in December 2020. Meanwhile, in most flood cases in Cibadak, the inundation duration was around 1 – 1.5 hours, which happened almost 9 – 10 times a year. The floods inundated a large area in the administrative village, covering Hamlet 5, 6, 7, and 8 causing a huge amount of material loss. Meanwhile, a flood with a depth of 2 meters caused several residents' houses damage, bridge damage and flooded irrigation [33].

4.3. Social Demographic

While other structural mitigation measures rely on government top-down projects, SUDS requires community's initiatives and responsibilities in the implementation since it is part of decentralized systems located in neighbourhood scale [34]. Moreover, challenges also found on the maintenance phase after the implementation of SUDS since the community might facing constraints to continuously engaging in the implementation of SUDS [35]. Therefore, social demographic characteristics such as demographic characteristics, public awareness and values, and social cohesion perform an essential role in the implementation and maintenance of SUDS.

From the survey, it was found that most the residents of Cibadak administrative village are traders whose income is relatively small. Riverbanks alongside the Citepus River become the place they choose as a place to live which might be an unfavourable location for higher socio-economic status groups. Furthermore, income and house price are related to the suitability of the SUDS types, where low-income groups of residents with low house prices are more feasible in implementing the micro approach of SUDS. On the other hand, population density also influences the suitability of SUDS implementation. Based on spatial analysis, it is found that Cibadak administrative villages is a high-density settlement with high population density. This kind of settlements typically related to higher levels of imperviousness.

The second variable of social demographic factor identified in this study is public awareness of the risk and their understanding about SUDS. From the interview, we found that the community is already aware with the flood that happened in their area because they have experienced it for years and they also have already known about what to do during and after the flood. Meanwhile for their understanding about SUDS, they said that they are quite familiar with SUDS, however, the implementation of SUDS is still limited in their area.

"Yes, that's right. They are used to flooding. It's normal for them to clean up the mud after the flood."
Ceppy Setiawan, Chief of Hamlet 2, Cibadak Administrative Village

Social cohesion, as the third variable in social demographic factor, illustrates the social capital that the community owns and uses to take care of other people and its surroundings which are impacted by the floods. The social cohesion in this community was found in volunteering activities and community organization membership such as *Karang Taruna* (youth community organization). The key informants mentioned that Fire and Disaster Management Agency in Bandung City has held a training about disaster response and recovery for the residents. Residents are also accustomed to doing *Gotong Royong* (cooperation and community service) after a flood occurs. Furthermore, the community organization membership affected how a community cooperate and take care of its surroundings.

4.4. Budget and Planning

The first point in budget and planning factor is political responsiveness, which means that SUDS requires strong, ongoing, and responsive commitment from the government. In Bandung City, there are several governmental regulations that were made to strengthen disaster management efforts, including floods. In a broad sense, there are two regulations. The first one is Regulation of the Mayor of Bandung City Number 626 of 2017 concerning the Establishment of the Organization and Work Procedure of the Bandung City Disaster Management Implementing Unit, and the second one is Bandung City Regional Regulation Number 9 Year 2009 concerning Regional Medium Term Development Plan (RPJM) 2009-2013. In the RPJM document, the government outlined the

IOP Conf. Series: Earth and Environmental Science **986** (2022) 012055 doi:10.1088/1755-1315/986/1/012055

guidelines for disaster mitigation strategies and plans for the classification of protected areas, such as Northern Bandung that is often hit by hazards. Furthermore, there is Bandung RTRW 2011-2031, which specifically mentioned the urgency to improve drainage infrastructure services in order to overcome the problems of flooding and inundation.

The second point is development opportunity, which can be seen from the organisational capacity in disaster management efforts. Based on the result of our interviews with Chief of Hamlet 2 of Cibadak Administrative Village and Chief of *Karang Taruna* of Cibadak Administrative Village, it can be understood that the local government of Cibadak Village is always cautious and on-guard whenever rain hits the area, and that they will know what to do if flood follows the rain and that the local government of Cibadak Village has a good coordination with the local community, the district-level government, and also the Social Office of Bandung City, to help them finding good shelters for the evacuated persons. The organisational capacity can also be seen from disaster management related trainings that have been done at Cibadak Administrative Village. Chief of Hamlet 2 of Cibadak Village stated that there was a training on how to handle flood risks held by Bandung City's Office of Fire and Disaster Management, and Chief of *Karang Taruna* of Cibadak Village stated that there is a regular training on how to handle floods, such as how to evacuate people trapped in the flooding, which safety tools should be used, how to provide public kitchen, and more. The training prepares the residents to be able to evacuate themselves during the floods, and a basic knowledge of post-disaster recovery such as collecting data on the affected houses, fatalities and material loss, and where to get help. Moreover, there are also several flood management related programs that have been implemented in Cibadak Village that involve the community

"There was a program called Citarum Harum, also a program by BRI Peduli in which a water filter was built and then taught the community on how to build green areas to filtrate water. There was also a program that built biopore, but it was not really effective to reduce floods in Hamlet 6 and 7."
M. Dio Wana Paratama, Chief of *Karang Taruna* of Cibadak Administrative Village

However, for the third point (economic cost), both Chief of Hamlet 2 and Chief of *Karang Taruna* of Cibadak Village stated that they mostly relied on voluntary funding for emergency posts and public kitchen, and that they must ask to the district government or the Social Office of Bandung City first to get the funding for the recovery of the affected community and for assistance in the form of the nine basic needs. If SUDS will be built and implemented in Cibadak, the local government should prepare a stable source of funding and can no longer rely on the voluntary funding, because SUDS will need not only initial cost, but also operational and maintenance costs throughout the year to ensure it functions and works effectively.

5. Conclusion: Implementation Challenge of SUDS in Bandung High Density Settlement

High-density settlements, especially that located alongside the riverbank, are known for the high level of flood risk. Many conventional structural measures have been implemented to reduce the risk, but that have not yet demonstrated the best result. For urban areas that are highly populated, such as Cibadak administrative village in Bandung, an innovative solution that refers to natrue-based is needed to mitigate flooding. As stated on the Bandung RTRW 2011-2031, Bandung City has implemented Sustainable Urban Drainage System (SUDS) which is one type of Nature-based Solution for mitigating the impacts of hydro-meteorological hazards. However, this implementation faced several challenges which is explored in this paper. In general, SUDS works by mimicking natural hydrological processes which includes harvesting, infiltrating, slowing, storing, conveying, and treating runoff on site [27]. Some examples of the SUDS that can be implemented in Cibadak admistrative village are green roofs, permeable surfaces, infiltration trenches, filter drains and filter strips, swales - shallow drainage channels, detention basins and purpose-built ponds and wetlands [36]. These examples can be implemented in every different landscape, nevertheless the implementation should be considering the characteristic of each area especially for high density settlement.

Based on the four factors explored in this study, it can be concluded that the adaptability of the scale, budget, and natural characteristic of urban settlement as the challenges in implementing SUDS in high-density settlement. From the scale perspective, various implementations of large-scale SUDS are more difficult to implement, due to limited public land and high land value in Cibadak. Nevertheless, micro or household scale SUDS would be an appropriate option to implement. Meanwhile, since SUDS (especially micro and household SUDS) requires individual or community initiatives, they also facing financial constraints because they mostly relied on voluntary funding. Moreover, based on physical characteristics of Cibadak administrative village, it is found that the site conditions such as flat slope, hydrological nature of riverbanks, climate variability, settlement density, and the lack of vegetation and open spaces make SUDS even more challenging to be implemented.

On the other hand, there are also opportunities for the implementation of SUDS in Cibadak administrative village in terms of its social cohesion and the existence of government support. The residents of Cibadak administrative village are aware of the impact of flooding and are ready to help those who are affected with Gotong Royong mechanism coordinating by *Karang Taruna*. Meanwhile, in terms of planning, there are several regulations of Bandung City that can enable and support the implementation of SUDS in Cibadak administrative village. Besides that, in order to reduce community vulnerability, the city government, along with community organisations have conducted several trainings related to flood management. Furthermore, this also indicate the challenges faced in implementing SUDS was in accordance with NBS implementation as a new approach in global south, such as the biophysical environment and the climate aspect, the risks in the environment, the relationships between human and nature, the policy and governance aspect[5].

Acknowledgments
We would like to thank to funding provided by Riset ITB (Institut Teknologi Bandung): Pengembangan Unggulan 2021 for this research. Besides that, we really appreciated the assistance of RDI staffs and interns who assisted us in providing data and analysing the results. Nonetheless, any errors and responsibilities remain at the authors.

References

[1] Indonesia S. Indonesia Population Projection 2015–2045. Statistics Indonesia, Jakarta 2018.
[2] Ellis P and Roberts M 2016 Leveraging Urbanization in South Asia. World Bank Group 207.
[3] Khan I and Zhao M 2019 Water resource management and public preferences for water ecosystem services: A choice experiment approach for inland river basin management. Sci Total Environ **646** 821–31. https://doi.org/10.1016/j.scitotenv.2018.07.339.
[4] Zhang S, Liu B, Zhu D, and Cheng M 2018 Explaining Individual Subjective Well-Being of Urban China Based on the Four-Capital Model. Sustainability **10** 1–14.
[5] Lechner A, Gomes R, Rodrigues L, Ashfold M, Selvam S, Wong EP, et al 2020 Challenges and considerations of applying nature-based solutions in low- and middle-income countries in Southeast and East Asia. Blue-Green Systems 2. https://doi.org/10.2166/bgs.2020.014.
[6] Wisner B, Blaikie P, Cannon T, Davis I 2004 At Risk: Natural Hazards, People's Vulnerability, and Disasters. Routledge.
[7] UNISDR 2017 Country Assessment Report for Indonesia: Strengthening of Hydrometeorological Services in Southeast Asia.
[8] Apip, Sagala S, Pingping L 2015 Overview of Jakarta Water-Related Environmental Challenges. United Nations University-IAS.
[9] Saut Sagala, Dodon Yamin, Lutfiana D, Ramanditya Wimbardana 2013 Alih Fungsi Lahan Rawa dan Kebijakan Pengurangan Risiko Bencana Banjir: Studi Kasus Kota Palembang. https://doi.org/10.13140/RG.2.1.2763.1206.
[10] BPS Kota Bandung 2020 Kecamatan Astanaanyar dalam Angka.
[11] Badan Pusat Statistik, Jakarta. 2021.

IOP Conf. Series: Earth and Environmental Science **986** (2022) 012055 doi:10.1088/1755-1315/986/1/012055

[12] Rianawati E and Sagala S 2014 Communal Based Flood Mitigation Measures in Bandung City.
[13] Bandung City Spatial Plan 2011-2031. n.d.
[14] Davis M, Naumann S 2017 Making the Case for Sustainable Urban Drainage Systems as a Nature-Based Solution to Urban Flooding 123–37. https://doi.org/10.1007/978-3-319-56091-5_8.
[15] Nurhikmah D, Akmalah E 2016 Pemilihan Metode Sistem Drainase Berkelanjutan Dalam Rangka Mitigasi Bencana Banjir Di Kota Bandung **12**.
[16] Jeffries C, Scotland & Northern Ireland 2004 Forum for Environmental Research. SUDS in Scotland: the monitoring programme of the Scottish Universities SUDS monitoring programme : final report. Edinburgh: SNIFFER.
[17] Uvini Srishantha, Upaka Rathnayake 2017 Sustainable urban drainage systems (SUDS) what it is and where do we stand today? Engineering and Applied Science Research **44** 4. https://doi.org/10.14456/EASR.2017.36.
[18] Jayawardena AW 2015 Hydro-meteorological Disasters: Causes, Effects and Mitigation Measures with Special Reference to Early Warning with Data Driven Approaches of Forecasting. Procedia IUTAM **17** 3–12. https://doi.org/10.1016/j.piutam.2015.06.003.
[19] UNISDR. Terminology on Disaster Risk Reduction 2009.
[20] IFRC. World Disasters Report 2012.
[21] Directorate-General for Climate Action (European Commission). Going climate-neutral by 2050: a strategic long term vision for a prosperous, modern, competitive and climate neutral EU economy. LU: Publications Office of the European Union.
[22] Sulasmi 2014 Analisis Kerentanan Kebakaran Permukiman Menggunakan Sistem Informasi Geografis Di Kecamatan Depok Kabupaten Sleman. skripsi. Fakultas Ilmu Sosial.
[23] Merz B, Thieken AH, Gocht M 2007 Flood Risk Mapping At The Local Scale: Concepts and Challenges. In: Begum S, Stive MJF, Hall JW, editors. Flood Risk Management in Europe: Innovation in Policy and Practice, Dordrecht: Springer Netherlands 231–51. https://doi.org/10.1007/978-1-4020-4200-3_13.
[24] Birkmann J 2006 Measuring Vulnerability to Natural Hazards: Towards Disaster Resilient Societies. United Nations University Press.
[25] Poleto C, Tassi R 2012 Sustainable Urban Drainage Systems. Drainage Systems, InTech. https://doi.org/10.5772/34491.
[26] Griffiths J 2016 Sustainable Urban Drainage. Reference Module in Earth Systems and Environmental Sciences. https://doi.org/10.1016/B978-0-12-409548-9.10203-9.
[27] Woods Ballard B, Wilson S 2015 Udale-Clarke H, Illman S, Scott T, Ashley R, et al. The SuDS Manual.
[28] European Commission 2012 Guidelines on best practice to limit, mitigate or compensate soil sealing.
[29] Goulden S, Portman M, Carmon N, alon mozes T 2018 From conventional drainage to sustainable stormwater management: Beyond the technical challenges. Journal of Environmental Management 219. https://doi.org/10.1016/j.jenvman.2018.04.066.
[30] Kellagher R, Laughlan C. Use of SUDS in high density developments. Guidance manual 2005. https://eprints.hrwallingford.com/550/ (accessed July 21, 2021).
[31] Jiménez Ariza, Martínez, Muñoz, Quijano, Rodríguez, Camacho, et al 2019 A Multicriteria Planning Framework to Locate and Select Sustainable Urban Drainage Systems (SUDS) in Consolidated Urban Areas. Sustainability;11:2312. https://doi.org/10.3390/su11082312.
[32] Barros Ramalho Alves P, Alexandra Alves Rufino I, Hermínio Cunha Feitosa P, Djordjević S, Javadi A 2020 Land-Use and Legislation-Based Methodology for the Implementation of Sustainable Drainage Systems in the Semi-Arid Region of Brazil. Sustainability **12** 661. https://doi.org/10.3390/su12020661.

[33] jabar.tribunnews.com 2020 Banjir Terparah di Astanaanyar Kota Bandung, Tinggi Air Capai 2 Meter di RW 07 Kelurahan Cibadak (Worst Flood in Astanaanyar, Bandung City, Water Height Reaches 2 Meters in RW 07 Cibadak Neighbourhood).
[34] Nóblega Carriquiry A, Sauri D, March H 2020 Community Involvement in the Implementation of Sustainable Urban Drainage Systems (SUDSs): The Case of Bon Pastor, Barcelona. Sustainability **12** 510. https://doi.org/10.3390/su12020510.
[35] Cotterill S, Bracken LJ 2020 Assessing the Effectiveness of Sustainable Drainage Systems (SuDS): Interventions, Impacts and Challenges. Water **12** 3160. https://doi.org/10.3390/w12113160.
[36] Zhou Q 2020 A Review of Sustainable Urban Drainage Systems Considering the Climate Change and Urbanization Impacts. Water 2014;6:976–92. https://doi.org/10.3390/w6040976.

Dynamics types of land use and land cover to determine time of the value erosion coefficient criticism in Keduang Watershed

Ahmad[1,2*] and R Utomowati[1,2]

[1] Environmental Research Center, Sebelas Maret University
[2] Lecturers of Geography Education, Faculty of Teacher Training and Education, Sebelas Maret University

*Email: ahmad-geo@fkip.uns.ac.id

Abstract. This research aims to: (1) Determine the trends in change of the type of plants planted by the community within a year for each type of land use in Keduang Watershed; (2) Understand the dynamics of land cover in each land-use unit in Keduang Watershed; (3) Determine the influential factors of the dynamics land use and land cover in Keduang Watershed; (4) Understand in what circumstances the criticality of erosion coefficient (C value) occurs in Keduang Watershed. The methods used are: (1) analysis of trends in changes of crop types are determined based on the rotation of the plant species planted by farmers in each land use unit; (2) The dynamics of land cover determined by the percentage (%) of the plant canopy area that covers each land use unit; (3) The affecting factor of the dynamics of land use and land cover are determined based on the causative factor of land-use change and land cover dynamics; (4) the time of occurrence of the critical value of the erosion coefficient is determined based on comparing the time of high rainfall and the lowest decrease of land cover. The result shows that: (1) In Keduang Watershed, crop rotation occurs in rainfed-rice fields and moorland; (2) Seasonal change is the main factor for change of land use and land cover; (3) The critical period of the erosion coefficient value in Keduang Watershed occurred in November.

1. Introduction

Land use which is very influential toward the erosive value (C), is the type of use and the percentage of land cover. Plants can intercept the rain that falls so that the raindrops do not directly hit the ground but hit the plants first before falling to the ground. The ability of plants to intercept raindrops is majority determined by the type of plant and the number of plants suitable for the land [1]. Further explained that land covered by forest and open fields would produce different levels of erosion which proved by the runoff produced by the two land uses [2]. Surface runoff on forest land is slightly clear than runoff on the open field. The results of previous research explained that land cover is closely related to the erosion coefficient and *run-off* coefficient [3–6].

Principally land use will reflect the vegetation type that can develop on the land unit, including settlement land with varied yard vegetation. This vegetation type can use as a guide for determining the land cover. Therefore land use is always used as the primary basis in determining land cover in every erosion study [7–9]. It is pretty interesting to note that the type of land use in a particular unit can be the same for one year, but land cover can change from time to time. Some previously mentioned researchers

IOP Conf. Series: Earth and Environmental Science **986** (2022) 012056 doi:10.1088/1755-1315/986/1/012056

assume that land cover for one year is always the same. So, that it is only done once in land cover data collection.

The results of observations made in the Keduang Watershed show that land use and land cover for one year are always dynamic. There are two main things that are always changing, namely: (1) Aspects of land use. This aspect is a form of crop rotation carried out by the community, especially on the use of seasonal cropland; (2) Land cover aspect. This is a form of plant response to seasonal changes, this generally occurs in all types of land use. The dynamics of land use and cover in turn can have an impact on the dynamics of the ability of vegetation to intervene in the rain that falls on the ground. If the first rain falls at the beginning of the rainy season, while the plant growth conditions are in very minimum conditions, then the erosion coefficient value of the plant will decrease from its normal condition. Therefore, even though the land use is the same every year, the land cover will decrease, due to the change of seasons.

As is known from the rainfall data, the climate in the Keduang Watershed is seasonal, namely the dry season and the rainy season. The two seasons change every six months. The rainy season occurs from December to April, while May is a transitional season. The dry season occurs from June to October, while November is a transitional season. The process of changing the two seasons can cause a critical value of the erosion coefficient. The criticality value of the erosion coefficient will occur when rainfall is high while vegetation growth is very minimal. In previous studies, researchers did not consider the dynamics of the erosion coefficient value, so it is not uncommon for erosion prediction results to deviate far from the actual erosion. Such as the result of erosion predictions made during the construction plan of the Wonogiri Multipurpose Reservoir to predict the volume of sediment that enters the reservoir. The forecast shows that it will be full of sediment at the age of 100 years but, it turns out within about 20 years. One of the factors suspected by the researcher is the deviation due to not paying attention to the dynamics of land use and land cover, so it does not consider the time of occurrence of the criticality of the erosion coefficient value. Therefore, this research trying to study the dynamics of land cover to determine the time of the criticality of the erosion coefficient in the Keduang Watershed.

2. Methods

2.1. Site Research

The research location is in Keduang Watershed. The area is selected because Keduang Watershed is one of the catchment areas of the Wonogiri Multipurpose Reservoir and is considered the highest contributor of sediment in the Wonogiri Multipurpose Reservoir. The materials in this research was land use maps and Landsat images, each with a scale of 1: 50.000.

2.2. Population

The population is the change of land cover in each type of land use. Land use in Keduang Watershed was categorized into 6 categories: forest, mixed gardens, irrigated fields, rain-fed rice fields, moorland, and settlements.

2.3. Sampling Technique

The sample in this research was:

2.3.1. A Land Use Sampling.

Land use sampling, used to determine the location of observations of land use dynamics and land cover measurement. The sampling method of land use locations is to classify the types of land use. Then, each type of land use is selected by purposive sampling; Measurement of land cover change did by making a 2 x 2-meter area for each selected land use unit. Then we measure the extent of the plot covered by plants.

2.3.2. Time Of Observation And Measurement Sampling
Sampling time of observation and measurement, used to determine the month of conducting research. Determination of time was base on the peak time of the rainy season, the end of the rainy season, and the beginning of the rainy season. Based on these criteria, the direct research was conduct in January, April, and November.

2.3.3. Collecting and Analyzing Data
Land use dynamics data is carried out directly through field observations. Rainfall dynamics data obtained through secondary data from rain measure stations in all sub-districts in Keduang Watershed. Determination of the time of the critical value of the erosion coefficient by comparing the time of high rainfall and the lowest decrease of land cover.

3. Result and Discussion

3.1. Change of Land Use Type
In general, land use in Keduang Watershed was categorized into 6 categories: forest, mixed gardens, irrigated fields, rain-fed rice fields, moorland, and settlements. The general of Forest is planted with various types such woody trees, such as teak, spruce, silk tree, mahogany, and others. A mixed garden, planted with few types of plants such as cashew, teak, mahogany, cloves. In the use of forest and garden land, there is no change in the types of plants in a year because it is a long-term plant species.

Usually, irrigated and rainfed rice fields used to grow rice. The difference is the irrigated rice fields do not only get water from the rain but also irrigation assistance from drilled wells built by the Wonogiri Regency Agriculture Service. In the use of rainfed rice fields, the water source is pure rainwater. The farming period for irrigated rice fields, usually done twice. The first planting is at the end of December, and the harvest period is from late March to early April. The second planting period is mid-April, and the harvest is in July. After the harvest in July, the irrigated rice fields were left empty until the first planting at the end of December. Therefore irrigated rice fields, there is no change of plants because both planting seasons prioritize rice cultivation. The rainfed rice fields entering the first farming period at the end of December, and the priority is to plant rice. In the second farming period in mid-April, the rainfed rice fields have replaced crops, such as corn and beans, which do not require much water.

The moorland, generally planted with intercropping corn and cassava. The planting period did in early December at the same time as the cassava planting period. The age of the corn plant is three months, so the corn harvest period is usually carried out in March, while the age of the cassava plant reaches 6 months, so the cassava harvest period is usually June/July. Therefore, in the moorland, the crop rotation system is in the first three months of growing corn plants, while in the second three months for cassava plants. Residential land use in rural areas is different from urban settlements. If in the urban areas, usually dense settlements with house buildings, but in the rural, the houses are sparse whereas the yards are wide. The yards are commonly for fruit and vegetable crops or dry fields such as corn and cassava, so it is often called the moor/yard.

Changes in land-use types in Keduang Watershed can be seen in changes of crop types each planting season. Based on the interviews with the community, crop replacement just did on seasonal crop farming activities, such as rain-fed rice fields and dry rice fields. There is no change in the forests and gardens because the plants are annual plants that tend to be long-term plants. In the irrigated ricefield, there is no change in the harvest because the prioritized is rice plants.

3.2. Land Cover Dynamics
Based on the results of field studies, the dynamics of land cover in 6 types of land use in the Keduang Watershed are affected by the change of seasons, namely the dry season and the rainy season. Rainfall and rainy day data show that the rainy season in Keduang Watershed occurs from December to April,

while May is a transition period from the rainy season to the dry season. The dry season occurs from June to October, then November is a transition from the dry season to the rainy season. These conditions can affect the dynamics of the land cover. In the dry season, there can be a decrease in land cover because the water supply in plants decreases, whereas, in the rainy season, the land cover will be maximum because the plants get sufficient water supply. Table 1 and Picture 1 show the dynamics of rainfall data at several stations in Keduang Watershed. Table 2 and Picture 2 show the dynamics of the rainy day at several stations in Keduang Watershed.

Table 1. The Dynamics of Rainfall Data at Several Station in Keduang Watershed

No	Station	Month											
		Jan	**Feb**	**Mar**	**Apr**	**May**	**Jun**	**Jul**	**Ags**	**Sep**	**Oct**	**Nov**	**Dec**
1	Ngadirejo	550	388	438	368	372	90	62	27	25	90	901	896
2	Sidoharjo	719	462	399	337	307	107	69	32	55	75	398	926
3	Jatiroto	738	484	447	440	351	82	0	0	0	60	341	756
4	Slogohimo	805	486	363	304	182	40	0	0	28	68	312	591
5	Jatisrono	1.406	546	407	316	233	202	91	73	0	120	350	577
6	Jatipurno	1.276	824	778	741	761	150	45	0	22	145	966	1.185
7	Girimarto	746	746	573	444	571	80	58	0	45	125	728	671

Source: Wonogiri District Agriculture Departmen, 2017

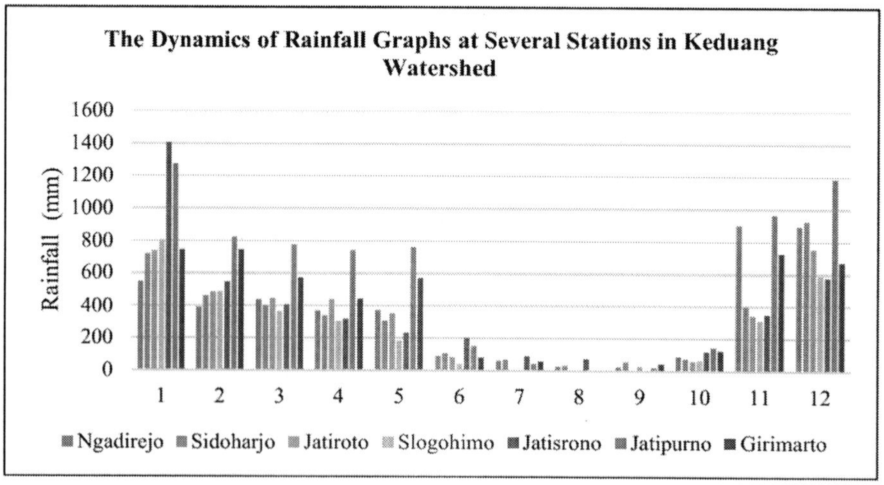

Figure 1. The Dynamics of Rainfall Graphs at Several Stations in Keduang Watershed.

Table 2. The Dynamics of Monthly Rainy Days/ Years at Several Stations in Keduang Watershed

No.	Stations	Month											
		Jan	Feb	Mar	Apr	May	Jun	Jul	Ags	Sep	Oct	Nov	Dec
1	Ngadirojo	11	12	9	5	5	2	1	1	1	3	8	12
2	Sidoharjo	15	16	14	9	6	4	1	1	2	3	14	18
3	Jatiroto	11	13	10	7	5	2	-	0	0	2	9	12
5	Slogohimo	13	13	11	7	5	1	-	0	1	2	5	12
6	Jatisrono	14	12	11	10	7	6	2	1	0	4	6	10
7	Jatipurno	13	13	11	10	7	3	1	0	1	4	11	14
8	Girimarto	13	16	12	9	9	2	1	0	2	4	12	16

Source: Wonogiri District Agriculture Departmen, 2017

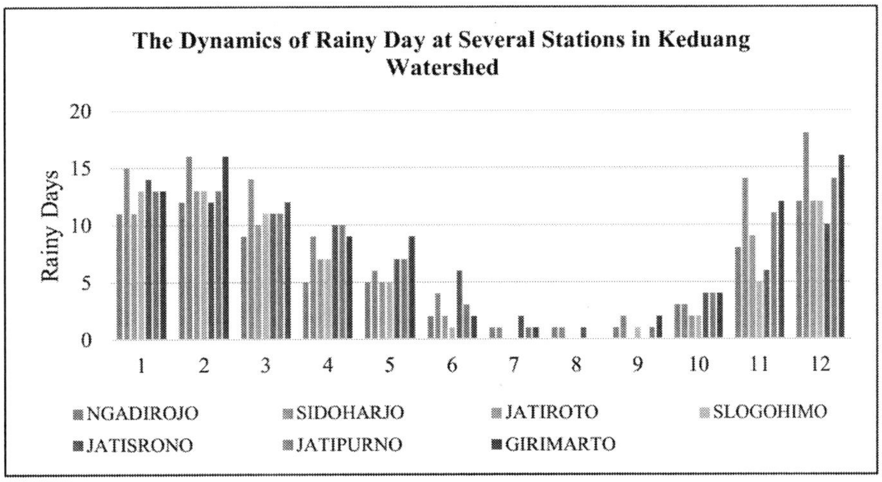

Figure 2. The Dynamics of Rainy Day Graphs at Several Stations in Keduang Watershed.

The results of observations on forest land and mixed garden showed that in January, it grew well because the supply of rainy water was pretty high, so the land cover is still 100% (great). In April is the end of the rainy season, the condition of the forest has not changed because it is still affected by the rainy season. So the land cover has not changed. In November, the circumstances of the forest affected by the dry season, many leaves fall as well as many small plants under it die so that the land cover only left by 80%.

IOP Conf. Series: Earth and Environmental Science **986** (2022) 012056 doi:10.1088/1755-1315/986/1/012056

Figure 3. The Condition of Forest at Keduang Watershed: a) In January, b) In April.

Figure 4. The Condition of Mixed Garden at Keduang Watershed: a) In January, b) In November.

In the first planting period in January, the condition of the land cover of the irrigated rice fields and the rainfed rice field is 100% due to the effect of the rainy season. In April, the land cover of the irrigated and rainfed rice field reduced to be 85% because March is the harvest season period until the planting period in April. In November, the land cover of the irrigated and rainfed rice field reduced to be 60% because, after the harvest season in June, the dry season has entered, so there is no planting until the beginning rainy season in early December.

Figure 5. The Condition of Irrigated and Rainfed Rice Field at Keduang Watershed: a) In January, b) In April, c) In November.

The land cover of moorland in January is 100%, but in April, there was a decrease of land cover to 85% due to the time of corn harvest makes the cassava hasn't grown optimally. The reduction in land cover occurred in November to 60% due to the after cassava harvest in June / July. It enters the dry season, so the land was left empty without plants until the rainy season in December.

Figure 6. The Condition of Moorland at Keduang Watershed; a) In January, b) In April, c) In November.

The land cover on settlement in January is 100%, as well as in April. Fruit and vegetable plants grow well due to the sufficient water availability that still supports them. There was a decrease in land cover in November to 80% because, in the dry season, some of the leaves on fruit trees fell while vegetable crops were not planting due to lack of water. The fruits grow well during the rainy season. At the same time, lower crops such as vegetables start to be replanted.

Figure 7. The Condition of Settlement at Keduang Watershed: a) In January, b) In November.

More clearly, the dynamics of land cover for some land uses in Keduang Watershed can be seen in Table 3. and Figures 8. below.

Table 3. The Dynamics of Land Cover / Month of Land Uses in Keduang Watershed

No	Land Use	Land Cover (%)		
		January	April	November
1	Forest	100	100	80
2	Mixed Garden	100	100	80
3	Irrigated Rice Field	100	85	60
4	Rainfed Rice Field	100	85	60
5	Moorland	100	85	60
6	Settlement	100	100	80

Source: Data Analysis, 2019

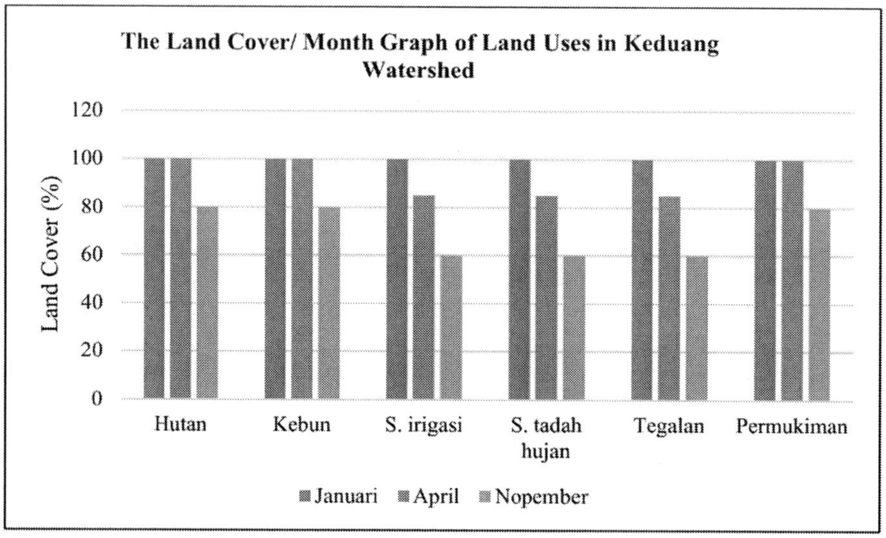

Figure 8. The Land Cover/ Month Graph of Land Uses in Keduang Watershed.

3.3. Time of Occurancer of Critical Value of The Coefficient of Erosion

The time of the critical occurrence of the erosion coefficient value is the indicated land condition by the decreasing ability of vegetation from ideal conditions to control erosion and at the same time rainfall as a crushing force, scraping and transporting soil particles in pretty high conditions. Therefore, the critical occurrence of the value of the erosion coefficient is to compare the month in which the erosion coefficient has declined with the month with high rainfall conditions. The results of the Soil Research Center (1984) and Abdurachman et al. (2002) determined that the value of the erosion coefficient of various land uses in Indonesia should ideally be as shown in Table 4.

Table 4. The Erosion Coefficient Value of Land Uses in Normal Conditions

No.	Land Use Type	C Values (Erosion Coefficient)
1.	Moorland	0,7
2.	Irrigated Rice Field	0,01
3.	Rainfed Rice Field	0,01
4.	Mixed Garden	0,2
5.	Forest	0,001
6.	Settlement	0,7

The determination of the coefficient value is based on the maximal condition of plants. It means that if plants grow well, the land cover will produce an erosion coefficient value for each type of land use, as shown in the previous table. Conversely, if plant growth is not optimal, the land cover will produce a coefficient value higher than the value listed in Table 5.

Table 5. The Erosion Coefficient Value When Occurs Dynamics Erosion Coefficient Value

No	Land Use	Normal C Value	Dynamics of C Value		
			January	April	November
1.	Forest	0,01	0,01	0,01	0,012
2.	Mixed Garden	0,2	0,2	0,2	0,24
3.	Irrigated Rice Field	0,01	0,01	0,0115	0,014
4.	Rainfed Rice Field	0,01	0,01	0,0115	0,014
5.	Moorland	0,7	0,7	0,805	0,98
6.	Settlement	0,7	0,7	0,7	0,84

Source: Data Analysis, 2019

According to the comparison of tabel 5 and table 4, data shows that the critically of the erosion coefficient value in the Keduang Watershed occurred in November. It is possible because in November there has been high rainfall. As is known in November, there was also decrease in the value of the erosion coefficient due to land cover in the dry season not growing maximally.

4. Conclusion

Based on the research as explained, the following conclusions are changes in land use in a year in Keduang Watershed occur only in rain-fed rice fields and dry fields. Seasonal change is a factor that causes land cover dynamics in each type of land use in the Keduang Watershed.Critical values of the erosion coefficient in Keduang Watershed occur in November due to just entered a period of change from the dry season to the rainy season with high rainfall, while vegetation on all land uses experiences a peak in growth decline due to the dry season.

Acknowledgement

The authors would like to thank LPPM UNS for funding the research, Wonogiri District Government Officials, and other parties who have participated in completing this study.

References

[1] Pertanian D I L 2007 PENGARUH MODELING MACAM TANAMAN TERHADAP NILAI EROSI DI LAHAN PERTANIAN Oleh : Pancadewi Sukaryorini 1) dan Moch. Arifin 1) **9** 96–100

[2] Berihun M L, Tsunekawa A, Haregeweyn N, Meshesha D T, Adgo E, Tsubo M, Masunaga T, Fenta A A, Sultan D and Yibeltal M 2019 Exploring land use/land cover changes, drivers and their implications in contrasting agro-ecological environments of Ethiopia *Land use policy* **87** 104052

[3] Chen H, Zhang X, Abla M, Lü D, Yan R, Ren Q, Ren Z, Yang Y, Zhao W, Lin P, Liu B and Yang X 2018 Effects of vegetation and rainfall types on surface runoff and soil erosion on steep slopes on the Loess Plateau, China *Catena* **170** 141–9

[4] Guzha A C, Rufino M C, Okoth S, Jacobs S and Nóbrega R L B 2018 Impacts of land use and land cover change on surface runoff, discharge and low flows: Evidence from East Africa *J. Hydrol. Reg. Stud.* **15** 49–67

[5] Aghsaei H, Mobarghaee Dinan N, Moridi A, Asadolahi Z, Delavar M, Fohrer N and Wagner P D 2020 Effects of dynamic land use/land cover change on water resources and sediment yield in the Anzali wetland catchment, Gilan, Iran *Sci. Total Environ.* **712** 136449

IOP Conf. Series: Earth and Environmental Science **986** (2022) 012056 doi:10.1088/1755-1315/986/1/012056

[6] Tadesse L, Suryabhagavan K V., Sridhar G and Legesse G 2017 Land use and land cover changes and Soil erosion in Yezat Watershed, North Western Ethiopia *Int. Soil Water Conserv. Res.* **5** 85–94

[7] Nut N, Mihara M, Jeong J, Ngo B, Sigua G, Prasad P V V and Reyes M R 2021 Land use and land cover changes and its impact on soil erosion in stung sangkae catchment of cambodia *Sustain.* **13**

[8] Yan R, Zhang X, Yan S and Chen H 2018 Estimating soil erosion response to land use/cover change in a catchment of the Loess Plateau, China *Int. Soil Water Conserv. Res.* **6** 13–22

[9] Rizeei H M, Saharkhiz M A, Pradhan B and Ahmad N 2016 Soil erosion prediction based on land cover dynamics at the Semenyih watershed in Malaysia using LTM and USLE models *Geocarto Int.* **31** 1158–77

IOP Conf. Series: Earth and Environmental Science **986** (2022) 012057 doi:10.1088/1755-1315/986/1/012057

The effect of rainwater injection into dug wells on the changing of ground water level in Kusumanegara Region of Yogyakarta City

A Maryono[1,*], A Prasetya[1], P T S Sembada[2], A S Basyarudin[2]
[1] Department of Civil Engineering, Vocational College, Universitas Gadjah Mada, Yogyakarta 55281, Indonesia
[2]Magister of Technology for the Sustainable Development of Postgraduate Schools, Universitas Gadjah Mada, Yogyakarta 55281, Indonesia

[*]Email: agusmaryono@ugm.ac.id

Abstract. The rapid growth of the Special Region of Yogyakarta makes the water shortage one of the problems that must be solved immediately. The study area is located between Semaki Gede and Muja Muju Village in the Umbulharjo district of Yogyakarta. Declining groundwater levels and droughts has caused an abnormal environmental behavior, as there are still many flood points when the rainy season arrives. Precipitation (mm) between 1976 and 2018 shows that the potential rainfall intensity of the Yogyakarta has not changed significantly. Therefore, it is possible to take advantage of the rainwater potential of an average of 7.23 mm per month, a maximum precipitation of 210.281 mm in November 2017 and a minimum precipitation of level 0 from May to October 2018. The method used is to load or recharge rainwater in the dug well to keep the groundwater level stable. In this study, GAMA Rainfilter rainwater injection technology was used and 10 excavated wells were used, of which 5 uninjected observation and research wells were excavated and 5 excavated injected observation and research wells. Observation of the groundwater level of 10 wells from 2018 to 2019 resulted in an additional production of 0.095 m or 9.5 cm. Changes in groundwater quality were successful and there are several indicators that need to be considered again in the testing methods process.

1. Introduction

The clean water crisis has hit almost all over the world due to climate change, exploration that does not pay attention to sustainability, and several other problems are among the common issues that must immediately find a solution. The lack and inadequacy of water supply which is used as the main medium in the daily life of every human being in developed to developing countries can have an impact on the rate of economic development. Currently, it is estimated that around 1.1 billion people have not had access to safe drinking water since 2002, so that 6,500 children die from diarrhea every day. In general, deaths due to diarrhea in children in the world reach 42,000 cases/week, 6,000 cases/ ay, 4 cases/minute, and 1 death/14 seconds. The number of urban residents has increased sharply, especially in developing countries due to rapid urbanization, this has created many slum areas and minimal access to clean water [1]. Data from BPS [2] from 2013 to 2018 shows that the development of clean water production capacity (lt/s) from Clean Water Companies in each Province continues to increase but is not commensurate with the rate of community growth which annually experiences such a large increase. In line with such a large increase, the construction of tourism facilities and proponent especially in the DI

Yogyakarta region has also increased rapidly. According to BPS data [3,4] In 2018 and 2019, there was an increase in the construction of hotels and other accommodations by 181 new places. This is what triggers the impact of lowering groundwater due to the large number of developments, especially hotels and other accommodation. The increase in development has resulted in a decrease in the quantity or level of groundwater levels in the public area, especially settlements on Jalan Kusumanegara, Yogyakarta City. Then came a movement called "Jogja Asat" which was coordinated by the Empowered Citizens Movement Association in 2014. The emergence of the association was based on a group of community members who refused to build a hotel because they paid attention to the impact that occurred after the hotel operated. Hotels that are supposed to make use of boreholes with deep groundwater sources are in fact converted into exploiting shallow or surface ground water sources. This is what causes public unrest in areas with quite high hotel growth. One of the areas that became the backdrop for the "Jogja Asat" movement was the construction of the Fave Hotel on Jalan Kusumanegara, Yogyakarta City.

On the other hand, widespread drought and a decrease in groundwater levels in many areas, especially in the Special Region of Yogyakarta, have resulted in the emergence of an anomaly of unbalanced environmental behaviour. This is no longer in line with the character of the seasons that occur in Indonesian territory. As a tropical country, Indonesia has a dry and rainy season. Over time, many areas experience drought during the dry season, while during the rainy season they experience flooding. Floods and droughts are currently a major problem in various regions in Indonesia. BNPB data records that in the 2018-2019 period there were 1,064 floods and drought disasters which reached 162 events [5]. Therefor we need alternative water resources, one of which is rainwater. The potential for rainwater in Indonesia is from 2000 mm/year to 4000 mm/year [6] and this potential is high compared to other countries, for example in Europe the average is only 1500 mm/year. Most of the potential rainwater falls in the rainy season from October to April. This results in a water surplus in the rainy season and a deficit in the dry season. The character of high rainfall in the Yogyakarta City area from 1976 to 2018 [7] is presented in Figure 1 below.

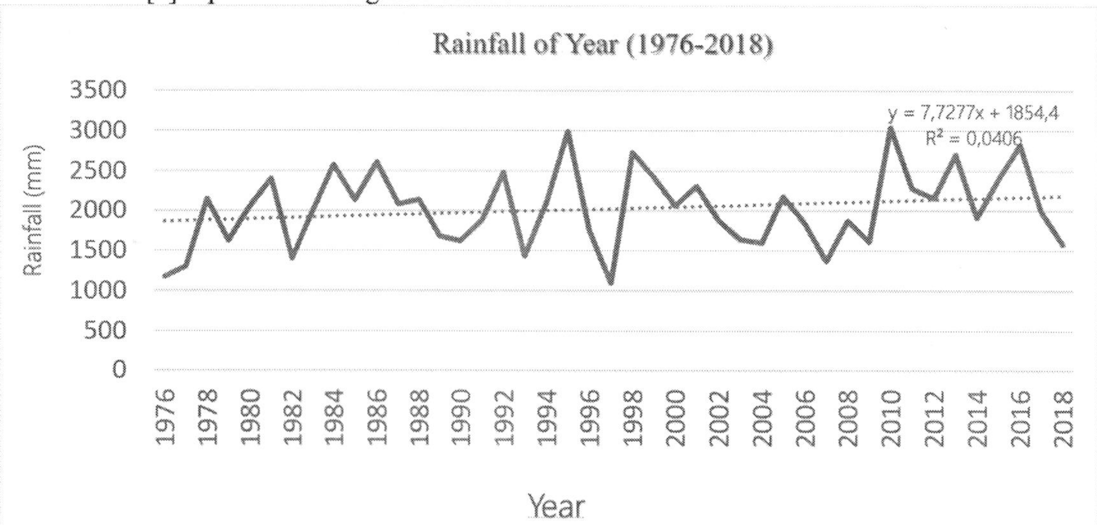

Figure 1. Rainfall intensity (mm) in Yogyakarta City between 1976-2018

The change in the significance of the intensity of rainfall (mm) that occurred during 1976 to 2018 did not show a major change with an average rainfall rate of 169.02 mm, this is indicated by the linear regression value, the results of the analysis of the rain data show the number 0,04 which is less than 1 so it can be said that the potential for rain in the Yogyakarta City area can be an alternative source of water besides groundwater. In addition, the potential magnitude of the availability of rainwater can also be optimized by inserting or recharging it into dug wells so that the stability of the ground water level can be maintained.

There are various ways to overcome the problem of water scarcity, especially in countries around the world by introducing actions to minimize water use. Apart from these actions, several countries have introduced methods of collecting rainwater as part of their water management plans [8–10] The uses that can be fulfilled by utilizing the potential of rainwater are as toilet water, hydroponics, washing vehicles, and other necessities. Rainwater can also be used as drinking water, water to support cooking, bathing, and other activities that often use water as a medium. The use applications of rainwater harvesting that are integrated with everyday life must have a standard of care that has been determined by the central government or local governments. This is done to prevent the unexpected event [11]. The process of applying rainwater harvesters to help provide water for daily needs cannot be separated from the quality that has been indicated by central government regulations and local governments so that it is suitable for use. The characteristics of the quality of rain water that is used as water for daily consumption are as follows:

1. Harvested rainwater is considered high quality water, but in its journey the water still contains contaminants. This happens due to the dissolving process in the air when the rain falls.
2. The resulting effect of contaminants occurs in the early weeks of rain due to the accumulation of sediment, roofing material, animal waste, and other contaminants in the air and media roof.
3. Contaminants with the highest concentration only occur in the first week of rain, when it rains the following week these contaminants have been carried away with the rain in the weeks early so that the quality produced is good [12].
4. Overall, the utilization of roofing media has absorption capacity which provides less contamination than other media. Therefore, the quality of rainwater that falls on the media can be maintained in comparison to the surface infiltration soil.

Several developed and water abundant countries have introduced rainwater harvesting measures because the application of rainwater harvesters has a good profit in helping to meet the needs of water availability for daily life. Germany has been promoting practices rainwater harvesting since the 1980s, and has conducted studies to assess the practicality and feasibility of collecting runoff from surface traffic in densely populated areas [9]. Singapore, on the other hand has focused on creating an extensive network of waterways and canals, aimed at controlling flooding and creating a drinking water supply [13]. The application of rainwater utilization as a support for domestic water for countries in the world is based on the profit and the impact obtained leads to a positive value. This prompted the local government to issue policies that had an impact on the mass application of rainwater utilization and organized. The profit generated from the rainwater utilization policy is based on an assessment of the following elements:

1. The use of the roof of the building as a medium for collecting rainwater which is then used as domestic water for households or industries and offices can reduce water use from 40% to 50% in the buildings.
2. Liaw and Chaing (2014) [14] estimate that the roof collecting approach optimally can supply 32% of the water consumption needs. According to estimates and analysis of water use according to its utilization, 10% will also affect the economic side.

2. Methods

This research was conducted in the Semaki Gede Village and Muja Muju Village, Umbulharjo District, Yogyakarta City. This is because the area is close to the location of the Fave Hotel which was the beginning of the issue of the problem of lowering ground water level (MAT) in Yogyakarta City on 2012. This research process was carried out from 2018 to 2019 with the help of public participation.

2.1 Research Area

This research design utilizes 10 points of research location with a division of 5 points utilizing non-injected wells and 5 Injection well points as well as the map presented in Figure 2 below. Then the types of data used are primary data and secondary data with primary data based on literature reviews or literature studies and data collection from agencies. As for secondary data, it uses observation methods that aim to determine the state of the 10 research points and objective or measurement methods that support the presentation of a map of groundwater flow patterns.

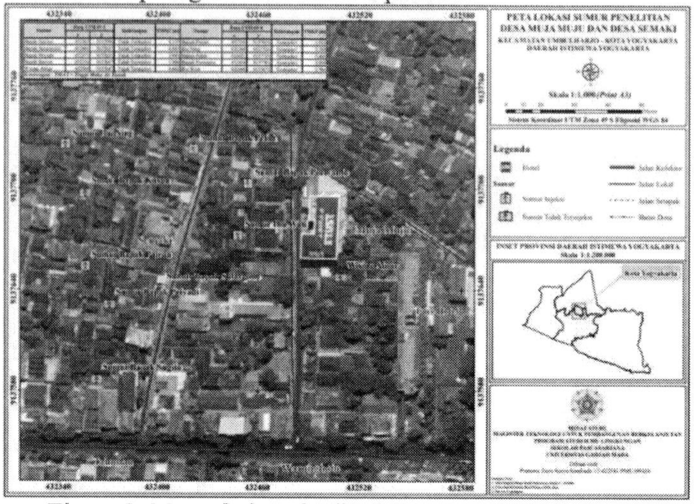

Figure 2. Map of The Distribution of Research Location

2.2 Analysis of Rainwater Availability and Injection

The initial analysis in this study utilizes the calculation of a rational formula used to calculate the amount of rainwater availability [6]. The amount of available rainwater that can be utilized will be correlated with changes in groundwater level (MAT). The following provides a rational formula for determining the volume of rainwater available and managed.

$$V = \alpha \; x \; \beta \; x \; A \; x \; I \; x \; t \tag{1}$$

Where,

V : volume of rainwater collected in a rainwater collecting pond (m^3)
α : run off coefficient
β : coefficient of rain distribution
v : water velocity (m/sec)
I : intensity of rain (mm/hour)
t : Duration of rain (hours), the unit can be converted into seconds
A : The area of the pipe section gutter (m^2)

In addition to utilizing the storage method as in equation (1), this research application utilizes the basis of calculations for the utilization of rainwater injected into the well using pressurized pipes in various models. Based on the Darcy formula, the method of utilizing injected rain water by utilizing a pressurized pipe (injection) can be divided into 5 conditions resulting in different water conservation systems, so that it can be seen which systems are suitable for the conditions and their effectiveness in reviewing the amount of water that can be infused [15]. The following is presented in Figure 3. the system and condition of the pressure pipe (injection) used in this study.

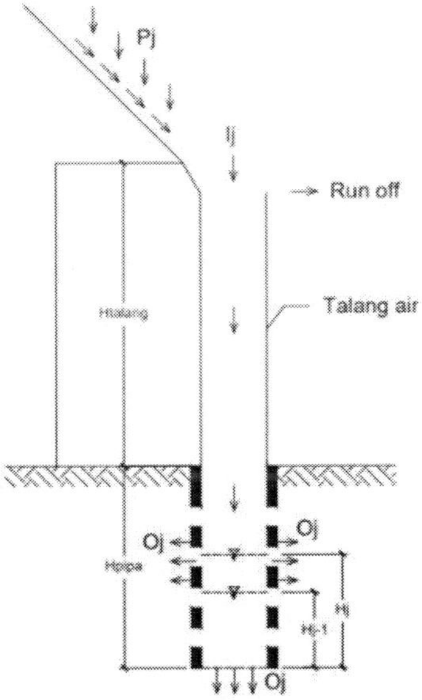

Figure 3. The Condition of the Pressurized Pipe System (Injection)

In this condition, the system has holes at the bottom and walls of the pipe. In addition, at the top of the infiltration system, it is connected to a gutter. With additional guttering, the height gradient hydraulic water and storage volume increase.

$$S_j = S_{j-1} + I_j - 0_j \tag{2}$$

Where,

S_j : infiltration building reservoir at the end of the period j (m^3)
S_{j-1} : infiltration building reservoir at the beginning of period j-1 (m^3)
I_j : volume of entry during period j (m^3)
O_j : volume of discharge during period j (m^3)

2.3 Analysis of Groundwater Quality Change

The method of measuring changes in groundwater quality after the application of the pressure pipe method (injection) uses 2 methods, namely laboratory tests and rational calculations. Water quality test parameters used are limited parameters from chemical and biological tests, including detergent (mg / L), iron (mg / L), calcium (mg / L), TDS (mg / L), pH, and E-Coli. The number of samples taken for tests was carried out laboratory 3 times with the division of time before injection, at injection, and after injection. In addition, the rational calculation of changes in water quality as presented in equation (3) below is used as a control medium in the testing process. So that it will get valid results or close to the calculation results.

$$X_3 = \frac{(V_1\ x\ X_1)+(V_1\ x\ X_2)}{(V_1+V_2)} \tag{3}$$

Where,

X_3 : change in chemical parameters
V_1 : Volume of Well Water (m^3)
V_2 : Volume of Rainwater (m^3)
X_2 : Chemical Parameters of Well Water

X_1 : Chemical Parameters of Rain Water

The combination of the three analysis above which were applied to 10 research points with 1year period of research showed changes in groundwater level fluctuations (MAT) and changes in groundwater quality which were influenced by rainwater utilization using pressurized pipe media (injection).

3. Result and Discussion

3.1 Analysis of Yogyakarta Rainfall (mm)

The amount of rainfall (mm) from 2017 to 2019 in Yogyakarta region averaged per month is 7.23 mm. The maximum value of the rainfall occurs in November 2017 of 210,281 mm and the minimum value occurs from May to October, especially in 2018 when climate change occured. In the dry season, there is no rainfall for average 7 months with rainfall data daily (mm) reviewed from 2017 to 2019. The data analysis comes from data daily rainfall (mm) of Lembah UGM rain station which is not far from the research location.

3.2 Rainwater Availability

Analysis of rainwater availability at 10 research points using equation (1) which is presented in Table 1 is as follows:

Table 1. Rainwater Availability (m3) in Research Area

No	*Name*	*Rainwater Availability (m^3) per week per year*	*Information*
1	Pak Sugito	1,21	non-injected well
2	Pak Sutargono	1,64	non-injected well
3	Pak Priyadi	2,27	non-injected well
4	Pak Kasun	1,08	non-injected well
5	Wisma Ainar	2,69	non-injected well
6	Pak Pairan	1,69	injected well
7	Bu Ning	1,26	injected well
8	Pak Peloh	0,83	injected well
9	Pak Purwanto	2,86	injected well
10	Bu Wati	0,86	injected well

The average magnitude of the potential rainwater by the 10 research observation points is 1.63 m3 per week per year. This becomes the basic data in the process of analyzing the effect of rainwater, especially when it is injected into 5 research points.

3.3 Evaluation of Groundwater Levels Changes

Utilization of pressurized pipe media (injection) in this study to injected into wells obtained the results of the magnitude of the effect on groundwater level (MAT) according to the application of the Managed Aquifer Recharge (MAR) method. Equation (2) shows that the roof reservoir (potential water availability) affects the volume of injection and discharge of water into the dug well, so that there is a change in the groundwater level. The effect of changes in groundwater level due to injection is influenced by several factors, namely:

1. The speed of pumping or the amount of ground water taken.
2. Permeability coefficient (soil passing), type, and soil conditions.
3. Type of pump well.
4. Distance of the pump well from the surface water source.
5. The natural direction of groundwater flow.

Changes in groundwater level in this study were obtained from the combination of the trend of daily rainfall (mm) with Lembah UGM station with groundwater level data taken manually from 2018 to 2019 as in the following graph.

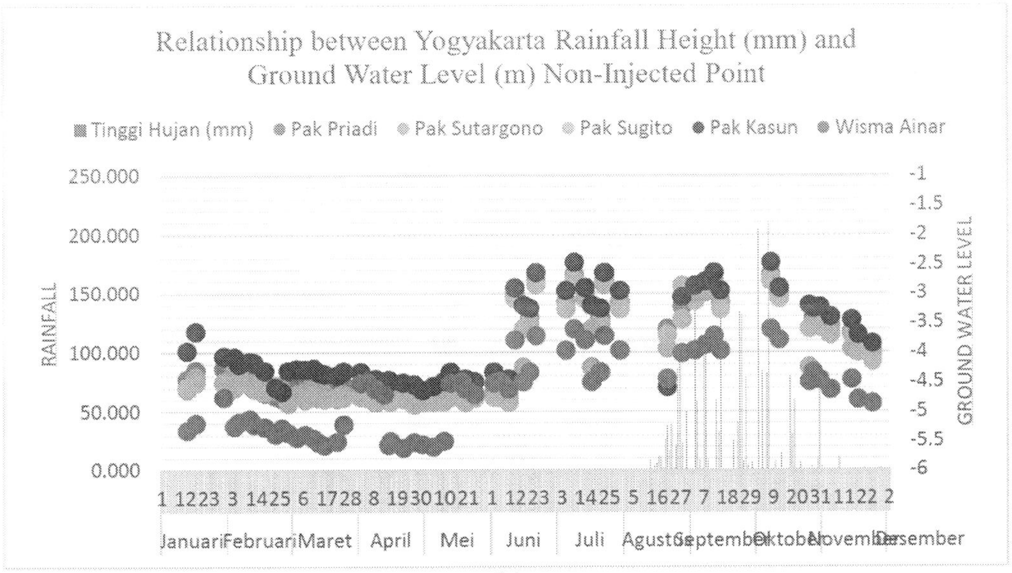

(a)

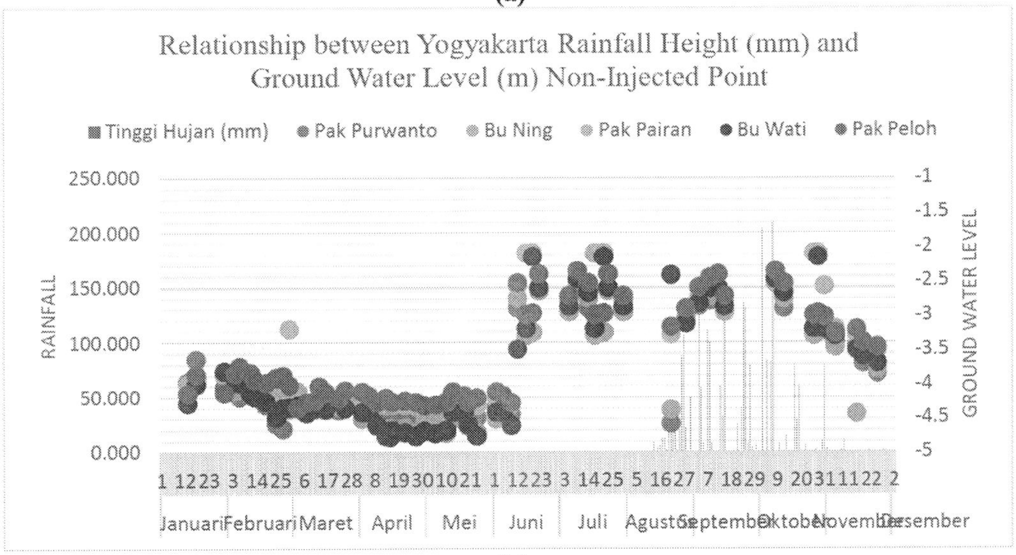

(b)

Figure 4. Pattern the amount of rainfall (mm) with Ground Water Surface (m) with non-injected research (a) and Injected (b)

The fluctuation pattern of the dug wells from the non-injected and injected research observations is presented in Figure 4. It appears that the amount of rainfall (mm) has an effect on increasing the quantity and the groundwater level (MAT). Groundwater flows in a uniform section with the force gravitational obtained from the slope of the groundwater contour [15]. The hydraulic gradient value of each different flownet segment can be used to determine the absolute slope of groundwater between onepoint groundwater to another [16]. Groundwater itself moves in the aquifer at a certain speed so that it has the potential for dynamic groundwater. So that the results of the map of ground flow patterns are presented in Figure 5 below.

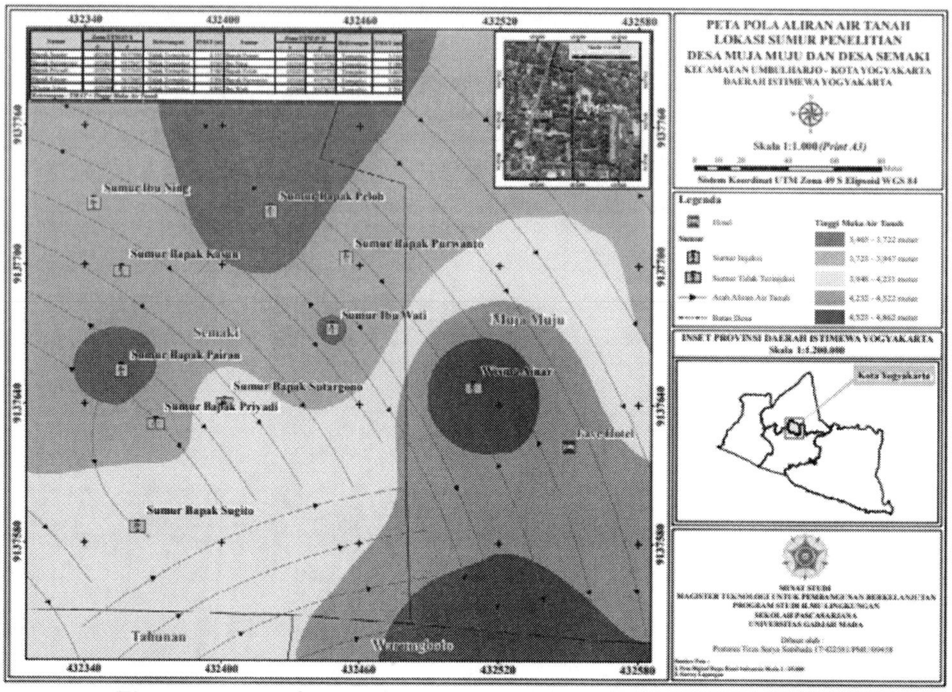

Figure 5. Map of Groundwater Flow Patterns in the Research Area

The groundwater level differences (MAT) that occurred between dug wells research observations in the non-injected with injected one was 0.095 m. The difference in groundwater level (MAT) that occurs is not large due to the short injection time starting from July 2018 to October 2019 and the flat contour of the ground. The selection of a comparison point, namely Pak Priyadi's Point 1 (not injected) with Pak Purwanto's Point 2 (injected) is based on the analysis of the ground flow pattern map which shows that the injected well (Pak Purwanto's well) has a lower contour height difference than the non-injected well. (Pak Priyadi's well) so there is a fluctuation of influence due to injection of rainwater into the well. This is indicated by the increase in the volume of water in the injected well than the volume of water in the non-injected well of 0.095 m as shown in Figure 6 below.

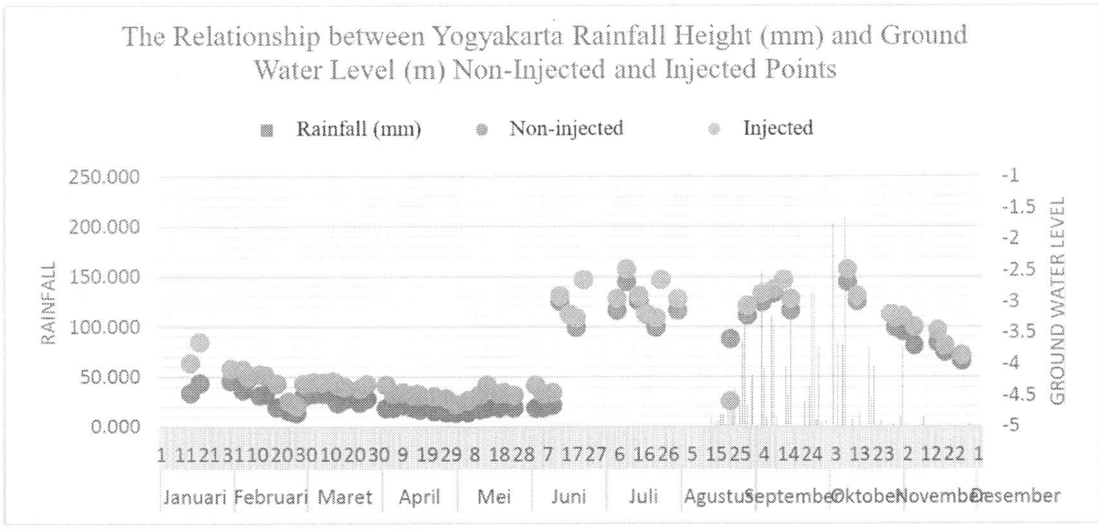

Figure 6. Fluctuation Patterns Between Dug Well Observation Research Non-Injected with Dug Well Observation Research Injected

In addition, this research area is located on two different types of geomorphology, namely volcanic slopes and alluvial plains with relatively gentle to flat topography. The type of soil formed in this research area is Regosol, the result of volcanic eruption material which has a high soil porosity with a high infiltration rate to pass rainwater into the groundwater flow. This is the basis for the phenomena presented in Figure 4 and 6, in which from December to mid-February there was an increase in the groundwater level in the study area while the high rainfall did not occur. This phenomenon is caused by local rain that occurs around the research area, thereby increasing the groundwater level, this is due to the presence of infiltrated rainwater into the groundwater flow and flowing according to the groundwater flow pattern leading to the research area.

3.4 Results of Changes in Groundwater Quality

The analysis of the results of changes in groundwater quality (well water) in the research area was obtained from several tests with limited parameters (chemical and biological) laboratory-based. The process and time testing are carried out before and after the application of the injection method. The following results shows the of the groundwater quality test in Table 2 and Table 3 below.

Table 2. Groundwater Quality Test Results at Non-Injected Points

No	Parameter	Satuan	Bu Ning		Pak Pur		Wisma Ainar		Pak Sugito		Bu Wati	
			Before Injection	After injection	Before Injection	After injection	Before Injection	After injection	Before Injection	After injection	Before Injection	After injection
1	Besi*	mg/L	< 0,0162	< 0,0162	< 0,0162	0,0251	< 0,0162	0,2164	0,0245	< 0,0162	< 0,0162	0,3698
2	TDS	mg/L	213	262	229	169	176	210	211	225	199	206
3	pH*	-	6,7	6,6	6,8	6,8	7,1	6,9	6,7	7,6	6,9	6,8
4	Deterjen	mg/L	0,0632	< 0,0002	0,0733	0,0048	0,0226	< 0,0002	0,0715	0,0178	0,0105	< 0,0002
5	Kalsium*	mg/L	40	43,78	43,2	27,86	32	35,82	35,2	38,21	34,4	33,43
6	E.Coli	Jumlah/ 100 ml	< 1,8	0	< 1,8	0	< 1,8	0	< 1,8	0	< 1,8	0

Table 3. Groundwater Quality Test Results at Injected Points

No	Parameter	Satuan	Pak Peloh		Pak Priadi		Pak Pairan		Pak Kasun		Pak Sutagrono	
			Before Injection	After injection	Before Injection	After injection	Before Injection	After injection	Before Injection	After injection	Before Injection	After injection
1	Besi*	mg/L	< 0,0162	< 0,0162	< 0,0162	0,0251	< 0,0162	< 0,0162	< 0,0162	0,0251	< 0,0162	0,0418
2	TDS	mg/L	236	237	226	227	257	229	267	273	219	271
3	pH*	-	6,8	6,9	6,8	6,8	6,8	7	6,8	7	6,8	6,9
4	Deterjen	mg/L	0,0612	< 0,0002	0,0167	< 0,0002	0,0576	< 0,0002	0,0037	0,0084	0,0671	< 0,0002
5	Kalsium*	mg/L	43.2	42,98	40	36,62	49,6	35,82	49,6	45,37	37,6	44,58
6	E.Coli	Jumlah/ 100 ml	< 1,8	0	> 1600	0	23	0	9,3	0	> 1600	0

There were some changes in the parameter shown in Tables 3.2 and 3.2 above for several elements. There are some elements that have changed for the better but there are also changes that are worse than before. This occurs due to several factors, including the following:

1. At the time of taking water samples, there was a certain moment of storage that was not known by the sample taker at the dug well observation research point.
2. Vulnerability of the testing time for the first and second stages too far so that the results that is expected not match the research design.
3. The sampling process utilizes less clean media, such as when washing the media to maintain the sterility of the media, there are elements or contaminants left in the media that affect the results of the water quality test.

Therefore, it is necessary to test the quality of ground water (well water) so that it is tested again to determine the actual treatment. The process of advanced groundwater quality testing using experimental methods and laboratory observations by utilizing a comparison of the size of the test media. The following shows the results of the groundwater quality test in Table 4.

Table 4. Compilation of Rainwater and Well Water Mixing Test Results

No	*Parameter*	*Satuan*	*Result Air Hujan 1 lt – Air Sumur 2 lt*	*Result Air Hujan 2 lt – Air Sumur 2 lt*	*Result Air Hujan 3 lt – Air Sumur 2 lt*	*Result Air Hujan 4 lt – Air Sumur 2 lt*	*Result Air Hujan 5 lt – Air Sumur 2 lt*	*Value of Clean and Drinking Water Standards*
1.	Zat padat terlarut (TDS) **	mg/L	12,4	10,6	10,5	10,3	10,1	500
2.	pH	-	7,83	7,8	7,77	7,72	7,65	6,5 – 8,5
3.	Kesadahan ($CaCO_3$)	mg/L	11,82	11,82	7,88	15,76	11,82	500
4.	Fosfat (PO_4)	mg/L	< 0,024	0, 420	< 0,024	0, 168	0, 153	-
5.	Nitrit (NO_2^- - N)	mg/L	< 0,001	< 0,001	< 0,001	< 0,001	< 0,001	3
6.	Nitrat (NO_3^- - N)**	mg/L	2,991	1,884	2,491	1,915	2,529	50
7.	Besi (Fe)	mg/L	0,127	0,124	0,125	0,126	0,127	0,3
8.	Deterjen **	mg/L	< 0,002	< 0,002	< 0,002	< 0,002	< 0,002	-

As a control media to determine trends in groundwater quality changes that occur due to mixing of rainwater with ground water (well water), it is necessary to have in-depth studies using theoretical methods to ensure the changes that will occur. The theoretical calculation used utilizes the rational formula equation in mixing 2 types of substances with differences in the quality results they have (3). The composition magnitude of the two test substances in supporting the modelling of the rational formula using the experimental design is shown in Table 3.4 above. The following results of modelling calculations by mixing rainwater with shows the groundwater in Tables 5 and 6 below.

Table 5. Modelling of Mixing Rainwater with Groundwater with Iron Parameter and Detergent

Information	*Besi*					*Deterjen*				
V1	1	1	1	1	1	1	1	1	1	1
V2	1	2	3	4	5	1	2	3	4	5
X1	0,0251	0,0251	0,0251	0,0251	0,0251	0,0048	0,0048	0,0048	0,0048	0,0048
X2	0,151	0,151	0,151	0,151	0,151	< 0,002	< 0,002	< 0,002	< 0,002	< 0,002
X3	0,09	0,11	0,12	0,13	0,13	0,0034	0,0029	0,0027	0,0026	0,0025

Table 6. Modelling of Mixing Rainwater with Groundwater TDS Parameter

Information	*TDS*				
V1	1	1	1	1	1
V2	1	2	3	4	5
X1	169	169	169	169	169
X2	12,9	12,9	12,9	12,9	12,9
X3	90,95	64,93	51,93	44,12	38,92

Theoretically, as shown in Tables 3.5 and 3.6 above, there is a change in the parameters of test results the groundwater quality after mixing it with rainwater using a predetermined composition. The changes shown by the result equation, namely X3, show the character that the greater the volume of rainwater mixed into groundwater at the research point of observing public wells, the greater the changes and the opposite applies. This is directly proportional to the amount of volume with changes in chemical parameters. As a comparison between the results of the quality of groundwater before it is mixed with rainwater, the graph is presented in Figure 7 below.

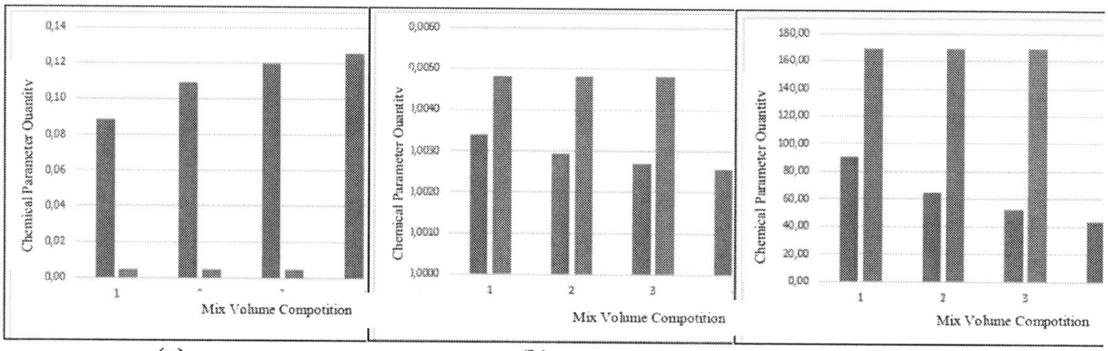

(a) **(b)** **(c)**

Figure 7. Pattern of Change after Mixing Rainwater with Well Water Parameters (a) Iron, (b) Detergent, and (c) TDS; blue for chemical parameters and orange for well parameter.

Theoretically, it is known that there is a change in the quality of well water which is influenced by the volume of rain water that can be mixed. This shows that the application of the injection method rainwater into public dug wells from a water quality perspective is good to be implemented widely and there is a need for rules in local regulations through the local and relevant central governments. Hopefully, flood disasters and water shortages during the dry season can be handled. In addition, the quality of water used by public for their needs is daily appropriate and in accordance with the required standards. A healthy environment will produce a productive society far from disease outbreaks

4. Conclusion

The potential for rain that occurs in the area of Yogyakarta City has a high average monthly rainfall of 7.23 mm with a maximum value of rainfall of 210.281 mm in November 2017 while the minimum value occurs in May to October with a magnitude of 0 due to climate change. The effect of rainwater recharge after rainwater injection technology is 0.095 m or 9.5 cm. Meanwhile, groundwater quality changes at the research location show good results and need to improve the actual testing method in order to obtain values validation that match the cues. This research also supports the achievement of SDG's number 6 concerning Clean Water and Sanitation in the provision of clean water to maintain the quantity of water availability, especially ground water.

Acknowledgments

The author expresses his gratitude for the support from the Ministry of Research, Technology and Higher Education of the Republic Indonesia and Magister of Technology for the Sustainable Development of the Environmental Science Study Program of UGM Graduate School for their support for the implementation of this research activity. Thank are also conveyed to the holding of the ICoDMC 2021 for providing space for the publication of this research.

Reference

[1] Dillon P, Toze S, Page D, Vanderzalm J, Bekele E, Sidhu J and Rinck-Pfeiffer S 2010 Managed aquifer recharge: Rediscovering nature as a leading edge technology *Water Sci. Technol.* **62** 2338–45
[2] BPS 2017 Statistik Air Bersih
[3] Badan Pusat Statistik Daerah Istimewa Yogyakarta 2019 Direktori Hotel dan Akomodasi Lain Daerah Istimewa Yogyakarta 2019
[4] Badan Pusat Stastika 2018 *Direktori Hotel dan Akomodasi Lainnya Daerah Istimewa Yogyakarta* (Yogyakarta)
[5] Badan Nasional Penanggulangan Bencana 2019 Data Informasi Bencana Indonesia (DIBI)
[6] Maryono A 2016 *Memanen Air Hujan* (Yogyakarta: Gama Press)
[7] Karlina A *Evaluasi Perubahan Iklim Tahun 1976-2018 dan Pemeliharaan Konstruksi Pemanen Air Hujan (Studi Kasus Di Dusun Miliran dan Dusun Semaki Gedhe Kelurahan Muja-Muju Kecamatan Umbulharjo Kota Yogyakarta* (Yogyakarta: Universitas Gadjah Mada)
[8] Eroksuz E and Rahman A 2010 Rainwater tanks in multi-unit buildings: A case study for three Australian cities *Resour. Conserv. Recycl.* **54** 1449–52
[9] Nolde E 2007 Possibilities of rainwater utilisation in densely populated areas including precipitation runoffs from traffic surfaces *Desalination* **215** 1–11
[10] Raksanagara A S, Santanu A M, Yusnita S, Sari I, Sunjaya D K, Farisya I, Arya D and Agustian D 2017 Faktor yang Memengaruhi Perilaku Penggunaan Air Bersih pada Masyarakat Kumuh Perkotaan berdasar atas Integrated Behavior Model Integrated Behavior Model : Factors Influencing Clean Water Use among Urban Slum Dwellers *Mkb* **49** 122–31
[11] Medina V 2016 *Rainwater Harvesting* (United State: Technical Report)
[12] Gould J E and McPherson H J 1987 Bacteriological quality of rainwater in roof and ground catchment systems in Botswana *Water Int.* **12** 135–8
[13] Zhang D, Gersberg R M, Ng W J and Tan S K 2017 Conventional and decentralized urban stormwater management: A comparison through case studies of Singapore and Berlin,

Germany *Urban Water J.* **14** 113–24
[14] Liaw C H and Chiang Y C 2014 Framework for assessing the rainwater harvesting potential of residential buildings at a national level as an alternative water resource for domestic water supply in Taiwan *Water (Switzerland)* **6** 3224–46
[15] Harjito H 2014 Metode Pumping Test sebagai Kontrol Untuk Pengambilan Airtanah Secara Berlebihan *J. Sains &Teknologi Lingkung.* **6** 138–49
[16] Sudarmadji, P H and M W 2016 *Pengelolaan Sumberdaya Air Terpadu* (Yogyakarta: Gama Press)

IOP Conf. Series: Earth and Environmental Science **986** (2022) 012058 doi:10.1088/1755-1315/986/1/012058

Food security for drought disaster adaptation in Cireundeu Traditional Village

E Maryani[1*]

[1] Universitas Pendidikan Indonesia

*Email: enokmaryani@upi.edu

Abstract. Food security has always been a crucial issue throughout the ages. Therefore, food security is a target that must be completed by 2030 in the sustainable development program launched by the United Nations. Kampung Cireundeu people have a habit on consuming cassava as a staple food, and it received an award for an independent food village in 2007. The purpose of the study is to determine the geographical factors that influence the people of Kampung Cireundeu to use cassava as food. The data collection method used were interviews, literature study, and documentation. The data analysed and described qualitatively. The results showed that this area had experienced drought almost every year. It also had undulating topography. This condition was not suitable for planting rice, except cassava. This finally prompted the people in Kampung Cireudeu to eat cassava rice, known as *rasi*. They also had several principles as their belief in consuming *rasi*. One of them was "Teu Nyawah Asal Boga Pare, Teu Boga Pare Asal Boga Beas, Teu Boga Beas Asal Bisa Nyangu, Teu Nyangu Asal Dahar, Teu Dahar Asal Kuat" (Don't cultivate rice fields as long as you have paddy. It is ok to not have paddy, as long as you have rice. It is ok to not have rice, as long as you can cook. Do not cook as long as you can eat. Then, do not eat as long as you are strong). Unfortunately, their tradition in consuming cassava began to weaken. It was because the increase of accessibility and higher mobility of the younger generation.

1. Introduction

Food security have been a crucial problem. The United Nations even declares that this problem must be resolved by 2030. Sustainable Development Goals, well known as SDGs, are the development to improve the socio-economic welfare of the community on an ongoing basis by maintaining the quality of the environment (SDGs) [1]. It is also the development that can ensure a justice and good management to improve the quality of life from one generation to the next. Population grows exponentially. On the other side, there is an imbalance between supply and need in food production. This creates a series of social problems, such as hunger, poverty, health, low education. Productivity also becomes low. As a result, circle of poverty occurred [2]. Land as the main asset in producing food, however, continues to narrow due to the increasing number of areas that are built for housing, offices or industry. In addition, the fertility of the land in each area is not the same. Agriculture is a system that is highly dependent on the availability of water, weather, technology and community. Food production becomes unequal, insufficient, and unreachable both in terms of distance and price. Thus, the World Summit on Food Security in 2009, G8 meetings, and the Madrid ministerial-level meeting proclaimed "Food Security for All" at the beginning of 2011 [3].

Foods is material consisting of protein, carbohydrate, and fat used in the body of an organism to sustain growth and to furnish energy [4]. The term food is used for food that has been processed and for food that is technically used, such as food technology, food production, food additives and so on [5]. Staple food is the most often food consumed by the population. It is generally most widely available food and is relatively easy to obtain or cultivate. The type of staple food consumed depends on the place, culture, and individual psychology. It is usually passed down from one generation to the next. Seventy-five percent of the global food supply comes from only 12 plant and five animal species, up nearly 60 percent of calories from just three plants (rice, maize, wheat) [3].

The culture of the community in consuming the staple food has an influence on food security. Indonesian people tend to consume rice. However, in Kampung Cireundeu, cassava is a staple food that has been consumed for generations. The type of cassava here is Karikil cassava, which is regarded as a poisonous cassava [6]. For this action, in 2007, Kampung Cirendeu was awarded "Food Independent Village" by President Bambang Yudhoyono and in 2008 received a food security award certificate [7]. Then, through Cireundeu Gastrodiplomacy in 2019, Kampung Cireundeu got Top 45 Public Service Innovation Information System (Sinovik) 2019 Public Service Innovation Competition [8]. Based on these facts, this paper tries to examine "What are geographical and historical factors supporting people of Kampung Cireundeu to consume cassava as a form of food security?

2. Methods

The purpose of this study is to find out geographical and historical factors that support the community in Kampung Cireundeu Village to consume cassava as a staple food. The approach used in this research was qualitative. There were three types of data. First is primary data, which was obtained from the interviews with traditional leaders (Mr. Y) and people who consumed cassava as a staple food. Second is secondary data. It is gained from map analysis, documentation study, and literature. The last is observation. It was used to see the natural conditions of Kampung Cireundeu and the process of making cassava into various foodstuffs. The collected data were then analyzed descriptively to answer the research objectives.

3. Results and Discussion

3.1. Geographical Condition of Kampung Cireundeu

3.1.1. Location

Administratively, Kampung Cireundeu is a part of Leuwigadjah, Cimahi. The distance from Kampung Cireundeu to urban village is ± 3 KM. It is also 4 KM to South Cimahi District and about 6 KM to the center of Cimahi City. Kampung Cireundeu is located in a valley between Mount Gajah Langu and Mount Jambul in the north. There is also Mount Puncak Salam in the east, Mount Cimenteng in the south, Pasir Panji and Mount Kunci in the west. To go to the settlement, we must pass a small road down for about 500 meters. This road can only be passed by two-wheeled vehicles.

Kampung Cireundeu comes from the words water (*ci*) and the *reundeu* plant (straurogyne elongate) which is widely grown as a medicinal plant and fresh vegetables for the Sundanese. The term "kampung" (village) actually refers to a group of settlements consisting of 50 families or 800 people, who most of them are farmer. The area is about 64 Ha, which are divided into 60 ha for agriculture and 4 ha for settlement. Kampung Cireundeu is placed in one area of 10 Hamlet. This hamlet has 5 neighborhoods. Not all the people in these five neighborhoods consume cassava as a staple food as there is no obligation to do it.

Figure 1. The location of Kampung Cireundeu, Cimahi
Sumber : https://goo.gl/maps/H92zn8zTgjmvgjKM9

3.1.2. Topography
The altitude of Kampung Cireundeu is between + 750 M to + 900 M above sea level. The topography is undulating. Kampung Cireundeu is located in part of the Bandung basin. The settlement is on a slope of 8 to 15%. It forms groups/clusters in relatively flat areas with narrow courtyards. The hilly area has a slope of 15% to 25%, thin soil layer, and a lot of rocks. This shows weathering that has not been perfect (not horizontal). The dominant rock types are Tufa Pumice (Qyt) and Breccia Tufaan Lava, Conglomerate Sandstone (Pb) (Geological Map).

3.1.3. Geohydrology
There are many rivers that flow in Cimahi, such as the Cimahi River, Cisangkan River, Cibabat River, Cibaligo River and Cibeureum River. However, there is no river that passes through Kampung Cireundeu. Aquifers in Cimahi are moderately productive. Shallow groundwater can be found at a depth of more than 10 m. Groundwater in the southern part is widely used for industry. Kampung Cirendeu has a groundwater with the depth of 25 to 35 meters. This groundwater is dug through wells, and is generally used for domestic needs.

3.1.4. Climate
Based on the classification of Smith and Ferguson, the climate in Cimahi is regarded as type A, which is a wet climate. The transition period of the two seasons occurs between April-November. The average rainfall is 40.5 mm/month. The highest rainfall is in April 74.6 mm, while the lowest is in August, November, and December. The average air temperature in Cimahi is 23.3°C. The maximum value occurs in October for about 24.4°C, while the minimum value reaches 22.9°C which occurs in February and December.

3.1.5. Land Use
There are 60 ha that commonly used for cultivation of cassava and other types of plants. Agricultural land spreads in the hills, and rely on rain fed. Pasir Panji, Jambul Mountain, Gajah Langu, Puncak Salam, Mount Cimenteng to the border of Batujajar District are agricultural land with a slope of 15% to 25%. Meanwhile, the area used for settlement is about 4 ha.

3.1.6. Population and Livelihood
Kampung Cireundeu contains of 200 families. There are 270 men and 256 women. Based on the age, there are 21 adults, 283 adolescents in the age of 7-20 years, and 83 children in the age under 7 years.

The educational background of the population is dominantly junior high school graduates. However, there are also 4 people who graduated from college, 65 high school graduates, and 24 elementary school graduates. The people's livelihoods are farming and animal husbandry (54.24%), civil servants/retirees (3.39%), labor/construction (14.12%), and private sector (28.25%). Most of the farmers plant cassava. This cassava is made as a raw material for making *Rasi* (cassava rice). Cassava starch is processed into raw materials for typical Cireundeu foods, such as *opak* and *ranginang* (kind of craker). Meanwhile, the breeders usually raise sheep. The sheep will be sold during Eid al-Adha.

Lingga, P., (1986) [9] says several advantages of planting cassava. Those are: 1) it can grow in dry and less fertile land; 2) its resistance to disease is relatively high; 3) the harvest period is not rushed so that it can be used as a living barn (the leaves and tubers can be processed into various foods when they are left in place for several weeks). These advantages make the availability of cassava more consistent. Thus, it can increase food security of the area. Moreover, Sundari, T (2010) [6] asserts that cassava will grow well in certain conditions, such as : 1) Temperature between 18° -35° C; 2) the optimum rainfall is between 760-1015 mm per year; 3) the types of soil are Alluvial, Latosol, and Podsolic, with an altitude between 10 to 1,500 meters above sea level; and 4) the soil texture is dominated by clay and silt, acidic pH (pH 4.2–4.3), very low organic matter, very low P content, low K content, high-very high Al saturation (35–80%). These condition is in line with the geographical condition of Kampung Cireundeu. Hence, planting cassava is the right choice for the farmers in Kampung Cireundeu.

a. Geographical Condition of Kampung Cireundeu

Cassava is grown from generations to generation. The type of cassava grown is a poisonous local species, called *karikil* [10]. This cassava has a bitter taste, due to the presence of cyanide acid (HCN) with HCN levels 50 mg/kg. The high HCN can cause poisoning for both humans and animals, so it is not recommended for direct consumption [11,12]. To remove the toxins, the cassava needs to be well processed. It has to be peeled and then washed. The clean cassava stems are then put into the grinding machine. After being crushed, the stem is put into a container and mixed with water to be filtered and squeezed. The cassava pulp is dried in order to make cassava rice, while the cassava extract is stored overnight. The water is then removed, and the remains is dried in the sun to be procceed into cassava starch. The starch is used as an ingredient to make crackers, cakes and other foods. This method is in accordance with the opinion of Winarno FG (2004) [13]. He states that to reduce and eliminate toxic levels in cassava, we can do traditional process by peeling the skin before processing, soaking before cooking, and fermenting for several days. Through this process, *linamarin* can be reduced and the hydrogen cyanide was also wasted, leaving about 10-40 mg/kg.

An elder from Cireundeu, Abah Emen, say that the *rasi* (cassava rice) tradition began in 1918. At that time, the agricultural land planted with rice experienced crop failures. The people were threatened food insecurity. The Dutch government did not care about this condition. In that condition, one of the Cireundeu community leaders, Haji Ali, initiated to do the conversion of land. He suggested to plant cassava instead of rice. Since then, the people of Cireundeu have started consuming cassava as a staple food. They have a principle that "*Teu Nyawah Asal Boga Pare, Teu Boga Pare Asal Boga Beas, Teu Boga Beas Asal Bisa Nyangu, Teu Nyangu Asal Dahar, Teu Dahar Asal Kuat* (Don't cultivate rice fields as long as you have paddy. It is ok to not have paddy, as long as you have rice. It is ok to not have rice, as long as you can cook. It is ok to not cook as long as you can eat. Then, it is ok to not eat as long as you are strong). This philosophy becomes a belief that continues to be passed down as a force to continue living even though they do not have rice. Another principle that they still adhere is "Ngindung ka Time, Mibapa ka Zaman" (Prenteding the time as a mother, and the era as a father). This means that a mother must be obeyed, and the customs must continue to be carried out even though the times and era are changing. Adaptation to the times, such as adopting the technology, are followed by the Cireundeu people. They use lighting, TV, radio, and household furniture. The other principle is living harmony with nature. They say that "Cadu Urang Sunda Ngarojong Anu

Ngaruksak Kabuyutan, Saha Anu Wani Ngarumpak Pek Tarima Binasana" (We should not damage anyone's environment. If the environment is damaged, then a lot of death will be happened). These three principles become the local wisdom of the Cireundeu people.

b. Social, Cultural, and Economic Value of Kampung Cireundeu
Kampung Cirendeu likes to be called "Dewitapa", which stands for Tourism Village of Cassava-based Food Security. The consumption of cassava is closely related to geographical conditions and the history of people who are traumatized by the failure of the rice harvest in the past. The trauma is gradually recovering. This is proven by the fact that:
1. Not all Cireundeu people consume cassava. Currently, there are only 2 neighborhoods left.
2. Consuming cassava as an experience-based food sources is passed down from generation to generation.
3. The choice of consuming cassava comes from each individual, self-awareness. There are no punishment if the people violates the custom.
4. Consuming cassava as a staple food is generally carried out if it is in the Cireundeu area. The people may combined it with any other food that is in accordance with their interests and tastes.
5. A free marriage system with outsiders makes the traditions increasingly lost. Some Cireundeu people move to the other area, then some outsiders come to Cireundeu because of marriage. This condition prompt them to consume rice.
6. The tradition are getting weaken. It is caused by the higher mobility and the intensive interaction with people outside Cireundeu. Eating *rasi* and rice is often combined, especially for the outsiders. However, if there are people who want to permanently consume *rasi,* a ceremony should be held. They also should ask the permission to a leader. If it is not done, they will feel constantly hungry.

Kinship system of Kampung Cireundeu is bilocal (may live with bride's and groom's family). This system is basically the same with Sundanese tradition. The difference is that the tradition in Kampung Cireunde still allows the marriage between families, such as the marriages between cousins from both the father's and the mother's side. There is also no prohibition for the adherents of religious beliefs to marry the people with the other beliefs.

Cireundeu people cultivate *karikil* with planting ages between 11-12 months. They use monoculture farming. Some of the harvest is used for family food needs, and the other is sold in the form of flour or various foods. According to one of Cireundeu people, the production of cassava from year to year is relatively stable at around 12.5 tons / hectare. This result is relatively the same as the average production of West Java cassava production in the dry land, which is 12.9 tons/ha-1 [14]. The stability of this production makes Cireundeu people who consume cassava have a guarantee of sustainable food needs. This is in line with the criteria for food needs where the benchmark for fulfilling the food needs includes various aspects, namely: (a) sufficient quantity, (b) good quality, which means safe for consumption, the availability of various types of food, good nutritional adequacy, (c)) spiritual food security, which means food should be appropriate with the religious principles, beliefs and culture of the community, and (d) affordable, which indicates that food is distributed evenly at affordable prices. Regarding the nutrition of cassava, its protein might be lower than rice [15]. However, protein needs can be gained from the other complementary food that are consumed with *rasi.*

People who consume cassava are generally Muslim, although some of them claim to be *Jatisunda* or *Sunda Wiwitan* (163 people). In Kampung Cireundeu, there is annual tradition, called 1 Syuro or 1 Muharram. This tradition is held at Bale Saresehan. On the 1st to 10th of Shura, peoples usually fast in order to be blessed and have a long life. The "Mubur Shura" ceremony which is held every 10th of Shura, has a special dish, red and white porridge accompanied by various fruits, meats and tubers. The ceremony is started by the speech from the elders. They give several advices including to remember and to respect the struggle of the ancestors.

Economically, the price of *rasi* is relatively cheaper than the price of rice. The price of *rasi* is only Rp. 500/kg, while the cassava starch is Rp. 1,500/kg. Cassava starch is an ingredient for making various processed foods such as pastries, pastels, *opak*, and *ranginang*. The price of pastries are varied from Rp. 10,000 to Rp. 15.000,-. One quintal of wet cassava can produce 30 kg of cassava starch, 25 kg of dried cassava, and 80 seeds of *opak*. The cheap price of *rasi* makes the cost of living cheaper. Meanwhile, the selling value of cassava starch and its processed products is relatively more expensive. This causes the consumption of *rasi* as a staple food is more profitable in terms of food security

The potentials contained in the local food system can overcome food problems at the community level. Community-based food systems offer an opportunity for the community to increase their capacity to producing food, and can fundamentally guarantee food security both now and in the future [16]. The diversity of staple foods of an ethnicity (food ethnography) aims to provide recognition of the food habits of certain ethnic groups that are associated with the nutritional conditions of the people [17]. Food security has always been a very crucial issue to human civilizations through the ages. In the last decades there is a growing concern about the impacts of economic development/crisis and population growth on global food security and food production, biodiversity is the key in maintaining food security [3,18].

4. Conclusion

Local wisdom in consuming staple foods in the surrounding environment is a form of adaptation to the natural environment. The sufficient quantity, the ease of access, and the diversity of types of food among communities will form food self-reliance and security. Adherence to local wisdom in the consumption of staple foods needs to be maintained. This is done not only to be involved to the customs, but also to solve the global problem of world food availability.

References

[1] Anon Sustainable Development Goals *https://www.sdg2030indonesia.org*
[2] Todaro M 2000 *Pembangunan Ekonomi di Dunia Ketiga* (Jakarta: Erlangga)
[3] FAO 2012 *WORLD AGRICULTURE TOWARDS 2030/2050*
[4] Dictionary W 2019 Food
[5] Ahdian D 2013 *Definisi Pangan*
[6] Sundari T 2010 Petunjuk Teknis Pengenalan Varietas Unggul dan Teknik Budidaya Ubi kayu (Materi Pelatihan Agribisnis bagi KMPH) *Balai Penelit. Kacang-kacangan dan Umbi-umbian*
[7] Radja A M 2014 Warga "kampung singkong" Cireundeu tak butuh nasi *antaranews*
[8] Febriani R N 2019 Inovasi Ketahanan Pangan Kampung Adat Cireundeu Diganjar Penghargaan *Pikiran Rakyat*
[9] Lingga P 1986 *Petunjuk Penggunaan Pupuk* (Jakarta: Penebar Swadaya)
[10] Anon https://balitkabi.litbang.pertanian.go.id
[11] Kompas 2016 Cara Menghindari Racun Sianida dari Singkong
[12] S S R 1993 *Bercocok Tanam Ketela Pohon* (Jakarta: CV. Yasaguna)
[13] G W F 2004 *Kimia Pangan dan Gizi* (Jakarta: Gramedia Pustaka Utama)
[14] Badan Pusat Statistik 2020 *Jawa Barat dalam angka 2000* (Bandung: Kantor Statistik Jawa Barat)
[15] Saleha Q 2005 Kajian pola dan kebiasaan makan masyarakat Cireundeu di Kelurahan Leuwigajah, Kecamatan Cimahi, Kabupaten Bandung *EPP* **2** 22–8
[16] Fadhilah A 2014 Budaya Pangan Anak Singkong dalam Himpitan Modernisasi Pangan : Eksistensi Tradisi Kuliner Rasi (Beras Singkong) Komunitas Kampung Adat Cireundeu Leuwi Gajah Cimahi Selatan Jawa Barat *Bul. Al-Turas* **20** 13–30
[17] D S 1991 Fungsi sosial dan budaya pangan *Maj. Pangan*
[18] Stefanis C 2014 Global Food Security: An Agricultural Perspective *J. Agric. Sustain.* **6** 69–87

IOP Conf. Series: Earth and Environmental Science **986** (2022) 012059 doi:10.1088/1755-1315/986/1/012059

Attitude and behavior of farmers on the slopes of mount lawu in adapting to climate change (a case study in the Jenawi District of Karanganyar Regency)

E Lestari[1], P Rahayu[2] and A Wibowo[1*]

[1] Faculty of Agriculture, Universitas Sebelas Maret Jl. Ir. Sutami 36 A, Surakarta, Indonesia

[2] Faculty of Engineering, Universitas Sebelas Maret, Jl. Ir. Sutami 36 A, Surakarta, Indonesia

*Email: agungwibowo@staff.uns.ac.id

Abstract. This paper describes the adaptation strategies of farmers on the slopes of Mount Lawu in the face of weather changes, their ways of dealing with the impact of weather changes on the farming sector, and the strategic efforts of farmers in handling the impact of weather changes to overcome the effects of unpredictable climate change on crop production. The research method is qualitative, with an explorative approach, to investigate the local knowledge that plays an important role in shaping the attitude and behavior of farmers in the face of unpredictable climate change. The research location is the district of Jenawi in Karanganyar Regency. The method of data analysis used for the research is a descriptive analysis. The results of the research show that the community living on the slopes of Mount Lawu have local knowledge that serves as a guideline for farmers in this area, and can be seen in their attitude and behavior in dealing with climate change, such as: (1) plant diversification; (2) organic farming; (3) water-saving irrigation; (4) healthy plant cultivation; (5) environmentally friendly pest and disease control; and (6) crop protection and improvement of crops yield.

1. Introduction

Global climate change is an environmental issue that has become a mainstream focus of the international community. The phenomenon of climate change has brought about a number of impacts that have ana significant effect on the sustainability of human life and that of other living creatures. In Indonesia, this phenomenon has caused various changes to the physical and biological systems of the environment, including an increase in intensity of tropical storms, changes in precipitation patterns, sea water salinity, wind patterns, reproductive period of animals and plants, distribution of species, and size of population, and an increase in the frequency of plant pest attacks [1].

One of the more predictable aspects of climate change is rainfall. Rain is an element of climate that is essential for all living creatures and their natural environment. Changes in wet and dry seasons, however, sometimes differ from previous predictions, and this has become a natural phenomenon that is difficult to ignore, and one of the reasons why it is important for this research to investigate the potential of existing human resources. If nature is not renewable, then it is hoped that humans can update technology and carry out necessary adaptations to the changing conditions of nature [2]. The most notable impact of climate change on plantation crops, in particular oil palm, rubber, and cocoa crops, is the decrease in production caused by changes in rainfall and extreme climate events. Drought has a huge effect on productivity and quality of yield of oil palm, rubber, cocoa, sugarcane, and coffee, so it is vital to initiate the implementation adaptation technology for crops. This kind of technology includes developing commodities that survive in drought and flooding stress conditions, implementing soil and crop management technology to improve crop adaptation, developing water saving technology, and implementing water management technology, especially on land that is prone to drought [3]. According to the results of research by [4], since most cultivated land is rainfed, the development of small-scale irrigation is an important step to ensure food security in the future, but farmers' affordability is still the main problem. Soil and water conservation measures are extremely important for maintaining crop productivity and building resilience to climate change.

Climate change is taking place as a consequence of global warming, caused by a global rise in carbon dioxide levels in the atmosphere. Most of the earth's CO2 is now carbon residue [5], primarily due to processes of weathering and biological activity [6]. It began with the increased use of fossil fuels associated with the industrial revolution, leading to a rise in CO2 concentration in the atmosphere of 270-280?? ppm (volume) to its current level of 364 ppm [7,8]. This is an increase of approximately 32% in a relatively short period of time, in terms of a geological and evolutionary time scale. Latest reports show that the potential effect of a twofold or threefold increase in atmospheric CO2 could be enormous. On a short time scale of 100-1000 years, higher CO2 in the atmosphere may affect several of the abiotic attributes of the earth, such as: (1) Temperature (higher average temperatures can lead to the greenhouse effect) [9]; (2) precipitation patterns [10]; (3) acidic precipitation [6]; (4) sea levels (higher sea levels will flood lowland areas); and (5) climate areas (a shift in climate zones will cause a change in the composition of the ecosystem and distribution of plants and animals [10,11].

Research by [2] in Africa shows that the problem of climate change has the potential to thwart efforts of sustainable development if steps are not taken to respond to the terrible consequences that are going to occur, so it is necessary to study the perception and adaptation of farmers to climate change in sub-Sahara Africa. Furthermore, research by [12] in Latin America illustrates that the risk of infection of Black Sigatoka has increased significantly in all banana producing areas of Latin America and the Caribbean, at a rate of 44.2% per pixel from the 1960s to the 2010s. This increased risk is due to climate change, which has caused a rise in temperature conditions for germination and spore growth and created wetter plant canopies.

Research by [13] demonstrates that temperature has a greater impact on banana yield in the Philippines, in comparison with rainfall. Among the temperature variables, seasonal temperature (ST) is shown to have the highest effect on banana yield, followed by annual mean daily temperature (AMDT), and finally annual mean temperature (AMT). With regard to future climate conditions, most countries will not experience a decline in land suitability for banana crops, with the exception of a number of countries that suffer from an attack of Fusarium, a wilt disease resulting from increased rainfall [14]. A study by [15] predicts that climate change will have the same impact on banana production as on other crops. Modern breeding tools, in particular genetic modification and genome editing, can supplement conventional breeding to develop climate smart varieties of banana crops. Research by [16] describes four social and economic factors that influence farmers' perception to climate change in the area of the study, specifically the gender of head of household, types of farming system, agricultural practices, and access to weather information.

On the subject of adaptation to climate change, [17] finds that the effect of general warming and regional climate drying requires the implementation of alternative management practices, such as irrigation, in order to counterbalance the impact of climate change. Research by [18] shows that wide areas of current plantation land will disappear in the future, most likely due to higher temperatures and a shift in agro-ecological zones. Although climate change is allowing new areas to be opened for coffee plantations, especially in the middle west and far west regions of Nepal, there is still the question of how to protect current areas of production and farmers' investments. In this research, we identify one of the possible measures of adaptation to the impact of climate change, using an agroforestry system such as banana intercropping. We identify locations where coffee can be intercropped with bananas. According to a study by [19], in order to adapt to climate change, banana farmers in Uganda must preserve moisture using conventional farming practices such as applying mulch, a shade system, agroforestry, and harvested water, to maintain moisture in banana plantations in most parts of Uganda. In addition, the application of animal manure and fertilizer is needed to ensure that the banana crops receive sufficient nutrition from the topsoil.

In another study by [20], adaptation through crop management and variety substitution, of whatever kind, is highest with a moderate or low level of warming, showing that deep systemic and/or transformational change may be needed in the event that a higher level of warming occurs. The main sources of greenhouse effect emissions on agricultural land associated with banana growing are fertilizer, packaging, and fossil fuels. Results of research by [21] illustrate that the Nigeria Agricultural Research Institution and NGOs are working together with agricultural organizations to find adaptation strategies that can be implemented by farmers through the various value chains (production, processing, packaging, and marketing). Planting drought resistant varieties is one way of dealing with climate change in the scenario of [22], but sometimes drought tolerant cultivars are not able to adapt completely to new environmental conditions, which can also be supported by inoculating them with potential plant growth promoting microbes [23].

A study by [24] demonstrates the effects of climate change, viewed from a social economic perspective, defining it as the primary factor that contributes to the fluctuation of banana prices on the international market, and providing a special evaluation about the importance of extreme weather events. We use a time progression that is no longer trending, for the volume of banana export and price of banana import in France, and apply a qualitative approach to connect the variations observed in this time progression with important events taking place in both the producing country and importing country. We place special emphasis on the relative role of extreme weather events in the price variations observed. The analysis shows that there are a large number of exogenous and endogenous events in various categories that may cause variations in price, under certain conditions. Intense weather events are the most frequent type of negative event reported, and have a direct impact on banana export volume in the producing country, and accordingly in the importing country. In addition, other exogenous events, such as geopolitical problems in the producing country or the importing region can significantly decrease the volume or flow or bananas. A volcanic eruption can drastically reduce export for a certain period of time, and currency changes can also impact banana trade. Endogenous events related to the banana market, such as changing regulations between 2006 and 2009, also have a significant effect on price.

The purpose of this article is to describe the adaptation efforts of individual farmers on the slopes of Mount Lawu in empowering their own potential as means of self-efficacy in utilizing all their skills and ability, which is manifested in the attitude and behavior of the farmers in adapting to climate change. Adapting to changing conditions of nature requires maximum capacity, both internally, of the individual, which is influenced largely by the background of the farmers' social identity, and also externally. The farmers' capacity is developed through a process of counseling, which provides information and knowledge, and teaches the attitudes, skills, and adaptation methods needed to adapt to changes in rainfall pattern, so that they are able to synergize in formulating a strategy. This strategy is to teach the appropriate behavior to farmers and their families, groups and communities, and decision-making institutions, so that farmers have the capacity to adapt.

2. Research Method

The research was carried out on the slopes of Mount Lawu, in Jenawi District, Karanganyar Regency. This area was chosen as the research location because it has been developed into a banana center in Karanganyar Regency, and also because research by [25] has already shown that Jenawi District is suitable for growing bananas. The agroclimate conditions in Jenawi District can be divided based on air temperature and rainfall. The differences in altitude in Jenawi District cause a difference in the agroclimate characteristics of each Land Map Unit (LMU). LMU 1, 2, 3, and 4 are at an average altitude of 2,056, 837, 593 and 399 MSL respectively, with an average temperature of 14.0, 21.3, 22.7 and 23.9 °C and an average rainfall of 3,750, 3,750, 3,250 and 3,000 mm/year. Geographically, the coordinates of Jenawi District are 111°4'54,063 – 111°11'40,01 EL and 7°37'14,103 – 7°31'39,311 SL. This district is situated on the western slope of Mount Lawu, with hilly to mountainous topography, and is dominated by 3 main geological volcanic formations, namely Qvl, Qlla, and Qvjb. The volcanic formation of Lawu Volcanic Rock (Qvl) consists of tuff and volcanic breccia, the Lawu Lava volcanic formation (Qlla) consists of andesite and basalt, and the Jobolarangan Breccia volcanic formation (Qvjb) consists of volcanic breccia, embedded with lava [26]. Jenawi District covers an area of approximately 5,608.28 ha, with 524.14 ha used for rice fields and 5,084.14 ha of dry land. Dry land includes 765.00 ha of private grounds and buildings, 1,991.00 ha of fields and gardens, 1,600.67 ha of state forest, 11.00 ha of grassland, and 611.00 ha of plantations [27].

The research method is qualitative descriptive, with a phenomenological approach, to determine the perception, attitude, and behavior of banana farmers on the mountain slopes in adapting to climate change. The area of study was intentionally chosen to include various locations where bananas had been grown and produced for the target period of 1980-2021. The research uses source triangulation and theory triangulation to collect data using guided interviews. Informants of this research are New Head of Menjing Village, banana farmer, banana traders at traditional market, small entrepreneurs, and external actors that building cooperation with the village. The data collection method included in-depth interviews with key informants, Focus Group Discussions (FGDs), and observation, to supplement the primary data and to capture key factors that determine the attitude and behavior of farmers in adapting to climate change. The data analysis was carried out systematically, with the following steps: (a) looking for data, in the field process with various pre-field preparations, (b) arranging systematically the results of findings in the field, (c) presenting the field findings, and (d) searching for meaning on a continual basis until no other deviation of meaning was found.

3. Research Results and Discussion

Adaptation to climate change is a combination of technical innovations and institutional innovations. In order to protect farmers from the effects of climate change, policies need to be formulated to develop information about climate change, and to share knowledge about management systems. In addition, cooperation needs to be strengthened between the Ministry of Agriculture and other stakeholders, in order to increase farmers' capacity to receive information about agro-meteorology and to ensure that there is an effective mechanism for delivering services, including climate smart extension services.

3.1. Strengthening the adaptive capacity of mountain slope farmers to overcome the impact of climate change

Although banana plants are quite adaptive to the environment, maintenance needs to be done so that they can grow optimally and produce quality fruit. The following are some of the maintenance carried out by farmers at the banana center: (a) Provision of biological agents to prevent wilt disease that generally attacks banana plants such as Fusarium wilt caused by Fusarium fungus. The biological agent that can be used is Trichoderma sp. and Gliocladium sp. (b) Prune dry leaves to prevent disease transmission. (d) Thinning tillers to keep production from decreasing. (e) Treatment of bunches by cleaning the dry leaves around the bunches so as not to injure the fruit and rapping the fruit bunches with plastic or paper wrap to avoid damaging the fruit from pests or scratches. (f) Garden sanitation by cleaning dry leaves, thinning tillers, and removing crop residues that have been harvested once every 45 days. (g) Integrated pest control that is adapted to the type of pest and disease that attacks and is controlled wisely by taking into account environmental conditions.

In principle, adaptation to climate change means minimizing vulnerability, building resilience, and developing the ability to take advantage of profitable opportunities that arise from the situations and conditions caused by climate change [28]. Vulnerability is the extent to which the existence of something is easily damaged, destroyed, or weakened. On the contrary, resilience refers to the ability to design something that will survive, recover, or even develop from a condition created by, or arising as a result of climate change [29]. The table below shows the adaptive capacity of banana farmers on the mountain slopes in overcoming climate change,

Table 1. Strengthening of Adaptive Capacity of Banana Farmers on the Mountain Slopes to Climate Change

Entity of Adaptive Capacity	Strengthening of Adaptive Capacity	Impact
Knowledge in the field of farming to adapt to climate change	Farmers join climate school	Able and willing to implement planting patterns that are adaptive to climate change No damage caused to the environment, and able to lead a decent life No conflict with social life
Farming technology	Adaptation of irrigation techniques Minimizes risk of crop failure due to flooding and or drought	Minimizes risk of crop failure due to flooding and or drought Better availability of water for irrigation enables farmers to implement adaptive planting patterns
Managerial farming skills	Integrated and sustainable farming management	Economically beneficial Increases productivity Farming sustainability
Ability to access climate information	Access to information through meetings with counselors, electronic media, and printed media	Encourages interest in innovation Increases understanding Encourages attempts at innovation Provides motivation to implement technology
Ability to access market	Adjustment of food supply systems and consumption patterns, and increase in planting intensity	Food availability Food stability Food access and utilization
Institutionalization	Builds collective action in activities of water irrigation management and plant pest control Group cohesiveness	Group cohesiveness Builds common perception Builds commitment Builds integration between subsectors Behavioral changes occur that lead to collective action

Data Source: Primary Data through In-Depth Interviews and Focus Group Discussions

The results of this research are relevant to research about the evaluation of the importance of farmers' perception and adaptation to climate change in sub-Sahara Africa conducted by [2] and [30]. Another research report about the coastal area of Bangladesh looks at four strategies of the main components of adaptation, which according to [31] are: (1) level of self understanding in adopting adaptive mechanisms; (2) priority of potential based on interest in the agricultural business; (3) identification of social and economic factors; and (4) recognition of potential obstacles from farmers' reports about how far they have managed to overcome these obstacles. The interaction of these various factors of climate change and the action response of farmers and farming policy makers will determine the future of Indonesian agriculture as well as the subsistence level of the community and the welfare level of the nation [31]. The strategies implemented by governments in Africa are to provide the resources needed to support farmers' adaptation, including credit services, information and counseling to train farmers about strategies and technology for adapting to climate change, and to improve building infrastructure of water, soil, and stations for climate monitoring and reporting [2].

Therefore, the government plays a vital role, especially in developing and accelerating the adoption of farming technology that is more productive and adaptive to climate change, providing agricultural infrastructure that is effective for supporting the application of this technology, developing a climate-farming information network, developing an institution to protect farmers from the negative effects of extreme climate in their farming practices, and to implement an input and output pricing policy that is conducive for improving farmers' income. Furthermore, effective communication about climate change is indicated by increasing awareness, improving knowledge, and encouraging action, so this research also shows the joint actions of farmers and the government, after training, as an endeavor to increase agricultural yield and improve irrigation systems, despite the threats of climate change [32,33].

3.2. Attitude and Behavior of Farmers in Adapting to Climate Change

Climate change has altered the behavior of the farming practices of the majority of farmers working on the mountain slopes. These changes in behavior include adaptive steps, as a way of coping, which aim to preserve their livelihood. The behavior changes are adapted within the cultivation system on the mountain slopes. Changes in behavior occur after being given information about how to increase production and income, and how to improve family welfare. A process of behavior change is essential so that the target of change is not merely the result of acquiring more knowledge, but it is hoped will also bring about a change in farmers' skills and mental behavior, leading to action or work that is better, more productive, and more profitable. The attitude and behavior of farmers on the slopes of Mount Lawu in adapting to climate change includes aspects of crop diversification, organic farming, water saving irrigation, healthy crop cultivation, environmentally friendly pest and disease control, as well as crop protection and increased crop yield, as shown in table 2 below.

Table 2. Attitude and Behavior of Farmers on the Slopes of Mount Lawu in Adapting to Climate Change

Activity	Description
Crop diversification	Farmers plant a variety of crops using an intercropping and or rotation system Use different planting dates Change crop combinations to drought tolerant and short season varieties
Organic farming	Farmers use materials available nearby for making organic fertilizer Farmers use organic pesticide made with local materials
Water saving irrigation	Farmers use a spinner to water crops more effectively and efficiently Banana crops are vulnerable to water logged or flooded soil conditions, so prevention is essential with proper drainage management Optimum water needed for banana crops ranges from 1,300 – 2,600 mm/year, so sufficient soil water content or proper irrigation management can be advantageous for banana crops to accelerate fruit ripening, increase yield, increase fruit weight, and increase number of fruit per bunch
Healthy crop cultivation	Farmers grow local crops that are resistant to pests and diseases, and are suitable for the topography of the area, so they are able to adapt to climate uncertainty Regulations about cultivation correspond to standards of fertilizer use, seeds, soil management, water regulation, and crop pest control More focus on growing bananas in areas that are at a lower altitude and have a higher temperature Proper fertilizer usage
Environmentally friendly pest and disease control	Plant drought resistant varieties Make use of microorganisms (make biofungicides and biocompost) Make use of plants that have potential as biopesticides Physical and mechanical control (heating, cooling, burning, drying, using light traps) Farming control (sanitation, soil management, water management, crop rotation, plant spacing, use of mulch and trap plants)
Crop protection and increased crop yield	Development of heat and drought resistant bananas Farmers are starting to use organic pesticides Use natural enemies to reduce the need for pest and disease control, and also to enhance the immunity of the crops

Data Source: Primary Data through In-Depth Interviews and Focus Group Discussions

In future conditions of climate change, the needs of communities that are vulnerable to the impact of climate change on food supply must be addressed. For example, crop breeding should perhaps place more emphasis on scientifically controlled development of heat and drought resistant banana crops, as well as other crops. Improved irrigation, disease control, and drought resistant banana crops will also increase production under weather conditions that are monitored properly in the area of the research and other similar places. In addition, according to [34], the storage and transportation of bananas from the production location (farm) to towns and supermarkets must be done properly, in controlled temperatures and humidity, to avoid damage that may spoil the quality of the fruit. However, the increasing dependence of the economy and individuals on farming, along with increasingly widespread poverty, inadequate technology, and lack of political support, tends to worsen the effects of climate change in tropical regions. Therefore, there is a huge and urgent need to implement mechanisms that will increase banana productivity and promote the adaptation of communities to conditions of climate change in the future.

According to the research of [35], in Kenya the main impact of climate change is to decrease banana productivity, where the decrease in yield can be viewed from two perspectives: first, low rainfall decreases banana productivity and lowers the quality of yield, producing smaller bunches, while high temperatures cause damage to the banana crops and an increase in pests and diseases, which leads to a lower yield. Second, high rainfall and storms cause damage to the banana crops and new outbreaks of pest attacks and diseases, which leads to a lower yield. However, the results of the research in the banana center in Jenawi show that banana crops are believed to have the ability to survive in conditions of uncertain climate change, and prices remain relatively stable. Therefore, farmers choose banana production as a fixed source of income and as food for their own households because of the perennial nature of the crops and the possibility of a year round harvest. Systematically, the flow of the problem from climate change to a change in attitude and behavior of banana farmers on the mountain slopes in adapting to climate change is presented in figure 1 below.

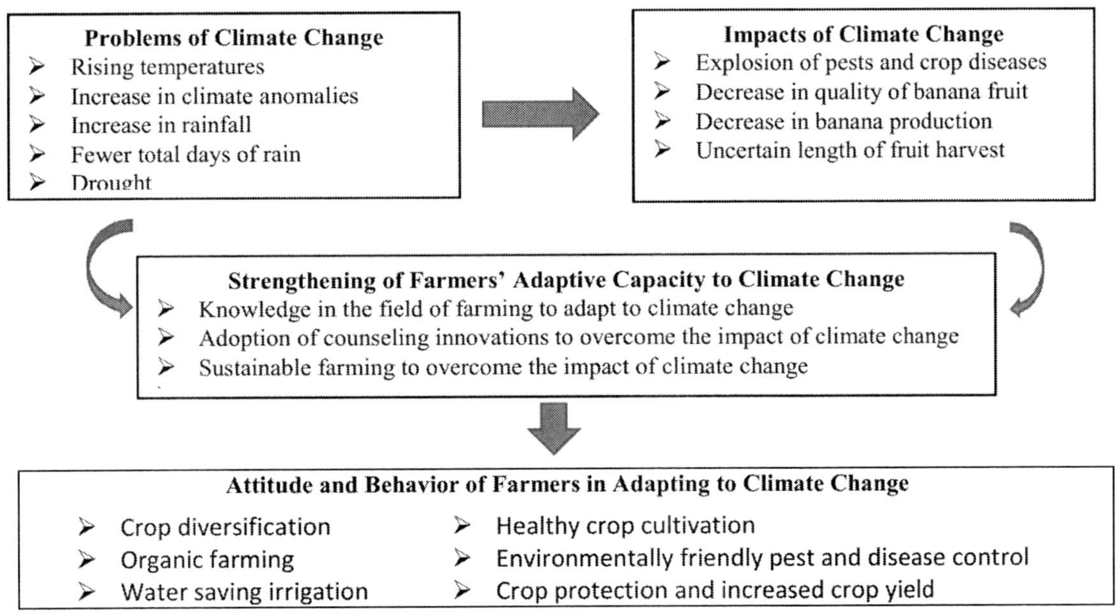

Figure 1. Attitude and Behavior of Banana Farmers on the Mountain Slopes in Adapting to Climate Change

4. Conclusion

The various problems that lead to climate change have a huge negative impact on banana farmers on the mountain slopes. For this reason, it is necessary to facilitate the strengthening of the individual capacity, community level, and enabliung environment level of farmers in adapting to climate change, which ultimately leads to a change in attitude and behavior of the farmers in their adaptation to climate change. This attitude and behavior is shown in: (1) crop diversification; (2) organic farming; (3) water saving irrigation; (4) healthy crop cultivation; (5) environmentally friendly pest and disease control; and (6) crop protection and increased crop yield. This research also proposes the following implications. First, in line with the development of banana crop cultivation in the local area, the local government needs to motivate and support farmers to carry out banana processing so that there is added value of the bananas in this area, by establishing processing plants for banana chips and flour, certification, and marketing of products to increase the welfare of the local community. It is also extremely important to increase the capacity of small farmers to make the most of their opportunities on the product value chain through training about value addition. Second, the government needs to facilitate intensive training associated with climate school, to develop technology for banana cultivation that is responsive to climate change. Third, access to credit facilities is essential for increasing capital and ensuring sustainability in banana cultivation and processing.

References

[1] IPCC 2013 *Climate Change 2013: The Physical Science Basis. Contribution of Working Group I to the Fifth Assessment Report of the Intergovernmental Panel on Climate Change* (United Kingdom: Cambridge University Press)

[2] Juana J S, Kahaka Z and Okurut F N 2013 Farmers' Perceptions and Adaptations to Climate Change in Sub-Sahara Africa: A Synthesis of Empirical Studies and Implications for Public Policy in African Agriculture *J. Agric. Sci.* **5**

[3] Kementrian Pertanian 2011 *Pedoman Umum Model Kawasan Rumah Pangan Lestari* (Jakarta)

[4] Adhikari U, Nejadhashemi A P and Woznicki S A 2015 Climate change and eastern Africa: A review of impact on major crops *Food Energy Secur.* **4** 110–32

[5] Hunten D M 1993 Atmospheric Evolution of the Terrestrial Planets *JSTOR* **256** 915–20

[6] Berner R A 1992 Weathering, plants, and the long-term carbon cycle *Geochim. Cosmochim. Acta* **56** 3225–31

[7] Watson R, Zinyowera M C and Richard H. M 1995 Climate Change 1995 Impacts , Adaptations and Mitigation of Climate Change *Sci. Anal.* **78** 2644

[8] Watkin B 1978 Rewards and penalties. *Nurs. Mirror* **147** 12

[9] Handel M D and Risbey J S 1992 An annotated bibliography on the greenhouse effect and climate change *Clim. Change* **21** 97–255

[10] Smith T M, Leemans R and Shugart H H 1992 Sensitivity of terrestrial carbon storage to CO2-induced climate change: Comparison of four scenarios based on general circulation models *Clim. Change* **21** 367–84

[11] Y K N, Kobak K I and Turchinovich I E 1993 Possible responses of terrestrial vegetation to increasing CO2 concentration in atmosphere and global warming *Lesovedenie* **4** 71–6

[12] Bebber D P 2019 Climate change effects on Black Sigatoka disease of banana *Philos. Trans. R. Soc. B Biol. Sci.* **374**

[13] Salvacion A R 2020 Effect of climate on provincial-level banana yield in the Philippines *Inf. Process. Agric.* **7** 50–7

[14] Salvacion A R, Cumagun C J R, Pangga I B, Magcale-Macandog D B, Cruz P C S, Saludes R B, Solpot T C and Aguilar E A 2019 Banana suitability and Fusarium wilt distribution in the Philippines under climate change *Spat. Inf. Res.* **27** 339–49

[15] Tripathi L, Ntui V O and Tripathi J N 2019 Application of genetic modification and genome editing for developing climate-smart banana *Food Energy Secur.* **8** 1–16

[16] Karienye D, Nduru G and Kamiri H 2019 Socioeconomic Determinants of Banana Farmers' Perception to Climate Change in Nyeri County, kenya David *J. Arts Humanit.* **08** 89–101

[17] Machovina B and Feeley K J 2013 Climate change driven shifts in the extent and location of areas suitable for export banana production *Ecol. Econ.* **95** 83–95

[18] Ranjitkar S, Sujakhu N M, Merz J, Kindt R, Xu J, Matin M A, Ali M and Zomer R J 2016 Suitability analysis and projected climate change impact on banana and coffee production zones in nepal *PLoS One* **11** 1–18

[19] Sabiiti G, Ininda J M, Ogallo L A, Ouma J, Artan G, Basalirwa C, Opijah F, Nimusiima A, Ddumba S D, Mwesigwa J B, Otieno G and Nanteza J 2018 Adapting agriculture to climate change: Suitability of banana crop production to future climate change over uganda *Clim. Chang. Manag.* 175–90

[20] Rippke U, Ramirez-Villegas J, Jarvis A, Vermeulen S J, Parker L, Mer F, Diekkrüger B, Challinor A J and Howden M 2016 Timescales of transformational climate change adaptation in sub-Saharan African agriculture *Nat. Clim. Chang.* **6** 605–9

[21] Elbehri A, Calberto G, Staver C, Hospido A and Skully D 2016 *An economic and biophysical assessment to promote a sustainable and climate-compatible strategy* (Roma: Food and Agriculture Organization of the United Nations)

[22] Adebisi-Adelani O, Adeogun M . and Zaka K . 2014 Farmers' perception of the effect of climate change on plantain and banana production in Ekiti State *Niger. J. Rural Sociol.* **15** 127–34

[23] Kaushal M 2019 *Climatic resilient agriculture for root, tuber, and banana crops using plant growth-promoting microbes* (Elsevier Inc.)

[24] Blake D, Dawson C, Loeillet D and Staver C 2018 Can global climate change affect prices in the world Banana market? *Ger. J. Agric. Econ.* **67** 48–64

[25] Mujiyo M, Widijanto H, Herawati A, Rochman F and Rafirman R 2018 Potensi Lahan Untuk Budidaya Pisang Di Kecamatan Jenawi Karanganyar *Caraka Tani J. Sustain. Agric.* **32** 142

[26] Sampurna and Samodra 1997 *Peta Geologi Lembar Ponorogo* (Bandung)

[27] Anon 2020 *Badan Pusat Statatistik Kabupaten Karangnayar*

[28] Burton I, Malone E and Huq S 2003 *Adaptation Policy Frameworks for Climate Change : Developing Strategis Polices and Measures* (United Kingdom: Cambridge University Press)

[29] ECA 2009 *Shaping Climate-Resilient Development: A Framework for DecisionMaking, A Report of the Economics of Climate Adaptati on (ECA) Working Group, ClimateWorks Foundation, Global Environment Facility, European Commission, McKinsey & Company, The Rockefeller Fou*

[30] Uddin M N, Bokelmann W and Entsminger J S 2014 Factors affecting farmers' adaptation strategies to environmental degradation and climate change effects: A farm level study in bangladesh *Climate* **2** 223–41

[31] Mutekwa V T 2009 Climate change impacts and adaptation in the agricultural sector: the case of smallholder farmers in Zimbabwe *J. Sustain. Dev. Africa* **11** 237–56

[32] Ter Mors E, Weenig M W H, Ellemers N and Daamen D D L 2010 Effective communication about complex environmental issues: Perceived quality of information about carbon dioxide capture and storage (CCS) depends on stakeholder collaboration *J. Environ. Psychol.* **30** 347–57

[33] Wirth V, Prutsch A and Grothmann T 2014 Communicating climate change adaptation. State of the art and lessons learned from ten OECD Countries *Gaia* **23** 30–9

[34] Salau O R, Momoh M, Olaleye O A and Owoeye R S 2016 Effects of Changes in Temperature , Rainfall and Relative Humidity on Banana Production in Ondo State, Nigeria *World Sci. News* **44** 143–54

[35] Karienye D K and Kamiri H 2020 Trends of Banana Production among Smallholders' Farmers Due to Rainfall and Temperature variations in Mount Kenya Region County, Kenya *Budapest Int. Res. Exact Sci. J.* **2** 213–27

IOP Conf. Series: Earth and Environmental Science **986** (2022) 012060 doi:10.1088/1755-1315/986/1/012060

Lora-based IoT sensor node for Real-time Flood Early Warning System

Yoeseph, N.M.[1,*], Purnomo, F.A.[1], Hartono, R.[1], and Nuryani[2]
[1] Informatics Engineering Departement, Sebelas Maret University
[2] Physisc Departement, Sebelas Maret University

[*]Email: nanang.my@staff.uns.ac.id

Abstract. Internet of Things (IoT) using LoRa technology has been widely used in many fields. One of them is in disaster mitigation and management. This paper presents sensor nodes as part of a Flood Early Warning System using LoRa technology. The node consists of a water level sensor and a water flow meter. Water level measurements were carried out by ultrasonic sensors HY-SRF05, and the speed of water flow is measured by water flow meter YF-S201. Communication between the nodes and the gateway is done by the BsFrance Lora32u4 II v1.4 Wireless Transceiver module. Data regarding the height of the water level and speed of water flow will be measured by the sensor and sent by the LoRa device to the gateway. The gateway then forwards that data to an open-source IoT platform for analysis and visualization. Testing ultrasonic sensor readings get an accuracy of 97.06%. Waterflow sensor testing obtained 91.96% accuracy. In addition to testing the measurement accuracy, data transmission consistency is also tested against the distance between the gateway and the node. The test results show that the data can still be sent at a space between the node and the gateway below 400 m. Above 400m, the data cannot be received by the gateway anymore. Testing delay time between reading data and receiving data on the IoT platform shows the average delivery delay of 0 ms.

1. Introduction

The flood phenomenon occurs in areas around rivers or urban areas, creating
an early warning system. A Flood Early Warning System was designed, consisting of sensor nodes, LoRA gateway, cloud server, and user interface application. Sensor nodes are used for measuring the water level and the water flow and can send data for analysis and alerts. LoRa is a proprietary spread spectrum modulation scheme derivative of Chirp Spread Spectrum modulation (CSS) [1].

LoRa is a technology module that is different from others, such as wifi, GSM, Zigbee, and Bluetooth. In addition, LoRa has less power than Mosul GSM/LTE and has a more extended range of up to 10 Km than wifi, Bluetooth, and Zigbee [2]. In Lora-based Internet of Things system, when the sensor on the node detects data, it passes through the gateway and server for data transmission [3]. The End-note, which has many sensors, is placed within the LoRa range to monitor an area. The acquired data is then transmitted wirelessly via a LoRa gateway device connected to a cloud server [2].

Due to those characteristics, Lora technology has been used in several fields, such as agriculture[4][5][6] and weather stations[2][7]. Several Flood Early Warning Systems using LoRa technology had been published [8][9]. This paper will focus on the construction of Lora Sensor Nodes as part of Flood EWS. The Sensor Nodes consist of a water level meter and water flow meter. The experiment is done by testing the accuracy of each sensor. In addition, a latency test is also carried out to test the consistency of data transmission against the distance between the gateway and the node.

2. Methods

The early warning system architecture consists of a LoRa node connected to several sensors, a LoRa gateway, a LoRAWAN network server, and a database. A LoRa gateway is used to bridge the LoRA wireless network to an IP network through wifi and ethernet [10]. The system uses The Things Network (TTN) as a LoRaWAN network server. The Things Network is an open-source infrastructure that provides open tools and networks for building IoT applications. Data from the sensors is then saved in the database through HTTP integration with The Things Network. The EWS system architecture is shown in Figure 1.

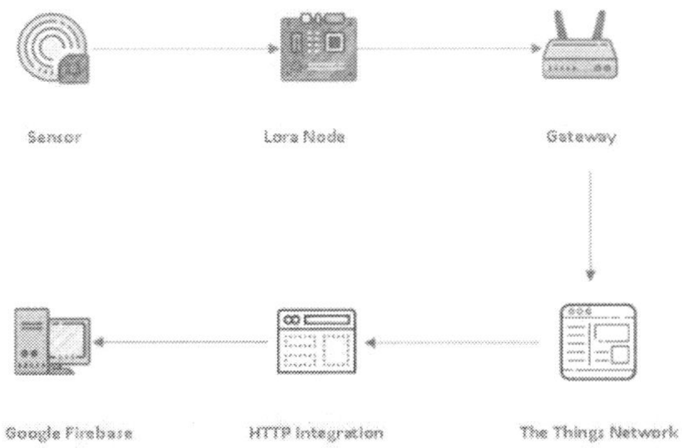

Figure 1. EWS System Architecture

LoRa nodes are built with BSFrance Lora32u4 v1.4 as LoRA wireless module to transmit data from ultrasonic sensors and water flow meters. The ultrasonic sensor is used to monitor water level, and the water flow meter is used to monitor the speed of water flow. The ultrasonic sensor used is HY-SRF05 module. The water flow sensor used is YF-S201.

Ultrasonic sensors work by utilizing waves ultrasonics. Ultrasonic is a wave whose frequency is above the sound wave. Ultrasonic sensors used in this research are the SRF05 module which is shown in Figure 2. This sensor works using a transmitter that emits ultrasonic waves and a receiver that receives back waves ultrasonics. This HY-SRF05 ultrasonic sensor works on a DC voltage of 5 Volts, a current load of 30 mA to 50 mA, with a frequency wave of 40 kHz. This sensor can read distances between 3 cm to 400 cm.

Figure 2. HY-SRF05 Ultrasonic Sensor

Waterflow is a sensor used to measure the flow of water flowing flow. This sensor has a pinwheel inside, connected with an integrated magnetic hall effect sensor that emits an electrical pulse for each round. By counting the vibration from the sensor output, the water flow can be estimated. The water flow sensor model used for this research is YF-S201 which is shown in Figure 3. This sensor model works on a voltage of 5 to 18V DC, with 5V output. This sensor can work with flow speed up to 30 liters/minute, at a temperature of -25 to 80°C and a humidity of 35% - 80% RH. The accuracy of this sensor ranges from 10%. The maximum acceptable water pressure is 2 MPa.

Figure 3. YF-S201 Water Flow Sensor

Ultrasonic sensor accuracy test is done by measuring the distance captured by the sensor with the existing accurate length. As for testing of the water flow sensor is done by testing the current water flow with the amount of water obtained in 60 seconds. Latency is measured by calculating the time between the data read by the sensor and the data saved in the database.

3. Result and Discussion

Testing on the node focuses on sensor accuracy and the time difference between sensor readings and data received by the database. The accuracy of the ultrasonic sensor is determined by comparing the distance data obtained by the sensor with the actual distance. The ultrasonic sensor SRF05 accuracy test result is shown in Figure 4. From the figure, it can be seen that the accuracy obtained from ultrasonic sensor testing is 97.06%.

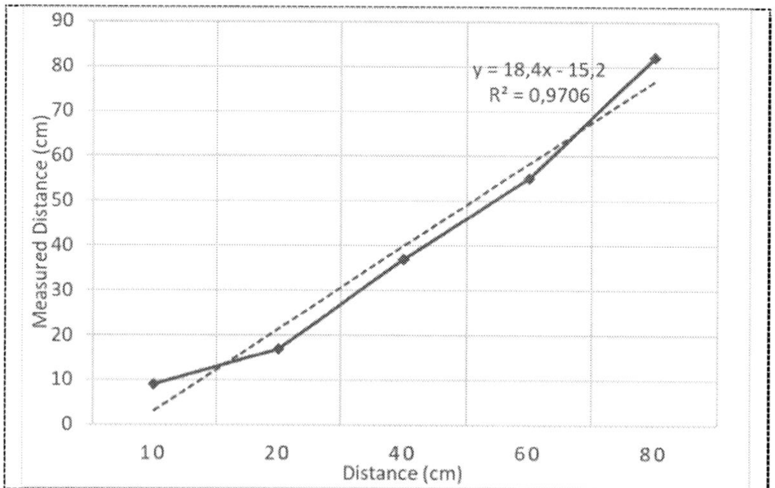

Figure 4. Ultrasonic sensor SRF05 accuracy test result

The testing of the water flow sensor is carried out by connecting it to a faucet that can be adjusted to the amount of water that flows out. The amount of water in 60 seconds is then calculated. The accuracy of the water flow sensor is obtained by comparing the value from the sensor with the volume of water obtained. The water flow sensor YF-S201 accuracy-test can be shown in Figure 5. From the figure, it can be seen that the accuracy obtained from water flow sensor testing is 91.96%.

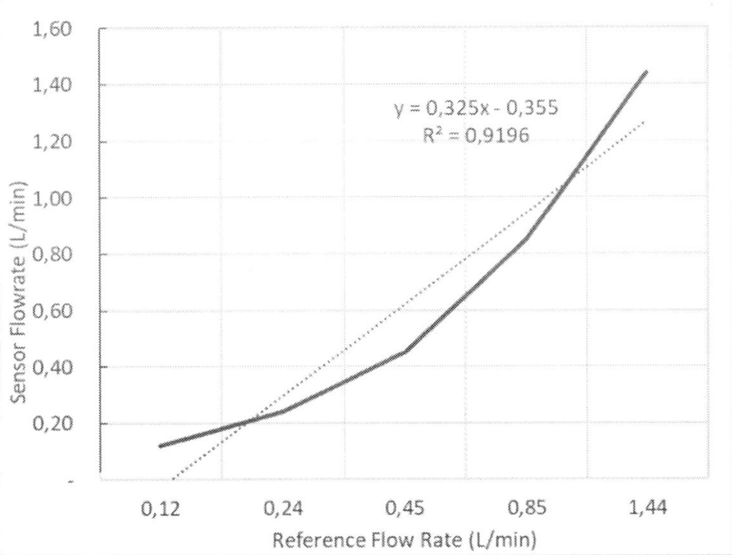

Figure 5. Water flow rate sensor YF-S201 accuracy test result

The latency test calculates the time between the data read by the sensor and sent by the nodes with the data received in the databases. This data latency is essential to know whether data arrive immediately at the database after the sensor has collected the data. Data transmission time is obtained from the time of sensor readings from the Serial monitor. The time of receiving the data is obtained from the timestamp of the data obtained in the Firebase console. Data latency is shown in Table 1. In Table 1, the latency between data transmits, and data receive 0 seconds. Zero-data latency can be called real-time data. The nodes for the Early Warning System can be used to monitor water levels and water flow rates. The transmitted data can be classified as zero-data latency or so-called real-time data.

Table 1. Data latency

Data Transmit (UTC+7)	Data Receive (UTC+7)	Latency (sec)
16:25:56	16:25:56	0
02:30:00	02:30:00	0
02:49:58	02:49:58	0
02:58:34	02:58:35	0
03:11:56	03:11:56	0

4. Conclusions

The conclusion of this paper presented is a manufacture of sensor nodes as part of a Flood Early Warning System using LoRa technology. The ultrasonic sensor readings test gets an accuracy of 97.06%, and water flow sensor testing obtained 91.96% accuracy. The delay time between reading data and receiving data on the IoT platform shows the delivery delay of 0 ms.

IOP Conf. Series: Earth and Environmental Science **986** (2022) 012060 doi:10.1088/1755-1315/986/1/012060

References

[1] Penkov S and Taneva A 2019 Algorithm development with LoRaWAN for monitoring system *IOP Conf. Ser. Mater. Sci. Eng.* **618**

[2] Murdyantoro E, Setiawan R, Rosyadi I, Nugraha A W W, Susilawati H and Ramadhani Y 2019 Prototype weather station uses LoRa wireless connectivity infrastructure *J. Phys. Conf. Ser.* **1367**

[3] Wu J, Tang X and Lei J 2020 Design of wireless monitoring system for greenhouse based on LoRa *J. Phys. Conf. Ser.* **1650**

[4] Purnomo F A, Yoeseph N M, Bawono S A T and Hartono R 2021 Development of air temperature and soil moisture monitoring systems with LoRA technology *J. Phys. Conf. Ser.* **1825** 1–5

[5] Jin J, Ma Y, Zhang Y and Huang Q 2018 Design and implementation of an Agricultural IoT based on LoRa *MATEC Web Conf.* **189**

[6] Tathod S 2018 Lora Based Smart Agriculture 1616–21

[7] Putra I, Rajibussalim R, Irwandi I and Muhammad S 2020 Utilizing LoRa for IoT based mini weather station as STEM learning media to support industrial revolution 4.0 *J. Phys. Conf. Ser.* **1470** 0–6

[8] Leon E, Alberoni C, Wister M and Hernández-Nolasco J 2018 Flood Early Warning System by Twitter Using LoRa *Proceedings* **2** 1213

[9] Ragnoli M, Barile G, Leoni A and Ferri G 2020 An Autonomous Low-Power LoRa-Based Flood-Monitoring System

[10] Salih T A and Noori M S 2020 Using LoRa Technology to Monitor and Control Sensors in the Greenhouse *IOP Conf. Ser. Mater. Sci. Eng.* **928**

IOP Conf. Series: Earth and Environmental Science **986** (2022) 012061 doi:10.1088/1755-1315/986/1/012061

GIS application for spatial analysis of public health centres in response to Covid-19 pandemic

Pranichayudha Rohsulina [1], Agung Hidayat [2*], MS. Khabibur Rahman [1], Talitha Rahmawati [1], Bayu Kurniaaji [1]
[1]Geography Education Study Program, Universitas Veteran Bangun Nusantara, Sukoharjo – Indonesia
[2] Universitas Sebelas Maret, Surakarta

*Email: agung.geographer@gmail.com

Abstract. The COVID-19 pandemic has made many people aware of the real condition of the national health system. The efforts to handle the pandemic occurred simultaneously throughout the territory of Indonesia that had never been thought of before. The Ministry of Health through the Health offices is working hard to tackle the Covid 19 outbreak. Public Health Centers are the technical implementing units in each region that have duties in carrying out promotive, preventive, curative, and rehabilitative efforts. Problems arise when the service standards of the public health center are not the same in all regions of Indonesia. On the other hand, the reachability of health facilities is still a problem in some places. In this study, we are interested in mapping the pattern and reachability of level one health facilities in Nguter Sub-District, Sukoharjo Regency, Central Java, Indonesia using a geographic information system (GIS). This study would be spatially analysed by using nearest neighbour method in ArcGIS. The evaluation of reachability was done by using a minimum service standard by the Decree of the Minister of Settlement and Regional Infrastructure No. 534/KPTS/M/2001. The results showed that the mapped health facilities in Nguter Sub-District were 2 Public Health Centers and 3 Sub-Public Health Centers. Based on the analysis of the nearest neighbors, the public health center in the Nguter Sub-District show a dispersed pattern. From the results of buffering, it is known that almost the entire area of Nguter Sub-District has covered by the Public Health Centers and Sub-Public health centers.

1. Introduction

For more than a year, the world including Indonesia has been struck by the Corona Virus Disease 19 (Covid 19) pandemic. The deadly virus has infected more than 183 million people and caused the deaths of nearly 4 million people (Figure 1). In Indonesia, Covid 19 has infected more than 2.3 million people and caused the death of more than 61,000 people (Figure 2) [1]. The impact of Covid 19 is very broadly felt by the population, both in health sector and other sectors such as economy and social.

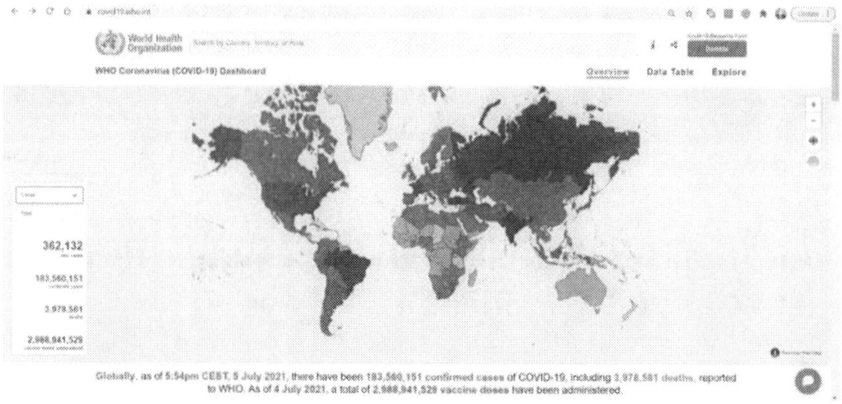

Figure 1. The Global situation of confirmed numbers and deaths from Covid 19 [2]

The spread of the virus is very fast due the transmission model between humans who have close interaction with the infected people of covid-19 virus (Figure 2). To date, epidemiologists have not succeeded in finding a vaccine that can prevent the spread of the corona virus. The only way to prevent the spread is to take preventive measures such as frequent washing hand, wearing masks, maintaining distance and avoiding crowds.

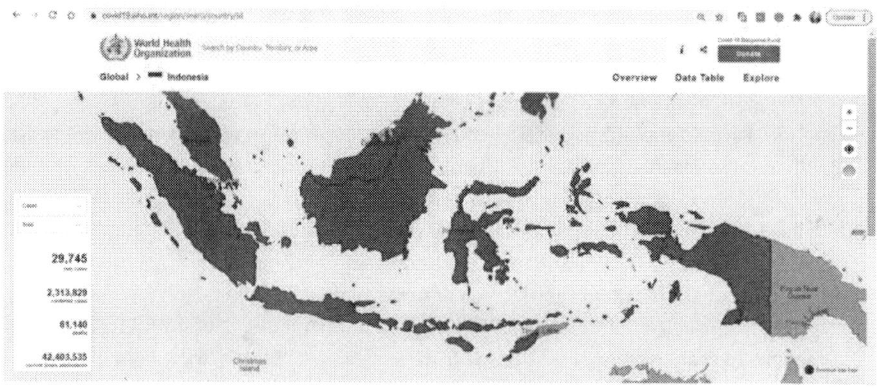

Figure 2. The cases of Covid 19 in Indonesia [3]

Since it was first announced by the Indonesian President where two Indonesian citizens were positively infected by covid-19, the government immediately took action to handle it. One of them was by activating the referral hospitals for covid 19 as well as the community health centres as the spearhead in handling covid 19 cases in sub-district level.

The community health centre (which is later called *Puskesmas*) as an extension of the District Health Office plays an important role as the Level 1 of Health facility,who has authorization to handle covid 19 outbreak. *Puskesmas* has many tasks, especially during this Covid-19 pandemic. According to the Technical Guidelines for Public Health Centers during Covid 19 Pandemic issued by the Ministry of Health, Health Center or *Puskesmas* has duties to provide education about COVID-19, to conduct tracing on the close contacts of affected people, to monitor self-isolated patients, to refer COVID-19 patients to referral hospitals, to play an active role in socializing the prevention efforts, and to collect the data of the suspected people of covid 19. If the suspects experience COVID-19 symptoms, they will be referred to referral hospital to get further treatment. Besides, the health center also becomes the first line in vaccinating people as the effort to tackle the outbreak. These additional functions can be interpreted as additional workload for them that makes the responsibility of health center is drastically increasing, and worsely some of their staffs are also affected by covid 19.

By considering the strategic function of the first-level health facilities in coping the covid 19 outbreak in both of preventive and curative measures, we are interested to conduct a study on the existence of health center and sub-health center in Nguter District, Sukoharjo Regency related to the distribution pattern and the reachability for the community. The mapping and the reachability analysis was carried out using Geographic Information System (GIS) technology [4] in which this technology has helped researcher in mapping and spatial analysis activity [5–12].

The evaluation of the reachability will refer to the Guidelines for Minimum Service Standards for Settlements through the Decree of the Minister of Settlements and Regional Infrastructure No. 534/KPTS/M/2001 for the coverage of the district/city area unit. Hopefully this evaluation can give information about the spatial distribution and the reachability of health center in Nguter Sub-district, Sukoharjo Regency.

2. Method

This section will discuss about the object of study area, research tool and material, variable, data collection, data analysis, and research output.

2.1 Study Area and Object of study

This study was conducted in Nguter sub-District, Sukoharjo District. Nguter consists of 16 villages namely Lawu, Baran, Nguter, Gupit, Pengkol, Jangglengan, Tanjungrejo, Serut, Juron, Celep, Plesan, Kedungwinong, Daleman, Kepuh, Pondok, and Tanjung (Figure 3). The object of this study was the service of health facilities in Nguter sub-district.

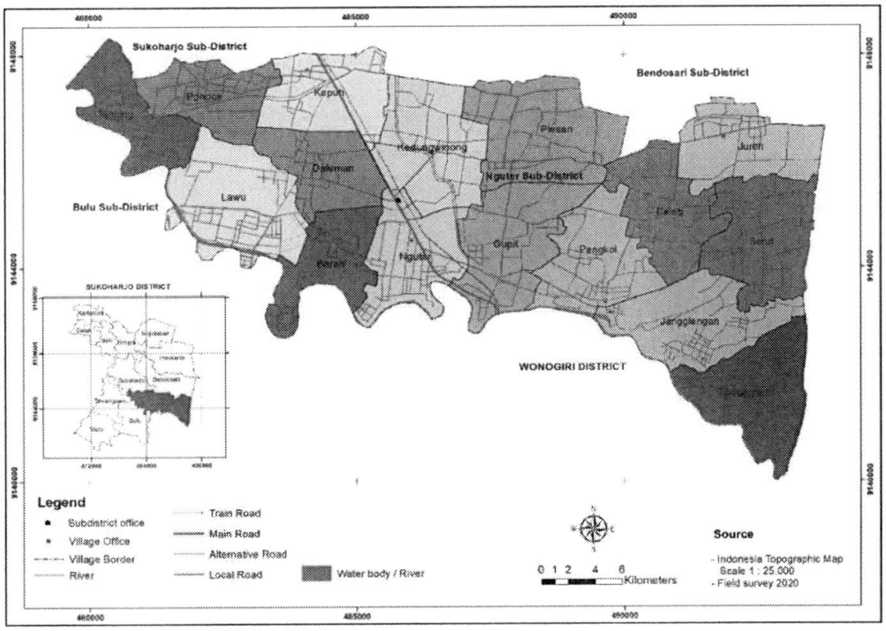

Figure 3. The administration Map of Nguter Sub-District in Sukoharjo

IOP Conf. Series: Earth and Environmental Science **986** (2022) 012061 doi:10.1088/1755-1315/986/1/012061

2.2 Research tool and material
This study used the application of Geographic Information System (GIS) to analyse the distribution and the reachability of health facilities including:

a. The administration map of Nguter that shows the boundaries areas
b. Topographical Map of Nguter sub-district
c. Road and river network data

The tools used in this study includes:

a. Global Positioning system (GPS) Garmin
b. Digital Camera
c. Field survey sheet
d. Stationary
e. Laptop installed with ArcGIS, Google Earth, Microsoft Office, and Corel Draw.

2.3 Data Collection
The information about the location of health facilities in study area was collected from Sukoharjo Statistical Bureau followed by field survey using GPS (Figure 4).

Figure 4. The field survey in collecting coordinate points of health facilities

2.4 Data Analysis
The data obtained from the field survey and imagery interpretation was then analysed in order to produce the map of health and education facilities. The steps were followed:

a. Extracting the data of road and rivet network as well as administrative boundaries from Indonesian topographical map
b. Input the location data based on field survey using GPS to ArcGIS
c. Processing the data through "reattributing" data of health facilities in Nguter sub district
d. Analyzing the spatial pattern of health facilities distribution using nearest neighbor analysis. This analysis is used to determine the pattern of a group of objects in certain region whether following random, clustered, or dispersed pattern that is showed from the value of Rn. Nearest neighbor analysis considers 3 aspects which are distance, the number of location points, and area.

The Rn value or the nearest neighbor distribution index is obtained from below formula:

$$Rn = \frac{D\ (obs)}{0{,}5\sqrt{\frac{a}{n}}} \tag{1}$$

Rn : Nearest Neighbor value
D (obs) : The average distance of neighbor observation
a : Area
n : The number of points (location)

Then, the result of Rn value will be interpreted using *Continum Nearest Neighbor Analysis* as pictured below:

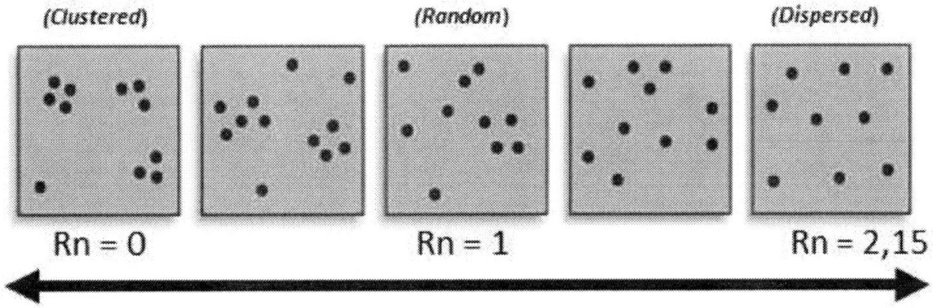

e. Analyzing the reachability of health facilities using buffering method and comparing with the number of populations within the scope of health facilities.

3. Result And Discussion

3.1 Description of study area

Nguter is one of sub districts in Sukoharjo Regency, Central Java [13]. It is located in the south east of the capital regency with approximately 8,2 km in distance. The area is around 5.488 Ha or about 11,76% of total area in Sukoharjo. To the north is bordered by Bendosari and Sukoharjo sub-districts and to the south by Bulu sub-district and Sukoharjo district. To the east is bordered by Karanganyar Regency and to the west is bordered by Sukoharjo, Tawangsari and Bulu Districts. Most of the region is plain with the average altitude 104 mdpl.

Administratively, Nguter is divided into 16 villages with the activity center is in Nguter Village as described in Table 1.

Table 1. The villages and the area in Nguter sub district

No	Village	Area (ha)	%
1	Lawu	436	7,94
2	Baran	252	4,59
3	Nguter	325	5,92
4	Gupit	392	7,14
5	Pengkol	365	6,65
6	Jangglengan	380	6,92
7	Tanjungrejo	356	6,49
8	Serut	390	7,11
9	Juron	320	5,83
10	Celep	292	5,32
11	Plesan	431	7,85
12	Kedungwinong	392	7,14
13	Daleman	268	4,88
14	Kepuh	395	7,20
15	Pondok	253	4,61
16	Tanjung	241	4,39
	Total	**5.488**	**100**

Source: Sukoharjo Statistical Bureau, 2020

3.2 Landuse

Nguter is entirely rural area where the land is generally used and classified as agriculture and non-agriculture area. The agriculture area includes 2.418 Ha while the non-agriculture area includes 3.070 Ha [13]. The non-agriculture area consists of yard, moor, and others. The land use of each village is presented in Table 2.

Table 2. The area of land use in each village in Nguter

No	Village	Paddy Fields (Ha)	Moor (Ha)	Yard (Ha)	Others (Ha)
1	Lawu	245	47	134	10
2	Baran	139	6	104	3
3	Nguter	148	41	123	13
4	Gupit	132	54	180	26
5	Pengkol	85	57	161	62
6	Jangglengan	104	141	68	67
7	Tanjungrejo	152	49	85	70
8	Serut	169	58	138	25
9	Juron	210	19	78	13
10	Celep	92	26	163	11
11	Plesan	115	52	239	25
12	Kedungwinong	158	102	103	29
13	Daleman	167	10	78	13
14	Kepuh	243	8	124	20
15	Pondok	154	13	76	10
16	Tanjung	105	66	66	4
	Total	**2418**	**749**	**1920**	**401**

Source: Agriculture Bureau of Nguter, 2020

The largest agriculture area is Lawu village with 245 Ha followed by Kepuh village with 243 Ha. While th smallest paddy field area is Pengkol village with only 85 Ha. On the other hand, the largest moor area is in Jangglengan village with 141 Ha while the smallest one is in Baran village with 6 Ha of moor area. Besides, the biggest yard is in Plesan village (239 Ha) while the least one is in Tanjung village (66 Ha).

3.3 Population

3.3.1 The population number and distribution

The total number of people in Nguter in 2019 is 41.811 individuals with 20.708 males and 21.103 females [13]. The highest population is in Nguter village with 3.791 inhabitants and the least number is in Baran village with 1.669 individuals. Table 3 presents the population distribution in Nguter.

Table 3. The number of people per village in Nguter

No	Village	Area (Km^2)	Male	Female	Total
1	Lawu	4,36	1359	1393	2752
2	Baran	2,52	834	835	1669
3	Nguter	3,25	1844	1947	3791
4	Gupit	3,92	1574	1517	3091
5	Pengkol	3,65	1279	1313	2592
6	Jangglengan	3,8	956	968	1924
7	Tanjungrejo	3,56	1030	1088	2118
8	Serut	3,9	1221	1200	2421
9	Juron	3,2	1243	1226	2469
10	Celep	2,92	1169	1195	2364
11	Plesan	4,31	1224	1307	2531
12	Kedungwinong	3,92	1411	1389	2800
13	Daleman	2,68	1290	1380	2670
14	Kepuh	3,95	1589	1654	3243
15	Pondok	2,53	1615	1570	3185
16	Tanjung	2,41	1070	1121	2191
	Total	**54,88**	**20708**	**21103**	**41811**

Source: Sukoharjo Statistical Bureau, 2020

3.3.2 Population Density

Population density is the number of populations per kilometer square [14]. In Nguter, the population density is 766/km^2 with the highest density in Pondok village (1.265/km2) and the lowest density in Jagglengan village (509/km^2). The data of population density is presented in Table 4

Table 4. Population Density in Nguter in 2019

No	Village	Area (Km^2)	Population Number	Population Density (individual/km^2)
1	Lawu	4,36	2752	631
2	Baran	2,52	1669	662
3	Nguter	3,25	3791	1166
4	Gupit	3,92	3091	789
5	Pengkol	3,65	2592	710
6	Jangglengan	3,8	1924	506
7	Tanjungrejo	3,56	2118	595
8	Serut	3,9	2421	621
9	Juron	3,2	2469	772
10	Celep	2,92	2364	810
11	Plesan	4,31	2531	587
12	Kedungwinong	3,92	2800	714
13	Daleman	2,68	2670	996
14	Kepuh	3,95	3243	821
15	Pondok	2,53	3185	1259
16	Tanjung	2,41	2191	909

Source : Sukoharjo Statistical Bureau, 2020

3.4 Health Facilities in Nguter

There are two levels of health facilities in Nguter which are *Puskesmas* (Community Health center) and *Puskesmas pembantu* (Sub Community Health center) and this study will focus on those facilities. The main reason Nguter has been chosen is because its highest population density among other sub districts, so it needs more accessible basic health facilities. In addition, those facilities represent the basic health facilities in a region, particularly in Nguter. Practically those facilities has been supported by their own buildings with professional technical support personnel. Meanwhile the *PKD* (Village Health Post) is united with the village office and do not serve public health checks every day. However, However, *PKD* remains little discussed in the description of the results of this study.

There are 2 health centers currently operating, namely Nguter 1 Health Center in Nguter Village and Nguter 2 Health Center located in Celep Village (Figure 5). In carrying out health services to community, the two existing health centers are assisted by 3 sub-health centers located in Lawu, Tanjungrejo, and Pondok villages.

(A) **(B)**

Figure 5. (A) Nguter Health Center I in Nguter Village and (B) Nguter II Health Center in Celep Village.

At the village level there is a Village Health Post (*PKD*) located in almost every village. *PKD* is usually a basic health service for the community coordinated by the local village midwife. It does not serve every day, but only on certain days. Based on field surveys, the *PKD* in some villages were in fairly good condition, because there is a regular schedule of health services and the health care facilities located in the local village hall. However, in some villages PKD was inactive either because it is close to the Puskesmas or indeed the village conditions do not allow it. The complete facilities of Puskesmas, Pustu, and PKD in Nguter District are presented in Table 5.

Table 5. The facilities of *Puskesmas, Puskesmas Pembantu (Pustu),* and *PKD* in Nguter

No	Village	*Puskesmas*	*Pustu*	*PKD*
1	Lawu	0	1	1
2	Baran	0	0	1
3	Nguter	1	0	0
4	Gupit	0	0	1
5	Pengkol	0	0	1
6	Jangglengan	0	0	1
7	Tanjungrejo	0	1	1
8	Serut	0	0	1
9	Juron	0	0	1
10	Celep	1	0	1
11	Plesan	0	0	1
12	Kedungwinong	0	0	1
13	Daleman	0	0	1
14	Kepuh	0	0	1
15	Pondok	0	1	1
16	Tanjung	0	0	1
	Total	**2**	**3**	**15**

Source: BPS, 2020 [15]

Figure 6. shows the spatially distribution of basic health facilities (*Puskesmas* and *Puskesmas Pembantu)* in Nguter.

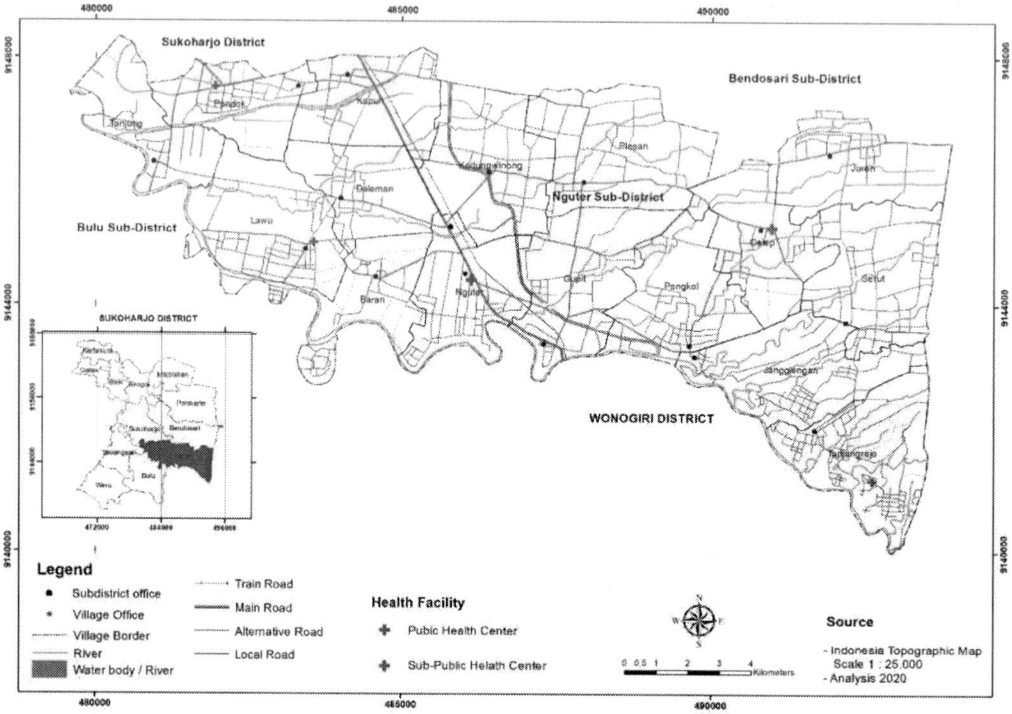

Figure 6. The distribution of basic health facilities in Nguter

3.5 Spatial Pattern of health facilities

The spatial analysis to see the pattern of a group of objects distribution in an area can be done by using nearest neighbor analysis [5,9]. This analysis is used to determine the pattern of a group of objects in an area whether it follows a random, clustered, or uniform pattern as indicated by the magnitude of the Rn value. This analysis considers 3 aspects namely distance, number of location points, and area.
The value of Rn (the nearest neighbor spread index) is obtained through the formula:

$$Rn = \frac{D\ (obs)}{0{,}5\sqrt{\frac{a}{n}}} \qquad (2)$$

Rn : Nearest Neighbor Value
D (obs) : The average distance of observed neighbor
a : Area
n : The number of points (location)

Then, the Rn value will be interpreted by using Continuum Nearest Neighbor Analysis which is pictured below:

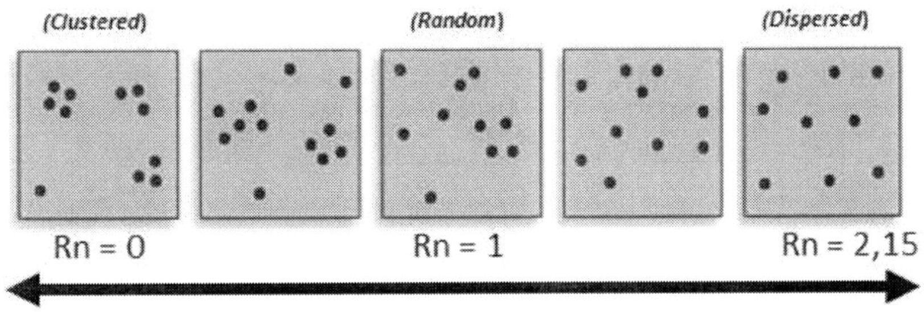

The result shows that the spatial distribution of health services in Nguter is following dispersed pattern. These are the results using ArcGIS 10.2 software:

Average Nearest Neighbor Summary
Observed Mean Distance: 3300,006943 Meters
Expected Mean Distance: 1852,025918 Meters
Nearest Neighbor Ratio: 1,781836
z-score: 2,991415
p-value: 0,002777

Dataset information

Input Feature Class	: puskesmas
Distance Method	: EUCLIDEAN
Study Area	: 54880000,00 Meters

The graphic that shows the result of nearest neighbor analysis is presented in Figure 7.

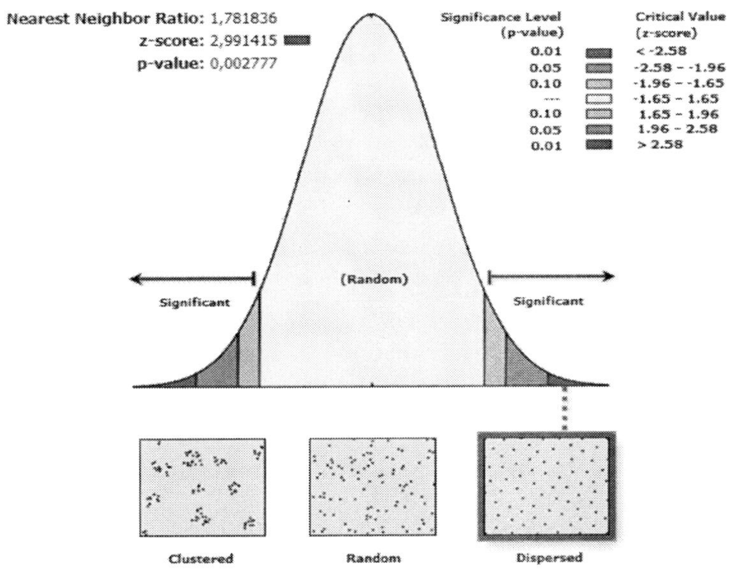

Figure 7. The result of nearest neighbor analysis of basic health facilities in Nguter

3.6 The Reachability of health facilities

The reachability analysis of health facilities in Nguter was done by doing the comparison between the real condition with the minimum standard of health facilities based on SNI 03-1733-2004. The radius range of one public health services is 3000 m and for one sub-public health services is 1500 m. To get this, buffering analysis was undertaken by pulling out the distance about 3000 m from the point of health facilities and 1500 m from the auxiliary health facilities (Figure 8). The public health services were symbolized by red plus and the sub-public health services were symbolized by blue plus. From buffering, it can be understood that almost the entire area of Nguter was covered by health facilities service. Only a small part of the area was not covered by health facilities service.

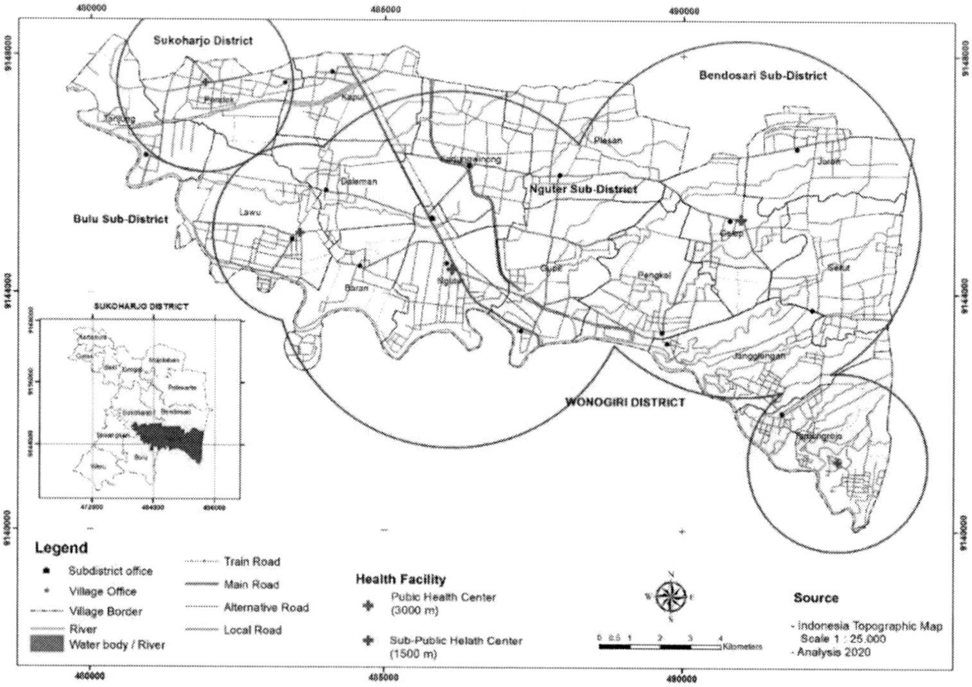

Figure 8. The radius of health facilities service in Nguter

Even though in terms of distance that small area is beyond the range of health facilities, but the good condition of road facilities which is categorized as arterial type makes the time travel become reachable for the community. The location reachability is not only seen from the proximity of the distance but also the road network. This road network will determine the closest alternative access to reach the location. As closer the location of health facilities and as better the road network, as more useful these facilities for the community. Table 6 proves this statement based on the field survey of Village Potency in 2019 that shows the ease of local community to reach health facilities such as hospital, health center, and pharmacy.

Table 6. The Reachability easiness of health facilities in Nguter in 2019

No	Village	Hospital	Health Facilities (*Puskesmas)*	Pharmacy
1	Lawu	Easy	Easy	Easy
2	Baran	Easy	Easy	Easy
3	Nguter	Easy	Very Easy	Very Easy
4	Gupit	Easy	Easy	Very Easy
5	Pengkol	Easy	Easy	Easy
6	Jangglengan	Easy	Easy	Easy
7	Tanjungrejo	Easy	Easy	Easy
8	Serut	Easy	Easy	Easy
9	Juron	Easy	Easy	Easy
10	Celep	Easy	Very Easy	Very Easy
11	Plesan	Easy	Easy	Very Easy
12	Kedungwinong	Easy	Easy	Easy
13	Daleman	Easy	Easy	Easy
14	Kepuh	Easy	Easy	Easy
15	Pondok	Easy	Easy	Easy
16	Tanjung	Easy	Easy	Very Easy
	Total	**2**	**3**	**15**

Source: PODES Survey [16]

3.7 The Sufficiency Evaluation of Health Facilities

To evaluate the sufficiency of health facilities service, the comparison with SNI 03-1733-2004 as the minimum standard of health facilities was done. The minimum service standard (threshold) for 1 health service is 120.000 people while the threshold for sub-health service is 30.000 people. The calculation of the sufficiency level of health facilities is presented in Table 7.

Table 7. The calculation of the sufficiency level of health facilities in Nguter

Health Facilities	Population	Threshold	Requirement	Availability	FCC
Sub health service	41811	30000	2	3	1,5
Health service	41811	120000	1	2	2

Based on above table, the carrying capacity of sub health service is 1,5 while for the health service is 2. It means that the existing facilities are able to support the needs of population. Facility Carrying Capacity (FCC) value > 1 can be interpreted as sufficient service level (surplus).

4 Conclusion

The Covid-19 pandemic has taught us about the importance of public health facilities as one of the spearheads in handling the pandemic. Based on the results of the mapping conducted in Nguter District, there are two public health centers and three sub-public health centers. The nearest neighbor analysis shows that health service facilities in Nguter indicate a dispersed pattern. This means that the existing of public health facilities are well spread in the Nguter District area. The buffering analysis also gives result that almost the entire area of Nguter is covered by public health centers. Only a small part of the area is not covered by their facilities. Finally, the calculation of the carrying capacity of public health services shows that the existing facilities have been able to support the needs of population.

Acknowledgement
This publication is part of the research grant given by Universitas Veteran Bangun Nusantara whom authors would like to thank for. We also thank to all assistants who help the research activities.

Reference

[1] Satuan Tugas Penanganan COVID-19 Republik Indonesia 2021 Statistik Kasus Covid 19 di Indonesia
[2] World Health Organization 2021 The Global situation of confirmed numbers and deaths from Covid 19
[3] World Health Organization The Indonesia Situation of confirmed numbers and deaths from Covid 19.
[4] Prahasta E 2002 *Konsep – Konsep Dasar Sistem Informasi Geografis* (Bandung: INFORMATIKA)
[5] Pujayanti J A D, Susilo B and Puspitaningrum D 2014 Sistem Informasi Geografis Untuk Analisis Persebaran Pelayanan Kesehatan di Kota Bengkulu *J. Rekursif* **2** 99–111
[6] Ekadinata A, Dewi S, Hadi D, Nugroho D and Johana F 2008 *Sistem Informasi Geografis untuk Pengelolaan Bentang lahan Berbasis Sumberdaya Alam. Buku I; Sistem Informasi Geografis dan Penginderaan Jauh Menggunakan ILWIS Open Source* (Bogor)
[7] Mutalazimah M, Handaga B and Sigit A A 2009 Aplikasi Sistem Informasi Geografis pada Pemantauan Status Gizi Balita di Dinas Kesehatan Kabupaten Sukoharjo *Forum Geogr.* **23** 153
[8] Waskito D Y, Kresnowati L and Subinarto S 2018 Pemetaan Sebaran Sepuluh Besar Penyakit Di Pusat Kesehatan Masyarakat Mojosongo Kabupaten Boyolali Berbasis Sistem Informasi Geografis *J. Ris. Kesehat.* **6** 7
[9] Budiman R and Cahyono A B 2017 Analisis Spasial Fasilitas Pelayanan Kesehatan Masyarakat Terhadap Pengunjung di Kota Blitar *J. Tek. ITS* **6** 2–5
[10] Kaswidjanti W, Santosa B and Wicaksono R S 2008 Sistem Informasi Geografis Pemetaan Fasilitas Kesehatan Di Kota Magelang Berbasis WEB *Seminar Nasional Informatika* pp 271–8
[11] Kurniawan P, Krisna A, Piarsa I N and Buana P W 2016 Sistem Informasi Geografis Pemetaan Penyebaran Penyakit Berbasis Web *Merpati* **2** 271–9
[12] Kurniasari N, astuti P, Widarsa T and Mulyawan H 2012 Aplikasi Sistem Informasi Geografis Untuk Memetakan Distribusi Sasaran Pemantauan Kesehatan Ibu Di Wilayah Kerja Puskesmas I Denpasar Selatan *Arch. Community Heal.* **1** 18–27
[13] Badan Pusat Statistik Kabupaten Sukoharjo 2020 *Kabupaten Sukoharjo dalam Angka Tahun 2020* (Sukoharjo: BPS)
[14] Muta'ali L 2015 *Teknik Analisis Regional untuk Perencanaan Wilayah, Tata Ruang, dan Lingkungan* (Yogyakarta: BPFG UGM)
[15] Badan Pusat Statistik Kabupaten Sukoharjo 2020 *Kecamatan Nguter dalam Angka Tahun 2020* (Sukoharjo: BPS)

IOP Conf. Series: Earth and Environmental Science **986** (2022) 012062 doi:10.1088/1755-1315/986/1/012062

Settlement land directions based on the tsunami disaster in the coastal of Purworejo District, Central Java

Fitria Dewi Kartika[1,2,*] **and Pipit Wijayanti**[1,2]

[1]Disaster Research Center, Universitas Sebelas Maret Surakarta
[2]Geography Education, Universitas Sebelas Maret Surakarta

[*]Email : fitriadewikartika@gmail.com

Abstract. The purpose of this study was to determine the direction of land use for settlements. The research location is on the coast of Purworejo Regency which includes Purwodadi District, Ngombol District, and Grabag District. This study uses a quantitative descriptive method. Data collection techniques include observation, interviews, and secondary data processing from related agencies. The analytical technique used in the Geographic Information System (GIS) analysis is to overlay tsunami hazard data, evaluate the suitability of residential land and area functions. The results of the overlay will show the zoning for the direction of the settlement. The results showed that the coast of Purworejo Regency has three directions of land use for settlements. The optimal direction of land use for settlements has an area of 427.6 Ha with the development of settlements with high density and can be developed as a Central Business District (CBD) area. The conditional direction of land use for settlements has an area of 3794.05 Ha. The law on the development of low-density residential areas and residential areas with structural and non-structural mitigation because this zone has a moderate level of tsunami vulnerability. The directive not to make settlements with an area of 1242.78 hectares is directed to the green belt area with limited community activities.

1. Introduction

The Population Growth Rate in Indonesia from 2010 to 2019 reached 1.31% [1]. At the same time, rapid socio-economic development has led to a high concentration of people in coastal areas, especially in big urban centers [2]. This happens in Indonesia because 60% of the population lives in coastal areas [3]. The very rapid development of infrastructure and high population density make urban areas very dynamic and become the goal of people's lives [4]. The coast is preferred by the community to be used as a place to live than inland areas because the beach is a dynamic area [5].

Moreover, the concentration of the population in Indonesia is centered on the island of Java. On the other hand, the coast of Java Island is prone to disasters. The Indian Ocean region south of Java Island is known as one of the areas with high tsunami risk; the region experienced tsunamis in 1994 and 2006, which resulted in significant loss of life [6]. The estimated time of arrival of waves has been predicted in coastal areas of Indonesia through tsunami modeling [7,8]. This research has been conducted in the Coastal District of Purworejo. Since 2009 the coast of Purworejo Regency has become a research area for GITEWS which produces a tsunami hazard map [9].

Indonesia has made a policy regarding the provision of land use for mitigation in spatial planning [10]. Spatial planning, whose function is to control land use in the long term, is instrumental in minimizing the potential for exposure of communities, socio-economic activities, and infrastructure to natural disasters. [11]. The development of housing and settlement areas must be by the provisions of responsible and sustainable development [12]. Information obtained from land suitability will provide clues about land suitability for final use [13]. identification of land suitability is needed for predicting future use of land resources [14]. In classifying land use, land for settlement must be by a conducive environment [15]. The aim is to ascertain the constraints and opportunities for sustainable land maintenance, planning, and development [13].

The coast of Purworejo Regency is an area affected by the construction of the New Yogyakarta International Airport. This development is the government's way of balancing growth in the coastal areas of southern Java and northern Java. The construction of Yogyakarta International Airport has an impact on the development of other infrastructure that is submerged, namely Jalan Deandles. The coast of Purworejo Regency is an area passed by Deandles Street, which is the access road to Yogyakarta International Airport from the west. This is one of the significant impacts for the Coastal region of Purworejo Regency. The development of the area around the New Yogyakarta International Airport is one of the plans of the Purworejo Regency government to build a border city built on an area of 2,000 hectares with the concept of an industrial city that is integrated with Aero City in Kulonprogo Regency. [16]. The development of this area attracts the attention of the community to live in coastal areas. On the other hand, the South Coast of Purworejo is prone to Tsunami disasters. If the wider area of land is converted to residential land, the risk of a tsunami disaster is higher.

Spatial planning is one form of mitigation in reducing disaster risk, one of which is the Tsunami disaster on the coast of Purworejo Regency. One of the most crucial lands used to be managed settlements. Settlements are centers of human activities and activities, and there is vital infrastructure in them, so they are very risky in the event of a Tsunami disaster. So that research on the direction of residential land use is vital. Many studies on disaster-based settlement directions have been carried out in Indonesia. As research conducted by Haryana et al. (2013) [17] in Bantul Regency with the direction of settlements based on the physical condition of the area and safe from the danger of earthquakes. Another similar study was conducted by Hermon (2019) [18] in the city of Padang with the direction of settlement using zoning maps of land suitability and landslide-prone maps.

In contrast to the two studies, the direction of settlements in the Coastal area of Purworejo Regency is based on the Tsunami disaster hazard zone, the suitability of residential land, and the area's function. The purpose of the study was to identify the direction of land use for settlements. This directive will produce a site that is used as a reference in deciding areas with optimal settlement development, conditional settlements, and not for settlements or relocation.

2. Methods

Research Locations in the Coastal District of Purworejo, Central Java Province, precisely in three sub-districts, including Purwodadi District, Ngombol District, and Grabbag District. Astronomically, the coast of Purworejo Regency is located at a position of 7°48'52.07" South Latitude - 7°51'16.11" South Latitude and 109°49'10.42" East Longitude - 110°2'32.89" East Longitude. Coastal as the area under study, the determination of the boundaries of the coastal area based on the landform map of Purworejo Regency, it is known that the coastal landforms in Purworejo Regency include the landforms of young coastal shelf, old coastal shelf, Lagoon, and Swale.

This research is a quantitative descriptive study with secondary and primary data analysis (field observations). To obtain spatial phenomena, a spatial analysis approach is used [19]. GIS analysis can be utilized for future coastal land use development [20]. Based on Weerasinghe et al. (2011), Habitable coastal zones are based on minor zoning according to high risk, low-risk zones, alert zones, and safe zones that can facilitate the development of early warning systems and evacuation procedures [21]. From the two statements, the direction of land use is based on zoning for the direction of optimal, conditional land use, and relocation for settlements. This zoning is made based on evaluating the suitability of residential land, area functions, and disaster-prone areas. Land suitability as a condition for planning [22]. While in the assessment of disaster-prone areas based on buffer analysis using GIS. Create a map by overlaying all score data using GIS [17], and then a zone is obtained to direct residential land use.

IOP Conf. Series: Earth and Environmental Science **986** (2022) 012062 doi:10.1088/1755-1315/986/1/012062

The directives on the use of residential land are in the form of maps with a scale of 1: 65,000. Parameter data in this study are shown in the table. 1.

Table 1. Variables, Parameters, Criteria, and scores in this study

No	Variable	Class	Parameter	Criteria	Score
1.	Physical Land Evaluation For Settlements	Suitable (S1)	Drainage	Good - Very Good	3
		Quite Suitable (S2		Medium	2
		Not Suitable (N1)		Quite Bad - Bad	1
			Flood	Without Flood	3
				Rarely Flood	2
				Often Flood	1
			Slope	0 – 15%	3
				>15 – 40%	2
				>40%	1
			Soil Texture	Rather Rough	3
				Rather Smooth	2
				Smooth	1
			Soil Depth	<50 Cm	3
				>50 – 90 Cm	2
				>90 Cm	1
			Erosion Hazard	No Erosion	3
				Light Erosion	2
				Heavy Erosion	1
			Elevation	>15 M	3
				5 – 15 M	2
				< 5 M	1
2	Area Function	Protected Area	Slope	0 -8 %	20
		Cantilever Area		8 – 15 %	40
		Seasonal Cultivation Areas And Settlements		15 – 25 %	60
		Annual Cultivation Area		25 – 40 %	80
				>40 %	100
			Type Of Soil	Alluvial, Planosol, Hydromorph Gray, Lateric	15
				Latosol	30
				Brown Forest Land, Mediterranean Land	45
				Andosol, Laterik, Grumusol, Podsol, Podsolic	60
				Regosol, Litosol, Organosol, Renzina	75
			Rainfall (mm/day)	0 – 13,6	10
				13,61 – 20,7	20
				20,71 – 27,7	30
				27,71 – 34,8	40
				> 34,8	50
			Big River Buffer	100 m	
			Small River Buffer	50 m	
3.	Disaster-Prone Areas	Buffer Zone		100 m	
		High Disaster Prone Area		100 – 300 m	
		Moderate Disaster Prone Area		300 – 1000 m	
		Low Disaster Prone Area		>1000 m	

(Source : [23–26]

IOP Conf. Series: Earth and Environmental Science **986** (2022) 012062 doi:10.1088/1755-1315/986/1/012062

3. Results and Discussion

3.1 Physical Land evaluation for settlement

The suitability evaluation is used to determine the housing area to be relocated and identify suitable places for relocation [27]. This evaluation of settlement land suitability only assesses the suitability of settlements from the physical aspect. This physical aspect is described from the parameters used.

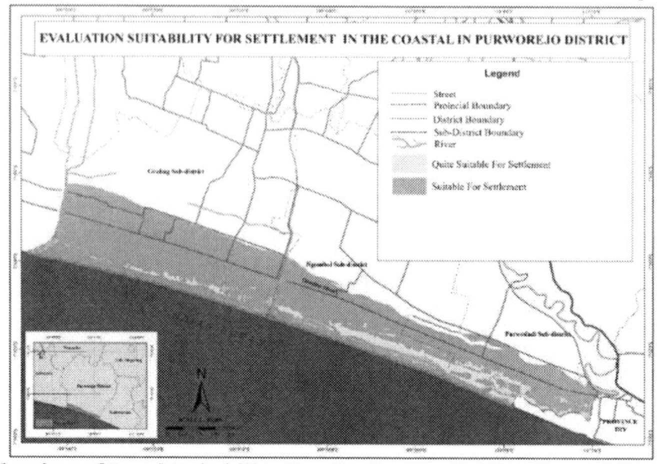

Figure 1. Evaluation of Land Suitability For Settlement In The Coastal Purworejo District

The coast of Purworejo Regency has two land suitability classes for settlements, namely the appropriate class and the quite comfortable class. Areas with suitable land suitability for settlement are scattered throughout the coastal sub-districts of Purworejo Regency, both in Grabag District, Ngombol District, and Purwodadi District. The appropriate class dominates in this area, around 5029.95 Ha. This is caused by various factors, such as the relatively flat slope, the rare occurrence of erosion, and the soil texture, which is suitable for residential areas. These factors make a high score in each land unit. In the areas that enter the appropriate class, the average height of the place is medium class and good class. This group is quite suitable in the coastal area of Purworejo Regency, covering an area of 438.84 hectares and is spread across all sub-districts in the coastal zone, both in Purwodadi District, Ngombol District, and Purwodadi District.

3.2 Disaster-prone areas

This disaster-prone area still uses buffer parameters from the coastline, not yet using tsunami wave modeling. The buffer is carried out from the shoreline, while the river buffer is carried out when calculating the area function. The coastal area of Purworejo Regency is dominated by low tsunami-prone areas with an area of 3230.44 Ha or 58.16%; this area has a distance of more than 1000 m from the coast. Low tsunami hazard areas are areas suitable for residential development areas with high-density levels. The moderate tsunami-prone zone, which is 300-1000 m from the coast, has an area of 1,688.28 hectares or 30.40%. This area is a conditional zone with the development of settlements with low population density. The next area is a high disaster-prone area with an area of 422.68 ha or 7.61%, and this area has a distance of 100-300 m from the beach. This area can be developed as a protected area, beach tourism, or created as a green open area. The coastal border buffer area is 100 m from the beach, this area on the coast of Purworejo Regency has an area of 212.51 Ha or 3.83%. The buffer zone area is an area that was developed as a protected area or green belt in coastal areas, and this area is the most critical in disaster mitigation to reduce waves when a tsunami comes.

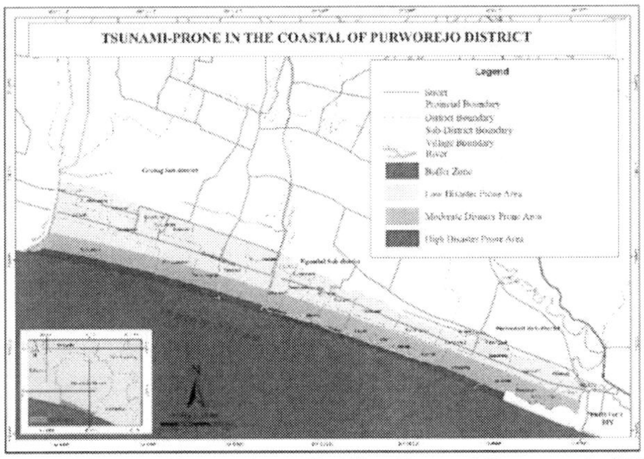

Figure 2. Disaster Prone Area In The Coastal Purworejo District

3.3 Area Function

The function of the cantilever area as the most significant area on the coast of Purworejo Regency is influenced by soil type factors. The soil type in the cantilever area is Regosol soil type with compassionate sensitivity to erosion. The average rainfall in the last ten years is 21.9 mm/day which is in the medium class, and the slope parameters in this area are class I. Based on the Decree of the Minister of Agriculture No. 837/Kpts/Um/11/1980 [26], a total score of 125-174 is included in the cantilever area function. Cantilever area with an area of 4,020.81 ha or equivalent to 72.39% of the entire coastal area of Purworejo Regency. The function of the protected area on the coast of Purworejo Regency has an area of 1,095.55 Ha or equivalent to 19.73% of the total coastal area. The function of this area includes the function of local protected areas, coastal borders, large and small river boundaries, each of which has a different distance. The function of this area is measured using buffer tools in the GIS application for a 100-meter long coastal border to the mainland, a 100-meter river border for a large river, and a small river border 50 meters from the right and left banks of the river. The function of the local protected area of the major river border is the function of the broadest protected area with an area of 599.03 Ha or equivalent to 10.79% of the entire coastal area of Purworejo Regency, and this is because this area has three major rivers or main rivers that empty into the sea—covering the Bogowonto River, Jati River, and Wawar River.

The function of the annual crop cultivation area and settlements is 437.73 hectares or equivalent to 7.88% of the entire coastal area of Purworejo Regency. This area has criteria for alluvial soil types that are not sensitive to erosion, and the slope is only class I, which is 0-8%, and the average rainfall for ten years is 21.9 mm/day, including medium type.

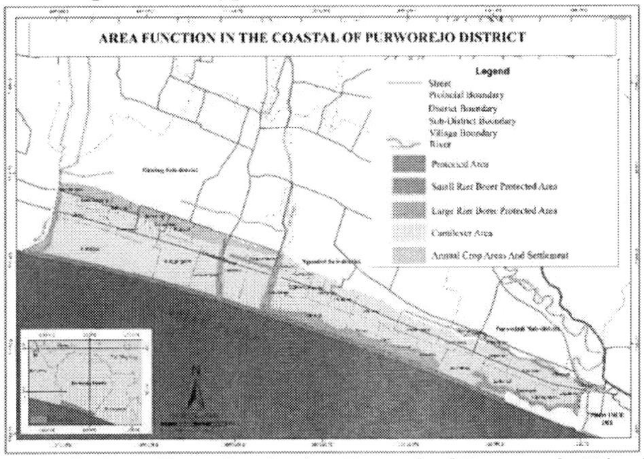

Figure 3. Area Function in the Coastal of Purworejo District

3.4 Directions for the use of Settlement Land

The overlay of land suitability evaluation maps for settlements, area function maps, and maps of disaster-prone areas show that the coast of Purworejo Regency has 13 zonings. Zoning is used as a reference for directing the use of settlement land. After obtaining zoning, then overlaying the zoning with settlements to determine zoning distribution in settlement land uses.

Table 2. Zoning of Settlement Directions on the Coast of Purworejo Regency

No	Suitability Class	Area Function	Tsunami-Prone Areas	Direction For Settlement	Area (Ha)
1	Quite Suitable	Seasonal Cultivation Areas And Settlements	Low	Conditional Settlement	10.09
2		Cantilever Area	Moderate	Conditional Settlement	195.97
3			High	Not For Settlement	37.47
4		Protected Area	Moderate	Not For Settlement	149.18
5			High	Not For Settlement	46.13
6	Suitable	Seasonal Cultivation Areas And Settlements	Low	Settlement	427.6
7		Cantilever Area	Low	Conditional Settlement	2499.12
8		Cantilever Area	Moderate	Conditional Settlement	1088.87
9			High	Not For Settlement	199.38
10		Protected Area	Low	Not For Settlement	245.54
11			Moderate	Not For Settlement	217.42
12	Suitable		High	Not For Settlement	135.15
13			Buffer Zone	Not For Settlement	212.51

The zone with optimal settlement direction has an area of 427.6 Ha. This zone can be developed as a high-density settlement area and a Central Business District (CBD) area. This is because this zone has suitable land characteristics for settlements and is included in the low tsunami-prone areas. Areas with optimal settlement development directions as far as more than 3 km from the coastline. Sites with conditional settlement directions covering an area of 3749.05 hectares located in this zone have different obstacles for each zone. Zones with barriers are in the moderate tsunami-prone regions, special requirements must be made for settlement areas. This requirement is in the form of mitigation, both structural mitigation and non-structural mitigation.

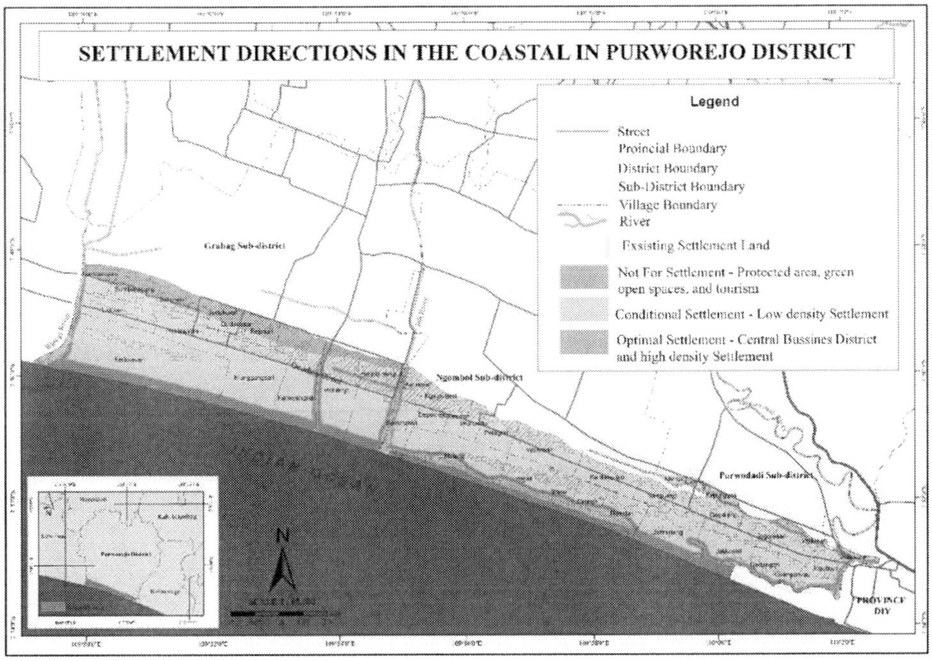

Figure 4. Settlement Directions in the Coastal In Purworejo District

Structural and non-structural mitigation in settlements with medium and high vulnerability levels include:

a) Residential areas within the tsunami hazard zone must be built on the elevated ground [28]
b) Planting broadleaf vegetation with large diameter branches near houses so that people can escape by climbing during a Tsunami disaster [29]
c) Improving existing public facilities buildings, targeted as tsunami evacuation buildings in tsunami-prone coastal areas to anticipate future tsunami threats [9].
d) Build housing behind the tsunami control forest or strengthen more significant buildings
e) Development of Formal/Informal Education for coastal communities, construction of evacuation routes, conducting socialization and training on disaster mitigation and preparedness, as well as increasing awareness of coastal communities about tsunamis [30]
f) We are placing the main access road outside the inundation area and the secondary access road perpendicular to the beach.
g) Creating mitigation facilities such as evacuation route signposts, temporary evacuation sites, and early warning systems

Areas located in the area not to be developed for settlements covering an area of 1242.78 hectares with a distance of 0 to 1 kilometer from the coastline. Areas located in the high tsunami-prone areas should not be developed as settlement areas because they can increase community risk. So it is recommended to relocate the community to a safer area. It is undeniable that moving people is not an easy thing. Therefore there is also a need for mitigation to reduce community vulnerability from disasters. evacuation, construction of embankments, evacuation routes, and planting of wave control vegetation. This zone can be developed as a protected area, green open space, and coastal tourism with limited land management. However, in the coastal area of Purworejo Regency, this zone is still widely used as a dry land agricultural area and ponds. The buffer zone in the tsunami-prone area is essential because this area is used as a protective belt. However, the coast of Purworejo Regency is still minimal, with mangroves and other vegetation that can dampen waves. Forests can serve to reduce the impact of tsunamis [31]. Coastal tourism development may be carried out if accommodation and tourism support are built with a maximum 15 (fifteen) meter high-rise structure design or a minimum of 3 (three) floors considering the history of the tsunami and tsunami inundation. [32].

4. Conclusion

The coast of Purworejo Regency has three directions of land use for settlements: the direction of optimal settlement development, the direction of conditional settlements, and the direction of not developing settlements or relocating settlements. The optimal direction of land use for settlements has an area of 427.6 Ha with the development of settlements with high density. It can be developed as a Central Business District (CBD) area. The direction of conditional land use for settlements has an area of 3794.05 Ha. The direction is to develop low-density residential areas and residential areas with structural and non-structural mitigation. The directive not to create settlements has an area of 1242.78 hectares directed at the green belt area with limited community activities such as parks or mangrove tourism. The government can consider the results of this study as a reference for structuring settlements on the coast of Purworejo. Moreover, the development of settlements in this area will be remarkably rapid in the future because of the New Yogyakarta International Airport, which attracts the public's attention to live. This settlement directive can also be used as a form of long-term Tsunami mitigation.

References

[1] Statistik B P 2020 *Statistik Indonesia 2020* ed Subdirektorat Publikasi dan Kompilasi Statistik (Indonesia: Badan Pusat Statistik)

[2] Merkens J L and Vafeidis A T 2018 Using information on settlement patterns to improve the spatial distribution of population in coastal impact assessments *Sustain.* **10** 10–2

[3] Handayani W, Rudiarto I, Setyono J S, Chigbu U E and Sukmawati A M awanah 2017 Vulnerability assessment: A comparison of three different city sizes in the coastal area of Central Java, Indonesia *Adv. Clim. Chang. Res.* **8** 286–96

[4] Liu J, Wen J, Huang Y, Shi M, Meng Q, Ding J and Xu H 2015 Human settlement and regional development in the context of climate change: a spatial analysis of low elevation coastal zones in China *Mitig. Adapt. Strateg. Glob. Chang.* **20** 527–46

[5] Rudiarto I, Handayani W and Setyono J S 2018 A regional perspective on urbanization and climate-related disasters in the northern coastal region of central Java, Indonesia *Land* **7**

[6] Mardiatno D 2013 A proposal for tsunami mitigation by using coastal vegetations: Some finding from southern coastal area of Central Java, Indonesia *J. Nat. Resour. Dev.* **3** 85–95

[7] Syamsidik, Rasyif T M, Suppasri A, Fahmi M, Al'ala M, Akmal W, Hafli T M and Fauzia A 2020 Challenges in increasing community preparedness against tsunami hazards in tsunami-prone small islands around Sumatra, Indonesia *Int. J. Disaster Risk Reduct.* **47** 101572

[8] Horspool N, Pranantyo I, Griffin J, Latief H, Natawidjaja D H, Kongko W, Cipta A, Bustaman B, Anugrah S D and Thio H K 2014 A probabilistic tsunami hazard assessment for Indonesia *Nat. Hazards Earth Syst. Sci.* **14** 3105–22

[9] Hakim F F, De Vries W T, Siegert F and Syahbana J A 2017 a Gis-Based Tsunami Evacuation Model Considering Land Cover and Spatial Configuration (Case of Purworejo Regency, Indonesia) *Geoplanning J. Geomatics Plan.* **4** 143

[10] Syamsidik and Fachrul F 2017 The Indian Ocean Tsunami and Land Use Changes in Indonesia

[11] Sutanta H, Rajabifard A and Bishop I D 2013 Disaster risk reduction using acceptable risk measures for spatial planning *J. Environ. Plan. Manag.* **56** 761–85

[12] Widodo B, Lupyanto R, Sulistiono B, Harjito D A and Hamidin J 2015 Analysis of environmental carrying capacity for the development of sustainable settlement in Yogyakarta urban area *Procedia Environ. Sci.* **28** 519–27

[13] Rusdi M, Roosli R and Ahamad M S 2014 A Conceptual Approach to Land Suitability Evaluation for Residential Settlement in Aceh Sing Geographic Information System (GIS)

Proceedings of the 2nd International Conference on Natural and Environmental Sciences (ICONES) pp 7–12
[14] Steiner F, McSherry L and Cohen J 2000 Land suitability analysis for the upper Gila River watershed *Landsc. Urban Plan.* **50** 199–214
[15] Masri R M 2012 Analisis Keruangan Kesesuaian Lahan untuk Permukiman di Kabupaten Bandung dan Bandung Barat *Forum Geogr.* **26** 190
[16] Ridarineni N and Assidiq Y 2017 Purworejo Berencana Bangun Kota Perbatasan dengan DIY
[17] Haryana I K, Fikriyah V N and Yulianti N V 2013 Application of Remote Sensing and Geographic Information System for Settlement Land use Classification Planning in Bantul based on Earthquake Disaster Mitigation (Case Study: Bantul Earthquake, May 27th 2006) *Procedia Environ. Sci.* **17** 434–43
[18] Hermon D 2019 Land Stability Model for Sustainable Spatial Planning in Padang City-Indonesia based on Landslide Disaster *J. Geogr. Earth Sci.* **7** 19–26
[19] Muta'ali L 2013 Environmental Carrying Capacity Based on Spatial Planning *Indones. J. Geogr.* **43**
[20] Pourebrahim S, Hadipour M and Bin M 2011 Landscape and Urban Planning Integration of spatial suitability analysis for land use planning in coastal areas ; case of Kuala Langat District , Selangor , Malaysia *Landsc. Urban Plan.* **101** 84–97
[21] Weerasinghe W K, Hokugo A and Ikenouchi Y 2011 Tsunami risk mitigation through strategic land-use planning and evacuation procedures for coastal communities in Sri Lanka *Sci. Tsunami Hazards* **30** 163–77
[22] Tian Y, Kong X and Liu Y 2018 Combining weighted daily life circles and land suitability for rural settlement reconstruction *Habitat Int.* **76** 1–9
[23] Hermon D 2015 *Geografi Bencana Alam* (Jakarta: PT Raja Grafindo Persada)
[24] Setyadi G P, Iskandarsyah T Y W M, Suganda B R, Barkah M N, Hidayat T and Hadian M S D 2018 Kesesuaian Lahan Untuk Pengembangan Perkotaan Berdasarkan Aspek Geologi Lingkungan Pada Daerah Pangandaran, Provinsi Jawa Barat *Padjadjaran Geosci. J.* **2** 355–63
[25] Umar I, Dewata I, Barlian E, Anwar S and Suarti Y 2019 Priority Selection Of Residental Development Areas With Flood Hazard In Limapuluh Kota District, West Sumatra *Int. J. GEOMATE* **16** 138–44
[26] Menteri Pertanian 1980 *Kriteria dan Tata Cara Penetapan Hutan Lindung* vol 837 (Jakarta)
[27] Tian Y, Kong X and Liu Y 2018 Combining weighted daily life circles and land suitability for rural settlement reconstruction *Habitat Int.* **76** 1–9
[28] Strusińska-Correia A 2017 Tsunami mitigation in Japan after the 2011 Tōhoku Tsunami *Int. J. Disaster Risk Reduct.* **22** 397–411
[29] Tanaka N 2009 Vegetation bioshields for tsunami mitigation: Review of effectiveness, limitations, construction, and sustainable management *Landsc. Ecol. Eng.* **5** 71–9
[30] Muhari A, Diposaptono S and Imamura F 2007 Toward an Integrated Tsunami Disaster Mitigation: Lessons Learned from Previous Tsunami Events in Indonesia *J. Nat. Disaster Sci.* **29** 13–9
[31] Nateghi R, Bricker J D, Guikema S D and Bessho A 2016 Statistical analysis of the effectiveness of seawalls and coastal forests in mitigating tsunami impacts in iwate and miyagi prefectures *PLoS One* **11** 1–21
[32] Yuniartanti R K 2021 Konsep Penataan Ruang Kawasan Rawan Bencana (KRB) Tsunami di Kabupaten Karangasem, Pulau Bali *J. Reg. Rural Dev. Plan.* **5** 1–14

IOP Conf. Series: Earth and Environmental Science **986** (2022) 012063 doi:10.1088/1755-1315/986/1/012063

Project-induced displacement: Rethinking the impact of spatial planning and disaster mitigation policies in Tambakrejo, Semarang City

S I Ayuni [*], **S Tsana, and N R Priyandianto**

Master of Urban and Regional Planning, Bandung Institute of Technology, Jl. Ganesa No.10, Lb. Siliwangi, Bandung 40132, Indonesia

[*]Email: susiyowati.indah21@gmail.com

Abstract. Development in accordance with disaster mitigation policies in urban areas sometimes involves relocation and forced eviction of people living in informal settlements. Semarang City has the same experience. In 2019 the normalization project of the Banjir Kanal Timur river as a government effort to control flooding had displaced the local communities in Tambakrejo. This paper aims to criticize the implementation of flood disaster mitigation projects that impact poor households and assess their livelihood assets after displacement. Utilizing spatial data and primary data collection using the purposive sampling method were carried out during the fieldwork. The findings highlight that, irrespective of potential opportunities to avoid local communities from flood risk and safer place to live, on the other hand, local communities experience impacts on their livelihood assets and unclear compensation for their assets. This phenomenon proves that there are still many challenges to creating safe, comfortable, and sustainable urban development mandated in Law No. 26 of 2007. In the name of national projects, this action has taken away the "right to the city", has increased marginalized urban communities, and the emergence of urban social issues such as the affected communities' dissatisfaction with local government authorities

1. Introduction

In general, developed countries have better capacity in disaster risk management, and developing countries are believed to have a higher vulnerability to hazard risks [1]. As a developing country, Indonesia has a heightened urgency to solve various problems, one of which is related to disasters. Data from the National Disaster Management Agency (BNPB) in 2020 shows that 4650 disaster events struck in Indonesia. The domination of these disasters is hydrometeorological disasters such as floods (1,518 disasters) and landslides (1,054 disasters) [2]. When referring to the distribution of the population, in 2020, 56.64% of Indonesia's population will live in urban areas [3]. This condition makes urban areas have higher social vulnerability than other areas based on population levels [4]. Furthermore, most urban areas currently face various challenges, especially concerning the rate of population growth, rapid land-use change, and environmental hazards [5,6]. In addition, most of Indonesia's urban areas are located in coastal and low-lying areas, and these locations increase risk vulnerability to disasters such as floods and tropical cyclones [6]. The existing trend shows that urban areas are the most vulnerable to damage and losses that may occur [1,6].

Issue related to the rate of population growth, urbanization, and rapid land-use change in urban areas is the existence of informal settlement areas in the city center. Generally, people living in poverty live in that area. The urban poor is seen as the group most vulnerable to disaster risk [6]. Furthermore, the urban poor also often lack assets to meet their daily needs and cannot afford land. They were forced to build houses on land unsuitable for settlement, such as coastal plains, riverbanks, railroad tracks, land close to pollution sources, and hazard sites [5,7,8]. So the existence of this area is a concern to be resolved immediately [6]. The Government's interventions are often carried out: rebuilding projects, renewal projects, and projects related to urban disaster management [9].

Government policies in urban planning and disaster mitigation sometimes cause evictions for certain groups affected by the program [9]. In the process of implementing development projects, infrastructure or disaster mitigation requires space. This need ultimately results in the dislocation or displacement of the population [10]. Vanclay [10] adds that displacement due to development projects can impact local people's livelihoods or sources of income, either temporarily or permanently. Furthermore, displacement can also affect social and environmental aspects, such as the emergence of social conflicts, changes in land use, and environmental degradation. Currently, the focus of development projects involving land acquisition is still limited to compensation for land and property assets owned by local communities [10,11]. However, the Government and developers sometimes ignore the plight of the poor, who are the development objects.

Displacement experienced by the poor in informal settlements is often dismissed as unavoidable, especially for reasons referred to as public interest [12]. The main goals of sustainable development and equitable, inclusive development for all urban areas seem challenging to obtain for marginal groups who inhabit these informal settlement areas. Therefore, this study aims to investigate more deeply and critically the displacement experienced by local communities living in informal areas by assessing changes to their livelihood assets. Livelihood assets are used to understand how urban people live in informal areas and assess their opportunities and weaknesses. Livelihood assets are also used to broaden the discussion regarding the impacts of displacement from an economic, social, environmental, and political perspective. Several researchers have shown that post-displacement is often accompanied by loss of livelihood sources and leads to the impoverishment of displaced communities. The risk of impoverishment is manifested through several interrelated phenomena: such as landlessness, loss of housing, unemployment, loss of access to shared resources, marginalization, food insecurity, social disarticulation, and other life uncertainties [13,14]. This research consists of 4 sections. The first section explains the background and the theory used, which is related to changes in urban spatial planning, urban vulnerability, and government projects as disaster mitigation policies. Section 2 describes the study area, methods, and analytical tools used. Section 3 describes the results of the analysis and discussion. The fourth or final section contains the conclusions of the research results.

1.1. The Changing of Urban Spatial Planning

The dynamics of the urbanization process and the increase in population are directly proportional to the increase in socio-economic activities that utilize space in urban areas [15]. The growth of diverse urban socio-economic activities has impacted land-use change [16], as well as increased pollution and decreased environmental quality [17]. In the last two decades, changes in land use in large urban areas, such as Jakarta, Bandung, Semarang, Surabaya, Makassar, and so on, have increased significantly. Surya et al. Urban areas are believed to have the most dynamic land change systems, with urbanization set to occur over the next few decades. Space utilization control in the form of spatial planning policies is needed to suppress the rapid changes in urban land and threaten distractions in social and environmental aspects [18]. Often, spatial planning policies prioritize economic aspects and ignore the other two aspects, making it challenging to balance the three [19]. For example, governments in developing countries control the use of slums to build the city's image by going through the process of eviction, cleaning, moving, or reoccupying slum areas in new locations or environments [9]. However, with a narrow conceptualization, these informal or slum settlements are practically marginalized in the policy, which ignores the social and economic elements of the people who live in it [20]. It is as if the governance actors are not paying attention to the effects of these policies. So that the impact of the socio-economic shift that occurred after the eviction policy was cleaning, moving, or reoccupying slum areas, especially related to people's livelihoods and the socio-cultural conditions of the communities.

1.2. Urban Vulnerability and Government Projects as Disaster Mitigation Policies

Urban areas in big cities face many challenges. Often these challenges are related to urbanization, population, land-use change, and environmental threats [6]. The World Bank predicts that 68% of Indonesia's population will live in urban areas in 2025 and an increase of 82% in 2045 [21]. In addition, most Indonesian cities are geographically located in the form of coastal areas, adding to their challenges. Population pressures, geographical conditions, and various other challenges can increase the level of vulnerability along with the threat of hazards and climate change. Although it is known that the area along the coast and canal is an area with a high risk of danger, some regulations regulate it. However, many urban residents do not comply with the rules and live in disaster-prone areas, especially the poor who are forced to live in these areas. This condition is quite complex, so efforts are needed from the Government to solve it, both related to disaster risk reduction and settlement area arrangement.

Semarang city, especially in coastal urban areas, is also experiencing the same problem. Tidal inundation caused by land subsidence and tidal waves has threatened the communities living in the area [22]. Areas below sea level become flooded during high tide and when it rains heavily due to waterlogging caused by several things, such as clogged drainage due to garbage. It's also caused by several massive development activities in the northern area of Semarang and riverbanks, beach reclamation, and other characteristics are related to the ratio of river length and elevation [23]. In disaster management, Semarang City has involved many actors in preparing various long-term and short-term policy plans and on a regional and local scale [24]. The Government has included the development plan and disaster management strategy in the Regional Medium-Term Development Plan (RPJMD) and Regional Spatial Plan (RTRW) through various activities such as the construction of water polders, water pumps, revitalization of flood canal systems and river normalization, development of drainage systems, and coastal embankment construction [23].

2. Methodology

2.1. Study Area

The study area of this research is in RT 05/RW 16, the fishing village of Tambakrejo which is in Tanjung Mas Village, North Semarang District, Semarang City. Tambakrejo Fishing Village is an informal and slum area located on the north coast of Semarang City. This area is located along the banks of the Banjir Kanal Timur (BKT) and frequent flooding due to the overflowing of the BKT river and sea tides. Most of the people of Kampung Tambakrejo work as fishermen, managing ponds, industrial workers/laborers, and the informal sector such as fish traders, market traders, odd jobs, and others. The Tambakrejo informal settlement has grown on government land since 1989 [25]. The place where they live is River Basin Center (BBWS) 's land belonging to the and has a function as an irrigation area or river border, which is not intended for residential areas. In other words, these settlements grow wildly.

In May 2019, this area experienced eviction due to the flood control project and the normalization of the BKT river. This project is a national project under the Ministry of Public Works and Public Housing (PUPR), implemented by the Pemali-Juana River Basin Center (BBWS) and collaborated with the Semarang City Government. Based on the BBWS Pemali-Juana Strategic Plan 2015 – 2019, in the context of disaster mitigation and improvement of a flood control infrastructure, the Government has included a program related to the preparation for the implementation of flood control and coastal protection under the name of normalization design and riverbank reinforcement of the BKT river, Semarang City along 14.37 km for 2015 budget allocation [26]. This activity is continued in the Pemali-Juana BBWS Strategic Plan 2020-2024 with the name of BKT control activities and the allocation of funds amounting to Rp. 101 billion. The physical normalization work will be carried out from 2020 to 2021, so cleaning the BKT watershed is necessary. Therefore, many people are affected by the project, which is hundreds of residents in Tambakrejo Village. In general, there are a total of 2,172 residents whose homes or businesses will be relocated. The details of the victims consist of 1374 street vendors, 666 house buildings, 38 public facilities, 94 other buildings [25]. Especially for Tambakrejo Village, there are 137 households, of which 40 households choose to move to rental flats (Rusunawa Kudu), Genuk District, Semarang City. Meanwhile, 97 other household heads survived until the forced transfer process took place in May 2019. This forced transfer process directly and indirectly affected aspects of their lives.

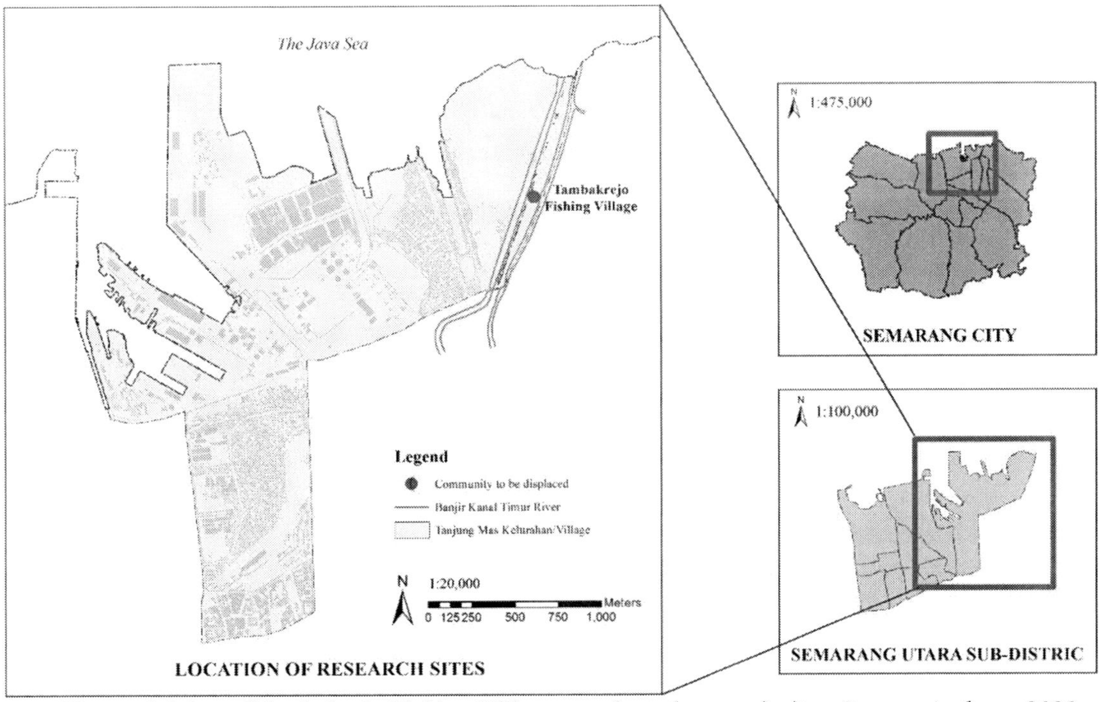

Figure 1. Map of Tambakrejo Fishing Village as selected research sites. Source: Authors, 2021

2.2. Research Methods and Data Collection

This study uses a quantitative research approach. Using this approach to describe and explain in more detail related to social phenomena into a more scientific explanation [27]. Furthermore, the data was collected primarily by using a questionnaire method to affected communities in July-August 2021. Two years after the forced displacement occurred in Tambakrejo Village. Field observations were also carried out to observe the environment around the settlements, both in temporary residential areas (Huntara) and rental flats (Rusunawa Kudu). The results of this observation are in the form of direct documentation. This study also uses mapping data obtained from government agencies and non-profit mapping service providers (Humanitarian OpenStreetMap) used for spatial analysis.

The research population is the local communities affected by the BKT river normalization project, where the method used for determining the sample is the purposive sampling method. This method is sometimes referred to as "judgment sampling". It is a non-random technique intended to select informants with the required criteria to provide information according to their knowledge to answer research questions [28]. Respondent criteria that need to be met are 1) the local communities of Kampung Tambakrejo, an adult who can represent each family head and is directly affected by the BKT river normalization project. In addition, 2) they currently live in temporary housing (Huntara) or Rusunawa Kudu. The total sample is 25 respondents, with 12 respondents living in the Huntara and 13 respondents living in Rusunawa Kudu. The determination of the sample and the criteria for respondents aim to make the identification of pre-and post-forced displacement conditions more reliable.

2.3. Variables and Analysis methods

Components and variables used to assess the impact of displacement due to the BKT river normalization project in this study refers to the framework of livelihood assets belonging to the Department for International Development (DFID). The indicator consists of five assets: social capital, natural capital, financial capital, physical capital, and human capital [29]. This approach using livelihood asset indicators helps assess the ability and availability of community resources to live. This indicator is further expanded to obtain the results of a more profound discussion by considering various topics such as land tenure, property rights, land compensation, and infrastructure services described in the situation both before and after displacement.

Table 1. Components dan sub-components of land tenure and livelihood assets

Component	Sub-component	Explanation	Source
Land tenure and property right	Land ownership	Identify the status of land ownership and property owned by the communities	[9,30]
	Property ownership		
Social Capital	Neighbor relationship	Refers to the social resources owned by the communities which are assessed from the individual's relationship with the social environment and the existence of forums for communities' participation	[9,29,31,32]
	Social networking		
	Social institutions		
	Participation		
Human Capital	Education level and access	Refers to the level of knowledge, skills, and abilities of the community so that it can support the life of the local communities	[9,29,31,32]
	Health level and access		
	Skills		
	Expertise		
Financial Capital	Type of jobs	Identify changes and effects on community's types of work, income, and expenditure	[9,29,31,32]
	Household income		
	Household expenses		
Physical Capital	Roads	Refers to the conditions and ease of access to basic infrastructure to support the lives of local communities	[9,29,31–33]
	Clean water		
	Sewage and sanitation		
	Electricity		
	Telecommunication		
	Drainage		
	Waste management		
Natural & Environment Capital	Land compensation	Refers to the existence of natural resources owned by the communities and those in the surrounding environment	[9,29,31–33]
	Land use		
	Land productivity		

To assess how the BKT river normalization project affects the livelihood assets of local people living in the Huntara and Rusunawa is to use a statistical analysis t-test. This test is used to identify significant differences between the two variables related to two cases of samples. Still, it also can be used to see the effects of behavioral variables [34].

$$t = \frac{\bar{x}_1 - \bar{x}_2}{\sqrt{\frac{s_1^2}{n_1} + \frac{s_2^2}{n_2}}} \tag{1}$$

Notes:

t = t-test value
$\bar{x}_1$ = mean group data 1
$\bar{x}_2$ = mean group data 2
s^2 = group sample standard deviation
n_1 = number of samples group 1
n_2 = number of samples group 1

The test utilizes the SPSS application, and the results of the statistical analysis of the independent t-test will be described further. So that it can be known the average of each livelihood asset (social, human, physical, financial, and natural) and how the characteristics of the different changes in these assets affect the affected communities after the forced displacement due to the BKT project in Tambakrejo Village, Semarang City.

3. Result and Discussion

3.1. Land Tenure Security of Local Community

The condition of the ownership status of the land parcels based on data BHUMI.atrbpn a plot that has not been officially registered and the value of the land zone per m squared in Tambakrejo is around IDR 500,000 – IDR 1,000,000 [35]. Although the status of the land has not to be listed on the system, according to local stakeholders' information, such land is designated as a border river and owned by BBWS Pemali-Juana. This fact makes the people do not have the right to land ownership. Tambakrejo communities began to settle in 1989 and buy land from the "landlords" at a low price, ranging from IDR 500,000 - IDR 9,000,000 per plot of land. Due to the absence of land rights, the Government does not provide compensation to their assets in the process. The local communities only give compensation in the so-called "tali-asih" to communities who choose to live in Huntara. Even this indemnity through a long process: the filing of protests, demonstrations, and advocacy assistance. Meanwhile, the communities who follow government directives and living in Rusunawa pleaded not given any compensation. These behavioral differences also raise an issue among local communities.

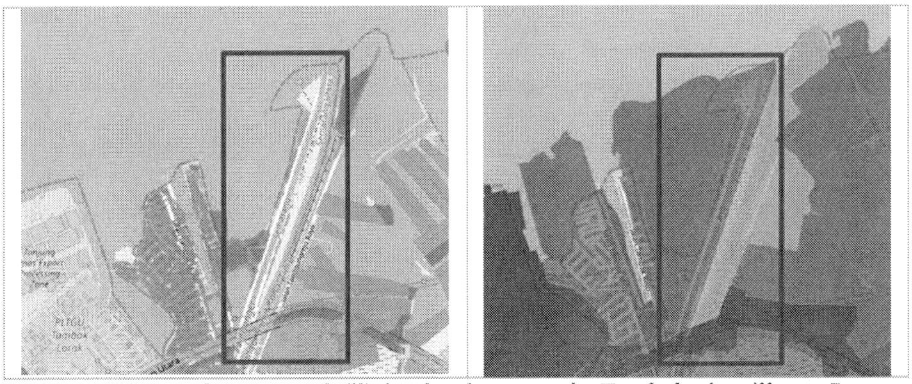

Figure 2. (i) Land status and (ii) land value zone in Tambakrejo village. Source: modified from [35]

Forced displacement directly impacts the form of loss of land assets that they have been living in - landlessness people [13]. Indirectly, this will also lead to the perception of "insecurity" due process that is considered unfair in compensation by the local Government. They were forced to move and lose assets on the public interest and flood disaster mitigation [9]. Although the project's purpose is better disaster mitigation, the process of land acquisition and displacement can also cause disturbances and provoke emotional pain on the experience of forced evictions experienced by the local communities [10].

3.2. Livelihood Impacts in the Pre and Post-Project Implementation

3.2.1. Characteristic of Livelihood Assets. Studies related to forced displacement experienced by the local communities are a dramatic phenomenon caused by government policies. It can influence and change the socio-economic life of the community [33]. Livelihood assets are used as an entry-point for understanding the phenomenon from various aspects and ways of life of local communities. The

IOP Conf. Series: Earth and Environmental Science **986** (2022) 012063 doi:10.1088/1755-1315/986/1/012063

assessment of each asset using a scale of 1-5. A score closer to 5 means that the ability and capability of communities on these assets is getting stronger.

- Social Capital
  This capital can be seen in the relationship between the individual and society. Although the local Tambakrejo communities live in an area of high hazard-prone, and almost always experience floods and tidal waves regularly. Before the forced displacement, they subjectively felt comfortable with these conditions. It is because they have been living and have a high degree of adaptation to the vulnerability. While living in the area tends to be informal slums, Kampung Tambakrejo has a social organization that actively embodies the aspirations and participation and routinely carries out social activities such as cooperation and community service. Social capital has not changed much and has even strengthened due to the relocation of the BKT river normalization project. Before displacement, local communities had social capital with a score of 4.67 (very strong) and strengthened after displacement with 4.92. In general, forced displacement will lead to social conditions of disarticulation, network destruction, and imbalance [9,13,36], because local communities must be forced to separate from their original communities. It does not occur entirely because the local people have strong family ties. However, the communities who chose to live in Rusunawa felt that the changes in the social organization system that occurred in their new neighbourhood had an intangible impact, such as the loss of the role of local champion and the existing organizational system.

- Human Capital
  The level of human resources of local communities can be identified by the level of education, health level, where education, skills, and the increasing expertise of local communities. Results of the analysis show that human capital has increased from a score of 3.20 (medium) to 3.47 (strong), strengthened 0.27 points. Basically, in Tambakrejo Village, many training activities are organized by both institutions and universities, such as seafood processing, utilization of mangrove plants for batik material, and training hydroponics to increase the capacity of human resources. Thus, even though the average level of the community's education is low (elementary-junior high school graduates), the training activities have increased the skills and knowledge of the local communities.

- Physical Capital
  Identification of physical capital is seen from the basic infrastructure conditions in the Tambakrejo communities, such as road conditions, access to water and sanitation, drainage networks, access to electricity, solid waste, telecommunications, and the presence of wifi/hotspots. The result shows that physical capital owned by the communities before the displacement was 2.15 (weak). It increased along with the rebuilding of settlements by the government, and the migration of some communities to Rusunawa with a score of 4.10 (strong). Direct impact occurred shortly after the eviction process. 97 Households lost their homes and were forced to settle in *Bedeng* (temporary buildings). The demolition of the house was the most significant loss, considering that the local communities had been living for a long time and regularly invested in the house they lived in [33]. The loss of housing will lead to chronic conditions such as alienation, deprivation, and feelings of loss of space - loss of attachment to housing [9,13,36]. However, this condition is getting better after the rebuilding of Huntara and the local community's adaptation in Rusunawa. They recognize that the basic infrastructure services are better than before relocation.

- Financial Capital
  Financial capital is believed to be the most affected due to displacement. Identification of the financial capital of communities is seen from the changes to the type of work, the amount of monthly income, and expenses. The average financial capital of the communities was 3.88 (strong) and weakened to 3.24 (medium) due to forced displacement. It happened because most local communities were forced to lose their jobs due to converting productive land (shrimp ponds, fish, and mangroves) into residential areas and displacement, which made it difficult for local

communities to access their former workplaces. Several researchers confirm that after relocation, people's incomes decrease, the majority of job losses and ultimately, this effect will increase the level of poverty of local communities [9,13,33].

- Natural Capital
  Drastic changes can be seen from natural resource assets. Original natural form such as riverbanks, ponds, and mangrove has changed with the redevelopment of temporary shelters (Huntara) in Tambakrejo Village. At first, the flood and tidal disaster mitigation activities were only in the form of normalizing the BKT river, but due to forced evictions, local communities demonstrated and negotiated, and the Government eventually rebuilt settlements for 97 households in the same location. Narrow canals, ponds, and mangroves were sacrificed for the redevelopment. This condition made the natural capital in Tambakrejo initially at 3.12 (medium) decreased to 2.48 (weak). Most communities eventually lose access to common assets (canals, ponds, and mangroves) as part of their source of income [13,36].

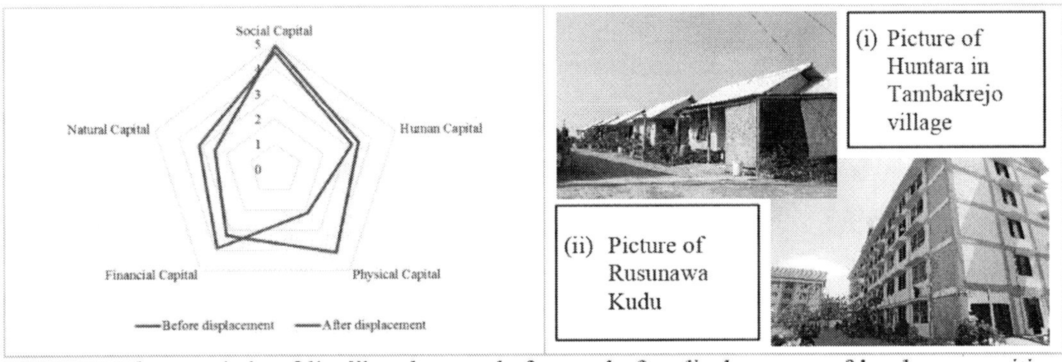

Figure 3. Characteristic of livelihood assets before and after displacement of local communities. Source: Author, 2021

3.2.2. The Difference Livelihood Conditions of Local Communities. When we try to analyze more deeply related to the living conditions of affected communities who choose to live in Huntara or Rusunawa. We found that there were different conditions. Before displacement, **Figure 4. (i)** shows that the communities have livelihood assets that are not much different from the physical capital values for the two communities with low scores and high social capital. However, after relocation, there is a significant difference, in **Figure 4. (ii)** visually, it can be seen that the natural capital of the local communities living in Rusunawa decreased, and the financial capital of the local communities living in the Huntara decreased. Further analysis was carried out by testing the independent t-test for the characteristics of changes in livelihood assets in both locations.

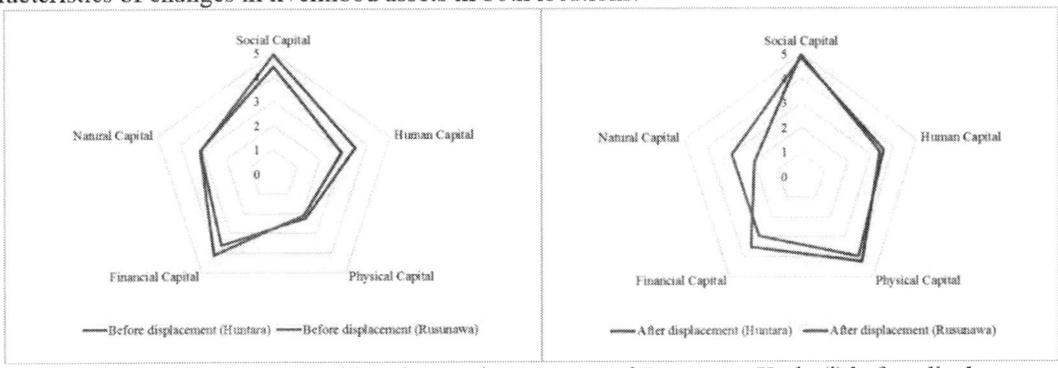

Figure 4. The difference of livelihood assets in Huntara and Rusunawa Kudu (i) before displacement and (ii) after displacement. Source: Author, 2021

Table 2. Results from independent t-test of livelihood assets in Huntara and Rusunawa

Livelihood Assets	Mean		Sig. (2-tailed)
	Huntara	Rusunawa	
Social Capital	-.056	.541	.000
Human Capital	.063	.462	.218
Physical Capital	1.750	2.122	.002
Financial Capital	-.667	-.615	.924
Natural Capital	-.142	-1.132	.020

*statistical significance level at 0.05

The test results show significant differences in conditions after the forced displacement of local communities living in Huntara and Rusunawa, which is found in social, physical, and natural capital with a sig value of < 0.05. 1) Social capital in the local Huntara community has decreased along with new settlements and reduced assistance programs. 2) physical capital tends to increase, and a significant increase occurs in the communities living in Rusunawa due to the ease of access to basic infrastructures such as water, sanitation, electricity, telecommunications, health, education, and security. 3) Natural resource capital in both locations tends to decrease, but the local community living in Rusunawa has experienced a drastic decline. They can no longer access shared resources such as ponds, canals, or the sea as part of their work due to the relatively remote location of the Rusunawa. The independent t-test also shows that the finances at both locations are similar for both locations are equally decreased. This phenomenon is proof that development projects involving forced displacement have social consequences. Although the goals of the development are reasonable, the consequences of the impoverishment of certain groups occurred.

3.3. Displacement: Dilemmas in The Implementation of Urban Planning

Development projects always have negative and positive side impacts, as well as the BKT normalization project policy, which results in changes in the local community's physical, social, and economic aspects. Conceptually, a development project will show a spatial nature, whether small or large, resulting in a 'reorganization of meaning and control of space,' meaning that it encourages natural displacement [37]. This project will directly restrict and alter some aspects of these locations, such as livelihood, social and cultural conditions. Forcing local communities to pull out and move, although no definite assurance is given by the government [37]. Cernea [38], explains that the impact of the development project will be different for each segment of the communities. This can be taken into consideration regarding the explanation of the opportunities that occur to analyse the intensity of the impact on each segment of the communities when the development is planned. This is due to what is perceived as a threat that could turn into opportunities for others. Differences in impact across age, gender, wealth, and other lines of differentiation should be carefully analysed and researched to arrive at conclusions about the nature and impact of changes that occur from development projects [37].

Based on several studies related to the impact of development projects, the economic aspect to be the riskiest is the risk of loss of personal property, employment, and access to public and social facilities [37]. Furthermore, social consequences in the form of marginalization, and social disarticulation as socio-cultural risks, while homelessness, food insecurity and increased morbidity and mortality as social welfare risks [13,38,39]. Analysis of the effect of displacement on livelihood assets has proven that there is an influence on the financial capital and natural capital of the Tambakrejo communities, which decrease in aggregate in the period before and after the forced displacement. Further found an interesting result that there is a difference between the condition of the communities who choose to live in Huntara or Rusunawa. The significant difference is seen in the aspects of social capital, physical capital, and natural capital. According to Burge [40], change in any form will have social costs and social benefits for others. The goal of both socio-economic benefits from disaster risk reduction activities is aimed at the whole communities. Regardless of that reason, the social costs that have a negative impact are always borne by the local-poor communities. In terms of regional structuring and disaster risk reduction involving displacement, the implementation of government policies seems to be a dilemma in the planning process itself. Cernea [13], emphasizes that the planning approach leading to displacement and only a portion of the "rehabilitated" planning process failed because the process cannot prevent the impoverishment of society. At the expense of marginalized communities, the planning approach showed a flaw in implementing policies and needs to be changed.

4. Conclusion

Urban planning activities, in this case, flood and tidal disaster mitigation projects, bring a noble goal in the public interest, disaster risk reduction, and the arrangement of slums. Often the Government still focuses on social-economic benefits, and the implementation of these activities often sacrifices the poor as objects of development. The study results on the implementation of the East Flood Canal (BKT) normalization project in Tambakerjo, Semarang City, which involved forced displacement, resulted in the following findings: 1) Communities do not have land rights where they live. Still, they have invested some of the money to build the house over the years. Therefore, it is difficult for them to get compensation. Ultimately, the communities experience "insecurity" in land tenure and property rights 2) In this case, livelihood assets are successfully utilized to assess the changes experienced by the communities. The implementation of projects involving forced displacement has significantly changed the situation before and after both in terms of social-human-physical-financial-natural capital. 3) Differences in local community response and post-displacement government treatment have led to different perceptions and new issues among the communities living in the Huntara and Rusunawa.

This research illustrates that the implementation of urban area redevelopment is still not fully able to achieve the spatial planning objectives in Indonesia contained in Law No. 26/2007. The objectives are to realize safe, comfortable, productive, and sustainability is still challenging to achieve. In addition, we found a study related to displacement can also be used to indicate the success-failure of the implementation of urban policy. Therefore, further research is needed, especially regarding the role of each actor and their interaction in developing policy planning. It is also necessary to study the participation of local communities in these activities to provide input on the development of science related to the dynamics of the urban population and the realignment of areas that involve forced displacement.

5. Acknowledgments

We express our deepest gratitude to the Master's Program in Urban and Regional Planning, School of Architecture Planning and Policy Development, Bandung Institute of Technology for providing material and non-material support. So, we can finish this research properly. We also thank the Semarang City government and the Tambakrejo communities for providing data and information during the field survey period.

References

[1] Hallegatte S, Green C, Nicholls R J and Corfee-Morlot J 2013 Future flood losses in major coastal cities *Nat. Clim. Chang.* **3** 802–6
[2] BNPB 2021 Kejadian Bencana Tahun 2020 *BNPB*
[3] World Bank 2021 Urban population (% of total population)
[4] Flanagan B E, Gregory E W, Hallisey E J, Heitgerd J L and Lewis B 2020 A Social Vulnerability Index for Disaster Management *J. Homel. Secur. Emerg. Manag.* **8**
[5] Tipple G 2006 Housing, Urban Vulnerability and Sustainability in Rapidly-Developing Cities *Built Environ.* **32** 387–99
[6] Wilhelm M 2011 The Role of Community Resilience in Adaptation to Climate Change: The Urban Poor in Jakarta, Indonesia *Resilient Cities* ed K Otto-Zimmermann (Springer Netherlands) pp 45–53
[7] de Sherbinin A, Schiller A and Pulsipher A 2007 The vulnerability of global cities to climate hazards *Environ. Urban.* **19** 39–64

[8] Kita S M 2017 Urban vulnerability, disaster risk reduction and resettlement in Mzuzu city, Malawi *Int. J. Disaster Risk Reduct.* **22** 158–66
[9] Nikuze A, Sliuzas R, Flacke J and van Maarseveen M 2019 Livelihood impacts of displacement and resettlement on informal households - A case study from Kigali, Rwanda *Habitat Int.* **86** 38–47
[10] Vanclay F 2017 Project-induced displacement and resettlement: from impoverishment risks to an opportunity for development? *Impact Assess. Proj. Apprais.* **35** 3–21
[11] Cernea M M 2003 For a new economics of resettlement: A sociological critique of the compensation principle *Int. Soc. Sci. J.* **55** 37–45
[12] van Noorloos F and Kloosterboer M 2018 Africa's new cities: The contested future of urbanisation *Urban Stud.* **55** 1223–41
[13] Cernea M M 1997 The risks and reconstruction model for resettling displaced populations *World Dev.* **25** 1569–87
[14] Claudianos P 2014 Out of Harm's Way; Preventive Resettlement of at Risk Informal Settlers in Highly Disaster Prone Areas *Procedia Econ. Financ.* **18** 312–9
[15] Prasetyo H B 2016 Perubahan Peruntukan Kawasan dalam Perspektif Penataaan Ruang (Studi di Kota Mojokerto) *Univ. Brawijaya Semarang* **2** 53–9
[16] Hersperger A M, Oliveira E, Pagliarin S, Palka G, Verburg P, Bolliger J and Grădinaru S 2018 Urban land-use change: The role of strategic spatial planning *Glob. Environ. Chang.* **51** 32–42
[17] Surya B, Ahmad D N A, Sakti H H and Sahban H 2020 Land use change, spatial interaction, and sustainable development in the metropolitan urban areas, south Sulawesi province, Indonesia *Land* **9**
[18] Brenner N and Schmid C 2015 Towards a new epistemology of the urban? *City* **19** 151–82
[19] Albrechts L, Healey P and Kunzmann K R 2003 Strategic Spatial Planning and Regional Governance in Europe *J. Am. Plan. Assoc.* **69** 113–29
[20] Plessis du D J 2013 A Critical Reflection on Urban Spatial Planning Practices and Outcomes in Post-Apartheid South Africa *Urban Forum* **25** 69–88
[21] Handayani W, Rudiarto I, Setyono J S, Chigbu U E and Sukmawati A M awanah 2017 Vulnerability assessment: A comparison of three different city sizes in the coastal area of Central Java, Indonesia *Adv. Clim. Chang. Res.* **8** 286–96
[22] Marfai M A, King L, Singh L P, Mardiatno D, Sartohadi J, Hadmoko D S and Dewi A 2008 Natural hazards in Central Java Province, Indonesia: An overview *Environ. Geol.* **56** 335–51
[23] Sembiring R P, Untung S H and Gading S A 2019 Pencegahan Dan Penanggulangan Banjir Dan Rob Menurut Peraturan Daerah Kota Semarang Nomor 14 Tahun 2011 Tentang Rencana Tata Ruang Wilayah Kota Semarang Tahun 2011-2031 *Diponegoro Law J.* **8** 664–87
[24] Bakti L M 2010 *Kajian Sebaran Potensi Rob Kota Semarang dan Usulan Penanganannya* (Universitas Diponegoro)
[25] Syofii M and Alfirdaus L K 2020 Koalisi Masyarakat Sipil Dalam Advokasi Kebijakan Relokasi Warga Tambakrejo Kota Semarang *J. Polit. Profetik* **8** 112–35
[26] BBWS Pemali Juana 2015 *Rencana Strategis Tahun 2015-2019* (Semarang, Indonesia)
[27] Franklin M 2008 Quantitative analysis *Approaches Methodol. Soc. Sci. A Plur. Perspect.* 240–62
[28] Tongco M D C 2007 Purposive Sampling as a Tool for Informant Selection *Ethnobot. Res. Appl.* **5** 147
[29] DFID 1999 *Sustainable Livelihoods Guidance Sheets* (London, UK: DFID)
[30] Cernea M M 2008 Compensation and benefit sharing: Why resettlement policies and practices must be reformed *Water Sci. Eng.* **1** 89–120
[31] Ashley C and Carney D 1999 *Sustainable Livelihoods Analysis: Lessons form early experience* (London, UK: Russell Press)
[32] Farrington J, Ramasut T and Walker J 2002 *Sustainable Livelihoods Approaches in Urban*

Areas: General Lessons, with Illustrations form Indian Cases (London, UK)

[33] Choi N 2015 Impacts of displacement on urban livelihoods: a railway project in Metro Manila *Dev. Pract.* **25** 643–54

[34] Kim T K 2015 T test as a parametric statistic *Korean J. Anesthesiol.* **68** 540

[35] Kementerian ATR/BPN 2019 Bidang Tanah dan Zona Nilai Tanah

[36] Jaysawal N and Saha S 2018 Impact of displacement on livelihood: A case study of Odisha *Community Dev. J.* **53** 136–54

[37] Gizachew A 2017 "Causes" of big development projects : Development - induced displacement and its socio-economic impacts on displaced people in Tekeze Dam *Int. J. Sociol. Anthropol.* **9** 114–30

[38] Cernea M M 2005 Concept and Method: Applying the IRR Model in Africa to Resettlement and Poverty *Displacement Risks in Africa: Refugees, Resettlers and their Host Population* ed I Ohta and G Yntsio (Tokyo, Japan: Tokyo University Press.) pp 195–258

[39] Koenig D 2001 *Toward Local Development and Mitigating Impoverishment in Development-Induced Displacement and Resettlement* (London, UK)

[40] Burdge R J 1987 The social impact assessment model and the planning process *Environ. Impact Assess. Rev.* **7** 141–50

Spatial analysis of damage due to the eruption of Mount Kelud in February 2014: lessons learned for better regional planning

S Tsana[1*], S I Ayuni[1] and N R Priyandianto[1]

[1] Master of Urban and Regional Planning, Bandung Institute of Technology, Jl. Ganesha No.10, Lb. Siliwangi, Bandung 40132, Indonesia

*Email: shafwatus.tsana@gmail.com

Abstract. The eruption of Mount Kelud in February 2014 released eruptive material that caused lava flooding from the sub-watershed. Lava floods cause land-use changes, land damage, and infrastructure damage. Ngantang Sub-district Malang Regency is the area most severely affected by the eruption of Mount Kelud. This paper aims as a preventive effort to reduce disaster risk through spatial evaluation of vulnerability due to lahar sediments and directives for infrastructure and settlement planning in volcanic areas. Using scoring and map overlay techniques for vulnerability analysis and planning directions. The findings highlight that villages along the river have a higher level of vulnerability and disaster risk and are not recommended to build and use land within a 100 meters radius of the river wall for various community activities. Another finding is the lack of strong indicators of disaster risk reduction capacity in Ngantang Sub-district. This phenomenon proves that there are still many challenges to realizing disaster risk reduction as mandated in Law No. 24 of 2007 and Regulation of the Head of BNPB No. 2 of 2012.

1. Introduction

Natural disasters are phenomena that threaten and cause harm to human life, both material and non-material, individuals, and communities. One type of natural disaster that has a significant impact is a volcanic eruption [1,2]. The impact of the disaster that occurs does not only happen directly (primary) when the volcano erupts but also occurs secondary (after the main event), namely lava floods that occur during heavy rains [3]. Areas affected by volcanic eruptions are generally agricultural areas and rural settlements around mountains and lava flows.

Mount Kelud is one of the active volcanoes in Indonesia, located in East Java Province [4]. The mountain, which has an explosive eruption type, showed its last volcanic activity in February 2014. The Kelud eruption that occurred was more severe than the eruption that occurred in 1990. It was caused by volcanic ash from the explosion spreading to Central Java, DI Yogyakarta, and parts of West Java. The next day, volcanic ash spread northward to Surabaya City, Banyuwangi Regency, and several cities/counties in East Java [5,6].

In addition to pyroclastic flows, the eruption of Mount Kelud, which occurred during the rainy season, caused several places to experience erosion, causing cold lava from the crater. Lava (*lahar*) has meaning as a sediment-laden mixture of water and rock debris flowing down the side of a volcano [7,8].

Lava is one of the processes capable of mobilizing large amounts of sediment and has the most damaging effects on human populations and even infrastructure [9].

The Kelud eruption in 2014 has forced people living within a 10 km radius of the crater to evacuate. The number of affected people who were displaced was 87,629 people from 36 villages [6]. When measured from the material side, the damage caused by the eruption occurred on a local and regional scale. On a local scale, a 3 km radius from the crater caused damage to tourism infrastructure, especially in Kediri Regency, such as damaged roads and bridges that were cut off due to thick sand and lava flows [6]. Previously the Center of Volcanology and Geological Hazard Mitigation (PVMBG) issued a delineation of Pandansari Village, Ngantang sub-district from a Disaster Prone Area (*Kawasan Rawan Bencana*). It was considered a safe distance from the Kelud crater, but in 2014, the village was also affected by an eruption in lava flows and volcanic material.

Furthermore, the damage to other facilities and infrastructure in the Ngantang sub-district includes clean water, road, and agricultural land [10]. On a regional scale, economic losses occurred in most Java islands due to volcanic ash's impact, which caused damage to transportation facilities, especially the aviation sector (e.g., airport closures). The eruption of Mount Kelud has had a direct or indirect impact on the surrounding community's physical, environmental, social, and economic conditions.

The current disaster management approach has undergone a paradigm shift from reactive to preventive actions [11]. The Government takes a preventative approach by building preparedness and mitigation in dealing with disasters and early warning activities to reduce disaster risk. In addition, to minimize casualties and damage caused by the disaster. Therefore, this study aims to spatially evaluate the vulnerability and impact of the lahar flood that occurs. It is expected to provide input as a preventive effort to local policymakers in dealing with disasters in Malang Regency, especially areas prone to volcanic eruptions in Kelud District. Ngantang by determining settlement planning and supporting infrastructure for disaster mitigation.

2. Method

2.1. Study Area

The study area of this research is in the Ngantang sub-district, Malang Regency. The Center for Volcanology and Geological Hazard Mitigation (PVMBG) in 2004 has eliminated Ngantang District on the Kelud Volcano Hazard Area Map. There are 3 (three) villages in the Ngantang sub-district within a 5 km radius of the Kelud Volcano crater, namely Pandansari, Ngantru, and Pagersari are included in disaster-prone areas II. On the map, these three villages are prone to volcanic ash, eruptive lava, cold lava flows, throwing stones (incandescent), and heavy ash rain [12]. Overall, the Ngantang sub-district should be included in the three classifications of areas prone to volcanic eruptions based on the KRB map of Mount Kelud.

In the eruption of Mount Kelud that occurred before 2014, the eruption's impact moved to the west, northwest, southwest, and south of the mountain so that Ngantang District, which is in the northeast of Mount Kelud, was too severely affected. However, on February 13, 2014, the eruption destroyed several hamlets, especially in Pandansari Village, including Wonorejo Hamlet, Putut Hamlet, Klangon Hamlet, and Sedaun Hamlet. The hamlets of Kelangon and Sedaun, which the Sambong River separates, are isolated because the bridge collapsed [13]. As a result of this eruption, residents within a 10 km radius of the crater fled to a safer place. The losses caused by the 2014 eruption were damage to residential facilities, public facilities, and infrastructure. In addition, it also caused damage to 5,146 ha of paddy fields, 1,792 ha of plantation agricultural land, and 260,060 ha of fruit plantations [14].

Administratively, the Ngantang sub-district has an area of 147.70 km^2 or about 4.96 per cent of the Malang Regency covering the Gunung Kelud National Park Area [10,15]. Furthermore, Mount Kelud, which has a height of 1731 MSL, is located at coordinates 07056'00" south latitude and 112018'30" east longitude. Mount Kelud is located in three regions, namely Kediri Regency, Blitar Regency, and Malang Regency, in East Java Province. Mount Kelud is one of the quarter volcanoes with a stratovolcanic type surrounded by several older volcanoes such as Mount Kawi and Mount Butak in the east, and Mount Anjasmoro in the northeast to form a rough morphology with steep hills and ravines [16,17].

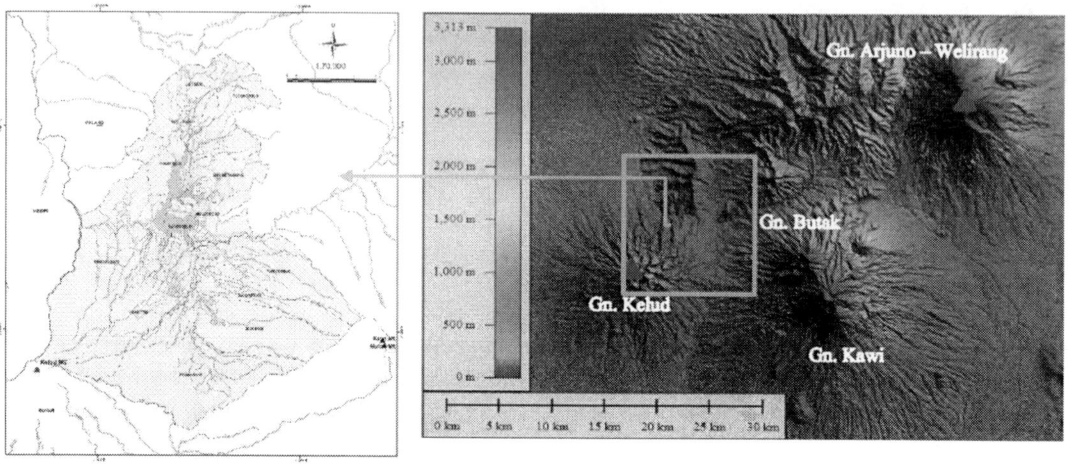

Figure 1. Map of Ngantang Sub-district as selected research site.
Source: DEMNAS BIG, 2021

Based on volcanic data, Mount Kelud has recorded nearly 33 eruptions and has experienced dozens of explosive eruptions since the 1000s. Explosive eruptions centred on crater lakes and preceded by phreatic eruptions with spilt lake water dominated this period. In addition, the condition of the crater is mixed with mud and old material. The mixture is also found in the crater and the area around the summit to the river's upper reaches around the summit. In contrast, effusive eruptions formed lava domes [15–17]. The largest eruption in Mount Kelud was the eruption that occurred in 1586, which killed 15,000 people, while the eruption that occurred on May 20, 1919, was the second-largest eruption of Mount Kelud. This eruption released volcanic material ejected from the mountain's crater, washed away 104 villages, 9,000 houses, and 135 km2 of agricultural land, and claimed the lives of 5,160 people [18,19]. Another feature of the late eruption of Mount Kelud is the formation of a lava plug in the crater vent [15,20].

Table 1. The Eruption of Kelud in the '20s

Date of Eruption	*Volcanic Eruption Index [VEI]*	*Tephra Volume [Million m³]*	*Eruption Duration [Hour]*	*Damage Radius [km]*	*Volcanic Eruption Range [km]*	*Vash Radius [km]*	*Fatalities [Person]*
May, 22-23 1901	3	200		6	27	-	exist
May, 20th 1919	4	190		5-7	37,5	10	5.160
Aug, 31th 1951	4	200	11,5	4-6,5	12	6,5	7
Apr, 24th 1966	4	90	7	2-5	31	9	210
Feb, 10th 1990	4	130	8	1-5	7,5	5	34
Nov, 3rd 2007	2	16,3					0
Feb, 13th 2014	4	219	3,2	1-7	18-32	4,9	7

Source: [6,10–12]

2.2. Data Collection

The research approach used in this research is descriptive quantitative using case studies, namely problem solving by describing a phenomenon based on data and facts in the field. Case studies are used in small areas to be more specific and not broad. In this study, the author uses a literature review approach that discusses the depth of the topic and the spatial method used to describe data and facts related to conditions in the field.

Furthermore, data collection was carried out using interviews with affected communities and local governments in April 2014. In addition, secondary data collection used data from journals, articles, books, government agencies, as well as spatial data issued by official agencies, such as data following map: (i) Administrative map of Ngantang District, Malang Regency with a scale of 1:25,000; (ii) Map of the Malang Regency road network with a scale of 1:25,000; (iii) Malang Regency spatial plan map; (iv) High-resolution satellite images for 2013-2015, 2017 and 2020 sourced from LAPAN; (v) Map of Volcanic Disaster Prone Areas issued by the Center of Volcanology and Geological Hazard Mitigation.

The population in this study was local communities affected by the eruption. Where the method used to determine the sample is the purposive sampling method. This method is sometimes referred to as "judgment sampling".

2.3. Variables and Analysis Methods

The variables used to assess the impact of the eruption of Mount Kelud in this study refer to the disaster risk framework derived from the Regulation of the Head of the National Disaster Management Agency (BNPB) with the variables of threat, vulnerability, and capacity to lahar flow.

Table 2. Component and Sub-component of Disaster Risk

Component	***Sub-component***	***Explanation***
Hazard	Areas affected	The affected area in an eruption of primary and secondary as a cloud heat, lava, material, and flood cold incandescent lava
Vulnerability	Physic	Built-up area, the density of buildings, distance to roads, distance to health facilities, quantities of health facilities
	Economic	Quantities of poor people
	Social	Population Density, Population growth rate, number of elderly and under-five people, education level
	Environment	Land Use (Protected Area such as conservation forest)
Capacity		It includes regulations and institutions for disaster mitigation, early warning, and disaster risk assessment, disaster education, and preparedness development.

Source: [10,13–15]

The analytical methods used in this study are scoring and overlay methods and map interpretation. In addition, this research uses scoring to determine the level of each component with low, medium, and high index units. Only the threat and vulnerability components will be analyzed in this study, while the capacity component will be the subject of discussion. The formulation of disaster risk is as follows:

$$r = h \times v \tag{1}$$

Explanation:
r = risk
h = hazard
v = vulnerability

The overlay technique is carried out to see changes in land use due to lava floods, determine residential areas that are safe from lava floods, and determine mitigation infrastructure in locations that have potential as disaster evacuation sites and community mitigation patterns.

3. Result

3.1. Spatial Assessment of Vulnerability Due to Lava Flood

Heavy rains that occurred for approximately 3.5 hours resulted in material deposits upstream being carried away by rain currents and entering the cross-section of the river carrying lava. Pyroclastic materials mobilized from the crater Kelud through Kali Konto and Kali Sambong up to 3 km. The material consists of stones, falling blocks, and gravel-sized material (2 cm to 15 cm in diameter), whereas volcanic ash spreads up to a distance of 10 km from Kelud Crater [5,16]
. The eruption of Kelud had a significant impact on the damage to village infrastructure, agricultural land, and settlements around the Mount Kelud area. The amount of loss due eruption of Kelud predicted at IDR 1.2 trillion, then Ngantang Sub-District is estimated at IDR 252 billion.

Tabel 3. The Value of Damage to Facilities and Infrastructure in Ngantang District Due to Eruption 2014

	Units of Million Rupiah					
	Residential	*Sanitation Facilities*	*Education Facilities*	*Health Facilities*	*Government Offices*	*Places of Worship*
Damaged	173.180	1.602,81	857,5	122,5	857,5	245
Moderately Damaged	53.235	-	180	135	335	360
Minor Damaged	20.335	-	87,5	37,5	75	262,5
Total	*246.750*	*1.602,81*	*1.125*	*295*	*1.267,5*	*867,5*

Source: [17]

Based on the results of disaster risk analysis, Pandansari, Pagersari, Sidodadi, and Banjarejo villages have the highest risk of volcanic disasters. Based on the results of the calculation of the scoring of the indicators of disaster threat and vulnerability. However, calculating a vulnerable index of each sub-component has different values. Meanwhile, areas classified as disaster areas with low classes are Banturejo, Kaumrejo, and Sumberagung village. The results of the disaster risk analysis as one of the considerations for optimizing early warning systems, disaster mitigation systems in regional spatial planning, empowering communities in disaster-prone areas, and optimizing the functions and services of infrastructure damaged by disasters [10].

Table 4. Component Scoring of Risk Disaster

Village	*Hazard*	*Vulnerability*	*Risk Level*
Banjarejo	High	Medium	Medium
Banturejo	Low	Medium	Low
Jombok	Medium	Low	Low
Kaumrejo	Low	Low	Low
Mulyorejo	Medium	Low	Medium
Ngantru	Medium	High	Medium
Pagersari	High	High	High
Pandansari	High	High	High
Purworejo	Medium	Low	Medium
Sidodadi	High	Medium	High
Sumberagung	Low	Low	Low
Tulungrejo	Medium	Low	Medium
Waturejo	Medium	Low	Medium

Source: Modified from [10,18,19]

Volcanic material carried by the river erodes land on the banks of the river, widens the river channel, and lowers the riverbed [4,12,20,21]. Lava flows cause river widening, threaten settlements and agricultural land on the banks of the river. The eruption in 2014 increased the width of Kali Konto by 5-10 meters [12]. The results of the overlay map of the slope and the map of the area prone to volcanic disasters, Pandansari has the highest level of risk. Konto River flows west while the Sambong River flows in the northern of Pandansari Village. Konto River has a river morphology with a U-shaped valley that causes horizontal erosion to be more dominant than vertical erosion and has a vulnerability to lahar material overflow [20]. The land uses of forests, shrubs, and local agriculture located around the walls of the Kali Konto and Kali Sambong rivers have the highest vulnerability to lava floods.

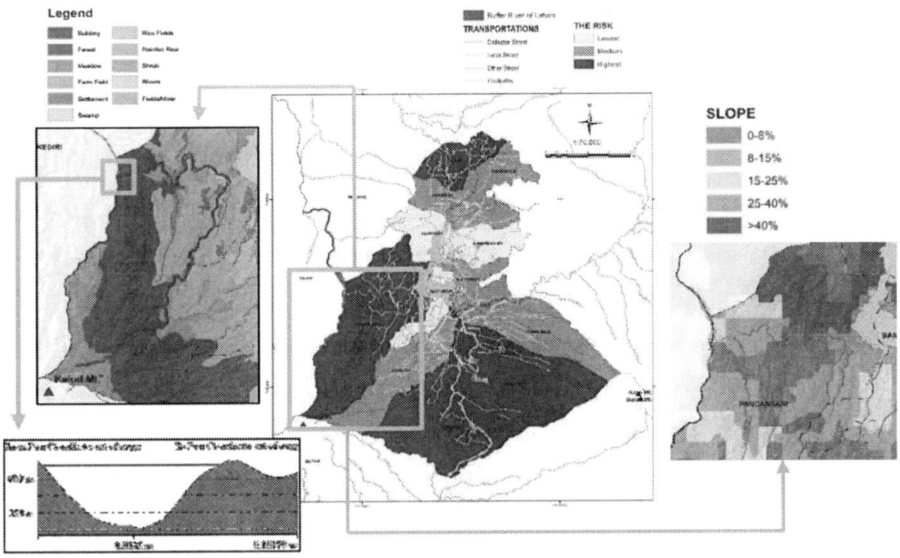

Figure 2. Risk Disaster Maps of Kelud and Areas vulnerable to lava flood
Source: Modified from [13,19] and LAPAN, 2021

3.2. Settlement and Mitigation Infrastructure Planning at Volcanic Areas

Based on land use data, the area of settlements and places of activity in Ngantang Sub-District is 6,796,504.23 m2. Residential areas are clustering with a pattern following a network of roads and river channels. Based on data from the Central Statistics Agency of Malang Regency, the population of Ngantang District in 2014 amounted to 59,278 people. The citizens of Pandansari Village were 5,528 people, Ngantru Village was 5,490 people, and Pagersari was 3,371 people. The population of the Ngantang Sub-District in 2020 is 65,894 people, and the projection of residents of the Ngantang sub-District is 78,990 people until 2035. Residents in Ngantang District were growing up and predicted to increase every year. The need for residential land will go hand in hand with population growth and allow finding space that suits the interests of community activities. In this case study of Ngantang Sub-District, the community's livelihoods are dominated by farmers, thus enabling the community to open land for settlements close to agricultural land.

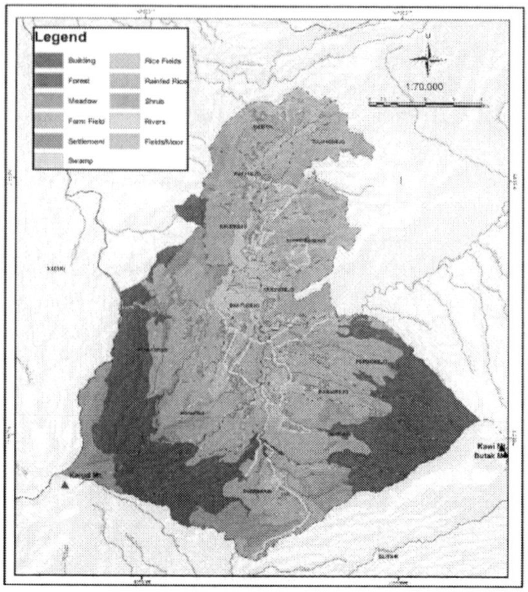

Figure 3. Land Use Map of Ngantang Sub-district.
Source: Modified from BIG and LAPAN, 2021

However, on the fact, residential areas are located in areas prone to volcanic disasters being Pandansari, Ngantru, and Pagersari villages [19]. The results of overlay analysis between land use in residential areas, slopes, and disaster-prone areas are disclosing residentials in Pandansari Village are 6,794.74 m2 or 1% of the built-up areas in Pandansari Village, which has the potential carried away by lava flows. Besides Pandansari Village, residential areas in Banjarejo and Sidodadi villages are also potentially to carried away by lava flows from Mount Kelud through the Laharkletek River flow.

Table 5. Population Projection of Ngantang Sub-District of 5 years 2020-2035

Tahun	*Proyeksi Penduduk (Jiwa)*
2020	65.894
2025	69.998
2030	74.359
2035	78.990

Source: Author, 2021

Road infrastructure in Ngantang Sub-District forms collector roads, local roads, and environmental roads. The type of road surface is asphalt/concrete pavement with good condition and welcomed by cars or more wheeled motorized vehicles [22]. Kali Sambong, one of the large rivers in Pandansari Village, has a bridge connecting separate hamlets (Klangon, Sedawun, Putut, and Wonorejo). In addition to roads, one of the supporting infrastructures for disaster mitigation is the availability of health facilities. Based on the Central Statistics Agency, the available health facilities in Ngantang Sub-District are Public Health Centers, Polyclinics, Sub-Puskesmas, Posyandu, and Polindes. The availability of these health facilities is still considered sufficient to meet the needs of the Ngantang Sub-District community until 2035 (SNI 03-1733-1989. Urban Housing Area Planning Procedures).

Table 6. Health Facilities at Ngantang Sub-district.

Village	*Maternity Hospital*	*Puskesmas*	*Polyclinics*	*Sub-Puskesmas*	*Posyandu*	*Polindes*
Pagersari					5	
Sidodadi	1				7	
Banjarejo					5	
Purworejo					4	1
Ngantru	1		1		5	1
Banturejo				1	4	
Pandansari					7	1
Mulyorejo					6	
Sumberagung			3		6	1
Kaumrejo		1	1		6	
Tulungrejo					4	1
Waturejo					6	1
Jombok				1	6	
Total	**2**	**1**	**5**	**2**	**69**	**6**

Source: [22]

From the results bout the carrying capacity and capacity of the environment in Ngantang District, the direction of settlement planning is directed to Kaumrejo Village and Sumberagung Village. In this case, the results of the disaster risk analysis are taken into consideration for determining the direction of development. It's by looking at physical, social, economic, and environmental vulnerability indicators and the availability of disaster mitigation infrastructure (road quality and availability of health facilities). It isn't recommended to build or use land within a radius of 100 meters from the river for community activities. It is necessary to strengthen the river walls so that they are not easily carried away by the flow of lava if there is another lava flood.

4. Discussion

According to Law no. 24/2007 concerning Disaster Management, implementing disaster management is a struggle with formulating development policies that create a disaster risk, disaster prevention activities, emergency response, and rehabilitation. Referring to the Regulation of the Head of the National Disaster Management Agency No. 1/2012 concerning General Guidelines for Disaster-Resilient Villages, the biggest disaster victims are the poor at the community level and those affected by the disaster. One of the assessment indicators for disaster risk reduction capacity is a strong institutional basis for implementing disaster mitigation through legal and policy frameworks [13]. On the other hand, the Regional Spatial Plan (RTRW) of Malang Regency only explains the existence of disaster-prone areas. It does not explain the supporting infrastructure for spatial planning for disaster mitigation even though Ngantang Sub-District is considered unsafe against volcanic eruptions. In this case, the capacity index in disaster risk reduction in Ngantang District is still not appropriate and needs attention.

In the Regional Spatial Plan (RTRW) document, Ngantang Sub-District, one of the main urban development areas in the area as a centre for local activities (PKL) and *agropolitan* development [23]. The RTRW also mentions efforts for natural disaster-prone areas because urban area development must be based on disaster mitigation and adaptation as stated in Presidential Regulation No. 28/2012 concerning Spatial Planning for the Java-Bali Island. Where efforts to manage volcanic eruption-prone areas include: (1) Volcanic eruption-prone areas must be protected by providing liquid lava flow channels. (2) In areas that are indicated to be affected by volcanic eruptions, such as prone I, prone II, and prone III, areas of lahar flow are also prohibited from being used for people's daily activities, especially for settlements. Volcanic disaster mitigation means a series of volcanic disaster risk reduction efforts, both through physical development and awareness and capacity building, to face the threat of volcanic disasters [20]. Several non-physical capacity-building efforts that can be implemented are regulatory arrangements such as disaster risk reduction policies and disaster risk studies. In addition, other efforts are implementing the implementation of information dissemination through education and culture and strengthening preparedness with effective responses at all levels.

Pandansari, Ngantru, Pagersari, and Sidodadi villages are included in disaster-prone areas III based on the Kelud Volcano KRB Maps. Meanwhile, refers the results of the research analysis, Pandansari, Pagersari, and Sidodadi villages have the highest disaster risk, both in terms of potential lava flows and volcanic material ejections. In the plan document, a disaster risk assessment must still pay attention to the actual conditions of disasters in disaster-prone areas [10,24]. The research of disaster risk analysis for optimizing the early warning system, empowering communities in disaster-prone areas, and optimizing the functions and services of infrastructure damaged by disasters. Disaster risk assessments must consider the formulation of disaster risk areas, environmental, physical, economic, and social factors. Disaster risk areas are needed to mitigate future disasters to reduce losses caused by disasters [10]. One of the physical efforts to reduce disaster risk, especially for lahar flows, is to build SABO DAM, allocate potential evacuation locations and evacuation routes.

4.1. Building Sabo DAM Kali Konto

In the rainy season, deposits of volcanic material that settle around peaks, slopes, and river channels mix with rainwater, turning into cold lava flows. Lava flow can be in the form of debris flow or mud flow with a large drag force so that the sediment transported downstream has a large volume and high speed. Not only to control debris flow, flash floods, and excess sediment flow, Sabo also has the function of conserving water resources, preserving the river environment and the area around the river, especially upstream of peaks and mountain slopes [25,26].

Sabo was first introduced in 1970 in volcanic areas for handling rain lava flows, such as in the areas of Mount Merapi, Kelud, Semeru, Agung, and Galunggung. However, sediment movement management has been used in Indonesia since the Dutch era [27,28]. The concept of controlling the flow of lahars in general from upstream to downstream has been recognized as a very environmentally friendly concept. The conception uses a structural approach by applying a non-structural approach, such as vegetation and contours of sediment source areas that can land landslides [28]. This concept also puts forward the balance of upstream-downstream sediment transport. Sabo also has a function to maintain the continuity of sediment supply in the river channel and maintain the river [29,30].

Comprehensively, areas that need to be wary of lahars are areas near valleys or downstream rivers. At the same time, their expansion often occurs, especially in bends of rivers with low cliffs (U-shaped valleys) such as Kali Konto and Kali Sambong.

4.2. Provision of Temporary and Final Evacuation Paths and Places

The provision of evacuation routes is another disaster mitigation effort. The determination of evacuation routes must pass through a safe route from dust caused by volcanic eruptions and avoid watersheds or lava pools if at any time it can become a cold lava flood disaster. In planning, this evacuation route must connect to the bunker or the location of public facilities/social facilities belonging to the local village government or an evacuation location [31]. Signage or unique signs are instructions for evacuation routes with clear and well-planned procedures for people living around the volcano. Therefore, during the Kelud eruption, people are not confused when evacuating and looking for a safe place.

In addition to the evacuation route, an evacuation site in a disaster event is significant. It is because natural disasters are challenging to predict. Comprehensive information and meticulous planning are needed for natural disaster mitigation. It is essential to form awareness and warning to the surrounding community. In this case, the Government must have determined potential locations that can be used as evacuation sites, both spatially as contained in the regional spatial plan also a detailed regional spatial plan. Determination of evacuation locations can use existing buildings such as school buildings/village halls. Moreover, the determination of evacuation location must be considered with other proponent indicators. The proponent indicators are the distance from the evacuation location to the main road, the availability of health facilities, and the completeness of clean water, electricity, and sanitation.

In this research, the authors directed the evacuation locations in Jombok Village, namely SD Negeri 3 Jombok and SMP Negeri 3 Ngantang. Based on previous analysis and research results, Jombok Village has the lowest risk disaster level [10]. The Jombok Village is linked with a collector road that is considered strategic. In addition, the distance between the two evacuation sites is also close to health facilities, namely the Ngantang Health Center where is approximately 2 km away.

Figure 4. Map of Disaster Evacuation Site for Eruption Kelud.
Source: modified from [25,26]

5. Conclusion

The results showed that villages along the Konto River, Sambong River, and their tributaries had a higher risk of flooding than villages that did not pass. The morphology of the river dramatically affects the rate of the lava flow. In addition, the direction of planning for settlements and supporting infrastructure for disaster mitigation is directed from the centre to the north of Ngantang District. It is based on the results of disaster risk analysis which show low disaster vulnerability. The availability of infrastructure that supports the mitigation process is more available in the region.

This research illustrates that the capacity index in Ngantang District is still not fully able to achieve the disaster risk reduction objectives contained in Law no. 24/2007 and Perka BNPB No. 2/2012. It is because the policy is still implicit and additional development.

Acknowledgments

We express our deepest gratitude to the master's program in Urban and Regional Planning, School of Architecture Planning and Policy Development, Bandung Institute of Technology, for providing material and non-material support so that this research can be completed properly. And the National Institute of Aeronautics and Space for providing multi-temporal satellite imagery of this research.

References

[1] Suparman T W 2015 Sabo untuk penanggulangan bencana akibat aliran sedimen

[2] Gschneidner K A, Bunzli J-C G and Pecharsky V K 2004 *Handbook on the physics and chemistry of rare earths* (Elsevier)

[3] Munir M D 2019 Bangunan Sabo Dam, Fungsi dan Potensinya Sebagai Pendukung Geowisata Gunung Merapi Sabo Dam Structure, Its Function and Potency as part of Geotourism in the Merapi Volcano

[4] Thouret J-C, Abdurachman K E, Bourdier J-L and Bronto S 1998 Origin, characteristics, and behaviour of lahars following the 1990 eruption of Kelud volcano, eastern Java (Indonesia) *Bull. Volcanol.* **59** 460–80

[5] Bachri S, Utaya S, Nurdiansyah farizki dwitri, Nurjanah A, Tyas lela wahyu ning, Purnama denny setia and Adillah akhmad amri 2017 Hidup Selaras Bersama Gunung Api : Kajian Dampak Positif dari Letusan Gunung Api Kelud Tahun 2014 sebagai Modal Pembangunan Berkelanjutan *Pros. Semin. Nas. Geogr. UMS 2017 PENGELOLAAN Sumberd. Wil. BERKELANJUTAN* **3** 147–58

[6] Heriwaseso, A; Andreastuti, S. D.; Budianto, A.; Santosa, I.; Kristiawan, Y. ; Natalia M C . K B K D 2017 Karakter Erupsi Kelud 2014, Pembelajaran Dalam Mitigasi Infrastruktur Di Kawasan Rawan Bencana *Seminar Nasional Kebumian Ke-10* (Yogyakarta: Departemen Teknik Geologi FT. UGM) pp 1314–32

[7] Iverson R M 2014 Debris flows: behaviour and hazard assessment *Geol. today* **30** 15–20

[8] Hadmoko D S, de Belizal E, Mutaqin B W, Dipayana G A, Marfai M A, Lavigne F, Sartohadi J, Worosuprojo S, Starheim C C A and Gomez C 2018 Post-eruptive lahars at Kali Putih following the 2010 eruption of Merapi volcano, Indonesia: occurrences and impacts *Nat. Hazards* **94** 419–44

[9] Franco-Ramos O, Castillo M and Muñoz-Salinas E 2016 Using tree-ring analysis to evaluate intra-eruptive lahar activity in the Nexpayantla Gorge, Popocatépetl volcano (central Mexico) *Catena* **147** 205–15

[10] Syiko S F, Rachmawati T A and Rachmansyah A 2014 Analisis Resiko Bencana Sebelum dan Setelah Letusan Gunung Kelud Tahun 2014 (Studi kasus di Kecamatan Ngantang, Malang) *Indones. J. Environ. Sustain. Dev.* **5** 22–9

[11] Pratomo I, Boudier J-L and Vincent P M 1991 The February 1990 eruption of Kelut volcano (East Java, Indonesia). Preliminary data **312** 777–83

[12] Dibyosaputro S, Dipayana G A, Nugraha H, Pratiwi K and Valeda H P 2015 Lahar at kali konto after the 2014 eruption of kelud volcano, east java: impacts and risk *Forum Geografi* vol 29

[13] BNPB 2012 *PERKA BNPB No. 2 Tahun 2012 Tentang Pedoman Umum Pengkajian Risiko Bencana*

[14] Sumekto D R 2018 Pengurangan Resiko Bencana Melalui Analisis Kerentanan dan Kapasitas Masyarakat dalam Menghadapi Bencana

[15] Kardono P and Pinudji S D E 2010 Penginderaan Jauh Untuk Penganggulangan Bencana *J. Dialog Penanggulangan Bencana* **1** 63–75

[16] Maulana E 2014 Pemetaan Daerah Rawan Jatuhan Material Piroklastik: Kasus Erupsi Gunungapi Kelud 2014 pp 27–51

[17] BPBD Kabupaten Malang 2014 *Data Kerusakan Akibat Erupsi Kelud 2014*

[18] Irawan I G A, Diara I W and Bhayunagiri I D A B P Pemetaan Potensi Lokasi Pengungsian Akibat Bencana Letusan Gunung Agung di Kabupaten Karangasem Berbasis Sistem Informasi Geografis (SIG) *J. Agroekoteknologi Trop. ISSN* **2301** 6515

[19] Pusat Vulkanologi dan Mitigasi Bencana Geologi Peta Kawasan Rawan Bencana Gunungapi Wilayah Jawa

[20] Wijaya I P K and Utama W 2016 Pemetaan Tingkat Kerawanan Bencana Lahar Hasil Erupsi

Gunungapi Kelud *J. Dialog dan Penanggulangan Bencana* **7** 31–41

[21] Thouret J C, Antoine S, Magill C and Ollier C 2020 Lahars and debris flows: Characteristics and impacts *Earth-Science Rev.* **201** 103003

[22] Badan Pusat Statistik Kabupaten Malang 2020 *Kecamatan Ngantang Dalam Angka 2020* (Malang)

[23] Anon 2010 *PERATURAN DAERAH KABUPATEN MALANG NOMOR 3 TAHUN 2010 TENTANG RENCANA TATA RUANG WILAYAH KABUPATEN MALANG*

[24] Ainun Y 2014 Gunung Kelud Meletus, Kota Batu "Mati" dan Jalur Malang-Kediri Ditutup *Kompas.com*

[25] Gunawan H, McCausland W A, Pallister J S, Iguchi M and Nakada S 2019 The eruptions of Sinabung and Kelud volcanoes, Indonesia *J. Volcanol. Geotherm. Res.* **382** 1–5

[26] Rustan R and Purqon A 2016 PEMODELAN ALIRAN DEBRIS FLOW UNTUK ANALISIS POTENSI LONGSORAN STUDI KASUS: PEGUNUNGAN FISHHWAK, CALIFORNIA *J. ONLINE Phys.* **2** 6–10

[27] Puslitbang Sumber Daya Air 2014 *Konsep Penataan Ruang Pada Daerah Rawan Bencana Sedimen: Teknologi Sabo Sebagai Elemen Pengendali Banjir Lahar Dalam Penataan Ruang di Kawasan Gunungapi Merapi* (Bandung)

[28] Sukatja C B and Alfianto A 2017 Revitalisasi Sabodam Sebagai Pengendali Aliran Lahar (Studi Kasus Di Daerah Gunungapi Merapi) *J. Tek. Hidraul.* **8** 29–42

[29] Dermawan V and Sisinggih D 2017 ANALISA POTENSI SEDIMEN DEBRIS DI DAS KONTO PASCA ERUPSI GUNUNG KELUD 2014 *J. Tek. Pengair. J. Water Resour. Eng.* **8** 231–40

[30] De Bélizal E, Lavigne F, Hadmoko D S, Degeai J-P, Dipayana G A, Mutaqin B W, Marfai M A, Coquet M, Le Mauff B and Robin A-K 2013 Rain-triggered lahars following the 2010 eruption of Merapi volcano, Indonesia: A major risk *J. Volcanol. Geotherm. Res.* **261** 330–47

[31] Supriatin L S 2017 KONSEP DAERAH ALIRAN UDARA ERUPSI GUNUNG API DAN APLIKASINYA *Ber. Dirgant.* **18**

IOP Conf. Series: Earth and Environmental Science **986** (2022) 012065 doi:10.1088/1755-1315/986/1/012065

The determinant factors of landslides in Lima puluh kota regency using bivariate statistical analysis 2015-2020

Ahyuni[1*], B B Susetyo[1], F Oktari[1], A Karim[1], A M Putriana[1] and H Nur[2]

[1]Department of Geography, Universitas Negeri Padang, Indonesia
[2]Department of Urban and Regional Planning, Universitas Bung Hatta, Indonesia

*Email: ahyuniaziz@fis.unp.ac.id

Abstract Regarding disasters, West Sumatra Province have multi-hazard risks. Lima Puluh Kota Regency as a part of the region was frequently hit by floods, landslides, and tornadoes. Especially, almost all of its districts are prone to landslides. The purpose of this study is to analyse the significant factors influencing or controlling landslides in susceptible areas. Eleven natural and human-related factors were supposed to influence were analyzed.This study used bivariate statistical analysis with weight of evidence (WOE) method. The 142 land slides points occurrence from 2015 to 2020 had collected from historical records and field survey. Data divided into 83 train and 59 test data set. The results show there are eight factors that influences landslide occurrence: geology condition, landform, land cover, elevation, density of vegetation greenness, slope, rainfall intensity, and proximity to stream. The other three factors such as soil type, curvature of the slope, and aspect of slope were not taken into account because they are not considered to have a significant influence on landslides.

1. Introduction

West Sumatra Province is called "the hazard supermarket" for its various susceptibility to disasters. Lima Puluh Kota Regency as a part of this region is frequently hit by floods, landslides, and tornadoes. Especially for landslides, almost all of its districts are prone to this hazard. There were recorded 127 landslide points based on data from the Local Disaster Management Agency (BPBD) in Lima Puluh Kota Regency for the last five years. From 2016 to 2019, respectively, there were 25 occurrance in 2016,10 in 2017, 22 in 2018, and 11 in 2019 that caused damage to roads, agricultural land, and houses.

Landslides are one type of soil or rock mass movement, or a mixture of both, down or out of the slope as a result of the disruption of the stability of the soil or rocks that make up the slope. Landslides are part of ground motion, the types such as fall, topple, slump, slide, flow, lateral spread, creep and landslides compound. The causes of landslides are due to two factors, natural and human. Landslide is directly related to natural properties such as geology, vegetation density, slope, slope direction, plan curvature, distance from the river, lithology type, distance from structure, and rainfall intensity. Also related to human factors in terms of land use and community activities in landslide-prone areas

The landslide susceptibility analysis carries out with three methods [1]. First, heuristic method is based on geomorphological analysis and weighting of each parameter based on expert judgment as a preliminary study that provides an overview on a small scale. Second, statistical method is based on the relationship between distribution landslide and control factors in each parameter class. This method is appropriate for the spatial planning level on a medium scale and the third, a deterministic method that based on slope safety factor analysis is used for the design of infrastructure development.

This study refers to statistical method aims to determine the factors of landslides in Lima Puluh Kota Regency based on bivariate statistical analysis using the WOE (Weight of Evidence) method to indicate the importance of the presence of the factor for the occurrence of the landslides.The advantage of this method is that its easiness and practicality to apply based on evidence. Researches on this topic have been widely applied in many places or compared with other methods [2–7].

2. Study Area

Lima Puluh Kota Regency is located between 0⁰25'28.71"N - 0⁰22'14.52" South Latitude and 100⁰15'44.10"E - 100⁰50'47.80" East Longitude. This region consists of 13 sub-districts with total area approximately 3,354.30 Km^2. The geological conditions of Limapuluh Kota Regency are varied, influenced by the dynamic Sumatra Fault System (SFS) [8] which is evidenced by the discovery of igneous rocks such as basalt, tuff, andesite, pumice, sedimentary rocks such as sandstone, limestone and metamorphic rocks such as slate, quartz, and philite. This geological formation is important to be studied because it is one of the main factors that cause landslides [9]. The topography of the area varies in the range of flat, undulating, and hilly. It has a height between 110 meters and 2,261 meters from sea level. In this area there are 3 inactive volcanoes namely Mount Sago (2,261 m), Mount Bungsu (1,253 m), and Mount Sanggul (1,495 m), and 13 large and small rivers. According to the 2020 national census, the regency population is 383.525. The research area is shown in figure 1.

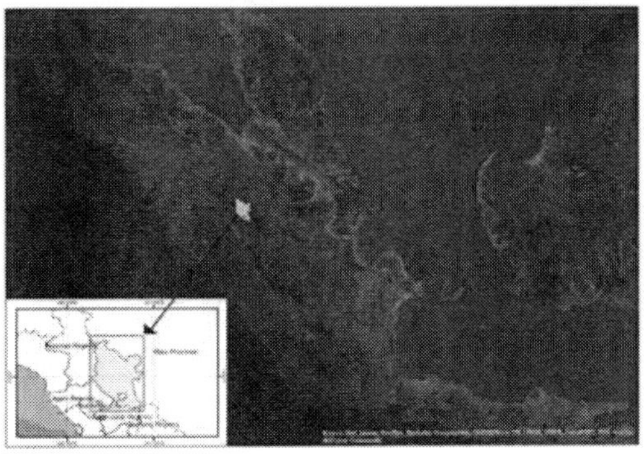

Figure 1. The research area

3. Material and Method

This research used bivariate statistics applying the measurement of Weight of Evidence (WOE) model to determine significance factors controlling landslides. The bivariate method assumes that landslide susceptibility factors were not inter-relatedand based on the Bayes' rule [3,10].Weight of Evidence (WOE) is calculated as follows:

$$Wij^{+} = \text{Log e} \frac{P(B/D)}{P(B/\bar{D})} \quad (1)$$

$$Wij^{-} = \text{Log e} \frac{P(\bar{B}/D)}{P(\bar{B}/\bar{D})} \quad (2)$$

Where:

W_{ij}^{+} = The positif weights of evidence of the j^{th}parameter class of ith landslide
W_{ij}^{-} = The negatif weights of evidence of the j^{th}parameter class of ith landslide
P = Probability
B = presence of the landslide evidential feature
D = The number of landslide belonging in the evidential feature
$\bar{B}$ = The total area on the map where the evidential feature is absent
$\bar{D}$ = The number of landslide not belonging in the evidential feature

The positive value of W_{ij} indicate the importance of the presence of the factor for the occurrence of the landslides otherwise the negative value indicates the factor is not favorable or absent [6]. The results of WOE calculation are presented in a probabilistic map. The greater the value, the stronger the predictive ability for susceptibility analysis. According to Bonham-Carter et al., 1989 [6], a value between 0.1-0.5 are middle predictive, 0.5-1 are moderately predictive, 1-2 are strongly predictive, and greater than 2 are extremely predictive.

The contrast (Cij) measures the spatial association between the analyzed parameter and landslide occurrence. Cij is the contrast value of the jth parameter class of ith evidential map with a positive or negative value. A class of parameters has a greater predictive power when $C > 1$ and very significant when the value is approximately 2 [6].

$$C_{ij} = W_{ij}^{+} - W_{ij}^{-} \tag{3}$$

The next step is a calculation of the value of the Area Under Curve (AUC) which measures the reliability of association of the parameters to landslides occurrence. This step is a model validation by comparing the susceptibility measurement of the training sets that were used for building the models and the test set that were not used during the model building process. The AUC in each parameter class measures by divide the percentage of the total area of the parameter class with the percentage of the total number of landslides. The AUC value of the factors prediction of landslides is 0.6 to 1. A factor regards as most predictive to the landslides analysis when its value is closer to 1. The classification value of AUC are: < 0.6 (as a poor model), 0.7-0.8 (as a medium or reasonable model), 0.8-0.9 (as a good model), 0.9 (very good model) [4,11]. Steps in the research are shown in the following flow chart diagram.

There was 142 landslides points occurrence collected from BNPBD historical record and field survey. The data was divided into 83 train and 59 test data set in proportion 60:40 (as shown in figure 2). Data of elevation, aspect of slope, curvature, and slope were processed from DEM Nas that obtain from Inageo portal BIG, with a resolution of 8.3 meters. The distance from the stream data is obtained from BIG by buffering the area. Lithology data were obtained from geological map. Landforms and soil types data from the BBSDLP semi-detailed soil map. Land cover data was obtained from Image Spot 7 in 2020. Rainfall intensity data was obtained from the West Sumatra PSDA and processed using the IDW method. NDVI data is processed from Landsat 8 imagery, and then pan sharpened to get a resolution of 15 meters. Data in pixel 25 x 25 meters and presented at a scale of 1:50,000.

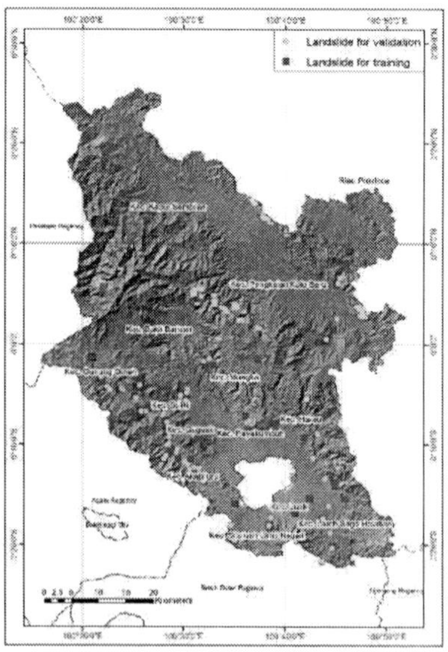

Figure 2. Landslide points occurence devide to train and test data set

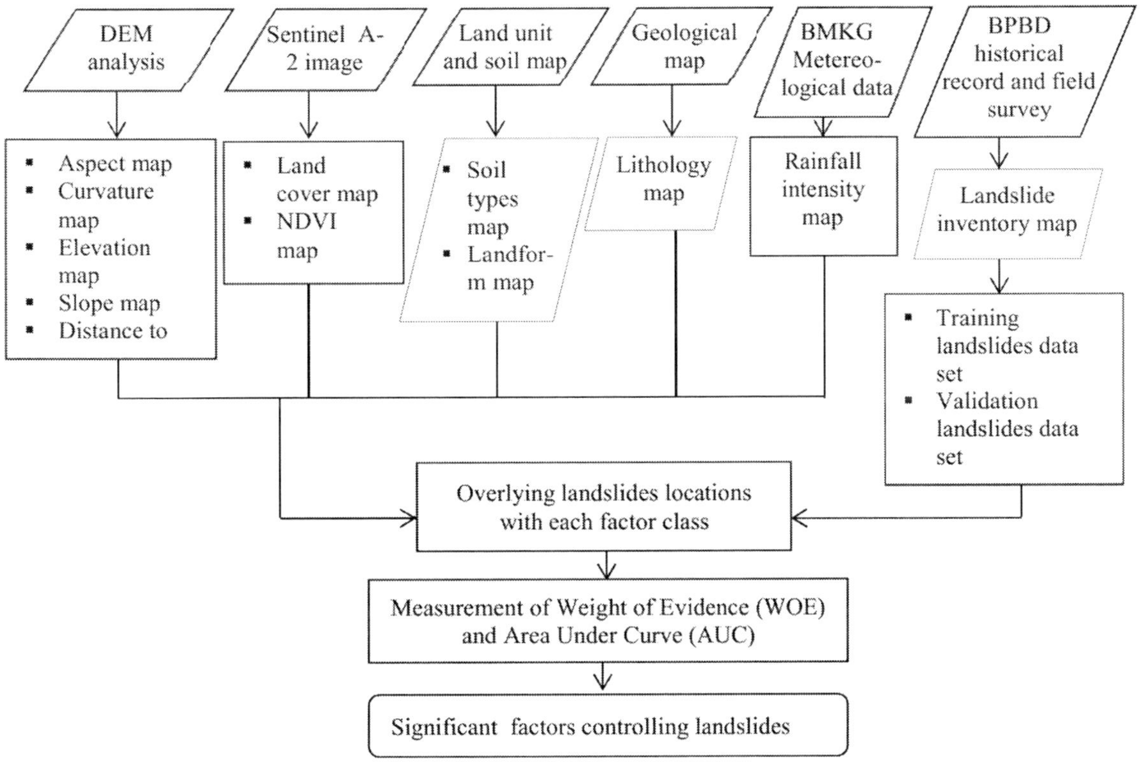

Figure 3. flow chart of the research

4. Results

From eleven parameters that had been analyzed, eight variables became the determining factors for landslides based on the AUC value (as shown in table 1). These factors were: geology (0.80); landform (0.74); land cover (0.68); elevation (0.65); NDVI (0.63); slope(0.63); rainfall intensity (0.62); and proximity to stream (0.62). Three parameters are not taken into account as significant factors for controlling landslides: soil type (0.60); curvature of the slope (0.57); and aspect of slope (0.42). Soil type is at the minimum threshold value as a determining factor for landslides but this study was considered this factor not significanct enough as controlling landslides .

Factors associated with landslides based on the calculation of Weight of Evidence are shown in Figure 4-11 and table 1.

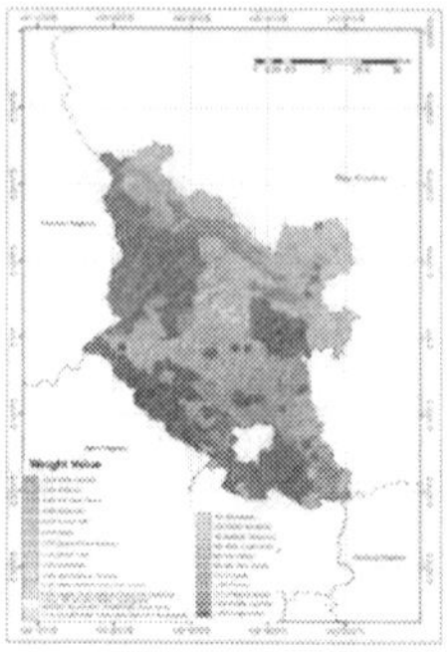

Figure 4.
WOE of geology conditon map

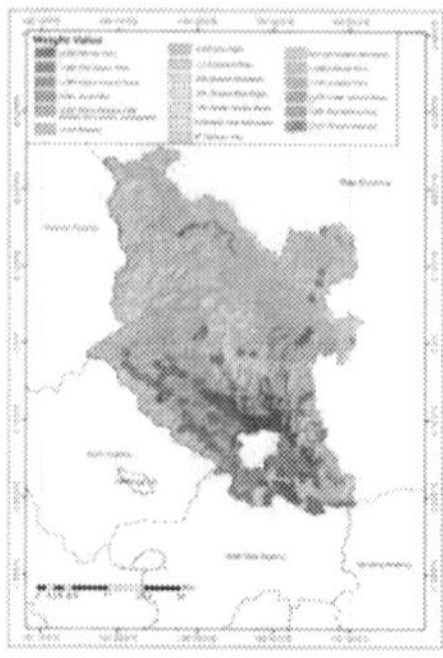

Figure 5.
WOE of landform map

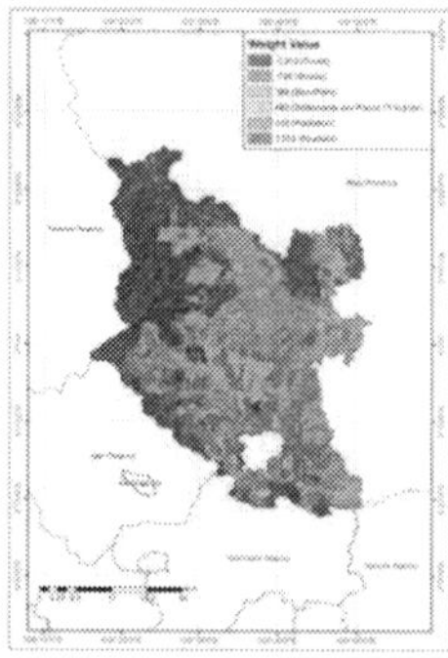

Figure 6.
WOE of land cover map

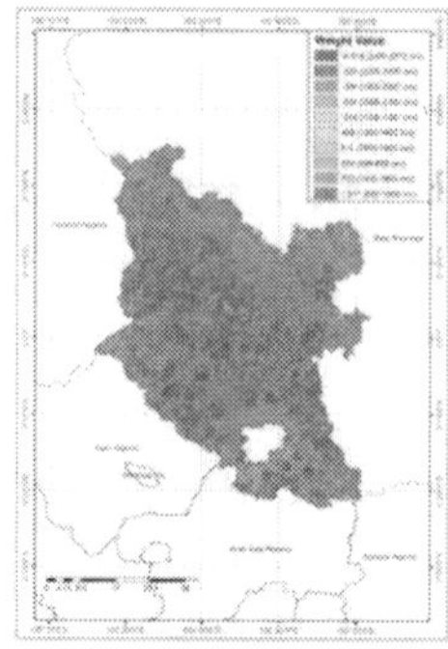

Figure 7.
WOE of elevation map

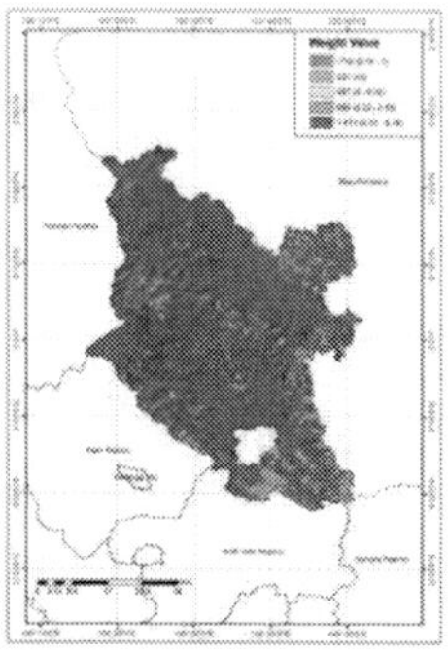

Figure 8.
WOE of NDVI map

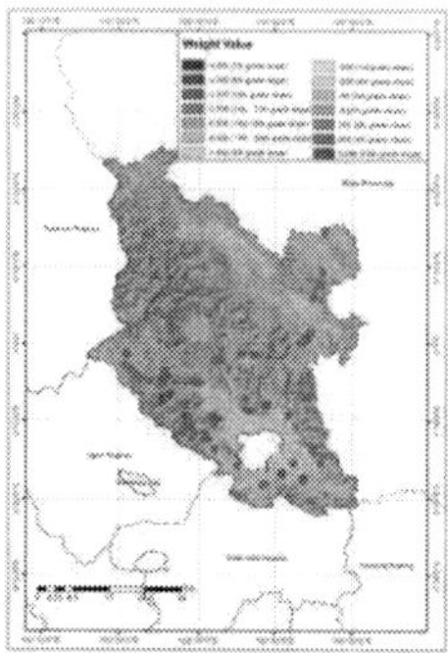

Figure 9.
WOE of slope map

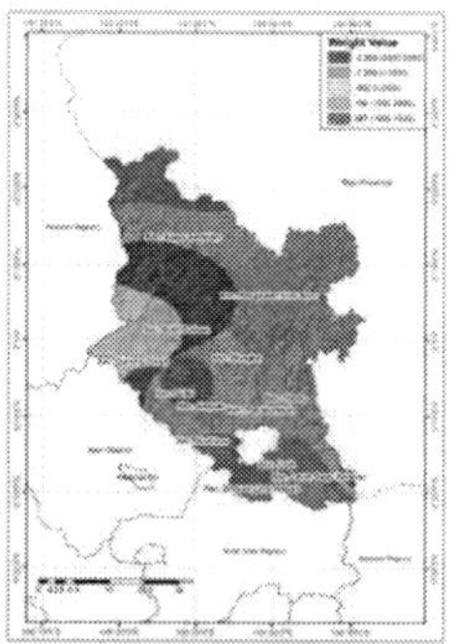

Figure 10.
WOE of rainfall intensity map

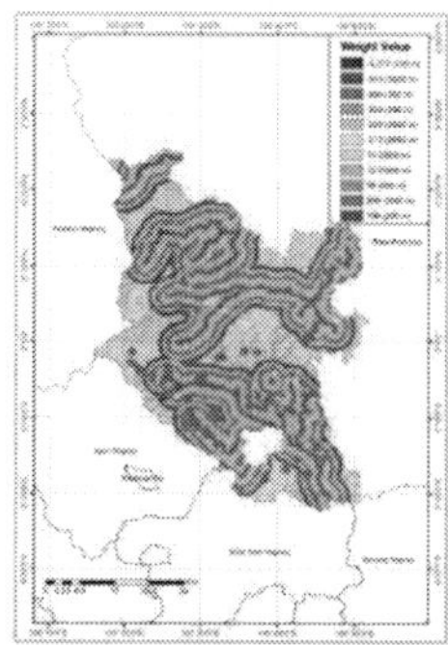

Figure 11.
WOE of proximity to stream

Table 1. Weights (W), contrast values (C) and Area Under Curve (AUC) of parameters classess

Classes	Class area from total area (%)	Landslide occurance	Weight +	Weight -	C	AUC
Slope (%)						0.63
0 – 12	21.4	9	-0.791	0.139	-0.930	
12 - 28	25.7	24	-0.007	0.002	-0.009	
28 – 42	19.9	18	-0.037	0.009	-0.046	
42 – 56	13.1	18	0.378	-0.072	0.449	
56 – 69	8.5	10	0.222	-0.023	0.245	
69 – 81	5.5	1	-1.604	0.046	-1.650	
94 - 106	1.4	1	-0.265	0.003	-0.269	
164 – 180	0.1	13	5.152	-0.146	5.298	
Geology						0.80
Sandstone	23.5	9	-0.747	0.150	-0.897	
Quartzite, sandstone	8.6	7	-0.013	0.001	-0.014	
Quartz sandstone	7.1	5	-0.146	0.010	-0.156	
Rude Conglomerate	2.5	3	0.389	-0.012	0.401	
Andesite, Basalt	12.6	29	1.021	-0.297	1.318	
Phylite, Quartzite	3.8	11	1.265	-0.104	1.370	
Pumice	3.3	9	1.190	-0.082	1.271	
Conglomerate	0.1	1	2.090	-0.011	2.100	

Classes	Class area from total area (%)	Landslide occurance	Weight +	Weight -	C	AUC
Granite	1.4	4	1.229	-0.035	1.264	
Phillyte, flakes	1.8	3	0.692	-0.019	0.710	
Silt, sand, gravel	0.4	1	0.997	-0.008	1.005	
Landform						0.74
Collovial plain	0.4	1	0.997	-0.008	1.005	
Tectonic plain	8.5	2	-1.215	0.063	-1.278	
Volcanic plain	3.4	9	1.156	-0.080	1.236	
Tectonic plate escarpment	1.3	1	-0.062	0.001	-0.062	
Volcanic intrusion	0.3	4	2.709	-0.046	2.756	
Lower Volcanic slope	1.9	6	1.356	-0.056	1.412	
Middle Volkanic slope	1.5	1	-0.193	0.003	-0.195	
Tectonic mountains	40.8	18	-0.609	0.273	-0.881	
Old volcanic mountains	8.9	14	0.647	-0.093	0.740	
Tectonic hills	20.7	17	0.002	0.000	0.002	
Old volcanic hills	1.7	8	1.743	-0.085	1.827	
Tectonic plate ridge	1.8	1	-0.369	0.006	-0.375	
Rainfall intensity (mm)						0.62
<1000	3.7	1	-1.081	0.025	-1.106	
1000-1500	17.1	24	0.532	-0.157	0.688	
1500-2000	55.1	50	0.100	-0.138	0.238	
2000-2500	15.2	2	-1.769	0.139	-1.908	
>2500	8.9	5	-0.371	0.030	-0.401	
NDVI						0.63
Non vegetation	4.4	4	0.108	-0.005	0.113	
Lower dense	4.1	4	0.181	-0.008	0.189	
Dense	5.9	7	0.362	-0.028	0.390	
Higher dense	13.8	28	0.898	-0.265	1.163	
Highest dense	71.8	39	-0.406	0.616	-1.022	
Land cover						0.68
Forest	35.4	6	-1.532	0.358	-1.890	
Plantation	29.2	34	0.345	-0.187	0.532	
Settlement	1.5	2	0.496	-0.010	0.506	
Paddyfield	7.9	16	0.903	-0.134	1.036	
Shrubs	7.2	3	-0.661	0.037	-0.698	
Dry agriculture	18.8	21	0.305	-0.086	0.392	
Elevation (m)						0.65
596-800	1.4	2	0.545	-0.010	0.555	
800-1000	2.8	6	0.971	-0.047	1.018	
1000-1200	4.5	3	-0.210	0.009	-0.219	
1200-1400	6.5	8	0.404	-0.035	0.439	
1400-1600	8.9	14	0.642	-0.092	0.735	
1600-1800	12.6	16	0.432	-0.081	0.513	
1800-2000	19.0	12	-0.257	0.052	-0.308	
2000-2200	21.0	14	-0.205	0.048	-0.253	
2200-2400	13.7	7	-0.463	0.057	-0.520	
Proximity to stream (m)						0.62
100-200	2.6	4	0.706	-0.028	0.735	
200-300	2.6	1	-0.654	0.013	-0.666	
300-400	2.5	2	0.085	-0.002	0.087	
400-500	2.5	1	-0.594	0.011	-0.606	
500-1000	11.7	9	0.036	-0.005	0.041	

Classes	Class area from total area (%)	Landslide occurance	Weight +	Weight -	C	AUC
1000-2000	21.0	16	0.031	-0.008	0.039	
2000-3000	17.5	22	0.526	-0.159	0.685	
3000-4000	14.5	9	-0.172	0.026	-0.199	
4000-5000	12.4	4	-0.811	0.076	-0.887	
5000-6000	9.6	6	-0.169	0.016	-0.185	

Note:

- The table not presented the parameter with AUC value ≤ 0.6
- The table not presented the clases of parameter with no existing landslide occurrence

Each significance factor controlling landslides that are found are described as follows:

a. Geological factor
Lithology types found as the most significant factor to predict landslides based on positive W and C and greater than 1 consist of: conglomerate; andesite, basalt; phylite, quartzite; pumice and granite. Lithological characteristic provide a condition of landslides. Sedimentary rocks such as conglomerate composed of a fraction of rounded to sub angular gravel-size clasts which are unconsolidated material. Andesite and basalt as igneous rocks are difficult to penetrate water but when surrounded of soil type with a lot of clay can formed a sliding plane so this can explain that several times there have been rockfall types of landslides. Phyllite and slate rocks contain clay that has a characteristic to store water with the rate of releasing water is small. In the case of pumice, its texture is hollow which can pass water easily, and can increase the rate of landslide susceptibility.

b. Land form factor
Volcanic intrusion landform is strongly associated with landslide, and then respectively folowed by old volcanic hills, lower volcanic slope, and volcanic plain. Volcanic intrusion landform area are unstable and give rise to landslide if there is vibration, heavy rain or changes in land use that can not maintain the stability of the soil aggregate. The majority of those landforms have alluvial soil composed of sand, clay and, dust deposited in several places such as riverbanks foothills.

c. Land cover factor
Landcover types are found associated with landslides are paddy field, plantation, and settlement. The characteristic of terraced paddy fields which are found largely in the regency is one of the explanation.

d. Elevation factor
Elevation of 600-1000 meters and 1400-1800 meters above sea level are classes associated with landslides, especially much strong association at elevation of 800-1000 meters.

e. Density of vegetation greenness (NDVI) factor
As expected, the class of the highest dense of vegetation index is not associated with landslides even though there is the most landslide point occurrence in this class. The other clases have a weak correlation with landslides.

f. Slope factor
The class of slope 164-180 % is a very significant factor for controlling landslides and the slope 42-69% is moderately significant but the other classes in not significant.

g. Rainfall intensity factor
Areas with rainfall intensity class between 1000-2000 correlated with the occurrence of landslides.

h. Proximity to stream factor
Landslide is associated with close proximity to the river (100-200 meters) and in a considerable distance (2000-3000 meters from the river)

In order to measure the validity of the prediction of the parameter model and data accuracy, a test was carried out using 59 (40%) landslide point data that had been recorded but were not used in the analysis process. It is obtained the value of AUC = 0.787. This validity test shows that the parameter model and inventory database are considered good.

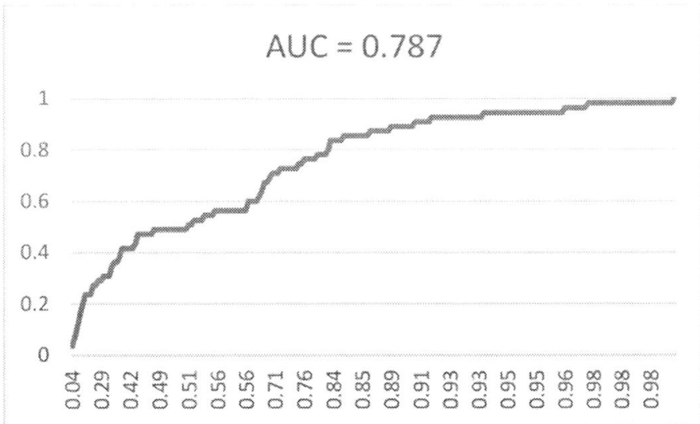

Figure 12. AUC value by using validation landslides data set

5. Conclusions

There are eight out of eleven factors that can be used to predict landslides in Limapuluh Kota Regency. Lithological types and landform parameters with AUC values > 0.7 are the most influencing factors in controlling landslides susceptible area. The other lower influencing factors than the two mention before are land cover, elevation, density of vegetation, slope, rainfall intensity, and proximity to stream.

Based on the significant or dominant factors controlling landslides in the region, landslide susceptibility zone map in the Lima Puluh Kota Regency can be derived. In next step, we can compare landslide susceptibility zone map based on statistical method analysis with susceptibility zone map in spatial planning document which is based on heuristics method analysis.

Acknowledgement

The author would like to thank Lembaga Penelitian dan Pengabdian Masyarakat (Research Institutions and Community Service) Universitas Negeri Padang for funding this work with a contract number: 969/UN35.13/LT/2021.

References

[1] Badan Standardisasi Nasional 2016 *Penyusunan dan penentuan zona kerentanan gerakan tanah* (Jakarta: BSN)

[2] Anis Z, Wissem G, Vali V, Smida H and Mohamed Essghaier G 2019 GIS-based landslide susceptibility mapping using bivariate statistical methods in North-western Tunisia *Open Geosci.* **11** 708–26

[3] Pamela, Sadisun I A and Arifianti Y 2018 Weights of Evidence Method for Landslide Susceptibility Mapping in Takengon, Central Aceh, Indonesia *IOP Conf. Ser. Earth Environ. Sci.* **118**

[4] Silalahi F E S, Pamela, Arifianti Y and Hidayat F 2019 Landslide susceptibility assessment using frequency ratio model in Bogor, West Java, Indonesia *Geosci. Lett.* **6**

[5] Pradhan A M S, Dawadi A and Kim Y-T 2012 Use of different bivariate statistical landslide susceptibility methods: A case study of Kulekhani watershed, Nepal *J. Nepal Geol. Soc.* **44** 1–12

[6] Sumaryono, Muslim D, Sulaksana N and DasaTriana Y 2015 Weights of Evidence Method for Landslide Susceptibility Mapping in Tandikek and Damar Bancah, West Sumatra, Indonesia

IOP Conf. Series: Earth and Environmental Science **986** (2022) 012065 doi:10.1088/1755-1315/986/1/012065

Int. J. Sci. Res. **4** 1283–90
[7] Wubalem A 2020 Landslide Susceptibility Mapping using Statistical Methods in Uatzau Catchment Area, Northwestern Ethiopia *Geoenvironmental Disasters* 1–21
[8] Darman H, Minarwan, Nasution F A, Bahesti F, Young W I H and Panjaitan J 2011 *Sedimentology Newsletter Sumatra* (Indonesia)
[9] Yin Y 2017 *Advancing Culture of Living with Landslides* vol 1
[10] Nandi A and Shakoor A 2010 A GIS-based landslide susceptibility evaluation using bivariate and multivariate statistical analyses *Eng. Geol.* **110** 11–20
[11] Pimiento E 2010 *Shallow Landslide Susceptibility Modelling and Validation*

IOP Conf. Series: Earth and Environmental Science **986** (2022) 012066 doi:10.1088/1755-1315/986/1/012066

An external event identification for Nuclear Power Plant (NPP) siting project at West Kalimantan, Indonesia

Heni Susiati [1], Pande Made Udiyani [2], and Muhammad Budi Setiawan [2]

[1]Center for Assessment of Nuclear Energy System, National Nuclear Energy Agency, Jalan Kuningan Barat, Mampang Prapatan, Jakarta Selatan 12710, Indonesia
[2]Center for Nuclear Reactor Technology and Safety, National Atomic Energy Agency, Building 80 PUSPIPTEK Area, South Tangerang, 15310

*Email: heni_susiati@batan.go.id

Abstract. The research aims to determine the existing external event for Nuclear Power Plant (NPP) Siting project at West Kalimantan. The identification of external hazard potential studies is carried out based on the IAEA safety guide and BAPETEN regulations regarding external potential hazards which include seismic aspects, surface faults, extreme meteorology, geotechnics, volcanoes which must be carried out in determining the NPP's site. The method used in this research is field ground check and desk work analysis with references related to identified external hazards. The results showed that the potential site of the NPP's in West Kalimantan Province is safe from potential external natural hazards. The NPP's site has also been determined to be safe from the dangers of flooding due to tsunamis and flood overflows, no capable faults are found, and far from potential volcanic hazards. Extreme meteorological potential is rare at site sites. To evaluate the oceanography aspect, the existence of physical oceanographic conditions must be carried out in-depth studies in the future, especially related to the sedimentation process that has occurred around the nuclear power plant site.

1. Introduction

In the West Kalimantan Province 2019 RPJMD Planning, it has been stated that the local government has programmed the benefits of nuclear power plants as a source of electrical energy in West Kalimantan[1]. In this context, the National Nuclear Energy (BATAN) as a promoter in implementing the nuclear power plant (NPP) site selection has coordinated to make West Kalimantan an energy barn. In selecting the NPP site, BATAN uses the IAEA Safety Standard Series Safety Guide No. 35[2] and BAPETEN Indonesian Nuclear Regulatory Body No. 4 of 2018 as a reference in selecting the NPP site[3]. It is recommended that in the survey and site selection of the NPP an external event evaluation, which includes evaluation activities to assess the aspects of volcanoes, seismicity, subsurface materials and meteorological hazards[4,5,6,7]

Kalimantan Island is an island with great potential for NPP sites, given the relatively stable geological conditions and far from the threat of natural external hazards, especially volcanoes and has a low level of seismicity. NPP is a power producer with a large enough capacity and can be built in accordance with the selected area with strict requirements, so that it will match the electricity needs of West Kalimantan Province. The long-term planning of West Kalimantan Province is to become a storehouse for future energy that will be fulfilled with the NPP program as an alternative energy that can be developed. The development of nuclear power plants as an alternative energy aims to meet national electricity needs in order to spur industrial growth, improve the regional economy and public welfare in general, and increase the capability of human resources in mastery of technology[7,8].

The pre-survey and site survey activities in the process of determining the location of nuclear power plants in the Coastal Zone of West Kalimantan were carried out in 2014-2016. In 2014-2015, data base line for the area of interest was obtained and in 2016 a potential site data base line was obtained along the Coastal area. West Kalimantan. These activities are based on general criteria. The results of these activities are the basis for the next survey stage, and in 2019, potential sites along the West Kalimantan Coastal Area were obtained including Salatiga District (Sambas Regency). Simping Beach and Tanjung Batubelah (Singkawang City), Tanjung Suak and Gosong Beach (Bengkayang Regency) and Temaju Island (Mempawah Regency), Kumbang Island (Kayong Utara Regency), Muara Pawan and Pagar Mentimun (Ketapang Regency)[7,8].

To support the selection of potential sites, external events both natural and man-made are an aspect to be considered so that NPP to be built can avoid phenomena or events that endanger the safety of the NPP[2,11]. In the site evaluation, every potential impact source from facilities and human activities around the site, which could potentially harm the power reactor over the life of the reactor is required to be identified and assessed. Estimates of the possibility of developing areas that are also potentially dangerous to safety, during the life of the reactor should also be anticipated.

The research objective was to identify external event sources to support potential site selection in West Kalimantan Province. The method used is literature review and secondary data collection related to external sources of hazards due to natural and man-made events. It is hoped that the results of the study will provide input for policy makers in choosing the site of choice for nuclear power plants in West Kalimantan Province.

2. Methods

In accordance with the safety principle of nuclear technology, the site selection for an NPP must meet various safety requirements, safety aspects that need to be studied and considered in selecting a nuclear power plant location include: (a) safety aspects caused by natural external factors (external natural events); (b) safety aspects caused by human activities (external human-induced events); (c) aspects of public and environmental safety (public and environmental safety).

Evaluation of nuclear power plant potential site acceptability on the coast of West Kalimantan is carried out by studying regional geological maps and geological conditions at the site supported by field observations, which are indicated as fault locations. The evaluation is carried out thoroughly on the cesarean activity that has been known from the research that has been done. The study areas along the coast of West Kalimantan are shown in Figure 1.

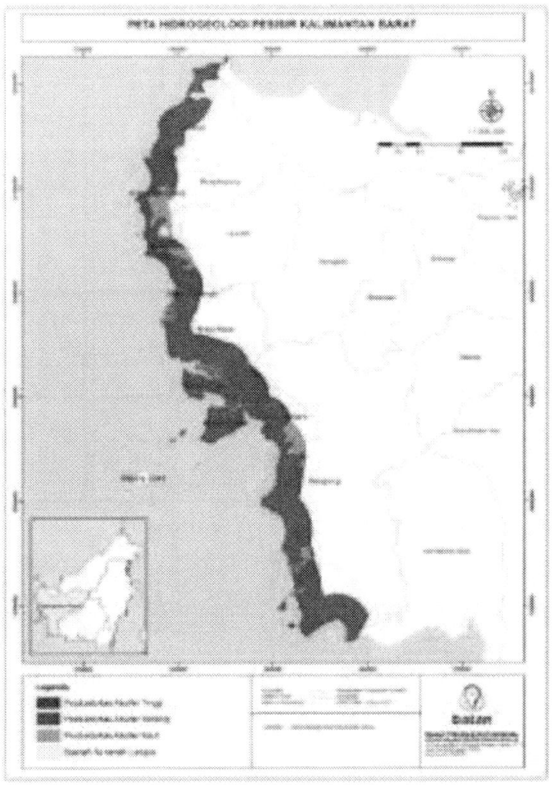

Figure 1. Study Area in Coastal West Kalimantan[7]

3. Results and Discussions

Various factors that influence the site evaluation are safety aspects, non-safety aspects and other considerations such as abrasion, gas emissions, sea water intrusion and others. Site selection related to safety aspects includes geology, seismicity, geophysics, volcanoes, hydrology, hydrogeology, meteorology, and man-made events. Based on the study of the criteria and distribution of active faults, active volcanoes, seismicity and peak acceleration maps in bedrock in the coastal area of West Kalimantan Province, 6 potential site locations were obtained, namely 1 (one) in Sambas Regency, 2 (two) in Singkawang City, 2 (two) in Bengkayang Regency, 1 (one) in Mempawah Regency, 1 (one) in Kayong Utara Regency and 2 (two) locations in Ketapang Regency[7,8]. The results of the NPP potential site suitability selection that have been produced are depicted in Figure 2.

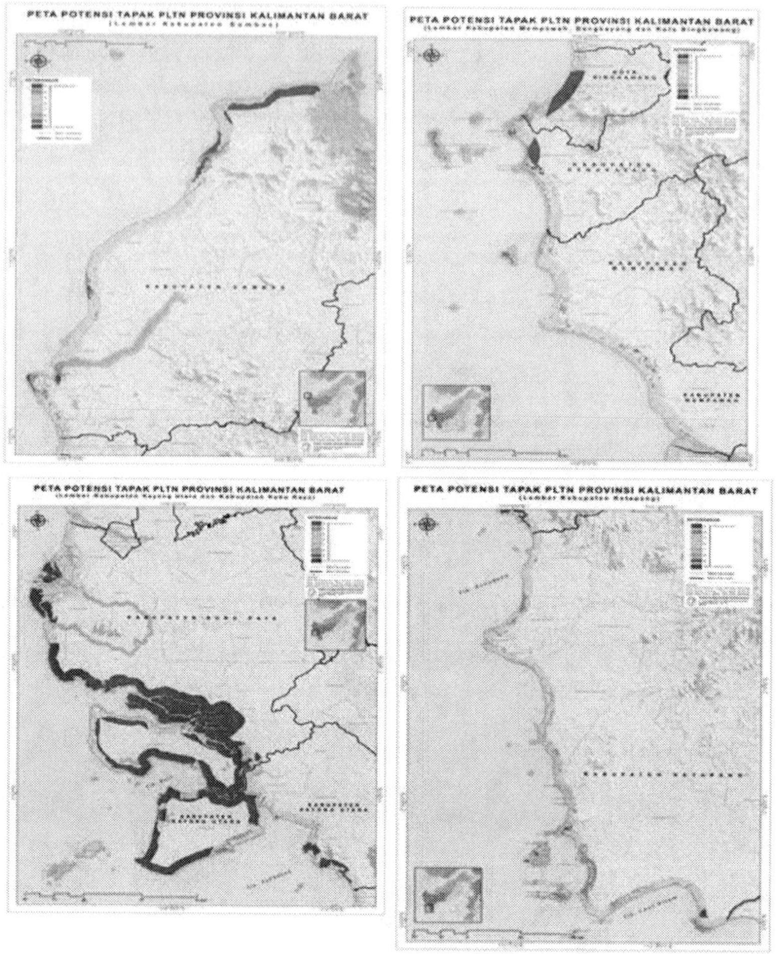

Figure 2. Distribution of NPP's Site Potential in West Kalimantan Province[9]

Identification of sources that can pose a potential hazard to the potential site of a nuclear plant is one of the steps in evaluating an external event, which includes the following aspects:

3.1 Seismotectonic

In general, the seismic condition in West Kalimantan is lower than that of other regions in Indonesia. The locations of the subduction earthquakes are along the island of Sumatra, Java, and Sulawesi. Meanwhile, in West Kalimantan there were only a few earthquakes (Figure 3).

IOP Conf. Series: Earth and Environmental Science **986** (2022) 012066 doi:10.1088/1755-1315/986/1/012066

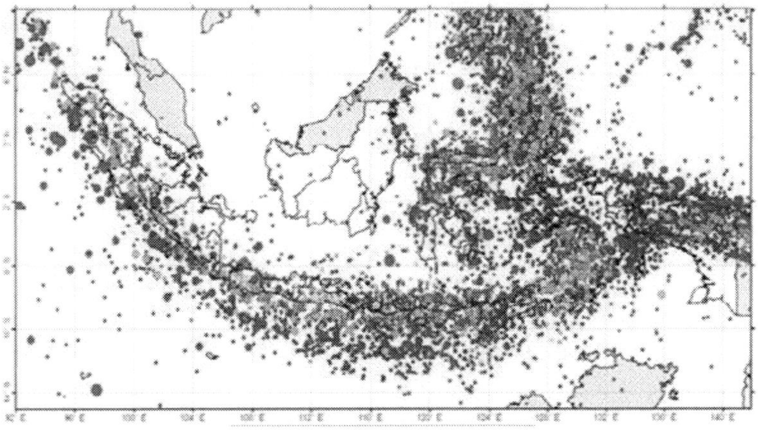

Figure 3. Map of Earthquake Distribution in Indonesia[11,12]

3.1.1 History of Seismicity in West Kalimantan

Historically, in the West Kalimantan region, a large and destructive earthquake has never been recorded [4]. However, there has been an earthquake in Bengkayang Regency and Singkawang City, which is thought to have been caused by a small earth plate fault about 37 kilometers from the west of Bengkayang. The first earthquake in Singkawang and Bengkayang occurred on August 23, 2011, at 08.40 WIB with a magnitude of 4.4 on the Richter scale; with the epicenter on the ground. Then followed by an earthquake on August 24, 2011 at 22.51 WIB with a strength of 4.6 on the Richter scale and the source of the earthquake was at sea. It did not last long, at around 23.33 WIB, an earthquake occurred again with a magnitude of 3.6 on the Richter scale, the epicenter of the earthquake came from the ground [5].

3.1.2 Seismotectonic Model

Based on a regional tectonic arrangement, Indonesia is located between the 3 main plates of the world, namely the Indo-Australian plate which is sinking under the Eurasian plate with a direction of movement relative to the northeast, while from the east the Pacific plate moves relative to the Eurasian plate to the west, and from the direction of south of the Australian continental plate pushing towards a relative north. This causes Indonesia to have a very complicated seismic activity, especially along the boundary of these plates.

The tectonic framework in Kalimantan is influenced by regional tectonic developments which are part of the interaction between the Philippine oceanic plate, the Indo-Australian plate and the Eurasian plate which produces a collection of oceanic basins and continental micro-blocks that are limited by subduction zones, distant movements between plates and faults[6]. Kalimantan Island is located quite far from the subduction zone. In addition, because of these collisions, fractures were formed on the island of Kalimantan, however, until now experts have not identified the existence of active faults on the island of Kalimantan.

Western parts of Indonesia such as Kalimantan, Sumatra and Java are composed of continental crusts, as is the shallow seabed between these islands. Under the Earth's crust there are zones where the rock is hotter and more plastic. Continental plates and oceanic plates float on top of this liquid material. In Kalimantan, there are four main geological units, namely first, rocks connected to plate edges; second, bedrock; third, young hardened, and non-hardened rocks; and fourth, alluvial rocks and shallow young deposits.

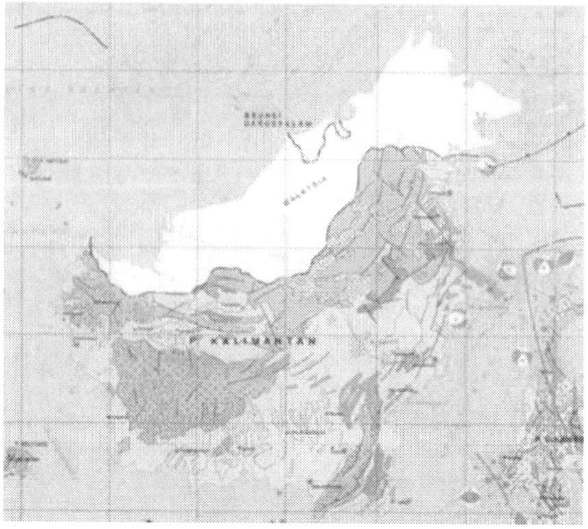

Figure 4. Seismotectonic Map of Kalimantan[8]

The regional seismotectonic of Kalimantan are strongly influenced by the submergence activity of the Philippine oceanic plate, the Indo-Australian plate, and the Eurasian plate. One way to determine the tectonic forces that act in an area is by analyzing the focal mechanism of the earthquake that occurred. Most of the earthquakes that occurred around the island of Kalimantan were around East Kalimantan which were shallow earthquakes with strike slip and thrusting faults. This indicated that the shallow earthquake was associated with subduction zone activity and fault activity that occurred on land, while in West Kalimantan there was no occurrence of destructive earthquakes with a magnitude of more than 5 SR. In the West Kalimantan region, there are several fault segments in the seismotectonic picture (Figure 4).

Based on earthquake catalog data from the United States Geological Survey (USGS), the number of earthquakes that occurred in West Kalimantan and its surroundings during the period 1973 to 2013 was 8 earthquakes (Table 1).

Table 1. Earthquake Occurrence in Radius 500 km from Pontianak[11]

No.	Date	Longitude	Depth (km)	Magnitude (Mw)
1	24/08/2011	108,831 BT	2,7	4,7
2	01/12/2007	105,577 BT	10	4,1
3	10/05/2005	109,381 BT	10	4,1
4	14/04/2003	109,249 BT	33	4,7
5	04/12/2001	111,971 BT	33	4,7
6	24/02/1996	109,947 BT	33	4,1
7	19/02/1994	112,710 BT	33	4,7
8	12/02/1994	112,765 BT	28,9	5,3

This condition indicates that the level of regional seismicity in the West Kalimantan region is very low. When compared with the seismic condition in East Kalimantan, the seismic condition in West Kalimantan is lower[3,9]. All earthquake events occurred at a depth of less than 70 km with a magnitude between 4.1 - 5.3 Mw. The earthquake was very felt and caused enough damage at the location of the incident which was about 100 km from the city of Pontianak. The source of the earthquake was estimated from a fault in the Bengkayang area.

The results of the earthquake distribution in Figure 5 show that in West Kalimantan Province there were only 3 earthquakes, most of which were shallow earthquakes. It is estimated that the occurrence of this earthquake was influenced by the presence of faults in the area.

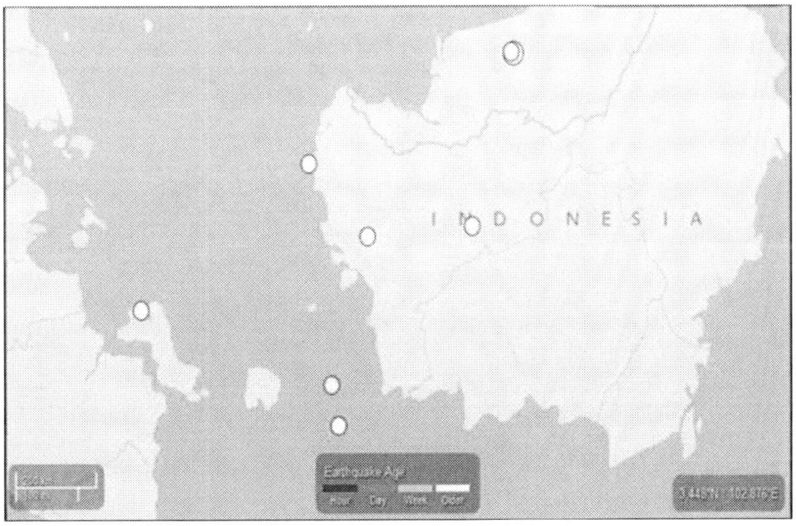

Figure 5. Map of Earthquake Distribution in 500 km Radius from Pontianak City[8]

3.1.3 Subsurface Materials

The area of West Kalimantan is mostly an area consisting of old lineages or faults. The indication of the developing fault is in the form of a relatively northwest-southeast alignment. Several locations show an alignment that cuts the basin into several segments and aligns parallel to the northwest-southeast. The alignments in the districts of Sintang and Sanggau are in the form of the Lupar Fault which cuts through the Cat area to Sekadau and the Adang Fault which intersects the Sambas, Sanggau, and Kapuas Hulu areas (Figure 4). In Singkawang City and Bengkayang Regency, the youngest fault can be identified regarding the Oligocene Sintang Formation (23.7–30 million years) in the form of a sinistral horizontal fault, namely the Adang Fault. The existing straight line of Melawi Regency is dominated by a relatively northwest-southeast direction. Based on the condition of the geological structure, suitable coastal areas include Mempawah, Pontianak, Kubu Raya, Kayong Utara and Ketapang.

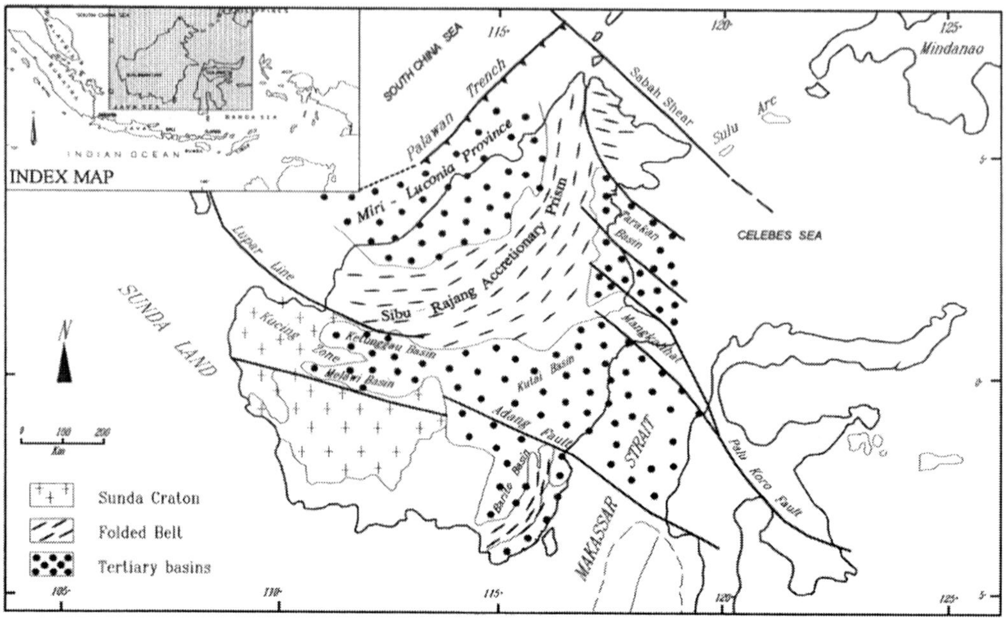

Figure 6. Map of West Kalimantan Geological Structure Distribution[8]

3.2 Maximum Ground Acceleration

Every earthquake that occurs will cause a value for the acceleration of the ground in a place. The value of ground acceleration that will be considered in building planning is the maximum ground acceleration value. The maximum ground acceleration is the value of the greatest ground acceleration in a place due to earthquake vibrations in a certain period. Although strong earthquakes do not occur frequently, they are still very dangerous to human life. One of the important things in seismological research is knowing the damage caused by earthquake vibrations to buildings in every place. This is necessary to adjust the strength of the buildings to be built in the area. Factors that affect the size of the ground acceleration in a place include earthquake magnitude, hypocenter depth, epicenter distance, and ground conditions. The greater the magnitude of an earthquake, it means that the greater the energy emitted from the source of the earthquake is greater, so that the ground surface acceleration that arises is also greater. The deeper the hypocenter and the farther the epicenter distance, the smaller the ground surface acceleration that occurs. Another factor that also determines the amount of ground surface acceleration is the level of soil density in that place. Thus, the surface acceleration that arises is directly proportional to the magnitude and inversely proportional to the epicenter distance, hypocenter depth, and soil density. Buildings with extraordinary strength can be made, so that if a very strong earthquake occurs, it will not have a response or reaction to the earthquake. The calculation of ground acceleration in an empirical way can be done with the approach of several formulas derived from earthquake magnitude and/or intensity data. This formulation is not always correct, even from one method to another it is not always the same, but it is sufficient to provide an overview of the maximum ground acceleration or Peak Ground Acceleration (PGA)[11]. This ground acceleration needs to be studied for every earthquake that occurs in an area and in a certain period of time.

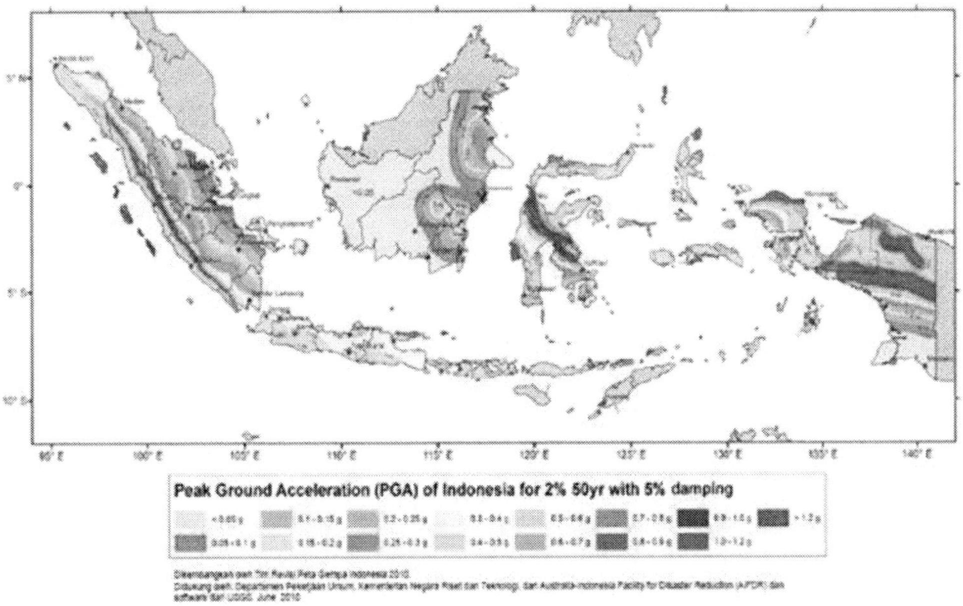

Figure 7. Map of Peak Ground Acceleration (PGA) in Indonesia[11]

Based on the map of the maximum ground acceleration, Kalimantan Island is only in the eastern part which has a high enough value, which is around 0.2 g. Especially in West Kalimantan Province, it has a value below 0.05 g, except in the southern part it is 0.1 g. This shows that most areas of West Kalimantan have a maximum acceleration value below 0.05 g. In general, the seismic condition in West Kalimantan Province is very low.

3.3 Stratigraphy

Based on the map of the maximum ground acceleration, Kalimantan Island is only in the eastern part which has a high enough value, which is around 0.2 g. Especially in West Kalimantan Province, it has a value below 0.05 g, except in the southern part it is 0.1 g. This shows that most areas of West Kalimantan have a maximum acceleration value below 0.05 g. In general, the seismic condition in West Kalimantan Province is very low. III.5 Stratigraphy In general, West Kalimantan is composed of old Triassic rocks (± 204 million years) in the form of layered sandstone, andesite, granite covered above by Quaternary sedimentary and alluvium rocks (<1.6 million years). The coastal area is composed of alluvium units, sedimentary rocks and several places in the form of breakthrough rocks including granite, granodiorite and dacite which occupy the beaches of Singkawang, Bengkayang, Sukadana, and Kendawangan (Figure 8).

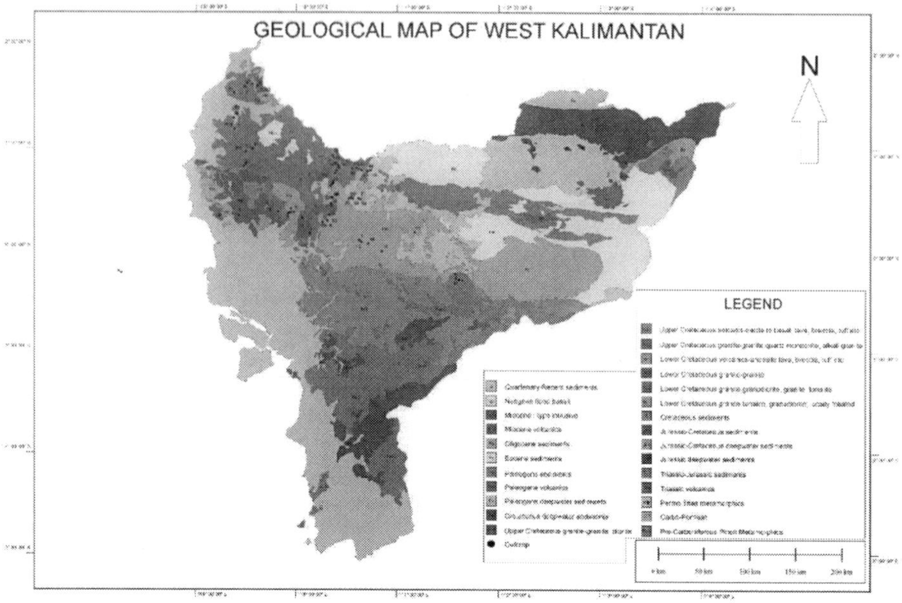

Figure 8. Geology Map of West Kalimantan[7, 11, 15]

4. Conclusions

Based on the results of the site safety and feasibility study which refers to the IAEA-Safety Standard Series safety requirement No. NS-R-3, 2003, it can be concluded that the evaluation results of the surface magnification aspect show that there are no active faults in the site area and site vicinity. The evaluation of the volcanic aspects shows that the potential site of the NPP in West Kalimantan is far from the existence of volcanic activity. The seismic aspect shows that the value of ground acceleration is still within the permissible limits based on the most conservative calculations. So that the overall evaluation of the external event aspect at the potential site of the NPP in West Kalimantan is a feasible, suitable and safe site.

Acknowledgment

The works are conducted under the Government research and funding by Center for Nuclear Energy Assessment, National Nuclear Energy Agency of Indonesia (BATAN), for the year of 2020 and partially supported by RISTEK/BRIN-LPDP project (14/E1/III/PRN/2020).

Reference

[1] GUBERNUR 2019 Rencana Pembangunan JangkaMenengah Daerah Provinsi Kalimantan Barat Tahun 2018-2023

[2] Dong H, Li G, Xu C, He J and Xu W 2019 *Dynamic quantitative allocation and prediction for multi-nuclear power projects resource based on Monte-Carlo simulation* (inis.iaea.org)

[3] Septiadi D, Yarianto Sugeng B S, Sriyana S, Anzhar K and Suntoko H 2018 An Extreme Meteorological Events Analysis for Nuclear Power Plant (NPP) Siting Project at Bangka Island, Indonesia *IOP Conf. Ser. Earth Environ. Sci.* **132**

[4] Alimah S, Alhakim E E, Suntoko H, Sunarko S and Mudjiono M 2020 Study on Hydrogeological Aspect of Galang Sub-District, Batam City for Acceptance of Prospective Npp Sites *Maj. Ilm. Pengkaj. Ind.* **14** 172–81

[5] Suntoko H 2014 Identifikasi Daerah Interes Calon Tapak Pltn Kalimantan Barat Berdasarkan Kriteria Umum *Eksplorium Bul. Pus. Teknol. Bahan Galian Nukl.* **35** 57–68

[6] Basri N A, Hashim S, Ramli A T, Bradley D A and Hamzah K 2016 No Title *J. Radiol. Prot.* **36** R96

[7] Newman A 2016 *Decision-making and radioactive waste disposal* (Nuclear Threat Initiative, Washington, DC, United States: Taylor and Francis)

[8] Susiati H, S.B.S. Y, Anzhar K, Kironi B and Mellawati J 2015 Aplikasi Data Penginderaan Jauh Dan Sig Dalam Pemilihan Tapak Potensial Pltn Kalimantan Barat *J. Pengemb. Energi Nukl.* **17** 121

[9] Euis Etty Al Hakim, Heni Susiati, Sunarko A H 2019 Analisis Spasial Awal Lokasi Calon Tapak PLTN di Kalimantan Barat *Pros. Semin. Nas. Infrastruktur Energi Nukl.* 173–9

[10] Song Y, Li H, Hu L, Hou Y and Liu H 2017 Portable gamma camera for nuclear power plants … *Electron. J. Nucl. Saf.* …

[11] Wu X and Ho D 1994 Bruce NGS a discharge channel flow simulation using MODTURC_CLAS vol 1 (Ontario Hydro, Toronto, Canada: Canadian Nuclear Assoc)

[12] Yuliastuti Y and Indrawati Y 2021 Seismic hazard disaggregation analysis of West Kalimantan NPP potential site: A precursory study *AIP Conf. Proc.*

[13] Masyhur Irsyam, Asrurifak M, Djazilus H and Latief H 2011 Geotechnics and earthquake geotechnics towards global sustainability *Geotech. Geol. Earthq. Eng.* **15**

[14] Heni S, Hadi S, Euis E A and Slamet S 2019 *Land surface temperature parameters and fire hazard potential considerations in NPP site candidate selection, West Kalimantan province* (inis.iaea.org)

Spatial analysis of the urban physical vulnerability using remote sensing and geographic information systems (case study: Yogyakarta City)

N Arif [1*], A Wardhana[2] and A Martiana[2]

[1] Department of Geography Education, Faculty of Social Science, Universitas Negeri Yogyakarta

[2] Department of Sociology Education, Faculty of Social Science, Universitas Negeri Yogyakarta

[*]Email: nursida.arif@uny.ac.id

Abstract. Vulnerability mapping in urban planning is very important to reduce damage caused and loss of life due to disasters. This paper discussed the mapping of urban physical vulnerability using remote sensing, and geographic information systems (GIS) approaches. Physical vulnerability parameters used were building density, distribution of public facilities, slope, and land use. Landsat 8 OLI satellite imagery used remote sensing data to obtain building density information using the Normalised Difference Built Index (NDBI) and land use interpretation approaches. With the use of GIS, the spatial distribution of urban vulnerabilities could be known through the overlay of each indicator. The results of the analysis obtained that the physical vulnerability classes of Yogyakarta City were very high (26%), high (29%), moderate (29%), low (16%), and very low (0%). Some areas categorized as vulnerable to very vulnerable are areas with high building density, namely the Sub-districts of Danurejan, Kraton, Gedongtengen, Jetis, Ngampilan, Pakualaman, Mantrijeron, parts of Gondomanan and Tegalrejo.

1. Introduction

Disasters that occur in humans can be caused by natural events or triggered by human activities. Natural hazards that exceed the response capacity cause natural disasters and significantly affect social economic development of a region [1]. The resulting impact can be in the form of loss of life and human injury [2,3]. Indonesia is a country that is vulnerable to various natural hazards, namely earthquakes, tsunamis, floods, droughts, landslides, volcanic eruptions, and disease outbreaks [4,5]. Therefore, Indonesia continues to make efforts by carrying out sustainable development planning that is responsive to disasters.

On the other hand, [1] stated that the conditions of urban development in developing countries are hazardous. Urban planning has an essential role in minimizing risk. The higher level of vulnerability of buildings in cities is triggered by an increase in population and unclear regulations [6]. Built-up land has a considerable risk of earthquake disaster [7,8]. Building users are vulnerable to injury or death from collapsing buildings. These facts are the reason to look at the level of physical vulnerability in a city by investigating several indicators contributing to disaster vulnerability, namely land use, building density, affordability of public facilities, and slopes. Yogyakarta City is an area that has a high multi-disaster threat, such as earthquakes, volcanic eruptions, floods, extreme weather, and other disasters. The findings of this paper would identify the areas that have high vulnerability to disasters. Implementing physical vulnerability is important in response to environmental conditions in the event of a disaster.

2. Methodology

2.1. Study Area

The research area covered the region of Yogyakarta City which is located at 7°48′5″N and 110°21′52″E (**Figure 1**). Physically, the development of Yogyakarta City is triggered by fairly high population growth. Administratively, the area of Yogyakarta City is 32.50 km2 which is divided into 14 districts. The city's residential land use increased by 0.06% from 2017 to 2020 (BPS, 2021). Based on the results of satellite imagery analysis, the distribution of urban heat island (UHI) areas is more dominant than non-UHI, which indicates that the energy needed by the city is very high due to a large amount of built-up land [9]. Topographically, Yogyakarta City is situated at an altitude of fewer than 200 meters above sea level.

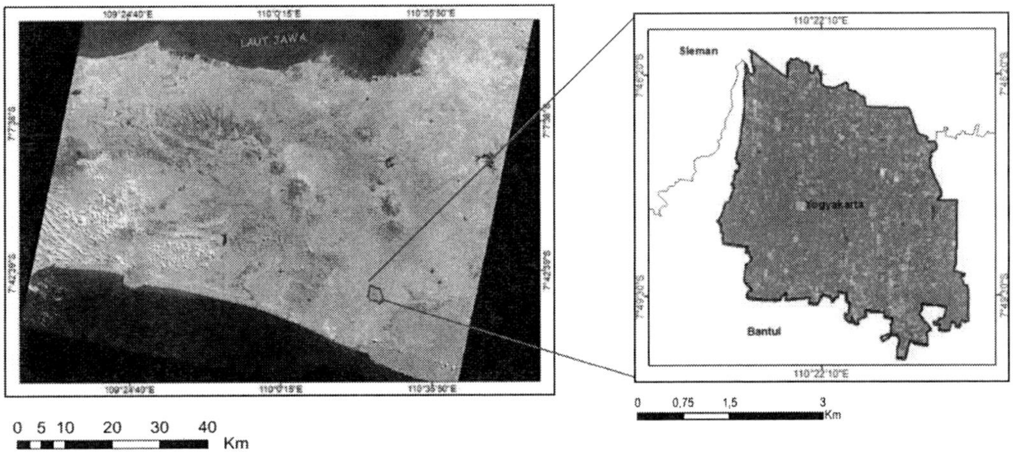

Figure 1. The study area (source: http://Landsat.USGS.gov/)

2.2. Physical Vulnerability Indicators

Determination of physical vulnerability indicators is an approach used in understanding urban disasters. Ahadnezhad Reveshty emphasizes building indicators which are divided into several criteria, namely the type of construction, the age of the building, the quality of the building, the number of floors of land cover, land use, and building density [1]. In Indonesia, physical vulnerability refers to general guidelines for risk assessment using indicators of houses, public facilities, and critical facilities [10]. In this study, the hands used were modifications of some of those indicators (**Table 1**). Each hand was analyzed using a geographic information system and overlaid according to their respective weights. The highest importance was building density. Adjacent buildings will have a higher vulnerability to disasters, such as fires and earthquakes because access to escape from the area will be more complex [11].

Table 1. Physical vulnerability indicators for urban area

Indicators	Weights
Building density	40
Affordability of public facilities	30
Slope	15
Land cover/land use	15

modified from BNPB, 2012; Reveshty et al., 2014

The building density indicator was obtained from the analysis of Landsat 8 OLI satellite imagery recorded on June 25, 2019, using the normalized difference built-up index (NDBI) approach. This index is used to see the reflectance of urban areas using shortwave-infrared (SWIR) compared to near-infrared (NIR). The number of building density classes was determined based on the number of pixels detected as buildings in one district region divided by the area of the district. Zha et al. (2003) developed the calculation of NDBI value as written in the following equation

$$NDBI = \frac{(SWIR-NIR)}{(SWIR+NIR)} \quad (1)$$

the range of SWIR band is 1.55-1.75 μm and NIR band is 0.76-0.9 μm

The availability of public facilities is an important indicator in urban disaster vulnerability as an institutional response to basic human needs, such as health, security, and recreation [13][13]. In this study, the distribution of public facilities was limited to hospitals, community health centers, and fields because they were assumed to be urgent facilities needed by the community in the event of a disaster. Other indicators used in this study were slope and land use, which are the most influential parameters in several natural disasters, such as erosion, landslides, and floods [14–17]. The vulnerability class interval of each indicator was determined using the equal interval classification method, namely by dividing the entire range of the highest value minus the lowest value by a predetermined number of classes [18][18].

3. Results and Discussion

Land use in Yogyakarta City is dominated by built-up area (94.19%), open field (1.88%), water body (0.08%), and vegetation (3.86%%) (**Figure 2 and Figure 4**). Built-up land has a complex relationship with urban development because it affects economic activity, social life and is physically vulnerable [19,20]. This study did not look specifically at the characteristics of urban buildings affecting physical vulnerability. But, in general, it is known that city buildings are identical to building density. Buildings with more floors than in rural areas make them vulnerable to several disasters, such as earthquakes and fires [3], [22 [3,21].

Figure 3 shows the building density of Yogyakarta City is relatively moderate to high. The results of NDBI analysis obtained a range of density values of 0.53 - 0.87, with the density class category for each district was calculated based on the average NDBI in each district divided by the area of the district. The area of each building density class based on Figure 3 is presented in **Table 2**. Although the NDBI method is fast and easy to implement, the results obtained are very dependent on the sensor and its weakness is that it cannot distinguish between built-up land and barren land [22]. So, these results could be different if done with different methods and approaches.

Figure 2. Land use map of Yogyakarta City

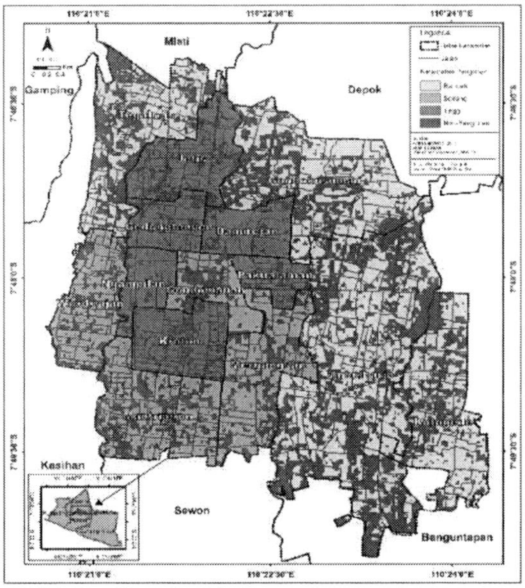

Figure 3. Building density City

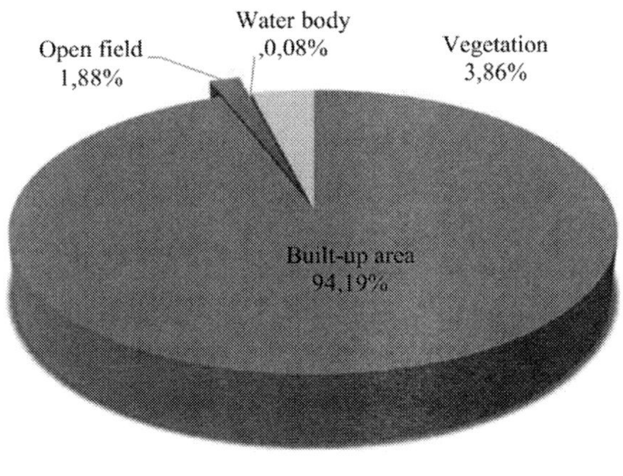

Figure 4. Percentage of land use in Yogyakarta

Tabel 2. The area of each building density class

Density class	Sub-districts	Area (Km2)	Density area (%)
High	Jetis	1.70	20,34
	Gedongtengon	0.96	
	Danurejan	1.10	
	Pakualaman	0.63	
	Kraton	1.40	
	Ngampilan	0.82	
Moderate	Wirobrajan	1.76	24
	Mantrijeron	2.61	
	Mergangsan	2.31	
	Gondomanan	1.12	
Low	Tegalrejo	2.91	55,66
	Gondokusuman	3.99	
	Umbulharjo	8.12	
	Kotagede	3.07	

Analysis, 2021

Almost half of the area of Yogyakarta City belongs to the moderate (24%) to high (20.34%) building density class in **table 2**. While the other half are included in the low building density class (55.66%). However, when viewed from the percentage of the number of districts, there were 6 districts or 42.86% of the total number of the districts that were classified into the high-density class, the moderate and low-density classes had the same number of 4 sub-districts or 28.57%, respectively. The same spatial pattern was shown by a previous researcher who stated that several districts in the middle of Yogyakarta City, namely Kraton, Danurejan, and Pakualaman districts had higher urban heat islands compared to the other districts [9]. This is inseparable from the existence of Malioboro which is administratively included in the Danurejan district. Malioboro is a constituent element of the north-south axis, is located in a strategic area that is the target of investors to build public and commercial facilities [23].

Figure 5 shows that, on average, the entire of Yogyakarta City has moderate to high access to affordability. The location of public facilities provided by the government ideally meets the needs of pedestrians at a comfortable speed and distance [24]. All districts in the city of Yogyakarta have the availability of accessible public facilities, in which 11 General Hospitals, 18 Community Health Centers, and several other service facilities are available [25]. Yogyakarta has the potential for physical regional development in the form of hierarchical urban system development, as well as quality and service improvement for a network of facilities and infrastructure [26]. So that the availability of public facilities will continue to increase and improve.

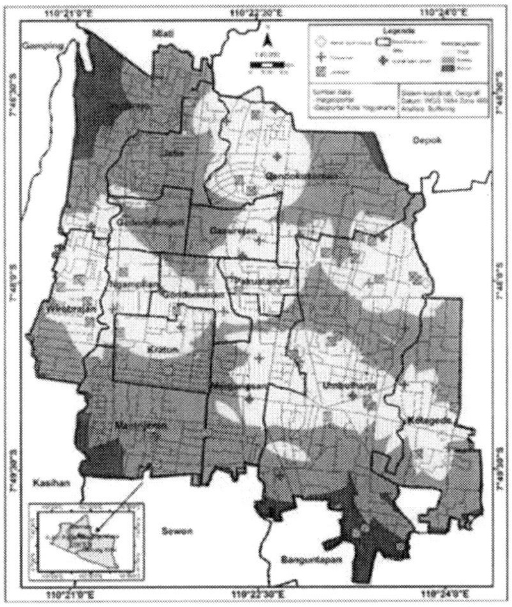

Figure 5. Affordability of public facilities facility

Topographically, Yogyakarta City is not prone to erosion and landslide disasters but is quite vulnerable to flooding when viewed from the results of the slope map analysis shown in **Figure 6**. The slope of Yogyakarta City is relatively flat (77%), sloping (19%), moderately steep (3%), steep (1%), and very steep (0.05%) (Figure 7). Decrease in slope and increase in flooding are directly proportional [27]. Three major rivers are flowing in the city of Yogyakarta, namely the Winongo river in the west, the Code river in the middle, and the Gajahwong river in the east. Most of the Code watershed was moderate and low levels of flood vulnerability [28].

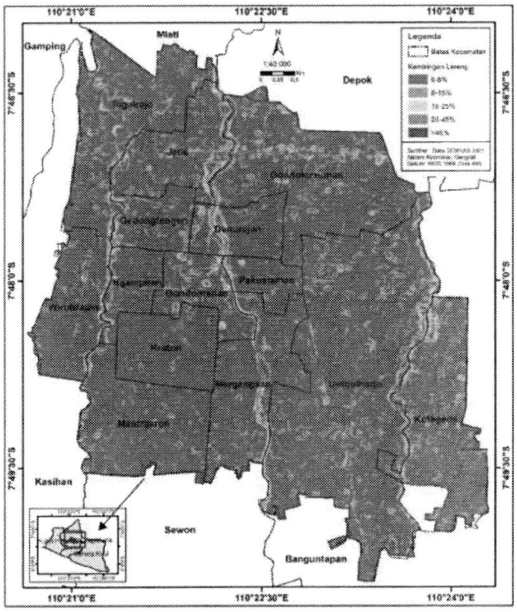

Figure 6. Slope Map of Yogyakarta

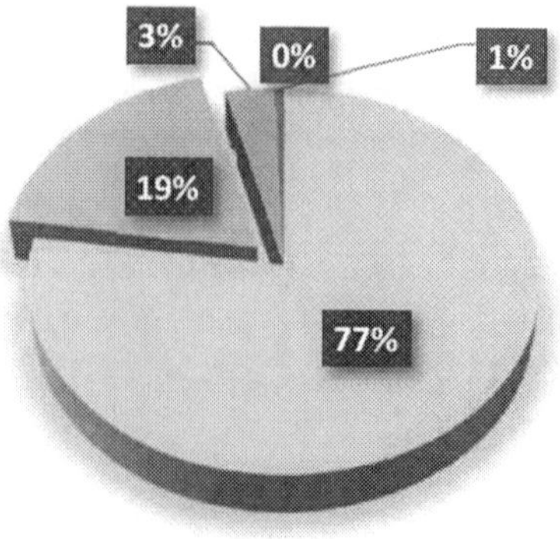

Figure 7. Percentage of Slope in Yogyakarta

A physical vulnerability map was obtained based on the analysis of all indicators, namely slope, land use, building density, and affordability of public facilities. **Figure 9** shows the percentage of each physical vulnerability class, namely very high (26%), high (29%), moderate (29%), low (16%), and very low (0%). From the map results in Figure 8, the percentage of vulnerability in each district is presented in **Table 3**.

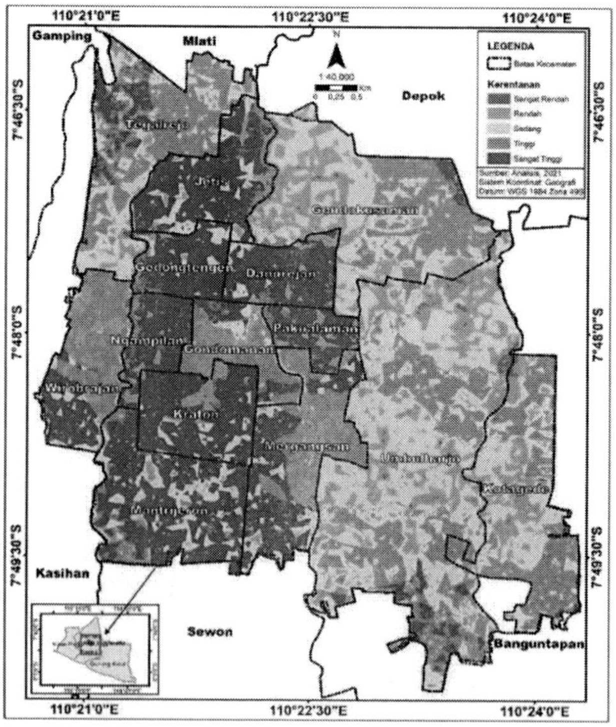

Figure 8. Physical vulnerability map of Yogyakarta City

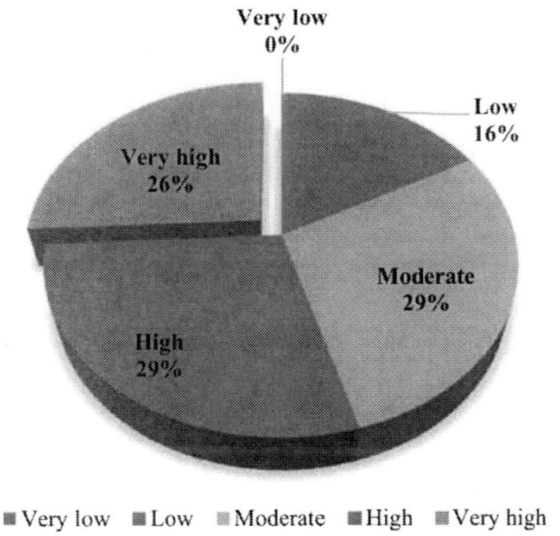

Figure 9. The percentage of each physical vulnerability class

Table 3. Percentage of Physical vulnerability area of each Sub-districts in Yogyakarta City

Sub-districts	Area (%)				
	Very Low	Low	Moderate	High	Very high
Danurejan	0,05	11,37	8,41	9,17	71,01
Gedongtengon	0,36	5,97	12,83	7,06	73,78
Gondokusuman	0,59	27,32	42,28	27,91	1,90
Gondomanan	0,10	23,78	10,58	48,59	16,94
Jetis	0,39	8,70	16,44	7,46	67,01
Kotagede	1,00	22,37	37,52	38,74	0,37
Kraton	0,01	10,51	10,00	2,96	76,52
Mantrijeron	0,04	1,60	19,20	13,57	65,58
Mergangsan	0,30	14,79	13,98	35,73	35,19
Ngampilan	0,78	14,79	0,36	16,11	67,96
Pakualaman	0,18	19,94	0,45	16,20	63,23
Tegalrejo	0,30	5,32	29,14	55,15	10,10
Umbulharjo	1,04	21,15	47,15	28,18	2,47
Wirobrajan	0,56	18,10	10,71	47,12	23,51

Analysis, 2021

Table 3 shows that areas with very high vulnerability are Danurejan (71%), Kraton (73.78%), Gedongtengen (73.78%), Jetis (67.01%), Ngampilan (67.96%), Pakualaman (63.23%), and Mantrijeron (65.58%) districts. Meanwhile, areas with high vulnerability are Gondomanan (48.59%) and Tegalrejo (55.15%). Judging from the population data, several areas with high and very high vulnerability, such as Danurejan, Gedongtengen, Gondomanan, and Jetis subdistricts have a higher population than other sub-districts [29].

4. Conclusion

This research is not specific to one of the disasters that are prone to occur in Yogyakarta City. However, it was generally analyzed using four indicators, namely building density, land use, slope, and affordability of public facilities. The approach used is easy to implement to assess the level of physical vulnerability of the city. The key indicator in this research is building density which was analyzed using the Normalized Difference Built-Up Index (NDBI). Land use in Yogyakarta City is dominated by built-up land, while the topography tends to be flat. Potential disasters that are prone to occur are earthquakes, fires, and floods. The analysis results showed that 9 of 14 subdistricts in Yogyakarta City were included in high and very high vulnerability classes, namely the subdistricts of Danurejan, Kraton, Gedongtengen, Jetis, Ngampilan, Pakualaman, Mantrijeron, some parts of Gondomanan and Tegalrejo. Where some of those areas were correlated with a larger population than other subdistricts. Urban planning needs to pay attention to the physical vulnerability of the city to create a city that is responsive to disasters.

Acknowledgments
The authors would like to thank the Institute for Research and Community Service and the Faculty of Social Science, Yogyakarta State University, for providing financial support for this research with contract numbers T/4.48/UN/34.21/PT.01.03/2021.

References

[1] Ahadnezhad Reveshty M, Kamelifar M J, Ranjbarnia B and Pashaiifar A 2014 A Comparative Study on Physical Vulnerability of Urban Area against Natural Hazards: Importance of Health Promoting Approach in Civil Engineering. *Heal. Promot. Perspect.* **4** 122–31
[2] Hassink R and Gong H 2020 *Regional Resilience* vol 11 (Elsevier)
[3] Hung H C, Yang C Y, Chien C Y and Liu Y C 2016 Building resilience: Mainstreaming community participation into integrated assessment of resilience to climatic hazards in metropolitan land use management *Land use policy* **50** 48–58
[4] Arif N, Danoedoro P and Mulabbi A 2020 Prediction Model using Fractional Vegetation *Indones. J. Sci. Technol. Eros.* **5** 125–32
[5] Mcdonald A, Wilcox T, Aslam P, Pannawadee S, Janne P, Animesh K, Iria T C and Omar A 2020 Disaster Risk Reduction in Indonesia Disaster Risk Reduction Status Report 2020 40
[6] Singh A, Kanungo D P and Pal S 2019 *Physical vulnerability assessment of buildings exposed to landslides in India* vol 96 (Springer Netherlands)
[7] Saputra A, Rahardianto T, Revindo M D, Delikostidis I, Hadmoko D S, Sartohadi J and Gomez C 2017 Seismic vulnerability assessment of residential buildings using logistic regression and geographic information system (GIS) in Pleret Sub District (Yogyakarta, Indonesia) *Geoenvironmental Disasters* **4**
[8] Yüce, G and Arun G 2012 Earthquake and Physical and Social Vulnerability Assessment for Settlements: Case Study Avcilar District
[9] Sumunar D R S, Arif N, Hadi B S and Endro K 2020 Urban Energy Modeling Using Remote Sensing Approaches *Int. J. GEOMATE* **19** 203–8
[10] Badan Nasional Penanggulangan Bencana 2012 Buku Saku Tanggap Tangkas Tangguh Menghadapi Bencana *Badan Nas. Penanggulangan Bencana* 62
[11] Nazmfar H 2019 An integrated approach of the analytic network process and fuzzy model mapping of evaluation of urban vulnerability against earthquake *Geomatics, Nat. Hazards Risk* **10** 1512–28
[12] Zha Y, Gao J and Ni S 2003 Use of normalized difference built-up index in automatically mapping urban areas from TM imagery *Int. J. Remote Sens.* **24** 583–94
[13] Anon 2012 Public Facilities and Services
[14] Arif N, Danoedoro P and Hartono 2017 Analysis of Artificial Neural Network in Erosion Modeling: A Case Study of Serang Watershed *IOP Conference Series: Earth and Environmental Science* vol 98

[15] Bera S, Guru B and Oommen T 2020 Indicator-based approach for assigning physical vulnerability of the houses to landslide hazard in the Himalayan region of India *Int. J. Disaster Risk Reduct.* **50** 101891
[16] Msabi M M and Makonyo M 2021 Flood susceptibility mapping using GIS and multi-criteria decision analysis: A case of Dodoma region, central Tanzania *Remote Sens. Appl. Soc. Environ.* **21** 100445
[17] Zhou W 2021 GIS for Earth Sciences *Encyclopedia of Geology (Second Edition)* ed D Alderton and S A Elias (Oxford: Academic Press) pp 281–93
[18] Stern B, Hurni L, Wiesmann S and Ysakowski Y 2011 Statistics for Thematic Cartography 16
[19] Surya B, Ahmad D N A, Sakti H H and Sahban H 2020 Land use change, spatial interaction, and sustainable development in the metropolitan urban areas, south Sulawesi province, Indonesia *Land* **9**
[20] Zhang Z, Xiao R, Yu W, Liu Y, Lin M and Wang M 2017 Characterizing factors associated with built-up land expansion in urban and non-urban areas from a morphological perspective *Sustain.* **9**
[21] Riyanto W, Irawan D, Joko T, Adi W and Iranata D 2020 Earthquake Vulnerability Assessment of High-Rise Buildings in Surabaya using RViSITS Android Application Earthquake Vulnerability Assessment of High-Rise Buildings in Surabaya using RViSITS Android Application
[22] Valdiviezo-N J C, Téllez-Quiñones A, Salazar-Garibay A and López-Caloca A A 2018 Built-up index methods and their applications for urban extraction from Sentinel 2A satellite data: discussion *J. Opt. Soc. Am. A* **35** 35
[23] Cahya G A, Mahendra Y K D and Damanik I I 2017 Malioboro as a value of Special District of Yogyakarta City *IOP Conf. Ser. Earth Environ. Sci.* **70**
[24] Ismail A S, Azili N S and Ahmad @ Taufik F 2019 The EFFECTIVENESS OF PUBLIC FACILITIES IN LOW COST HOUSING FLATS FROM RESIDENTIAL SATISFACTION TOWARDS SUSTAINABLE LIVING *Asia Proc. Soc. Sci.* **5** 1–6
[25] Dinkes Kota Yogyakarta 2020 Profil Dinas Kesehatan Kota Yogyakarta tahun 2020 *Profil Dinas Kesehat. Kota Yogyakarta tahun 2019* 1–234
[26] Yogyakarta P K 2010 PERATURAN DAERAH KOTA YOGYAKARTA NOMOR 2 TAHUN 2010 TENTANG RENCANA TATA RUANG WILAYAH KOTA YOGYAKARTA TAHUN 2010-2029 634
[27] Padulano R, Rianna G, Costabile P, Costanzo C, Del Giudice G and Mercogliano P 2021 Propagation of variability in climate projections within urban flood modelling: A multi-purpose impact analysis *J. Hydrol.* **602** 126756
[28] Nursetiawan and Faizah R 2019 Identification of flood risk zones in the region of Yogyakarta *IOP Conf. Ser. Mater. Sci. Eng.* **650**
[29] BPS-Statistics of Daerah Istimewa Yogyakarta 2020 *Daerah Istimewa Yogyakarta Province in Figures 2020* (Yogyakarta)

IOP Conf. Series: Earth and Environmental Science **986** (2022) 012068 doi:10.1088/1755-1315/986/1/012068

Urban spatial temporal analysis and the urban heat island potential

P Wijayanti [1,2*]

[1] Geography Education, Universitas Sebelas Maret
[2] Disaster Research Center Universitas Sebelas Maret Jl. Ir. Sutami 36A Kentingan Jebres Surakarta 57126, INDONESIA

*Email: pipitwijayanti@staff.uns.ac.id

Abstract. The rate of urbanization has in changes in the urban spatial development, including in the city of Kediri. This study aims to analyze the distribution of urban spatial development in Kediri Regency/City, 2) Analyze the factors the factors that influence the spatial development, and 3) Analyze the UHI phenomenon and its relationship with urban spatial development. This study used Landsat 5 (1996) and 8 (2016, 2021) OLI to analyze Normalized Difference Vegetation Index (NDVI), Normalized Difference Built-up Index (NDBI) and land surface temperature (LST). Ndbi and ndvi are used to determine the urban spatial development, meanwhile lst and ndvi are to analyse surface temperature. The analysis was conducted from 1996 to 2021. Urban spatial development in study area has a linear type along with the development of roads. Built-up land in the study area has increased rapidly in the last 5 years (2016 – 2021) by 83.8% compared to changes that occurred in 20 years (1996 – 2016). This is in line with the increase in urbanization, especially since the advent of the shopping centers. Temperature is increasing in line with the increase in built-up land. There is a potential UHI phenomenon in the study area as indicated by a significant temperature difference between urban and sub-urban. This makes it a challenge for urban spatial planning so that the study area remains a comfortable place to live.

1. Introduction

Climate change and urbanization are today's issues in sustainable urban planning (1), and they've become a global concern about the different unintended repercussions of these human activities ((2); (3)). Human activities have caused a temperature rise of around 1°C from pre-industrial times, which will reach 1.5°C by 2030. This risk is created by the increasing severity of disasters, which will have an impact on all facets of human life. As a result, adaptation and mitigation are required in order for residents to continue to live properly and comfortably. Urbanization is fueled by population growth, which necessitates more space to carry out its activities, resulting in the conversion of land functions ((4), (5)). The city has experienced extraordinary development as a result of the activities and services of its residents. ((6)).

Population growth in tandem with urbanization has hastened land conversion ((7)), altered energy balances, and contributed to environmental degradation and a reduction in quality of life ((8); (5)). The reduction in the quality of life in urban areas will lead to the city's inability to sustain the lives of its citizens, necessitating a proper reaction. The impact of urbanization is evident in the UHI phenomenon, which has an impact on urban planning and quality of life ((9), (10); (11)).

IOP Conf. Series: Earth and Environmental Science **986** (2022) 012068 doi:10.1088/1755-1315/986/1/012068

UHI is the main product of urban human activities that affect temperature ((12), (13)), where urban temperatures are higher than surrounding non-urban areas ((14); (15); (5); (16)). Land cover is a key component in determining the intensity of UHI, while urban characteristics are also important, though not the most important. The most common factors influencing UHI are vegetation cover (44%), season (33%), building area (28%), day/night (25%), population density (14%), and air body (12%) ((17)). The heat absorption properties of the soil surface are influenced by the micro topography of the soil surface, which affects the soil surface temperature (LST). Thus, LST analysis is very important to understand the UHI phenomenon ((18)).

This study was conducted to determine urbanization as indicated by changes in soil surface temperature in Kediri Regency/City. The specific goals are to 1) analyze the distribution of urban spatial development in Kediri Regency/City, 2) analyze the factors that influence spatial planning development, and 3) analyze the UHI phenomena and its link with urban spatial development.

2. Method

2.1. Studi Area

This research was conducted in the district and city of Kediri, East Java during the period 1996-2021. The research area is a closed city located on the slopes of Mount Wilis with a height of 67 amsl and an area of 152233.07 Ha and 6915.32 Ha. Closed cities are necessary for developing a clear UHI model and claiming that the UHI phenomenon in the study area is driven by urbanization and land use changes. Kediri Regency/City has a population of 1673569/286796 people (in 2020), with a population density of 1048/4524 people/km2, with the majority of them being of working age (59.53 percent/60.03 percent) and working in the service sector (41.74 percent/75.68 percent) ((19)). The large number of persons employed in the service sector is a characteristic of urban areas. Kediri has a lot of potential locations and cities that could be developed in the industrial sector. Industry expansion has resulted in fewer vegetation, more buildings, and an increase in land surface temperature (LST) in these two places.

2.2. Method

This study uses 3 years, namely 1996, 2016 and 2021. The data used are Landsat 5 images in 1996, path 118 row 065 and path 119 row 065, Landsat 8 in 2016 and 2021, path 118 row 065 and path 119 row 065 and the administrative map of the Regency/City of Kediri. The radiometric image correction was performed on all image years, while the geometric correction was performed on the 1996 Landsat 5 image.

2.2.1. Analyze

The spatiotemporal development of the city is carried out by analyzing the Normalized Difference Vegetation Index (NDVI) and Normalized Different Built-up Index (NDBI) classifications. UHI phenomenon by analyzing NDVI, LST, while the correlation between building development and temperature is done by overlaying NDBI and LST

a. NDVI

NDVI classification on Landsat images using Red Band and NIR Infrared Band. Using the following equation:

NDVI = float (Band NIR-Band Red)/(Band NIR+Band Red)

NDVI classification is divided into 5 (Table 1)

IOP Conf. Series: Earth and Environmental Science **986** (2022) 012068 doi:10.1088/1755-1315/986/1/012068

Table 1. Classification of NDVI

Kerapatan Vegetasi	**NDVI value**
Very high	>0.72
High	0.42-0.72
Moderate	0.22-0.42
Low	0.12-0.22
Very Low	<0.12

(Franklin, 2001)

b. *NDBI*

NDBI classification on Landsat images using Red Band and NIR Infrared Band. Using the following equation:

NDVI = float (Band SWIR-Band NIR)/(Band SWIR+Band NIR)

NDVI classification is divided into 5 (Table 2)

Table 2. Classification of NDBI

Kerapatan Bangunan	Nilai NDBI
Very high	>0.2
Moderate	0.1-0.2
Low	0-0.1
Non Built	<0

(Handayani et, al, 2017)

c. *LST*

NDBI classification on Landsat images using the following equation::

LST= Trad/Emissivity0.25

Trades are obtained from:

(K2 Constant B10/(alog((K1 Constant B10/B10)+1))))

The values of K1 and K2 are obtained from the image metadata and B10 is replaced by the Thermal Band. Meanwhile, the emissivity is obtained from NDVI processing.

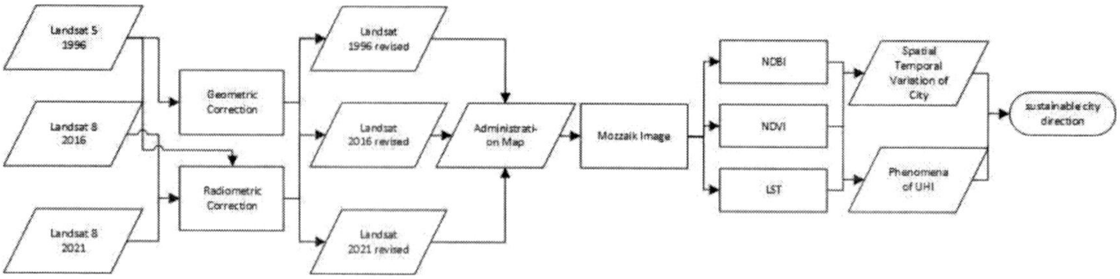

Figure. 1 Flowchart of Research

3. Result and Discussion

3.1. Distribution of urban spatial development in Kediri Regency/City,

From time to time, the research area is subjected to spatial urbanization. This can be seen in land cover changes as reflected in the NDVI (Figure 2) and NDBI (Figure 4) values. The distribution pattern in Kediri Regency is dominated by the high vegetation density class (HDV), but in 2021 it shows a decrease in area in that class. The increase in area occurs in areas that are not vegetated and have low vegetation every year.

The city of Kediri has a slightly different distribution, where the distribution of vegetation is dominated by medium and high density vegetation classes, and there is no very high density vegetation class. However, there is a trend of decreasing area every year which allows a shift in the spatial distribution of vegetation density in the future. Non-vegetation and low-frequency vegetation classes appear to be increasing. The increase in non-vegetation classes from 2016 – 2021 shows almost 4 times that of the previous year. The rate of increase is apparently much higher than in the previous 20 years (1996 – 2016). This shows the high demand for space as a consequence of urban development. Two aspects of urban ecosystems are closely related to land-vegetation and water changes ((5)).

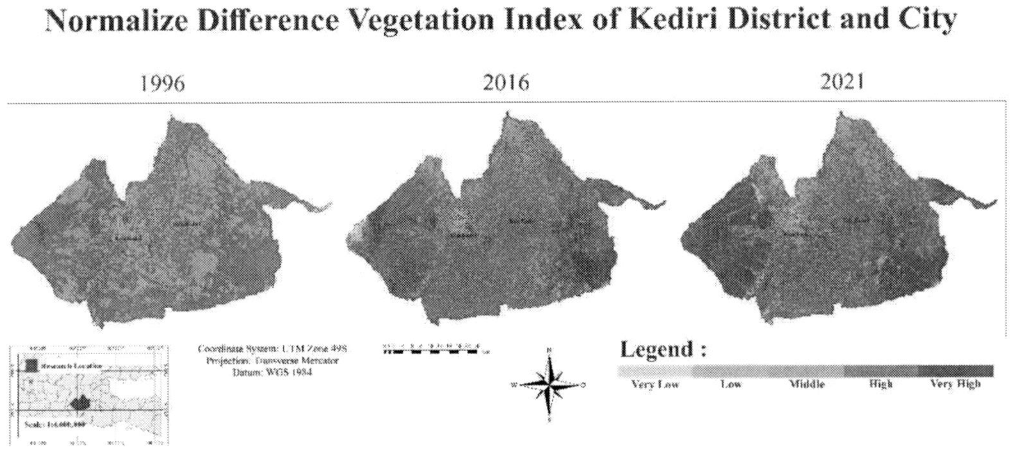

Figure 2. Spatial temporal of NDVI

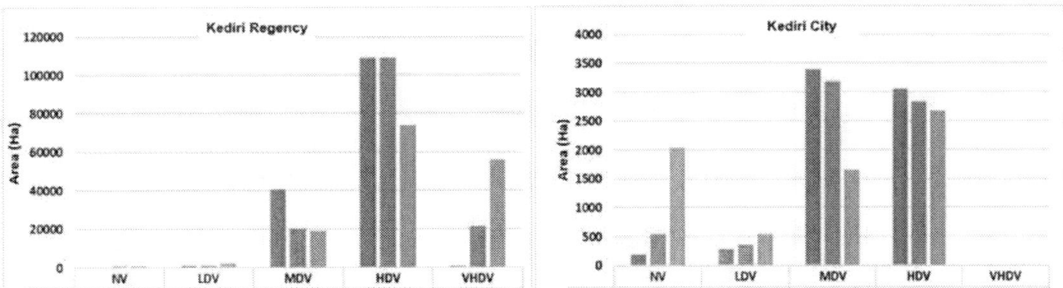

Kediri Regency	NV	LDV	MDV	HDV	VHDV
1996	540,5	1292,8	40579,3	109123,4	697,1
2016	756,7	1033,2	20105,8	109049,7	21287,7
2021	871,8	2286,3	19189,6	73803,3	56082,0

Kediri City	NV	LDV	MDV	HDV	VHDV
1996	188,0	279,1	3393,9	3054,4	0,0
2016	543,2	361,8	3180,6	2829,7	0,0
2021	2041,4	541,6	1653,1	2679,2	0,0

Figure 3. Graph of the distribution of the area of vegetation density in the Regency /City of Kediri

The spatial distribution of NDBI shows an inverse trend to NDVI. NDBI is an assessment of urban development based on built-up areas. Based on Fig. 4 and 5, it can be seen that in the research area, the trend of NDBI is the same. The awakened area increases over time, while the unbuilt area decreases. However, there are differences in the characteristics of the two research areas. Kediri Regency experienced an increase in built area about twice in a period of 20 years (1996–2016). The increase was even greater in the last 5 periods (2016-2021), which reached almost 3 times the 2016 level. The increase in the last 5 years increased by 83.8% compared to the increase in the previous 20 years. Although in 2021 the spatial distribution will still be dominated by non-built areas, given the very high increase, it is possible to shift the spatial distribution to be dominated by built-up areas.

Kediri City follows the same pattern as Kediri Regency, though not to the same extent. In 1996, non-built regions still dominated Kediri Regency (65%).The distribution of non-built areas declined between 1996 and 2016, despite the fact that they still contributed to the spatial distribution. In 2016, the non-built area accounted for 54% of total land area.The distribution alters over time, with built-up areas dominating in 2021 (52 percent). When you consider that the developed area had only increased by 10% in 20 years, but by 6% in just five years, this is a huge increase.Given the increasing trend of growth, if steps to manage land utilization are not implemented, it will be a threat for years to come.

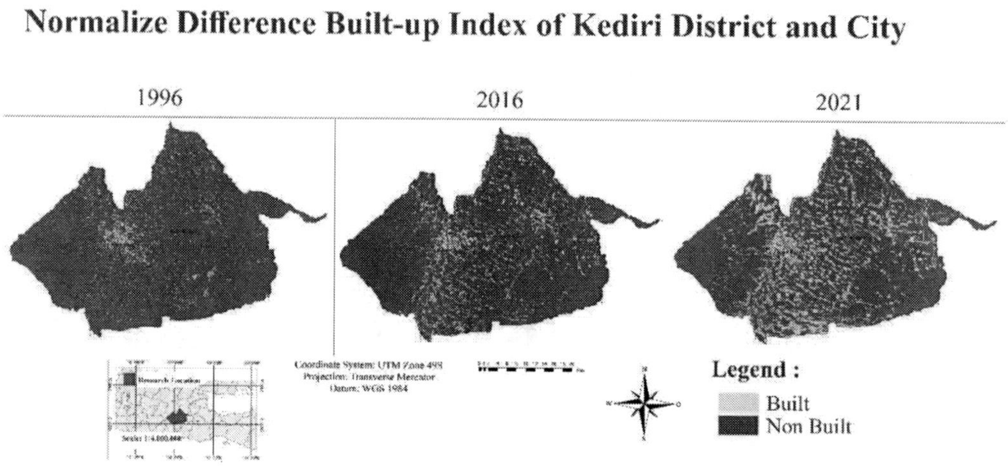

Figure 4. Spatial temporal of NDVI that develops over time

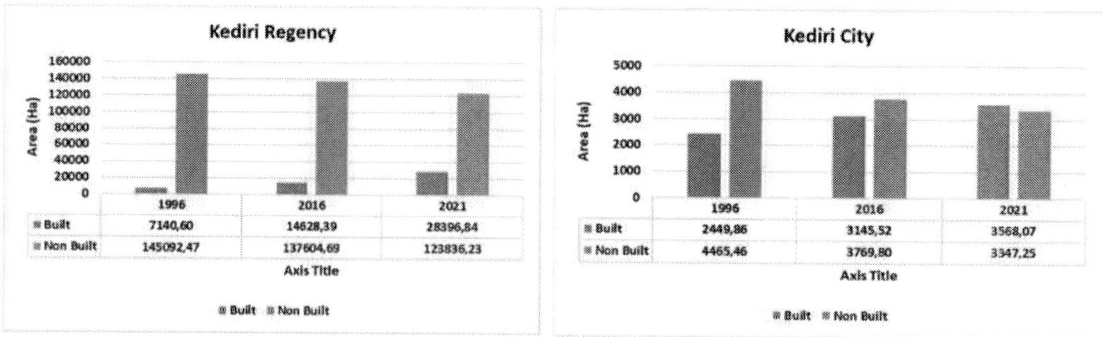

Figure 5. The NDBI distribution graph demonstrates that it is increasing in the built-up area over time.

3.2. Factors affecting the development of the city

The development of urban spatial planning in the study area is linear, with the development of roads leading to the district/suburbs. City (Fig. 4). The presence of road access helps in this regard, as it makes various daily tasks easier. The increase in population activity in urban areas appears to be a significant factor in the decline in the number of people migrating outside the region. In 2010, the number of people moving out exceeded the number who moved. Kabupaten Kediri was responsible for 79% of all migration, while Kota Kediri was responsible for 48%. (Table 3). This shows that the residents of Kediri Regency/City prefer to work outside the region. Districts/cities are considered less attractive areas for improving their welfare.

In 2015, the situation changed, with an increase in in-migration and a decrease in out-migration. The decrease in out-migration indicates that the local area is attractive enough to improve living conditions, so there is no need to look elsewhere. This increase is attributed to the expansion of job prospects in the city.

Table 3. Kediri Regency and City Migration

Kabupaten/Kota	Tahun 2010			2015		
	In migration	Out migration	Netto	In migration	Out migration	Netto
RegencyKediri	24711	93850	-69139	32624	42699	-10075
City Kediri	14868	13765	-1103	13175	10165	3010

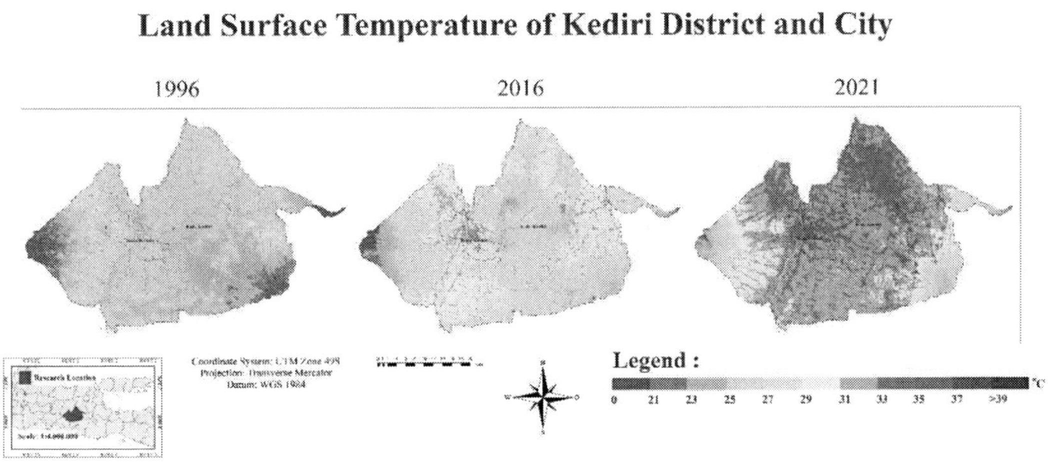

Figure 6. The spatial and temporal distribution of LST has increased rapidly in the last 5 years.

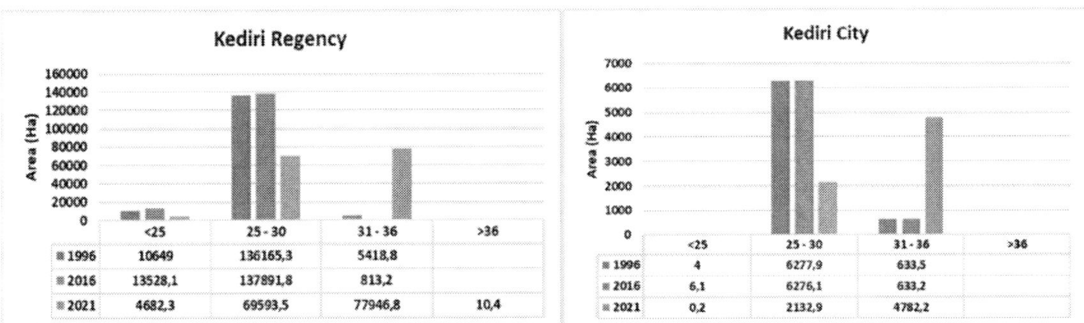

Figure. 7 The LST distribution graph shows that there has been an increase in temperature, especially in the last 5 years.

The surface temperature indicated by the LST value has a varied distribution. Figure 6 shows that the distribution of surface temperatures in the two study areas in 1996 and 2016 has the same pattern. The distribution of surface temperature is dominated by a temperature of 25-30 ° C, with an area of 90%. Kediri Regency, which covered 6-8% of the land area in 1996 and 2016, whereas Kediri City is much smaller, covering around 0.05-0.008% of the land area.This trend changes in 2021, where there is an increase in surface temperature with the largest distribution being at a temperature of 31-36° C (Fig 7). The increase was very significant, especially in Kediri Regency which reached 51% of its land area (an increase of 9400% compared to 2016). The distribution of surface temperature at a surface temperature of 31 – 36 ° C in Kediri City reaches 69% of the land area with an increase of 600%. This phenomenon is certainly very worrying, especially in the last 5 years.

Based on the transect analysis conducted from West to East across the center of Kediri City (Fig. 8), it can be seen that there is a spatial correlation between the increase in temperature and the increase in the built-up area. The increase is very significant, especially in the last 5 years (Fig. 9). The highest temperature is in the city of Kediri with a value of 36 °C, while the lowest is < 25 which is at the peak of Mount Wilis. Higher temperature values in built-up areas and lower in undeveloped areas indicate the UHI phenomenon.

The two research cities have relatively similar patterns related to the phenomenon of reducing green vegetation, increasing built-up area and increasing surface temperatures. These results are in line with research by (8)), (20)), (18), Lu et al. (2009), and (21), which investigated the cities of Vila Velha (Brazil), Noida (India), Hong Kong (China), and Nanjing (China), respectively. In these studies, it was found that there was a negative correlation coefficient between LST and NDVI and a positive one between LST and NDBI. (21) also emphasized that LST always builds a positive correlation with NDBI. Moreover, many studies have also proved the positive correlation between LST and NDBI ((22); (23); (24)). (25)

The UHI phenomenon, which is the main product of urbanization ((13)), indicates that urbanization has been and is ongoing in the research location. The very high speed of increasing the built-up area, especially in the last 5 years (2016-2021), followed by rising temperatures, has become a real threat to a sustainable city. If there is no mitigation and adaptation as well as good planning for the city, it will accelerate the increase in temperature and the affected area. Moreover, climate change with a projected increase of 1.5 C by 2040 (IPCC, 2018) will further magnify the threats facing the city. UHI is increasingly becoming a concern for policy makers and urban planners due to natural degradation, health risks, and economic losses ((26)).

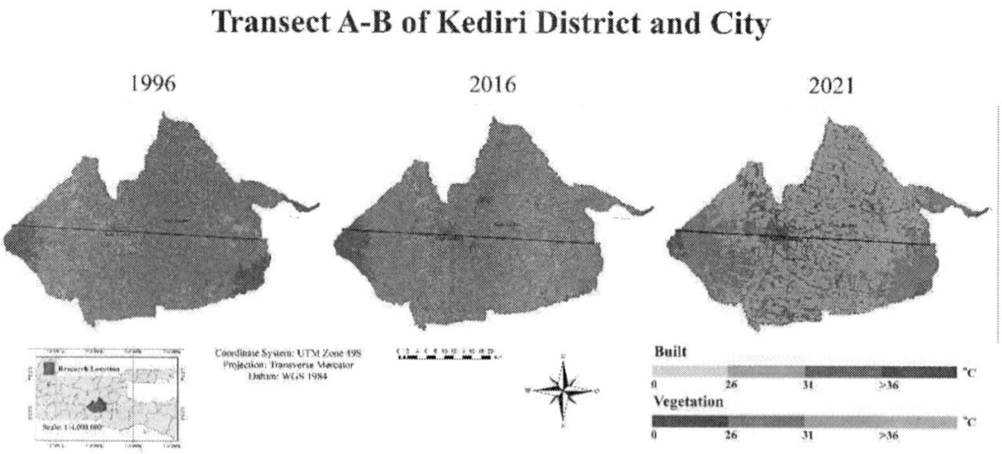

Figure. 8 The image shows a transect traced across the city of Kediri from point A to point B.

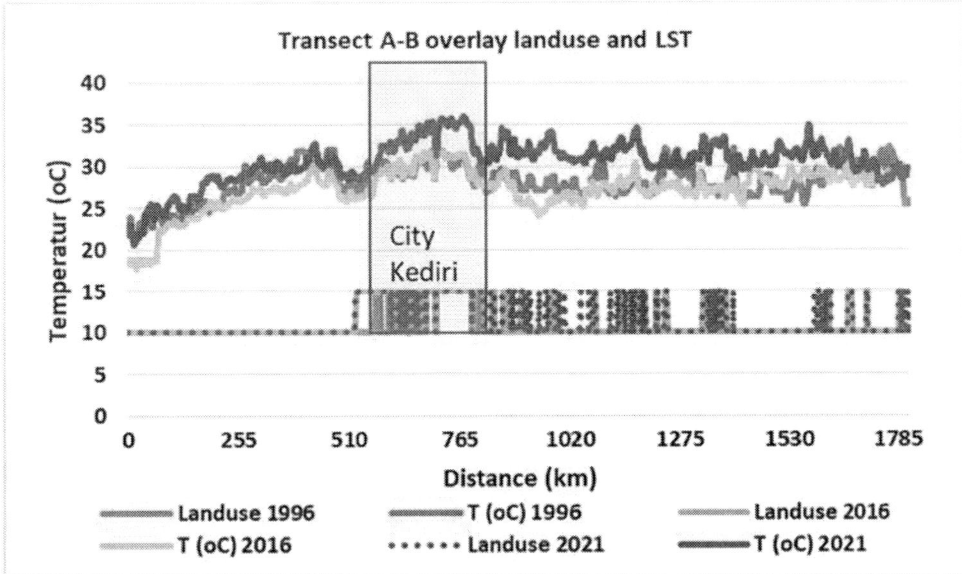

Figure. 9 As shown in this graph, the surface temperature has risen significantly in the last five years. The graphic clearly shows a positive relationship between surface temperature and building area.

Environmentalists and urban planners should consider the remaining open space for better ecological conditions ((21)). In addition, for the success of realizing a sustainable city, it is necessary to overcome three challenges in urban development. : The three problems of urban development today are: (1) active participation in urban development planning; (2) city management and government action; and (3) public participation and engagement ((27).

4. Conclussion

The development of urban spatial planning in the study area is linear, with the development of roads leading to the district/suburbs. The spatial development of the Regency and City of Kediri experienced a decrease in the amount of non-building land/vegetation and an increase in the amount of built-up land. Built-up land in the last 5 years increased by 83.8% compared to the increase in the previous 20 years. This is in line with increasing urbanization, especially the increase in population, the development of shopping centers, and ease of accessibility. The temperature increases with the increase in the built-up area. This shows that urbanization is happening. There is a potential UHI phenomenon in the study area as indicated by a significant temperature difference between urban and sub-urban. This is a challenge for urban spatial planning so that the study area remains a comfortable place to live.

References

[1] Van Hove L W A, Jacobs C M J, Heusinkveld B G, Elbers J A, Van Driel B L, Holtslag A A M 2015 Temporal and spatial variability of urban heat island and thermal comfort within the Rotterdam agglomeration. Build Environ [Internet] 83:91–103.

[2] Golden J S 2004 The Built Environment Induced Urban Heat Island Effect in Rapidly Urbanizing Arid Regions – A Sustainable Urban Engineering Complexity. Environ Sci. **1**, 4 321–49.

[3] Rasoolimanesh S M, Badarulzaman N, Jaafar M 2011 City Development Strategies (CDS) Contribution Toward SUstainable Urban Development in Developing Countries. Journal of

the Malaysian Institute of Planners Volume IX (2011), Page 1 - 18
[4] Iyengar S 2003 Environmental Damage to Land Resource: Need to Improve Land Use Data Base. Econ Polit Wkly **38** 34 :3596–604.
[5] Pickett STA, Cadenasso ML, Grove JM, Boone CG, Groffman PM, Irwin E, et al. 2011 Urban ecological systems: Scientific foundations and a decade of progress. J Environ Manage [Internet] **92** 3 : 331–62
[6] Baroroh N, Pangi P 2019 Perubahan Penutup Lahan Dan Kerapatan Vegetasi Terhadap Urban Heat Island Di Kota Surakarta. Semin Nas Geomatika. **3** 641.
[7] Xu Y, Qin Z, Wan H 2010 Spatial and Temporal Dynamics of Urban Heat Island and Their Relationship with Land Cover Changes in Urbanization Process: A Case Study in Suzhou, China. J Indian Soc Remote Sens. **38** 4:654–63.
[8] dos Santos AR, de Oliveira FS, da Silva AG, Gleriani JM, Gonçalves W, Moreira GL, et al 2017 Spatial and temporal distribution of urban heat islands. Sci Total Environ [Internet]. **605–606**:946–56
[9] Chen Y, Du M, Dong R 2007 Correlation Between Urban Heat Island Effect and the Thermal Inertia Using Aster Data in Beijing , China. Archives 1723–8.
[10] Chen Y C, Chiu H W, Su Y F, Wu Y C, Cheng K S 2017 Does urbanization increase diurnal land surface temperature variation? Evidence and implications. Landsc Urban Plan [Internet] **157**:247–58
[11] David R. Streutker 2017 A Remote Sensing Study of the Urban Heat Island of Hong Kong. Geomatics Sci Technol **5** 02:23–33.
[12] Xu Y, Zhou D, Li Z 2016 Research on Characteristic Analysis of Urban Heat Island in Multi-scales and Urban Planning Strategies. Procedia Eng [Internet] **169**:175–82.
[13] Levermore G, Parkinson J, Lee K, Laycock P, Lindley S 2018 The increasing trend of the urban heat island intensity. Urban Clim. **24**:360–8.
[14] Oke 1982 The energetic basis of the urban heat island p. 1–24.
[15] Rizwan A M, Dennis L Y C, Liu C 2008 A review on the generation, determination and mitigation of Urban Heat Island. J Environ Sci. **20** 1:120–8.
[16] Santamouris M 2013 Using cool pavements as a mitigation strategy to fight urban heat island - A review of the actual developments. Renew Sustain Energy Rev. **26**:224–40.
[17] Deilami K, Kamruzzaman M, Liu Y 2018 Urban heat island effect: A systematic review of spatio-temporal factors, data, methods, and mitigation measures. Int J Appl Earth Obs Geoinf [Internet]. **67** (September 2017):30–42
[18] Liu B, Jiang G, Zhang Y, Li J, Li X, Yue J, et al 2011 Analysis of transcriptome differences between resistant and susceptible strains of the citrus red mite panonychus citri (acari: Tetranychidae). PLoS One. **6** 12.
[19] BPS 2021 Kabupaten Kediri dalam Angka. **148**:148–62
[20] Kikon N, Singh P, Singh SK, Vyas A 2016 Assessment of urban heat islands (UHI) of Noida City, India using multi-temporal satellite data. Sustain Cities Soc. **22**:19–28.
[21] Guha S, Govil H, Gill N, Dey A 2021 A long-term seasonal analysis on the relationship between LST and NDBI using Landsat data. Quat Int [Internet] **575–576**(April 2020):249–58.
[22] Ibrahim G R F 2017 Urban land use land cover changes and their effect on land surface temperature: Case study using Dohuk City in the Kurdistan Region of Iraq. Climate **5** 1
[23] Yang H, Xi C, Zhao X, Mao P, Wang Z, Shi Y, et al 2020 Measuring the urban land surface temperature variations under Zhengzhou city expansion using landsat-like data. Remote Sens **12** 5
[24] Zhang Y, Odeh IOA, Han C 2009 Bi-temporal characterization of land surface temperature in relation to impervious surface area, NDVI and NDBI, using a sub-pixel image analysis. Int J Appl Earth Obs Geoinf. **11** 4:256–64.
[25] Aslam A, Rana I A, Bhatti S S 2021 The spatiotemporal dynamics of urbanisation and local climate: A case study of Islamabad, Pakistan. Environ Impact Assess Rev.

91(August):106666.
[26] Estrada F, Botzen W J W, Tol R S J A 2017 global economic assessment of city policies to reduce climate change impacts. Nat Clim Chang. **7** 6:403–6.
[27] Parsaee M, Joybari M M, Mirzaei P A, Haghighat F 2019 Urban heat island, urban climate maps and urban development policies and action plans. Environ Technol Innov. **14**:100341.

Assessing urban heat island in Jakarta, Indonesia during the pandemic of Covid-19

A Saputra[1*], M H Ibrahim[2], S Shofirun[3], A Saifuddin[4] and K Furoida[1]

[1] Geography Faculty, Universitas Muhammadiyah Surakarta, Indonesia
[2] Universiti Pendidikan Sultan Idris, Perak, Malaysia.
[3] Universiti Utara Malaysia, Sintok, Malaysia
[4] Geography Faculty, Universitas Gadjah Mada Yogyakarta, Indonesia.

*Email: sholehsofyan@gmail.com, yasinyusup@staff.uns.ac.id

Abstract. Jakarta, the capital city of Indonesia, is the most populated area in Indonesia, with a population of 10.56 million people or around 3.91% of the total Indonesian population. Jakarta has also become an important city in Indonesia as around 80% of Indonesian economic activity is located in this area. No doubt, the urban activity caused the air pollution to infuriate its microclimate, including the urban heat island phenomena. During the Covid-19 pandemic, Indonesia applied the partial lockdown for several areas to decrease virus transmission. Most of the transportation and commercial areas were closed, and most people worked from home. This significant shifting will affect the temperature and urban heat island conditions. Thus, the main objective of this research was to analyse the urban heat island during the early stage of the pandemic in Jakarta. We used Landsat 8 imagery to extract the land surface temperature (LST) and generated the urban heat island (UHI). The correlation was used to determine the relationship between the distribution of covid-19 cases with the UHI distribution. From the four different recording times (May, July, September, and December 2020) of Landsat 8 imagery, the highest UHI, 7.76^{0}C was found in December 2020 (the late first year of the pandemic). The lowest UHI, 4.91^{0}C was found in May 2020 (The early stage of the first year). Furthermore, the UHI hot spot moved from almost evenly in East Jakarta in May 2020, a tiny spot in East Jakarta in July and September 2020, and evenly distributed in Southeast Jakarta in December 2020. In addition, we found that the increase of covid-19 cases had a positive correlation with the UHI in Jakarta, which means the areas with the high UHI value have high new covid-19 cases. These results showed that the area with high activities is very prone to covid-19 transmission.

1. Introduction

The population is one of the causes of climate change. But more than that, the activities carried out by the population have a major impact on climate change such as industrial activities that release greenhouse gases and population migration on a large scale [1]. These activities can be found in urban areas. The World Population Data Sheet 2020 shows that 40% of twenty-six countries have urban populations of more than 1 million people. The reason why is many people choose to live in urban areas is that cities have a higher standard of living than villages so that residents can generate higher incomes to fulfil consumptive desires [2]. A large number of residents in urban areas will certainly be directly proportional to the burden borne by urban areas, especially on natural resources so that urban areas have the vulnerability to the risks and impacts of climate change [3]. One of the risks of climate change is the urban heat island (UHI) phenomenon. UHI is a condition where the city has a higher temperature in the surrounding area because the city experiences the fastest and most urbanization process [4,5]. UHI has 2 types, namely surface UHI which can be measured using Land Surface Temperature (LST), and atmospheric UHI which is measured based on air temperature [6].

In 2019, there was a new outbreak that attacked the world's population, namely the coronavirus disease (COVID-19), which was first reported to appear in Wuhan, China on December 31. The outbreak has a rapid spread of cases, resulting in a rapid increase in COVID-19 cases even on a global scale. In May 2021, WHO reported that more than 170,000,000 cases of COVID-19 had occurred worldwide. Even in serious cases, COVID-19 can result in death. The process of spreading COVID is through small fluid particles from the mouth or nose of an infected person when they cough, sneeze, talk, sing or breathe. In addition, the WHO also reported that COVID-19 would have a very large risk of spreading in the interaction of several people at a distance of less than 1 meter if they were in a room with minimal ventilation. The risks and impacts of COVID-19 have caused many countries to take decisions to reduce the spread of COVID-19 more widely in their countries. One of the decisions taken was to impose a lockdown. The Lockdown in Wuhan, as the first place where COVID-19 emerged, was carried out by restricting travel to and from Wuhan, closing public places including schools and universities, banning social activities and having to adhere to strict quarantine at home [7]. The Lockdown was also carried out by several other countries, including Italy, Poland, Spain, England, France, Malaysia, and many other countries.

The State of Indonesia, especially in Jakarta, has also implemented restrictions on activities for its people to reduce the spread of COVID-19 with the existence of Large-Scale Social Restrictions (PSBB). PSBB includes restrictions on religious activities, schools, workplaces, public facilities, socio-cultural activities, modes of transportation, and other activities specifically related to defence and security aspects, but in terms of essential services and public transportation, there are no total restrictions but only capacity reduction and strict implementation of health protocols. The PSBB, which restricts Jakarta residents from doing activities outside the home, will certainly make a difference, not only in terms of habits but also environmental effects. Anugerah et al. (2021) [8] in their research showed that PSBB can reduce the concentration of air pollution, namely CO, NO2, and SO2. Another thing that can be predicted by PSBB is the impact of the urban heat island phenomenon. Previously, several studies showed that Jakarta experienced an urban heat island. Jakarta had temperatures reaching 31 – 34 °C in 2013 and 34.1 °C in 2014 [8,9]. While the UHI comparison research in Putra et al. (2021) [10] shows that the 3 sample years studied, namely 2008, 2013, and 2018, have several areas with temperatures over 30 °C.

This recent study, researchers aimed to examine the relationship between UHI and COVID-19 in Jakarta. This study uses surface UHI as a parameter to determine UHI so that image data for obtaining LST values is carried out using LANDSAT-8. Several studies of the same study have also been carried out, such as Hadibasyir et al. [11] showing that the average LST value in built-up areas experienced a significant difference in the form of a decrease in the LST value of 1.19^{0}C. Research in the United Arab Emirates on the implementation of lockdown was able to reduce the value of the surface urban heat island intensity (SUHII) by 19.2% [12].

2. Material and study area

The study area is located in DKI Jakarta, the capital city of Indonesia. The data used in this research was taken in May, July, September and December of 2020. Those selected months represent two seasons in Indonesia. July and September is the dry season, while May & December is the wet season. The use of remote sensing imagery gives an excess to actual comprehensive in the case of the city region [4]. This research uses imagery remote sensing of Landsat 8 thermal infrared sensors (TIRS). Data (for the first 4 months of 2020) was obtained from the United States Geological Survey (USGS) website. There is no radiometric and geometric correction because this research uses data type L1TP that is already corrected by the system. The Covid 19 data case in Jakarta was obtained by the government website of DKI Jakarta. The data used cumulative cases of covid 19 in 1 district in 1 month.

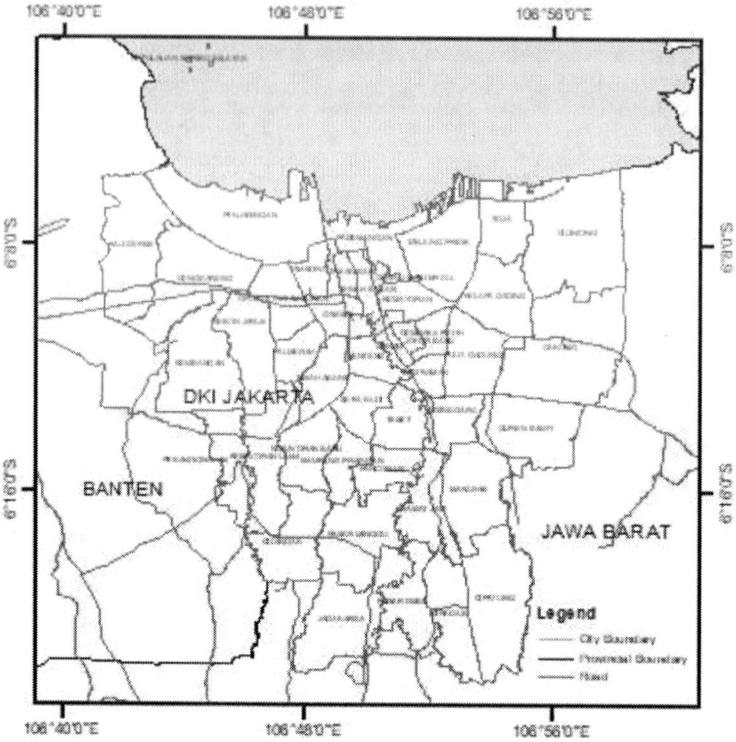

Figure 1. Map of District DKI Jakarta

3. Method

The data process to generate the urban heat island (UHI) use derivatives from land surface temperature (LST) value. Then, UHI value in every district has been correlated with cumulative cases Covid 19. Here is the method process:

3.1. Radiance (Lλ)

Land surface temperature value got by process that refers to Landsat 8 data handbook [13]. Digital value or digital number (DN) pixel changed to radiance value (Lλ) with algorithm

$$L\lambda = (ML)*(DN) + AL - Oi \quad (1)$$

ML & AL is factor value radiance from metadata; DN is pixel value; Oi is constant value band 10 viz 0.29.

3.2. Conversion radiance (Lλ) to Brightness Temperature (BT)

Radiance value is not represented temperature value so it needs to convert to Brightness Temperature (BT) value. Brightness Temperature is real temperature recorded by remote sensing. BT value obtained by equation 2. K1 & K2 obtain from metadata Landsat imagery is constant value. The result of equation is minus by value -273.15 to get Celsius temperature.

$$BT = K2/\ln (K1L\lambda + 1) - 273,15 \quad (2)$$

3.3. Land surface emissivity (LSE)

Emissivity is capacity of city material that influenced thermal condition to determine solar energy reflectance and absorption [14]. Land emissivity value are varied by heterogeneity of vegetation, land humidity, roughness, view angle with energy. LSE value evaluation by Normalized Difference Vegetation Index (NDVI) and Proportion Vegetation (PV) method [15].

$$E = 0.004 * Pv + 0.98 \quad (3)$$

PV is proportion vegetation value. If pv value show lower it is built land or open land and value upper show the vegetation. Pv equation use the relation between NDVI and percent covered land [15].

$$Pv = ((NDVI- NDVI_{max})/(NDVI_{max}-NDVI_{min}))^2 \quad (4)$$

NDVI give green vegetation illustration in raster pixel and give land composition to get emissivity value. NDVI calculated by Equation 5. NDVI result value from -1 to 1, value near -1 showing less density vegetation, water or soil and near 1 showing more density vegetation.

$$NDVI\ (NIR\ band - Red\ band)/(NIR\ band+ Red\ band) \quad (5)$$

3.4. Land Surface Temperature (LST)

Land surface temperature (LST) calculated by use of corrected land emissivity value [8] use Equation 6 to calculated.

$$LST = BT/\ (1+\lambda*BT/c2)*\ln\ (\varepsilon) \quad (6)$$

BT is Brightness temperature value from Equation 2; λ is effective wavelength (Band Thermal); C2 is $h*c/s= 1.4338*10^{-2}mK = 14338MK$; s is Boltzmann constant value (1,38 ×10− 23JK− 1); h is Planck constant value (1,38 ×10− 23JK− 1); and c is light speed in vacuum (2,998 × 10− 8 Ms− 1).

3.5. Urban Heat Island (UHI)

UHI phenomenon is divided into two namely Land UHI & atmospheric [14] . Remote sensing will result only the land UHI. Land UHI is used for visualization difference temperature condition cause land temperature related with air temperature [14]. UHI value calculated from LST value used threshold calculation [16].

$$UHI = Ts - \ (\mu + 0{,}5\ \alpha) \quad (7)$$

UHI is urban heat island; Ts is land surface temperature value; μ is mean value Ts; α is standar deviation value Ts. UHI calculated used threshold calculation if positive value so represented the UHI and if lower than 0 represented no UHI in area [15].

3.6. Corelation Analysis

Pearson correlation is used in this research to know the relation among two variables between UHI value at district in DKI Jakarta and cumulative increase positive cases covid 19 in one month. The Pearson value result can represent the relation value of the variables.

4. Result and discussion

4.1. Land Surface Temperature (LST)

Land surface temperature calculation in DKI Jakarta generates a different value each month. This result is affected by the cloud cover that covers the land, so it has become the main problem with LST calculation. The maximum LST value at December is 34.36 °C and the minimum value looks similar every month between 13 – 14 °C. It is due to the cloud cover in imagery being less clean, so some areas in study location have a noise value from that. Figure 2 shows that the dry season (July & September) looks hotter than the wet season (May & December). From that, we know that the LST depends on the season. The great temperature of LST occurred in July, which shows that almost the whole area in DKI Jakarta is yellow to red, which shows the high symbol of LST. But July is not the hottest month of the year. The hottest month is September. In September, the temperature reached 35,73 °C. Otherwise, May become the coldest month in which the temperature is 30,65 °C. Most of the LST results show the same pattern of each area which has the higher temperature each month. But the temperature is dynamically shown depending on each month, such as in northeast DKI Jakarta in May, September, and December show the green colour which shows the lowest temperature, but in July it shows the yellow to orange colour that fits with the fact that July has a larger area of high temperature.

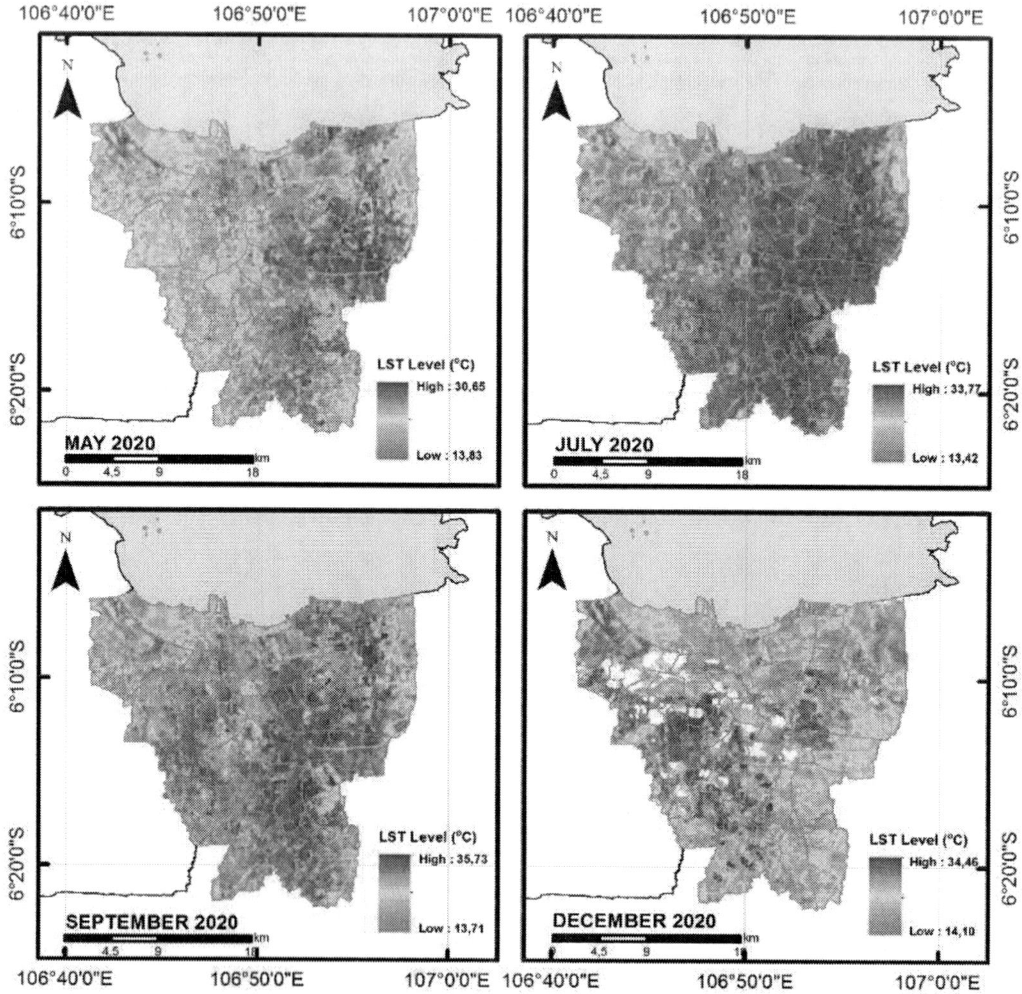

Figure 2. Land Surface Temperature in DKI Jakarta

4.2. Urban Heat Island (UHI)

The urban heat island value is calculated from the land surface temperature value. The UHI value represents the LST value in each area. UHI detected in the east area of DKI Jakarta in May, July and September, while in December there is a high UHI value detected in the east area of DKI Jakarta. The UHI value in each month has increased from May, which has a maximum value of 4.92 °C, July at 5.64 °C, September at 6,97 °C and December at 7,76 °C. UHI spatial distribution can be related to a lot of people's activities in that area. The more people who have activity outside, the more it will affect the expansion of spatial distribution and increase the UHI value in that area. The result of this research has applied to *Pembatasan Sosial Skala Besar* (PSBB) that limits people's activities outside. May is the early month of PSBB in DKI Jakarta. The rules must be people decrease the activity outside. PSSB affected schools, companies, public facilities, and shopping centres to reduce activity. This factor represented UHI value with UHI distribution due to people's activity. May is the first PSBB and UHI that can be affected by human activity. May shows the lowest UHI value compared to the other 3 months. Figure 3 shows that May, July and September have the same UHI pattern that shows that the UHI value is mostly in the east area of DKI Jakarta. Whereas, the UHI pattern in December is very different compared to the other 3 months, where December has many larger UHI phenomenon. In December, most of UHI is in the west area of DKI Jakarta, which is mostly the urban areas of this city. Related to the lowest and highest UHI value, this research shows that each month the UHI value always increases. The increase in UHI value can be affected by the first PSBB in May, but the next month, the PSBB is loosened up so that people do activities outside and increase the cause of the UHI phenomenon. Figure 3 also shows that not all areas of DKI Jakarta have a UHI value, although the LST value in figure 2 shows that DKI Jakarta mostly has high LST. But, the different patterns of LST and UHI mainly in July that have the larger area of LST do show the larger area of UHI. The larger area of UHI occurred in September, which has the higher LST. Those results mean that the larger area of LST does define the UHI phenomenon, but the higher LST can define the larger UHI phenomenon.

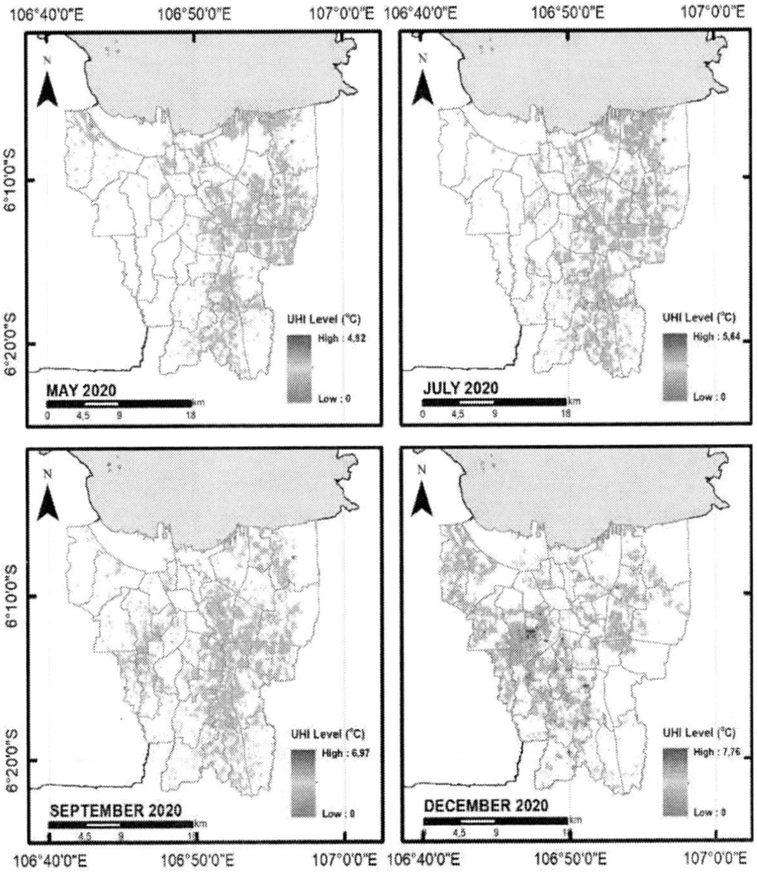

Figure 3. Urban Heat Island in DKI Jakarta

4.3. Correlation of UHI and COVID-19 Cases

The correlation results between urban heat islands and cumulative cases covid 19 show that there is a positive relationship between them. It means that higher UHI value will have higher positive cases of covid-19 in that area. The Pearson correlation value shows that it increased every month from May at

0,147, July at 0,228 and December at 0,514. Yet, the correlation in September doesn't have a significant result because the calculated UHI in September has over cloud cover.

Table 1. Correlation UHI Value with Covid 19 Data Cases May

		UHI	DATA COV19
UHI	Pearson Correlation	1	**,147***
	Sig. (2-tailed)		,016
	N	268	268
DATA COV19	Pearson Correlation	,147*	1
	Sig. (2-tailed)	,016	
	N	268	268

*. Correlation is significant at the 0.05 level (2-tailed).

Table 2. Correlation UHI Value with Covid 19 Data Cases July

		UHI	DATA COV19
UHI	Pearson Correlation	1	**0,228***
	Sig. (2-tailed)		0,000
	N	263	263
DATA COV19	Pearson Correlation	0,228**	1
	Sig. (2-tailed)	0,000	
	N	263	263

*. Correlation is significant at the 0.05 level (2-tailed).

Table 3. Correlation UHI Value with Covid 19 Data Cases September

		UHI	DATA COV19
UHI	Pearson Correlation	1	**0,069**
	Sig. (2-tailed)		0,265
	N	81	81
DATA COV19	Pearson Correlation	0,069	1
	Sig. (2-tailed)	0,265	
	N	81	81

*. Correlation is significant at the 0.05 level (2-tailed).

Table 4. Correlation UHI Value with Covid 19 Data Cases December

		UHI	DATA COV19
UHI	Pearson Correlation	1	**,514***
	Sig. (2-tailed)		,000
	N	265	265
DATA COV19	Pearson Correlation	,514**	1
	Sig. (2-tailed)	,000	
	N	265	265

*. Correlation is significant at the 0.05 level (2-tailed).

Significancy of the correlation can be affected by the data quality of imagery. On September cases of covid 19 keep increase in every month but availability of remote sensing imagery cannot be found in good imagery because its depend on the temporal radiometric of Landsat-8 that in each month may only have 2 recorded imagery so that in this research use the best recorded imagery but still have the cloud cover that affected the result. Table 1, Table 2, Table 3 and Table 4 show that the correlation result is based on the district data in Jakarta that was selected only at the area which have UHI phenomenon to look the relation between UHI and covid 19 cases. December has the higher value of correlation related to figure 3 that show that UHI phenomenon is also larger and higher than the other 3 months. It means that on December, when PSBB is loosened up and most of people back to do activity outside, the covid 19 cases is also increase because the spread of covid 19 is based on the people activity. From that correlation, it can be concluded that rules is not affective in Jakarta at 2020.

Signify the LST results of Hadibasyir et al [11] research, this result found the similar phenomena in Wuhan China, although they used different resolution of satellite imagery. During the lock down policy both areas, Wuhan and Jakarta, experienced the same phenomena. The LST of both areas decreased significantly. This condition led to the shifting of urban heat island phenomena which usually occur in industrial area to the settlement area. During the pandemic, the government force the lock down policy in particular area in Jakarta. Most of workers work from home and used more electricity to support their activity at home. As a result, the LST also affect the distribution of UHI. In term of Covid transmission, we found high positive correlation of Covid-19 accumulative case with the UHI value in several observed months. We assumed that the area which has high UHI value has more activities, therefore the transmission of covid increase as the developing country tend to have more problem in maintaining the social distancing or partial lock down in particular area [17].

5. Conclusion

During the implementation of partial restriction or lockdown policy, there were shifting of population concentration in Jakarta. In normal time people tend to concentrated in central business district and industrial area. Meanwhile, during the lock down policy, most of people were work from home. As a result, there were shifting of LST temperature. The results showed that the LST decreased significantly on industrial area and increase in settlement area. Furthermore, the UHI pattern also match with the LST pattern. In addition, we found that the prone areas to covid-19 (have high accumulative covid-19 case) were have positive correlation between the high UHI value. Further research need to be conduct in different places signify this result.

Acknowledgments

The authors are thankful to Universitas Muhammadiyah Surakarta, Universiti Pendidikan Sultan Idris, Malaysia, and Universiti Utara Malaysia for providing excellent collaboration research. The author is also most thankful to Universitas Muhammadiyah Surakarta who provide the research funding.

References

[1] Stephenson J, Newman K and Mayhew S 2010 Thesis Population dynamics and climate change : what are the links ? *J. Public Health (Bangkok)*. **32** 150–6

[2] Kahn M E 2009 Urban Growth and Climate Change *Annu. Rev. Resour. Econ.* **1** 333–50

[3] Sharma R, Pradhan L, Kumari M and Bhattacharya P 2021 Urban Climate Assessing urban heat islands and thermal comfort in Noida City using geospatial technology *Urban Clim.* **35** 100751

[4] Chakraborty T C, Sarangi C and Lee X 2021 Reduction in human activity can enhance the urban heat island : insights from the COVID-19 lockdown OPEN ACCESS Reduction in human activity can enhance the urban heat island : insights from the COVID-19 lockdown *Environ. Res. Lett.* **16**

[5] Ibrahim M H, Latiff N A A, Ismail K and Isa N K M 2018 Effect of urbanization activities towards the formation of urban heat island in Cameron Effect of urbanization activities towards the formation of urban heat island in Cameron Highlands , Malaysia *IOP Conf. Ser. Earth Environ. Sci. 148* **148**

[6] Mukherjee S 2020 Correlation between Land Surface Temperature and Urban Heat Island with COVID-19 in New Delhi , India 1–9

[7] Lau H, Khosrawipour V, Kocbach P, Schubert J, Bania J, Khosrawipour T, Protection C H and Khosrawipour T 2021 The positive impact of lockdown in Wuhan on containing the COVID-19 outbreak in China *J. Travel Med.* **27** 1–7
[8] Rizky A, Suksessanno P and Adi D 2021 Effect of large-scale social restriction (PSBB) during COVID-19 on outdoor air quality : Evidence from five cities in DKI Jakarta Province , Indonesia *Environ. Res.* **197** 111164
[9] Rushayati S B, Shamila A D and Prasetyo L B 2018 The Role of Vegetation in Controlling Air Temperature Resulting from Urban Heat Island *Forum Geogr.* **32** 1–11
[10] Putra C D, Ramadhani A and Fatimah E 2021 Increasing Urban Heat Island area in Jakarta and it ' s relation to land use changes Increasing Urban Heat Island area in Jakarta and it ' s relation to land use changes *IOP Conf. Ser. Earth Environ. Sci. 148* **737**
[11] Hadibasyir H Z, Rijal S S and Sari D R 2020 Comparison of Land Surface Temperature During and Before The Emergence of Covid-19 Using Modis Imagery in Wuhan City , *Forum Geogr.* **34** 1–15
[12] World Population Data Sheet 2020 Demographic Trends May Make Us Vulnerable to Pandemics Data Table. 22
[13] Department of the Interior U.S. Geological Survey 2019 Landsat 8 (L8) Data Users Handbook
[14] H A, R B, T B, D C, M E, G H, D H, B J, M L, G M and T O 2008 *Reducing Urban Heat Islands : Compendium of Strategies Urban Heat Island Basics*
[15] Ngie A, Abutaleb K, Ahmed F, Darwish A and Ahmed M 2014 Assessment of urban heat island using satellite remotely sensed imagery : a review *South African Geogr. J.* **96** 37–41
[16] Jatmiko R H 2016 *Penggunaan Citra Saluran Inframerah Termal untuk Studi Perubahan Liputan Lahan dan Suhu sebagai Indikator Perubahan Iklim Perkotaan di Yogyakarta* (Gadjah Mada University)
[17] Crimmins E M 2020 Age-Related Vulnerability to Coronavirus Disease 2019 (COVID-19): Biological , Contextual , and Policy-Related Factors *Public Policy Aging Rep.* **30** 142–6

IOP Conf. Series: Earth and Environmental Science **986** (2022) 012070 doi:10.1088/1755-1315/986/1/012070

Spatial analysis of air surface temperature using M-AST model in complex sub-urban area

N F Fauzan[1*], A Wibowo[1] and I P A Shidiq[1]

[1] Department of Geography, Faculty of Mathematics and Natural Science, Universitas Indonesia, Depok, Indonesia

*Email: nur.fadhilah72@ui.ac.id

Abstract. The distinction between urban and sub-urban becomes faded nowadays. The urban sprawl changes and transforms the sub-urban into more stand-alone cities in many ways, even more with the emergence of various activity centers spread over the area. Intense and complex human activities within the area will affect the micro-climatic condition, such as surface temperature and relative humidity. This study aims to identify changes in air surface temperature in a complex sub-urban area, like the IPB University campus and its surroundings in Cibanteng, Cihideung Ilir, Babakan, and Dramaga, of Bogor Regency. The changes in land-cover over some times also needed to understand, as it will raise the surface temperature, especially when natural vegetation turns into the built-up area. M-AST was generated from regression analysis between ground-based air surface temperature (AST) and image-based land surface temperature (LST). The $y = 0.9988x + 4.3082$ from the M-AST has an acceptable result with 0.70008 in the R^2. The highest air surface temperature was found on the built-up area (35.6 °C), and the lowest was on the non-agricultural area (32.9 °C). This study concluded that the increase in air surface temperature of more than 2 °C was caused by changes in land-cover from vegetation to the built-up area.

1. Introduction

From time to time development of urban areas is dynamically changing and shaping the land-cover of its surroundings. Land-cover change is one of the most visible outcomes of ecosystem modification and significantly impacts the local, regional, and global environment [1]. Changes in land-cover are impacting two aspects, natural variations of the earth system and human activities [2]. Some changes in land cover caused by human activities include changes in vacant land and vegetated land into impermeable surface land such as concrete, concrete, and cement or changes in urban morphology such as building areas with varying heights and densities [3]. Such kind of circumstances can lead to changes in temperature on the Earth's surface [4].

Urban areas are characterized by reduced vegetation cover and increased buildings compared to rural areas, and this often results in the city being warmer than the surrounding area [5]. This phenomenon is known as Urban Heat Island (UHI) [6]. UHI is a unique phenomenon of urban ecosystems and has the main characteristics as a result of anthropogenic activities on air temperature that have an impact on global climate warming [7]. Based on the research that has been done, there is a dominant physical relationship between air temperature and ground surface temperature [8].

The university can be considered as a micro-scale urban area because of its size, population, and various activities that occur in it [9]. it can dynamically change the land cover in the campus area.

Moreover, since urban areas and campuses have similar features, changes in the land-cover will impact land surface temperature variations [10,11]. The existence of the campus causes changes in the use of the surrounding land to meet the needs of students, such as housing, restaurant/cafe, and other supporting places such as mini markets, photocopy stations, internet cafes, and laundry facilities [12]. This activity will cause changes in land-cover around the campus. The surface temperature inside the campus and the surrounding area can describe conditions in urban areas, especially with temperature differences based on different land-cover types.

IPB University (IPB) seeks to realize the dream of becoming an environmentally friendly green campus [13]. Green Open Space (RTH) is the focus of IPB as one of the variations of land-cover, which is the theme of IPB as a green campus [14]. However, over time, IPB carried out developments to support student needs that unfortunately changes their land-cover [10]. Because of the rapid change in land cover, it is essential to look at the dynamics of temperature changes around a large education center in Bogor, such as IPB University.

Changes in land cover will impact changes in surface temperature and air temperature [15]. This study aims to identify changes in air surface temperature in a complex sub-urban area, like the IPB University campus and its surroundings in Cibanteng, Cihideung Ilir, Babakan, and Dramaga, of Bogor Regency. Due to the nature and extent of the phenomenon, this study uses a remote sensing approach. Remote sensing is considered to be better at identifying the relationship between land cover and land surface temperature in tropical areas, such as Indonesia.

2. Methods

Monitoring land-cover change and dynamic land surface temperature (LST) are challenging due to the nature and extent of the events. This study utilizes a medium-resolution Landsat 8 OLI-TIRS to extract land-cover features and generate temporal LST as part of the remote sensing approach to address those challenges. Remotely sensed data processing includes land-cover classification and LST generation.

2.1. Image processing

The satellite image used in this study is the Landsat 8 OLI-TIRS acquired in the dry months with 20% in the cloud cover.

Table 1. List of Landsat 8 OLI-TIRS imageries in this study

Year	Path/Row	Acquisition Date	Remote Sensing Imagery
2013	122/064	08 July	Landsat 8 OLI/TIRS
2017	122/064	17 June	Landsat 8 OLI/TIRS
2021	122/064	11 May	Landsat 8 OLI/TIRS

2.1.1. Land-cover classification

Land-cover were categorized using the supervised classification method on the Google Earth Engine. The Classification and Regression Tree (CART) technique is applied to generate several land-cover classes. The first step in CART is image processing by taking several areas of interest (AOI) as samples to obtain the pixel characteristics of each object/class, which are then grouped based on the characteristics of the pixel values [16]. The determination of AOI is based on observations of Landsat imagery and verified based on Google Earth, especially for 2013, 2017, and 2021.

The land-cover classification follows the 2010 national standard from the *Standar Nasional Indonesia* (SNI), which is generalized to land-cover in the IPB University area and its surroundings. The classification is divided into five classes: water body, built-up land, open space land, non-agricultural vegetation, and agricultural vegetation. The area of each class will be calculated every year, and then the change in area will be seen over the years.

The Khat Kappa method is a statistical calculation method used in research to test the accuracy of the data between land-cover classification resulting from satellite image processing and the testing ROI from Google Earth. The Khat Kappa method calculates every error in each form of land-cover from the results of the image classification process [17].

2.1.2. Land Surface Temperature (LST)

LST is obtained from the calculation of the satellite brightness value. It is assumed that the water content in the atmosphere is constant for a small area.

In identifying the land surface temperature, a thermal band (Band 10) from Landsat 8 OLI-TIRS image is being calculated with a formula according to [18].

$$L\lambda = (M \times DN) + A \tag{1}$$

The first step is to change the image pixel value (DN) into spectral radiance, where Lλ is spectral radiance (wm-2SR-1µm-1), M is the multiplicative digital number at thermal band, DN is the digital number at thermal band, and A is the additive value at the thermal band.

After the spectral radiance value is obtained, it will be converted to the soil surface temperature value with the following formula:

$$T = \frac{K2}{In\left(\frac{K1}{L\lambda} + 1\right)} \tag{2}$$

Where T is temperature from satellite sensor (Kelvin), K1 is calibration constant 1 for Landsat OLI (774853 K), K2 is calibration constant 2 for Landsat OLI (1321.0789 K), and Lλ is spectral radiance at the thermal band.

After obtaining the Kelvin temperature value, it is converted into Celsius with the following formula:

$$LST(Celcius) = T - 272{,}15 \tag{3}$$

Temp-Kelvin is the LST approximation of the satellite sensor in Kelvin. Temp-Celsius is the International Standard for LST in degrees Celsius, where one Kelvin is equal to 272.15 °C.

2.2. Air Surface Temperature Model (M-AST)

The M-AST is determined from a simple linear regression between the LST obtained from the image processing, and the AST obtained from field survey measurements. The simple linear regression formula can be seen in the following equation:

$$Y = a + bX \tag{4}$$

Y is the dependent variable, X is the independent variable, a is constant (value of Y if X = 0), and b is the regression coefficient (positive or negative effect). Then, the M-AST map can be developed by applying the model in the ArcGIS 10.3 raster calculator. The air surface temperature model that has been obtained will then be tested for accuracy. The Root Mean Square Error (RMSE) method is a statistical calculation method used in research to test the accuracy of the data between the M-AST value and the air surface temperature value from the ground truth. The RMSE formula that has been determined by [19] can be seen in the following equation:

$$RMSE = \sqrt{\Sigma_{i=1}^{N} \frac{(x_{esti,i} - x_{meas,i})^2}{N}} \tag{5}$$

IOP Conf. Series: Earth and Environmental Science **986** (2022) 012070 doi:10.1088/1755-1315/986/1/012070

Where $x_{esti,i}$ is the estimated value, $x_{meas,I}$ is the measured value, N is the amount of data used for validation, and $\sum$ is the summation (the total number of scores).

3. Results and Discussions

3.1. Land-cover change

The land cover on the IPB University campus and its surroundings is classified into five classes: water body, built-up land, open space land, non-agricultural vegetation, and agricultural vegetation. Based on Table 2, it can be seen that in 2013, 2017, and 2021 built-up land, agricultural vegetation, and non-agricultural vegetation are the dominant land-cover. The agricultural vegetation cover was primarily found in 2013 (38.28%), but it was being transformed to built-up land where it was primarily found in 2017 (44.09%) and 2021 (55.72%). The built-up is developing in the form of settlements, apartment buildings, and campus buildings. In comparison, the decreasing agricultural vegetation is rice fields, plantations, fields, and dry fields. On the other hand, the non-agricultural vegetation is in the form of forests and shrubs. Meanwhile, water bodies are land cover with the lowest proportion each year, less than 1%.

Table 2. Land Cover Area in IPB University and Its Surroundings

Land Cover	2013		2017		2021	
	Area (Ha)	(%)	Area (Ha)	(%)	Area (Ha)	(%)
Water Body	1.80	0.20	1.02	0.11	1.53	0.17
Build-up Land	308.84	34.38	396.54	44,09	500.86	55.72
Open Space Land	7.75	0.86	2.16	0.24	7.87	0.88
Non-Agricultural Vegetation	236.12	26.28	240.90	26.78	173.73	19.33
Agricultural Vegetation	343.91	38.28	258.84	28.78	214.86	23.90

Each land-cover class on the IPB University campus and its surroundings from 2013, 2017, and 2021 continues to change. Based on Table 2, in 2013, the land cover in the IPB University campus and its surroundings was dominated by agricultural vegetation land with an area of 343.91 Ha. In eight years (2013-2021), the area of agricultural vegetation decreased by 14.38% (129.05 Ha) from 343.91 Ha to 214.86 Ha. In addition to agricultural vegetation, non-agricultural vegetation also has a large enough area. However, similar to agricultural vegetation, the area of non-agricultural vegetation also tends to decrease every year.

In 2013-2021 the area of non-agricultural vegetation decreased 6.96% (62.39 Ha) from 236.12 Ha to 173.73 Ha. The decline in the area of agricultural vegetation and non-agricultural vegetation was followed by a significant increase in built-up land area by 21.34% (192.02 Ha) from 308.84 Ha in 2013 to 500.86 Ha in 2021. As presented in Figure 1, spatially land-cover changes occur rapidly in the area around the IPB University campus, which is dominated by changes in agricultural and non-agricultural vegetation into built-up land.

Based on the results of the calculation of the accuracy test of the Khat Kappa method, the results of image classification and Google Earth validation of land cover of the IPB University campus and its surroundings in 2013, 2017, and 2021 have an Overall Accuracy value of 89%. And the Kappa Coefficient value is 0.84, based on the Khat Kappa classification a value of> 0.80 indicates a very good accuracy value.

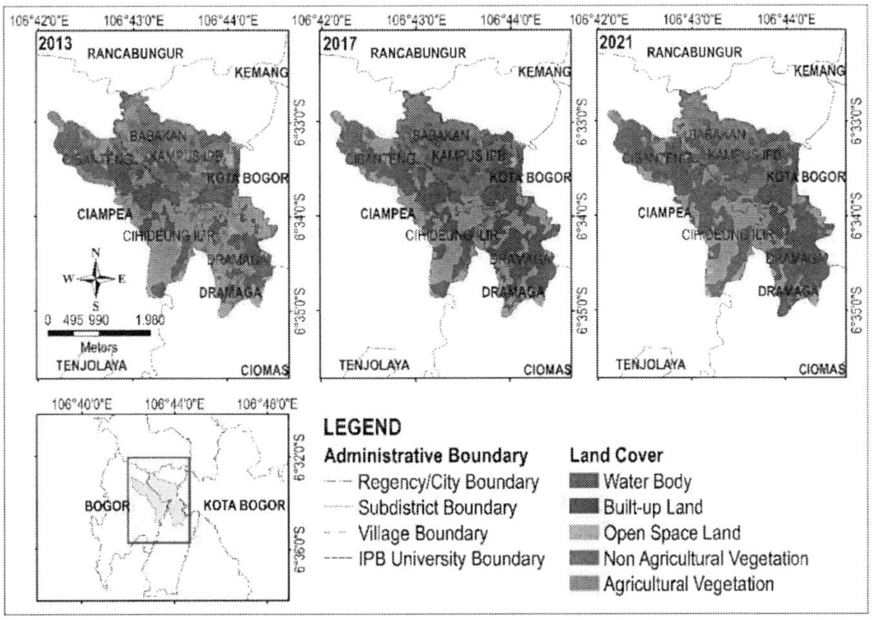

Figure 1. Land-cover of IPB University and its surroundings in 2013, 2017, and 2021

3.2. Spatial-temporal analysis of LST

Landsat 8 OLI-TIRS image processing results in 2013, 2017, and 2021 show that the land surface temperature varies. In 2013 the lowest temperature value was 25.13 °C, and the highest temperature was 31.55 °C. In 2017 the lowest temperature value was 25.82 °C, and the highest temperature was 32.14 °C. While in 2021, the lowest temperature value is 27.32 °C, and the highest temperature is 33.83 °C. Every year, the temperature difference is caused by several factors, such as the month of recording on the Landsat 8 OLI-TIRS imagery used. In addition to the month of recording, the climate is also the cause of the difference in temperature every year. LST values on the IPB and its surroundings are classified into five classes.

Table 3. LST in IPB University

Land Surface Temperature (°C)	2013		2017		2021	
	Area (Ha)	(%)	Area (Ha)	(%)	Area (Ha)	(%)
25-27	244.78	27.21	56.26	6.26	0	0
27-29	503.67	56.00	488.73	54.35	235.76	26.23
29-31	149.30	16.60	327.35	36.40	449.01	49.96
31-33	1.72	0.19	26.97	3.00	204.75	22.78
>33	0	0	0	0	9.24	1.03

In Table 3 it can be seen that the land surface temperature of the IPB University campus and its surroundings continues to increase in terms of distribution, number of grids, and area of each temperature class in 2013, 2017, and 2021. Overall from 2013-2021 the value of land surface temperature is higher than 29 °C their area increases. Especially the land surface temperature of 29-31 °C in 2021, their area will increase compared to 2013. An increase in area of 33.36% from 149.30 Ha in 2013 to 449.01 Ha in 2021. Meanwhile, areas with land surface temperatures less than 29 °C have a reduced area every year.

The land surface temperature has a pattern caused by differences in land cover in the IPB University campus area and its surroundings from time to time. Land surface temperature with a value of less than 29 °C is generally a land cover of water body and non-agricultural vegetation. While the land surface temperature with a value of more than 29 °C in general is a land cover of agricultural vegetation, open space land, and built up land. The highest land surface temperatures tend to be in built-up areas and the lowest land surface temperatures are in forest areas [10].

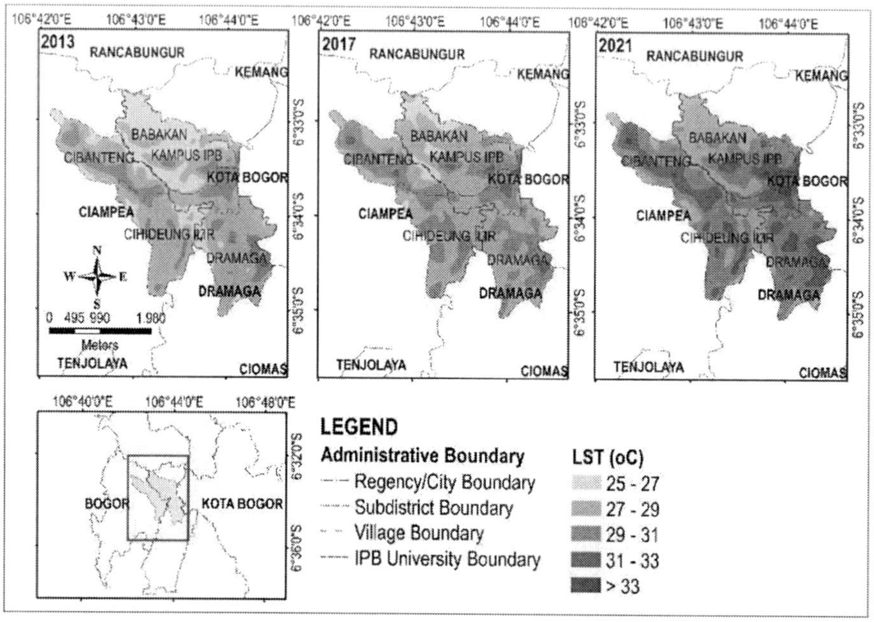

Figure 2. Land Surface Temperature in IPB University and Its Surroundings

Changes in land cover that occur every year are one of the causes of changes in land surface temperature. The land surface temperature of the IPB University campus and its surroundings has generally increased from 2013-2021. The increase in temperature generally occurs due to an increase in development in the IPB University campus area and its surroundings. Increased development such as settlements, parks, and other supporting facilities generally converts vegetated land into built-up land. Therefore, every year the LST on the IPB University campus and its surroundings increases along with the increase in built-up land and the decrease in vegetative land.

The pattern of LST changes follows changes in land cover that occur in the IPB University campus area and its surroundings. Land cover change will affect LST spatially and temporally [20]. LST on the campus of IPB University and its surroundings continues to increase every year. ESG occurs as a result of a significant increase in built-up land. The highest increase in LST occurred in areas that experienced changes in land use for vegetative development. [10].

3.3. Spatial-temporal Analysis of AST

The air surface temperature model can be obtained by a simple linear regression method, by regressing the land surface temperature value that has been processed using satellite imagery with the air surface temperature value obtained in the field. LST is the result of processing the Thermal Band of Landsat 8 Satellite Imagery. Meanwhile, the air surface temperature is obtained from a field survey of temperature measurements at 46 sample points which are divided by land cover class in the IPB University campus and its surroundings.

In the regression equation between the land surface temperature value and the air surface temperature value, the AST model equation (M-AST) is obtained, namely y = 0.9988x + 4.3082 with a coefficient of determination (R^2) of 0.70.

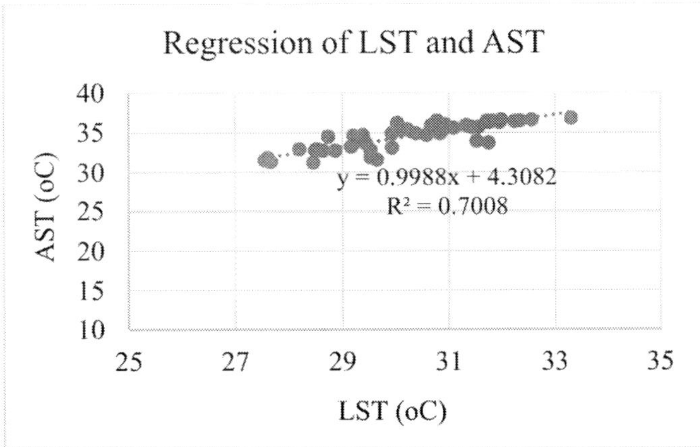

Figure 3. Regression of LST and AST Graphic

Based on the regression results, it is known that the land surface temperature can explain the air surface temperature model by 70%. There are positive results linearly between the land surface temperature value and the air surface temperature model on the IPB University campus and its surroundings. So that the following equation can explain the AST Model (M-AST):

$$\text{M-AST} = 0.9988\,(\text{LST}) + 4.3082 \quad (6)$$

The air surface temperature modeling on the IPB University campus and its surroundings will be carried out with the existing equations. Air surface temperature modeling was carried out based on land use at the IPB University campus and its surroundings in 2013, 2017, and 2021. Based on Table 4 shows the model of ground surface temperature and air surface temperature, there is a temperature difference of about 4 °C.

The RMSE test was carried out on the air surface temperature model in 2021 to determine the level of accuracy and correctness of the modeling air surface temperature data on the air surface temperature data from field measurements. The RMSE test results produce an RMSE value of 0.93, meaning that there is an error of 0.93 in the air surface temperature model data. This means that the largest possible deviation value is 0.93 from the model value of the air surface temperature to the air surface temperature in the field.

Table 4. LST and M-AST by Land Cover in IPB University

Land Cover	2013		2017		2021	
	LST (°C)	M-AST (°C)	LST (°C)	M-AST (°C)	LST (°C)	M-AST (°C)
Water Body	27.5	31.8	28.5	32.8	29.4	33.7
Build-up Land	28.9	33.1	29.9	34.1	31.3	35.6
Open Space Land	28.4	32.6	29.0	33.3	30.3	34.5
Non-Agricultural Vegetation	26.5	30.8	27.7	31.9	28.7	32.9
Agricultural Vegetation	27.7	31.9	28.7	32.9	29.7	33.9

Based on the results of air surface temperature modeling in 2013, 2017, and 2021, the IPB University campus and surrounding areas are divided into five temperature classes, namely 29-31 °C, 31-33 °C, 33-35 °C, 35-37 °C, and >37 °C.

Table 5. Air Surface Temperature in IPB University

Air Surface Temperature (°C)	2013		2017		2021	
	Area (Ha)	(%)	Area (Ha)	(%)	Area (Ha)	(%)
29-31	166.51	18.60	22.06	2.46	0	0
31-33	534.96	59.77	445.67	49.69	177.86	19.99
33-35	189.32	21.15	379.69	42.34	453.21	50.95
35-37	4.25	0.48	49.44	5.51	241.15	27.11
>37	0	0	0	0	17.32	1.95

In Table 5 shows the air surface temperature of the IPB University campus and its surroundings continues to increase, seen from the distribution and the area of each temperature class in 2013, 2017, and 2021. Overall from 2013-2021 the value of air surface temperature smaller than 33 °C their area continues to decrease. This is in line with the increase in the area of temperature more than 33 °C.

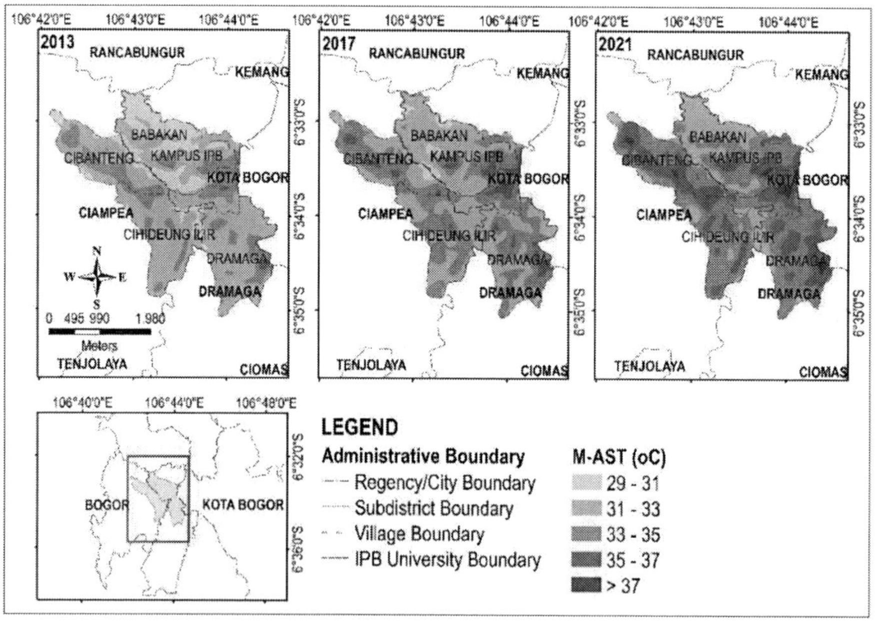

Figure 4. Air Surface Temperature Model in IPB University and Its Surroundings

Air surface temperature has a pattern caused by differences and changes in land cover in the IPB University campus and surrounding areas. Land cover produces land surface temperatures that affect air surface temperatures [21]. Air surface temperature with a value of less than 33 °C is generally a land cover of water bodies and non-agricultural vegetation. While the air surface temperature with a value of more than 39 °C in general is a land cover of agricultural vegetation, open space land, and built-up land.

4. Conclusions

The air surface temperature in the IPB University campus and its surroundings is obtained from the modeling results. M-AST was generated from regression analysis between ground-based AST and image-based LST . The y = 0.9988x + 4.3082 from the M-AST has an acceptable result with 0.70008 in the R^2. Between the land surface temperature and the air surface temperature model there is a temperature difference of 4 °C. Air surface temperature has a pattern caused by differences and changes in land cover in the IPB University campus and surrounding areas. The highest air surface temperature was found on the built-up area (35.6 °C), and the lowest was on the non-agricultural area (32.9 °C). This study concluded that the increase in air surface temperature of more than 2 °C was caused by changes in land-cover from vegetation to the built-up area.

Acknowledgments

The authors are deeply grateful to the supervisor and lecturers from the Departement of Geography, Faculty of Mathematics and Natural Science, Universitas Indonesia, and support from the IPB University.

References

[1] Xiao H and Weng Q 2007 The impact of land use and land cover changes on land surface temperature in a karst area of China *J. Environ. Manage.* **85** 245–57
[2] Dong S, Yan X and Xiong Z 2013 Varying responses in mean surface air temperature from land use/cover change in different seasons over northern China *Acta Ecol. Sin.* **33** 167–71
[3] Adulkongkaew T, Satapanajaru T, Charoenhirunyingyos S and Singhirunnusorn W 2020 Effect of land cover composition and building configuration on land surface temperature in an urban-sprawl city, case study in Bangkok Metropolitan Area, Thailand *Heliyon* **6** e04485
[4] Wang L, Tian F, Wang X, Yang Y and Wei Z 2020 Attribution of the land surface temperature response to land-use conversions from bare land *Glob. Planet. Change* **193** 103268
[5] Alexander C 2020 Normalised difference spectral indices and urban land cover as indicators of land surface temperature (LST) *Int. J. Appl. Earth Obs. Geoinf.* **86** 102013
[6] Sheng L, Tang X, You H, Gu Q and Hu H 2017 Comparison of the urban heat island intensity quantified by using air temperature and Landsat land surface temperature in Hangzhou, China *Ecol. Indic.* **72** 738–46
[7] Wang R, Cai M, Ren C, Bechtel B, Xu Y and Ng E 2019 Detecting multi-temporal land cover change and land surface temperature in Pearl River Delta by adopting local climate zone *Urban Clim.* **28** 100455
[8] Benali A, Carvalho A C, Nunes J P, Carvalhais N and Santos A 2012 Estimating air surface temperature in Portugal using MODIS LST data *Remote Sens. Environ.* **124** 108–21
[9] Alshuwaikhat H M and Abubakar I 2008 An integrated approach to achieving campus sustainability: assessment of the current campus environmental management practices *J. Clean. Prod.* **16** 1777–85
[10] Madakarah N Y, Supriatna, Wibowo A, Manessa M D M and Ristya Y 2019 Variations of Land Surface Temperature and Its Relationship with Land Cover and Changes in IPB Campus, Dramaga Bogor 2013-2018 *E3S Web Conf.* **125**
[11] Wibowo A, Yusoff M M and Salleh K O 2020 Monitoring urban heat signature and profiles of localized urban environment in the University of Malaya *IOP Conf. Ser. Earth Environ. Sci.* **481**
[12] Kolawole O A and Boluwatife A R 2016 Assessment Of The Factors Influencing Students' Choice Of Residence In Nigerian Tertiary Institutions *Sains Humanika* **8** 39–47
[13] IPB 2019 IPB University Rebranding
[14] GreenCampus 2016 Green Space
[15] Jiang J and Tian G 2010 Analysis of the impact of Land use/Land cover change on Land Surface Temperature with Remote Sensing *Procedia Environ. Sci.* **2** 571–5

[16] P U F, Y P and A S 2016 Analisis spasial perubahan luasan mangrove akibat pengaruh limpasan sedimentasi tersuspensi dengan metode penginderaan jauh *J. Geod. Undip* **5** 305–315
[17] H N N, B S and B S 2013 Analisis Pengaruh Perubahan Lahan Pertanian Terhadap Hasil Produksi Tanaman Pangan di Kabupaten Patu Tahun 2001-2011 *J. Geod. Undip* **2** 127–40
[18] USGS 2013 Using the USGS Landsat Level-1 Data Product
[19] Jaelani L M, Setiawan F and Matsushita B 2015 Uji Akurasi Produk Reflektan-Permukaan Landsat Menggunakan Data In situ di Danau Kasumigaura , Jepang *Pertem. Ilm. Tah. Masy. Penginderaan Jauh Indones.* 9–16
[20] Shidiq I P A, Ismail M H, Supriatna S and Wibowo A 2020 Spatial-temporal analysis of changes in land-cover and land surface temperature (LST) within Universiti Putra Malaysia campus area *IOP Conf. Ser. Earth Environ. Sci.* **561**
[21] Wibowo A, Yusoff M M, Adura T A, Supriatna and Hakim Zaini L 2020 Spatial model of air surface temperature using Landsat 8 TIRS *IOP Conf. Ser. Earth Environ. Sci.* **500**

A geographical review of urban farming and urban heat island in developing countries

T A Ladan[1], M H Ibrahim[2*], S S B S Ali[3] and A Saputra[4]

[1] Kashim Ibrahim College of Education Maiduguri, Nigeria.
[2] Department of Geography and Environment, Faculty of Science Humanities, Universiti Pendidikan Sultan Idris, Malaysia.
[3] School of Government, College of Law, Government and International Studies, Universiti Utara Malaysia.
[4] Geography Faculty, Universitas Muhammadiyah Surakarta, Indonesia

*Email: hairy@fsk.upsi.edu.my

Abstract. Cities in developing countries are heavily involved in urban agricultural activities, and a lot of research has been done to address the challenges in this regard. However, several studies have proposed taking advantage of them in reducing the impact of urban heat islands (UHIs), which are manifested as a consequence of stringent changes in energy use, human settlement patterns, industry, transportation, and population growth. Data for this study were obtained from a literature search of existing academic journals, the domain of agriculturally related organisations, and books; the investigation was mainly based on developing nations. Best practices of scholarly work did the review. Authors stand posited. Urban farming has been employed in cities mainly as a food source while also catering to the changing needs of the urban dweller's lifestyles. There is an urgent need to investigate the potentials of urban farming not just as food crops but as green infrastructure in ameliorating urban heat island intensities in our cities.

1. Introduction

Today, urban centres have a higher population density than their rural counterparts; by 2050, two-thirds of the world's population is predicted to live in cities, with this number projected to increase by 1.44 per cent over the next decade [1– 4]. Also, Africa and Asia will account for 90% of urbanisation until 2050 [5]. Crop production in and around cities is gaining popularity in many places worldwide, with a range of forms, different backgrounds, and passions. There has also been an upsurge in literature and debate about the growing interest in urban farming in academics and the media. (UF) [6], maintain that, like Urban Agriculture, Urban farming is regarded as an agricultural practice that primarily involves growing crops on the outskirts of cities, primarily for economic gain, and is motivated and operated by experts with medium to large distribution channels ranging from direct marketing to global supply chains. Many scholars use different terms to explain the agricultural practice in towns and cities, like 'urban agriculture ', professional urban agriculture' 'professional farms' 'urban agriculture and growing crops on the outskirts of cities [6,7]. In their work, the role of a community toward urban farming activities, [8] described urban farming as food or livestock production or processes carried out within the metropolitan area or around the urban centres to generate income. Thus, urban agriculture in the same vain refers to the intensive plant cultivation and livestock farming used in the production, processing, and delivery of food and other products., horticulture, aquaculture and other practices, for producing fresh food in and around cities for feeding local population [8–10].

The term "urban heat island" refers to the phenomenon of high air temperature in urban areas compared to rural counterparts. [11–19]. This, according to [9], include small-scale urban farming, land sharing, residential food production, land sharing, allotment, schoolyard gardens, rooftop gardens and beehives, restaurant-supported salad gardens, guerrilla gardening, food production in public spaces, balconies and vegetable growing in window sills and other activities". There has been a lot of research on urban agriculture and the effects of urban heat islands, and this study aims to explore the impacts of urban heat islands, their causes and highlight how urban agriculture can be used in mitigating some of its consequences and to revive awareness about the importance of agriculture further. Urban areas and the impact of the urban heat island effect on urban populations in developing countries.

2. Methods

The paper is a desk review. It examines urban heat island studies conducted in developing nations and highlights how urban farming can alleviate some of its impacts on urban dwellers by citing existing literature. Data used for this review is obtained or sourced from scholarly work. The academic work searched for this review spans through the field of Geography, Agricultural science, and another Environmental discipline in the last five years. About 30 articles were explored. The database majorly used was google scholar using the search word "urban farming", "urban heat island", and "developing nations". In addition, to the above, the information used is linked to the developing countries irrespective of the continent. Historical data were compared with recent academic contributions. All sources are acknowledged in accordance with best practices. The conclusion was drawn based on the author's disposition derived from the perusal of past articles/publications.

3. Results And Discussion

3.1. Urban Farming

Urban farming is defined as food or livestock production or processes carried out within or near urban areas to generate income [8]. Similarly, [6] opined that the term "urban farming" refers to "small- to large-scale agriculture that takes place in cities."

It is primarily a farming practice on the outskirts of cities. It is primarily driven by financial factors and is handled by experts with medium- and large distribution channels ranging from direct marketing to global supply chains. But, [20] considers urban farming to grow crops to produce entrepreneurial food for sustainable urban neighbourhoods. A study by [21] reveals that Urban farming is a rigorous and crop production activity that utilises space effectively, with most shorter-cycle, higher-yield products. It makes good use of both horizontal and vertical areas by using multi-cropping and integrated farming practices (through practices such as shelf-mounted chicken coops, different kinds of fish farming, and growing crops in a container). Urban farming uses water more efficiently than rural farming since water is expensive and generally in short supply.

3.1.1. Types of Urban Farming

The use of urban farming differs depending on the location and ecosystem of the city. Several studies have reported that urban farming has already been implemented on grounds, balconies and rooftops of buildings, as well as vertical walls, in a variety of cities in various climatic environments, including high-density cities such as the United States of America, Japan, Singapore, and other developing countries [8]. They further highlighted urban farming to include community farming, rooftop farming, and vertical farming. Each farm has similar techniques and goals, yet each has its own unique qualities due to space organisation and boundaries.

Traditionally, city dwellers employed various methods to cultivate food crops or flowers in their homes, such as in pots. Breeding and raising animals, bees, aquaculture (fish farming), aquaponics (fish farming alongside vegetables), and non-food goods such as planting flowers and growing seeds are all examples of urban farming. [22].

3.1.2. Urban Farming Technology

Many urban farming technologies have been implemented in a variety of urban centres. [23], identified the following, aquaponics, aeroponics, hydroponics, and vertical farming are four prominent urban farming technologies employed by Malaysian city dwellers. Aquaponics is the practice of growing crops in water in a symbiotic relationship. Both hydroponics and fertigation use a similar technique to ensure that nutrients and fertilisers are delivered directly to the plant's roots, preventing root infections. One of the most common methods for quick and easy cultivation is hydroponics. Crops that are grown vertically are referred to as vertical farming [22].

3.1.3. Benefits of Urban Farming

Many experts have emphasised the potentials of urban farming. [24], categorised the benefits into four: environmental management, economic dependence, social improvement, and health and nutrition. Similarly, Reduce the cost of household expenditure, Improve landscape, food supply, community engagement. Also, [24] highlighted the benefits of urban food production, including improved food access, improved health, higher incomes, increased skills, sustainable community development, and mitigating structural inequalities. Urban farming is also a viable means of increasing green areas, vegetative fields, and gardens for leisure and urban rejuvenation. It is an approach to "shifting cultivation" by modifying unutilised land to create new open spaces. By maintaining the city's landscapes, the city's ecological balance and environment will be improved and preserved. [25]. For people of all income levels, urban farming is an easy-in, easy-out entrepreneurial venture. It ensures that the urban impoverished have easy access to food. It provides a source of income as well as low-cost, high-quality food for the steady poor. It allows middle-income families to save money and get a return on their investment in urban real estate. It is a profitable enterprise for both small and large business owners [22]. It cannot be overemphasised that urban farming is an essential strategy for ensuring the long-term viability of cities.

3.2. Urban Heat Island

The urban heat island effect is a phenomenon that refers to an increase in air temperature in urban centers compared to rural areas [1,11–18]. This according [7] " itu melibatkan pertanian perkotaan skala kecil, pembagian tanah, produksi makanan di perumahan, taman halaman sekolah, taman atap dan sarang lebah, produksi makanan di ruang publik, kebun salad yang didukung restoran, jatah, berkebun gerilya, balkon dan penanaman sayuran di ambang jendela dan kegiatan lainnya".

3.2.1. Types of Urban Heat Island

Three types of urban heat islands were recognised by [26]. The Boundary Layer Heat Island (BLHI), Canopy Layer Heat Island (CLHI), and Surface Layer Heat Island (SLHI) were identified. The BLHI is the layer above the average height of a building. It begins at the rooftops and extends up to the point where the urban landscape has no influence on the atmosphere. The CLHI is the layer just above the surface to the top of trees and roots. The SLHI are the actual surface temperatures that can be measured directly on the surface or indirectly by satellite or aircraft.

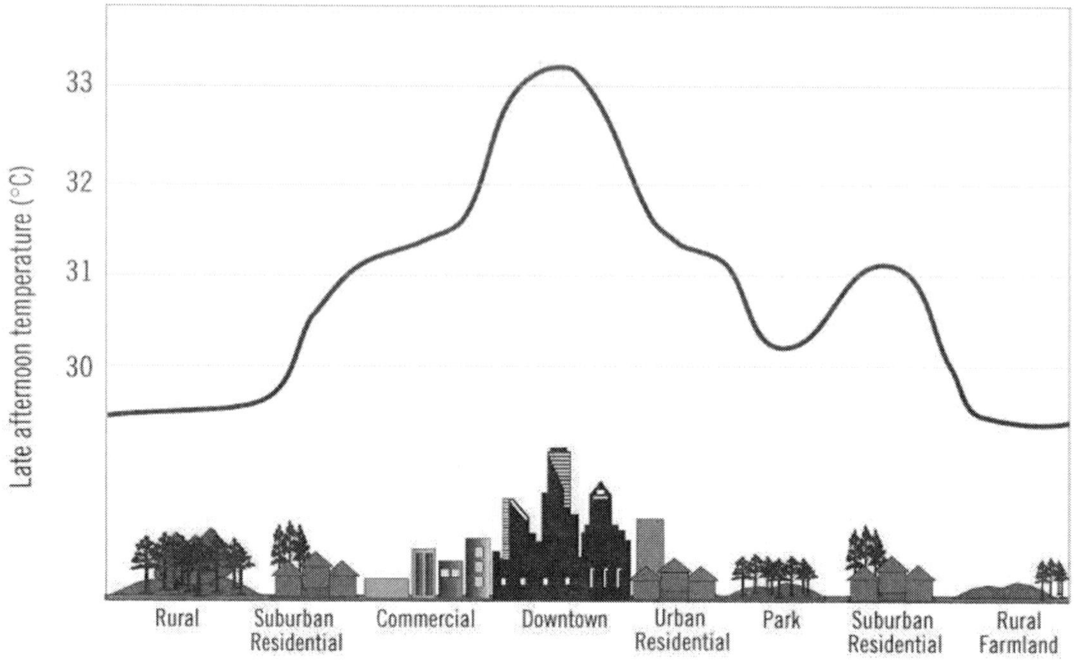

Figure 1. Showing the profile urban heat island.
Source: [27]

3.2.2. Causes of Urban Heat Island

There are many reasons why urban centres are warmer than their rural counterparts, which are well documented [28]. These include absorption of incoming solar radiation as a result of the expansion of tall buildings; solar radiation is deposited as incoming heat due to higher heat ingress from the hard surface of urban surfaces; loss of vacant land as a result of urban sprawl, radiation loss due to limited sky views; reduced evapotranspiration from water bodies as well as soil, and vegetation, as they are replaced by harsh urban structures [29]. Similarly, the urban heat island intensity is further excavated by the following factors, [1] these include: increased heat storage; reduced albedo; reduce the sky view factor; heat release from buildings, cars and people; increased challenging surface area; loss of soil and plant water content; and backscattering of air pollution are all contributing factors to UHI. In addition to urban design, the weather impacts UHI intensity (temperature difference between rural and urban land cover), with the highest intensity occurring at night with steady and moderate wind speeds. Indeed, "anthropogenic heat production (traffic, residential and industrial heating, etc.) in cities contribute to the formation of UHI, and, in addition to physical aspects of the urban landscape (canyon streets, dark materials such as asphalt or pavements) that contribute to trapping a consistent portion of this heat, meteorological conditions (wind, humidity, sky view factor, radiation, precipitation, etc.) that contribute to trapping a consistent portion of this heat. However, it is worth noting that UHIs are neither the cause nor the consequence of global warming" [30].

3.2.3. Impact of Urban Heat Island

According to growing scientific data, climate variability and change pose severe threats to urban dwellers around the world, in both developing and developed countries. The negative consequences of the UHI on cities are widely acknowledged. Exposure to high ambient temperatures is widely accepted as a severe health hazard [32]. It has also been observed by [33] that Human health is enormously affected by climate and weather. [34] observes that higher air pollution levels reduced nighttime cooling and increased daytime surface temperature related to urban heat islands can negatively influence human health, causing general discomfort, exhaustion, respiration problem, headaches, non-fatal heat stroke, heat cramps, and heat-related mortality. Similarly, [35] highlights that many studies have shown that this affects a person's psychology, namely depression, anxiety, and aggressive behaviour. It also affects physical health such as respiratory disease, heart problems, and heat stroke associated with the heat island effect [36], also stressed that heat-related sickness and mortality will be more common among vulnerable population groups across all social strata and geographical regions, especially the very young, the frail old, and those with a pre-existing physical or mental condition.[30], also indicated that, When compared to plant growth in rural areas, urban heat islands can hurt their growth phase in cities.

4. Conclusion

Urbanisation is a continual phenomenon, and as such more people will continue to migrate and live-in urban areas. Thus, there will be more people to feed and expose them to urban heat island effects. Therefore, in today's urbanising world, urban farming makes increasing sense. It's a practical and necessary practice in the twenty-first century. In today's urbanising world, urban agriculture makes increasing sense. It's a good and necessary practice in the twenty-first century. Urban farming's full potential will become more reachable as it obtains recognition as an industry with a role to play in the sustainability of cities and the sustenance of their population

Reference

[1] Bassett R, Young P J, Blair G S, Cai X-M, Chapman L 2020 Urbanisation's contribution to climate warming in Great Britain. *Environ Res Lett* **15**:114014
[2] Vandercasteelen J, Beyene S T, Minten B, Swinnen J 2018 Big cities, small towns, and poor farmers: Evidence from Ethiopia. *World Dev* **106**:393–406.
[3] World Bank. World Urban Population 2016:SP @ data.worldbank.org. (n.d.). Retrieved from ht.
[4] Pölling B, Sroka W, Mergenthaler M 2017 Success of urban farming's city-adjustments and business models—Findings from a survey among farmers in Ruhr Metropolis, Germany. *Land Use Policy* **69**:372–85
[5] Opitz I, Berges R, Piorr A, Krikser T 2016 Contributing to food security in urban areas: differences between urban agriculture and peri-urban agriculture in the Global North. *Agric Human Values* **33**:341–58
[6] Yusoff NHB, Hussain MRM, Tukiman I 2017 Roles of community towards urban farming activities. *Plan Malaysia* **15**:271–8
[7] Tornaghi C 2014 Critical geography of urban agriculture. *Prog Hum Geogr* **38**:551–67.
[8] Game I, Primus R 2015 GSDR 2015 Brief: Urban Agriculture. *Urban Agric* 1–13.
[9] Karlessi T, Gaitani N, Synnefa A, Santamouris M 2013 Cool pavements. *Adv Dev Cool Mater Built Environ* 104–19
[10] Ohwo O, Abotutu A 2015 Environmental Impact of Urbanization in Nigeria. *Br J Appl Sci Technol* **9**:212–21
[11] Ishola K A, Okogbue E C, Adeyeri O E 2016 A Quantitative Assessment of Surface Urban Heat Islands Using Satellite Multitemporal Data over Abeokuta, Nigeria. *Int J Atmos Sci 2016*:1–6.
[12] Folorunsho F, Balogun I A, Adediji A T, Olumide A, Abdulkareem S B, State O, et al 2017 Assessment of Urban Heat Island over Ibadan Metropolis Using Landsat and Modis. *Int J Environ Bioenergy* **12**:62–87.
[13] Abur A A, Ako T, Yusuf I T 2018 Mitigating Urban Heat Islands in Abuja , Nigeria : Adopting Cool Pavement Technology – A Review. *Am J Environ Policy Manag* **4**:60–6.
[14] Ogun L G A, Adewara M B, Oyewole A M 2019 Dynamic Effects of Urban Heat Island in Ilaro Town , *Yewa South* **13**:51–60
[15] Santamouris M 2020 Recent progress on urban overheating and heat island research. Integrated assessment of the energy, environmental, vulnerability and health impact. Synergies with the global climate change. *Energy Build* **207**.

[16] Ulpiani G 2021 On the linkage between urban heat island and urban pollution island: Three-decade literature review towards a conceptual framework. *Sci Total Environ* **751**:141727.
[17] Ibrahim M H 2017 Impact of Urbanisation on the Formation of Urban Heat Island in Kota Bharu, Kelantan. *J Mech Contin Math Sci* sp **l9** :63–72.
[18] Poulsen M N, Neff R A, Winch P J 2017 The multifunctionality of urban farming: perceived benefits for neighbourhood improvement. *Local Environ* **22**:1411–27.
[19] Mahmood R, Jia S, Mahmood T, Mehmood A 2020 Predicted and projected water resources changes in the chari catchment, the Lake Chad Basin, Africa. *J Hydrometeorol* **21**:73–91.
[20] Muhammad R M, Masdek N R N M, Haimid M T, Ponari S Z, Sayuti Z 2020 Impact of urban farming technology on urban community in Malaysia (Impak teknologi pertanian bandar kepada komuniti bandar di Malaysia). *Econ Technol Manag Rev* **15**:37–49.
[21] Friday Nkwocha K, Oyilieze Akanwa A, Abba Jimme M, Oji Iheukwumere S, Joe-Nkechebelu N 2020 Perception and Impacts to Climate Emergency in Maiduguri Urban, North-East Nigeria: A Case for Community Based Adaptation Approach. *Rev Environ Earth Sci* **6**:24–33.
[22] van Veehuizen R 2006 *Urban Agriculture for Green and Productive Cities.*
[23] Carolan M 2020 "Urban Farming Is Going High Tech": Digital Urban Agriculture's Links to Gentrification and Land Use. *J Am Plan Assoc* **86**:47–59
[24] Ibrahim M H, Isa N K M, Ismail K, Adnan M, Ali S S S, Saputra A 2019 Assessment of urban heat Islands in Rawang small cities, Malaysia. Int J Adv Sci Technol **28**:184–93.
[25] Bahi H, Mastouri H, Radoine H. M*aterials Today : Proceedings Review of methods for retrieving urban heat islands*. Mater Today Proc 2020
[26] Taslim S, Parapari D M, Shafaghat A 2015 Urban design guidelines to mitigate urban heat island (UHI) effects in hot-dry cities. *J Teknol* **74**:119–24
[27] Yusof N S, Mohd Hussain N H, Rusli N 2017 The Relationship of Heritage Trees in Urban Heat Island Mitigation Effect At Taiping, Perak, Malaysia. *Malaysian J Sustain Environ* **3**:157.
[28] Waffle A D, Corry R C, Gillespie T J, Brown R D 2017 Urban heat islands as agricultural opportunities: An innovative approach. *Landsc Urban* Plan **161**:103–14
[29] Berry P, Enright P M, Shumake-Guillemot J, Prats E V, Campbell-Lendrum D 2018 Assessing health vulnerabilities and adaptation to climate change: A review of international progress. *Int J Environ Res Public Health* **15**
[30] Kim K H, Jahan S A, Kabir E, Brown R J C 2013 A review of airborne polycyclic aromatic hydrocarbons (PAHs) and their human health effects. *Environ Int* **60** 71–80.
[31] World Bank. 2015 Climate and Health country Profile - 2015 Nigeria. World Bank Glob Heal Expend Database United Nations Dev Program Hum Dev Reports **8**.
[32] Cole D, Dietsch N, Gero G, Hitchcock D, Lewis M, Magee J, et al 2008 Heat Island Reduction Activities. Reducing Urban Heat Islands Compend Strateg 1–23.
[33] WONG L P, Alias H, Aghamohammadi N, Aghazadeh S, Sulaiman N M N 2018 Physical, Psychological, and Social Health Impact of Temperature Rise Due to Urban Heat Island Phenomenon and Its Associated Factors. *Biomed Environ Sci* **31**:545–50.
[34] Block A H, Livesley S J, and Williams N S G 2012 Responding to the Urban Heat Island : A Review of the Potential of Green Infrastructure. *Vic Cent Clim Chang Adapt* 1–62.

IOP Conf. Series: Earth and Environmental Science **986** (2022) 012072 doi:10.1088/1755-1315/986/1/012072

LST-based threshold method for detecting UHI in a complex urban landscape

M S Labib[1,*], A Wibowo[1] and I P A Shidiq [1]

[1] Department of Geography, Faculty of Mathematics and Natural Sciences, Universitas Indonesia, Depok, Indonesia

*Email: muhammad.syauqi72@ui.ac.id

Abstract. Urban Heat Island (UHI) explains the occurrence of excessive heat that spread unevenly in urban areas. Intense and complex human activities in urban areas are mainly causing the formation of UHI. This study aims to identify the spatial characteristics of UHI in a complex urban feature, like the Universitas Indonesia campus area and its surroundings in Depok City. The campus and its surroundings might represent a complex urban area and can be used as a reference to understand the widespread impact of UHI. UHI was determined based on the threshold value, calculated from the average and standard deviation of the surface temperature. This study utilized Landsat 8 OLI-TIRS imagery to extract Land Surface Temperature (LST) value. The LST observations were collected for three different periods, including 2013, 2017, and 2021. The LST value tends to be increased during these times, with the LST_{max} recorded at 34.68 °C, 35.47 °C, and 36.4 °C for 2013, 2017, and 2021, respectively. Based on the number of the accommodated grid, the UHI was getting larger every year. Started from 313 grids in 2013, the UHI has been expanded to more than 550 grids in 2021.

1. Introduction

The development of urban areas occurs on a large scale, which is characterized by widespread land conversion into buildings [1]. Land conversion is the changing process of natural land into built-up land [2]. Very intensive land conversion can increase Urban Heat Island/UHI phenomenon [3]. The UHI phenomenon is an increase in temperature unevenly in urban areas compared to surrounding areas [4] due to the high percentage of built-up land especially in urban areas compared to surrounding areas, it's basically known as rural areas [5], even in metropolitan cities, the temperature gap with the surrounding rural areas can reach 12 °C [6].

The UHI phenomenon is generally influenced by various factors, such as anthropogenic factors related to human activities [7] and the development of urban areas such as urbanization, especially in big cities [3]. Big cities have a history of massive urbanization due to overcrowding [8], and will move towards the surrounding area which is a suburban area because its support for the metropolitan area known as buffer area [9,10]. The existence of urbanization is one of the benchmarks in understanding the causes of the UHI phenomenon, especially in big cities.

Indonesia has a downtown in DKI Jakarta, which is a metropolitan city and surrounded by its buffer areas known as Bodetabek or Bogor, Depok, Tangerang, and Bekasi [11]. All of the buffer areas have their respective interactions with the metropolitan city, especially Depok city which has a very massive interaction, with 87% of commuters going to DKI Jakarta every day [12]. Depok city is also known as an area with a large population growth, about 9% per year [13] with a land conversion rate about 29% in the last 20 years [14]. Land conversion will affect the increase in land surface temperature/LST [15], which will have big impact on increasing UHI [16]. When UHI increases, mitigation is needed to reduce the negative effects of UHI for humans such as heat stroke or heat stress [3]. The proper mitigation of UHI is add the vegetation cover, especially natural vegetation and also artificial vegetation [17,18]. According to Depok City is a city that has a lot of vegetation due to the University of Indonesia (UI) campus, also located in Depok which there is an urban forest in UI Campus [19]. Urban forests can improve the microclimate quality to reduce LST and UHI [20,21]. The UI Campus area is also an educational area which is representing a complex urban landscape due to its characteristics [22,23].

In addition to the UI area with all the complexity of the landscape, the area around the UI campus is also complex landscape with a lot of development. Many development projects spread around UI Depok Campus area in the last few years such as the development of UI Hospital (RS UI), TOD Pondok Cina, and new Cijago toll road construction [24,25]. The rapid development will certainly contribute to UHI around the complex landscape, especially in the UI campus area and its surroundings.

Many researches related to the impact of UHI and LST on complex areas, especially the campus environment [3]. Manifestation of remote sensing data will simplify extraction about UHI [26] because the main role of remote sensing data is to utilize electromagnetic radiation energy information [27] which actually UHI detection mainly depends on it [28], especially the Landsat 8 OLI/TIRS satellite image which has a thermal band that is very suitable for UHI detection [29]. Nowadays, UHI research is not only generated in representative scale with the complexity of the landscape, but now has developed in several methods of data processing [30]. Pratiwi and Jaelani (2021) asserted that there are at least two methods for determining UHI, such as the hot spot analysis method with Getis Ord-gi and also the LST-based threshold method [31].

Some development of these methods is an extension of the initial method of determining UHI: classification-based LST, when the LST values in tropical areas above 30 °C it will be designated as UHI because it has been categorized as high temperature [32–34]. The developed method then compared, which the LST-based threshold method has the advantages of flexibility of use and efficiency of application in complex urban areas [31,35]. In addition, this is the background of this research, to analyze spatial and temporal distribution of UHI changes in complex landscape using the LST-based threshold method.

2. Methods

2.1. Study Area

Research covers the UI Campus area and its surroundings, which is located at coordinates -6,359, 106,827. Geographically, the northern part of the study area is the UI Student Dormitory which is directly adjacent to Jl. Srengseng Sawah. To the south is the Cijago toll road which borders the Beji Timur Village. The western part of the study area is bordered by South Jakarta City with the outermost area located at Kukusan Village. The eastern part is bordered by the Ci Liwung river. The area around the campus is an area that is assumed to be relatively close, in accordance with its proximity to educational support areas, such as the type of housing (boarding house), facilities, access to transportation, and commercial centre [36].

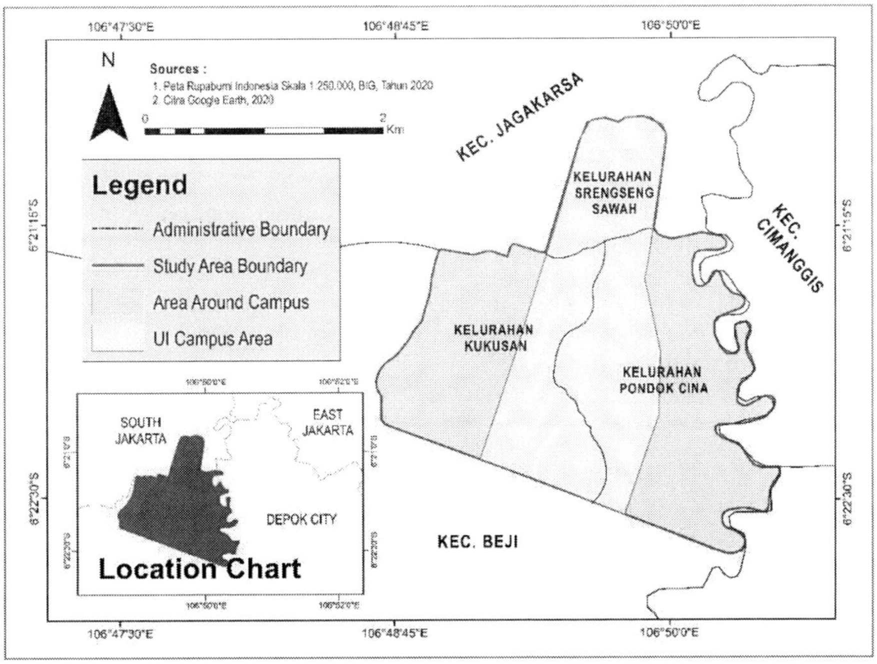

Figure 1. Study Area

The study area is focused in Kukusan and Pondok Cina Village which is represents of the closest region from the UI Campus with the highest number of boarding houses. This study is chosen because the area's characteristics around the Campus are marked by the number of paid residences (such as apartments and boarding houses) and places of business (such as culinary businesses and service industries), housing, and worship facilities [37].

2.2. Materials

The following materials were used in this study:

- Landsat 8 OLI/TIRS satellite imagery, path/row 122/164 divided into three years of data processing; 2013, 2017, dan 2021. Satellite imagery is selected with a cloud cover range less than 20% and freely downloaded from the USGS platform
- Digital thermometer for measure air surface temperature
- Digital map, for validation purpose

2.3. Methods

The method applied in this research is LST algorithm with emissivity parameter as popularized by Artis and Carnahan (1982) [38] by going through stages such as radiometric correction, temperature brightness conversion, NDVI calculation, vegetation proportion, epsilon valuation, and LST formulation [39]. Radiometric correction is applied the following equation:

$$L = \beta \times DN + \alpha \tag{1}$$

L = TOA spectral radiance/corrected spectral value (W/m^2sradμm)
β = number of multi band radiance
DN = pixel value
α = number of additive band radiance

The image that has been applied radiometric correction and brightness temperature then formulated to the following equation:

$$Tb = \frac{K2}{ln\,(\frac{K1}{L\lambda}+1\,)} \tag{2}$$

Where Tb is the brightness temperature (in degrees Kelvin), Lλ is the spectral radiance value (W/m2sradμm) which means the corrected spectral radian value, K2 and K1 are calibration constants, which are obtained from the metadata file, which represent the thermal constant values of Band 10 images. Landsat 8 OLI/TIRS satellite. The value of K1 itself is 774.8853, and the value of K2 is 1321.0789. After the Tb value is obtained, the NDVI value is calculated to determine the NDVI value along with the minimum and maximum according to equation (3) [26], which is then calculated by equation (4) the proportion of vegetation.

$$NDVI = \frac{B5-B4}{B5+B4} \tag{3}$$

$$PV = \left(\left(\frac{NDVI-NDVI\,min}{NDVI\,maks-NDVI\,min}\right)\right)^2 \tag{4}$$

Prior to the LST formulation, epsilon calculations were carried out to find out the ln of the epsilon before it was entered into the LST algorithm according to equation (5). The resulting Epsilon is then entered into equation (6) to find LST.

$$\varepsilon = 0.0004 \times PV\; + 0.986 \tag{5}$$

$$LST = \frac{Tb}{\left(1+\left(\frac{\lambda Tb}{\alpha}\right)ln\,\varepsilon\right)} \tag{6}$$

Where $\alpha = {}^{h \times C}/_{K}$

The formulation contains several parameters. means the wavelength of the emitted radiance, = 1,438 x 10-2 mK, for h = Planck's constant (6,626 x 10-34 Js), C is representing the speed of light (2,998 x 108 m/s), and K refers to the Boltzmann's Constant (14388).

The result of LST value then applied to the threshold method to see form of the urban heat island. The requirements for the formulation of the threshold that must be met are as follows [40]:

$$T > \;\mu + 0{,}5\;\alpha \tag{7}$$

$$0 < T \;\leq\; \mu + 0{,}5\;\alpha \tag{8}$$

The T symbol represents the LST value, while μ and α are LST average value and LST standard deviation, respectively. This symbolization also applies to equation (9), which is the threshold algorithm [31,35,40] to mathematically calculate ESG that is in the UHI range, where a negative result (-) to 0 represents UHI, while those greater than zero represent non-UHI values [31]. The threshold equation can be seen in equation (9) below:

$$UHI = LST - (\mu + 0{,}5\;\alpha) \tag{9}$$

Field measurements were also carried out, in the term of a survey for validation. This is important considering that information on satellites must be basic truth to ensure that there is little bias that may arise [41]. Ground truth is carried out in the form of taking the field temperature, namely AST. The results of field temperature measurements were carried out using a linear regression mechanism to see the level of representation between processed LST and AST field measurements. Interpretation was carried out to see the correlation between AST and LST. Regression result information is also used as a reference to see the RMS Error [42], which contains the amount of bias that occurs between LST and AST.

$$RMSE = \sqrt{\sum_{i=1} n \frac{(X\ est,i - X\ mea,i)^2}{N}} \quad (10)$$

2.4. Data Analysis

The analysis used in this study consisted of spatial, temporal, statistical, and descriptive analysis. Spatial analysis explains the spatial-temporal distribution of UHI which is described in three years of data processing; 2013, 2017, and 2021. The unit of analysis based on a 100 x 100 m grid, which represents the local climatological zone [43], the zone of micro-UHI growth [44].

Statistical analysis was also performed using simple linear regression to see the level of data representation between satellite imagery and AST field measurements. The RMS Error statistical test is also carried out to see the margin of errors and biases that may occur.

3. Results and Discussions

3.1. Field Validation

Simple linear regression processing was carried out to see the correlation between the temperature of LST processing and the temperature of field observations. If a positive result is obtained, it means that the relationship can be said to be strong so that it influences each other simultaneously if there is an increase or decrease in value

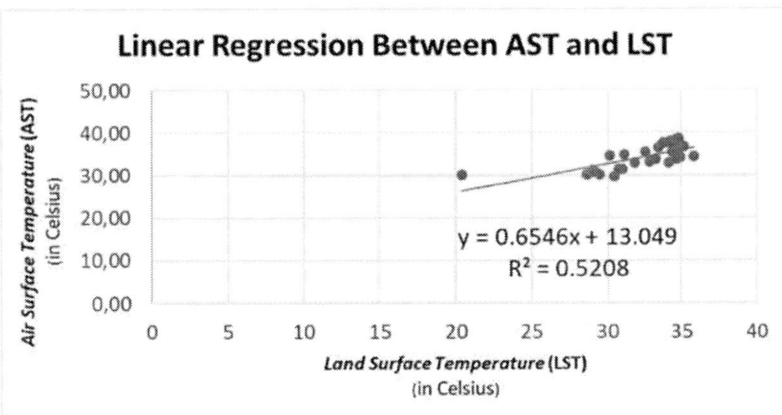

Figure 2. Linear Regression between AST and LST

The results of a simple linear regression test show a line equation y = 0.6546x + 13.049 with a correlation coefficient of 0.721 with a positive correlation which means that according to [31] shows a strong correlation, indicating AST will be directly proportional to LST if the value of the LST variable increases by 1%, the AST variable will increase by 13,049. The regression coefficient obtained is 0.654, which is included in the category of moderate correlation. The coefficient of determination of 0.5208 indicates that the LST variable can explain the AST variable by 52.08%, the rest may be influenced by other variables. The significance value is 0.000003142 which is used for comparison with an alpha value of 0.05 resulting in the conclusion that the significance value < alpha value, so it can be concluded that there is a significant effect between LST and AST.

IOP Conf. Series: Earth and Environmental Science **986** (2022) 012072 doi:10.1088/1755-1315/986/1/012072

In addition to linear regression, the RMS Error Test value is also formulated then produces a value of 1.92 from 32 data, which means that there is a large change (error) of 1.92 in the estimated total value of the entire test carried out, this means that the largest possible deviation value occurs is 1.92 different from the LST value to the AST.

3.2. Land Surface Temperature

The LST temperature value contains the lowest class information temperatures below 27 which are symbolized by blue colouring. Temperatures with a class of 27 to 30 C are represented in green, while yellow is chosen to visualize an ambient temperature of 30 to 33 C. The highest temperature values are those categorized in a temperature class above 33 C, defined in red.

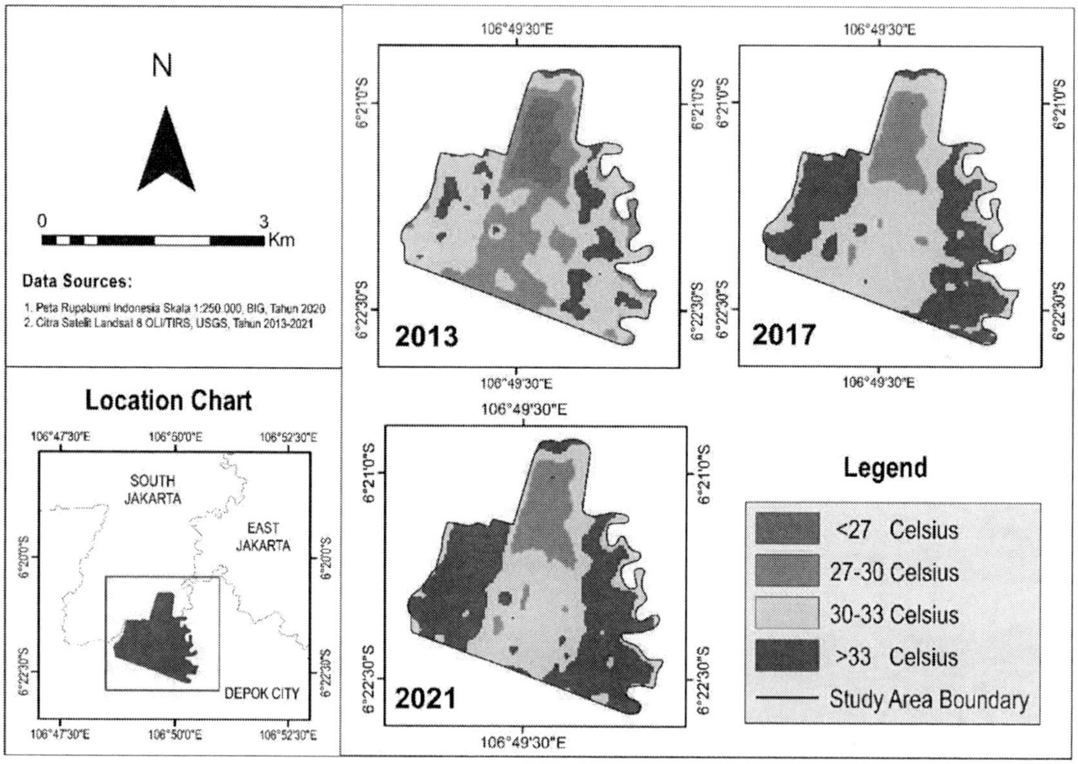

Figure 3. Temporal Map of Land Surface Temperature

In 2013, the LST value was visually dominated by temperature which included the representation of yellow colour, which was between 30 to 33 °C, then visually green colour also dominated in the middle where the green colour showed a relatively lower temperature than the yellow colour which is represent interval 27 to 30 °C. The uniqueness is in the blue colour with an LST description below 27 C, which in comparison between other LST maps, no longer looks blue, both on the 2017 and 2021 maps.

The 2017 map itself, apart from the absence of temperatures below 27 °C (blue), the LST value is also dominated by yellow, but the difference that appears visually with the previous year's LST map is that temperatures above 33 °C are increasingly widespread, and expansion occurs on the east and west sides (east is more dominant and longitudinal). Other information on the 2017 map that is no less interesting is the shrinking of the green colour to yellow, meaning that the area that was originally in the temperature range of 27 to 30 °C has now more or less turned warmer, at 30 to 33 °C.

The next information is a map with the processed year 2021 or the most recent, a very contrasting difference can be seen with the previous maps where if the previous maps are visually dominated by yellow with a temperature range of 30 to 33 °C, then it will be different with the 2021 map. Temperatures above 33 °C are increasingly widespread, although compared to 2017 they are not too different, but the expansion is in the western part which tends to widen.

Table 1. Temporal LST Value

Years	LST-max	LST-min	Lst-avg	Std-Deviation
2013	34,68	25,95	30,58	2,08
2017	35,47	28,87	32,12	1,53
2021	36,54	28,31	32,47	1,87

The value of LST processing results the highest LST values in 2013, 2017, and 2021 are 34.68 °C, 35.47 °C, and 36.54 °C, respectively. All the whole situation of high value and can endanger humans if not mitigated the hazard that will arise [3,34]. The average value every year continues to increase, in 2013 the average LST value was 30.58 °C, then increased in 2017 to 32.12 °C, and then reached its peak in 2021 with an average of 32.47 °C. The standard deviation is used to find the threshold, in 2013 the standard deviation was 2.08, then it fell to 1.53 in 2017, and again rose in 2021 to 1.87.

The minimum LST value which represents the lowest surface air temperature can be interpreted as the area with the coolest conditions. The coldest point in 2013 reached 25.95 °C, then in 2017 the value of the coldest point rose to 28.87 °C, a significant increase considering the increase of 2.92 °C. Finally, in 2021 the coldest point decreased again, although not significantly, which was 28.31 °C.

3.3. Urban Heat Island Based on Threshold

The spatial-temporal distribution of UHI is determined based on the LST threshold has the meaning of the existence of UHI areas that occur every year. The reprocessed LST is formulated based on its threshold to see the area that contains UHI. The division of the UHI area is arranged based on the temporal scale (time scale), in which the presentation of the UHI map is visualized based on the year of UHI processing, namely in 2013, 2017, and 2021.

Table 2. UHI Area Comparison in 2013

Classification	Area (Ha)	Percentage (%)	Accumulated Grid
UHI	297.5	44.27	313
Non-UHI	374.3	55.71	451

In 2013, the study area was dominated by non-UHI areas (55.71% of the total area) which was still larger at around 11.44% compared to UHI areas which was only around 44.27%, indeed the difference was not big enough, but from the grid side, it can be seen that there is a considerable difference, namely 313 grids for areas with UHI, and 451 grids for non-UHI regions. This can happen because of the nature of the grid which only takes general information, and its representation is limited to the relative shape of the research area; not geometric, sometimes there are those containing the size of one hectare (100 x 100) grid, sometimes there are also truncated ones, but in general, the smallest grid in the analysis is still considered as representative of each value or class that is counted as dominant on the grid.

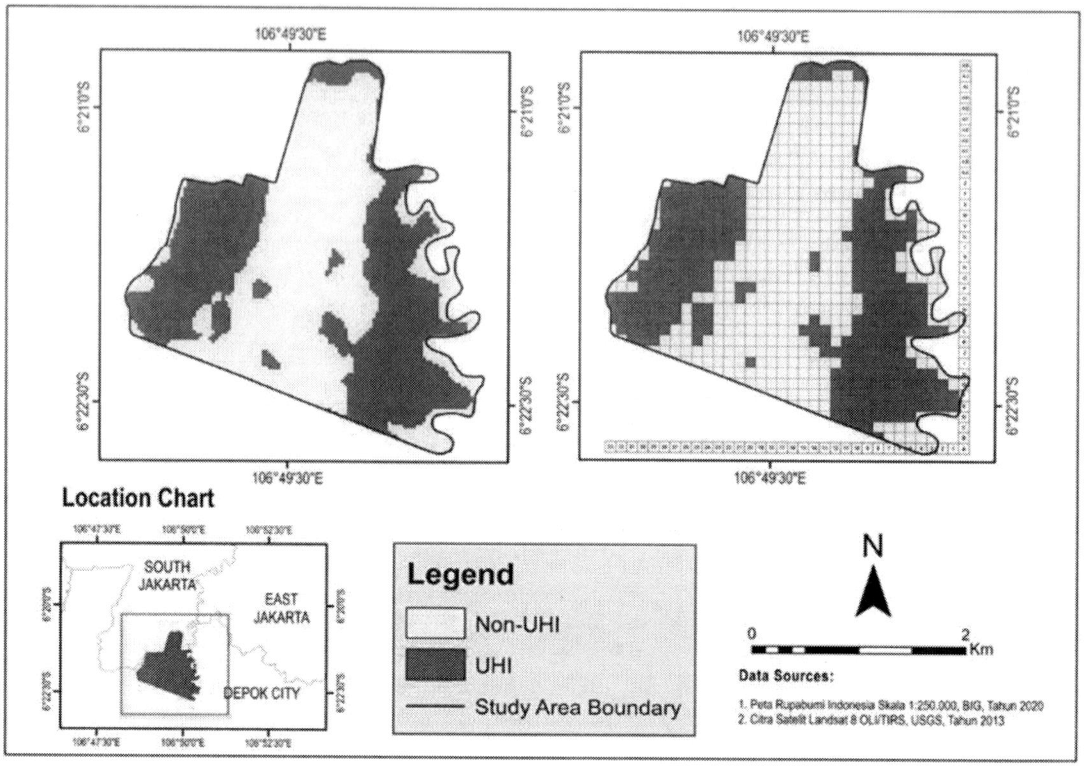

Figure 4. Spatial Distribution of UHI in 2013

In the next processing year, 2017, the area of UHI increased to 361 grids from the original 313 grids, accompanied by a reduction in non-UHI areas to 405 grids which were previously 451 grids. Comparison of UHI and non-UHI areas in 2017 can be seen in table 3 below:

Table 3. UHI area Comparison in 2017

Classification	Area (Ha)	Percentage (%)	Accumulated Grid
UHI	331.09	49.27	361
Non-UHI	340.73	50.71	405

The area of UHI in 2017 is smaller than that of non-UHI. The UHI area itself has an area of up to 331.09 ha while the non-UHI area reaches 340.737 ha. The area only has a very small difference, namely 9.647 ha the difference between non-UHI and UHI areas, with the difference in the percentage area of only 1.44% of the total area. The percentage of each region itself is 49.27% for regions with UHI, and 50.71% for regions with non-UHI categories. The resulting grid is also different, for the UHI category there are 361 grids, while for the non-UHI category there are 405 grids.

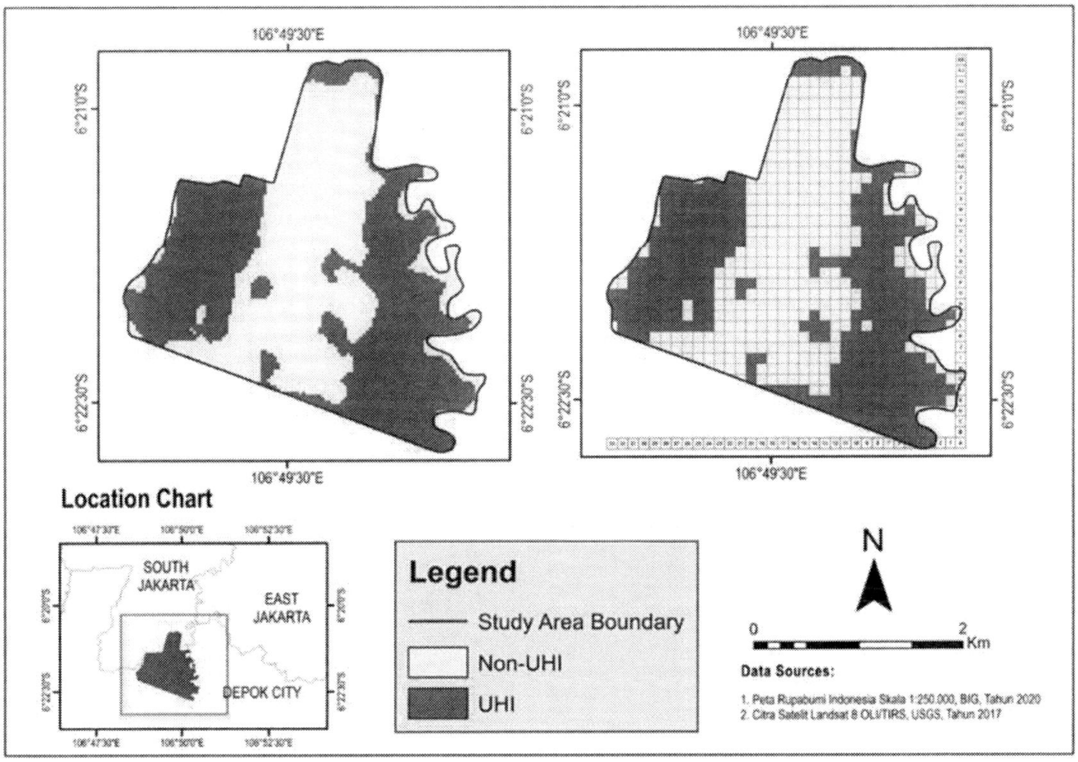

Figure 5. Spatial Distribution of UHI in 2017

Visually, the UHI area in the 2017 was identical, occurring in the east, west, south, and north regions relatively, while in the central region, areas showing UHI were rare. This indicates that the factors causing UHI are not in that area.

Further information is related to the existence of UHI in the 2021 processing year, in which there is a significant addition of UHI from 361 grids to 553 grids accompanied by a reduction in non-UHI areas from 405 grids to 211 grids.

Table 4. UHI area Comparison in 2021

Classification	Area (Ha)	Percentage (%)	Accumulated Grid
UHI	469.15	69.82	553
Non-UHI	202.67	30.16	211

The presentation of UHI information in the table above is sufficient to explain that the UHI area in the 2021 processing year dominates the research area, even the cover area of the UHI affected area itself reaches more than half of the research area, which is around 69.82%. Meanwhile, non-UHI areas can only cover the total area of 30.16%. The area affected by UHI itself reaches 469,155 ha, which has a difference of 266,483 ha from the non-UHI area with an area of 202,672 ha.

The number of grids produced is also very different, even if non-UHI areas are given a duplication of the number of grids (multiplied by two), it is not enough to give the same number of grids as UHI areas, which is 553 grids, while for non-UHI areas only the range 211 grids. That's quite a comparison. The comparison is interesting to examine the cause so that it gets a figure that is quite far away and even doubles. It is alleged that there is an x factor that makes UHI in 2021 able to make a lot of duplication. In order to clarify the causality, comparisons can be made between UHI each year, in order to see the changes that occur before further analysis is projected. The spatial distribution of UHI in 2021 can be seen in the following figure:

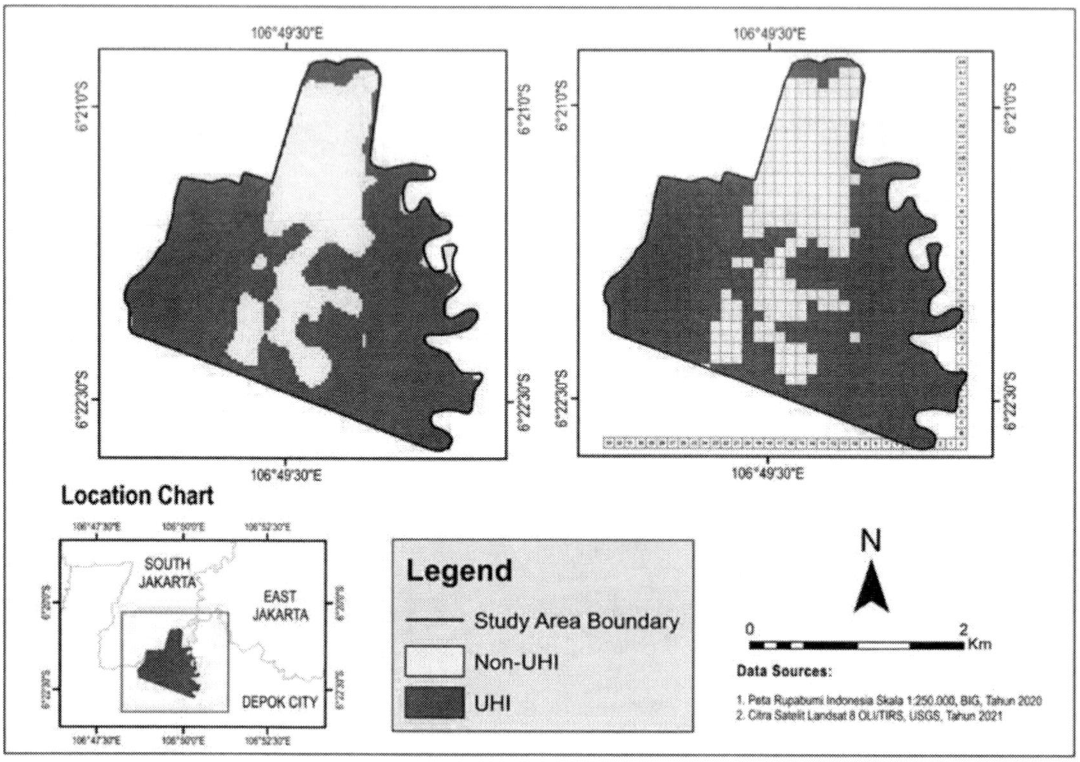

Figure 6. Spatial Distribution of UHI in 2021

4. Conclusions

Urban heat island extraction using the LST-based threshold method in the UI Depok Campus and its surrounding area is suitable for use in complex urban landscapes due to the result of a strong correlation with a few error margins from the actual air surface temperature. Research also explain information related to the spatial and temporal distribution of UHI, which always increases and expands every year. UHI began to spread very intensively in 2021, accompanied by many non-UHI areas decrease in area. In the last year, UHI has covered 69.82% of the research area based on the calculation of the LST threshold.

Acknowledgments

The authors are deeply grateful to the Department Geography, Faculty of Mathematics and Natural Sciences, Universitas Indonesia. This research and publication support by a Research Grant from the Directorate of Research and Development Universitas Indonesia, based on contract number: NKB-1945/UN2.RST/HKP.05.00/2020.

References

[1] S Y H 2008 *Dinamika wilayah peri-urban: determinan masa depan kota* (Yogyakarta: Pustaka

belajar)

[2] Ke X, van Vliet J, Zhou T, Verburg P H, Zheng W and Liu X 2018 Direct and indirect loss of natural habitat due to built-up area expansion: A model-based analysis for the city of Wuhan, China *Land use policy* **74** 231–9

[3] Wibowo A, Yusoff M M and Salleh K O 2020 Monitoring urban heat signature and profiles of localized urban environment in the University of Malaya *IOP Conf. Ser. Earth Environ. Sci.* **481**

[4] Senanayake I P, Welivitiya W D D P and Nadeeka P M 2013 Remote sensing based analysis of urban heat islands with vegetation cover in Colombo city, Sri Lanka using Landsat-7 ETM+ data *Urban Clim.* **5** 19–35

[5] Voogt J A and Oke T R 2003 Thermal remote sensing of urban climates *Remote Sens. Environ.* **86** 370–84

[6] Aguado E and Burt J E 2001 Understanding Weather and Climate (2nd Edition) (New Jersey: Prentice Hall Inc)

[7] Fabrizi R, Bonafoni S and Biondi R 2010 Satellite and ground-based sensors for the Urban Heat Island analysis in the city of Rome *Remote Sens.* **2** 1400–15

[8] Zhang Y, Su Z, Li G, Zhuo Y and Xu Z 2018 Spatial-temporal evolution of sustainable urbanization development: A perspective of the coupling coordination development based on population, industry, and built-up land spatial agglomeration *Sustain.* **10** 1–20

[9] Pribadi D O 2006 MODEL PERUBAHAN TUTUPAN LAHAN DAN FAKTOR-FAKTOR YANG MEMPENGARUHINYA *J. Teknol. Lingkung.* **7**

[10] Vioya A 2010 Tahapan Perkembangan Kawasan Metropolitan Jakarta *J. Reg. City Plan.* **21** 215–26

[11] Harahap F R 2013 DAMPAK URBANISASI BAGI PERKEMBANGAN KOTA DI INDONESIA Fitri Ramdhani Harahap, S.Sos., M.Si □ *J. Soc.* **I** 35–45

[12] Sitanala F 2010 Pergerakan Penduduk Kota Depok Menuju Ke Tempat Bekerja Tahun 2001 *MAKARA Sci. Ser.* **9** 41–4

[13] Sulistyaningrum S and Sumabrata J 2018 Transit Oriented Development (TOD) index at the current transit nodes in Depok City, Indonesia *IOP Conf. Ser.: Earth Environ. Sci.* pp 4–15

[14] Ristianingrum N and Osly P J 2017 ANALISIS URBAN SPRAWL TERHADAP PERUBAHAN TUTUPAN LAHAN KOTA DEPOK *J. Infrastruktur* **3** 31–9

[15] Madakarah N Y, Supriatna, Wibowo A, Manessa M D M and Ristya Y 2019 Variations of Land Surface Temperature and Its Relationship with Land Cover and Changes in IPB Campus, Dramaga Bogor 2013-2018 *E3S Web Conf.* **125**

[16] Nadira C, Saraswati R and Wibowo A 2019 Pengaruh Perubahan Tutupan Lahan Terhadap Fenomena Urban Heat Island di Kecamatan Cikarang Utara , Kabupaten Bekasi Tahun 2007 – 2018 Menggunakan Citra Landsat 5 dan 8 The Effect of Land Cover Change to Urban Heat Island Phenomenon in 2007 – 2018 at North *Seminar Nasional Penginderaan Jauh ke-6 Tahun 2019* pp 82–98

[17] Bretz S E and Akbari H 1997 Long-term performance of high-albedo roof coatings *Energy Build.* **25** 159–67

[18] Gober P, Brazel A, Quay R, Myint S, Grossman-Clarke S, Miller A and Rossi S 2010 Using watered landscapes to manipulate urban heat island effects: How much water will it take to cool phoenix? *J. Am. Plan. Assoc.* **76** 109–21

[19] Emiyati ., Kusratmoko E and Sobirin . 2017 Spatial Pattern of Hydrologic Response Unit (Hru) Effect on Flow Discharge of Ci Rasea Watershed Using Landsat Tm in 1997 To 2009 *Int. J. Remote Sens. Earth Sci.* **13** 39

[20] Dwi Sulistyana M I C, Yuwono S B and Rusita R 2017 Kenyamanan Hutan Kota Linara Berbasis Kerapatan Vegetasi, Iklim Mikro Dan Persepsi Masyarakat Di Kota Metro *J. Sylva Lestari* **5** 78

[21] Tauhid T 2008 *KAJIAN JARAK JANGKAU EFEK VEGETASI POHON TERHADAP SUHU*

UDARA PADA SIANG HARI DI PERKOTAAN (Studi Kasus: Kawasan Simpang Lima Kota Semarang) (Universitas Diponegoro)

[22] Wong N H, Kardinal Jusuf S, Aung La Win A, Kyaw Thu H, Syatia Negara T and Xuchao W 2007 Environmental study of the impact of greenery in an institutional campus in the tropics *Build. Environ.* **42** 2949–70

[23] Wibowo A, Semedi J M and Salleh K O 2017 Spatial temporal analysis of urban heat hazard on education Area (University of Indonesia) *Indones. J. Geogr.* **49** 1–10

[24] Yunika and Putro H 2021 ANALISIS PENERAPAN SISTEM K3 TERHADAP KINERJA PROYEK JALAN TOL CIJAGO SEKSI 2B kecelakaan kerja . Anas Zaini Z Iksan selaku Ketua Umum Asosiasi Ahli Keselamatan dan Kesehatan Kerja dan Pengurus yang tidak memenuhi dikenakan sanksi sesuai dengan Undang- pe *Politeknologi* **20** 53–63

[25] Rahmawati 2018 Penerapan Fuzzy Linear Programming Pada Optimasi Pembangunan Rumah Susun (Rusun) Di Kawasan Pondok Cina Provinsi Jawa Barat *J. Sains Mat. dan Stat.* **4** 79–89

[26] Sobrino J A, Oltra-Carrió R, Sòria G, Jiménez-Muñoz J C, Franch B, Hidalgo V, Mattar C, Julien Y, Cuenca J, Romaguera M, Gómez J A, de Miguel E, Bianchi R and Paganini M 2013 Evaluation of the surface urban heat island effect in the city of Madrid by thermal remote sensing *Int. J. Remote Sens.* **34** 3177–92

[27] Pradhan A M S, Dawadi A and Kim Y-T 2012 Use of different bivariate statistical landslide susceptibility methods: A case study of Kulekhani watershed, Nepal *J. Nepal Geol. Soc.* **44** 1–12

[28] Weng Q 2009 Thermal infrared remote sensing for urban climate and environmental studies: Methods, applications, and trends *ISPRS J. Photogramm. Remote Sens.* **64** 335–44

[29] Tsou J Y, Li X, Tsou K, He J and Pan D 2018 *Detect relationship between urban housing development and urban heat island dynamic in hyper-density Hong Kong by integrating GIS and RS techniques* vol 10863 LNCS (Springer International Publishing)

[30] Prasasti I, Sari N M and Febrianti N 2015 Analisis Perubahan Sebaran Pulau Panas Perkotaan (Urban Heat Island) di Wilayah DKI Jakarta dan Hubungannya dengan Perubahan Lahan , Kondisi Vegetasi dan Perkembangan Kawasan Terbangun Menggunakan Data Penginderaan Jauh *Pros. Pertem. Ilm. Tah. XX 2015* 383–91

[31] Pratiwi A Y and Jaelani L M 2021 Analisis Perubahan Distribusi Urban Heat Island (UHI) di Kota Surabaya Menggunakan Citra Satelit Landsat Multitemporal *J. Tek. ITS* **9** C48–55

[32] Tursilowati L, Sumantyo J T S, Kuze H and Adiningsih E S 2012 The integrated WRF/urban modeling system and its application to monitoring Urban Heat Island in Jakarta, Indonesia *J. Urban Environ. Eng.* **6** 1–9

[33] Wibowo A and Salleh K O 2018 LAND COVER TYPES AND THEIR EFFECT ON THE URBAN HEAT SIGNATURE OF UNIVERSITY CAMPUSES USING REMOTE SENSING Adi Wibowo 1,2* , Khairulmaini Osman Salleh 1 1 *Int. J. Technol.* **3** 479–90

[34] Liviona C D A, Saraswati R and Wibowo A 2020 The Effect of NDVI and NDBI on Land Surface Temperature in Cirebon City 2015 and 2019 *E3S Web Conf.* **202**

[35] Ma Y, Kuang Y and Huang N 2010 Coupling urbanization analyses for studying urban thermal environment and its interplay with biophysical parameters based on TM/ETM+ imagery *Int. J. Appl. Earth Obs. Geoinf.* **12** 110–8

[36] Kolawole O A and Boluwatife A R 2016 Assessment Of The Factors Influencing Students' Choice Of Residence In Nigerian Tertiary Institutions *Sains Humanika* **8** 39–47

[37] Anggraeni O, Astuti W and Mukaromah H 2020 Hubungan pemanfaatan lahan dengan volume pergerakan di Jalan Margonda Raya, Kota Depok *Reg. J. Pembang. Wil. dan Perenc. Partisipatif* **15** 16

[38] Mehta A, Shukla S and Rakholia S 2021 Vegetation Change Analysis using Normalized Difference Vegetation Index and Land Surface Temperature in Greater Gir Landscape *J. Sci. Res.* **65** 01–6

[39] Alsaady W F and Mohammed R 2021 Detecting of climatic drought by combination geo-information system and remote sensing in semi-arid zones: A case study *Mater. Today Proc.*

[40] El-Hattab M, Amany S M and Lamia G E 2018 Monitoring and assessment of urban heat islands over the Southern region of Cairo Governorate, Egypt *Egypt. J. Remote Sens. Sp. Sci.* **21** 311–23

[41] F S F 2007 *Remote Sensing: Principles and Applications* (Waveland Press)

[42] Vanhellemont Q 2020 Combined land surface emissivity and temperature estimation from Landsat 8 OLI and TIRS *ISPRS J. Photogramm. Remote Sens.* **166** 390–402

[43] Wang R, Cai M, Ren C, Bechtel B, Xu Y and Ng E 2019 Detecting multi-temporal land cover change and land surface temperature in Pearl River Delta by adopting local climate zone *Urban Clim.* **28** 100455

[44] Roth M 2013 Urban Heat Islands *Handbook of Environmental Fluid Dynamics* (CRC Press)

Urbanization rate in SUBOSUKAWONOSRATEN

Rita N[1*], Ahmad[1], Sarwono[1], Sugiyanto[1], and Singgih P[1]

[1] Faculty of Teacher Training and Education, Sebelas Maret University, Surakarta 57126, Indonesia

*Email: ritanoviani@staff.uns.ac.id,

Abstract. Urbanization has changed landscape patterns and ecological functions, resulting in a decline in service ecosystem services and environmental problems, especially in areas experiencing rapid Urbanization. The increase in population will affect the development process and affect land needed to support daily activities, and affect the temperature that triggers the Urban Heat Island phenomenon. A better understanding of the relationship between Urbanization and ecosystem services is critical to future regional sustainability and relevant policy development. This study aims to assess the relationship between Urbanization and ecosystem services in the urban area of Solo Raya, consisting of 7 districts/cities, namely Surakarta, Boyolali, Sukoharjo, Wonogiri, Sragen, and Klaten. The research method used is spatial quantitative descriptive analysis to analyze the impact of Urbanization on ecosystem services in 2020 in an isolated Solo Raya urban area. Ecosystem services are calculated from food production and Urbanization rate measured from four aspects: population, population growth, population density, and land expansion. The results show that areas with a high urbanization rate are automatically viewed in terms of Ecosystem Services. The region will be in deficit or not a self-sufficiency area. In other words, it shows that there is a relationship between the level of urbanization and ecosystem services.

1. Introduction

In recent years, the world's attention has focused on the increasingly intense anthropogenic impacts. Anthropogenic, especially on the environment, can have harmful effects on the environment and humans themselves. Rapid population growth is the cause of new symptoms, one of which is Urbanization. Urbanization contributes significantly to social improvement and economic development. But on the other hand, Urbanization can be a significant threat to the health of non-urban ecosystems. The trend generally indicates a decrease in ES supply due to changes in land cover in peri-urban areas caused by nearby Urbanization [1].

The way Urbanization affects ecosystems is a worldwide concern. Ecological risks (e.g., potential water crises, air pollution) generated by Urbanization are significant contributors to global change [2]. Urbanization has a strong influence on the regional climate. As core areas of human activity, cities affect restricted land use and land cover changes and affect global climate change [3]. Rapid Urbanization and large-scale Urbanization can lead to drastic land-use conversions. As a result, rapid Urbanization can have an impact on ecosystem services. Among others have turned into human-dominated ecosystems of unprecedented scale and speed. These conditions can result in the effects of the loss of ecosystem services.

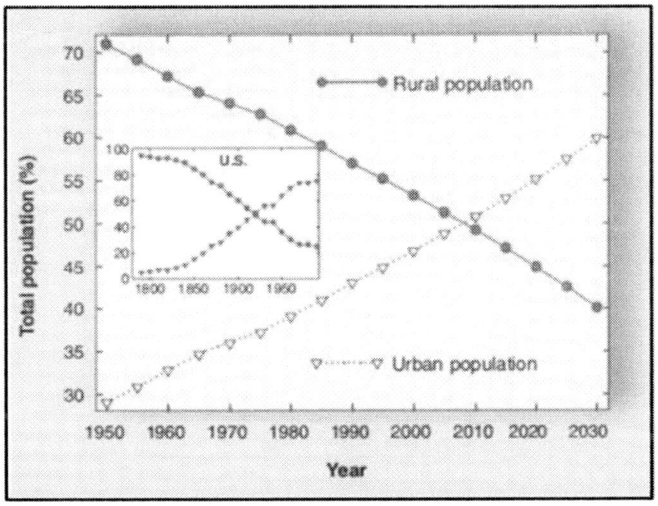

Figure 1. Changes in The World Population of Rural and Urban Areas 1950-2030

Source:[4]

Urbanization also has other impacts, such as Urban Heat Island. Urban heat island refers to high temperatures in urban areas compared to rural environments (U.S. Environmental Protection Agency, 2008: 6). The problem faced today is the increasing phenomenon of urban heat islands as the land is increasingly built. Urban heat islands can have some negative impacts and new issues on air quality, the human environment and even affect the region's local climate. This will undoubtedly have a significant effect on human life and can endanger the lives of humans or other living things. Therefore, management of urban spatial planning is needed to reduce the urban heat island phenomenon. Control of built land and open land will be very influential in reducing the negative impact of urban heat islands.

Urban sustainability is one of the most pressing problems and must have a solution to the problem. Understanding the relationships or interrelationships between Urbanization and ecosystem services can provide informative advice in urban planning. The relationship between Urbanization and ecosystem services depends not only on the characteristics of the study area but also on the ecosystem services or benefits that humans derive from ecosystems.

The research was conducted in the Surakarta, Boyolali, Sukoharjo, Karanganyar, Wonogiri, Sragen, and Klaten (SUBOSUKAWONOSRATEN) metropolitan area to support sustainable regional planning. Investigating the Impact of Urbanization on ecosystem services in urban megaregions is excellent for achieving a double-win economic development and environmental protection situation. Many studies have been conducted to investigate the impact of Urbanization and ecosystem services in urban megaregions. The leading cause is land-use change, a unique feature of airflow dynamics within cities and artificial heat sources, the most critical land-use change (Du et al., 2020).

2. Methods

The research was conducted in 7 (seven) districts/cities, namely Surakarta City, Boyolali Regency, Sukoharjo, Wonogiri Regency, Sragen Regency, and Klaten Regency SUBOSUKAWONOSRATEN and now better known as Solo Raya. Solo Raya area consists of 124 subdistricts and 1565 villages, with an area of 5,722.4 Km^{2}. The research site is the entire administrative area in the Greater Solo region. The location of Solo Raya was chosen with several considerations, namely: Geographically, the Greater Solo area is part of central Java Island which is located adjacent to the provinces of Yogyakarta Instimewa Region (DIY) and East Java Province, which makes its strategic position, as a driving force for regional development, so that it can become a center of growth and driver of economic growth for the surrounding region. The Greater Solo Region in this study was designated as a strategic area from the point of interest of economic development of Central Java Province in the Central Java Provincial Spatial Plan 2009-2029. The Spatial Plan of SUBOSUKAWONOSRATEN or Solo Raya urban areas in this research will be developed as a metropolitan area with a service center in Surakarta City.

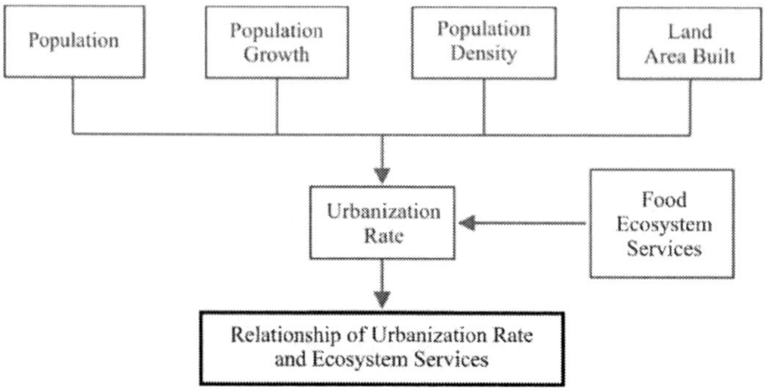

Figure 2. Research Mindset

2.1. Mapping Rate Urbanization

In this study, we characterized the Urbanization rate of SUBOSUKAWONOSRATEN urban areas from four variables: population number, population growth, population density, and the land area built up. Each indicator gives a score categorized into three classes to know which regions have low, medium, and high Urbanization rates.

2.2. Measuring Ecosystem Services

Ecosystem services focused on food were based on rice production and land needs. Rice production is measure from the annual rice production in each sub-district in SUBOSUKAWONOSRATEN collected by library study method. For land needs are measured by the following formula:

$$DL = N\,x\,KHLL \quad (1)$$

Source: Regulation of the Minister of Environment No. 17 of 2009
Information:

- DL: Total rice equivalent land needs (ha)
- N: Population (soul)
- KHLL: The area of land needed for decent living needs per resident

Calculation of the land needs side there are several provisions that the need for life
Per resident is:

a) The land area needed for decent living per population is a proper living requirement divided by local productivity.
b) Decent living needs per population are assumed to be 1 ton equivalent of rice/capita/year.
c) Areas with local rice productivity

2.3. Relationship of Rate Urbanization and Ecosystem Services

Urbanization with Ecosystem Services can be seen by overlaying on both maps, namely the Urbanization rate and ecosystem services maps. It can be seen how the alignment between the two in SUBOSUKAWONOSRATEN.

2.4. Data

In this study, there are several data used to find land production rates and land needs and measure the rate of Urbanization:

Table 1. Data of the Study

No.	Criterion	Data
1	Rate of Urbanization (Population)	Population
2	Rate of Urbanization (Population)	Population density 2020
3	Rate of Urbanization (Population)	Growth rate 2015-2020
4	Rate of Urbanization (Land Built)	Land area awakened
5	Land Needs	Rice Production
6	Land Needs	Rice Needs

3. Results and Discussion

3.1. Urbanization Rate

This research was conducted by calculating the rate of Urbanization in the development area of SUBOSUKAWONOSRATEN. First, the calculation of Urbanization rates is carried out based on population data in 2020, the rate of population growth in 2015-2020, population density, and the land area built in each sub-district in the SUBOSUKAWONOSRATEN development area. Then, the results of the calculation of the four data are carried out analyses using the scoring method. After overlay, so that the classes are formed. In this study, the rate of Urbanization was divided into three categories: low, medium, and high Urbanization rates. Areas that tend towards cities will have a rate of Urbanization from the Medium-High rate. At the same time, areas that are typically towards the countryside, in general, have a rate of Urbanization than the Low-Medium rate. Here are the calculations of the rate of Urbanization:

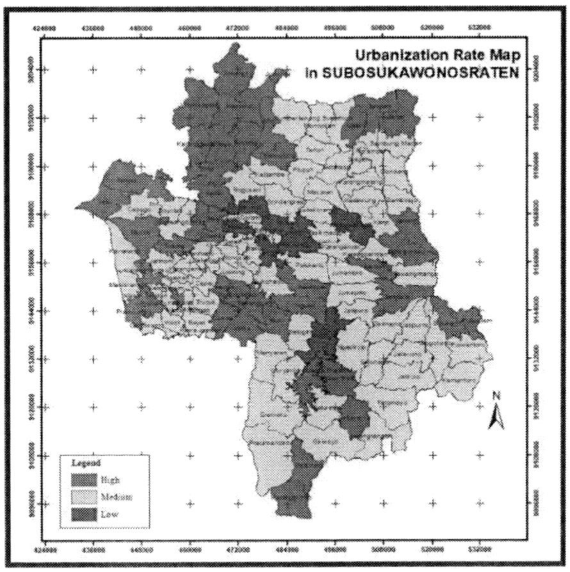

Figure 3. Urbanization Rate Map

Table 2. Urbanization Rate of SUBOSUKAWONOSRATEN by Number of Sub-Districts

No	District/City	Urbanization Rate		
		Low	Medium	High
1	Surakarta City	0	3	2
2	Boyolali	14	4	1
3	Sukoharjo	5	4	3
4	Karanganyar	4	11	2
5	Wonogiri	6	18	1
6	Sragen	5	15	0
7	Klaten	8	18	0

Based on **Table 2**, it can be known that subdistricts in Surakarta City have a rate of Urbanization of Medium-High. Subdistricts that have high Urbanization rates are the Banjarsari subdistrict and Jebres Subdistrict. Then, sub-districts with a medium Urbanization rate are in Laweyan District, Kliwon Market, and Serengan. This is because the city of Surakarta is a growth center in the SUBOSUKAWONOSRATEN area, which also encourages immigrants to enter the Surakarta area. Factors that dominate the rate of Urbanization in Surakarta are the factors of population and population density. The rate of Urbanization in Boyolali Regency is in the Low-High range. However, there is one subdistrict that has a high rate of Urbanization, namely the Ngemplak subdistrict. Meanwhile, the medium's Urbanization rate is in Boyolali, Cepogo, Mojosongo, and Nogosari districts. Then the most dominating is the low Urbanization rate found in Ampel, Andong, Banyudono, Juwangi, Karanggede, Kemusu, Klego, Musuk, Sambi, Sawit, Selo, Simo, Teras, and Wonosegoro. Sukoharjo Distritct is in the Low-High range. Subdistricts with low Urbanization rates are in Bendosari, Bulu, Neuter, Tawangsari, and Weru districts. The subdistrict is farthest from the urban impact of Surakarta City or located to the south. Meanwhile, Sub-district with a medium Urbanization rate is in Baki, Gatak, Polokarto, and Sukoharjo districts. The rate of Urbanization High is in Grogol, Kartasura, and Mojolaban districts that are heavily affected by the urban area of Surakarta City.

Based on **Table 2**, Based on the table above, it can be known that Karanganyar Regency has a range of Urbanization rates from Low-High. Karangnyar regency is dominated by the level of urbanization of Medium located in Colomadu, Gondangrejo, Jumantoro, Jumapolo, Karanganyar, Karangpandan, Kebakkramat, Kerjo, Tasikmadu, and Tawangmangu. While for the level of Urbanization, High is in Jaten and Mojogedang districts. For low Urbanization rates, there are many factors, one of which is the existence of several protected areas in the Karanganyar Regency. Sub-districts that have low Urbanization rates are Jatiyoso, Jenawi, Matesih, and Ngargoyoso. Wonogiri Regency has a range of Low to High. Districts that have a high level of Urbanization are only found in Wonogiri Subdistrict. At the same time, sub-districts with low urbanization rates are Batuwarno, Bulukerto, Giritontro, Ngutoronadi, Paranggupito, Puhpelem. While for sub-districts that have a level of medium Urbanization are in Baturetno, Eromoko, Girimarto, Giriwoyo, Jatipurno, Jatiroto, Jatisrono, Karangtengah, Kismantoro, Manyaran, Ngadirejo, Precimantoro, Purwantoro, Sidoharjo, Tirtomoyo, and Wuryantoro. The level of Urbanization in the Sragen Regency is near low-medium. Low urbanization rates are in Gemolong, Gesi, Jenar, Miri, and Tangen districts. Meanwhile, the level of urbanization of Medium is in Gondang, Kalijambe, Karangmalang, Kedawung, Masaran, Mondokan, Ngrampal, Plupuh, Sambirejo, Sambung Macan, Sidoharjo, Sragen, Sukodono, Sumberlawang, and Tanon.

Klaten Regency has a level of urbanization in the Low-Medium range. For sub-districts that have low urbanization rates are in Cawas, Jogonalan, Karangdowo, Karangnongko, Klaten Tengah, Pedan, Prambanan, and Tulung. Then the subdistrict that has a level of medium urbanization is in Bayat, Ceper,

Delanggu, Gantiwarno, Jatinom, Juwiring, Kalikotes, Karanganom, Kemalang, South Klaten, North Klaten, Manisrenggo, Ngawen, Polanharjo, Trucuk, Tulung, Wedi, and Wonosari.

3.2. Ecosystem Services

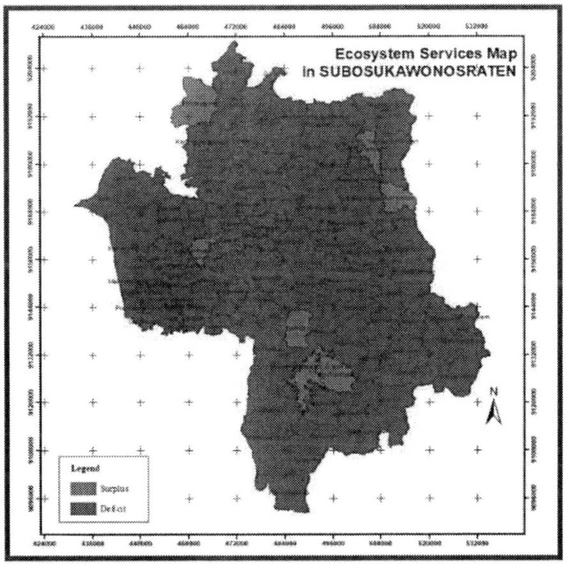

Figure 4. Ecosystem Services in the SUBOSUKAWONOSRATEN

The next research stage is the calculation of ecosystem services in the development area of SUBOSUKAWONOSRATEN based on food production. This calculation is done to determine whether each sub-district has been able to suffice or exceed its Ecosystem Services. If in the analysis of obtaining surplus results, the subdistrict can be said to be self-sufficiency because the need for food (especially rice) has been met even more than needed by the population. Meanwhile, if the calculation is Deficit, the subdistrict has not fulfilled its residents' food needs (especially rice). Therefore, in the SUBOSUKAWONOSRATEN development area in 2021, most sub-districts can still not meet the standard rice needs of 1000 kg/population/year.

Table 3. Ecosystem Services of SUBOSUKAWONOSRATEN by Number of Sub-Districts

No	District/City	Ecosystem Services	
		Decifit	**Surplus**
1	Surakarta City	5	0
2	Boyolali	18	1
3	Sukoharjo	12	0
4	Karanganyar	17	0
5	Wonogiri	23	2
6	Sragen	18	2
7	Klaten	25	1

Based on **Table 3**, It can be known that no subdistrict is a subdistrict of self-sufficiency in the city of Surakarta. All sub-districts in Surakarta city have not been able to meet the food needs of its residents. This is possible because of the number of residents who are too much or small rice production. In addition, the city of Surakarta is a growth center in the SUBOSUKAWONOSRATEN area, which also encourages immigrants to enter the Surakarta area. Therefore, exceeding the limits of ecosystem services needs special attention because it will affect food security in Surakarta.

Based on **Table 3,** It can be known that the subdistrict, which is a sub-district of self-sufficiency where it can be sufficient, even exceeds the need for food in Boyolali Regency only located in Wonosegoro Subdistrict. While other sub-districts have not been able to meet the population's food needs, this is possible because of the overcrowding, or the rice production is small. Sub-district of self-sufficiency where it can be sufficient, even exceeds the need for food in Sukoharjo Regency only located in Wonosegoro Subdistrict. While other sub-districts have not been able to meet the population's food needs, this is possible because of the overcrowding, or the rice production is small. It can be known that no subdistrict is a self-sufficiency subdistrict in Karanganyar Regency. All sub-districts in Karanganyar Regency have not been able to meet the food needs of their residents. This is possible because of the overcrowding or small rice production. So, this needs special attention because it will affect food security in a region.

Based on **Table 3,** It can be known that the subdistrict, which is a sub-district of self-sufficiency where it can be sufficient, even exceeds the need for food in Wonogiri Regency only located in Ngutoronadi District and Selogiri Subdistrict. While other sub-districts have not been able to meet the population's food needs, this is possible because of the overcrowding or the small production of rice.

Based on **Table 3,** It can be known that the subdistrict, which is a self-sufficiency subdistrict where it can be sufficient, even exceeds the need for food in Sragen Regency only located in Ngrampal District and Sambirejo Subdistrict. While other sub-districts have not been able to meet the food needs of their residents, this is possible because of the overcrowding or small rice production.

Based on **Table 3,** It can be known that the subdistrict, which is a self-sufficiency subdistrict where it can meet even exceeds the need for food in Klaten Regency only located in Polanharjo Subdistrict. While other sub-districts have not been able to meet the population's food needs, this is possible because of the overcrowding, or the rice production is small.

3.3. Relationship of Urbanization Rate and Ecosystem Services

Based on the results of both calculations, it can be concluded that areas with a high urbanization rate are automatically viewed in terms of Ecosystem Services. The region will be in deficit or not a self-sufficiency area. In other words, it shows that there is a relationship between the level of urbanization and ecosystem services. Then the region that has surplus Ecosystem Services or is a self-sufficiency area with surplus Ecosystem Services in some regions has low and medium urbanization rates. Nine districts have a high level of urbanization with ecosystem services that are in deficit. This condition will cause various city problems so that there is a need for attention to food needs in the region. Here is the relationship between the Level of Urbanization with Ecosystem Services:

Table 4. Relationship of Urbanization with Ecosystem Services

No	Sub-Disctrict	District	Urbanization Rate	Ecosystem Services
1	Ngemplak	Boyolali	High	Deficit
2	Jaten	Karanganyar	High	Deficit
3	Mojogedang	Karanganyar	High	Deficit
4	Banjarsari	Surakarta	High	Deficit
5	Jebres	Surakarta	High	Deficit
6	Grogol	Sukoharjo	High	Deficit
7	Kartasura	Sukoharjo	High	Deficit
8	Mojolaban	Sukoharjo	High	Deficit
9	Wonogiri	Wonogiri	High	Deficit

Urbanization is one of the most important factors in ecosystem services. Previous studies have also widely stated that ecosystem services are influenced by physical factors and social and even economic factors. Ecosystem Services were tightly correlated with social-ecological conditions [6]. Therefore, mapping related to urbanization and ecosystem services is important in the development and regional spatial planning. Some aspects can be one of the considerations for policymaking, especially related to land use. Identifying low- and high-value ecosystem services areas is crucial for land use management [7]. This indicates that urbanization and ecosystem services have become important things for the future. The relationship between urbanization and ecosystem services has received widespread attention in recent years (Cao et al., 2017). Districts in SUBOSUKAWONOSRATEN have a wide variety of urbanization rates and ecosystem services. Surplus ecosystem services do not support Districts with a high urbanization rate but will cause a lack of food carrying capacity. More attention is needed to address problems related to urbanization in urban areas, especially in districts close to urban development areas. Therefore, mapping urbanization and ecosystem services will be important. Without mapping about urbanization and ecosystem services, it will likely cause various problems where this is also supported by urbanization closely related to Land Surface Temperature. The application of urbanization mapping and ecosystem services will be very useful to prevent this. In the future, many effective measures should be implemented to realize the harmonious development of urbanization and the ecological environment (Peng et al., 2017).

4. Conclusion

Mapping the relationship between urbanization and ecosystem services needs to be considered in regional development and planning. Special attention to urbanization and ecosystem services will support sustainable development where natural resources will not be massively exploited. The medium-high level dominates the urbanization rate of the SUBOSUKAWONOSRATEN area. This condition indicates that the city of Surakarta has a considerable impact on urban development in the surrounding area. Urbanization also causes almost all sub-districts in SUBOSUKAWONOSRATEN has been exceeded for its carrying capacity, causing ecosystem services in the region to become deficient or have not been able to meet the needs of their communities. Nine districts have a high level of urbanization with ecosystem services that are in deficit. This condition will cause various city problems so that there is a need for attention to food needs in the region. Of course, this condition is not a good thing in running the development of a city. Policymakers must take appropriate steps to reduce urbanization rates or at least control the level of urbanization in a region. With controlled urbanization, ecosystem services in the area will also be in good condition or surplus.

IOP Conf. Series: Earth and Environmental Science **986** (2022) 012073 doi:10.1088/1755-1315/986/1/012073

References

[1] García-Nieto A P, Geijzendorffer I R, Baró F, Roche P K, Bondeau A and Cramer W 2018 Impacts of urbanization around Mediterranean cities: Changes in ecosystem service supply *Ecol. Indic.* **91** 589–606

[2] Li B, Chen D, Wu S, Zhou S, Wang T and Chen H 2016 Spatio-temporal assessment of urbanization impacts on ecosystem services: Case study of Nanjing City, China *Ecol. Indic.* **71** 416–27

[3] Cui Y, Xu X, Dong J and Qin Y 2016 Influence of urbanization factors on surface urban heat island intensity: A comparison of countries at different developmental phases *Sustain.* **8**

[4] Grimm N B, Faeth S H, Golubiewski N E, Redman C L, Wu J, Bai X and Briggs J M 2008 Global change and the ecology of cities *Science (80-.).* **319** 756–60

[5] Du H, Zhou F, Li C, Cai W, Jiang H and Cai Y 2020 Analysis of the impact of land use on spatiotemporal patterns of surface urban heat island in rapid urbanization, a case study of Shanghai, China *Sustain.* **12**

[6] Yang Y, Zheng H, Kong L, Huang B, Xu W and Ouyang Z 2019 Mapping ecosystem services bundles to detect high- and low-value ecosystem services areas for land use management *J. Clean. Prod.* **225** 11–7

[7] Hamann M, Biggs R and Reyers B 2015 Mapping social-ecological systems: Identifying "green-loop" and "red-loop" dynamics based on characteristic bundles of ecosystem service use *Glob. Environ. Chang.* **34** 218–26

[8] Cao W, Li R, Chi X, Chen N, Chen J, Zhang H and Zhang F 2017 Island urbanization and its ecological consequences: A case study in the Zhoushan Island, East China *Ecol. Indic.* **76** 1–14

[9] Peng J, Liu Y, Liu Z and Yang Y 2017 Mapping spatial non-stationarity of human-natural factors associated with agricultural landscape multifunctionality in Beijing–Tianjin–Hebei region, China *Agric. Ecosyst. Environ.* **246** 221–33

Assessment of Water Quality Status of Cengklik Reservoir due to Eutrophication

A Kurniawan[1*], F N M Jayatri[2], and K Khasanah[2]

[1] Department of Civil Engineering, Vocational College, Universitas Gadjah Mada, Yogyakarta 55281, Indonesia
[2]Magister of Technology for the Sustainable Development of Postgraduate Schools, Universitas Gadjah Mada, Yogyakarta 55281, Indonesia

[*] Email: adhy_kurniawan@ugm.ac.id

Abstract. Cengklik Reservoir is one of the water sources that has the main function for irrigation and aquaculture. Based on satellite imagery of the Cengklik Reservoir 2000-2020, about 30% of the waters of the Cengklik Reservoir have been covered by water hyacinth. The use of chemical fertilizers, pesticides and fish farming activities using floating net cages are the main factors causing eutrophication. Therefore, the purpose of this study was to analyze the water quality of the Cengklik Reservoir at the inlet and outlet points of the Cengklik Reservoir and then determine the level of pollution using the Pollution Index (PI) method. The water quality parameters tested were temperature, pH, TSS, DO, BOD, COD, TSS, Total-P, and Total N. Based on the results of the PI analysis, the water quality of Cengklik Reservoir was categorized as lightly polluted and does not exceed the quality standard value of Government Regulation No. 22 years 2021 other than phosphate. Phosphate parameter values at inflow 1, inflow 2, inflow 3 and outflow locations are 0.166 mg/L; 0.166 mg/L; 0.723 mg/L and 0.166 mg/L which are estimates of the cause of eutrophication in the Cengklik Reservoir.

1. Introduction

Reservoirs used by the community as irrigation, fisheries and water tourism destinations [1,2]. Increased activity in the communities around the dam has the potential to produce waste that can cause problems, such as water pollution and eutrophication. Eutrophication occurs due to natural factors (climate change, seasonal factors and meteorological conditions) and human activities [3]. Cengklik Reservoir is located in Kabupaten Boyolali, where the majority of the population are farmers and fish farmers. The Cengklik Reservoir is used as an irrigation facility for agriculture around the reservoir and aquaculture using the floating net cage method. Based on the appearance of the satellite image in Figure 1, about 30% of the Cengklik Reservoir has been covered by water hyacinth. In addition, there is an increase in the number of floating net cages in the Cengklik Reservoir during 2000 - 2020. Therefore, Pollution index assessment method for water quality analysis describes the condition and severity of water pollution, so that reservoir management policies can be planned to overcome these problems. . The aims of this research were to assess the water quality of the Cengklik Reservoir using the Pollution Index method. These results can be used to propose the technical and environmental recommendations to improve the integrated and sustainable management of the Cengklik Reservoir.

IOP Conf. Series: Earth and Environmental Science **986** (2022) 012074 doi:10.1088/1755-1315/986/1/012074

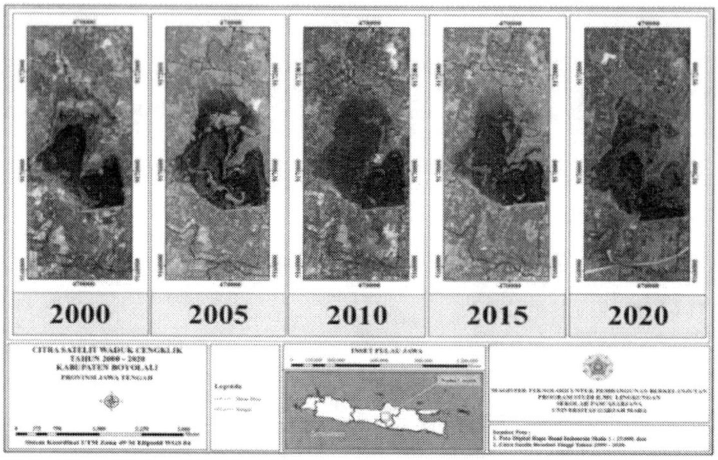

Figure 1. The appearance of satellite imagery in the Cengklik Reservoir

2. Method

2.1. Study Area

Cengklik Reservoir is located in Ngargorejo Village, Ngemplak District, Boyolali Regency with coordinates 7°30'26.1" LS 110°43'29.7" east longitude. Water sampling points were selected using a purposive sampling method (based on considerations of ease of access and time by researchers) at 4 location points, namely at 3 inflows and 1 outflow in Cengklik Reservoir as shown in Figure 2. Water samples were taken by means of grab samples (only describes the conditions when the sample was taken) on June 21, 2021 at inflow 1, inflow 2, and outflow and July 6, 2021 at inflow 3. The water sample that has been taken is tested at the Yogyakarta Health and Calibration Laboratory using the standard Government Regulation no. 22 of 2021. The limited parameters used in testing water quality are temperature, Total Suspended Solid (TSS), pH, Chemical Oxygen Demand (COD), Biological Oxygen Demand (BOD), Dissolved Oxygen (DO), Total-P, and Total N.

Figure 2. Aerial Map of Sampling Locations and Inflow Locations of Cengklik Reservoir (June 2021)

Based on the study [4] explained that the input of organic matter carried through the waste produced by human activities will enter the waters and under certain conditions will disrupt the existing waters.

IOP Conf. Series: Earth and Environmental Science **986** (2022) 012074 doi:10.1088/1755-1315/986/1/012074

Therefore, the location selection was chosen with the assumption that the incoming pollutant sources affect the eutrophication of the Cengklik Reservoir. Inflow locations 1 and 2 come from agricultural drainage, while inflow 3 locations come from the Pepe River, which is a non-point source of pollution.

2.2. Pollution Index Method

In this research using the Pollution Index Method in determining the status of water quality. This method is in accordance with [5] about Guidelines for Determining Water Quality Status. PI analysis is carried out with the formula:

$$PI = \sqrt{\frac{(C_i/L_i)^2{}_R+(C_i/L_i)^2{}_M}{2}} \quad (1)$$

where L_i is water quality standart, C_i is the result of water quality test and PI is pollution index. Calculation of pollution index (PI) for parameters: pH, TSS, DO, BOD, COD, Phosphat, and Total N. Furthermore, the determination of the Water Quality Index based on the value is categorized as follows **Table 1.**

Table 1. Pollution Index

PI_j value	Category
$0 \leq PI \leq 1$	Good conditions
$1 \leq PI \leq 5$	Lightly polluted
$5 \leq PI \leq 10$	Moderately polluted
$PI \geq 10$	Heavily polluted

3. Results and Discussions

3.1. Water Quality

Water quality assessment is needed to describe the suitability of water based on the reservoir water quality standard regulated in Government Regulation No. 22 years 2021. Based on its designation, the Cengklik reservoir is used for irrigation and fisheries activities so that it is classified according to [6], it is classified as water class III. The results of water quality testing are described in **Table 2** and **Figure3.**

Table 2. Cengklik Reservoir Water Quality Test Results

No.	Parameter	Units	Water Quality Standart	Outflow (21/6/21)	Inflow 1 (21/6/21)	Inflow 2 (21/6/21)	Inflow 3 (6/7/21)
1	Air Temperature	^{0}C	±3 from air temperature	24	22,9	22,7	22,6
2	Sample Temperature	^{0}C		24,1	25,2	28,7	24,1
3	pH	-	6-9	7,96	7,92	7,89	8,00
4	DO	mg/L	minimum 3	5	5,7	6,42	5,11
5	TSS	mg/L	100	51	17	18	19
6	COD	mg/L	40	34,441	25,258	28,473	12,075
7	BOD	mg/L	6	4,96	2,71	3,27	0,86
8	Phospat	mg/L	0,1	0,166	0,166	0,166	0,723
9	Total-N	mg/L	1,9	0,495	0,499	0,469	0,44

Based on **Table 2** and **Figure 3**, it can be seen that at the outflow point the water pollution is higher than the inflow. This is indicated by an increase in the value of TSS, COD and BOD, as well as a decrease in DO. At the inflow point, it is known that inflow 2 has more polluted water quality compared to inflow 1 and 3. However, inflow 3, water pollution contains the highest phosphate. Phosphate level of 0.025 – 0.1 mg/L is a high level of water fertility and affects algae growth, while the maximum acceptable phosphate level in waters so as not to cause eutrophication of waters is 0.1 mg/L [7]. Based on this theory, inflow 3 has a phosphate value that exceeds 0.1 mg/L so it is known as the cause of eutrophication in the Cengklik Reservoir.

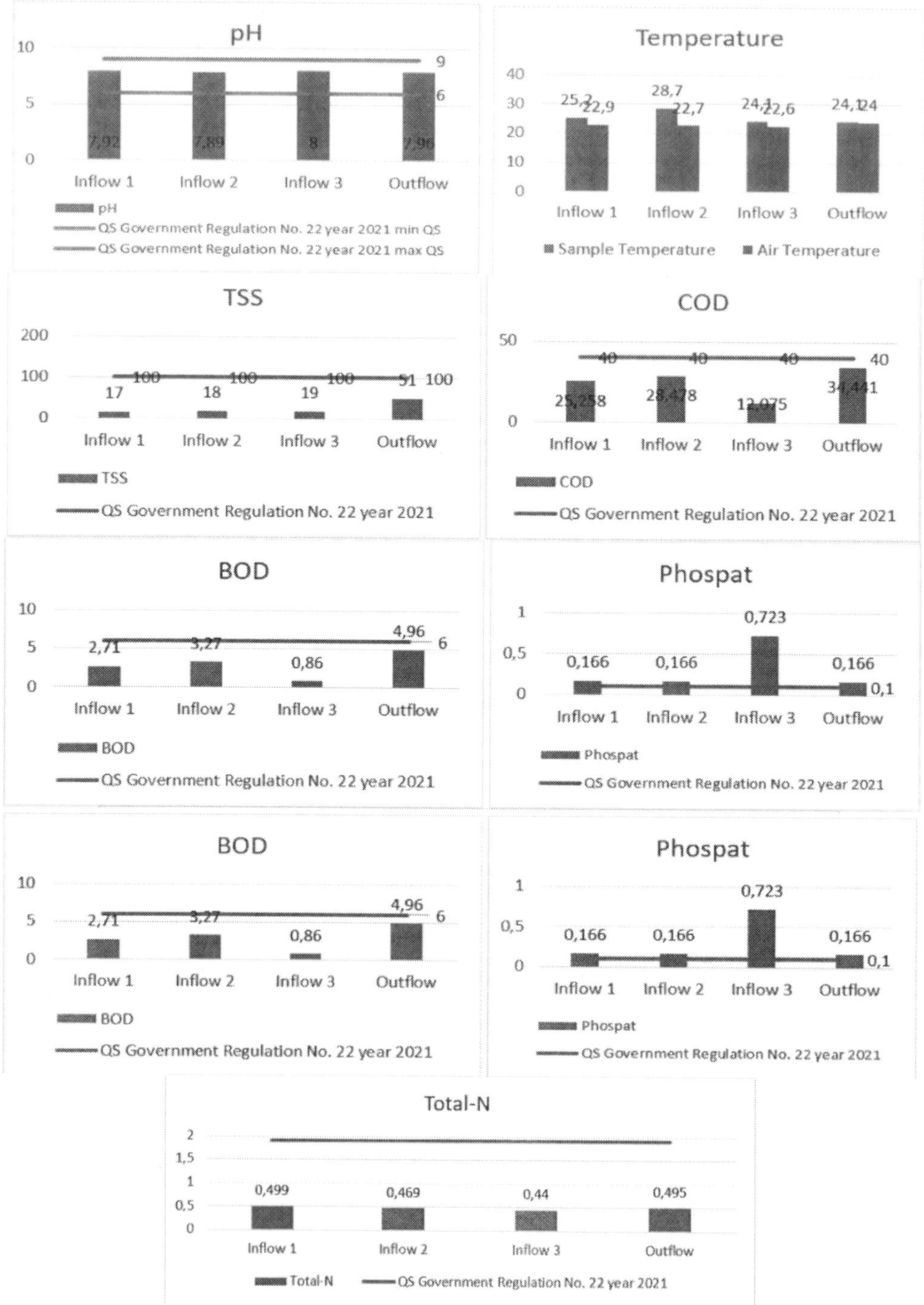

Figure 3. Results of Water Quality Test

Table 3. Calculation of Pollutant Index

(a) Pollution Index

Location	*PI*	Polluted
Inflow 1	1,426	lightly
Inflow 2	1,423	lightly
Inflow 3	1,442	lightly
Outflow	1,392	lightly

(b) Inflow 1

Parameter		Ci	Li	Ci/Li	Ci/Li new
pH	-	7,92	6-9	0,389	0,389
DO	mg/L	5,7	3	0,108	0,108
TSS	mg/L	17	100	0,17	0,17
COD	mg/L	25,258	40	0,631	0,631
BOD	mg/L	2,71	6	0,452	0,452
Phospat	mg/L	0,166	0,1	1,66	2,101
Total-N	mg/L	0,499	1,9	0,263	0,263
				(Ci/Li)2R	0,345
				(Ci/Li)2M	4,412
				PI	1,426

(c) Inflow 2

Parameter		Ci	Li	Ci/Li	Ci/Li new
pH	-	7,89	6-9	0,351	0,351
DO	mg/L	6,42	3	0,048	0,048
TSS	mg/L	18	100	0,18	0,18
COD	mg/L	28,473	40	0,712	0,712
BOD	mg/L	3,27	6	0,545	0,545
Phospat	mg/L	0,166	0,1	1,66	2,101
Total-N	mg/L	0,469	1,9	0,247	0,247
				$(Ci/Li)^2R$	0,364
				$(Ci/Li)^2M$	4,412
				PI	1,423

(d) Inflow 3

Parameter		Ci	Li	Ci/Li	Ci/Li new
pH	-	8	6-9	0,500	0,500
DO	mg/L	5,11	3	0,158	0,158
TSS	mg/L	19	100	0,190	0,190
COD	mg/L	12,075	40	0,302	0,302
BOD	mg/L	0,86	6	0,143	0,143
Phospat	mg/L	0,723	0,1	7,230	2,101
Total-N	mg/L	0,44	1,9	0,232	0,232
				$(Ci/Li)^2R$	0,268
				$(Ci/Li)^2M$	4,412
				PI	1,442

(e) Outflow

Parameter		Ci	Li	Ci/Li	Ci/Li new
pH	-	7,96	6-9	0,442	0,442
DO	mg/L	5	3	0,167	0,167
TSS	mg/L	51	100	0,51	0,51
COD	mg/L	34,441	40	0,861	0,861
BOD	mg/L	4,96	6	0,827	0,827
Phospat	mg/L	0,166	0,1	1,66	2,101
Total-N	mg/L	0,495	1,9	0,261	0,261
				$(Ci/Li)^2R$	0,534
				$(Ci/Li)^2M$	4,412
				PI	1,392

3.2. Pollution Index
Analysis of the PI is intended to determine the level of pollution from waters based on water quality standards. This method can provide recommendations for decision making water body quality management according to its designation. The calculation of the water PI is in Table 3. Based on Table 3. it is known that the pollutant index value in the inflows and outflow is included in the lightly polluted category. The PI value in inflow 3 is the largest value compared to other flows, this is due to the high phosphate content.

3.3. Cengklik reservoir management recommendations
The source of nutrient pollutant load entering the reservoir causes excessive algae population growth and significantly affects the dissolved oxygen concentration and thus inhibits the growth of aquatic organisms [8,9]. Tested water quality analysis,, the phosphate value in all inflows is very high, especially inflow 3 which comes from non-point sources so that the source point is unknown. Fink 2018 [10] said the source of nutrient phosphate that enters the waters comes from the use of inorganic fertilizers, livestock waste and domestic waste (septic tanks, latrines, and others). Nutrient loads that enter the waters can be reduced in various ways, such as treating waste before it is discharged into the reservoir stream, for example by providing communal WWTPs (Waste Water Treatment Plants) and phytoremediation.

4. Conclusion and Suggestion
Phosphate content in the waters of the Cengklik Reservoir is >0.1 mg/L at all location points and exceeds the maximum allowable total phosphate limit in the waters, thus indicating the cause of rapidly growing eutrophication in the Cengklik Reservoir. However, based on calculations using the PI method, it is stated that the water quality of the Cengklik Reservoir is classified as lightly polluted and is still in accordance with its utilization. Further research on Trophic State Index analysis needs to be carried out to determine the severity of Cengklik Reservoir eutrophication so that recommendations for management can be formulated appropriately.

Acknowledgments
This research was funded by Universitas Gadjah Mada, with a Recognision Final Project (RTA) grant.

References

[1] Sharip Z, Zaki A T A, Shapai M A H M, Suratman S and Shaaban A J 2014 Lakes of Malaysia: Water quality, eutrophication and management *Lakes Reserv. Res. Manag.* **19** 130–41
[2] Nurhayati A, Herawati T, Nurruhwati I and Handaka A 2018 STRATEGI PENGELOLAAN ADAPTIF BERBASIS KELEMBAGAAN LOKAL SUMBER DAYA PERAIRAN (Studi Kasus di Perairan Umum Waduk Cirata Kab Cianjur , Jawa Barat) *J. Kebijak. Sosek* **8** 63–75
[3] Lin S S, Shen S L, Zhou A and Lyu H M 2021 Assessment and management of lake eutrophication: A case study in Lake Erhai, China *Sci. Total Environ.* **751** 141618
[4] Rolle Longley K, Huang W, Clark C and Johnson E 2019 Effects of nutrient load from St. Jones River on water quality and eutrophication in Lake George, Florida *Limnologica* **77** 125687
[5] Menteri Negara Lingkungan Hidup 2003 *Keputusan Menteri Negara Lingkungan Hidup Nomor 115 Tentang Pedoman Penentuan Status Mutu Air*
[6] Anon 2021 *Peraturan Pemerintah Nomor 22 Tahun 2021 tentang Pedoman Perlindungan dan Pengelolaan Lingkungan Hidup* (Indonesia)
[7] Wimalawansa S A and Wimalawansa S J 2015 Protection of Watersheds, and Control and Responsible use of Fertiliser to Prevent Phosphate Eutrophication of Reservoirs *Int. J. Res. Environ. Sci.* **1** 1–18
[8] Liu W wen, Zhao E, Kuo Y M and Jang C S 2017 Identifying the relationships between trophic

states and their driving factors in the Shihmen Reservoir, Taiwan *Limnologica* **64** 38–45
[9] Costa J A da, Souza J P de, Teixeira A P, Nabout J C and Carneiro F M 2018 Eutrophication in aquatic ecosystems: a scientometric study *Acta Limnol. Bras.* **30**
[10] Fink G, Alcamo J, Flörke M and Reder K 2018 Phosphorus Loadings to the World's Largest Lakes: Sources and Trends *Global Biogeochem. Cycles* **32** 617–34

Study on the Application of Phytoremediation of Phosphate Content to Eutrophication in Cengklik Reservoir, Boyolali Regency

A Kurniawan[1,*] K Khasanah[2], F N M Jayatri[2]

[1] Department of Civil Engineering, Vocational College, Universitas Gadjah Mada, Yogyakarta 55281, Indonesia
[2]Magister of Technology for the Sustainable Development of Postgraduate Schools, Universitas Gadjah Mada, Yogyakarta 55281, Indonesia

[*] Email: adhy_kurniawan@ugm.ac.id

Abstract. Cengklik Reservoir is a reservoir for water from agricultural irrigation flows that use chemical fertilizers so that there is an increase in nutrients in the waters. Eutrophication is an environmental problem caused by an increase in nutrients such as nitrogen and phosphorus in aquatic ecosystems. Phytoremediation is a technique that is more efficient and economical than conventional techniques in degrading excess nutrients in polluted air. In phytoremediation, plants absorb air pollutants such as $NO3^-$, $NH4^+$ and $PO43^-$ directly. The purpose of this study was to examine the potential of aquatic plants in wastewater phytoremediation and provide recommendations for phytoremediation potential for the waters of the Cengklik Reservoir. The method used in this research is Systematic Literature Review (SLR). Based on the research that has been done, the Pistia stratiotes plant has been widely used as agricultural and domestic wastewater treatment. Pistia stratiotes has the potential to remove 80-95% of the phosphate content. So that the Pistia stratiotes plant is a suitable plant to be applied in the waters of the Cengklik Reservoir. The phytoremediation application was carried out in an inflow reservoir with a plant cover area of 75% of the phytoremediation area, and harvesting was carried out on the 14th day of treatment. The harvest can be utilized or managed by the community or related agencies to become a bioenergy generator.

1. Introduction

Cengklik Reservoir, which is located in Ngargorejo Village, Ngemplak District, Boyolali Regency, Central Java Province, is a water where most of its utilization is used to support agricultural and fishery irrigation. According to [1,2] agricultural activities in Indonesia generally use pesticides and chemical fertilizers, which in the long term can result in a buildup of nitrogen and phosphate in the soil. Surface runoff will bring nutrient content to reservoir waters through irrigation water. This causes the water quality to be disturbed and accelerates the occurrence of Eutrophication. Disturbed aquatic ecosystems due to algae blooms can inhibit utilization activities around the waters [3].

Several studies on water quality control have been carried out, [1,2,4] explained that phytoremediation is a biological method that aims to reduce water pollutants in an effective and environmentally friendly manner. This method utilizes plants as bioaccumulation of contaminants by absorbing NO3-, NH4+ and PO43−. Currently, more and more people are using phytoremediation for the treatment of polluted water because it is economical and has less impact than physical or chemical treatment. Based on the problems described previously, this study aims to examine the potential of aquatic plants in wastewater phytoremediation and provide recommendations for phytoremediation opportunities for the Cengklik Reservoir.

2. Method

The method that will be used in this research is Systematic Literature Review (SLR). Literature review is important and is the foundation in the academic field. The SLR is used to map and evaluate the literature, to identify potential research gaps, and to identify knowledge limits. This is usually done through an iterative cycle to determine the appropriate search terms, search the literature, and then complete the analysis. SLR has three stages, namely planning, implementation, and reporting. The planning stage in this research is carried out by ensuring what is needed in the literature study including formulating problems, validating the formulated problems that are still relevant for discussion, analyzing research trends over the last five years by looking for research related to the keywords "eutrophication", "phytoremediation" , "phosphate", and "wastewater" via ScienceDirect, Scopus, and others. Then proceed with developing according to the provisions set to conduct a review. At the implementation stage, it is done by searching the literature through a title review, then getting the core of the literature by reviewing abstracts, assessing the quality of the contents of the literature, extracting data, and analyzing and synthesizing data. Furthermore, at the reporting stage, results are reported in the form of research mapping and recommendations for future research opportunities [5].

3. Results and Discussions

3.1. Potential of aquatic plants in phytoremediation

The treatment of polluted water using phytoremediation was chosen because of several advantages including using natural resources, namely plants so that it only requires low capital and energy. But on the other hand, this method works slower than conventional water treatment and contaminants that are absorbed by plants have the potential to enter the food chain (contain toxic) [6] Several studies have tested the effectiveness of plants in aquatic phytoremediation processes and are described in Table 1. Nitrogen and phosphate are the main ingredients that cause eutrophic waters carried by domestic wastewater, wastewater, river water, lakes and laundry waste [7]. Based on the table, it can be seen that the plant Pistia stratiotes (shown in **Figure 1**). has the highest level of effectiveness to reduce pine phosphate and nitrate, this is also mentioned by [6,8]. One of the phytoremediation models carried out by Pistia stratiotes with vertical and horizontal models with a process time of 3 months [9].

Figure 1. *Pistia stratiotes* plant

Table 1. Potential of Aquatic Plants for Wastewater Phytoremediation

Source Waste	Plants	Effectiveness of the absorbed Nutrient	Treatment Time and Process	Writer
Domestic Waste	*Hydrilla verticillata*	44,69% TP; 39,45% NH_3-N; 50,35% NO_3-N; 47,59% NO_2-N; 36,14% COD; 37,16% Mg; 24,39% Ca	1 year (artificial culture system)	[10]
	Alternanthera (Joyweed)	36% TP; 45% TKN; 87% BOD; 78,9 – 83,9% COD; 93% SS; 43% Cu	10 days / sheet flow root level (shefrol bioreactor)	[11]
	Salvinia natans	37% PO4-P; 40% NO2-N; 85,2% TKN; 79% NH4-N; 96,9% BOD; 95% COD	8 months (tanks)	[12]
Laundry	*Azolla microphylla*	19,30% P; 37,19% COD	12 days (batch system)	[13]
Waste Water	*Pistia stratiotes*	95,4% P; 82% BOD; 98,8% TSS; 51% Cl	Vertical and horizontal system	[10]
	Pistia stratiotes	81,6% P; 83,3% $NO3^-$; 84,8% NH3	10 days (batch system)	[14]
River water	*Myriophyllum aquaticum*	93,6 TP; 88,3% TKN; 75% BOD; 67,4% COD	30 day (glass tanks)	[6]
	Pistia stratiotes	88% TP; 77% TN; 93% NH_4^+-N; 61,70% COD	6 months (pvc water tanks)	
	Eichornia crassipes	84% TP; 87% TN; 96% NH_4^+N; 68,20% COD		
	Myriophyllum spicatum (Eurasian watermilfoil)	80,55% TP; 77% TN; 93% NH_4^+N; 62,21% COD		

3.2. Recommended Phytoremediation Potential for Cengklik Reservoir Waters

Reservoir is a reservoir for water from a watershed that flows water continuously and gets input of organic and inorganic materials that can pollute the water quality of the reservoir. Therefore, a water quality test was carried out in a preliminary study to analyze the water quality condition of the Cengklik Reservoir. The water quality standard in Indonesia is regulated in Government Regulation Number 22 of 2021, which explains that the water quality standard of the Cengklik Reservoir is classified as group III because it is used for agricultural irrigation, aquaculture, and recreational areas. Table 2 below describes the results of testing the water quality of the Cengklik Reservoir.

Based on table 2, the phosphate values are 0.166 mg/L, respectively; 0.166 mg/L; 0.166 mg/L; and 0.723 mg/L, which levels exceeded the permitted phosphate limit of 0.1 mg/L. The results of this water quality test can be identified as a source of eutrophic waters and environmental quality conditions. High levels of phosphate cause the growth of moss and microalgae in excess which is called eutrophication, so that the water becomes cloudy and smells due to the decay of dead moss. Plants in a eutrophic state can use oxygen in rivers or ponds at night or when plants die and are digesting and during the day the sunlight entering the waters will decrease, so that the photosynthesis process that can produce oxygen is also reduced [10]. In addition, [11] also mentions that environmental factors are important factors in the phytoremediation process, such as pH. The pH factor is important in phytoremediation because it affects the solubility of nutrients that cause plant growth. A good pH condition for the absorption of phosphate by plants is between 6-7. Below or above this number, the absorption of elemental phosphate will be disturbed.

Table 2. Cengklik Reservoir Water Quality Test Results

No.	Parameter	Units	Water Quality Standart	Outflow (21/6/21)	Inflow 1 (21/6/21)	Inflow 2 (21/6/21)	Inflow 3 (6/7/21)
1	Air Temperature	^0C	±3 from air	24	22,9	22,7	22,6
2	Sample Temperature	^0C	temperature	24,1	25,2	28,7	24,1
3	pH	-	6-9	7,96	7,92	7,89	8,00
4	DO	mg/L	minimum 3	5	5,7	6,42	5,11
5	TSS	mg/L	100	51	17	18	19
6	COD	mg/L	40	34,441	25,258	28,473	12,075
7	BOD	mg/L	6	4,96	2,71	3,27	0,86
8	Phospat	mg/L	0,1	0,166	0,166	0,166	0,723
9	Total-N	mg/L	1,9	0,495	0,499	0,469	0,44

The application of phytoremediation in the waters of the Cengklik Reservoir is carried out in the inflow section of the reservoir, which can be seen in Figure 2. The determination of the phytoremediation area is based on the water source point of the Cengklik Reservoir, where the phosphate value content exceeds the water quality standard. Pistia stratiotes plants are stocked directly on the inflow that enters the reservoir. To reduce the distribution of Pistia stratiotes to the reservoir waters, an isolation type waste filter will be made. The equipment functions as an insulator so that the distribution of Pistia stratiotes is limited to certain areas. In this phytoremediation process, the discharge aspect is a very important factor because it will affect the length of stay of water. Where the length of stay of water will affect the length of time the plant absorbs pollutants in the water. Based on [11] the longer the absorption time, the greater the pollutants absorbed by aquatic plants. However, this factor applies if the aquatic plants have not reached the saturation point. When it has reached the saturation point, aquatic plants will no longer be able to absorb pollutants. This can be used as a guide in determining when aquatic plants should recover.

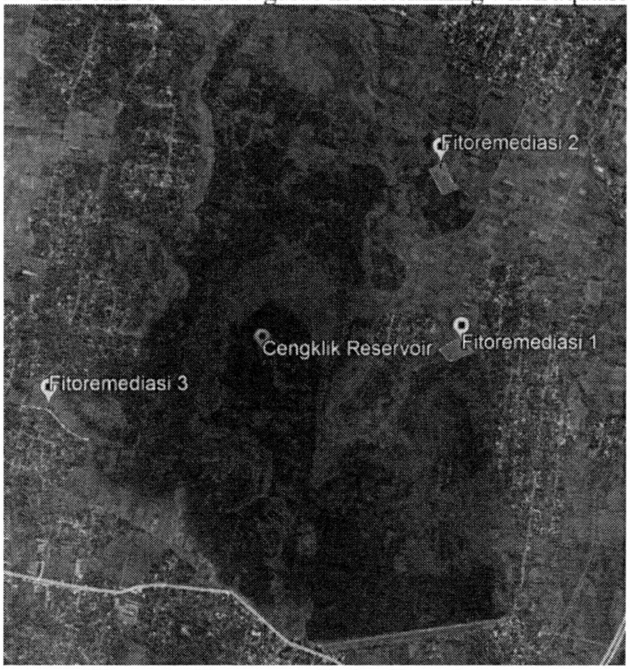

Figure 2. Phytoremediation Location Area

The recovery period or harvest of Pistia stratiotes plants is carried out after 14 days. Based on research conducted by [12] after 14 days of treatment, the phosphate concentration increased, it was suspected that the phosphate precipitated and then dissolved back into the water, resulting in an increase in the phosphate concentration. In addition, the increase in phosphate concentration is thought to be because plants are no longer effective in decomposing organic matter so that phosphate cannot be absorbed by plants. Changes in aquatic plant parts such as leaf wilting of aquatic plants, stem rot, and the growth of new shoots are thought to be caused by a lack of nutrients in wastewater and toxic substances. Changes in leaf color to yellow can be caused by contamination of organic matter. The growth of new shoots is considered a way for plants to survive. According to [13] 100% plant cover causes many plants to turn yellow. This is because there is no room to grow, so the ability of plants to survive and photosynthesis is reduced and eventually turns yellow. Therefore, the application of Pistia stratiotes as a phytoremediation agent in the Cengklik Reservoir is carried out by treating 75% of the area overgrown with aquatic plants, so that there is still room for plants to grow. The results from the recovery or harvesting of the Pistia stratiotes plant in the Cengklik Reservoir can be utilized or managed by the community or related agencies. This has been proven by [14], which produced a total of 9667.33 ml of biogas for 45 days through the biogas processing process from Pistia stratiotes.

4. Conclusions

Pistia stratiotes has an effectiveness level of 80-95% to reduce the phosphate content in wastewater and river water through a phytoremediation process. The application of phytoremediation was carried out in inflow with a plant cover area of 75% of the phytoremediation area, and harvesting was carried out on the 14th day of treatment. The advantages of this process are its availability, it is more economical and easy to implement by distributing it. Harvested from Pistia stratiotes can also be used as biogas. So that this plant can be used as a recommendation to be planted at the inflow point of the Cengklik reservoir to reduce the nitrate and phosphate content that enters the waters.

Acknowledgments

This research was funded by Universitas Gadjah Mada, with a Recognision Final Project (RTA) grant.

References

[1] Retnaningdyah C 2017 The improvement of the quality of polluted irrigation water through a phytoremediation process in a hydroponic batch culture system *AIP Conf. Proc.* **1908**
[2] Lu Q, He Z L, Graetz D A, Stoffella P J and Yang X 2010 Phytoremediation to remove nutrients and improve eutrophic stormwaters using water lettuce (Pistia stratiotes L.) *Environ. Sci. Pollut. Res.* **17** 84–96
[3] Chrismadha T, Suryono T, Magfiroh M, Mardiati Y and Mulyana E 2019 Phytoremediation of Maninjau Lake water using Minute Duckweed (Lemna perpusilla Torr.) *IOP Conf. Ser. Earth Environ. Sci.* **308**
[4] Chen C, Wang F, Hong Y, Liu R and Luo L 2019 The biomass accumulation and nutrient storage of five plant species in an in-situ phytoremediation experiment in the Ningxia irrigation area *Sci. Rep.* **9** 1–9
[5] Xiao Y and Watson M 2019 Guidance on Conducting a Systematic Literature Review *J. Plan. Educ. Res.* **39** 93–112
[6] Lu B, Xu Z, Li J and Chai X 2018 Removal of water nutrients by different aquatic plant species: An alternative way to remediate polluted rural rivers *Ecol. Eng.* **110** 18–26
[7] Ali H, Khan E and Sajad M A 2013 Phytoremediation of heavy metals-Concepts and applications *Chemosphere* **91** 869–81
[8] Rezania S, Taib S M, Md Din M F, Dahalan F A and Kamyab H 2016 Comprehensive review on phytotechnology: Heavy metals removal by diverse aquatic plants species from wastewater *J. Hazard. Mater.* **318** 587–99

[9] Yasar A, Zaheer A, Tabinda A B, Khan M, Mahfooz Y, Rani S and Siddiqua A 2017 Comparison of Reed and Water Lettuce in Constructed Wetlands for Wastewater Treatment *Water Environ. Res.* **90** 129–35

[10] Padmaningrum R T, Aminatun T and Yuliati 2014 Pengaruh Biomasa Melati Air (Echinodorus paleafolius) Dan Teratai (Nyphaea firecrest) Terhadap Kadar Fosfat *Penelit. Saintek* **19** 64–74

[11] Rizky N, Budiyono and Setiani O 2017 PENGARUH VARIASI LAMA KONTAK TANAMAN AZOLLA MICROPHYLLA TERHADAP PENURUNAN KADAR FOSFAT DAN COD PADA LIMBAH LAUNDRY *J. Kesehat. Masy.* **5** 465–72

[12] Audiyanti S and , Zahidah Hasan, Herman Hamdani dan H H 2019 Efektivitas Eceng Gondok (Eichhornia crassipes) dan Kayu Apu (Pistia stratiotes) Sebagai Agen Fitoremediasi Limbah Sungai Citarum *J. Perikan. dan Kelaut.* **10** 111–6

[13] Ni'ma N, Widyorini N and Ruswahyuni 2014 KEMAMPUAN APU-APU (Pistia sp.) SEBAGAI BIOREMEDIATOR LIMBAH PABRIK PENGOLAHAN HASIL PERIKANAN (SKALA LABORATORIUM) *DIPONEGORO J. MAQUARES* **3** 257–64

[14] Pantawong R., Chuanchai A., Thipbunrat P., Unpaprom Y. R R 2015 Experimental Investigation of Biogas Production from Water Lettuce, Pistia stratiotes L. *Emer Life Sci Res* **1** 41–6

Water quality during the rainy seasons and drought seasons in the Garang River Basin (Semarang, Indonesia)

M Hashim[1*], D L Setyowati[2], Suroso[2] and K D I A P Yohanes[2]

[1] Department of Geography and Environment, Faculty of Human Sciences, Universiti Pendidikan Sultan Idris

[2] Department of Geography, Faculty of Social Sciences, Universitas Negeri Semarang

*Email: mohmadisa@fsk.upsi.edu.my

Abstract. The Garang River is the city of Semarang's primary source of water, particularly for drinking water and other domestic needs. This research attempts to examine the current water quality status of the Garang River Basin by using the Malaysian Water Quality Index (WQI). Six water quality parameters were selected and analysed; DO, BOD, pH COD, NH_3-N and TSS. Water sampling was conducted at eight stations based on the Governor Regulation of Central Java No. 156 (2010) on Water Allocation and Water Quality Management of Garang River. Water quality was sampled twice, once during the rainy season (2 January 2020) and once during the drought season (1 September 2020). The findings of the study showed that the water quality status of the Garang River Basin during the rainy season recorded a WQI value of 81 and 92 during the drought season. Both values belong to Class II, which represents clean waters. Three water quality parameters were identified to have exceeded the set standards: the DO, BOD, and NH_3-N. Changes in landuse, particularly the clearance of forest areas for agriculture, plantation activities, domestic waste disposal, and wastewater directly from industrial activity and local inhabitants, have all contributed to the Garang River Basin's deterioration of water quality. However, in the context of water classes, it is shown that the water of the Garang River is suitable to be used as a domestic water supply for local residents but requires further treatment prior to consumption.

1. Introduction

Rivers area main source of water on the earth. They are essential for human and physical environments. Rivers are recognised for their importance to human activities since they are needed for a variety of activities such as home use, agricultural irrigation, and industry. Concurrently, debates over water quality and river quantity are common, as river resources have been declining and becoming polluted due to a variety of human activities [1–4]. Thus, it is unsurprising that water resources issues are becoming a priority in achieving the Sustainable Development Goals (SDGs), specifically in item six on Clean Water and Sanitation.

Generally, issues concerning river quality are prioritised in the tropics because they affect the quality of human life. Although river resources are primarily used for agricultural irrigation, they are still important for domestic human needs such as water to drink, cooking, and washing. Many studies have shown that land development activities cause significant river pollution in these tropical regions through the development new agricultural areas, population settlements [2,5–7], industrial waste discharge and heavy metals disposal [3,7,8] landfill areas [9,10], mining activities [1], and domestic waste disposal such as rubbish [5,11,12].

The Garang River gets its water from a drainage basin, considered an open system that is constantly threatened unless it is managed properly and effectively. Furthermore, a number of human activities that have an impact on the amount and quality of water in the Garang River Basin are constantly threatening the Garang River Basin. The Garang River provides a significant source of domestic water supply for Semarang City residents. Many previous studies from various disciplines and research interests have studied the Garang River for its water quantity, as well as quality and flood issues[13–17]. The aim of this article is to examine the level of water quality in the Garang River during the rainy seasons and drought seasons; the Malaysian Water Quality Index (WQI) was used to indicate the water quality index of the Garang River. This study was conducted in collaboration between the Sultan Idris Education University, Malaysia and Universitas Negeri Semarang, Indonesia.

2. Methods

2.1. Sampling stations and area

The Garang River Basin situated in the Central Java Province. This basin covers an area of approximately 21,277 hectares (212.7km^2). The Garang River covers three main districts: Semarang City, Semarang District, and Kendal District. The Garang River Basin contains four sub-basins: Garang Hulu, Kreo, Kripik, and GarangHilir (West Flood Canal). Mount Ungaran provides water to the Garang River. Meanwhile, the river flow originates from the Kreo, Kripik, and Garang Hulu rivers, which merge into the Garang River in the basin's downstream part [18].

The region of Central Java has a tropical monsoon climate. As a result, this region has two different seasons: the rainy season and the drought season. According to the Climatology and Geophysics Meteorological Agency, the rainy season lasts from November to April, and the drought season lasts from May to October [19]. These rainy and drought seasons indirectly impact the quality and quantity of water in the Garang River.

Furthermore, the downstream region of the Garang River Basin is frequently flooded, and previous research has indicated that floods are caused by heavy rainfall and that rivers have flash flood characteristics [5]. These situations occur due to the short distance between the upstream and downstream parts (about 35 km). Floods are extremely dangerous because of the increased speed and force of their flow. Due to the river's proximity to the upstream and downstream areas, water flows very swiftly downstream, and if it rains heavily and for an extended period in the upstream area, water reaches the downstream part in less than two hours.

Moreover, the Garang River Basin has been divided into seven (7) sections by the Governor Regulation No. 156 of 2010 on Water Allocation and Water Quality Management of the Garang River in Central Java Province.These segments were created to ascertain permissible landuse activities. Moreover, eight water sampling stations were designated based on the aforementioned rule (See Table 1 and Figure 2). This study scrutinises the Garang River Basin area in its upstream and downstream parts. The study sampled eight stations representing river segments in four distinct sub-catchments, namely the Garang Hulu sub-catchment (Station 1-4), the Kreo sub-catchment (Station 5), the Kripik sub-catchment (Station 6), and the Garang Hilir sub-catchment (Station 7-8). Rapid sampling was carried out to ensure the accuracy of samples taken for immediate laboratory analysis.

Table 1. Names of the quality monitoring stations based on their latitudes and longitudes.

Stations	Location	Latitude	Longitude
1	Hulu Sungai Garang di Dusun Lempuyangan	7°11'16.36"	110°22'36.90"
2	Sungai Garang, Jembatan Sungai Garang, Jalan Pramuka	7°06'32.15"	110°24'05.54"
3	Sungai Garang, Jembatan Sungai Garang Tinjomoyo, Perbatasan Kelurahan Bendan Duwur & Kelurahan Tinjomoyo, Perbatasan Kec. Gajahmungkur & Kec. Banyumanik	7°01'40.87"	110°24'07.19"
4	Sungai Garang, Jembatan Tugu Soeharto - Menoreh Tugu	7°01'01.11"	110°23'15.96"
5	Kelurahan Sikopek	7°05'47.1"	110°20'25.4"
6	Jembatan Desa Gisik Sari	7°01'15.5"	110°22'30.8"
7	Bendung Simongan	6°59'27.01"	110°24'08.77"
8	Jembatan Arteri Yos Sudarso	6°57'18.4"S	110°23'53.73"

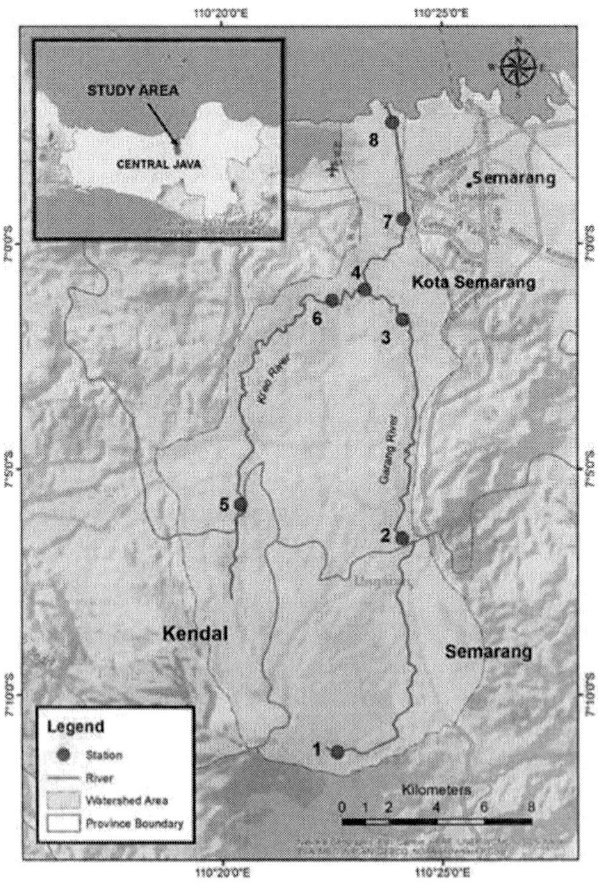

Figure 1. Monitoring station's location in the Garang River Basin.

2.2. Sampling activities and analysis

Water samples of the Garang River were collected twice, once during the rainy season on 2 January 2020 and once during the drought season on 1 September 2020. The American Public Health Association (APHA) advocated standard techniques in this field. [20]. Six water quality parameters were used, namely Dissolved Oxygen (DO), Chemical Oxygen Demand (COD), Total Suspended Solids (TSS), Ammoniacal Nitrogen (NH_3-N), Biochemical Oxygen Demand (BOD) and pH. Water samples from eight monitoring stations were gathered and analysed at the Health Laboratory and Medical Equipment Testing Hall, Health Department, Central Java Province.

The purpose of this study was to assess the level and state of water quality in the Garang River using Malaysian water quality standards known as the DOE Water Quality Index Classification (DOE WQIC). This index was based on the National Water Quality Standard for Malaysia (NWQSM). It was conducted to compare with the Indonesian Government's water quality standards, specifically the Governor Regulation Number 82 of 2001dated 14 December 2001 on Water Quality Management and Water Pollution Control. The Pollution Index (PI) classifies water quality into four categories: Good, Lightly Polluted, Moderately Polluted, and Heavily Polluted. Meanwhile, the status of river quality in Malaysia is classified into three categories: Clean, Slightly Polluted, and Polluted. Based on a study conducted by Ujianti et al. [21], the water quality of Garang River was PI 1.23, which was considered Lightly Polluted.

Six water quality characteristics were chosen because they correspond to those used by the Malaysian DOE to determine river quality status, also known as the NWQSM. As stated in Table 2, the NWQSM was utilised to classify water quality into five categories and assess the current status of the water. Table 2 indicates the range of subindex values (SI) for each water quality index (WQI) parameter. The WQI was derived using the Formula 1 established by the Malaysian DOE, which is as follows:

$$WQI = (0.22xSIDO) + (0.19xSIBOD) + (0.16xSISS) + (0.16xSICOD) + (0.12xSIpH) + (0.15xSIAN) \quad (1)$$

where:

SIDO = subindex DO (%saturation)
SIBOD = subindex BOD
SISS = subindex SS
SICOD = subindex COD
SIpH = subindex pH
SIAN = subindex NH_3-N
$0 \leq WQI \leq 100$

To simplify the computation of the WQI, the SI values were obtained from the River Water Quality Index Calculator via a dedicated website. Five WQI classifications were used: Very Good - Class I, Good - Class II, Moderate - Class III, Polluted - Class IV, and Heavily Polluted - Class V. Based on these water classes, the use of water resources for various human activities could be determined as shown in Table 3. At the same time, the Malaysian DOE established the WQI for a body of water, particularly rivers, and classified it as Clean (81-100%), Slightly Polluted (60-80%), and Polluted (0-59%) by the DOE, WQC, and WQI. All water quality parameter data were analysed using descriptive statistics to determine the mean and standard deviation of each water quality parameter. The data for each parameter is presented in the form of tables and graphs to aid in the explanation of the data and to illustrate whether the parameters exceed or do not exceed the established standards.

Table 2. DOE Water Quality Index Classification.

Parameter	Unit	Class				
		I	II	III	IV	V
NH_3-N	mg/l	<0.1	0.1-0.3	0.3-0.9	0.9-2.7	>27
COD	mg/l	<10	10-25	25-50	50-100	>100
BOD	mg/l	<1	1-3	3-6	6-12	>12
pH	-	>7.0	6.0-7.0	5.0-6.0	<5.0	>5.0
DO	mg/l	>7	5-7	3-5	1-3	<1
TSS	mg/l	<25	25-50	50-150	150-300	>300
WQI		>92.7	76.5-92.7	51.9-76.5	31.0-51.9	<31.0

Source: DOE [24]

Table 3. Water classes and water uses.

Class	Uses	
Class I	Conservation of natural environment. Water Supply I – Practically no treatment necessary.	Fishery I – Very sensitive aquatic species.
Class IIA	Water Supply II – Conventional treatment required.	Fishery II – Sensitive aquatic species.
Class IIB	Recreational use with body contact.	
Class III	Water Supply III – Extensive treatment required.	Fishery III – Common, of economic value and tolerant species; livestock drinking.
Class IV	Irrigation.	
Class V	None of the above.	

Source: DOE [22]

3. Results and Discussions

The mean and standard deviations of the Garang River parameters are shown for each monitoring station. Table 4 and Figure 2 show the concentrations of six water quality parameters observed at eight monitoring stations along the Garang River during the rainy seasons and drought seasons. Table 5 compares the water quality of the Garang River during the rainy seasons and drought seasons. Figure 4 depicts information about the water quality index of Garang River during the rainy seasons and drought seasons.

3.1. Analyses of the GarangRiver's water quality based on parameters

3.1.1. Dissolved Oxygen (DO)

The DO is a term that refers to the total amount of oxygen contained in a body of water, such as a river, at any given time and is typically proportional to the river's current velocity [23]. This parameter indicates the current level of water in the reservoir and the requirements of aquatic living ecosystems. The higher the DO value, the better the water quality. Fundamentally, the required DO content of clean water is between 8 mg/l and 10 mg/l. According to the DOE [22], the DO value that is suitable for any use is between 3 and 5 mg/l in Class III. The findings showed that the average DO value during the rainy season was 4.81 mg/l, compared to 6.70 mg/l during the drought season. A low DO value in the water indicates that the aquatic environment is being harmed. It was found that during the rainy season, the DO content was lower than during the drought season. Station 4 recorded the lowest DO value with 2.01 mg/l during the rainy season and 6.14 mg/l at Station 8 during the drought season. Overall, DO values were recorded as Class III during the rainy season and Class II during the drought season.

3.1.2. Biochemical OyxgenDemand (BOD)

The BOD is a parameter that indicates the quantity of oxygen needed to sustain the stability of organic matter during decomposition [24]. A high BOD value indicates that a body of water has a high organic matter content, which decreases DO levels, which eventually leads to the death of some aquatic life. The DOE [22] has established Class III water quality standards for BOD parameters ranging from 3 mg/l to 6 mg/l. The study's findings explained that the average BOD values during the rainy and drought seasons were suitable for any use. The average BOD value during the rainy season was 3.61 mg/l (Class III), while the average BOD value during the drought season was 0.92 mg/l (Class I). These results showed that the BOD parameter performed better during the drought season, registering as Class I rather than Class III during the rainy season. Station 4 recorded a high BOD value of 7 mg/l (Class IV) during the rainy season.

3.1.3. Chemical Oxygen Demand (COD)

Meanwhile, the COD parameter used to determine the amount of oxygen required for the decomposition process. In contrast to BOD measurements, the COD measurement entails the chemical decomposition of organic matter while simultaneously producing carbon dioxide. A high concentration of organic matter and pollutants in a body of water represents a high COD value. The term "organic matter" refers to material that microbes cannot decompose, such as domestic sewage wastes, agricultural activities, rainy markets, and industrial waste discharge [1,22]. The DOE [24] also specifies a COD value of between 25 mg/l and 50 mg/l for Class III. A high COD value indicates that the water is highly polluted. The study's findings indicated that the COD values were 11.09 mg/l (Class II) during the rainy season and 3 mg/l (Class I) during the drought season. Regardless of the season, all stations recorded COD values that were within Class I and Class II ranges.

3.1.4. pH

The pH value is measured on a body of water to determine whether it is acidic or alkaline. The pH scale ranges from 0 to 14, with a value closer to 0 indicating acidic water; the more the value approaches the pH of 14, the more the water becomes alkaline. The neutral pH value is 7. According to Abdul Maulud et al. [1], the solubility and availability of nutrients required by aquatic organisms may be influenced by pH value. The DOE [22] has established that the appropriate value for the pH parameter in water contents must be between 5.0 and 6.0 in Class III. Aquatic life can survive in a pH range of 6.5 to 9 [22]. The study's findings showed that the average pH readings throughout the rainy seasons and drought seasons were 7.42 and 8.05, respectively, in Class I. These values indicate that seasonal influences on the pH value of Garang River's water quality are negligible.

3.1.5. Total Suspended Solids (TSS)

The TSS is a measure to determine the concentration of a solid in water composed of organic or inorganic particles, is insoluble in water, and has a minimum particle size of 0.001 mm. This TSS may be composed of various dissolved materials, including silt, dead plants, sewage effluent, industrial waste, and other waste discharges. Water with a low TSS value is critical for all types of aquatic life. The primary source of TSS in the river is surface water runoff from landuse changes such as the clearing forest areas for agricultural, settlement, and logging activities [1,25,26]. The DOE [24] has developed a standard for the TSS parameter that should be present in rivers ranging from 50 mg/l to 150 mg/l (Class III). The study's findings showed that the TSS value in the Garang River was in a good condition, with all stations recording Class I conditions during the rainy and drought seasons, except for Station 4 (25mg/l, Class II). The average readings reported during the rainy season were 17.13 mg/l and 1.63 mg/l during the drought season. These results indicate that TSS water quality is better during the drought season than during the rainy season.

IOP Conf. Series: Earth and Environmental Science **986** (2022) 012076 doi:10.1088/1755-1315/986/1/012076

3.1.6. NH_3-N

The amount of NH3-N in a river is measured to determine whether it has been contaminated by sewage effluent, agricultural activities, or livestock consumption [24]. A high NH_3-N concentration value in the river indicates the presence of pollutants in the river. The presence of NH_3-N in water can harm human health and aquatic life [26]. The DOE [22] has developed the standard NH_3-N value for Class III ranging from 0.3 mg/l to 0.9 mg/l. The study's findings showed that the NH_3-N value of the rainy and drought seasons belonged to Class III, making the water suitable for any general use. During the rainy season, the NH_3-N value was 0.34 mg/l (Class III) and during the drought season it was 0.26 mg/l (Class II). Station 6 recorded a high value of 1.07 mg/l (Class IV) during the rainy season, and Station 8 recorded the highest value of 1.05 mg/l (Class IV) during the drought season.

Table 4. Six water quality parameters were measured in the Garang River Basin.

Stations	DO (mg/l)		BOD (mg/l)		COD (mg/l)	
	Rainy Season	Drought Season	Rainy Season	Drought Season	Rainy Season	Drought Season
1	6.67	7.38	2	1	6.2	3
2	5.80	7.14	3	1	9.3	3
3	6.93	7.69	1	1	3.09	3
4	2.01	6.44	7	0.8	22	3
5	6.76	6.29	1	0.9	3.09	3
6	3.09	6.29	6	0.7	18	3
7	3.86	6.21	4	0.98	12	3
8	3.38	6.14	4.9	1	15	3
Mean±SD	4.81±1.95	6.70±0.61	3.61±2.26	0.92±0.11	11.09±6.95	3.00±0.00

Stations	pH		TSS (mg/l)		NH_3-N (mg/l)	
	Rainy Season	Drought Season	Rainy Season	Drought Season	Rainy Season	Drought Season
1	7.17	7.89	1	1	0.03	0.01
2	7.32	8.17	3	1	0.098	0.05
3	7.83	8.01	14	1	0.03	0.004
4	7.56	8.29	25	2	0.1	0.01
5	7.7	7.98	21	2	0.03	0.03
6	7.17	7.58	7	2	1.07	0.4
7	7.32	8.12	22	2	0.153	0.5
8	7.28	8.37	44	2	1.18	1.05
Mean±SD	7.42±0.25	8.05±0.25	17.13±14.12	1.63±0.52	0.34±0.49	0.26±0.38

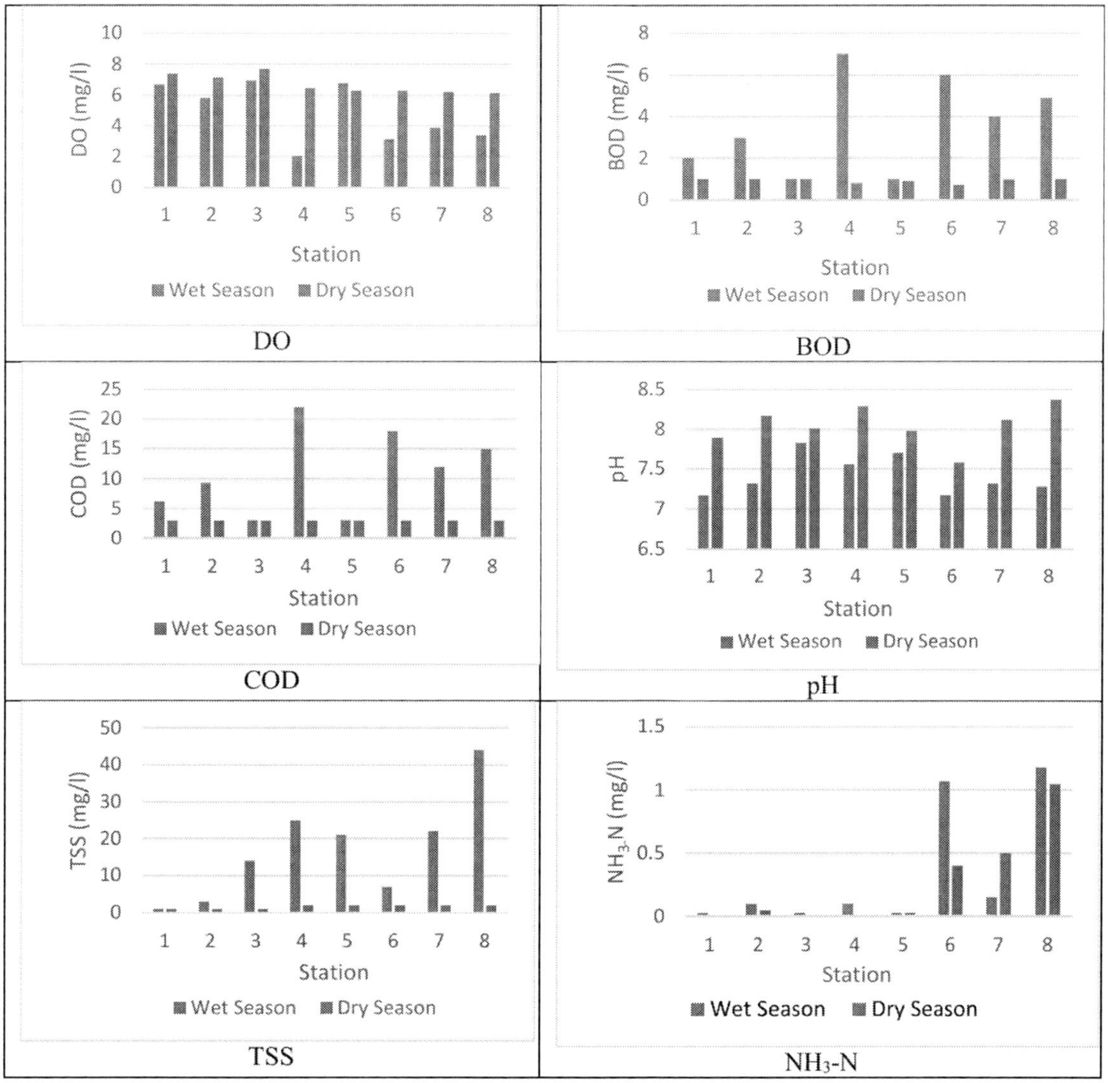

Figure 2. Comparison of quality parameters during the rainy and drought seasons in the Garang River Basin

3.2. Water quality index of the Garang River Basin

The WQI values were calculated for each station and season to determine the Garang River's water quality index. Using Formula 1, the WQI value was determined at each station with reference to the DOE Water Quality Classification based on WQI, which has range indices for a Clean (81-100), Slightly Polluted (60-80), and Polluted (0-59) status. These range indices were developed based on six parameters that had been used before. According to Table 5 and Figure 3, the WQI in the Garang River Basin during the rainy season was 81 (Class II), corresponding to a clean status. Only Stations 1, 2, 3, and 5 had WQI values classified as Clean. Meanwhile, Stations 4, 6, 7, and 8 were classified as Slightly Polluted. Indeed, Stations 1, 3, and 5 recorded WQI values greater than 90, classifying them as Class I.

During the drought season, the WQI of the Garang River Basin was higher than during the rainy season. It was found that theWQI value of the Garang River Basin during the drought season was classified as Clean with a value of 92 (Class II). Many stations recording the WQI values were within

IOP Conf. Series: Earth and Environmental Science **986** (2022) 012076 doi:10.1088/1755-1315/986/1/012076

Class I, specifically Stations 1 to 5. Class II waters were detected in stations in the downstream areas, particularly at Stations 6 to 8. However, all stations were recorded as Clean during the drought season. The findings of the study show that the water quality in the Garang River Basin was superior during the drought season compared to the rainy season. While natural elements such as weather and climate cannot be used to improve river's quality, the diversity of human activities, regardless of the rainy or drought season, has a significant impact on the river quality.

Table 5. Water quality index of the Garang River Basin by station and season.

	Rainy Season			Drought Season		
Stations	**WQI**	**Class**	**Status**	**WQI**	**Class**	**Status**
1	94	I	Clean	96	I	Clean
2	88	II	Clean	95	I	Clean
3	94	I	Clean	97	I	Clean
4	67	III	Slightly Polluted	94	I	Clean
5	93	I	Clean	93	I	Clean
6	67	III	Slightly Polluted	89	II	Clean
7	77	II	Slightly Polluted	88	II	Clean
8	66	III	Slightly Polluted	84	II	Clean
Mean	**81**	**II**	**Clean**	**92**	**II**	**Clean**

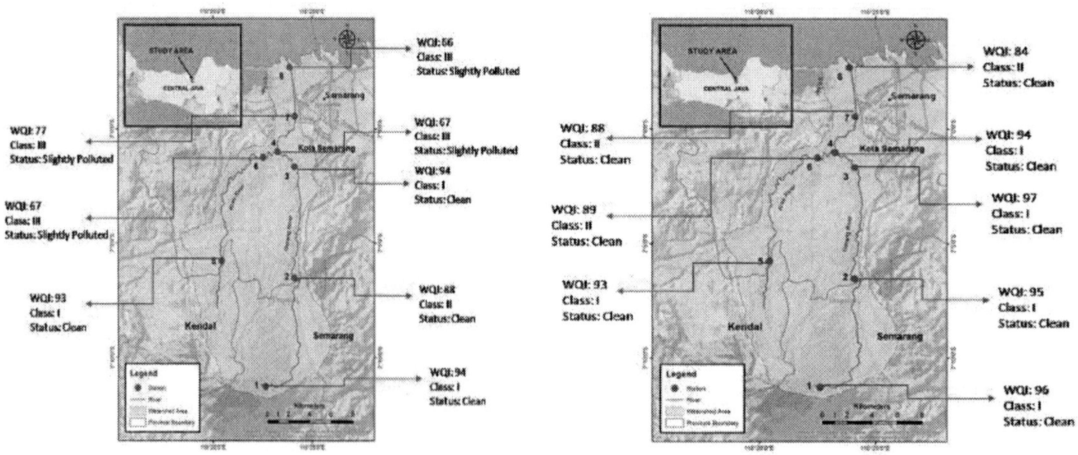

Rainy Season Drought Season

Figure 3. Water quality status of the Garang River Basin during the rainy and drought seasons.

4. Conclusions

According to the DOE's Water Quality Classification based on Malaysian WQI, the water quality status in the Garang River Basin was Clean, with a value of 81 (Class II) during the rainy season and 92 (Class II) during the drought season. In terms of Class II, these indicators show that the Garang River's water quality is suitable for human consumption but requires initial conventional treatment. Furthermore, the Garang River's water quality is ideal for fisheries, particularly those involving sensitive fish species, and recreational activities involving physical contact. Although the Garang River's water quality is deemed to be clean in this study, authorities cannot simply disregard any landuse development activities occurring in the Garang River Basin. Domestic waste discharge, as well as agricultural and industrial activities were identified as the primary contributors towards the deterioration of Garang River's water quality, and these activities should undergo close and regular monitoring.

Acknowledgements
The author conducted this study while on sabbatical leave at the Department of Geography, Faculty of Social Sciences, Universitas Negeri Semarang, Indonesia. The author wishes to thank the Sultan Idris Education University for providing him with a sabbatical leave in Indonesia from 1 December 2019 to 31 March 2020. Appreciation to Dr Koh Liew See for the assistance provided.

References
[1] Abdul Maulud K N, Fitri A, Wan Mohtar W H M, Wan Mohd Jaafar W S, Zuhairi N Z and Kamarudin M K A 2021 A study of spatial and water quality index during dry and rainy seasons at Kelantan River Basin, Peninsular Malaysia *Arab. J. Geosci.* **14**
[2] Xu J, Liu R, Ni M, Zhang J, Ji Q and Xiao Z 2021 Seasonal variations of water quality response to land use metrics at multi-spatial scales in the Yangtze River basin *Environ. Sci. Pollut. Res.* **28** 37172–81
[3] Ujianti R M D, Anggoro S, Bambang A N and Purwanti F 2018 Water Quality and The Heavy Metal Occurence of Fish in Polluted Watershed *Indones. Food Nutr. Prog.* **15** 21
[4] Rezagama A, Sarminingsih A, Sariffudin S, Hariyanto H, Daniswara C R A and Febbyany D G 2021 Prediction of spatial pollution load using the PLOAD model *IOP Conf. Ser. Earth Environ. Sci.* **623**
[5] Setyowati D L, Arsal T, Hardati P, Suroso and Prabowo K Z 2019 Morphoconservation analysis on Kali Garang as a river conservation effort *IOP Conf. Ser. Earth Environ. Sci.* **243**
[6] G P I A and W O W H 2020 Water Air Soil Pollut
[7] Fan X, Cui B, Zhang K, Zhang Z and Shao H 2012 Water Quality Management Based on Division of Dry and Wet Seasons in Pearl River Delta, China *Clean - Soil, Air, Water* **40** 381–93
[8] Rahman et al. 2010 A Study on Selected Water Quality Parameters along the River Buriganga, Bangladesh *Iran. J. Energy Environ.* **1** 81–92
[9] Adi Saputra W, Setiani O and Rahardjo M 2020 Water Quality and Pollution Index of Kreo and Garang River from Jatibarang Landfill in Semarang City *E3S Web Conf.* **202** 1–12
[10] Edhisono S 2015 Jatibarang Final Solid Waste Dumping Site of Semarang City and Its Problems *Waste Technol.* **3** 25–9
[11] Asocadewi G, Oktiawan W and Hadiwidodo M 2015 PENENTUAN STATUS MUTU AIR DENGAN MENGGUNAKAN METODE INDEKS PENCEMARAN (STUDI KASUS: SUNGAI GARANG, SEMARANG) Gessy Asocadewi, Wiharyanto Oktiawan, Mochtar Hadiwidodo *) *J. Tek. Lingkung.* **4** 1–7
[12] Hardati P and Setyowati D L 2019 Population Growth in the Upper Garang Watershed Semarang Regency, Central Java Province, Indonesia *IOP Conf. Ser. Earth Environ. Sci.* **256**
[13] Octaviana A, Prasetyo Y and Amarrohman F J 2020 Analisis Perubahan Nilai Total Suspended Solid Tahun 2016 Dan 2019 Menggunakan Citra Sentinel 2a (Studi Kasus : Banjir Kanal Timur, Semarang) *J. Geod. Undip* **9** 167–76
[14] Haeruddin, Supriharyono, Rahman A, Ghofar A and Iryanthony S B 2020 Spatial distribution and heavy metal pollution analysis in the sediments of garang watershed, Semarang, Central Java, Indonesia *AACL Bioflux* **13** 2577–87
[15] Ujianti R M D and Androva A 2020 Heavy metal toxicity and the influence of water quality in watershed for enhancing fisheries food security *IOP Conf. Ser. Mater. Sci. Eng.* **846**
[16] Setiawan M A and Masduqi A 2019 Environmental carrying capacity of Garang River basin in Central Java Province *IOP Conf. Ser. Earth Environ. Sci.* **259**
[17] Ujianti R M D and Androva A 2019 Analysis of Water Quality and River Waters Microbology for Manifestation of Food Safety *Adv. Sustain. Sci. Eng. Technol.* **1** 1–7

[18] L S D and E S 2011 *DAS Garang Hulu: Tata Air Erosi dan Konservasi* (Semarang: Widya Karya)
[19] Badan Meteorologi Klimatologi & Geofisika 2020 *Prakiraan Musim Hujan 2020/2021 di Indonesia* (Jakarta)
[20] America Public Health Association 2005 *Standard Methods for Examination of Water and Wastewater21nd edition* (Washington)
[21] Ujianti R M D, Anggoro S, Bambang A N and Purwanti F 2018 Water quality of the Garang River, Semarang, Central Java, Indonesia based on the government regulation standard *J. Phys. Conf. Ser.* **1025**
[22] Department of Environment 2019 *Environmental Quality Report 2019* (Malaysia)
[23] Hashim M, Nayan N, Saleh Y, Mahat H, Mat Said Z and Fhei Shiang W 2018 Water quality assessment of former tin mining lakes for recreational purposes in Ipoh city, Perak, Malaysia *Indones. J. Geogr.* **50** 25–33
[24] Shaharudin N, Nayan N, Hashim M, Ngah M S Y C, Saleh Y and Ismail K 2013 Pemetaan Kualiti Air Tasik Untuk Aktiviti Rekreasi Di Negeri Perak : Kajian Kes Taman Tasik Taiping *J. Perspekt.* **5** 31–44
[25] Yusoff W A, Jaafar M, Mohd Khairul A K and TorimaN M E 2015 Land exploration study and water quality changes in Tanah Tinggi Lojing, Kelantan, Malaysia *Malaysian J. Anal. Sci.* **15**
[26] Environment Protection Authority 2017 *Environmental Quality Criteria Reference Document for Cockburn Sound* (Perth)

IOP Conf. Series: Earth and Environmental Science **986** (2022) 012077 doi:10.1088/1755-1315/986/1/012077

Use of drones to determine potential locations of reservoir to anticipate water crisis in Weninggalih Village, Jonggol District, Bogor Regency, West Java

Cahyadi Setiawan[1,*], Ode Sofyan Hardi[1], Fauzi Ramadhoan A'Rachman[1], Oktavia Ariyanti[2], Rio Maulana Fattah[2], Zainul Abidin[2], Mush'ab Baihaqy[2]

[1] Lecturer of Geography Education, Faculty of Social Sciences, Universitas Negeri Jakarta, Indonesia

[2]Student of Geography Education, Faculty of Social Sciences, Universitas Negeri Jakarta, Indonesia

*Email: cahyadi-setiawan@unj.ac.id

Abstract. Weninggalih village experiences drought that keeps recurring every year which has an impact on the people in the area. The purpose of this study was to determine the potential location for the construction of a reservoir to anticipate water crisis in Weninggalih Village, Jonggol Districts, Bogor Regency. The research method used is a survey method to determine the sampling location and quantitative methods are used for analysis. The field survey used a drones for aerial photo data recording. Sampling was determined purposively based on the Digital Elevation Model in three dimensions from the resulting aerial photos. Quantitative methods are used to determine the potential location of the reservoir. The results showed that there were 6.609 photos taken with the drone flying at a height of 150 meters above ground level with 80% overlap. Based on the results show that there are three potential locations of reservoirs. The first potential sites of reservoir is in the area of 423 m^2. The second one potential sites of reservoirs 1.415 m^2 and the last one potential site of reservoirs are in area three land area of 1.191 m^2.

1. Introduction

Water is one of the main needs of living things, especially humans, which need to be fulfilled for their survival. As much as 0.3% of the water found on earth can be used and utilized. In addition, water use increased and at the same time some areas experienced water shortages [1]. In 2015 there were more than 90% of deaths occurred in developing countries caused by diarrheal diseases, this figure is certainly a consideration for the world government in reducing child mortality by improving water and sanitation services [2]. Meanwhile, the average daily need for drinking water is 1.5 liters [3].

Seeing climatic conditions that have changed significantly, of course, has a direct impact on low extreme rainfall levels, causing drought events. According to Kim, et al., (2020)[4] global warming that occurs shows the existence of different seasonal features, so that the distribution of rainfall is uneven in the Asian region. In addition, according to [5], Asia is one of the regions that is very vulnerable to drought due to the monsoon climate and many areas are affected. Based on BNPB data, 28 villages in 11 sub-districts in Bogor Regency experienced a drought in 2019, one of which occurred in Weninggalih Village. [6] The need for water will always increase every year, it follows the rate of population growth. The United Nations states that the increasing demand for water, damage to the soil surface, and damage to water quality can affect the level of conventional water availability [7]. There are three solutions to overcome the global water crisis, namely water infrastructure, market-based regulations and instruments by setting prices and regulating the water market, as well as the use of technology in the form of desalination, precision agriculture, and reuse of water that has been used [8]. The recommended solution to overcome the problem of drought during the dry season in Lembata is by building reservoirs in several watersheds [9]. This solution can also be applied in Weninggalih Village, namely by determining the potential location of the reservoir as a water reservoir. Utilization of reservoirs is used as a water reservoir to anticipate water scarcity [10]. Other functions of water reservoirs are as agricultural water distribution stations, micro hydro power plants, and temporary flood diversion systems.

Generally, the determination of potential locations for the construction of reservoirs is only done traditionally, namely using conventional decision-making or with political interests [11]. The digital era that continues to develop certainly makes it easier for analytical techniques to determine existing potential locations, namely with a geographic information system (GIS). The study uses several parameters to determine potential locations with geological formations, geomorphology, drainage flow density, slope, elevation, rainfall, surface water runoff, total dissolved solids, and large fractures [11]. In another study on the island of Boda Kalvsik, Sweden, to determine the potential location for the construction of a reservoir using the parameters of rainfall, evapotranspiration, recharge potential, piezometric level, geological formation, and geomorphology. The method used in this study is the method of topographic validity index and groundwater balance. The results of the study are favorable for areas that have a humid climate with limited water storage capacity [12].

This research was conducted using the Spatial Multi Criteria Evaluation (SMCE) analysis method to determine the weight value for each parameter and the overlay to determine the exact location for the construction of the reservoir. Three areas were identified as suitable locations for dam construction. The parameters used are rainfall data, drainage streams density, land cover, elevation, slope map, and geology. The purpose of this study was to determine the potential location for the construction of a reservoir in Weninggalih Village by utilizing remote sensing using drones, GIS, and SMCE.

2. Method

Weninggalih Village is located in Jonggol District, Bogor Regency with astronomical lines -6°.464'312'', 107°.102'556''. Bordered by Sirnajati Village (north), Sirnagalih Village and Karang Indah Village (south), Babakan Raden Village and Sukagalih Village (east), as well as Jonggol Village and Sukamanah Village (west) the total area owned is 444,506 Ha, with most of the area being agricultural land. Weninggalih has an average rainfall between 2,700 mm to 2,900 mm based on data for the period 2011 to 2020 [13]. The population of Weninggalih Village in 2020 reached 4,893 people, with a population growth of around 0.97% per year. This of course causes an increase in domestic water demand. According to the World Health Organization (2003)[14], the domestic water requirement of each person is 50 to 100 liter per day (l/day).

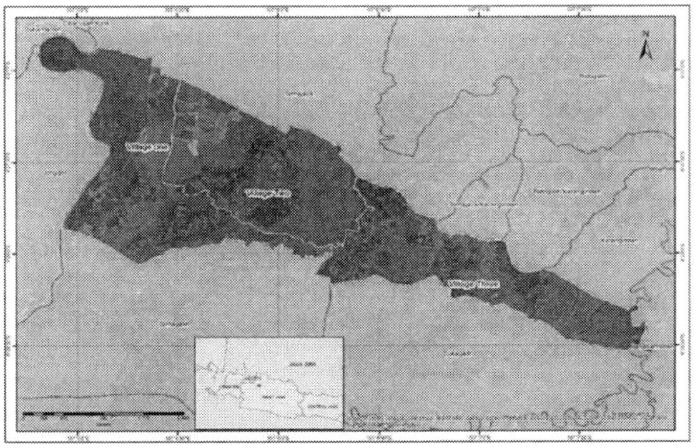

Figure 1. Research sites

The data analysis process uses the Spatial Multi Criteria Evaluation (SMCE) method and the overlay based on the research design to identify the best location for the construction of the reservoir is in Weninggalih Village, using drone data. The use of drones is used to see the conditions of the Weninggalih area in detail. The results of maximum aerial photos of course need to consider the range of objects and the distance between targets when determining the height of the drone [15]. The initial stage is carried out by collecting data from the meteorological, climatological, and geophysical agency, mapping drone, and ministry of energy and mineral resources. The data used are precipitation map, drainage streams density, slope map, elevation, land use, and geology.

The second stage, the data that is already available is then analyzed using the Spatial Multi Criteria Evaluation (SMCE) method which will produce the weights of the parameters and values for each category. The next step after the weighting and classification of the assessment is spatial analysis with the overlay technique. In similar research, it is known that the analytical method used is Multi-criteria Evaluation (MCE) with the AHP approach [16]. The result of the final analysis is a potential location suitable for the construction of a reservoir in Weninggalih Village.

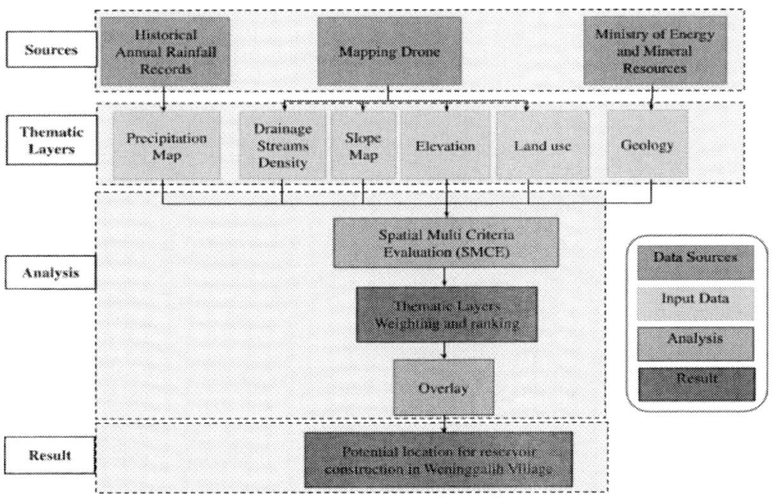

Figure 2. Research Design

The weight determination of each parameter is carried out using the eigenvalues and the results of the SMCE correlation. Spatial Multi Criteria Evaluation (SMCE) is an alternative method in spatial analysis. SMCE is a decision-making tool in a complex socio-ecological context [17]. The SMCE approach also combines spatial and non-spatial analysis which produces data output in the form of maps [18]. The available class in each parameter is given a value in the ranks column (**Table 1**).

Table 1. Weights and Ratings for Each Parameter and their Characteristics

Thematic Layers	Weight	Classes	Ranks
Precipitation	16%	276 – 285 mm	1
		267 – 276 mm	2
		258 – 267 mm	3
		249 – 268 mm	5
		241 – 249 mm	7
Drainage Stream Density	18%	Very low	1
		Low	2
		Medium	3
		High	5
		Very High	7
Geology	15%	Food Plain Deposits	1
		Unit Sandstone	2
		Cihoe Formation	3
		Older Alluvium	5
		Alluvium	7
Land Cover	14%	Settlement	1
		Industry	2
		Field	3
		Rice Field	4
		Undeveloped Land	5
		Water bodies	7
Elevation	17%	105 m - 140 m	3
		85 m - 105 m	5
		16 m - 85 m	7
Slope	20%	> 55^0	1
		35 - 55^0	2
		16 - 35^0	3
		8 - 16^0	4
		4 - 8^0	5
		2 - 4^0	6
		0 - 2^0	7

3. Results and Discussion

Based on the results of the analysis, there are three locations that can be considered for the construction of reservoirs based on spatial analysis. The location for determining the construction of

IOP Conf. Series: Earth and Environmental Science **986** (2022) 012077 doi:10.1088/1755-1315/986/1/012077

the reservoir certainly involves areas that are higher up. The process of taking aerial photo data is obtained by using the survey method. The number of aerial photos used is 6,609 units with an area of 444,506 hectares Weninggalih Village. The process of taking aerial photos with drones uses a flying height of 150 meters above sea level with 80% overlap. Aerial photos that have been processed get a GSD resolution of 21.9 cm/pixel. In previous studies, the determination of the location of the reservoir utilized the use of UAVs (Unmanned Aerial Vehicles) which resulted in reconstruction in 3D [19]. The results of this study indicate that the use of UAVs in determining the location of reservoirs has high accuracy [19].

This study uses several factors, including precipitation, drainage stream density, geology, land cover, elevation and slope [11]. In a similar study, the determination of potential locations for dam construction in Mozambique used the parameters of elevation, slope, the distance reservoir from road, rainfall, lineament, the distance of reservoir from village, land use, land cover, soil type, and stream density [16]. Based on the results of the weighting, in the Weninggalih area the dominant factor is influenced by the slope, while in the study of Luis & Cabral (2021)[16] the factor that has a significant influence is the density of the river. In other cases, using different parameters to determine the location of rain harvesters, namely in the form of runoff depth, slope, land cover, soil texture, and stream order [20]. The study at the Sharjah location used AHP analysis with the dominant factors produced were precipitation and drainage stream density with a weight of 20% [11]. The difference in the dominant factors is certainly influenced by the low density of the river in Weninggalih, as well as the geological structure in the form of white rock which is not able to absorb water, the parameters used for analysis of the location of the reservoir construction:

a. **Precipitation**, the data used is the average rainfall during five years in the 2016 to 2020 period. Rainfall data were obtained from the meteorological, climatological, and geophysical agencies in the form of daily rain data. Rainfall stations used are stations that are close to the study area, one of which is the Dayeuh Sukanegara (Jonggol) rainfall stations. Analysis of the data used is Inverse Distance Weighted (IDW), which is an interpolation method to find out a value at a location that is not sampled. According to Junior, et al. (2019)[21] IDW interpolation assumes that objects that are close to each other have more similar interpolation values than those that are farthest away. Based on **Figure 3** can be seen that the average amount of rainfall in Weninggalih Village varies from a minimum rainfall of 241 mm to 285 mm of rainfall as the maximum annual total.

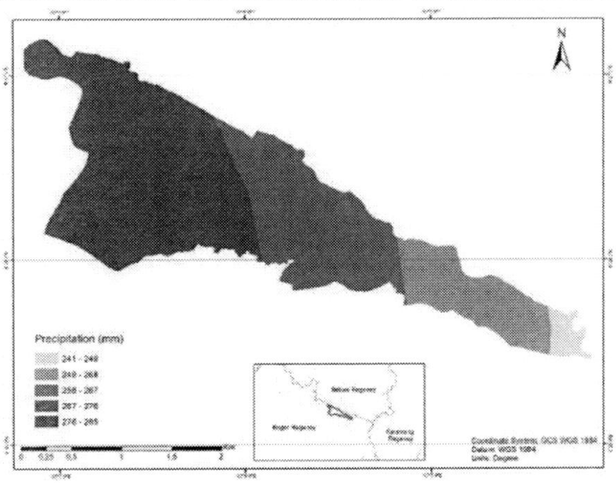

Figure 3. Distribution of the Level of Rainfall in Weninggalih Village

IOP Conf. Series: Earth and Environmental Science **986** (2022) 012077 doi:10.1088/1755-1315/986/1/012077

b. **Drainage Density**, have a good role of knowing the availability of water surface or subsurface [11]. The water flowing in the river in Weninggalih comes from two rivers, namely the Cipamingkis River (west) and the Cihoe River (east). The drainage density at the research site which is included in the highest category is in area two and area three Weninggalih. Drainage density is one of the important aspects in determining the location of the construction of the reservoir, because the location of the reservoir is closely related to surface runoff flow and permeability as shown in **Figure 4.** However, Adham's research (2018)[20] uses stream order as one of the parameters used to show the hierarchical relationship between flow and size in the existing drainage basin.

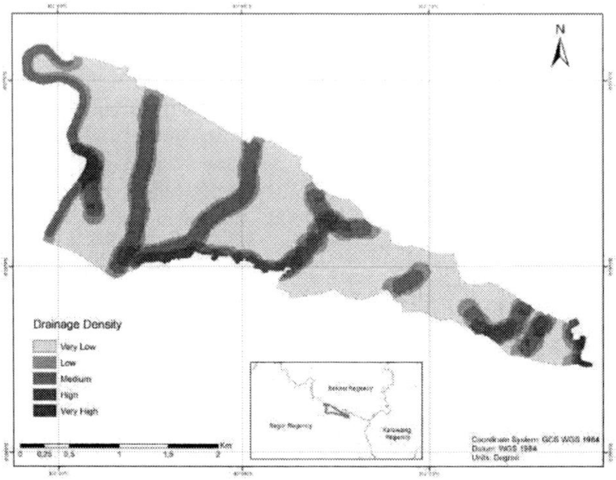

Figure 4. Level of Drainage Density in Weninggalih Village

c. **Slope Map**, this parameter affects the velocity of surface ground water. The higher the slope, the faster the surface runoff, and vice versa. The slope is directly proportional to the speed of the water flowing. Areas with slope values below 5% are very suitable as water reservoirs [22]. Most of the area in Weninggalih Village is classified as sloping (see **Figure 5**).

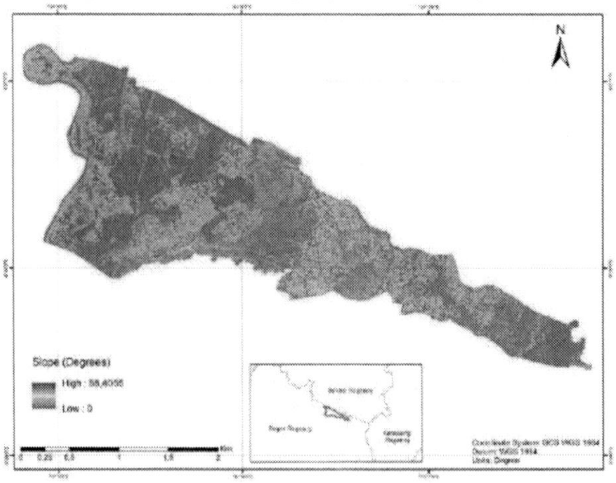

Figure 5. Distribution of Slope Levels in Weninggalih Village

IOP Conf. Series: Earth and Environmental Science **986** (2022) 012077 doi:10.1088/1755-1315/986/1/012077

d. **Elevation**, using the results of drone mapping data in the form of DEM with an aerial photo taking elevation of 150 meters above ground level. The height of Weninggalih Village ranges from 16 to 140 meters above sea level (see **Figure 6**). DEM is used to determine the potential location of the reservoir because it affects the movement of water [23]. The location of the reservoir can be considered suitable if the height of the study area is low, but higher than the residential area because it is possible that the quantity of water that is deposited is greater in the lower area and is easy to channel. Based on the **Figure 6,** it can be seen that the area that is classified as a higher land is in area two, with a maximum height of 140 meters above sea level.

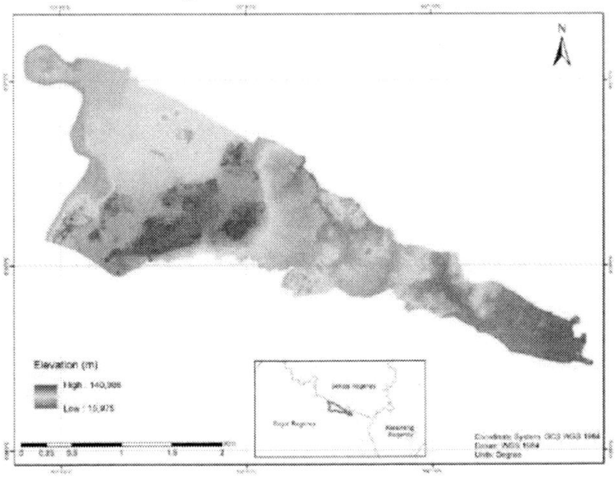

Figure 6. Distribution of Elevation in Weninggalih Village

e. **Land Cover area**, is needed to determine the undeveloped land in Weninggalih Village, so that later it can be used as a potential area for the construction of a reservoir. Undeveloped land is very suitable for construction of reservoir, by ensuring land ownership. This data was obtained from taking aerial photographs with drones which were carried out with spatial analysis in the form of maps. Weninggalih has various land covers, rice fields, field, settlement, water bodies and undeveloped land. Based on **Figure 7** It is known that almost 80% of the area in Weninggalih is used as rice fields.

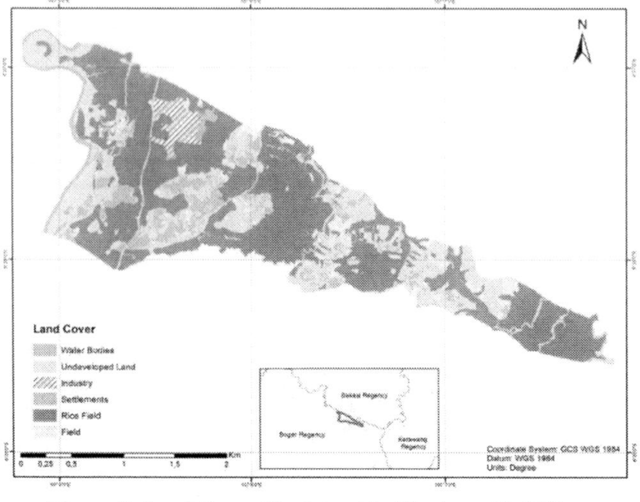

Figure 7. Land Cover Contained in Weninggalih Village

f. Geology, these data are used to determine the geological structure and spreading the Weninggalih village. There are five geological formations available at the study site, namely Alluvium, Cihoe Formation, Flood Plain Deposits, Older Alluvium, and Sandstone Units as shown in **Figure 8**. Most of Weninggalih have older alluvium geological structures, which have good shear strength, low compressibility, and are good for construction materials [24].

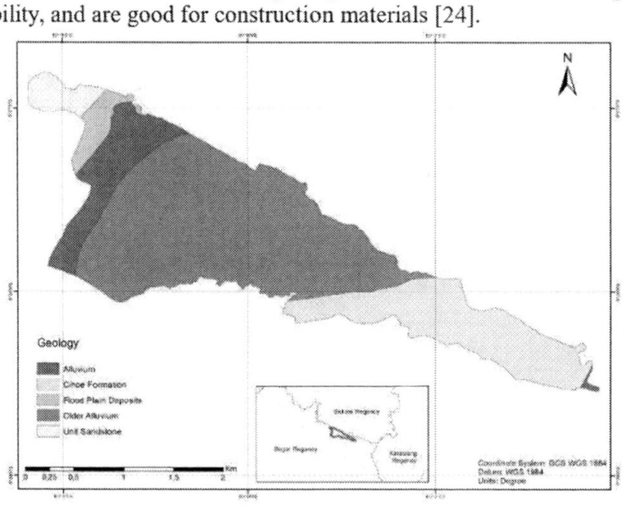

Figure 8. Rock Formation and Distribution of Rock Types in Weninggalih Village.

Very high zoning is located at the reservoir at point B. The zoning has parameter criteria including slope 0 - 2^0 at an altitude of 77 meters above sea level, geological structure of Older Alluvium. At these locations will have a water storage capacity of 4,185 m^3 or can meet the needs of 69.750 people. The **Figure 9** show that the middle part of Weninggalih, especially area two, is classified as very suitable for building a dam, because the drainage density and elevation of the area are very suitable. On the other hand, the eastern part of Weninggalih (area three) is not suitable for the construction of a reservoir, because the location tends to be lower and sloping than area two. Weninggalih does not have a reservoir, while in some places with water shortage conditions there are have reservoirs, for example in Sharjah [11] and Iraq [20]. Sharjah has 11 dams already owned, while the results of the AHP analysis show that there are three dam construction sites [11]. The Iraq region has 39 potential sites selected based on the estimated runoff available and capable of holding water [20].

The first location has a potential area of 423 m^2 with a water storage capacity of 1,296 m^3, while at the third location in the eastern part of Weninggalih bordering Sukagalih Village, the potential area of the reservoir building is 1,191 m^2, with a water storage capacity of 4,680 m^3. The third location gets its water supply from the Cihoe River, where the geological structure of this area is the Cihoe formation. The third location has many parameters. The parameters consist of the form of slope, drainage density, and land use. The third land is the most suitable for the construction for the reservoir. This area has a very high drainage density because a reservoir really needs a water source as the main supplier. The slope is the main determining factor in the location for the construction of the reservoir, in Weninggalih. Because, there are many differences slope. The finding is different with Luis & Cabral (2021)[16] and Ruzouq, et al. (2019)[11]. They found that river density, rainfall, and straightness are critical factors in determining the location of the reservoir construction.

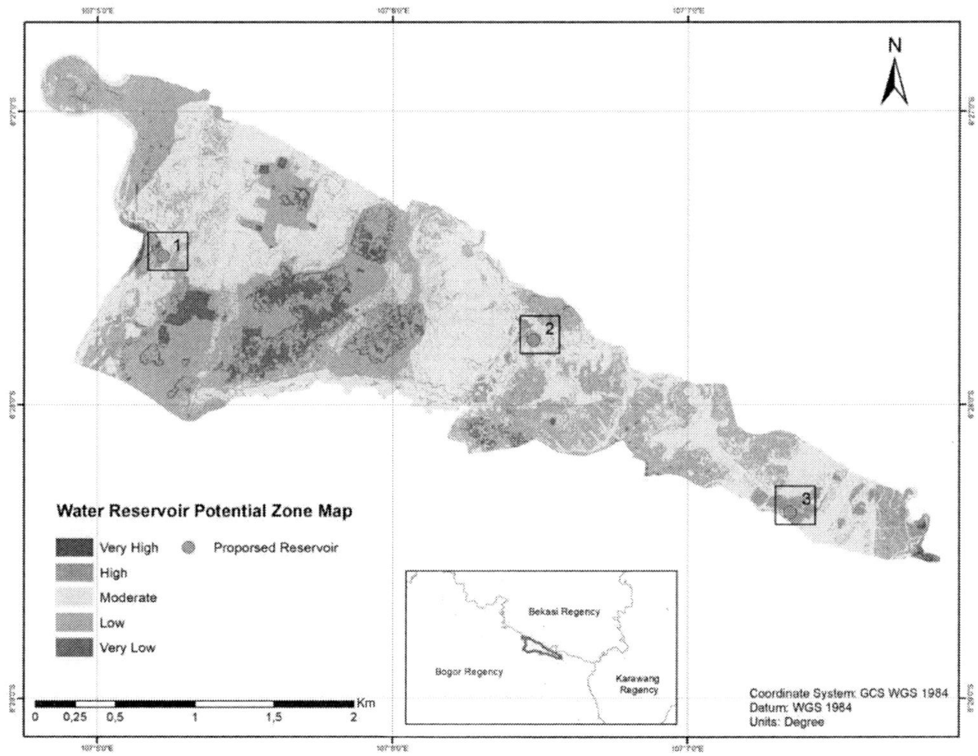

Figure 9. Water Reservoir Potential Zone Map in Weninggalih

Based on results of the research known that the potential land area for reservoir construction are categorizes into three points. The first potential site of reservoir is in the area of 423 m^2. The second one potential site of reservoirs 1.415 m^2 and the last one potential site of reservoirs are in area three land area of 1.191 m^2 of the study area. While in the Mozambique region, there are three categories of suitability identified, namely not suitable (15%), quite suitable (78%), and suitable (7%) of the total areas [16]. The existence of differences in development potential is influenced by land characteristics, in Weninggalih Village which has a geological structure in the form of rock making water unable to control groundwater flow patterns properly. Apart from differences in physical form, another factor is the use of data. The data obtained in this study using DEM from aerial photographs with a high level of accuracy and detail considering the research location has a small scope. In another study, the use of drones can also be used to monitor the health of coral reefs by differentiating the type of substrate [25].

4. Conclusion

Identifying potential locations for the construction of reservoirs is one of the strategic plans for water conservation in various areas, especially in areas that often experience drought every year. In this study, the analysis process utilizes GIS using drone data to identify potential locations for the construction of a reservoir in Weninggalih Village. This study uses six thematic layers as consideration for spatial analysis, including rainfall, drainage density, land cover, geology, slope, and topography. The slope factor is the most influential factor in spatial analysis, with a weight of 20%. Based on these parameters, it can be seen that the area of the reservoir which is in the very suitable category has a total area of 43,762.53 meters or 9.8% of the total area of Weninggalih Village.

Based on the analysis using the SMCE method, it is known that there are three locations that have the potential to build a reservoir in Weninggalih Village. These three potentials can be a reference for the Bogor Regency government to overcome the water crisis in Weninggalih Village by building a small dam or reservoir. The planning of the location of the construction of the reservoir that borders with other villages certainly needs to be considered in terms of elevation and slope. This is done to ensure that later the water flow from the reservoir is focused on the domestic needs of Weninggalih Village. The results of this study can then be used as material for consideration by the Bogor Regency government, to be able to follow up on priority locations for the development of village reservoirs in terms of the level of village water availability in meeting the community's domestic water needs.

Acknowledgments

We would like to thank to Mr. Mamat Rahmat, S.T. as the village head and all Weninggalih Village officials who have given us the opportunity to carry out research. After that, we would like to thank to all parties who contributed to the completion of this research.

Reference

[1] Kılıç Z 2020 The importance of water and conscious use of water *Int. J. Hydrol.* **4** 239–41
[2] WHO/UNICEF 2005 Water for life: make it happen. 40
[3] Jéquier E and Constant F 2010 Water as an essential nutrient: The physiological basis of hydration *Eur. J. Clin. Nutr.* **64** 115–23
[4] Kim J B, So J M and Bae D H 2020 Global warming impacts on severe drought characteristics in Asia monsoon region *Water (Switzerland)* **12**
[5] IPCC 2013 *Climate Change 2013: The Physical Science Basis. Contribution of Working Group I to the Fifth Assessment Report of the Intergovernmental Panel on Climate Change* (United Kingdom: Cambridge University Press)
[6] Badan Nasional Penanggulangan Bencana 2019 *Katalog Desa/Kelurahan Rawan Kekeringan*
[7] United Nations 2009 *ECONOMIC AND SOCIAL COMMISSION FOR WESTERN ASIA (ESCWA) ESCWA*
[8] CITIGROUP 2017 *Solutions for a Global Water Crisis*
[9] Masduqi A, Nugroho A R and Wilujeng S A 2020 Solution to water scarcity in the eastern indonesia: A case study of the lembata regency *Int. J. GEOMATE* **19** 69–76
[10] Abdullah F 2019 Studi Perencanaan Embung Sawahsumur Di Pulau Kangean Kabupaten Sumenep *J. Rekayasa Sipil* 85–94
[11] Al-Ruzouq R, Shanableh A, Yilmaz A G, Idris A E, Mukherjee S, Khalil M A and Gibril M B

A 2019 Dam site suitability mapping and analysis using an integrated GIS and machine learning approach *Water (Switzerland)* **11**
[12] Jamali I A, Olofsson B and Mörtberg U 2013 Locating suitable sites for the construction of subsurface dams using GIS *Environ. Earth Sci.* **70** 2511–25
[13] BMKG Climate Analysis. Retrieved from Monthly Precipitation
[14] World Health Organization 2003 *The Right to Water* (France: World Health Organization)
[15] Saitoh T and Kobayashi M 2021 Appropriate drone flight altitude for horse behavioral observation *Drones* **5**
[16] Luís A dos A and Cabral P 2021 Small dams/reservoirs site location analysis in a semi-arid region of Mozambique *Int. Soil Water Conserv. Res.* **9** 381–93
[17] Etxano I and Villalba-Eguiluz U 2021 Twenty-five years of social multi-criteria evaluation (SMCE) in the search for sustainability: Analysis of case studies *Ecol. Econ.* **188**
[18] Meena S R, Mishra B K and Piralilou S T 2019 A hybrid spatial multi-criteria evaluation method for mapping landslide susceptible areas in Kullu valley, Himalayas *Geosci.* **9**
[19] Langhammer J, Janský B, Kocum J and Minařík R 2018 3-D reconstruction of an abandoned montane reservoir using UAV photogrammetry, aerial LiDAR and field survey *Appl. Geogr.* **98** 9–21
[20] Adham A, Sayl K N, Abed R, Abdeladhim M A, Wesseling J G, Riksen M, Fleskens L, Karim U and Ritsema C J 2018 A GIS-based approach for identifying potential sites for harvesting rainwater in the Western Desert of Iraq *Int. Soil Water Conserv. Res.* **6** 297–304
[21] Oliveira Júnior A J de, Souza S R L de, Dal Pai E, Rodrigues B T and Souza V C de 2019 Aurora: Mobile application for analysis of spatial variability of thermal comfort indexes of animals and people, using IDW interpolation *Comput. Electron. Agric.* **157** 98–101
[22] Sayl K N, Muhammad N S, Yaseen Z M and El-shafie A 2016 Estimation the Physical Variables of Rainwater Harvesting System Using Integrated GIS-Based Remote Sensing Approach *Water Resour. Manag.* **30** 3299–313
[23] Mura J C, Gama F F, Paradella W R, Negrão P, Carneiro S, de Oliveira C G and Brandão W S 2018 Monitoring the vulnerability of the dam and dikes in Germano iron mining area after the collapse of the tailings dam of fundão (Mariana-MG, Brazil) using DInSAR techniques with terraSAR-X data *Remote Sens.* **10** 1–21
[24] Wong K S and Li W 2001 Old Alluvium : Engineering Properties and Braced Excavation Performance 223–33
[25] Cornet V J and Joyce K E 2021 Assessing the potential of remotely-sensed drone spectroscopy to determine live coral cover on heron reef *Drones* **5** 1–20

Assessment of water quality and its effect on the health of Selong District Resident, East Lombok, West Nusa Tenggara Province

B L Widiyanti[1,*]

[1] Environmental engineering, Hamzanwadi Universitas University

*Email: leea91819@gmail.com

Abstract. Water is the basic resource necessary for sustaining all human activities, so its provision in desired quantity and quality is of utmost importance. Water pollution affects drinking water, which consequently harms human health and the natural environment. The study focuses on measuring the quality of drinking water from dug wells in rural areas of Selong District, East Lombok and its effects on human health. Various analyses, including physical, chemical and microbiological assessment, carried out on the water samples collected from the dug wells in the villages. The result of samples analysis shows high phosphate, showed groundwater pollution has occurred from domestic waste. Microbiological quality was also questionable in most of the cases with high content of total coliform. On the contrary to these findings, most people living in these areas were not suffering from various water-borne diseases. So the study argues about the need and importance of water management systems in current times.

1. Introduction

Water is an important substance in the life of living things, including humans [1,2], but not everyone has adequate access to it because what needed is clean and healthy water. The amount of water available in the environment is actually quite adequate for all living things, but because of quality requirements, not all water sources can fulfill it. This is because of many factors that cause changes to the quality and quantity of water sources, both natural and artificial.

Access to clean water is a major health problem in the world [3,4]. WHO emphasizes global efforts to prevent the transmission of water-borne diseases. Clean water and proper sanitation are basic human needs. Fulfillment of clean water and proper sanitation is still a problem worldwide [5–7]. The need for clean water, drinking water and sanitation is a sustainable development goal. In the period 2015 to 2019, the level of household access to adequate sanitation and the level of household access to the availability of safe drinking water. At household level, access to proper sanitation, both for rural and urban from 2015 to 2019 has increased [8–10].

Water in natural get impurities from the surrounding environment. Impurities formed from exchange and dissolution reactions in the soil, such as organic/inorganic matter and solid or dissolved living organisms [11–13]. Drinking water must be free from the main types of contaminants and water pollutants [14].

In Indonesia, village contributions reach 74 percent of the national SDGs. This means, in fact, the role of the village is very dominant in the achievement of the SDGs. However, the village not included in the list of action plans or measures for calculating the national SDGs. If disaggregated, the village contribution realized because the 74,953 villages administration area covers 91 percent of the Indonesian government area. This means that fulfillment of village development goals contribute 91 percent to the ten regionally oriented national SDGs: clean energy, economic growth, industry and innovation, inequality reduction, climate mitigation, ocean preservation, land conservation, institutions and justice, and development cooperation networks. Meanwhile, 118 million villagers cover 43 percent of Indonesia's population. Thus, meeting the needs of villagers contribute 43 percent to the five national SDGs related to citizenship: eradicating poverty, eliminating hunger, access to health, access to education, access to clean water, and anti-gender discrimination.

Village SDGs (Sustainable Development Goals) refers to Presidential Regulation Number 59 of 2017, concerning National Sustainable Development Goals. In order for the Village SDGs to be realized, the Ministry issued a Regulation of the Minister of Villages for Development of Disadvantaged Regions and Transmigration (PerMen Desa PDTT) Number 13 of 2020, concerning Priorities for the Use of Village Funds in 2021, which refers to Economic Improvement and Community Empowerment. SDGs Desa is a global action plan agreed by world leaders, to reduce inequality, poverty and protect the environment. The Village SDGs are a role model for Sustainable Development which will be have included in the Priority for Use of Village Funds for the 2021 Fiscal Year [15–17]. The problem of access to clean water and health [18] is a concern in the MDGs, SDGs and even up to the Village SDGs [19], so the research aims to examine the quality of village water in Selong District, East Lombok Regency, West Nusa Tenggara Province and its effect on the health of the population against water-borne diseases.

2. Methods

The area used as the research site is several villages with a high population density in Selong District. They are: Sekarteja, Pancor, Khusus Kota Selong, Sandubaya, Kelayu Utara and kelayu Selatan. The area of the Selong sub-district is 31.68 sq km which consists of 11 sub-districts and a village. Selong sub-district height ranges from 74-192 meters above sea level. The absence of heavy industry causes the possibility of contamination or pollution of groundwater through dug wells were limited to the risk of pollution originating from household waste.

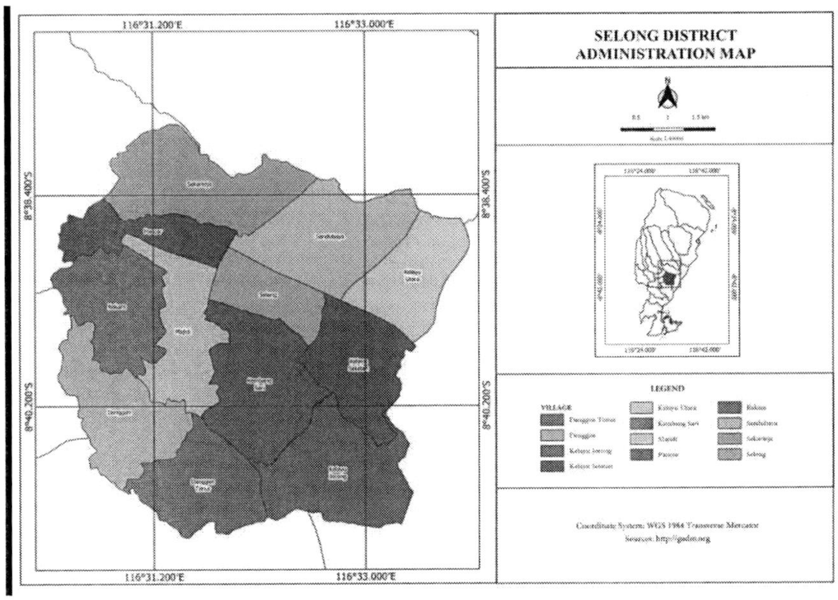

Figure 1. Selong District Administration Map (*source: Selong District in Figures, 2020*)

The first step of the research is to take water samples. Samples were tested in the laboratory in test tubes and analyzed for 24 hours. Samples were taken from all villages that had water supply. eleven water samples represent the conditions of the selected villages, in order to obtain information on the physical, chemical and micrological quality of drinking water. The household sites where the water sample taken chosen by purposive sampling technique based on groundwater flow direction Map. Physical parameters measured at the field qualitatively. Some chemical parameters chosen, based on domestic waste composition, measured in laboratory. Assessment of microbiological parameter, using most probable number techniques. Self -designed pre-tested semi-structured questionnaire interviewed a hundred participants. The questionnaire describes information about the respondent's life regarding water use, water storage, hygiene, and disease. The research was conducted for 2 months, from May 2021 to July 2021.

3. Results and Discussions

3.1. Water Quality

From the results of measurements both directly in the field and testing in the laboratory, the quality of groundwater originating from dug wells in the research area are as seen in Table1.

Table 1. Water Quality in Selong District

Parameters	**Village Location**					
	Sekarteja	**Pancor**	**Sandubaya**	**Selong**	**Kelayu Utara**	**Kelayu Selatan**
Temperature	27.8	27.6	28.1	27.8	27.3	27.3
EC	332	302	358	358	232	310
Turbidity	0.53	0.81	21.5	0.42	0.5	0.32
pH	6.64	6.64	7.03	6.81	7.5	7.39
BOD	< 2	< 2	< 2	< 2	24.5	< 2
COD	< 4	< 4	< 4	< 4	32.6	< 4
Phosphate	0.92	0.85	0.53	1.07	1.42	1.18
Nitrate	2.62	0.43	0.63	1.88	0.18	1.34
Nitrite	< 0.008	<0.008	< 0.008	< 0.02	< 0.008	< 0.008
Ammonia	0.055	0.24	0.15	< 0.008	0.04	< 0.02
Coliform	>24000	>24000	2815	3500	80	5400

Source: primary data, 2021.

The data in Table 1 shows the results of water quality inspections from the most commonly used sources by the community. The most prominent being the bacteriological content which is quite high in presentation, and also the phosphate content. Phosphate is found in natural water or wastewater as orthophosphate compounds, polyphosphates, and organic phosphates. Most of the phosphates come from cleaning materials that contain phosphate compounds. The sodium triphosphate compound used in detergents aims to bind calcium/magnesium ions from hard water. This is what causes phosphate waste easily found at the waters. The high content of coliform bacteria in groundwater is due to the proximity of the septic tanks and drainage channels to the wells, which are located in densely populated areas, where the use of on-site sanitation is still used.

3.2. Health Status of Resident

Demographic characteristics of respondents and their health status can be seen in the following table.

Table 2. Respondent Characteristic

Respondent Characteristic	**Amount of KK (head of family)**	**Percentage (%)**
1. Ages:		
- 20 – 37	32	32
- 38 – 53	51	51
- 54 – 71	17	17
2. Number of dependent family member:		
- 0 – 3	47	47
- 4 - 7	50	50
- 8 - 10	3	3
3. Type of work:		
- farmer	10	10
- Carpenter and builder	9	9
- Cleaning Services	2	2
- Self-Employee	25	25
- Government employees	9	9
- Housewife	4	4
- Driver	2	2
- Trader	16	16
- Labour	4	4
- Others	12	12
- Un-employee	5	5
4. level of income:		
- Low	18	18
- Medium	32	32
- High	50	50
2. Level of education:		
- Elementary	29	29
- Medium	10	10
- High School	33	33
- College	3	3
- University	15	15
- Undergraduate	2	2
- Un-educated	8	8
Total	**100**	**100**

Currently, morbidity and mortality caused by environmental-based diseases such as diarrhea, dengue fever and skin diseases are still prominent types of disease. Diarrhea is one of the environmental-based diseases, and the dominant cause is inadequate clean water facilities and indiscriminate disposal of feces. This factor will interact together with human behavior. An unhealthy environment and unsafe water quality are the main causes of the development of various types of diseases caused by water pollution [20,21].

Types of water-related diseases reported that the respondent has suffered: typhus, diarrhea, cholera, dysentery, itching. Duration of illness not too long, just about 3 days to 1 week. They feel the illness is a disturbance in daily activities, but quite disturbing. No special allocation of funds for health. More than half of respondent do not have special funds (90%). Special funds for obtaining health services, not specifically budgeted (85%)

3.3. Discussion
To create a conducive living environment for efforts to improve the degree of public health as well as the empowerment of individuals, families and community in the health sector, the main program for a healthy environment has the following goals and objectives:

3.3.1. The healthy environment
Program aims to create a healthy environment so that can protect the community from all possible disturbances that can pose a health hazard to health status better families and communities. The targets to be achieved in this program include:

a) An increase in the percentage of families occupying houses that meet health requirements to 70%, the rate of families having access to clean water to 75%, the percentage of families using latrines meeting the health requirements to 65%, and the portion of public places (TTU).) qualified up to 80%.
b) Availability and socialization of policies and guidelines as well as laws that support distributed programs, handling Extraordinary Events (KLB) to the village.
c) Availability of tools, materials and reagents to control risk factors and supporting the implementation of a healthy environment program. Its main activities include the provision of water facilities, basic sanitation, maintenance and supervision of environmental quality, impact control the risk of environmental pollution and the development of the East Lombok region healthy ones.

3.3.2. Healthy Behavior and Community Empowerment
Community empowerment through *Desa Siaga* in East Lombok aims: to create a healthy, caring and responsive village community to health problems in the area. What is meant by Desa Siaga is a village or sub-district whose population is ready in terms of capabilities and resources to deal with a disaster, health, and health emergencies independently. The target is that all individuals and families in the village are expected to lead a healthy life and care about health problems in the village area. Parties who influence individual and family behavior changes that can create a conducive environment for behavior change are community characters, including religious leaders, women and youth leaders, cadres, and health workers.

Poor drainage systems also caused poor water quality in the environment and inadequate waste water disposal facilities [22–24]. Currently, most people in Selong District use an on-site processing system with latrines built in each house or in certain places that are shared. People simply released the waste water into environment through drainage system and rivers. The villages, where the research is done, do not have a special system for treating domestic waste, judging by the treatment at the household level or at the community level. At some place, scattered trash interspersed around houses and also on the streets, that could affect the quality of groundwater, especially at rainy season. Over 50 percent of respondents reported that with such water quality, they did not feel symptoms related to water-related diseases or water borne disease. This is because they have increased immunity in the communal sphere against various water-borne viral and bacterial infections through their daily habits. Another reason may be that people are not aware of these symptoms, so they are responsible for under-reporting.

4. Conclusions
The results of field tests and laboratory analysis of groundwater quality in the research area show that the value is still included in the bad category because the values of all parameters are not in the same class for the suitability class based on the designation as raw material for drinking water.

People's behavior towards environmental sanitation included in the good category, with a fairly high level of knowledge, attitudes that are also included in the good category and the belief that a clean environment can make their lives better. At the level of practical action, many community members have not carried out activities that lead to the category of clean and healthy living behavior.

Acknowledgments
I express its gratitude and sincere appreciation to all the respondents at research sites and my student, who have contributed their time to help this research.

References

[1] Everard M 2019 Meeting global drinking water needs *Nat. Sustain.* **2** 360–1
[2] Payen G 2011 *Worldwide needs for safe drinking water are underestimated: billions of people are impacted* (Aquafed)
[3] Martínez-Santos P 2017 Does 91% of the world's population really have "sustainable access to safe drinking water"? *Int. J. Water Resour. Dev.* **33** 514–33
[4] Cassivi A, Johnston R, Waygood E O D and Dorea C C 2018 Access to drinking water: time matters *J. Water Health* **16** 661–6
[5] Moriarty P, Batchelor C, Fonseca C, Klutse A, Naafs A, Nyarko K, Pezon C, Potter A, Reddy R and Snehalatha M 2010 Ladders for assessing and costing water service delivery *Int. Water Sanit. Centre, available http//www. washcost. info/page/196*
[6] Kalbermatten J M, Middleton R and Schertenleib R 1999 Household-centred environmental sanitation *Swiss Fed. Inst. Environ. Sci. Technol. Ueberlandstrasse* **133**
[7] Supply W J W and Programme S M 2014 *Progress on drinking water and sanitation: 2014 Update* (World Health Organization)
[8] Rhodes B and McKenzie T 2018 To what extent does socio-economic status still affect household access to water and sanitation services in South Africa? *J. Econ. Financ. Sci.* **11** 1–9
[9] Komarulzaman A, Smits J and de Jong E 2017 Clean water, sanitation and diarrhoea in Indonesia: Effects of household and community factors *Glob. Public Health* **12** 1141–55
[10] Patunru A A 2015 Access to Safe Drinking Water and Sanitation in I ndonesia *Asia Pacific Policy Stud.* **2** 234–44
[11] Fawell J and Nieuwenhuijsen M J 2003 Contaminants in drinking water: Environmental pollution and health *Br. Med. Bull.* **68** 199–208
[12] Akple M, Keraita B, Konradsen F and Agbenowu E 2011 Microbiological quality of water from hand-dug wells used for domestic purposes in urban communities in Kumasi, Ghana *Urban Water J.* **8** 57–64
[13] Bain R E S, Gundry S W, Wright J A, Yang H, Pedley S and Bartram J K 2012 Accounting for water quality in monitoring access to safe drinking-water as part of the Millennium Development Goals: lessons from five countries *Bull. World Health Organ.* **90** 228–35
[14] Onda K, LoBuglio J and Bartram J 2012 Global access to safe water: accounting for water quality and the resulting impact on MDG progress *Int. J. Environ. Res. Public Health* **9** 880–94
[15] Afifah T, Nuryetty M T, Cahyorini, Musadad D A, Schlotheuber A, Bergen N and Johnston R 2018 Subnational regional inequality in access to improved drinking water and sanitation in Indonesia: results from the 2015 Indonesian National Socioeconomic Survey (SUSENAS) *Glob. Health Action* **11** 31–40
[16] Sirajuddin T 2020 Rural development strategies in Indonesia: Managing villages to achieve sustainable development *IOP Conference Series: Earth and Environmental Science* vol 447 (IOP Publishing) p 12066
[17] Elson S 2019 Engaging people prior to problems: providing drinking water for villages in South Kalimantan, Indonesia *Wacana J. Soc. Humanit. Stud.* **22**
[18] Holmes J, Jones B and Heap B 2015 Smart villages
[19] Odagiri M, Cronin A A, Thomas A, Kurniawan M A, Zainal M, Setiabudi W, Gnilo M E, Badloe C, Virgiyanti T D and Nurali I A 2020 Achieving the Sustainable Development Goals for water and sanitation in Indonesia–Results from a five-year (2013–2017) large-scale effectiveness evaluation *Int. J. Hyg. Environ. Health* **230** 113584
[20] Wang Q-F, Shen W-L, Liu C, Mu D-L, Wu X-F, Guo N-G and Zhu J-Q 2017 Effects of multi-

environmental factors on physiological and biochemical responses of large yellow croaker, Larimichthys crocea *Chemosphere* **184** 907–15
[21] Adimalla N and Wu J 2019 Groundwater quality and associated health risks in a semi-arid region of south India: Implication to sustainable groundwater management *Hum. Ecol. risk Assess. an Int. J.* **25** 191–216
[22] Juneja T and Chaudhary A 2013 Assessment of water quality and its effects on the health of residents of Jhunjhunu district, Rajasthan: A cross sectional study *J. public Heal. Epidemiol.* **5** 186–91
[23] Yassin M M, Amr S S A and Al-Najar H M 2006 Assessment of microbiological water quality and its relation to human health in Gaza Governorate, Gaza Strip *Public Health* **120** 1177–87
[24] Mohsin M, Safdar S, Asghar F and Jamal F 2013 Assessment of drinking water quality and its impact on residents health in Bahawalpur city *Int. J. Humanit. Soc. Sci.* **3** 114–28

IOP Conf. Series: Earth and Environmental Science **986** (2022) 012079 doi:10.1088/1755-1315/986/1/012079

Identification of groundwater depth in the Northern Sukoharjo Regency using Vertical Electrical Sounding (VES) method to overcome agricultural irrigation drought

Sorja Koesuma* and Aldi
Department of Physics, Faculty of Mathematics and Natural Sciences, Sebelas Maret University
Jl. Ir. Sutami 36A Kentingan Jebres Surakarta 57126. INDONESIA

*Email: sorja@uns.ac.id

Abstract. Groundwater use in Sukoharjo increases every year. This study aims to identify the depth of groundwater in the northern part of Sukoharjo district as a solution to fulfill agricultural irrigation using the Vertical Electrical Sounding (VES) method. This study used the OYO resistivity meter. Research data processing uses uses Non-Linear Optimization method to obtain the true layer resistivity value vertically. The research was conducted in 5 different locations using the *Vertical Electrical Sounding (VES)* method with *Schlumberger* configuration. The length of the research site is 250 - 300 meter with a target depth of more than 120 meters. Based on the value of rock resistivity can be determined the depth of groundwater level in the northern Sukoharjo district. The results of the interpretation showed the groundwater level is at a depth of 15.2 - 82.21 meter consisting of a layer of top soil, sand, gravel and clay rocks.

1. Introduction

Sukoharjo regency has an area of 466.66 km2 with a population of 907,587 people in 2020 [1]. In terms of fulfilling the needs of clean water, Sukoharjo people mostly utilize shallow groundwater resources, namely pump wells and wells. Shallow groundwater above the first layer that is waterproof is usually not too deep beneath the surface soil. Groundwater occurs due to the presence of water infiltration from the ground [2]. Groundwater is stored in a or so-called aquifer which is a geological rock formation that is saturated with water and able to store and pass water in sufficient quantities and economically [3]. The conversion of agricultural land into industry in Sukoharjo Regency makes a very pronounced impact when the dry season arrives, because hydrologically this event reduces the water catchment area so that the storage of water into the soil tends to be reduced and the water resources concerned will be disrupted circulation [4].

Some of the complaints from Sukoharjo people are difficult to get clean water, especially during the dry season. This is due to several factors including the presence of shallow groundwater that is very deep from the ground level. In addition, the use of groundwater for rice field irrigation is also a factor in the rapid reduction of shallow and deep groundwater. Based on what has been stated above, the authors conducted related research to find out the depth of groundwater in the northern part of Sukoharjo area to be a solution that can be done by the community in groundwater extraction.

IOP Conf. Series: Earth and Environmental Science **986** (2022) 012079 doi:10.1088/1755-1315/986/1/012079

Geoelectricity utilizes the electrical properties of a medium on earth and is carried out on the earth's surface [5]. The identification of the medium is carried out based on the resistivity properties of a rock. Resistivity is a material parameter that indicates the level of the material's ability to inhibit electric current. If the material has a low resistivity value, it can easily drain the electric current. One of the uses of geoelectric is the Vertical Electrical Sounding (VES) method, where this method has been widely used such as looking for groundwater potential [6]. The VES method is well used to determine the resistance of each layer of subsurface conditions, where the method is based on the concept that the earth has only one layer but in reality the earth consists of several layers [7]. The earth is assumed to be an isotropic homogeneous medium, but it is not in accordance with the actual earth conditions so that the resistivity value obtained is not the actual resistivity value [8].

2. Methods

This method of resistivity, the current is inserted into the ground using two or more electrodes, then the potential difference is measured between two previously determined probes with current electrodes [9]. Pseudo-resistivity values measured on the earth's surface that are not homogeneous are given by the following equation [10] :

$$\rho = K\frac{\Delta V}{I} \tag{1}$$

where K is a geometry factor, depending on the current configuration and measuring electrode. ΔV is the difference in electrical potential, I is the electric current.

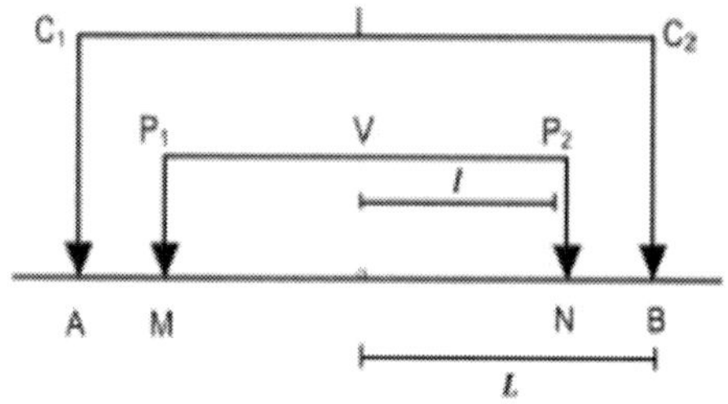

Figure 1. Schlumberger Configuration

The distance between MN potential electrodes is always fixed while at ab distances that will change. When the AB distance will change to relatively large, then the MN distance will be changed. Measurement using variations in current electrode position (AB) and potential electrodes (MN) [11] can be seen in figure 1. Schlumberger's configuration is that it can detect the presence of non-homogeneity of rock layers and can detect deep enough rock layers.

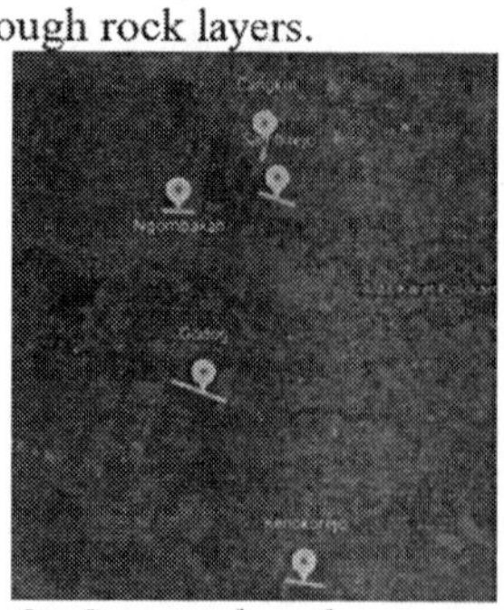

Figure 2. Survey location map (yellow pin), survey trajectory direction (yellow line) using google earth

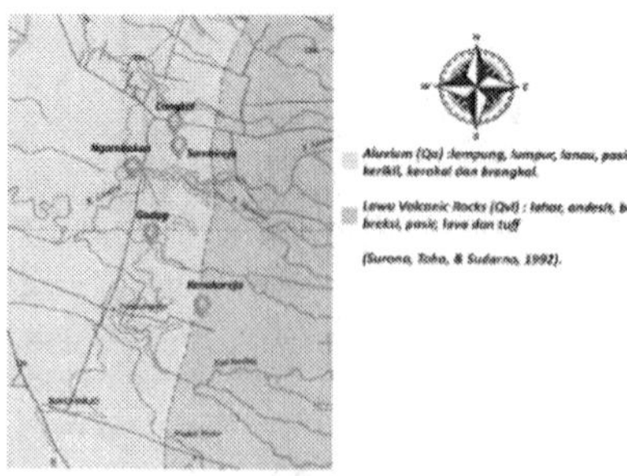

Figure 3. Survey location map (yellow pin) (geological map modification from [12])

3. Result and Discussion

Resistivity measurement is done with 5 different research locations with 1 dimensional method *(Sounding)* (Figure 2). The 1-dimensional method uses Schlumberger's configuration, MN potential electrodes are always placed between two AB current electrodes (Figure 1). This study aims to find out the depth of groundwater in northern Sukoharjo Regency using Vertical Electrical Sounding (VES) method. Resistivity measurement using OYO resistivity meter. The results obtained from the measurements are the magnitude of current (mA) and the potential magnitude (mV), using ohm hokum obtained large pseudo-type prisoners that were first multiplied by the distance factor (K). The length of the measuring instrument is 250 – 300 meters, with a target depth of 120 meters from the ground. This research was conducted in 5 different measurement points in North Sukoharjo Regency, i.e. Ckl, Sbrj, Nbk, Gdg and Knrj.

Data processing uses curve matching method by comparing or matching field measurement data and parent curve data to get errors between the two. With this method will be known the value of resistivity [13]. The inversion process is performed to obtain the actual resistivity value to the depth reflecting subsurface conditions based on the electrical properties of the rocks [14]. The inversion process is done with software and then obtained the actual resistivity value to the depth to be used as a reference to determine the distribution of aquifers based on the resistivity value obtained. The result of inversion in the form of resistivity log data can be seen in figure 4 below and table 1 log resistivity Y axis is resistivity and X axis is the distance of the current electrode [15]:

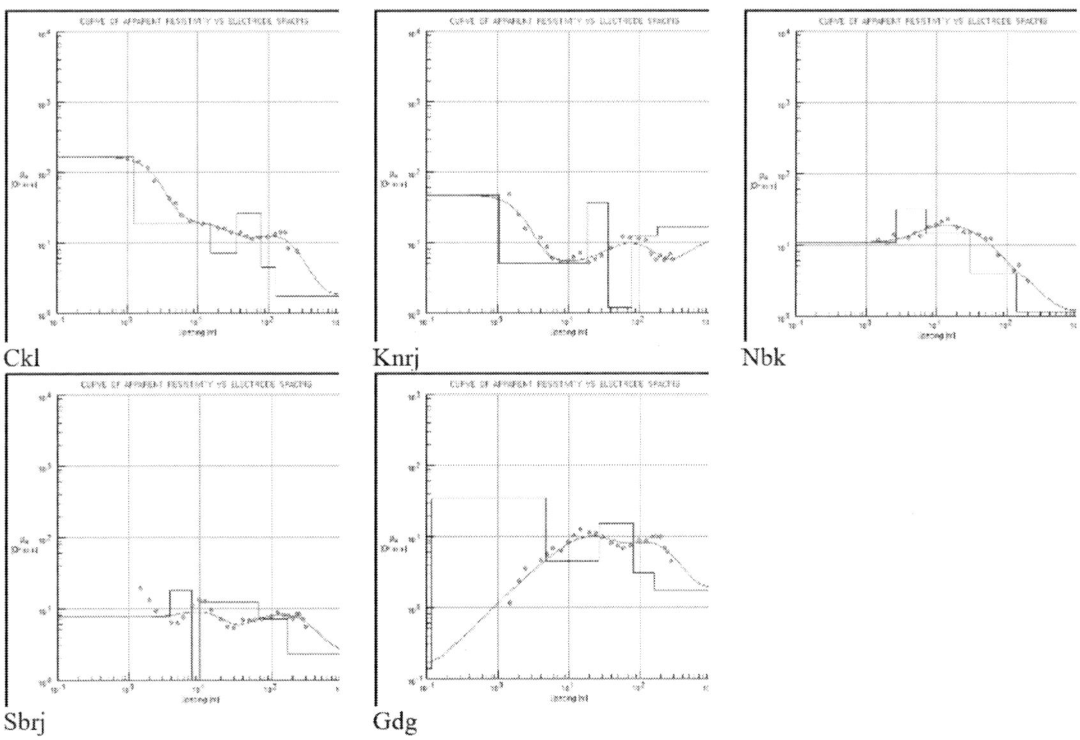

Figure 4. Results of processing 1-D geoelectric data.

Data processing is done using Non-Linear Optimization method. This method works as to match the measurement data with the model so that the smallest error is obtained. The results of the resistivity log will be matched with the resistivity table of rock layers and geological formations of the measurement site and can be indicated as aquifer zones [16]. Table 1 shows the value of rock resistivity of some type of rock material [17].

Table 1. Value of Rock Resistivity

Material	Resistivitas (Ωm)
Air	~
Pirit	3×10^{-1}
Galana	2×10^{-3}
Kwarsa	4×10^10 s.d. 2×10^14
Kalsit	1×10^12 s.d. 1×10^13
Salt Rocks	30 s.d. 1×10^13
Mika	9×10^12 s.d. 1×10^14
Granit	10^2 s.d. 1×10^6
Gabro	1×10^3 s.d. 1×10^6
Basalt	10 s.d. 1×10^7
Limestone	50 s.d. 1×10^7
SandStone	1 s.d. 1×10^8
Shale Rocks	20 s.d. 1×10^3
Dolomit	10^2 s.d. 10^4
Sand	1 s.d. 10^3
Clay	1 s.d. 10^2
Groundwater	0.5 s.d. 3×10^2
Sea water	0.2

The geological formations of the measurement area can be seen in the Figure 3, i.e., alluvium formation (Qa), consisting of clay, mud, lanau, sand, gravel, kerakal and brangkal, picture of 3, i.e., lawu volcanic rocks formation (Qvl) consisting of lava, andesite varies in size, basalt, breksi, sand, lava and tuff [12]. The geomorphological condition of measurement in Mojolaban and Polokarto subdistricts is generally mainly in the northern region is basically the morphology of the foothills of Mount Lawu. Most of the area is occupied by residential and rice fields. Stratigraphy and lithology in the area of measurement based on regional geology and direct observation in the field are alluvium formations – Lawu Volcanic Rocks consisting of clay, sand, gravel, andesite and basal rocks.

Based on the inversion results from field measurements in figure 4 obtained resistivity log, then interpreted with table 1 the value of rock resistivity to indicate the presence of aquifers below ground level. When measuring resistivity in the field is conducted direct observation by knowing the drill well in the settlement of the measurement area, it also aims to determine the existence of aquifers. The results of the interpretation indicate that the measurement location consists of topsoil, gravel sand, sand, clay. Gravel sand and clay spread from the measurement site in Ckl village to Knrj village.

Table 2. Results of measuring groundwater depth (meters) from ground level

No	Measurement Location	Ground Water Depth Measurement Location (Meter)
1	Ckl	15.42 Meter
2	Knrj	37.23 Meter
3	Nbk	29.46 Meter
4	Sbrj	65.43 Meter
5	Gdg	82.21 Meter

The measurement results (table 2) in Ckl village showed the presence of shallow aquifers at a depth of 15.42 meters. In Knrj village shows the presence of aquifers at a depth of 37.23 meters. In Nbk village shows the presence of aquifers at a depth of 29.46 meters. In the village of Sbrj shows the presence of aquifers at a depth of 65.43 meters. In Gdg village shows the presence of aquifers at a depth of 82.21 meters. The depth of the aquifers is spread differently from each measurement site, followed by the spread of geological conditions at each measurement point that has a different depth.

4. Conclusion

Based on measurements made to determine the existence of aquifers in northern Sukoharjo Regency using the Vertical Electrical Sounding (VES) method of Schlumberger configuration, it was concluded that the existence of shallow aquifers was at a depth of 15.42 meters to deep aquifers at a depth of 82.21 meters from the ground surface. The existence of this aquifer can be used as a reference in making boreholes for agriculture in the northern part of Sukoharjo. Types of rock layers known in the form of top soil, gravel sand, sand and clay.

Acknowledgment

The authors would like to thank members of the Geophysical Laboratory for their support of the field work required to obtain the data used in this study. This research is expected to help the community around the research location to overcome drought and get water for irrigation of farmers' fields.

References

[1] BPS Statistics of Sukoharjo Regency, Sukoharjo Regency In Figures 2021, Sukoharjo Regency ISSN: 2715-2847: BPS-Statistics of Sukoharjo Regency.

[2] H. H. A. Matin, S. Rachmawati, L. Kusumaningrum, I. Y. Septiariva and Purwono 2020 Analysis of Shallow Groundwater Quality in Joho Village *International Journal of Engineering Education,* vol. 2, no. 2, pp. 99-102.

[3] B. Sadjab, As'ari and T. Adey, "Pemetaan Akuifer Air Tanah di Kecamatan Prambanan Kabupaten Sleman Daerah Istimewa Yogyakarta Dengan Metode Geolistrik Tahanan Jenis," *Jurnal MIPA,* **1** (1) 37-44, UNSRAT Online.

[4] A. N. Anna, R. W. Kaeksi and W. A. Astuti 2010 Analisis Karakteristik Parameter Hidrologi Akibat Alih Fungsi Lahan Di Daerah Sukoharjo Melalui Citra Landsat Tahun 1997 Dengan Tahun 2002 *Forum Geografi,* **24** (1) 57-72.

[5] W. Lowrie, Fundamentals of Geophysics Second Edition, Cambridge: Cambridge University Press, 2007.

[6] L. Marfuatik, L. Koesuma, B. Legowo and Darsono 2018 Identification of aquifer potential in Karanganyar City by using vertical electrical sounding method *Journal of Physics: Conference Series*, pp. 1-7.

[7] M. A. Khalil, A. M. Abbas, F. M. Santos, U. Masoud and H. Salah 2013 Application of VES and TDEM techniques to investigate sea water intrusion in Sidi Abdel Rahman area, northwestern coast of Egypt *Arabian Journal of Geosciences,* **6** 3093-3101.

[8] S. Koesuma, S. D. Sutarno and A. H. Ramelan 2021 Groundwater level in the southern part of surakarta city," *Journal of Physics,* vol. 1825, pp. 1-6.

[9] P. K. Bhattacaharya and H. P. Patra 1968 Direct Current Geolectric Sounding, Amsterdam: ELSEVIER PUBLISHING COMPANY.

[10] F. Waskito, T. Yulianto and D. J. Suprapto 2016 Analisis Rembesan Pada Bendungan Cengklik Menggunakan Metode Resistivitas Konfigurasi Schlumberger Dan Uji Permeabilitas Untuk Menghitung Debit Rembesan *Youngster Physics Journal* **5** (4)425-432.

[11] S. Darmawan, U. Harmoko and S. Widada 2014 Identifikasi Struktur Bawah Permukaan Menggunakan Metode Geolistrik Konfigurasi Schlumberger Di Area Panasbumi Desa Diwak Dan Derekan Kecamatan Bergas Kabupaten Semarang, *Youngster Physics Journal* **3** (2) 159-164.

[12] Surono, B. Toha and I. Sudarno 1992 Peta Geologi Lembar Surakarta, Giritontro, Jawa. Pusat Penelitian dan Pengembangan Geologi.

[13] S. Koesuma and Sulastoro 2016 Identifying of ground water level by using geoelectric method in Karanganyar, Central Java, Indonesia *Journal of Physics: Conference Series 776,* pp. 1-6.

[14] Y. Yuniardi, Hendarmawan, Abdurrokhim, . V. Isnaniawardhani, F. Mohammad, M. K. Alfadli and P. Ridwan 2019 Pendugaan Akifer Airtanah Dengan Metode Geolistrik Konfigurasi Schlumberger Di Lereng Utara Gunungapi Tangkubanparahu, *Jurnal RISET Geologi dan Pertambangan,* vol. **29** (2) 239-253.

[15] L. Marfuatik, S. Koesuma, B. Legowo and D 2017 Identification of aquifer potential in Karanganyar City by, *Journal of Physics,* vol. Conference Series 983 1-6.

[16] F. W. Maulana 2013 Potensi Akuifer Daerah Desa Karangmojo Kecamatan Weru Kabupaten Sukoharjo Propinsi Jawa Tengah Berdasarkan Data Geolistrik, *Jurnal Teknologi Technoscientia,* **6** (1) 55-64.

[17] K. Nisa, T. Yulianto and S. Widada 2012 Aplikasi Metode Geolistrik Tahanan Jenis Untuk Menentukan Zona Intrusi Air Laut Di Kecamatan Genuk Semarang, *Berkala Fisika* **15** 7-14.

IOP Conf. Series: Earth and Environmental Science **986** (2022) 012080 doi:10.1088/1755-1315/986/1/012080

An environmental monitoring system for the availability of drinking water for the population as an ecosystem service in areas exposed to covid-19

S Samadi[1,2*], S Suhardjo[1,3] and A Munandar [1,4]

[1] Department of Geography Education, State University of Jakarta, Jalan Rawamangun Muka, Jakarta Timur, 13220, Indonesia
[2] Lecture of Environmental Geography
[3] Lecture of Cartography
[4] Lecture of Population Geography

*Email: samadi@unj.ac.id

Abstract. An effectiveness of the protected or rehabilitated natural assets will provide information on the availability of natural resources to respond to the impacts of climate change and ecosystem health. The emergence of the COVID-19 pandemic disaster increasingly shows that if humans destroy biodiversity, it also destroys human life-support systems. The purpose of the research to obtain a dynamic-empirical description of the availability of clean water for drinking water for residents in areas exposed to the Covid-19 pandemic as an implementation of an environmental monitoring system for ecosystem services. The research method uses quantitative methods. The results showed that: (1) although the low pressure area affected the number of customer complaints, it was not entirely the main factor that most dominantly affected the number of complaints of small water, dead water, and water quality in the studied area. (2) The results of the R square regression test show the percentage of the influence of low pressure areas on complaints of small water complaints, dead water, and water quality; is below 100%. The conclusion shows that although low water pressure occurs in areas exposed to Covid-19 in DKI Jakarta, it does not always affect customer complaints. This is because low water pressure is not entirely caused by small water, dead water, or water quality. Therefore, the environmental monitoring system for the availability of drinking water for the population as an ecosystem service pays more attention to other parameters besides small water, dead water, and water quality.

1. Introduction

Efforts to meet the needs of clean water in DKI Jakarta have been carried out by the Regional Drinking Water Company (PAM JAYA) since early 1998, and until now it has collaborated with two private partners, namely PT. Aetra Air Jakarta (AETRA) in the eastern region of the City of Jakarta and PT. Pam Lyonnaise Jaya (PALYJA) in the West region. And one of the service indicators in terms of quantity is that PAM Jaya together with the two private partners have agreed to meet a minimum water pressure of 0.75 atm at the customer point. However, according to the monthly reports received by BRPAM DKI Jakarta from the two private partners, there are still areas that do not meet the minimum standard of 0.75 atm; this is called a low-pressure water receiving area.

So far, detailed data has not been found regarding the impact of the Covid-19 crisis on the clean water industry, which is the domain of regional owned enterprises (BUMD). The decline in demand for clean water from the non-household sector is a derivative effect of the weakening of the national economy. Various industries and commercial businesses reduce the amount of production, reduce the number of hours worked, and even reduce the number of workers. The potential for decreased income in clean water supply system management enterprises is due to reduced demand from the non-household sector (industrial and commercial) as well as increased consumption in the household sector with an increase in income revenue that is not commensurate with the loss of revenue from the non-household sector. The curve for clean water demand from household groups is an implication of work, schooling and worshiping at home policies. However, this increase cannot compensate for the loss of revenue due to lower demand for the non-household sector due to the high disparity in water rates between customer groups.

The research questions are: (1) How do we see the consumption of clean water in areas exposed to Covid-19, while those areas experience low water pressure (small water); (2) Is it true that the community cannot access other clean water sources (due to pandemic conditions / PPKM policies, self-isolation, etc.); (3) Is it true that the community did not complain because of the policy during the pandemic, even though the customer community experienced low water pressure (small water).

2. Literature Review

The emergence of the COVID-19 (Coronavirus) pandemic disaster has increasingly shown that if humans destroy biodiversity, it will also destroy human life support systems [1,2]. The vast ability of water to purify and re-dissolve various pollutants in nature has earned it the nickname "universal solvent" [3], and it is this ability that makes water an invaluable life-sustaining force [4] and its role in social-human life. and closely related hydrology [5], known as ecosystem service capability [6–8]. Coronavirus has exposed the fragility at the heart of the clean water supply system as an ecosystem service [9–11] as well as proving the strengthening of the driving factors that contribute to the emergence of zoonotic diseases to humans, thus driving high demand for clean water individually and communally.

The increase in demand for clean water, especially in household groups during the pandemic, is an implication of work, schooling and worshiping at home policies. This is in contrast to the decreasing demand for clean water from the non-household sector where various industries and commercial businesses are doing at least three things, namely reducing the amount of production, reducing the number of working hours, and even reducing the number of workers [12]. This symptom has resulted in decreased people's incomes [12–14] even at the macro level the United Nations Development Program (UNDP) has released two new data dashboards that highlight the big differences in the ability of countries to cope with and recover from the COVID-19 crisis [15] where the pandemic is more than just a global health emergency but also a systemic human development crisis, which has affected the economic and social dimensions of development in an unprecedented way.

The increase in demand for clean water by household groups cannot compensate for the loss of revenue from clean water management companies [16] due to the decline in demand for the non-household sector due to the high disparity in water rates between customer groups. However, in the midst of increasing demand for clean water in household groups, the capacity of clean water supply is still experiencing various obstacles including inadequate volume of water received (quantity), disruption of the smooth distribution of clean water (continuity), and the quality of water arriving at the household. . However, from the temporary facts found in the field, as the uniqueness of the research, the above problems do not make the community who customers clean water respond negatively. Usually, the customer community will complain and even protest when the clean water supply is disrupted.

Clean water as one of the protected or rehabilitated natural assets will provide information on the availability of natural resources to respond to the health impacts of ecosystems during a pandemic. And therefore, this research is important to build an environmental monitoring model for the availability of clean drinking water for the population as an effort to improve ecosystem services in areas affected by the Covid-19 pandemic.

3. Research Methodology

With quantitative methods, the research aims to obtain a dynamic-empirical environmental monitoring model for the availability of clean drinking water for the population as an effort to improve ecosystem services in areas exposed to the Covid-19 pandemic.

The process of collecting secondary data series 2018, 2019, 2020 and limited observations were carried out in several areas affected by the pandemic in DKI Jakarta during January to February 2021. The limited observations referred to are about: 1) the times/hours of water flowing to customers; 2) adequate water needs; 3) received water quality; 4) peak hours of water usage; 5) other sources of drinking water; 6) customer complaints related to water supply services in the last 1 year; 7) types of complaints (water does not flow 24 hours, turbidity, small water supply, inappropriate prices); and 8) PAM's response to customer complaints. All of the above data is intended to strengthen the empirical analysis of the availability of clean water for drinking water for the population.

Data analysis used the correlational method with Pearson's product moment correlation test formula and continued with dynamic system analysis. A dynamic system with its black box analysis diagram is needed to help measure the availability of clean water for drinking water for residents in low pressure areas against small water quantities, continuity of dead water, and water quality.

4. Findings and Discussion

The water supply system (water supply systems) is one of the most important and expensive core public infrastructure. The main objective of a water supply utility is for this valuable asset to operate at the maximum possible efficiency at minimum cost during its design period. Therefore, the aspect of customer satisfaction is one of the performance indicators that must be considered especially as a reinforcement for empirical analysis of the availability of clean water for drinking water for the population [17].

Based on the data series obtained, the analysis results illustrate that there is a relationship between areas of low pressure to quantity (small water), continuity (dead water), and water quality in the East and West areas of DKI Jakarta from 2018, 2019 and the first semester of 2020; collectively shows different results when compared with the analysis on each complaint variable. For the eastern region of DKI Jakarta, showing the results of the analysis of the relationship between low pressure areas and the combination of all complaint aspects shows a diversity of relationship patterns from 2018, 2019, and the first semester of 2020. In 2018, the relationship between low pressure areas and complaints had a low level of relationship. 2019 has a strong relationship level, and the first semester of 2020 has a very strong relationship level category. The relationship of low pressure areas in each complaint variable shows that the low pressure area variable with complaints of small water has the highest correlation in 2018 and 2019, while in the first semester of 2020 low pressure with complaints of dead water has the highest correlation in the eastern area of DKI Jakarta.

In the western region of DKI Jakarta, the results of the analysis of the low pressure area relationship to all complaints from 2018, 2019 and the first semester of 2020 show the same level of relationship, namely at the level of a strong category relationship. Meanwhile, the low pressure relationship with each of the complaint variables from 2018, 2019, and the first semester of 2020 in the West Jakarta area is that the problem of low pressure areas has the highest correlation with water quality complaints.

The relationship between the number of customer complaints and the presence of low pressure areas for drinking water supply in DKI Jakarta can be seen in the following table.

IOP Conf. Series: Earth and Environmental Science **986** (2022) 012080 doi:10.1088/1755-1315/986/1/012080

Table 1. Relation of Number of Complaints to Low Pressure Areas in DKI Jakarta

Year	East Jakarta	West Jakarta
2018	0.254 Low	0.485 Strong
2019	0.564 Strong	0.611 Strong
2020 (first semester)	0.806 Very strong	0.498 Strong

Source: BR PAM DKI Jakarta, 2020-2021.

In the table above which explains the relationship between the number of customer complaints and the existence of low pressure areas for drinking water supply, it is followed by a further correlation analysis, thus showing that although low pressure areas affect the number of customer complaints regarding the quantity-continuity-quality of water received, it is not is the main factor that dominates the number of complaints of urination, dead water, and water. quality in the East and West. DKI Jakarta.

The following table further explains the correlation between customer complaints in areas of moderately high low pressure; This means that there is a very significant relationship between complaints in areas with low pressure water conditions.

Table 2. The Relationship with The Highest Correlation between Complaints to Low Pressure in DKI Jakarta

Year	East Jakarta	West Jakarta
2018	0.121 Low pressure - Small water	0.237 Low pressure - Water quality
2019	0.392 Low pressure - Small water	0.503 Low pressure - Water quality
2020 (first semester)	- 0.627 Low pressure - Water is off	- 0.374 Low pressure - Water quality

Source: BR PAM DKI Jakarta, 2020-2021.

From the two tables above, dynamic analysis steps (system dynamics) are then carried out which show that in the drinking water service area with the low pressure category, it turns out that all customers do not complain about low water conditions, dead water, and water quality. This can be seen from the data from the regression test in the two regions, where the R square number which shows the percentage of the influence of low pressure areas on complaints of water quality continuity received (complaints of small water, dead water and water quality) is still below 100% .

And based on the results above, a flow diagram of the system dynamics was obtained regarding the model for developing an environmental monitoring system (EMS) instrument for the availability of drinking water as a provider of ecosystem services, especially in areas exposed to Covid-19, DKI Jakarta.

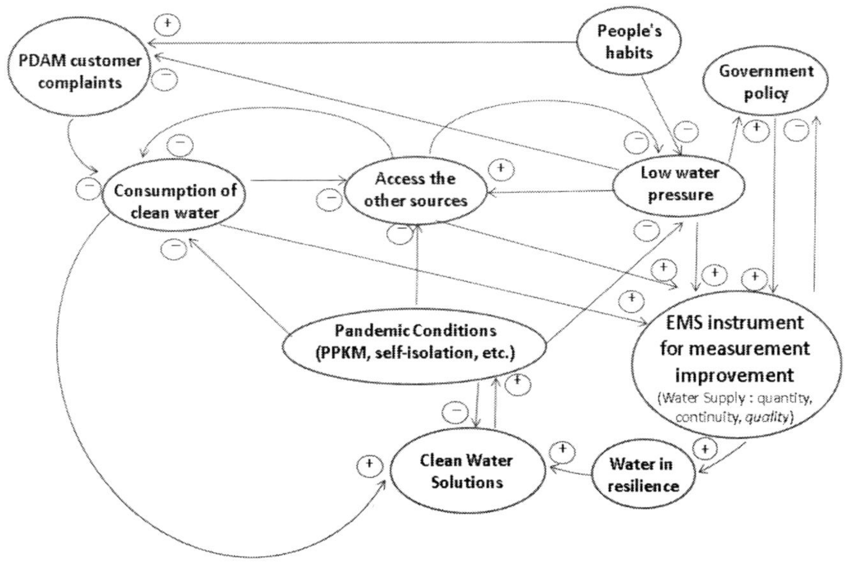

Figure 1. Environmental Monitoring System (EMS) Instrument Development Model for Availability of Drinking Water as an Ecosystem Service in Covid-19 Exposed Areas

The following diagram shows that there are other factors beyond the scope of the problem being studied at this time, which also affect the high level of complaints of urination, dead water, and water quality in the area. And finally obtained a black box analysis which produces the following black box diagram.

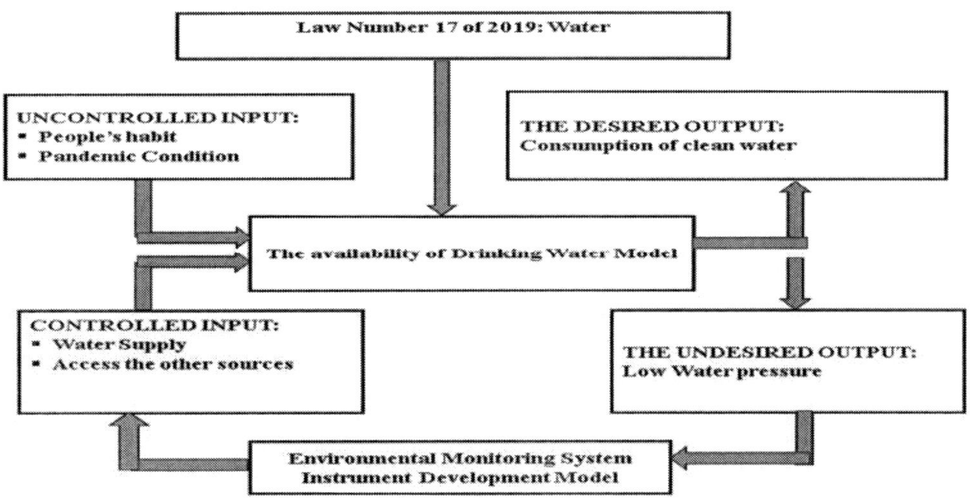

Figure 2. Black-box Diagram to Environmental Monitoring System (EMS) Instrument Development Model for Availability of Drinking Water as an Ecosystem Service in Covid-19 Exposed Areas

From this black box analysis diagram, it can finally be seen the efforts that can be made in measuring the availability of clean water for drinking water for residents in low pressure areas with respect to the small amount of water, continuity of dead water, and water quality in DKI Jakarta Province.

5. Conclusion and Further Research

The conclusion shows that although low water pressure occurs in areas exposed to Covid-19 in DKI Jakarta, it does not always affect customer complaints. This is because low water pressure is not entirely caused by small water, dead water, or water quality. Therefore, the environmental monitoring system for the availability of drinking water for the population as an ecosystem service pays more attention to other parameters besides small water, dead water, and water quality.

Furthermore, this study is still limited due to the problem of obtaining data directly from the source (PDAM customers), which in "normal" circumstances primary data will be more effective. The uniqueness and novelty of this research is that it obtains a description of the correlation between low-pressure service areas and low customer complaints regarding small water, dead water, and water quality. Usually, if the PDAM water pressure is low, customer complaints will be high due to small water, dead water, and water quality.

And in the future, as a follow-up to the findings of this study, it is necessary to carry out continuous analysis related to the following matters.

1. There needs to be an additional team and equipped with training development, especially to solve technical problems.
2. Cooperate with the community to supervise cases of water theft and illegal connections.
3. Technically and geographically; it is necessary to make relevant efforts in order to minimize the distribution, especially for the low pressure area.
4. Carry out water pressure control through a transmission control center in each zoning area at the same geographical height, so that can be controlled the water distribution process.

References

[1] Wall G 2002 Introduction To Life Support Systems and Sustainable *UNESCO - Encycl. Life Support Syst.*
[2] Iosef I. G and Genry M. L 2008 Creation of Closed Ecological Life Support Systems: Results, Critical Problems and Potentials *J. Sib. Fed. Univ. Biol.* **1** 19–39
[3] Sargen M 2019 Biological Roles of Water: Why is water necessary for life?
[4] Phillips F M, Hall G E and Black M E 2015 *Reining in the Rio Grande: People, Land, and Water* (United State: University Of New Mexico Prees)
[5] Krause F and Strang V 2016 Thinking Relationships Through Water *Soc. Nat. Resour.* **29** 633–8
[6] Keeler B L, Polasky S, Brauman K A, Johnson K A, Finlay J C, O'Neille A, Kovacs K and Dalzell B 2012 Linking water quality and well-being for improved assessment and valuation of ecosystem services *Proc. Natl. Acad. Sci. U. S. A.* **109** 18619–24
[7] Bateman I J, Mace G M, Fezzi C, Atkinson G and Turner K 2011 Economic analysis for ecosystem service assessments *Environ. Resour. Econ.* **48** 177–218
[8] Charles G, Klemick H, Massey M, Moore C, Newbold S, Simpson D, Walsh P and Wheeler W 2015 US Environmental Protection Agency valuation of surface water quality improvements vol 6816
[9] Liu S, Costanza R, Troy A, D'Aagostino J and Mates W 2010 Valuing New Jersey's ecosystem services and natural capital: A spatially explicit benefit transfer approach *Environ. Manage.* **45** 1271–85
[10] Costanza R, D'Arge R, Groot R de, Farber S, Grasso M, Hannon B, Limburg K, Naeem S, O'Neil R V, Paruelo J, Raskin R G, Sutton P and Belt M Van den 1997 The Value of the World's Ecosystem Services and Natural Capital *Nature* **387**
[11] Brauman K A, Daily G C, Duarte T K eo and Mooney H A 2007 The nature and value of ecosystem services: An overview highlighting hydrologic services *Annu. Rev. Environ.*

Resour. **32** 67–98
[12] Dodds W K, Bouska W W, Eitzmann J L, Pilger T J, Pitts K L, Riley A J, Schloesser J T and Thornbrugh D J 2009 Eutrophication of U. S. freshwaters: Analysis of potential economic damages *Environ. Sci. Technol.* **43** 12–9
[13] Goolsbee A and Syverson C 2021 Fear, lockdown, and diversion: Comparing drivers of pandemic economic decline 2020 *J. Public Econ.* **193** 104311
[14] Heuveline P and Tzen M 2021 Beyond deaths per capita: Comparative COVID-19 mortality indicators *BMJ Open* **11** 1–8
[15] UNDP 2020 COVID-19: New UNDP data dashboards reveal huge disparities among countries in ability to cope and recover. COVID-19: New UNDP data dashboards reveal huge disparities among countries in ability to cope and recover | UNDP
[16] Naidoo R and Fisher B 2020 Goals : pandemic reset *Nature* **583** 198–201
[17] Haider H, Sadiq R and Tesfamariam S 2014 Performance indicators for small-and medium-sized water supply systems: A review *Environ. Rev.* **22** 1–40

IOP Conf. Series: Earth and Environmental Science **986** (2022) 012081 doi:10.1088/1755-1315/986/1/012081

Determining River Absorption Capacity Using Simple Mathematical Model: A Case Study in Palur River, Sukoharjo Regency, Central Java Province, Indonesia

Setya Nugraha [1,*], Candra Purnawan [2], and Muhammad Habib Rizki [1]
[1]Department of Geography Education, Universitas Sebelas Maret, Indonesia
[2]Faculty of Mathematics and Natural Sciences, Universitas Sebelas Maret, Indonesia

[*]Email: setya.nug@gmail.com

Abstract. The increase in population causes a decrease in the quality of river water. Rivers are the most vulnerable ecosystem due to human activities. Humans throw the results of daily activities into rivers both from agricultural activities, domestic waste, and industrial activities. This waste results in a decrease in the quality of river water. Therefore the river needs to be managed properly. One of the efforts to manage the river is to determine the absorption capacity of the river itself. However, not many studies have been conducted regarding the absorption capacity of rivers, especially in small rivers. This study aims to determine the absorption capacity of the Palur River using a simple mathematical model called the Mass Balance (Neraca Massa). The results of calculations using the Mass Balance model obtained the average concentration value (Cr) from each segment of the Palur River. The worst river segment is Segment II with an average concentration (Cr) of TSS 54.22 mg/L, BOD 8.12 mg/L, and COD 26.21 mg/L. The calculation results show that the Palur River Segment II no longer has the capacity for TSS, BOD, and COD parameters. Waste that is dumped directly into the river needs to be processed first so that the load entering the river meets quality standards. Furthermore, it is hoped that the results of this study can be used as the first step in efforts to control water river pollution.

1. Introduction

Humans are very dependent on water. Humans need water for various commercial purposes. Water is the most essential substance needed by living things [1–4]. Freshwater as a basic human need occupies no more than 1% of the total available water in the world, where 97% of the world's water is found in the form of saltwater originating from the sea and only 3% is freshwater [5,6]. Water as an environmental component will affect and be influenced by other environmental components so it needs good management. Inefficient water management allows water quality to decline and result in a bad environment that will affect human health conditions and the lives of other living things [7]. In other words, the decline in water quality will threaten the quality of water resources [8] including biodiversity [9].

One of the water media on the earth's surface is in the river area. The river is a complex and orderly water system [10]. Rivers as ecosystems have a high level of biodiversity as well as being the most vulnerable habitat due to disturbances from human activities [9]. Palur River is one of the rivers whose upstream is in Karanganyar Regency and its downstream is in Sukoharjo Regency. However, in this study, the study of the absorption capacity of the Palur River is determined only to enter the administrative area of Sukoharjo Regency because it receives a lot of input from the pollutant load. The Palur River which passes through Mojolaban District, Sukoharjo Regency has a length of 7.6 km whose upstream point is in Jatisari Hamlet, Sapen Village and its downstream is located in Jogobondo Hamlet, Palur Village. The flow of the Palur River passes through 4 villages, namely: Sapen Village, Joho Village, Triyagan Village, and Palur Village which are included in one sub-district, namely Mojolaban District. The location of the Mojolaban sub-district which borders the city of Surakarta has resulted in this area getting a lot of direct influence from the city of Surakarta so that its development is quite rapid [11] and is accompanied by very rapid land conversion [12].

The Palur River was chosen as the object of the study of the absorption capacity because the Palur River is one of the rivers in the Bengawan Solo watershed which receives quite a lot of pollutant load discharges, both from agricultural, domestic, and industrial activities. The quality of the Palur River which flows down the Bengawan Solo River affects the quality of the Bengawan Solo River which is the raw material for clean water managed by PDAM. The pollutant load that enters the Palur River will enter the Bengawan Solo River flow and affect the quality of the supply of clean water raw materials.

Determining the status of the environment carrying capacity is important and effective in the use of resources for sustainable development [13]. However, information related to the environment, especially the status of the environment carrying capacity and absorption capacity of the river is not yet available properly. Limited information related to the environment is one of the obstacles in making policy so that the impact of pollution on the population cannot be estimated [14]. Therefore, this study seeks to determine the absorption capacity of the Palur River.

2. Method

2.1. Study Area
This research was conducted in the Palur River which is located in Mojolaban District, Sukoharjo Regency. Astronomically, the Palur River is located at a position of 07°35'33.55" South Latitude & 110°54'07.2" East Longitude (upstream) to 07°34'04.9" South Latitude & 110°51'47.5" East Longitude (downstream). The flow of the Palur River through Mojolaban District has a length of 7.6 km. The Palur River originates in Jatisari Hamlet, Sapen Village, while its downstream is in Jogobondo Hamlet, Palur Village. The flow of the Palur River passes through four villages, namely: Sapen Village, Joho Village, Triyagan Village, and Palur Village, with a total area of 11.52 km^2. Palur River is one of the rivers in the Bengawan Solo River Basin.

2.2. Data Collection
This study used descriptive-quantitative analysis. The data used in the preparation of this study are secondary data (S) and primary data (P). The secondary data used is an environmental document belonging to the Department of Environment (DLH) of Sukoharjo Regency. Secondary data obtained from DLH Sukoharjo Regency is data that describes the condition of the river during the dry season. In addition, primary data is also used by making direct measurements in the field during the rainy season.

Table 1. Data Collection

No.	Data	Type	Source
1.	Temperature (dry season)	S	
2.	Water discharge	S	
3.	TSS	S	
4.	pH	S	
5.	COD	S	
6.	BOD	S	DLH Sukoharjo Regency
7.	DO	S	
8.	*Total Coliform*	S	
9.	River length	S	
10.	River width	S	
11.	River depth	S	
12.	Rainfall	S	BPSDA Bengawan Solo
13.	Population	S	BPS
14.	Land use	P	Google Earth and field observation
15.	Temperature (rainy season)	P	Field observation

2.3. River Absorption Capacity

Determination of the carrying capacity of the pollution load can be determined using the mass balance method. The mass balance can be used to determine the average concentration of downstream flow originating from pollutant sources, both point sources and non-point sources. Determination of the carrying capacity of river pollution loads using the mass balance method based on the Keputusan Menteri Lingkungan Hidup No. 110 Tahun 2003 concerning Guidelines for Determining the Capacity of Water Pollution Loads in Water Sources.

This method assumes that several streams that meet at a final flow can be analyzed by mass balance. Mathematically it can be formulated as follows:

$$Cr = \frac{\Sigma CiQi}{\Sigma Qi} = \frac{\Sigma Mi}{\Sigma Qi} \tag{1}$$

After obtaining the Cr value, then a comparison is made with the quality standard value of each parameter listed in the applicable regulations. If the Cr value is below the quality standard, then the area is still able to accept the pollution load, whereas if the Cr value is above the quality standard, the area is no longer able to accept the pollution load. The calculation of the carrying capacity of the pollution load is carried out on the receiving water body, namely the Palur River by dividing the Palur River into several parts or segments. River segmentation is carried out by considering the pollutant load that enters the receiving water body representing the upstream, middle, and downstream sections.

3. Results and Discussion

3.1. Water Quality of Pollutant Source

The water quality in the main river is determined by comparing the concentration value of each parameter with the river water quality standard. The value of river water quality standards is regulated in Government Regulation (PP) no. 82 of 2001 concerning Water Quality Management and Water Pollution Control. The parameters used to determine the water quality of the Palur River include TSS,

BOD, and COD. The following presents the results of the comparison of the value of each parameter with river water quality standards.

3.1.1. TSS

The results of the comparison of TSS concentration with water quality standards are presented in the table below.

Table 2. TSS Concentration of Pollutant Source

No.	Sample	Code	Standard	TSS (mg/l)
1.	Sapen Village Farm	P1	100	14
2.	Triyagan Village Domestic	D1	30	9
3.	Joho Village Farm	P2	100	54
4.	Triyagan Village Milkfish Industry	I1	60	168
5.	Palur Village Domestic	D2	30	42
6.	Zaini Showroom Farm	P3	100	198
7.	Palur Village Tannery Industry	I2	60	507

3.1.2. BOD

The results of the comparison of BOD concentration with water quality standards are presented in the table below.

Table 3. BOD Concentration of Pollutant Source

No.	Sample	Code	Standard	BOD (mg/l)
1.	Sapen Village Farm	P1	50	2,82
2.	Triyagan Village Domestic	D1	30	19,12
3.	Joho Village Farm	P2	50	4,47
4.	Triyagan Village Milkfish Industry	I1	50	414
5.	Palur Village Domestic	D2	30	35,10
6.	Zaini Showroom Farm	P3	50	5,28
7.	Palur Village Tannery Industry	I2	50	479,37

3.1.3. COD

The results of the comparison of COD concentration with water quality standards are presented in the table below.

Table 4. COD Concentration of Pollutant Source

No.	Sample	Code	Standard	COD (mg/l)
1.	Sapen Village Farm	P1	100	<3,91
2.	Triyagan Village Domestic	D1	100	38,5
3.	Joho Village Farm	P2	100	16,69
4.	Triyagan Village Milkfish Industry	I1	100	1.306,90
5.	Palur Village Domestic	D2	100	65,27
6.	Zaini Showroom Farm	P3	100	26,22
7.	Palur Village Tannery Industry	I2	100	1.425,84

3.2. Water Quality of Palur River

The water quality in the main river is determined by comparing the concentration value of each parameter with the river water quality standard. The value of river water quality standards is regulated in Government Regulation (PP) no. 82 of 2001 concerning Water Quality Management and Water Pollution Control. The parameters used to determine the water quality of the Palur River include TSS, BOD, and COD.. The following presents the results of the comparison of the value of each parameter with river water quality standards.

3.2.1. TSS

The results of the comparison of TSS concentration with river water quality standards are presented in the table below.

Table 5. TSS Concentration of Palur River

No.	Sample	Code	Standard	TSS (mg/l)
1.	Headwater	S1	50	6
2.	Colo irrigation	S2	50	11
3.	Zaini Showroom Bridge	S3	50	38
4.	Downstream	S4	50	14

3.2.2. BOD

The results of the comparison of BOD concentration with river water quality standards are presented in the table below.

Table 6. BOD Concentration of Palur River

No.	Sample	Code	Standard	BOD (mg/l)
1.	Headwater	S1	3	<2
2.	Colo irrigation	S2	3	2,85
3.	Zaini Showroom Bridge	S3	3	<2
4.	Downstream	S4	3	2,66

3.2.3. COD

The results of the comparison of COD concentration with river water quality standards are presented in the table below.

Table 7. COD Concentration of Palur River

No.	Sample	Code	Standard	COD (mg/l)
1.	Headwater	S1	25	<3,91
2.	Colo irrigation	S2	25	4,80
3.	Zaini Showroom Bridge	S3	25	<3,91
4.	Downstream	S4	25	4,24

3.3. Palur River Absorption Capacity

Based on the consideration of the input pollutant load, the Palur River is divided into three segments from upstream to downstream. The following shows the results of the calculation of the status of the pollution load capacity in the Palur River using the Mass Balance method.

3.3.1. Segment I

The results of the calculation of the pollution load capacity in Segment I of the Palur River are shown in the following table.

Table 8. Segmen I

No.	Sample	Discharge (m/dtk)	TSS (mg/l)	BOD (mg/l)	COD (mg/l)	C_R (mg/l)		
						TSS	BOD	COD
1.	P1	0,21197	14	2,82	3,91			
2.	D1	0,02109	9	19,12	38,52	11,91	3,37	5,60
3.	S2	0,41675	11	2,85	4,80			
Standard		Class II				50	3	25
		Class III				400	6	50

3.3.2. Segment II

The results of the calculation of the pollution load capacity in Segment II of the Palur River are shown in the following table.

Table 9. Segmen II

No.	Sample	Discharge (m/dtk)	TSS (mg/l)	BOD (mg/l)	COD (mg/l)	C_R (mg/l)		
						TSS	BOD	COD
1.	P2	0,02232	54	4,47	16,69			
2.	I1	0,00014	168	414	1.306,9	54,22	8,12	26,31
3.	D2	0,00091	42	35.1	65,27			
Standard		Class II				50	3	25
		Class III				400	6	50

IOP Conf. Series: Earth and Environmental Science **986** (2022) 012081 doi:10.1088/1755-1315/986/1/012081

3.3.3. Segment III

The results of the calculation of the pollution load capacity in Segment III of the Palur River are shown in the following table.

Table 10. Segmen III

No.	Sample	Discharge (m/dtk)	TSS (mg/l)	BOD (mg/l)	COD (mg/l)	C_R (mg/l)		
						TSS	BOD	COD
1.	S3	0,43974	38	2	3,91			
2.	P3	0,00923	198	5,28	26,22	42,63	3,45	8,47
3.	I2	0,0013	507	479,37	1.425,84			
Standard		Class II				50	3	25
		Class III				400	6	50

Determination of the pollution load capacity of the Palur River is carried out by dividing the river into three segments. Determination of river segmentation based on the consideration of the input pollutant load. The status of the carrying capacity of the river pollution load is obtained from the comparison of the average concentration value of the key parameters of each segment with the class II and III water quality standards regulated in PP. 82 of 2001 concerning Water Quality Management and Water Pollution Control. The class II water quality standard was chosen because the Palur River is a river whose water quality class has not yet been determined. Based on PP No. 82 of 2001 it is stated that rivers that have not yet determined the water quality class are grouped into class II rivers. If the existing concentration value of each parameter is below the quality standard value, it is considered that the river still has the capacity for that parameter, in other words the river is still able to accept the input load without causing the river to become polluted. On the other hand, if the existing concentration value of each parameter is above or beyond the quality standard, it is considered that the river no longer has the capacity for that parameter. This means that efforts need to be made to reduce the pollutant load that enters the river in order to meet quality standards.

The calculation results show that according to the water quality standard class II, Segment I of the Palur River has a capacity for the TSS and COD parameters, but it no longer has the capacity for the BOD parameter and has to reduce the load that enters the river by 0.37 mg/l. Segment II of the Palur River no longer has the capacity for all parameters of class II quality standards, even it still exceeds the BOD class III quality standards. Thus, it is necessary to reduce the pollutant load entering the river, which is 4.22 mg/l for TSS class II; 5.12 mg/l and 2.12 mg/l for BOD, class II and III, respectively; and 1.31 mg/l for COD class II. Segment III of the Palur River still has capacity for TSS and COD parameters, but it no longer has capacity for class II BOD parameters and must reduce the load that enters the river by 0.45 mg/l.

Viewed from upstream – downstream, the status of the absorption capacity of the Palur River shows the same trend pattern as the status of the water quality of the Palur River, which goes up and then down. Segment I of the Palur River receives input from the pollutant load from agricultural and domestic waste. Segment I of the Palur River has one parameter that exceeds the class II water quality standard, namely BOD. Segment II of the Palur River receives input in the form of agricultural, industrial and domestic waste. In Segment II there was an increase in the average concentration of the three parameters, namely TSS, BOD, and COD to exceed the class II water quality standard. Segment III of the Palur River receives input from the pollutant load from agricultural and industrial waste. In Segment III there is a decrease in the average concentration of three key parameters, but there is one parameter that exceeds the class II water quality standard, namely BOD. Based on this description, it can be seen that the upstream and downstream areas of the Palur River still have capacity for key parameters, while the middle region of the Palur River no longer has capacity for key parameters.

Segment II of the Palur River, which is in the middle region, is the worst segment in terms of its capacity status. This means that the segment receives input of a large enough pollutant load so that it is no longer able to accommodate the pollutant load, both TSS, BOD, and COD parameters. Based on the results of the research, it is known that in Segment II of the Palur River, the three key parameters have exceeded the quality standard for Class II water quality, and have even exceeded Class III water quality for BOD parameters. This can happen because Segment II of the Palur River is the middle part of the river that receives various inputs of pollutant loads. Segment II of the Palur River receives input from three types of pollutant sources which are located close to each other, namely agricultural (P2), industrial (I1), and domestic (D2) pollutant sources. The accumulation of these three pollutant sources and the relatively close distance between pollutant sources results in the average concentration values of the three key parameters, all of which exceed the Class II water quality standard values and make the Palur River Segment II the worst segment that no longer has the capacity to accommodate three different sources. key parameters.

4. Conclusion

The absorption capacity of the Palur River according to class II water quality standards based on PP no. 82 of 2001 concerning Water Quality Management and Water Pollution Control shows that segments I and III no longer have the capacity for BOD parameters, while segment II which is the worst segment has no longer capacity for all parameters, namely TSS, BOD, and COD.

Aknowledgement

The authors would like to thank the Environmental Office of Sukoharjo Regency for the opportunity to work together.

References

[1] Garg S K 1979 *Water Resources and Hydrology [For Civil Engg. Degree, M. Tech., AMIE (Section B), Other General Courses in Hydrology Or Agriculture Engg: And for Professionals].* (New Delhi: Khanna Pub.)

[2] Kodoatie R J and Sjarief R 2010 *Tata Ruang Air* (Yogyakarta: Penerbit Andi)

[3] Pepper LL; Gerba CP; Brusseau ML 1996 *Pollution Science* (California: Academic Press)

[4] Pimentel D, Berger B, Filiberto D, Newton M, Wolfe B, Karabinakis E, Clark S, Poon E, Abbett E and Nandagopal S 2004 Water Resources: Agricultural and Environmental Issues *Bioscience* **54** 909–18

[5] Asdak C 2009 *Hidrologi dan Pengelolaan Daerah Aliran Sungai* (Yogyakarta: Gadjah Mada University Press)

[6] Gerba C P, Pepper I L, Gerb C P and Brusseau M 2006 *Risk Assessment. Environmental and Pollution Science* (California: Academic Press, San Diego, CA)

[7] Chen B, Wang M, Duan M, Ma X, Hong J, Xie F, Zhang R and Li X 2019 In Search of Key: Protecting Human Health and the Ecosystem from Water Pollution in China *J. Clean. Prod.* **228** 101–11

[8] Zhao C S, Yang S T, Sun Y, Zhang H T, Sun C L, Xu T R, Lim R P and Mitrovic S M 2018 Estimating River Accommodation Capacity for Organic Pollutants in Data-scarce Areas *J. Hydrol.* **564** 442–51

[9] Xiong W, Ni P, Chen Y, Gao Y, Li S and Zhan A 2019 Biological Consequences of Environmental Pollution in Running Water Ecosystems: A Case Study in Zooplankton

Environ. Pollut. **252** 1483–90
[10] Maryono A 2018 *Restorasi Sungai* (Yogyakarta: Gadjah Mada University Press)
[11] Setiawan A R 2015 *Perkembangan Wilayah Kabupaten Sukoharjo Antara Tahun 2008 dan 2013* (Universitas Muhammadiyah Surakarta)
[12] Desianingtyas M 2015 *Dampak Pertumbuhan Penduduk Terhadap Alih Fungsi Lahan di Kabupaten Sukoharjo Tahun 2000 - 2013* (Universitas Muhammadiyah Surakarta)
[13] Lu Y, Xu H, Wang Y and Yang Y 2017 Evaluation of Water Environmental Carrying Capacity of City in Huaihe River Basin Based on the AHP Method: A Case in Huai'an City *Water Resour. Ind.* **18** 71–7
[14] Garg T, Hamilton S E, Hochard J P, Kresch E P and Talbot J 2018 (Not so) Gently Down the Stream: River Pollution and Health in Indonesia *J. Environ. Econ. Manage.* **92** 35–53

IOP Conf. Series: Earth and Environmental Science **986** (2022) 012082 doi:10.1088/1755-1315/986/1/012082

Hydrologic response analysis to calculate water yield of Cileungsi River

A P Nugraha[1], M P Hadi[1,2*] and N Christanto[1]

[1] Faculty of Geography, Universitas Gadjah Mada, Yogyakarta, Indonesia
[2] Center for Environmental Studies, Universitas Gadjah Mada, Yogyakarta, Indonesia

*Email: mphadi@yahoo.com

Abstract. The study of watershed hydrological response is essentially to understand the hydrological characteristics produced by specific rainfall on the watershed area. These flow parameters also rely on watershed factors such as land use, soil type, and watershed morphometry. The research was performed in the Cileungsi watershed, Bogor Regency, West Java. This research aimed to determine the hydrological response and to investigate the water yield of the Cileungsi watershed. Distributed hydrological modelling was used. Curve Numbers (CN) were estimated using SCS-UH triangular hydrograph technique, while the baseflow was determined using the exponential recession method. Muskingum-Cunge flow routing techniques were used to calculate the flow hydrograph. Yearly water yield was estimated based on measured rainfall data. The simulated hydrograph of several flood event shows similarities to the measured hydrograph. It is based on calibration and optimization of the model parameters. The objective function values are less than 15 percent, with peak discharge range from 654-1,729 m^3/second. The hydrograph form tends to feature a rising limb that climbs rapidly with a sloping recession limb. The result of this research shows that the Cilengsi water yield are 4,520.33 mm.

1. Introduction

Hydrological response characteristics are essential for understanding the status of watershed resources using the study of the hydrograph. The hydrograph had a strong relationship with the transformation of land use, watershed morphometrics, and soil type characteristics. Rainfall is an essential stimulant of hydrological processes, although its effects are unknown [1]. Due to the differences in the physical properties of each soil type, land uses have a significant effect on soil physical features [2].

The Cileungsi watershed, situated in Bogor Regency, West Java Province, was chosen for this research. Industries, villages, rice farms, moors, mixed plantations, shrubs, woods, and open fields exist in the Cileungsi watershed. Watershed land use and soil types were classified as part of the hydrograph tracking study. As a result, we wrap off the hydrograph reactions by following the rainfall as it transforms into a runoff.

The objectives of this study are: (1) determine the hydrological responses of the Cileungsi watershed and (2) comprehend the water yield characteristics of the Cileungsi watershed. The hydrograph, which was formed by rain that fell inside the watershed region, was used to evaluate the characteristics of hydrological reactions. The HEC-HMS program was used to simulate the hydrograph.

2. Study Area

The Cileungsi watershed has an area of 210.13 km². The Cileungsi watershed is located in the 48s zone on the Universal Transverse Mercator (UTM) at latitudes 703621-719791 mE and longitudes 9266192-9286653 mN on the Universal Transverse Mercator (UTM). The Cileungsi watershed has many sub-regencies, including Babakanmadang, Citeureup, Gunungputri, Jonggol, Klapanunggal, and Sukamakmur. Babakanmadang is a sub-regency of the Cileungsi watershed. It is located on a slope on the northside of Pangrango Mountain in West Java, and it is the upstream watershed portion of the Cileungsi River. This watershed was subdivided into 13 sub-watersheds. Furthermore, the Cileungsi watershed only had one outlet, which was located in the Gunungputri sub-district.

3. Methods

The hydrograph at the outlet of watershed was generated from rainfall base on conceptual hydrological distributed modeling of HEC-HMS. Research on hydrological responses uses modelling concept to understand waterflow characteristics. This research concept was shown in the Figure 3.1.

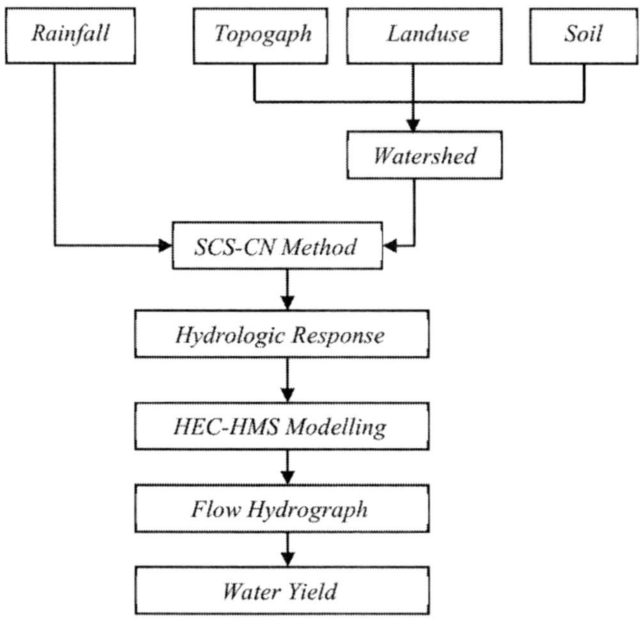

Figure 3.1 Research Framework

3.1. Tools and Materials

The SRTM DEM and Bogor Regency RBI Map (scale 1:25.000) are used in determining watershed segmentation and land use classification. Soilgrids from Geoportal data are used to classify the soil hydrologic groups. The CHRS Geoportal data is used as the source of hourly interval rainfall data. Water discharge data were recorded directly from the field using AWLR.

This research uses ArcGIS software, HEC-GeoHMS, and HEC-HMS as tools. ArcGIS Software and HEC-GeoHMS are used in determining the watershed boundary, while HEC-HMS software are used in building the flood hydrograph model.

3.2. Watershed Segmentation

Using sequential Terrain Processing, which includes DEM reconditioning, fill sinks, flow direction, flow accumulation, stream definition, stream segmentation, catchment delineation, catchment polygon processing, and drainage line, the boundaries of the Cileungsi watershed segmentation can be determined and mapped. To identify the watershed, it is necessary to investigate topographic boundaries and watershed flow characteristics.

3.3. Defining the Curve Number (CN)

The CN value is used to define the characteristics of river flow depending on the watershed land area. As part of this research, the Soil Conservation System-Curve Number (SCS-CN) classification technique was used to categorize CN numbers. This method is used to calculate direct runoff from rainfall and the potential for erosion that occurs [3]. SCS-CN can be used to determine the infiltration that occurs and can be used to calculate the sediment rate in the watershed [4]. The SCS-CN technique calculates the amount of rainfall that has collected and is being discharged as direct runoff.

The categorization of the watershed was based on the soil types and land-use characteristics of the watershed. Hydrological Soil groups are classified based on the characteristics of the surface conditions that affect the infiltration rate and the soil profile that affects the value of the transmission rate [5].The hydrological soil group (HSG) classification system is used to classify soil types, then further categorized by soil textures (Table 3.1)

Table 3.1 Classification of Soil Hydrology based on the Soil Texture

Texture	Minimum Infiltration Rate (mm/hour)	Classification based on SCS
Sand	210	A
Loamy Sand	61	A
Sandy Loam	26	B
Loam	13	B
Silty Loam	6,9	C
Sandy Clay Loam	4,3	C
Silty Clay Loam	2,3	D
Clay Loam	1,5	D
Sandy Clay	1,3	D
Silty Clay	1,0	D
Clay	0,5	D

Source: Triatmodjo, 2014

Landuses are classified by overlaying the data of landuse with HSG classification (Table 3.2).

Table 3.2 Classification of SCS-CN Land Use

Landuse Type	Hydrological Soil Group			
	A	B	C	D
Treated and Planted Soil				
- With Conservation	72	81	88	91
- Without Conservation	62	71	78	81
Meadow				
- Bad Condition	68	79	86	89
- Good Condition	39	61	74	80
Meadow: Good Condition	30	58	71	78
Forest				
- Infrequent plant, bad coverage	45	66	77	83
- Good coverage	25	55	70	77
Open field, lawn, golf field				
-Good condition: grass coverage 75% or more of surface area	39	61	74	80
-Medium condition: grass coverage 50% -75% of surface area	49	69	79	84
Commercial and Businness area (85% impermeable)	89	92	94	95
Industrial area (72% imprmeable)	81	88	91	93
Settlement				
Surface area % impermeable				
1/8 acre or less 65	77	85	90	92
1/4 acre 38	61	75	83	87
1/3 acre 30	57	72	81	86
1/2 acre 25	54	70	80	85
1 acre 20	51	68	79	84
Parking lot, roof, carpot (yard)	98	98	98	98
Road				
- Pavement with drainage	98	98	98	98
- Gravel	76	85	89	91
- Soil	72	82	87	89

Source: Triatmodjo, 2014

Watershed or sub-watershed $CN_{\text{-}Composite}$ are measured from the overlayed of landuse classification and HSG data (1).

$$CN_{Composite} = \frac{\sum A_i CN_i}{\sum A} \quad (1)$$

Where A_i is the surface of every landuse which equal to (Ha), CN_i is the CN value of the same landuse and A is the sub-watershed surface area (Ha).

After calculating the $CN_{\text{-}Composite}$ Value, the Antecedent Moisture Condition is determined (AMC). AMS was computed using rainfall data beginning five days prior to the examined date. AMC was classified into three water content situations based on SCS: The typical AMC (AMC II) state is defined as one in which the values range between 36 and 53 mm. Second, the dry state (AMC I) occurs when values are less than 36 mm, while the moist condition (AMC III) occurs when values exceed 53 mm. AMC II was computed using the CN value under normal condition, while AMC I and AMC III were determined using the following formulas:

$$AMC\ (I) = \frac{4{,}2\ CN\ (II)}{10-0{,}058\ CN\ (II)} \tag{2}$$

$$AMC\ (III) = \frac{23\ CN\ (II)}{10+0{,}13\ CN\ (II)} \tag{3}$$

3.4. Modeling Flood Hydrograph using HEC-HMS

Hydrology responses was analysed by using HEC-HMS software as the flood hydrograph modeller. The HEC-HMS model is used to analyze the hydrological conditions of a watershed. HEC-HMS includes a soil moisture algorithm to be a requirement in the consideration of the analysis of the conversion of rain to runoff [6]. This model uses SCS-CN as loss methods, CN lag as hydrograph transform method, Recession Constant as baseflow calculation method, and Muskingum-Cunge for hydrograph routing method. Validation and calibration are conducted using multiple trial and error by comparing the modeled with observed field data.

Measured hydrograph or the observed field data hydrograph was calculated using slope area method. The hydrograph discahrge was calculated using the formula as below (4):

$$Q = \frac{AR^{2/3} \times S^{1/2}}{n} \tag{4}$$

Where, Q is the water discharge (m^3/s), A is the surface area (m^2), R is the hydraulic radius (m) acquired from the result of dividing the river width with wet perimeter, S is the hydrolic gradient, and n is the Manning roughness coefficient as described in the Table 3.3 below.

Table 3.3 Rouhness Manning Coefficient

Channel Condition			n Value
Base Material	Soil	n0	0.020
	Rock		0.025
	Soft Gravel		0.024
	Rough Gravel		0.028
Channel ununiformity level	Soft	n1	0.000
	Less soft		0.005
	Medium		0.010
	Rough		0.020
Channel cross-section variaton	Gradually Change	n2	0.000
	Occasionally change		0.005
	Often change		0.010 – 0.015
The impact of the building, narrowing, and others to the cross section	Ignored	n3	0.000
	Less impactful		0.010 – 0.015
	Impactful enough		0.020 – 0.030
	Very Impactful		0.040 – 0.060
Plant	Low	n4	0.005 – 0.010
	Medium		0.010 – 0.025
	High		0.025 – 0.050
	Very High		0.050 – 0.100
Meander Level	Low	m5	1.000
	Medium		1.150
	High		1.300

Note : n = (n0 + n1 + n2 + n4) m5

Source: Chow, 1959

IOP Conf. Series: Earth and Environmental Science **986** (2022) 012082 doi:10.1088/1755-1315/986/1/012082

Modelling process was repeated until the modeled hydrograph result is similar with the measured hydrograph. Similiarity criteria was seen from the peak discharge, time to peak, and the shape of hydrograph. Objective function result must be less than 10%. Each parameter models are calibrated in each of the sub-watersheds.

3.5. Water Yield Calculation
Water yield is the amount of water aggregate that leaves and enters a unit of hydrological response [7]. Water yield itself consists of the accumulation of output from the watershed for an annual period. Water yield of Cileungsi river was calculated base on accepted rainfall-runoff model. Base on the rainfall data of the whole year, and inputting to the model, the hydrograph for a year was generated. Base on these hydrograph, then the volume of water yield can be estimated.

4. Results and Discussion

4.1. Watershed Characteristic
The Cileungsi watershed was located in the Bekasi upstream river section, which was densely populated with industries and settlements but was still surrounded by green space in forests, plantations, and sh rubs. At the outlet of the Cileungsi watershed has a river flow monitoring station (Stasiun Pengamat Aliran Sungai/SPAS). The outlet lay on 66 m above sea level terain while the most upper area of the cathment on the highes of 1554 m.

The Cileungsi watershed was segmented into seven slope classes: flat, undulating, undulating-rolling, rolling-hilly, hilly-steeply dissected, steeply dissected-mountainous, and mountainous. The geology of the area was dominated by many major formations, including the Jatiluhur and Klapanunggal formations. As shown in Table 4.1, the most prevalent soil types in the area are acrisols, andosols, and nitisols.

Table 4.1. Soil Hydrology Group of Cileungsi Watershed

Soil Types	HSG	Texture	Surface Area (Ha)
Andosols	B	Loam	1571
Acrissols	C	Sandy Clay Loam	17406
Nitisols	D	Clay	2036
		Total	21013

Source: Data processing results, 2020

Landuses within Cileungsi watershed was classified by using topographic map. Landuse classification shown in landuses map in the Figure 4.1 and Table 4.2.

IOP Conf. Series: Earth and Environmental Science **986** (2022) 012082 doi:10.1088/1755-1315/986/1/012082

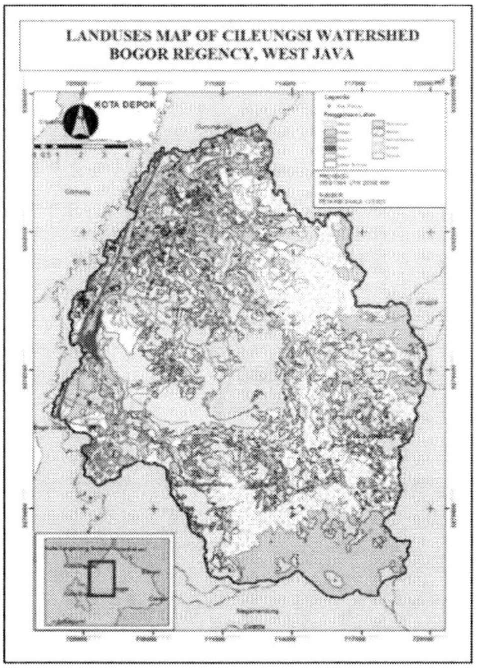

Figure 4.1 Landuses Map

Table 4.2. Landuses of Cileungsi Watershed

Landuses	Surface Area (Ha)	Percentage (%)
River	189	0.90
Lake	21	0.10
Roadway	126	0.60
Industry/building area	205	0.98
Settlement	2854	13.58
Rice Field	1972	9.38
Moor	3830	18.23
Mixed Plantation	4121	19.61
Open Spaces	1009	4.80
Bushes/shurbs	4544	21.62
Forest	2142	10.19
Total	21013	100.00

Sources: Interpretation of topo map, 2020

The surface area of the Cileungsi watershed is 210.13 km². Bushes/shurbs (21.62 %) and mixed plantations occupy this Watershed (19.61 %). As shown in the table, other landuses exist, including moors (18.23 %) and rice fields (9.38 %). Then there are woods (10.19 %) in the upstream portion of the Watershed. On the other hand, there are barren fields (4.8 %) and lakes (0.1 %). Among the others, as shown in Figure 4.1, are settlements (1.58 %), industrial structures (0.98 %), and highways (0.6 %). The proportion was determined using the Watershed's surface area.

4.2. Rainfall Characteristic

Apart from the watershed shape factor, duration and rainfall intensity have a significant effect on the shape of hydrographs [8]. Rainfall is a component of the hydrology cycle, and it is critical for hydrological study within a watershed. Topographic variables influenced the differences in rain properties. Different intensities in different locations combined with the effect of orographic rain result in variations in rain characteristics. Orographic rain occurs when there is a significant variation in air pressure and temperature between a high and low elevated site and is affected by the direction of the wind that brought vaporized water [9].

The rainfall data for the Cileungsi Watershed were collected using CHRS Geoportal data and processed in line with Figure 4.2, which depicts the rainfall activities in the Cileungsi Watershed from 14 March 2020 to 11 August 2020. The rainfall data gathered over six months has a one-hour time interval. The highest rainfall of 99 mm was recorded on 16 March 2020, with a rain thickness of 166 mm during five hours and a rain intensity of 33.2 mm/hour.

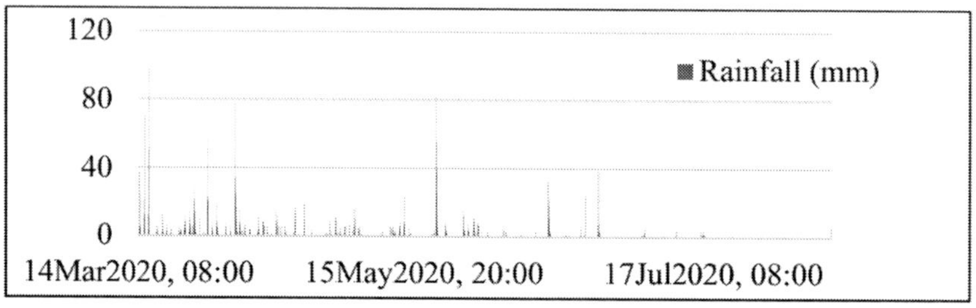

Figure 4.2 CHRS Cileungsi Watershed Rainfall Data, Period of 14 March 2020 - 11 August 2020

Source: CHRS data processing result, 2021

4.3. CN Value Analysis

Soil hydrology and landuses classifications are used in analyzing the watershed CN value and HEC-HMS model construction. CN value used in the model are shown as in the Table 4.3.

Table 4.3. CN Optimalization

Sub-Watershed	Initial Value	Optimalized Value					
	Normal	15-Mar-20	16-Mar-20	26-Mar-20	4-Apr-20	18-May-20	11-Jun-20
A	84.87	81	83	60	80	55	81
B	74.95	65	84	75	85	62	71
C	79.45	67	83	75	80	63	72
D	85.49	67	85	76	81	62	71
E	87.53	67	85	74	81	62	71
F	83.34	70	84	74	82	62	72
G	79.39	67	85	74	82	66	72
H	80.74	69	85	72	82	66	72
I	79.25	67	85	71	82	65	72
J	72.19	58	83	71	80	65	70
K	71.26	57	87	71	80	64	80
L	77.11	67	81	63	78	58	68
M	73.55	67	84	60	75	63	68
N	71.35	61	85	71	80	64	70

Source: Data processing result, 2020

Analysis of CN classification was based on sub-watershed segmentation units. CN Normal value variation in each Sub Watershed within Cileungsi watershed is ranging from 71.26 – 87.53, while for the calibration is ranging from 55-87 as corresponded to the table above. CN is the function of watershed model characteristics based on the soil type, plant coverage, landuses, humidity and cultivation [3].

4.4. Flood Hydrography HEC-HMS Model Analysis

Hydrograph analysis was done by using HEC-HMS model. Waterflow model scheme as shown in Figure 4.3 had 13 subbassins, 7 junctions, 6 reachs and 1 outlet. The watershed morphometric parameters used in this modeling are the area of each Sub-watershed in the Cileungsi watershed. The rain model used in the hydrograph model is the specified hyetograph. In this case, the transform parameter is that the lag time is influenced by the initial abstraction, besides that the length of the river and the slope of each sub-watershed greatly affect the size or length of the lag time. In the baseflow model, there are several recession parameters, namely initial discharge, recession constant, and ratio to peak. In addition to the previous parameters, there are other parameters that build a flood hydrograph model using routing on each section of the river. Muskingum and Cunge methods in the HEC-HMS model are used for flow tracking based on channel characteristics at each reach.

Model parameter was divided into parts including; watershed morphometry, loss, transform and baseflow. Those parameters are shown in the Table 4.4.

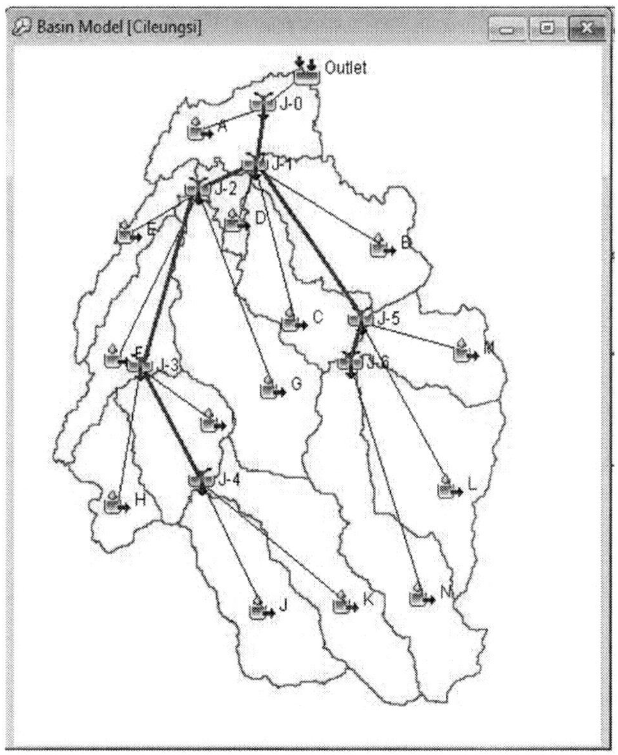

Figure 4.3 Cileungsi Watershed Waterflow model scheme

Table 4.4. HEC-HMS Model Parameter

Sub-watershed	Watershed Morphometry	Loss Method: SCS CN			Transform Method	Baseflow Method		
	Surface area	Initial Abstraction (mm)	CN AMC II	Impervious (%)	Lag Time (min)	Initial Discharge (m^3/s)	Recession Constant	Ratio to peak
A	13.52	2.26	84.87	5.6	60	1.39	0.15	0.150
B	18.48	4.24	74.95	4.2	46	1.13	0.17	0.180
C	12.75	3.29	79.45	3.7	39	2.32	0.18	0.145
D	3.38	2.16	85.49	5.2	23	0.82	0.17	0.150
E	6.88	1.81	87.53	4	31	2.79	0.172	0.160
F	13.25	2.54	83.34	4	40	1.96	0.18	0.154
G	24.74	3.30	79.39	4	52	2.45	0.15	0.150
H	8.46	3.03	80.74	4	33	1.27	0.19	0.180
I	15.9	3.32	79.25	4	53	1.97	0.2	0.150
J	16.41	4.89	72.19	5	65	0.78	0.15	0.160
K	19	5.12	71.26	3	69	3.12	0.12	0.130
L	28.06	3.77	77.11	3	59	1.53	0.17	0.160
M	6.16	4.57	73.55	5	59	1.53	0.16	0.130
N	23.88	5.10	71.35	4	51	1.99	0.18	0.160

Source: Data processing result, 2020

The hydrology model was build by the characteristics, elements, and paramaters that comes from Cileungsi Watershed. CN Value for 6 flood activity models have different value compaerd to the value of normal conditions. Flood activity model of 4 April 2020 has the closest CN Value to CN AMC II (normal). The other five have different CN Values compared to normal conditions.

This research adopt Muskingum-Cunge routing model. Muskingum-Cunge method on HEC-HMS used in tracking based on the characteristics of each channel in each river segment. Calibration process in routing was done by trial and error. Parameter of the process are length, slope, index flow, shape, and invert as shown in Table 4.5.

Table 4.5 Munskingum-Cunge Routing Parameter

Reach	R1	R2	R3	R4	R5	R6
Length (m)	6512.84	3011.48	6828.43	4971.01	9001.45	2272.92
Slope (m/m)	0.0057	0.0057	0.0041	0.0079	0.0044	0.0079
Manning	0.017	0.021	0.021	0.021	0.021	0.021
Index flow (m^3/s)	150	150	150	150	150	150
Width (m)	50	40	20	16	30	15
Invert (m)	6	4.5	4	4	5	5

Source: Data processing result, 2020

Hydrograph model which match with the measured flood hydrograph happen within 15 March 2020, 16 March 2020, 26 March 2020, 4 April 2020, 18 May 2020 and 11 June 2020 and as shown in the Figure 4.4. Based on these 6 flood models, there are 3 good objective function results that had value of less than 10% they are the flood model of 18 May 2020 (1.39%), 4 April 2020 (7.56%) and 11 June 2020 (8.85%). In the other side, the rest models have objective function above 10% they are the flood model of 15 March 2020 (10.8%); 26 March 2020 (12.49%) and 16 March 2020 (13.98%). This happen because of inacuratte CHRS rainfall data input which resulted in the increase of objective function value to go beyond 10%. The result from 6 flood models were shown in the Table 4.6.

Table 4.6 Objective Function Result

Flood incident date	Peak Discharge (m^3/s)		Volume (mm)		Peak time		Objective Function Percentage (%)
	Model	Measured	Model	Measured	Model	Measured	
15 March 2020	797.5	798	73.96	66.73	18:50	18:50	10.8
16 March 2020	1727.4	1728.6	148.27	130.02	19:20	19:20	13.98
26 March 2020	996.2	994.8	84.61	96.52	20:30	20:30	12.49
4 April 2020	992.1	993.3	75.51	70.17	18:20	18:20	7.56
18 May 2020	1374.4	1373.2	119.7	117.93	17:10	17:00	1.39
11 June 2020	655.1	654	66.02	60.62	21:40	21:40	8.85

Source: Data processing result, 2020

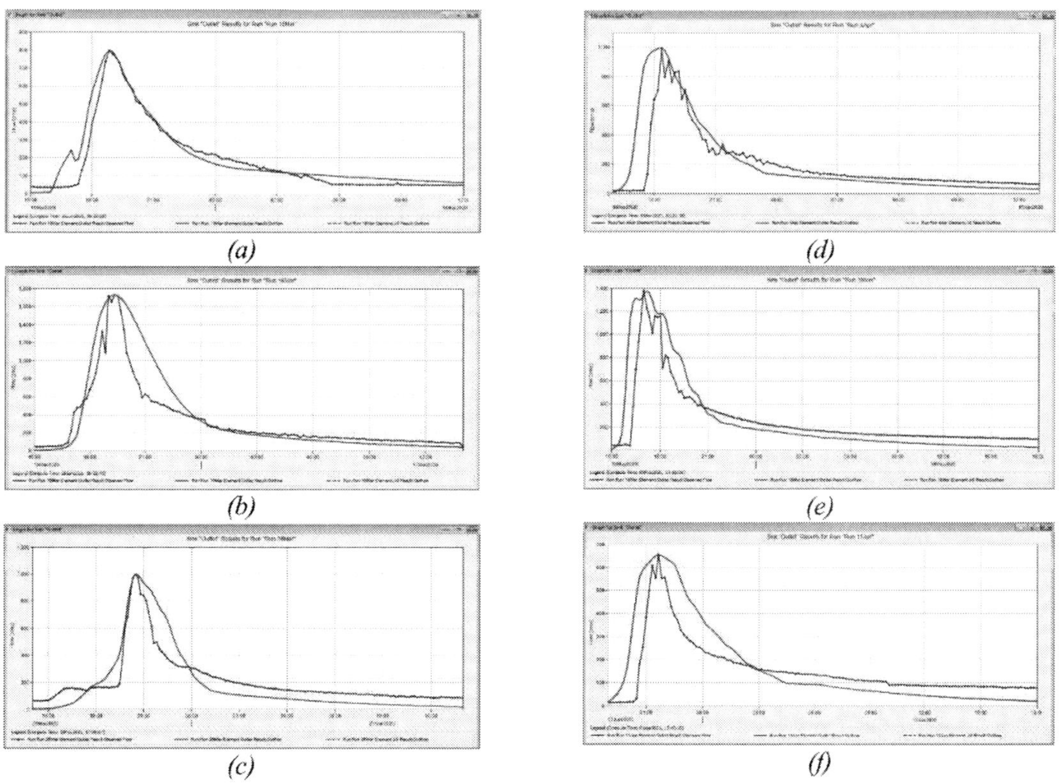

Figure 4.4 Result of Modeled and Measured Hydrography; (a) 15 Mach 2020, (b) 16 March 2020, (c) 26 March 2020, (d) 4 April 2020, (e) 18 May 2020, (d) 11 June 2020

Source: Data processing result, 2020

According to the six modeled hydrographs, the maximum discharge occurred on 16 March 2020 at a rate of 1727.4 m^3/s and a volume of 148.27 mm. While the hydrographs tested had a maximum discharge rate of 1728.6 m^3/s with a volume of 130.02 mm. Both occur at the exact moment, 19:20.

The predicted flood occurrence with the lowest objective function value occurred on 18 May 2020. With an outflow volume of 119.7 mm, the highest peak value is 1374.4 m^3/s. While the maximum discharged value recorded was 1373.2 m^3/s with an outflow volume of 117.93 mm. The hydrograph model's peak time was 17:10, which was ten minutes later than the actual data. This may occur due to the time lag value affecting the peak time; a change in the time lag may result in a more significant mistake since the peak discharged value will also change.

According to the hydrology responses study performed using HEC-HMS, the most sensitive factors for modeling a hydrograph are the CN value and time lag. Apart from that, the rainfall data utilized in this model must be pretty precise to minimize the potential of mistakes. Additional studies on CN and time lag must account for the scarcity of rainfall data to enhance the quality of the same research.

4.5. Water Yield Analysis

Water yield analysis was performed using HEC-HMS on one of six selected flood events from 4 April. This flood case has the most optimum CN value and is closest to the observation. From 1 January 2020 at 00:00 to 31 December 2021 at 23:00, a year of water yield analysis was conducted. Figure 4.5 illustrates the hourly output interval. Water yield is critical for examining runoff and the impact of environmental changes on ecosystem services. So many ecosystem services and flood-prone areas are affected by water yield [10].

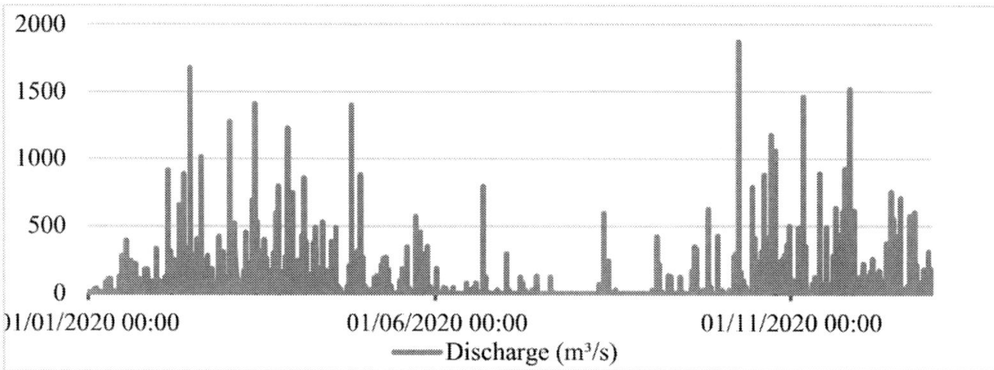

Figure 4.5 Hydrograph of Cileungsi Watershed flow from 1 january 2020 to 31 December 2020.

Soruce: Data processing result, 2021

Thus hydrograph are describing discharged flow from Cileungsi watershed outlet from 1 January 2020 to 31 December 2020. Through hydrograph above, the rain with the highest intensity happen in January – May and October – December, while June – August had lower rain intensity. The highest discharge happened in 9 October 2020 with the volume of 1,871.7 m^3/s.

The result from 1 year accumulation of water yield from Cileungsi watershed was shown at Table 4.7.

Table 4.7 Water Yield Result of Cileungsi Watershed 2020

1 year of Cummulative Rainfall (mm)	Volume of 1 Year Hydrograph (mm)
4662	4520.33

Sources: Data processing result, 2021

From the table above, we could understand the result of water yield in Cileungsi watershed. If we accumulated the result from the entire rainfall within a year, it would reach 4622 mm and the water yield could reach 4520 mm. Thus result was modeled using HEC-HMS.

5. Conclusions

We can conclude from the results of this research that the hydrology responses of the Cileungsi watershed can be simulated using HEC-HMS because the modelled and measured hydrographs are similar. By converting rainfall to runoff, the hydrograph depicts the Cileungsi watershed's water flow characteristics. Peak discharges resulting from the model of the six most significant flood episodes range between 655.1 and 1727.4 m^3/s, the volume ranges between 66.02 and 148.27 mm, and objective function results range between 1,39 and 13,98 per cent. While the peak discharge values from the six most significant flood episodes ranged from 654 to 1728.6 m^3/s with a volume of 60.62 to 130.02 mm. Base on the shape of the hydrograph is comparable along the fast-ascending limb but becomes slowed down as it approaches the recession limb.

Hydrology responses and water yield were analyzed using a one-year rainfall-runoff model with rainfall as input. The model was developed in response to the most similar flood event. One year's rainfall totaled 4662 mm, with a water yield of 4520 mm.

References

[1] Christanto N, Sartohadi J, Setiawan A, Hadi M P and Jetten V 2019 Investigating the Role of Rainfall Variability on the Hydrological Response of Small Tropical Upland Watershed *J. Geogr. Media Inf. Pengemb. dan Profesi Kegeografian* **16** 72–9
[2] Masnang A, Sinukaban N, Sudarsono and Gintings N 2014 Kajian Tingkat Aliran Permukaan Dan Erosi, pada Berbagai Tipe Penggunaan Lahan di SUB DAS Jenneberang Hulu *J. Agroteknos* **4** 32–7
[3] Mishra S K, Tyagi J V., Singh V P and Singh R 2006 SCS-CN-based modeling of sediment yield *J. Hydrol.* **324** 301–22
[4] Tyagi J V., Mishra S K, Singh R and Singh V P 2008 SCS-CN based time-distributed sediment yield model *J. Hydrol.* **352** 388–403
[5] USDA 2007 National Engineering Handbook: Chapter 7 Hydrologic Soil Groups, Natural Resources Conservation Service *Natl. Eng. Handb.* 7-1,7-5
[6] Laouacheria F and Mansouri R 2015 Comparison of WBNM and HEC-HMS for Runoff Hydrograph Prediction in a Small Urban Catchment *Water Resour. Manag.* **29** 2485–501
[7] Ayivi F and Jha M K 2018 Estimation of water balance and water yield in the Reedy Fork-Buffalo Creek Watershed in North Carolina using SWAT *Int. Soil Water Conserv. Res.* **6** 203–13
[8] Triatmodjo B 2014 *Hidrologi Terapan* (Yogyakarta: Beta Offset Yogyakarta)
[9] Hadi M P 2006 Pemahaman Karakteristik Hujan Sebagai Dasar Pemilihan Model Hidrologi (Studi Kasus di Das Bengawan Solo Hulu) *Forum Geogr.* **20** 13–26
[10] Redhead J W, Stratford C, Sharps K, Jones L, Ziv G, Clarke D, Oliver T H and Bullock J M 2016 Empirical validation of the InVEST water yield ecosystem service model at a national scale *Sci. Total Environ.* **569–570** 1418–26

Ecosystem service measurement model for urban population clean water monitoring

S Samadi[1,2*], S Suhardjo[1,3], A Munandar [1,4] and A Husen[5]

[1] Department of Geography Education, State University of Jakarta, Jalan Rawamangun Muka, Jakarta Timur, 13220, Indonesia
[2] Lecture of Environmental Geography
[3] Lecture of Cartography
[4] Lecture of Population Geography
[5] Department of Environmental Management, Universitas Negeri Jakarta, Jalan Rawamangun Muka, Jakarta Timur, 13220, Indonesia

*Email: samadi@unj.ac.id

Abstract. Clean Ecosystem service measurement model for urban population clean water monitoring Clean water as one of the ecosystem services and also a protected natural asset provides important information about the availability of natural resources for humans in response to the impacts of climate change and overall ecosystem health. The emergence of the COVID-19 pandemic which was exacerbated by the emergence of low-pressure areas for piped clean water services (PAM water) increasingly shows that if humans destroy biodiversity, it will have implications for other life support systems including clean water. This study identifies aspects of quantity, continuity, and quality (called 3K) of clean water from the piping system in the eastern part of DKI Jakarta, especially in areas with low pressure water supply. Through the Research and Development method, the aim of the study is to build a new model for measuring ecosystem services in monitoring clean water for residents in urban areas. Until the end of the first semester of 2020, the results of the study showed that it was true that there were areas with low pressure, but statistically it was not proven to significantly affect the emergence of complaints in terms of quantity, continuity, and quality of water supply to the population, so more in-depth parameters are needed. regarding the parameters of small water complaints (quantity), dead water (continuity), and piped clean water quality.

1. Introduction

Nowadays, many urban areas are increasingly facing the dynamics of climate change, demographic aging, and depletion of natural resources. On the other hand, the ecosystem still plays an important role as a facilitator of the transformation needed to overcome the challenges above. Urban communities are increasingly being driven to understand how urban ecosystems work, how ecosystems change, and what limits ecosystem performance; so everything will add understanding of ecosystem change including urban governance is required [1].

The COVID-19 (Coronavirus) pandemic increasingly shows that if humans destroy biodiversity, it will also destroy human life support systems [2,3]. Coronavirus has exposed the fragility at the heart of the water supply system as an ecosystem service [4] while at the same time proving the strengthening of the driving factors contributing to the high demand for clean water individually and communally.

The research questions are: First, how the pattern of clean water demand by household groups should be balanced with low low pressure areas, but on the contrary; and (second) what is the potential loss of revenue for the clean water management company [5]. From the number of non-commercial groups of 743555 customers, household customers are the largest number of clean water customers (99.5%) or about 739944 of the total non-commercial group customers. The data above explains that the increasing need for clean water in household groups should be directly proportional to the capacity of providing clean water. However, the reality on the ground shows that there are still service constraints, including inadequate water volume (quantity) which is indicated by not meeting the minimum standard of 0.75 atm at the customer point or can be termed low pressure on water that reaches households.

The problem of clean water services less than the minimum standard of 0.75 atm highlighted in this study did not make the community of clean water customers respond negatively. Usually the customer community will complain and even protest when the clean water supply is interrupted. This is what makes this research new and unique.

1.1. Literature Review

Ecosystem services (ES) are parts of ecological functions (physical, chemical, and biological processes) that are directly relevant or beneficial to human well-being [6]. In urban communities who are increasingly encouraged to understand the dynamics of urban ecosystems that function and undergo changes, and have environmental limitations; all of which will add to the increasing understanding of the community about ecosystem change including urban governance is needed [1].

Ecosystem services are generated by ecological functional attributes community; this is a characteristic of the ES indicator and service provider units consisting of population components, species, functional groups, food webs, and habitat types collectively provide services in a particular area [7].

The water supply system as an ecosystem service also has an important role in strengthening the factors driving the increasing supply of clean water [8,9], both individually and communally. Referring to Daily et al. [10] who suggested that ecosystem conditions and functions to become ES require interdisciplinary and user-oriented research, including collaborating with stakeholders to define public services [11,12] and therefore this research collaborates with the DKI Jakarta Provincial Government in this case the Drinking Water Regulatory Agency. ES is also assumed to be able to develop transparent and flexible models of ecological production function [13]. And this study applies the R and D method of the ADDIE model, which in the process tries to show the theoretical and empirical aspects of a relationship between ecosystem function and ES in the urban context so as to produce an ES measurement model for monitoring the population's clean water.

2. Methods

This study identifies aspects of quantity, continuity, and quality (called 3K) of clean water from the piping system in the eastern part of DKI Jakarta, especially in areas with low pressure water supply. Through the Research and Development method, the aim of the study is to build a new model for measuring ecosystem services in monitoring clean water for residents in urban areas. The data used in this study uses secondary data from piped water customers in the Eastern region of DKI Jakarta as many as 469534 customers (as of December 2019) with categories: (1) area data for water pressure < 0.75 atm; (2) there are customer complaints related to 3K services.

The selection of the East region of DKI Jakarta Province as the study locus is because this area has a larger number of piped water customers (469534 customers; as of December 2019) compared to customers in the western region of DKI Jakarta (408449 customers; as of December 2019), so that the resulting model simulation is expected can be representative and relevant for DKI Jakarta.

IOP Conf. Series: Earth and Environmental Science **986** (2022) 012083 doi:10.1088/1755-1315/986/1/012083

3. Results and Discussion

3.1. Results

Based on a series of data obtained throughout 2019 to 2020 the first semester in both the East and West areas of DKI Jakarta, the results of the analysis illustrate that:

The number of customers in the Eastern Region of DKI Jakarta in December 2019 was 469,534 connections. The customer complaints related to 3K during January to December 2019 were 385 for quantity problems, 1,317 complaints about water continuity, and 58 complaints about water quality.

Complaint data related to the problem of water supply quantity received by customers shows a dynamic number of complaints every month. In January it had the highest number of complaints, reaching 422 complaints, then in February it decreased to 349 complaints. This means that there has been an improvement in water intensity. The lowest water supply quantity complaints were in November with 224 complaints. However, it is known that the data on complaints of water supply continuity throughout 2019 is the highest complaint compared to complaints of water quantity and quality.

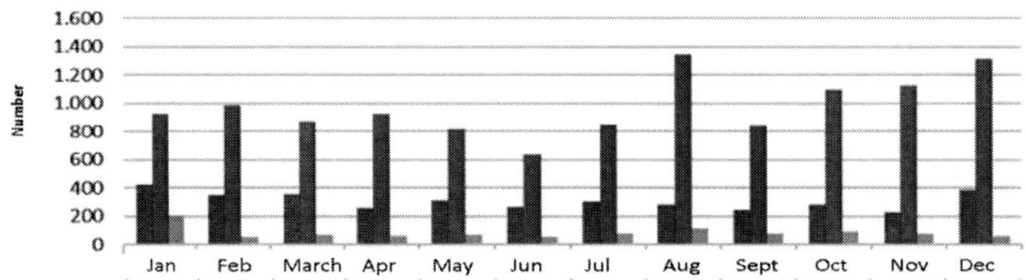

	Jan	Feb	March	Apr	May	Jun	Jul	Aug	Sept	Oct	Nov	Dec
Quantity	422	349	358	260	309	269	300	282	246	282	224	385
Continuity	926	982	868	923	816	639	850	1344	837	1094	1128	1317
Quality	204	56	69	64	70	57	73	111	79	90	76	58

Figure 1. Number of Complaints regarding 3K

Complaints about the continuity of water supply from January to May reached the range of 800-900 complaints. In June there was a significant decrease and became the lowest month regarding continuity complaints, namely 639 complaints, but increased to 850 complaints. The increase in complaints continued until December. The months of August, October, November, and December reached more than 1,000 continuity complaints. This shows that the occurrence of low water supply continuity for customers has increased in August, October, November, and December.

Complaint data related to water quality throughout 2019 is the smallest compared to complaints of continuity and quantity of water supply. The low number of complaints regarding water quality compared to other complaints indicates that the water quality service is quite good. The number of water quality complaints in 2019 is in the range of 50-200 complaints. In January there were 204 complaints and became the highest month regarding water quality complaints. The next highest number of complaints was in August with 111 complaints. In February to July and September to December the number of water quality complaints was below 100 complaints.

Complaint data related to 3K (quantity, quality, and continuity) in 2020 is only in semester 1, which is from January to June. Complaint data is obtained from the results of the contact center recapitulation for 24 hours every day this Eastern region supervisory company (PT. Aetra). The complaint data can be used to evaluate performance or prepare for a better management system.

The number of customers in the eastern area of DKI Jakarta until June 2020 was 473910 customers. Customer complaints related to 3K recorded in June 2020 were 397 for water supply quantity complaints, 926 water continuity complaints, and 38 water quality complaints. The total 3K complaints in June 2020 were 1,361 complaints.

Data on water quantity complaints in semester 1 of 2020 shows a dynamic number of complaints every month. January had the highest number of complaints in the early semester of 2020, which reached

616 complaints, but decreased in February to 464 complaints and continued to decline until April. And the lowest complaints occurred in April with 52 complaints. However, the increase in complaints occurred in May with 192 complaints and June with 397 complaints. Data on complaints of low water quantity in semester 1 of 2020 is the highest complaint compared to complaints of continuity and water quality.

Complaints against continuity from January to April continued to experience a significant decline, namely January 2017 complaints, February 1229 complaints, March 997 complaints, and April 107 complaints. April had the lowest number of complaints in semester 1 of 2020. However, the increase in complaints occurred from May to June. There were 479 complaints in May and 926 complaints in June.

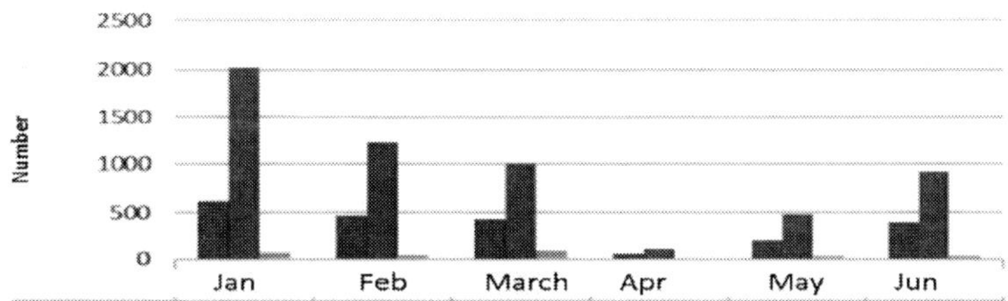

	Jan	Feb	March	Apr	May	Jun
Quantity	616	464	435	52	192	397
Continuity	2017	1229	997	107	479	926
Quality	70	48	90	1	31	38

Figure 2. Number of Complaints related 3K in Semester 1 of 2020

Water quality data in semester 1 of 2020 is the smallest complaint data compared to complaints of water quantity and water continuity problems. The number of water quality complaints in semester 1 of 2020 is no more than 90 cases of complaints. March had the highest number of complaints with 90 complaints. April was the month with the lowest complaint cases, which was only 1 complaint. However, there was an increase in complaints in May (31 complaints) and June (38 complaints).

3.1.1. Low pressure area data for 2019 and early semester 2020

Related to low pressure area data in 2019 which was below 0.75 atm every month it fluctuated. The area of high water pressure (≥0.75 atm) is mostly in the southern part, and the further north the piped water service coverage area is, the lower the water pressure area.

Table 1. Result of the 3K Complaint Correlation Test with Low Preasure Areas in 2019

Correlation	**2019**
	Pearson correlation
Low Preasure - Quantity	0,392
Low Preasure - Continuity	-0,359
Low Preasure - Water Quality	0,17

Meanwhile, at the beginning of the 2020 semester, the highest high pressure area (≥0.75 atm) was in the southern part of the study area. The further to the north, the lower the pressure area.

Table 2. Results of 3K Complaint Correlation Test with Low Preasure Areas Semester 1 of 2020

Correlation	Semester 1 2020
	Pearson correlation
Low Preasure - Quantity	-0,737
Low Preasure - Continuity	-0,627
Low Preasure - Water Quality	-0,661

3.2. *Multiple Linear Regression Test results*
The following analysis is a test of all variables simultaneously between independent variables (quantity, continuity, and water quality complaints) to the dependent variable (low pressure area) using multiple linear regression test.

Table 3. Multiple Linear Regression Test Results 3K Complaints with Low Pressure in 2019

Model	R	R Square	Adjusted R Square	Std. Error of the Estimate
1	,564[a]	,318	,063	3,236

Based on the results of multiple linear regression, R with 0.564 was obtained. This shows that the combined analysis produces a strong relationship as reflected by the R^2 (R Square) value of 0.318 or (31.8%). Thus, it can be explained that the relationship between low pressure areas and the number of 3K complaints in 2019 is categorized as strongly related.

Meanwhile, at the beginning of semester 1 of 2020, the results of the multiple linear regression test obtained R with 0.806.

Table 4. Multiple Linear Regression Test Results 3K Complaints with Low Pressure Area in Semester 1 of 2020

Model	R	R Square	Adjusted R Square	Std. Error of the Estimate
1	,806[a]	,650	,125	3,924

This shows that the low pressure area is very strongly related to all independent variables, namely the number of complaints 3K; with an indication of the value of R^2 (R Square) of 0.65 or (65%). From these results, it is assumed that the remaining 35% may be influenced by other variables not included in the study.

3.3. *Model Analysis*
There are several pre-conditions needed to build a new model for measuring ecosystem services in monitoring clean water for residents in urban areas, namely:

1. The factual condition of clean water consumption for the community and piped water customers in areas exposed to Covid-19,
2. The service area also experiences low pressure water,
3. Specifically for customers who cannot access other sources of clean water (due to family economic conditions, PPKM policies, self-isolation, etc.); In this case, it is closely related to government policy,

4. Piped water customers cannot protest further; and if this is allowed to happen, it is feared that it will form new behavior from customers who are apathetic to the factual conditions of clean water services while in fact they do not have their need for clean water fulfilled.

So that a new model is needed as a solution for piped clean water services, especially for urban communities affected by the Corona-19 pandemic.

The following is an overview of the mechanism for complaints service for piped clean water that has been running from the supervisor/company to the customer.

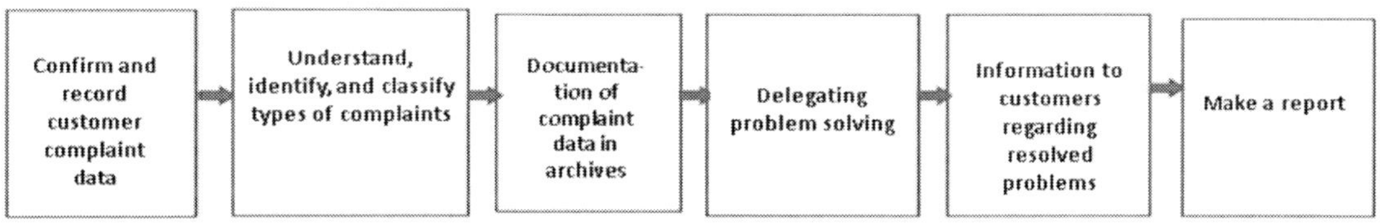

Figure 3. Scope of Handling Customer Complaints

The above mechanism describes the effort to record customer complaints and then confirm the existence of customer data at the clean water supervisory agency / company. The company then carries out an identification process and attempts to group the types of complaints which are then archived to be forwarded to the relevant department as a follow-up to resolving complaints from customers. And the estuary of this process is the preparation of a written report from the company.

The construction of a new model to measure ecosystem services in monitoring clean water for residents in urban areas, especially piped water customers affected by Corona-19, is to innovate piped clean water services. The following is a flow chart adapted to the findings/facts of this study.

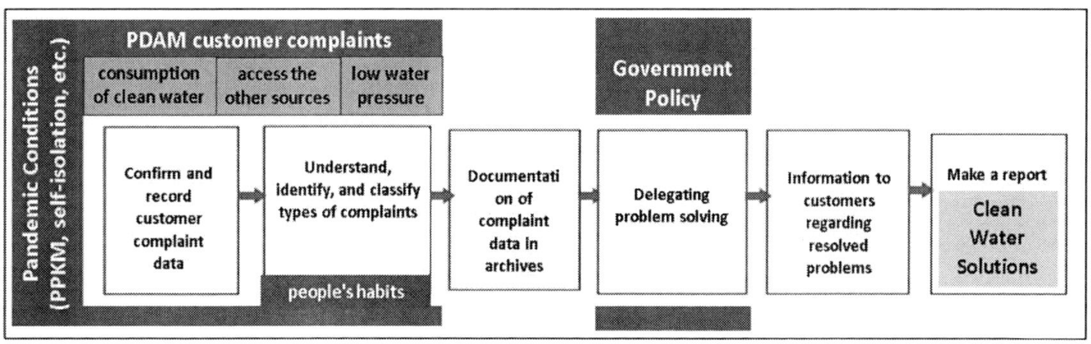

Figure 4. New model for measuring ecosystem services in monitoring clean water for residents in urban areas

In this newly developed model, customer complaint data is used as a baseline for the process of identifying and classifying types of complaints; where the complete complaint data has informed how much customer consumption of clean water is, customer accessibility problems to resources (clean water including other sources besides piped water) as well as information related to the position of customers in the category of being in low pressure areas or not. The last two categories need to be known in order to create the character of customers who are increasingly concerned about the piped water service mechanism. This is especially important during the Corona-19 pandemic.

From the data construction process in the end the company has a better and more accurate data system so that it is more targeted in the process of delegating further problem solving. And in the end, customers have certainty about the in-process-out mechanism for problems/complaints about clean water/piped water services. And the estuary is not just a file to be reported, but also a concrete solution to customer problems/complaints.

4. Conclusion and Further Research

The results of the analysis of the relationship between low pressure areas and all aspects of 3K complaints show that there is a diversity of relationship patterns in 2019 and semester 1 of 2020. In 2019, the relationship between low pressure areas and 3K complaints has a strong relationship level, and semester 1 of 2020 has a strong relationship. category level relationship is very strong. The relationship between low pressure areas in each 3K complaint variable shows that low pressure area variables with complaints of low water supply quantity have the highest correlation in 2019, while in semester 1 of 2020 low pressure water supply and water continuity complaints have the highest correlation.

The results of the correlation analysis show that although low pressure areas affect customer complaints regarding the quantity of water supply that is still low, the continuity of water flow is still not optimal, and water quality (3K); however, these three problems are not entirely a factor that is considered important by customers. This is indicated by the existence of (several) negative correlation results in the correlation test; The negative value in the results of this statistical calculation shows that there is a condition of low water quantity (water under standard pressure) for example, it really does happen but the customer doesn't really care.

As a solution, this research indicates that customer complaint data has covered data on clean water consumption, customer accessibility to water resources including other sources besides piped water and customer positioning in low pressure areas or not.

Acknowledges

This research was funded by BLU funds from the Faculty of Social Sciences, State University of Jakarta in the 2021 fiscal year.

References

[1] T E, Alfsen C and Colding J 2008 *Urban systems. In Encyclopedia of Ecology* (Oxford: Elsevier)
[2] Wall G 2002 Introduction To Life Support Systems and Sustainable *UNESCO - Encycl. Life Support Syst.*
[3] Iosef I. G and Genry M. L 2008 Creation of Closed Ecological Life Support Systems: Results, Critical Problems and Potentials *J. Sib. Fed. Univ. Biol.* **1** 19–39
[4] Liu S, Costanza R, Troy A, D'Aagostino J and Mates W 2010 Valuing New Jersey's ecosystem services and natural capital: A spatially explicit benefit transfer approach *Environ. Manage.* **45** 1271–85
[5] Naidoo R and Fisher B 2020 Goals : pandemic reset *Nature* **583** 198–201
[6] Groot R S de, Wilson M A and Boumas R M J 2002 A typoloy for the classification, description and valuation of ecosystem functions, goods and services *Ecol. Econ.* **41** 375–92
[7] Kremen C 2005 Managing ecosystem services: What do we need to know about their ecology? *Ecol. Lett.* **8** 468–79
[8] Costanza R, D'Arge R, Groot R de, Farber S, Grasso M, Hannon B, Limburg K, Naeem S, O'Neil R V, Paruelo J, Raskin R G, Sutton P and Belt M Van den 1997 The Value of the World's Ecosystem Services and Natural Capital *Nature* **387**
[9] Brauman K A, Daily G C, Duarte T K eo and Mooney H A 2007 The nature and value of ecosystem services: An overview highlighting hydrologic services *Annu. Rev. Environ. Resour.* **32** 67–98
[10] Daily G C and Matson P A 2008 Ecosystem services: From theory to implementation *Proc. Natl. Acad. Sci. U. S. A.* **105** 9455–6
[11] Carpenter S R, Bennett E M and Peterson G D 2006 Scenarios for ecosystem services: An overview *Ecol. Soc.* **11**

[12] Cowling R M, Egoh B, Knight A T, O'Farrell P J, Reyers B, Rouget M, Roux D J, Welz A and Wilhelm-Rechman A 2008 An operational model for mainstreaming ecosystem services for implementation *Proc. Natl. Acad. Sci. U. S. A.* **105** 9483–8

[13] Ricketts T H, Regetz J, Steffan-Dewenter I, Cunningham S A, Kremen C, Bogdanski A, Gemmill-Herren B, Greenleaf S S, Klein A M, Mayfield M M, Morandin L A, Ochieng' A and Viana B F 2008 Landscape effects on crop pollination services: Are there general patterns? *Ecol. Lett.* **11** 499–515

IOP Conf. Series: Earth and Environmental Science **986** (2022) 012084 doi:10.1088/1755-1315/986/1/012084

Microplastic pollution in the water and sediment of Krukut River, Jakarta, Indonesia

A Azizi[1]*, N Maulida[1], W N Setyowati[1], S Fairus[1] and D A Puspito[1]

[1] Environmental Engineering Department, Universitas Bakrie, H.R. Rasuna Said Road, Kuningan, South Jakarta, 12920, Indonesia

*Email: aqil.azizi@bakrie.ac.id

Abstract. One of Jakarta's drinking water sources, the Krukut River, has been contaminated by a variety of industrial and domestic wastes, including microplastic pollution. To investigate the abundance and identify the of microplastics in the Krukut River, water and sediment from 3 sites, upstream, middle and downstream were collected using a water sampler and Ekman sediment grab sampler, respectively. The microplastic from water and sediment was extracted using density separation method. The microplastics were counted and categorized according to the shape under a microscope. The differences in microplastic abundance in three different sites were determined using a one-way ANOVA. The result shows fiber dominated microplastic shape in the water across the sampling sites, whereas pellet dominated microplastic shape in the sediment from upstream to middle- stream. The downstream water has the highest abundance of microplastics (265 ± 5.0 particle/100 mL), which is significantly different ($p<0.05$) from the middle (221 ± 3.0 particle/100 mL) and the upstream (215 ± 1.5 particle/100 mL). Similarly, the downstream sediment had the largest abundance of microplastics (150 ± 5.0 particle/kg dry weight), which was significantly different ($p<0.05$) from the middle (125 ± 6.0 particle/kg dry weight) and the upstream (112 ± 3.5 particle/kg dry weight).

1. Introduction

Indonesia, the world's fourth-largest country, produces an estimated 123 million discarded facemasks and over 21 million tonnes of plastic garbage per day as a result of the Covid-19 outbreak [1]. If those plastics are not adequately managed and dumped into the environment, they may disintegrate into tiny pieces called microplastics due to a mixture of physicochemical and mechanical variables such as UV exposure, temperature, hydrophobicity, wind, and current [2].

Recently, several studies have shown microplastics found in drinking water, prompting many questions about the quality of the raw water supply and the ability of microplastic particles to be removed from drinking water treatment plants (WTPs). Despite the fact that the ecological and toxicological effects of microplastics are still largely unknown, microplastics are regarded an emerging pollutant, with concerns about their potential impact on human health [3].

The Krukut River, one of the primary rivers running through Jakarta, plays a significant role in supporting community activities and is designated as raw water supply for the Cilandak Water Treatment Plant (WTP), which is operated by a private company and offers drinking water services to Jakarta's western area. The Krukut River's water quality is deteriorating in conjunction with the growing population of Jakarta. Various human activities, including as industrial, residential, and agricultural activities, will produce waste, which will contribute to the deteriorated quality of river water [4].

IOP Conf. Series: Earth and Environmental Science **986** (2022) 012084 doi:10.1088/1755-1315/986/1/012084

The presence of microplastic particles is most likely to affect water suppliers who use water river as a direct or indirect raw water source. It is most likely that larger particles will be trapped through membrane filtration, medium filtration, bank filtration, artificial recharge, or underground passages. At this stage, however, there is a scarcity of information on the presence of microplastic particles in rivers as sources for drinking water especially in Jakarta, which is concerning due to the potential for human exposure to microplastics. This study was conducted to identify and quantify the types and abundance of microplastic in the water and sediment of the Krukut River from three sampling stations; upstream, middle and downstream.

2. Materials and Methods

2.1. Time and Place
The water and sediment sample of Krukut River were collected in 2020 from three sampling locations: upstream (Cipedak village), middle (Cilandak Timur village), and downstream (Karet Semanggi village) (see Figure 1). The upstream is approximately 7.1 km apart from the middle, while middle is about 9.9 km apart from downstream. Further sample analysis was carried out at Environmental Engineering Laboratory, Bakrie University.

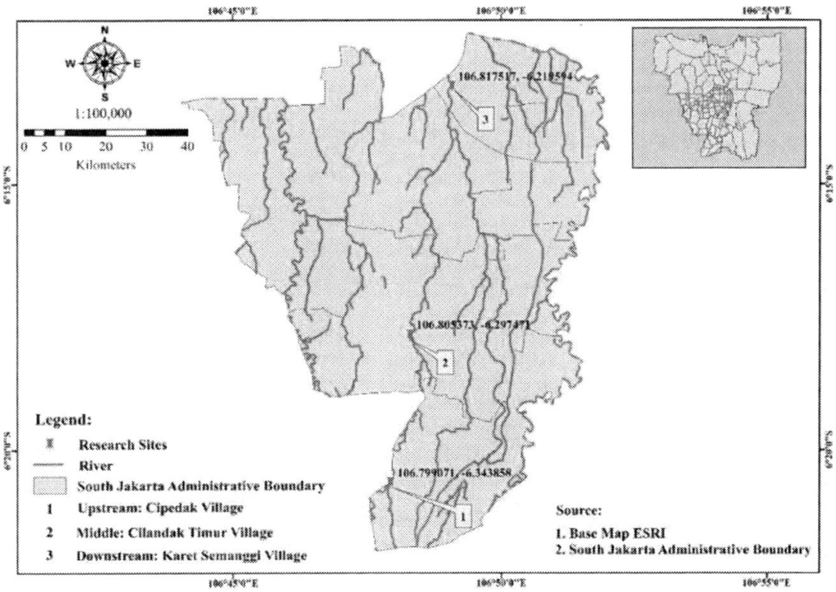

Figure 1. The location of sampling points on the map

2.2. Sampling Procedure
A total of 15 liters of water were gathered in triplicate at each sampling site using a water sampler. A plankton net with a size of 0.40 millimeters was used to filter these samples. After that, the filtered water samples were placed in a sterile glass bottle, sealed with a ParaFilm® sealing film, and stored at 4 °C for analysis. Sediment was collected in triplicate (0 – 30 cm depth) at each sampling station using an Ekman grab sampler. Before further examination, the sediment samples were maintained in an ice box, transported to the laboratory, and stored in a -2 °C freezer.

2.3. Isolation of Microplastics
In order to prevent organisms from interfering with the samples, the wet peroxide oxidation (WPO) procedure was used to pre-treat water samples. In a beaker, 20 mL of water samples were inserted. The mixture was then mixed at room temperature for 5 minutes with 40 mL of aqueous 0.05 M Fe (II)

solution and 40 mL of 30% hydrogen peroxide. The beaker was swirled and cooked on a hotplate at 75°C with a watch glass on top. The beaker was placed in the fume hood after being taken from the hotplate as soon as gas bubbles emerged, where it remained until the boiling stopped. When there were any natural organic compounds present, 20 mL of 30% hydrogen peroxide was added. The density separation was accomplished using sodium chloride to raise the aqueous solution's density (1.5 gcm^{-3}). For 15 minutes, the mixture was heated on a hotplate with 6 g sodium chloride added. The mixture was covered in aluminum foil for two days, after being removed from the hotplate. After two days, filtration of the mixture was performed using Whatman no. 4 filter paper and a vacuum pump. It was necessary to dry out the filter paper before it could be examined further.

The sediment samples were first dried using a drying method at a temperature of 60°C for one night or until totally dry. The dried sediment was then weighed and placed in a beaker using an analytical balance. Following the pre-treatment, sediment samples were subjected to a density separation technique. The amount of microplastics in sediment samples was determined using density separation, which involved raising the aqueous solution's density using sodium chloride (1.5 gcm^{-3}). For 15 minutes, the samples were heated in a hotplate with 6 grams of sodium chloride. After that, the mixture was taken off the hotplate and wrapped in aluminum foil for three days. After that, filtration of the mixture was performed using Whatman no. 4 filter paper and a vacuum pump. It was necessary to dry out the filter paper so that it could be studied further.

2.4. Sorting and Identification of Microplastics

Microplastic particles were observed and counted under a stereo Olympus SZH1O microscope at a magnification of 10 × 10 and classified into four groups: films, fibers, fragments, and pellets. The mean and standard deviations of microplastic abundance are given in n/100 mL ± (SDs) for water samples and n/kg dry weight ± (SDs) for sediment samples. One-way ANOVA, Tukey's test, and Post-Hoc analysis with SPSS 19 software were used to examine the differences in microplastic abundance between sampling stations.

3. Results and Discussion

3.1. Abundance of Microplastics

The downstream water had the largest abundance of microplastics (265 ± 5.0 particle/100 ml), which was significantly different ($p<0.05$) from the middle (221 ± 3.0 particle/100 ml) and the upstream (215 ± 1.5 particle/100 ml), as shown in Figure 2 (a). The downstream sediment of Krukut river had the highest microplastic abundance (150 ± 5.0 particle/kg dry weight), which was significantly different ($p<0.05$) from the middle (125 ± 6.0 particle/kg dry weight) and the upstream (112 ± 3.5 particle/kg dry weight), as shown in Figure 2 (b).

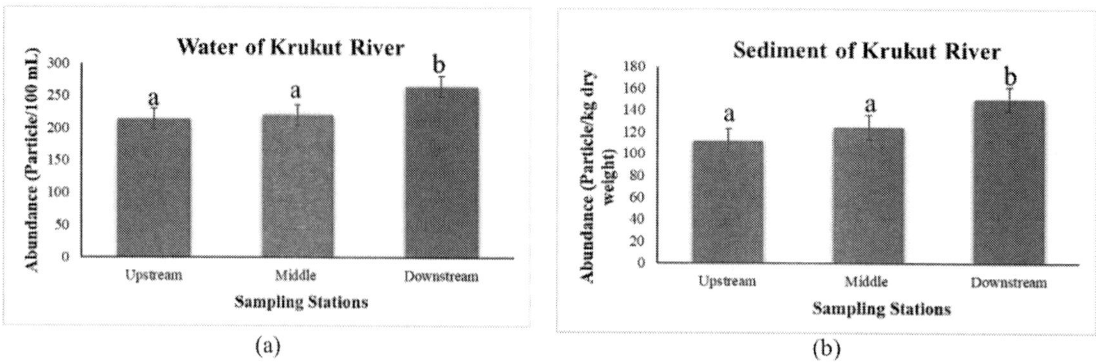

Figure 2. The mean abundance of microplastic in the water (a) and the sediment (b) from three different sampling stations of Krukut River

The highest microplastic abundance in downstream for both water and sediment samples has been indicated as the confluence of commercial, public facilities, office buildings, and residential districts, where are all within close proximity to the region. In sample stations that have been influenced by anthropogenic activities, microplastic pollution is more widespread. There has been research showing a link between microplastic pollution and population density in numerous areas [5]. Previous study also reported that microplastic concentration in downstream was higher than upstream [6]. The number of microplastic concentration in the water of Krukut river is much higher than other microplastic in the water of Surabaya River [7] and Cimandiri Watershed [8].

3.2. Shape of Microplastics

Figure 3 shows the identification of microplastics collected from all three sampling stations, water (a) and sediment (b), demonstrating four different types of microplastics: fiber, film, fragment, and pellet. Figure 4 shows the mean percentage of microplastic shapes found in the water and sediment samples from upstream (a); middlestream (b), and downstream (c) of Krukut river.

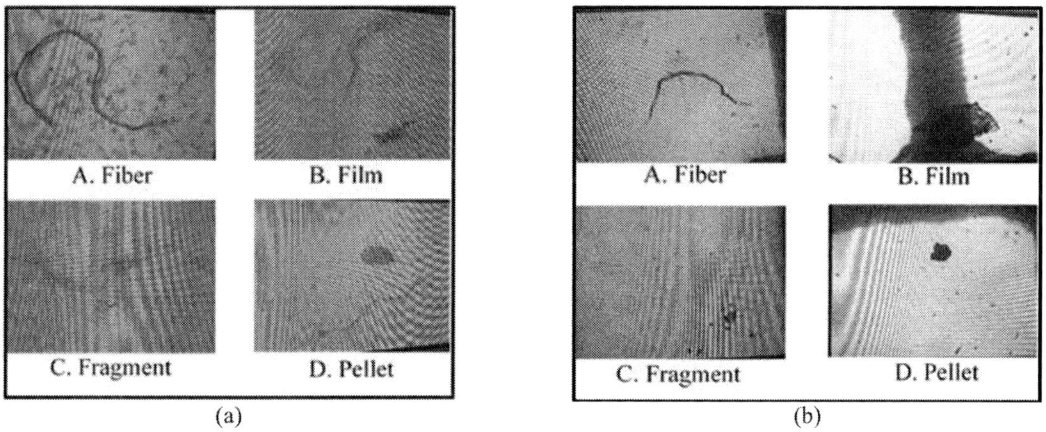

Figure 3. The shape of microplastic found in the water samples (a) and sediment sample of Krukut river A: Fiber, B: Film, C: Fragment, D: Pellet

Research shows fiber dominated microplastic shape in the water from upstream to downstream, accounting for 35.4%, 43.5%, and 44.3% respectively, whereas pellet dominated microplastic shape in upstream and middlestream sediment. Fragment and film had the lowest percentage of microplastics shape across all locations.

Many factors influenced the shapes of microplastics found, including human activities, hydrodynamics, the weather, and topography [9]. Fiber is most typically seen in water due of the fragmentation of monofilament nets, ropes, and synthetic fabrics. Residents' laundry waste is assumed to be the source of the fiber, which is discharged directly into the river. Pellets are found in sediments because they are part of the primary microplastics (such as polypropylene and polyethylene) that enter water bodies through household waste. The pellets discovered were made from microbeads found in beauty and cleaning products in household waste that was dumped directly into rivers.

IOP Conf. Series: Earth and Environmental Science **986** (2022) 012084 doi:10.1088/1755-1315/986/1/012084

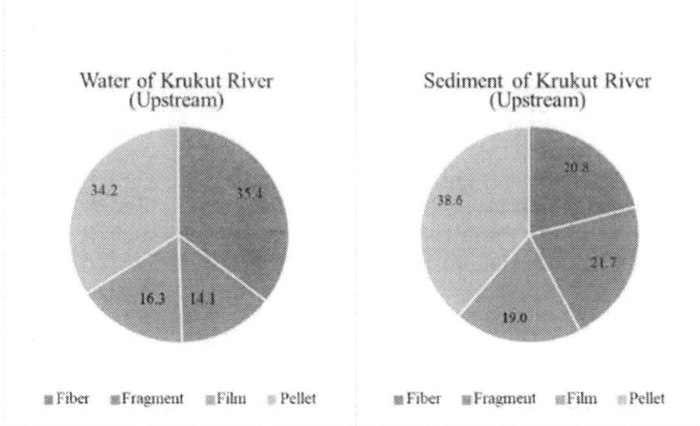

(a)

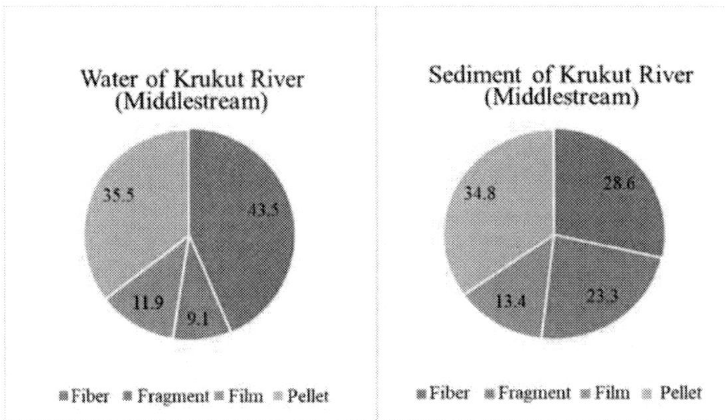

(b)

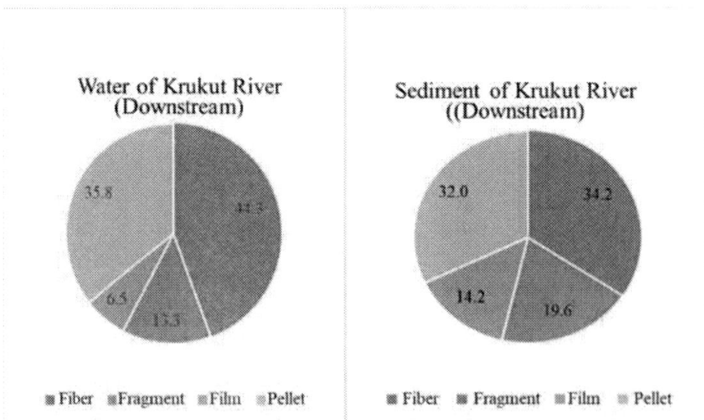

(c)

Figure 4. The mean percentage of microplastic shapes found in the water and sediment samples from upstream (a); middlestream (b), and downstream (c) of Krukut river

4. Conclusion
The presence and properties of microplastics in the Krukut river supplying water treatment plants (WTP) for Jakarta's western area was determined. The average number of MPs in raw water ranged from 215 ± 1.5 particle/100 ml in the upstream to 265 ± 5.0 particle/100 ml in the downstream. Generally, the downstream water and sediment of Krukut river contain significantly higher microplastic compared to the middle and the upstream. Fiber was predominated microplastic shape in the water across the sampling sites, while pellet dominated microplastic shape in the sediment from upstream to iddlestream. Further research should be focus on determination of microplastic in the treated water from WTPs.

Acknowledgments
This research was supported by The Grant of Directorate General of Higher Education (DIKTI) and Universitas Bakrie Research Grant 2021.

References
[1] Benson N U, Bassey D E and Palanisami T 2021 COVID pollution: impact of COVID-19 pandemic on global plastic waste footprint *Heliyon* **7** e06343
[2] Khoironi A, Hadiyanto H, Anggoro S and Sudarno S 2020 Evaluation of polypropylene plastic degradation and microplastic identification in sediments at Tambak Lorok coastal area, Semarang, Indonesia *Mar. Pollut. Bull.* **151** 110868
[3] Pivokonsky M, Cermakova L, Novotna K, Peer P, Cajthaml T and Janda V 2018 Occurrence of microplastics in raw and treated drinking water *Sci. Total Environ.* **643** 1644–51
[4] Rachmawati I pramudita, Riani E and Riyadi A 2020 Status Mutu Air Dan Beban Pencemar Sungai Krukut, Dki Jakarta *J. Pengelolaan Sumberd. Alam dan Lingkung. (Journal Nat. Resour. Environ. Manag.* **10** 220–33
[5] Browne M A, Crump P, Niven S J, Teuten E, Tonkin A, Galloway T and Thompson R 2011 Accumulation of microplastic on shorelines woldwide: Sources and sinks *Environ. Sci. Technol.* **45** 9175–9
[6] McCormick A R, Hoellein T J, London M G, Hittie J, Scott J W and Kelly J J 2016 Microplastic in surface waters of urban rivers: Concentration, sources, and associated bacterial assemblages *Ecosphere* **7**
[7] Lestari P, Trihadiningrum Y, Firdaus M and Warmadewanthi I D A A 2021 Microplastic pollution in Surabaya River Water and Aquatic Biota, Indonesia *IOP Conf. Ser. Mater. Sci. Eng.* **1143** 012054
[8] Pe E O L, Mashar A, Taryono and Wardiatno Y 2020 Microplastic distribution and abundance in cimandiri watershed flowing to Palabuhanratu Bay, Sukabumi, West Java, Indonesia *AACL Bioflux* **13** 657–68
[9] Febriani I S, Amin B and Fauzi M 2020 Distribusi mikroplastik di perairan Pulau Bengkalis Kabupaten Bengkalis Provinsi Riau *Depik* **9** 386–92

AUTHOR INDEX

Institute of Physics
Dirac House, Temple Back
Bristol BS1 6BE UK

ISSN: 1755-1307
ISBN 978-1-7138-4967-4

CURRAN ASSOCIATES INC.
proceedings.com